McGRAW-HILL INTERNATIONAL SERIES
IN THE EARTH AND PLANETARY SCIENCES

FRANK PRESS, Consulting Editor

AGER • Principles of Paleoecology
BERNER • Principles of Chemical Sedimentology
BROECKER and OVERSBY • Chemical Equilibria in the Earth
DE SITTER • Structural Geology
DOMENICO • Concepts and Models in Groundwater Hydrology
EWING, JARDETZKY, and PRESS • Elastic Waves in Layered Media
GRANT and WEST • Interpretation Theory in Applied Geophysics
GRIFFITHS • Scientific Method in Analysis of Sediments
GRIM • Applied Clay Mineralogy
GRIM • Clay Mineralogy
HEISKANEN and VENING MEINESZ • The Earth and Its Gravity Field
HOWELL • Introduction to Geophysics
JACOBS, RUSSELL, and WILSON • Physics and Geology
KRAUSKOPF • Introduction to Geochemistry
KRUMBEIN and GRAYBILL • An Introduction to Statistical Models in Geology
LEGGET • Geology and Engineering
MENARD • Marine Geology of the Pacific
MILLER • Photogeology
OFFICER • Introduction to the Theory of Sound Transmission
RAMSAY • Folding and Fracturing of Rocks
ROBERTSON • The Nature of the Solid Earth
SHROCK and TWENHOFEL • Principles of Invertebrate Paleontology
STANTON • Ore Petrology
TURNER • Metamorphic Petrology
TURNER and VERHOOGEN • Igneous and Metamorphic Petrology
TURNER and WEISS • Structural Analysis of Metamorphic Tectonites
WHITE • Seismic Waves: Radiation, Transmission, and Attenuation

FRANCIS BIRCH
Sturgis Hooper Professor of Geology
Harvard University

DEDICATION

This volume was written to honor Francis Birch on the occasion of his retirement from full-time teaching duties at Harvard University. Birch's work on the composition, structure, origin, and thermal character of the solid earth constitute a large portion of our current understanding of those subjects. His solid, forthright mode of attack on problems of the earth, utilizing high-pressure experimentation and relevant data and theory from geology and physics, have set an exemplary standard for his colleagues and students.

THE NATURE OF THE SOLID EARTH

Dedicated to Francis Birch

Editor

EUGENE C. ROBERTSON
U.S. Geological Survey
Silver Spring, Maryland

Associate Editors

JAMES F. HAYS
Harvard University
Cambridge, Massachusetts

and

LEON KNOPOFF
University of California
Los Angeles, California

Papers presented at a Symposium held at Harvard University
Cambridge, Massachusetts, 16-18 April 1970

THE NATURE OF THE SOLID EARTH

Library of Congress Catalog Card Number 76-168753

07-053165-X

2 3 4 5 6 7 8 9 0 MAMM 7 9 8 7 6 5 4 3 2

This book was set in Press Roman by Scripta-Technica, Inc., and printed and bound by The Maple Press Company. The designer was Scripta-Technica, Inc. The editor was Bradford Bayne. Sally Ellyson supervised production.

CONTENTS

LIST OF CONTRIBUTING AUTHORS

Anderson, Don L.
Seismological Laboratory
California Institute of Technology
PO Box 2, Arroyo Annex
Pasadena, California 91109

Anderson, Orson L.
Institute of Geophysics and
Planetary Science
University of California, Los Angeles
Los Angeles, California 90024

Blackwell, David D.
Department of Geology
Southern Methodist University
207 Heroy Hall
Dallas, Texas 75222

Bullard, Edward
Department of Geodesy and Geophysics
Cambridge University
Madingley Road
Cambridge, CB3 OEZ England

Clark, Sydney P., Jr.
Department of Geology and Geophysics
Yale University
New Haven, Connecticut 06520

Cox, Allan
Department of Geophysics
Stanford University
Stanford, California 94305

Decker, Edward R.
Department of Geology
University of Wyoming
Laramie, Wyoming 82070

Diment, William H.
Department of Geological Sciences
University of Rochester
Rochester, New York 14627

Doell, Richard R.
U.S. Geological Survey
Regional Geophysics Branch
345 Middlefield Road
Menlo Park, California 94025

Garrels, Robert M.
Scripps Institute of Oceanography
University of California, San Diego
La Jolla, California 92037

Gast, Paul W.
National Aeronautics & Space Administration
Manned Spacecraft Center
Houston, Texas 77058

Gilbert, Freeman
Institute of Geophysics and
Planetary Physics
University of California, San Diego
La Jolla, California 92038

Gilluly, James
975 Estes Street
Lakewood, Colorado 80215

Griggs, David T.
Institute of Geophysics and
Planetary Physics
University of California, Los Angeles
Los Angeles, California 90024

Grossman, Lawrence
Department of Geology and Geophysics
Yale University
New Haven, Connecticut 06520

Hales, Anton L.
Geosciences Division
University of Texas at Dallas
Dallas, Texas 75230

Hays, James F.
Department of Geological Sciences
Harvard University
Cambridge, Massachusetts 02138

Herrin, Eugene
Department of Geology and Geophysics
Southern Methodist University
Dallas, Texas 75222

Jordan, Tom
Seismological Laboratory
California Institute of Technology
Pasadena, California 91109

Kaula, William M.
Institute of Geophysics and
Planetary Physics
University of California, Los Angeles
Los Angeles, California 90024

Knopoff, Leon
Institute of Geophysics and
Planetary Physics
University of California, Los Angeles
Los Angeles, California 90024

Mackenzie, Fred T.
Department of Geological Sciences
Northwestern University
Evanston, Illinois 60201

McKenzie, Dan P.
Department of Geodesy and Geophysics
Cambridge University
Madingley Rise
Madingley Road
Cambridge, CB3 OEZ England

Menard, H. W.
Institute of Marine Resources and
Scripps Institute of Oceanography
University of California, San Diego
La Jolla, California 92037

Press, Frank
Department of Earth and Planetary
Sciences
Massachusetts Institute of Technology
Cambridge, Massachusetts 02139

Revetta, Frank A.
Department of Geology
State University of New York, Potsdam
Potsdam, New York 13676

Ringwood, A. E.
Department of Geophysics and
Geochemistry
Australian National University
Canberra, Australia

Robertson, Eugene C.
Office of Earthquake Research and
Crustal Studies
U.S. Geological Survey
Silver Spring, Maryland 20910

Roy, Robert F.
Department of Geosciences
Purdue University
Lafayette, Indiana 47907

Sammis, Charles
Seismological Laboratory
California Institute of Technology
Pasadena, California 91109

Shapiro, J. N.
Department of Geophysics
College of Geosciences
Texas A&M University
College Station, Texas 77843

Siever, Raymond
Department of Geological Sciences
Harvard University
Cambridge, Massachusetts 02138

Slichter, Louis B.
Institute of Geophysics and Planetary Physics
University of California, Los Angeles
Los Angeles, California 90024

Turekian, Karl K.
Department of Geology and Geophysics
Yale University
New Haven, Connecticut 06520

Urban, Thomas C.
Department of Geological Sciences
University of Rochester
Rochester, New York 14627

Woollard, George P.
Hawaii Institute of Geophysics
University of Hawaii
Honolulu, Hawaii 96822

PREFACE

The perspective of this book reflects the research that earth scientists are doing now on the interior of the earth. Subjects of the chapters were chosen to cover the broadest possible range of topics on the nature of the solid earth, with stress on recent developments. In their treatment the authors combined, as appropriate, geophysical, geochemical, and geological aspects of their topics, implications for the future, and review of recent results. The emphasis is on interpretation of experimental and field observations.

The content of this book includes discussions of the distribution of geochemical elements in the earth at its origin and at the present time, the structure and gross physical properties of the earth from seismology and magnetism, the kinematics of continents as drifting plates and the consequent mountain building and deep-focus earthquakes, the significance of the patterns of heat generation and of gravity in the earth, and the elastic, melting, and strength properties of selected rocks and minerals. Methods of analysis vary from rigorous mathematical development for working out physical models of the earth to thoroughly empirical study of the results of heat flow and gravity measurements. The sources of data vary from modern field techniques, such as, geodesy by satellite, and free oscillations of the earth from strain seismometers and gravimeters, to modern laboratory methods of high-pressure measurements of elastic properties and experiments on mineral equilibria.

The symposium at which these papers were presented orally was held at Harvard University, Cambridge, Massachusetts, 16-18 April 1970. The symposium and this book were conceived to honor Professor Francis Birch on his retirement from active teaching at Harvard and to set forth the achievements of the United States program for the Upper Mantle Project. Sponsors of the symposium were a Harvard Committee composed of former students of Birch and faculty members, the Department of Geological Sciences at Harvard, and the U.S. Upper Mantle Committee. A description of the activities of the Upper Mantle Project during the last decade follows this preface.

Students were encouraged to attend the symposium as the sponsors believed that the next generation of earth scientists would benefit from the comprehensive and conclusive program of recent research, as well as from personal contacts with the speakers. Interest was aroused around the United States and Canada, and as a consequence, about one-half the total number of registrants (750) were students. Grants from NSF funds were awarded as "Travel Scholarships" to 75 advanced graduate students.

A reader's guide of a new type is provided at the back of this book to simplify finding discussions of specific topics. A listing of broad, key words and the ideas subsumed under them in a "Topic Index" substitutes for the usual, detailed subject index. This innovation should improve access to the actual coverage of the book; the titles of the papers tell about their content well enough for a first identification, and the "Topic Index" should help readers find their way to less obvious general subjects or concepts of interest.

Acknowledgments are due to the following organizations for financial support to the symposium and thus indirectly to this volume: The National Science Foundation, The National Academy of Sciences (Geophysics Research Board and the U.S. Upper Mantle Committee), Harvard University (Committee on Experimental Geology and Geophysics, Center for Earth and Planetary Physics, and Committee on Oceanography), and the U.S. Geological Survey. In addition, the sponsors and editors wish to express their appreciation to Charlotte Isaacs and Pembroke J. Hart for their encouragement and unselfish effort in helping with the symposium and this volume. We are grateful to the many persons whose reviews of the manuscripts have certainly improved the quality of the book. The authors themselves deserve our grateful acknowledgment for their cooperation, because they literally make this book.

Eugene C. Robertson

UPPER MANTLE PROJECT

The Upper Mantle Project was proposed in 1960 as a program of international cooperation in research in solid-earth geophysics, sponsored by the International Council of Scientific Unions (ICSU) through the International Union of Geodesy and Geophysics and the International Union of Geological Sciences. The period until 1964 was primarily a planning period, and the program was fully under way in the United States and other countries during the interval 1965-1970. The guidance and coordination of the Upper Mantle Project (UMP) was provided by the International Upper Mantle Committee, of which Professor V. V. Beloussov (USSR) was chairman, Professor Leon Knopoff (USA) general secretary. Corresponding national committees for the UMP were established in the participating countries.

The Panel on Solid-Earth Problems of the Geophysics Research Board prepared a "Proposed U.S. Program for the UMP," which was issued at the fall meeting of the National Academy of Sciences, Austin, Texas, November 1962. Review of this proposed program by the President's Science Advisory Committee led to endorsement of the principle of the U.S. participation in the UMP and to creation of an Interagency Committee for the UMP to coordinate activities of the federal agencies. To develop and coordinate the U.S. Program for the UMP, a U.S. Upper Mantle Committee was established in the National Academy of Sciences under the Geophysics Research Board with the following membership: L. Thomas Aldrich, Francis Birch, Charles L. Drake, Eugene Herrin, Harry H. Hess, Leon Knopoff (chairman), Jack E. Oliver, Frank Press, John Verhoogen, and George P. Woollard. Three subcommittees were established: Committee on the Transcontinental Geophysical Survey, Committee on Deep Drilling for Scientific Purposes, and Committee on Data Exchange.

Progress reports on the U.S. program were issued in 1965 and 1967. When the period of the UMP was extended to 1970, the U.S. Upper Mantle Committee prepared a supplementary report "Upper Mantle Project: Proposed U.S. Program for 1966-1970." This proposed program stressing international cooperative aspects of the U.S. program, especially with Latin American countries, led to a telegram from the President of the United States to the President of the American Geophysical Union at the spring meetings in 1967, encouraging continued U.S. participation in the UMP.

Two symposia sponsored by the International Upper Mantle Committee were held in the United States: Symposium on the Origin of Andesite Magma, Eugene, Oregon, July 1968 (convened by A. R. McBirney); and Symposium on Mechanical Properties and Processes of the Upper Mantle, Flagstaff, Arizona,

June 1970 (conveners: Orson L. Anderson and Lynn R. Sykes). In addition, three symposia devoted to the UMP were sponsored domestically in the United States: in Austin, November 1962 (at the time of the announcement of the proposed U.S. program); at the spring meetings of the American Geophysical Union in 1968; and in Cambridge in 1970.

This volume, based on the symposium in Cambridge, in addition to being dedicated to Francis Birch, also presents a report on the scientific achievements of the U.S. program for the Upper Mantle Project. The U.S. program consisted of two main parts: (1) pertinent research already planned in the academic community and government agencies; (2) specially planned contributions. The National Science Foundation, Advanced Research Projects Agency, Office of Naval Research, Naval Oceanographic Office, U.S. Geological Survey, and U.S. Coast and Geodetic Survey were major supporters of the program.

The first U.S. UMP symposium in 1962 presented a rather different image of the earth than we have today. There now appears to be strong evidence favoring models of the upper mantle of the earth in which horizontal motions of rather large magnitudes occur among parts of the earth relative to one another, in particular the model of plate tectonics. This kinematic description of the surface and near-surface regions appears to be well-correlated to many of the physiographic features of the surface: the occurrence of mountains, oceanic trenches, oceanic and continental rifts, and even the ocean basins themselves. The model is correlated with geographic distributions of earthquakes and earthquake focal mechanisms, and it is further elaborated by the nature of heterogeneities in the mantle, which are evidence of dynamic instabilities in the earth's interior.

During the period of the UMP, many new developments have been brought forth in addition to those of plate tectonics. These include the study of reversals in the magnetic field, the role of free oscillations in the determination of the interior structure of the earth, the generalizations and limitations of the inverse problem, studies of equations of state including information derived from shock-wave experiments, the differences between crust and mantle structure in oceanic and continental areas, the nature of the low-velocity channel, and the major discontinuities in the mantle at about 400- and 600-km depth.

The papers in this volume present reviews of most of these developments, many of which were pioneered by Francis Birch and many were developed under the aegis and encouragement of the UMP. They point the way to the topics that should be explored in the next decade; and in furtherance of this, ICSU and the scientific unions have initiated a long-range program, the Geodynamics Project, which will bring together geology and geophysics in concerted study of the dynamics of the earth's interior.

Leon Knopoff

THE NATURE OF THE SOLID EARTH

CHEMICAL MODELS OF THE EARTH

1

MODEL FOR THE EARLY HISTORY OF THE EARTH

by **Sydney P. Clark, Jr., Karl K. Turekian, and Lawrence Grossman**

A model of inhomogeneous accretion of the earth is considered in some detail. The process is rapid, most of the earth forming in no more than 10^5 years. Accretion of the earth accompanied condensation of the planetary nebula, and the earliest major condensate–nickel-iron–began accumulating before the condensation of the major constituents of the mantle. During the main stage of accretion an appreciable part of the energy of accretion was retained because of the short time scale, and temperatures within the earth were high. The outer part of the earth was added more slowly at low temperatures. Conditions at this time were favorable for the retention of volatile elements which comprised part of the late-stage veneer.

INTRODUCTION

Speculating about the origin of the earth is one of mankind's oldest intellectual pastimes. Credit for initiating its modern European version is usually given to Descartes (1644). Since then theories of the origin of the solar system have fallen into two categories called *monistic* if they envisage its evolution from a single body, be it a star, a protostar or a stellar nebula, and *dualistic* if the interactions of two or more stellar bodies are assumed to be involved in the process of formation.

Theories of the origin of the solar system have been extensively reviewed, and we see no merit in adding a lengthy contribution at this time. The reader is referred to Spencer Jones (1956), and ter Haar and Cameron (1963) for a more detailed treatment. A popular account of the subject has been given by Berlage (1968); it is biased in favor of the author's ideas, which are subject to numerous objections. Here we simply note the principal contributions to the subject.

The first scientific theory of the solar system is found in the work of Descartes (1644), and examples of early monistic theories are the nebular hypotheses of Kant (1755) and Laplace (1796). These theories were generally discarded because in the actual solar system the sun contains practically all the mass and practically none of the angular momentum, and the theories seemed incapable of explaining these observations. As is so often true in this field, the gain of a little knowledge led to a simple way of removing these objections, and modern monistic theories bear a certain resemblance to these early ideas.

Dualistic theories were initiated by Buffon in 1745. He sought to explain the planets as a result of a comet hitting the sun, but once the properties of comets were better determined, this idea lost all force. Dualistic theories involving close approach of two stars, collision of two stars, a supernova explosion in one member of a binary stellar system or collisions or close encounters in binary or ternary stellar systems were proposed by Chamberlain (1901), Moulton (1905), Jeans (1917, 1919, 1928), Jeffreys (1916, 1918, 1929), Hoyle (1945, 1946), and Lyttleton (1938, 1941). Their none-too-satisfactory agreement with observation and their strong dependence on most improbable circumstances led Urey to complain that "The subject seems always beset with miracles" (Urey, 1952, p. 5).

Urey sought to remove the miraculous foundation of theories of the origin of the solar system by approaching the problem in a new way, already attempted by Whipple (1948). Instead of trying to arrive at a quantitative theory, Urey sought to establish as many initial and boundary conditions as could be provided by geochemical and cosmochemical measurements, and then to deduce the qualitative features of a model of origin that was forced to satisfy these constraints. A number of workers, notably Ringwood and Anders and his coworkers have followed Urey's path and have added immensely to the number of constraints that a successful theory must satisfy. Quantitative models must satisfy these constraints and any additional ones that arise from our growing knowledge of the chemistry of planetary bodies and of stellar evolution.

It might be supposed that we are hardly nearer to a proper, quantitative theory of the origin of the solar system than was Descartes. This view is overly pessimistic, however; a number of approximate, incomplete theories have recently been advanced, and these show sufficient promise to suggest that major improvement in our theoretical understanding is imminent. Progress has been slow, and this fact is in part a result of the basically difficult nature of the problem. It is also a result of the fact that although the subject is almost certainly not beset with Urey's miracles, it is nonetheless beset with a number of problems that behave as evanescent phantoms. We give two examples. The mass and moment of inertia of the solar system are quantities that we can observe with high precision. But what fraction of the mass of the body that evolved into the solar system is represented by the mass that we detect today? There are reasons to suppose that this fraction is small, perhaps less than 10%. Similarly, how much angular momentum has the sun lost as a result of hydromagnetic interaction with gas now spun off into space? This too may have been a large fraction of the original angular momentum of the sun and of the solar system; it is this fact that makes the rejection of early monistic theories of the solar system seem premature. These early theories are probably not acceptable today, but for entirely different reasons than are historically given for their rejection.

Before walking into a field in which we might conclude that angels fear to tread, a general disclaimer concerning the accuracy of our results is in order. In this book, dedicated to Francis Birch, the geophysical point of view predominates. In geophysics we are used to situations where we have a fair to good idea of the order of magnitude of the quantities involved in a particular problem. This is not always true, but the statement holds much of the time. In astrophysics, however, a great deal more uncertainty may prevail, and we cannot always be confident that we know the order of magnitude of the logarithm of the quantity in question. Part of this uncertainty is a result of the sort of evanescent phantoms that we have already encountered; the remainder can be laid at the door of imperfect and incomplete physical measurements and imperfect assessment of their applicability to a given astrophysical situation. Any discussion of the origin and early history of the earth is clearly subject to much larger uncertainties than we are used to in geophysics.

We shall adopt a monistic view of the origin of the solar system in keeping with most modern thought. We consider the solar system to originate essentially in a single event, the sun forming with the planets. Such a view is in complete contrast to the dualistic theories proposed to date, which suppose that the planets are formed from pre-existing suns. It is consistent with many monistic theories, and seems to subject cosmochemical constraints to minimal strain.

CONFIGURATION OF THE SOLAR NEBULA AND THE TIME SCALE OF ITS EVOLUTION

The collapse of a protostellar gas cloud to a star has been examined in detail only in the case where the cloud is not rotating. With this simplification Larson (1969) was able to supplement the earlier results of Ezer and Cameron (1963, 1965) and give a reasonably satisfactory quantitative theory of stellar synthesis. For our purposes, however, these results must be extended, since they can lead to nothing more than a single central body. The needed extensions require consideration of an initially rotating cloud, which complicates the problem enormously from the mathematical point of view. We suppose, however, that it would be possible to have the initial cloud split in two, resulting in a central core that contracts into a star in much the same way as a nonrotating cloud, leaving behind a hollow nebular disk. The ratio of mass of the core to that of the disk is highly variable, depending on the angular velocity of the original protostellar cloud.

This result, tentative though it is, gives strong encouragement to the idea that the formation of a sun and a planetary system is dynamically reasonable. Further evolution depends on the behavior of the nebular disk since the central core seems inevitably destined to become a single star. The disk has more possibilities; it might condense to a single body that would become the binary companion of the star formed from the central core. This appears to be the most likely outcome, since about half of the visible stars are binary. There is also the possibility, however, that the nebular disk might be unstable with respect to the formation of rings. It is risky to predict this sort of instability because one must be careful to distinguish true physical instabilities from instabilities that are wholly a result of the mathematical methods used to solve the relevant differential equations. Keeping this warning in mind, we are nonetheless encouraged to suspect that there are suitable circumstances under which the fragmentation of the nebular disk into rings may occur. The best evidence that this is true may be the very existence of the solar system. The formation of a planetary system may be much rarer than the formation of a binary companion, and, if true, this circumstance will vastly complicate the quantitative investigation of the matter.

Larson (1969) derived the very important result that during the collapse of a gas cloud the time that a particle takes to fall close to the center of the cloud is closely approximated by its gravitational free-fall time. The reason is that during gravitational collapse the bulk of the gas moves toward the center of the cloud with nearly constant velocity. Pressure and density build up only in a small central core. Larson (1969) found asymptotic relations indicating that the condensation time of a star in its main collapse stage should not exceed the free-fall time by more than a factor of 2. This result was anticipated by Cameron (1962).

The basic arguments hold equally well for the collapse of a cylinder of gas towards its axis. This fact enables us to use the free-fall time as an estimate of the time of condensation in the cylindrical configuration, with fair confidence that the results thus obtained will be in error by no more than a factor of 2.

Results by Larson (1969) and by Ezer and Cameron (1963) show that star formation is well under way by 10^7 years after collapse is initiated in the protostellar gas cloud. The main collapse phase is completed in roughly 10^6 years, but subsequent evolution onto the main sequence is slowed to a time scale of 10^7 to 10^8 years, partly because heightened opacity causes the star's temperature to rise, and the resulting increased internal pressure inhibits collapse. Additionally, the higher internal temperature lights the star's nuclear fuel source, which ultimately results in the burning of hydrogen and a star on the main sequence. These effects (resulting from high internal temperatures) which we expect to cause the greatest departures from a free-fall time scale, may be important in the evolution of the sun, but they seem likely to be far less important in the history of the planets. Planetary bodies have never attained the temperatures necessary to initiate nuclear energy sources, and the energy derived from their Helmholtz contraction phases is far less than is the case of the sun. We may conclude that the time of accretion estimated by the free-fall time is more likely to be valid for a planet than for a star and that it yields a reasonably close estimate of the actual time in question.

The time scale of 10^6 to 10^7 years is based on what is known of the dynamics of a nonrotating stellar nebula. We extrapolate this time scale into the rotating case by assuming that the effects of rotation are purely Keplerian. That is, we assume that the orbital velocity of a particle in the planetary disk is sufficient to counterbalance centripetal gravitational forces. Nevertheless, a ring formed out of an initial nebular disk is unstable with respect to its self-gravitation, and it may be expected to contract into a ring of negligible cross section, or more likely but less predictably, into a number of bodies moving in orbits that roughly coincide with the axis of the ring. Considering the first possibility first, we can calculate a free-fall time assuming that the orbital radius of the ring greatly exceeds its cross section. This assumption enables us to neglect the curvature of the ring in the orbital sense and to treat the problem of gravitational collapse as a two-dimensional collapse onto a cylinder. The result is that if the ring had the mass of the earth, augmented by a factor of 300 to account for the loss of light elements (mainly H and He), the free-fall time of collapse would be about 3 years. For an unaugmented earth the free-fall time of a ring is about 50 years.

The time required to accumulate objects in the orbital sense is probably not well given by the preceding argument. The accumulation of objects already accompanying each other on common Keplerian tracks is going to be far slower than is the time of condensation onto a common line. As a consequence, we would expect two time scales for the solar system, one appropriate to the accumulation of protoplanets from the condensing rings, and the other for the coalescing of protoplanets to form the planets that we observe today.

The time of accumulation of the earth can also be inferred from geophysical arguments as was done by Hanks and Anderson (1969). They made the assumption that the formation of the earth's core must predate the oldest known rocks, and must therefore have occurred within about one billion years of the origin of the earth itself. Such an assumption is plausible because if the core formed subsequently to the earth, energy sufficient to cause widespread melting in the upper mantle would be released in the process. No crustal rock could be expected to survive such a catastrophe. Hanks and Anderson found that the core could form within the allotted time only if the earth retained an appreciable part of the energy of accretion. This requirement sets a limit to the time of accretion, which was found to be about 2×10^5 years. Hanks and Anderson considered this figure to be an upper limit, and they did not exclude the possibility of a shorter time scale with core formation accompanying, rather than following, accretion.

The geophysical arguments in favor of rapid accretion of the earth are in harmony with modern astrophysical thinking. The reasons for adopting a short time scale of accretion are perhaps not altogether compelling, but the consequences show promise of accounting for observations that were not well explained by previous theories. We find no valid arguments to contradict the idea of rapid accretion.

TEMPERATURES IN THE PLANETARY NEBULA

A considerable body of thought concerning the origin of the earth starts from the presumption that the planetary nebula passed through a high-temperature stage before coalescence of the planets took place. This idea is found in the work of Cameron (1962), Larimer (1967), Larimer and Anders (1967), Wood (1967), and Lord (1965), among others. By high, we mean temperatures in the neighborhood of 2000°K, sufficient to ensure that all the material is in the gaseous form. In spite of the prevalence of assumptions of high pre-accumulation temperatures, they are not the automatic consequence of any and all models of stellar evolution. It would seem easy to create ones that did not possess this feature. It is physically reasonable, however, to suppose that high temperatures were produced early in the evolution of the planetary nebula (Cameron, 1962).

Starting with a hot planetary nebula losing energy by radiation, we first estimate the time required for cooling. Predictably, this is highly uncertain. It seems to be well established that radiative transfer is the dominant cooling mechanism, and that its efficiency is much more restricted by the opacity contributed by solid particles than it is by molecular absorption processes at higher temperatures where only gasses are present. The cooling of the nebula is controlled by its shape and by the inhibition of radiative transfer by particulate matter. We assume that the nebular disk is considerably flattened compared to the mean orbital radius of Pluto. This latter figure is 6×10^{14} cm, and we might plausibly consider thicknesses of the disk of 6×10^{12} to 6×10^{13} cm. Evidently, it is the thickness in the direction of the axis of the disk that limits the cooling rate.

Once condensation begins and particulate matter appears, the opacity of the planetary nebula rises markedly. Unfortunately, its magnitude is extremely sensitive to the size of the condensed particles. At a temperature of 2000°K, the Wien's law maximum on the blackbody curve is at about 1.5μ, still in the infrared. Much of the energy is at wavelengths of 10μ or more. At these long wavelengths it is likely that the physical dimension of the scattering particle is considerably less than the wavelength, particularly in the case of interstellar dust. Gaustad (1963) has investigated this problem and has estimated the opacity produced by, among other materials, "silicate" and iron grains. The result is proportional to the cube of the mean grain size, and since the number density of grains is inversely proportional to this quantity (for a fixed total density), we arrive at an opacity that is formally independent of grain size and that is numerically about 10^{-7} cm^{-1}. This estimate is only valid if the particle size is greatly exceeded by the mean wavelength of the radiation.

If we consider particles whose size may equal or exceed the predominant wavelength of the radiation, we must abandon Gaustad's treatment and consider a more complete one, due originally to G. Mie and described in the books by Goody (1964) and Middleton (1952). For our purpose we may consider an asymptotic form of the theory in which it is supposed that the real part of the refractive index of the grains has gone to infinity and thereby produced black particles. Under these circumstances, the opacity is proportional to the number of the grains and to their cross-sectional areas. For a fixed total density, the number density of grains is again inversely proportional to the cube of the grain size, leading to an opacity inversely proportional to the first power of the grain size. For an assumed composition of the solar nebula, discussed below, the opacity is of the order of $6 \times 10^{-11}/a$ cm^{-1}, where a is the mean grain radius. This differs by about an order of magnitude from the estimate in the last paragraph of grains 1μ (10^{-4}cm) in radius. Thus this is somewhat larger than the particle size at which we may expect the small-particle theory to fail.

We have made no attempt to solve the radiative transfer problem in detail, because of the large uncertainties contained within it. It seems clear, however, that the planetary nebula is optically thick as long as the mean radius of the particulate matter does not exceed a few centimeters. We calculate the order of magnitude of the cooling time of the nebula by setting it equal to $x^2/4\alpha$, where x is half the thickness of the nebula and α is the thermal diffusivity, which is calculated from the radiative thermal conductivity, heat capacity, and density. Taking an opacity of $10^{-7}cm^{-1}$ as given by small-grain theory, we derive a cooling time of the order of 10^4 years.

Actually, a case can be made for believing that the time scale of cooling changed progressively from fast to slow. This is partly a result of the decline in temperature, which makes radiative transfer progressively less efficient. Also we can expect that the opacity will rise enormously at least during the early stages of the cooling process. As a consequence, rapid cooling will occur until condensation of a large number of small grains increases the opacity and decreases the cooling rate by two orders of magnitude. With the maximum opacity estimated above, the time scale of cooling exceeds 10^5 years once the temperature has fallen to 600°K. This estimate is unreasonably large for two reasons. First, collapse of the nebula has set in, reducing its thickness and, consequently, its cooling time. Second, the number density of dust grains will decline as accumulation into protoplanets takes place, thereby reducing the opacity. By the time the nebula begins to clear up appreciably, however, the accumulation of larger bodies will be well under way, and we can shift our attention from conditions in the nebula to conditions in the planets. Their internal environment is set by the accumulation process, not by conditions in the planetary nebula.

In summary, it seems entirely possible that the time scale of cooling and condensation of the planetary nebula is comparable with the gravitational free-fall time: that is, less than 100 years. We expect that 10^4 years represents an upper limit to the cooling time, but calculation is highly uncertain because it is sensitive to the grain size of the condensing material. We suppose that the beginning of planetary accumulation overlaps the condensation process in time. Before considering accumulation, however, we must examine the chemical details of the condensation.

CONDENSATION OF THE PLANETARY NEBULA

Calculation of the order and temperatures of condensation of various constituents in the planetary nebula has been done at the higher temperatures in essentially the same way as Larimer (1967) with one important exception. As each crystalline phase becomes stable, it is allowed to equilibrate with the vapor, the composition of which, in many cases, undergoes considerable alteration. This in turn changes the projected condensation temperatures of lower temperature condensates. Larimer's work represents an extension and improvement of the earlier work of Lord (1965), which in turn supplanted previous calculations by Latimer (1950), Urey (1952), and Eucken (1944). Although the method is essentially the same in all these calculations, the comprehensiveness of the thermodynamic data, the degree of refinement of the solar system abundance table, and the completeness of the calculations have increased with time.

We have calculated the partial pressures of 364 gaseous species in a gas made up of 22 elements. The total pressure was assumed to be 10^{-3} atm, and since this pressure is mainly produced by hydrogen, which does not condense in significant amounts at the temperatures under consideration, it can be taken to be constant. The composition of the gas was taken from Cameron's (1967) abundance table.

In principle, the calculation is straightforward. Given the necessary thermodynamic data and assuming that the Ideal Gas Law and Dalton's law hold, the partial pressure of any molecular species assumed to exist in the planetary nebula can be calculated. Difficulty may be caused by lack of thermodynamic data for important compounds and by failure to recognize others and include them in the list of compounds to be considered. The possibility that we may have overlooked important molecular species for which data are available is not attributable wholly to lack of scholarship. The number of combinations of chemical elements is very large, probably in excess of 10^5, and the rules determining which chemical formulas are possible, and also interesting, are so complicated that no automatic procedure to search for relevant compounds has been developed. We must still rely on personal judgment and mechanically unaided search of the standard compilations of thermodynamic data to provide us with a complete list of important compounds.

As an example, we have found that gehlenite ($Ca_2Al_2SiO_7$), merwinite ($Ca_3MgSi_2O_8$), and akermanite ($Ca_2MgSi_2O_7$) are among the highest temperature condensates. None of these even appeared in earlier lists of the order of condensation, simply because they are exotic compounds that earlier authors did not consider.

In addition to the Ca, Mg, Al silicates, several of the refractory metals may be expected to condense early in the cooling of the nebula. These include Mo, Nb, Ru, Ir, Re, Ta, and W. It is found by a direct calculation that ignores the formation of compounds and alloys that these metals are supersaturated in the vapor phase by factors ranging from 10^2 to 10^9 at the temperature at which condensation of iron commences. ZrO_2 is supersaturated by a factor of 10^6 and gehlenite by 10^4. Even allowing for considerable supercooling, it seems likely that these early condensing elements and compounds will provide seeds upon which later, more abundant phases may grow. We may expect many nuclei per cubic centimeter if the degree of supersaturation is high enough to initiate homogeneous nucleation among the early condensates, which we believe to be the case.

The order in which the more abundant compounds condense is as follows: corundum at 1760°K, perovskite at 1645°K, and gehlenite at 1625°K. At 1515°K, corundum is destroyed by reaction with the gas to form spinel, and merwinite forms from gehlenite at 1475°K. Metallic iron first appears at 1473°K and contains 12.5 mole % Ni. As the gas cools and more alloy condenses, the equilibrium Ni content decreases, reaching 5 mole %, the solar Ni/Fe ratio, by 1350°K. Meanwhile, merwinite has reacted to form akermanite at 1455°K. Forsterite first appears at 1444°K, at which temperature 44% of the total iron is already condensed. Diopside then forms from akermanite at 1433°K, and at 1360°K, anorthite forms by the reaction of spinel with diopside. At 1350°, enstatite appears by reaction of forsterite with the gas, and the forsterite to enstatite ratio falls smoothly with decreasing temperature. The high temperature olivines and pyroxenes are practically Fe-free, but they gradually increase their Fe content with decreasing temperature so that there may be up to 3.0 mole % fayalite in the olivine by 700°K. The first-formed plagioclase is nearly Na-free but will contain 2 mole % albite by 1200°K. Pure alkali feldspars are probably stable at 1000°K. The condensation calculations have not been pursued in detail to lower temperatures, but it seems clear that by the time the temperature has fallen to around 1000°K, condensation of all major condensable elements except sulfur will have taken place.

THE ACCRETION OF THE EARTH AND MOON

The planets start to accumulate during condensation of the planetary nebula. At any given time, the rate at which gravitational energy is released is proportional both to the mass that has already accumulated and to rate at which mass is being added. Thus, we expect the rate of energy release to be small at first because little mass has had time to accumulate into a planet, to pass through a maximum at an intermediate planetary size, and to decrease again to low values as the rate of addition of matter declines in response to the exhaustion of accretable material in the planetary nebula. These features of the accumulation process were previously noted by Hanks and Anderson (1969).

We shall confine our attention to the earth and moon for the remainder of this paper, since these are the only planetary bodies about which we have enough chemical information to justify further speculation. The gravitational potential energy (GPE) of the earth is about 2.5×10^{39} ergs (Birch, 1965); retention of all of this energy would raise the temperature by about 3×10^{4}°K on the average. For the moon the GPE is about 1.3×10^{36} ergs, sufficient to raise the mean temperature by about 1400°K if all of it is retained. In fact, neither the earth nor the moon can be expected either to acquire or to retain all its GPE upon accumulation. Material falling on the earth or moon has been accelerated relative to the gas of the planetary nebula, but the buildup of relative velocity is opposed by viscous drag in the nebula. Thus, part of the GPE goes towards retarding the cooling of the nebula rather than heating the earth or moon. The fraction of the GPE that is dissipated in this fashion is difficult to estimate, since it depends critically on the sizes of the infalling particles. Small ones are slowed more than large ones. We expect the infalling material to have a broad spectrum of sizes, ranging from dust to objects large enough to produce the lunar maria by impact. A sizable but unknown fraction of the GPE will be lost to the nebula by the smaller particles. The fraction that is retained by the earth and moon depends on this factor and on the rate of accumulation as well.

As accretion of the earth begins, the temperature of the growing body will not depart sensibly from the temperature of the planetary nebula at the time, since the amount of GPE released is small because of the body's small mass. This temperature is above 1200°K, but the first accumulate seems certain to be solid. As mass is added, however, the rate of release of GPE ceases to be negligible; the earth heats, and eventually a molten zone is formed around the initial solid core. At this stage, the infalling material is almost entirely nickel-iron; with a density in excess of 7 gm/cm^3. The resulting earth is fairly massive, with a fairly high GPE as a consequence. We estimate that the molten zone could form while the radius was decidedly smaller than the present radius of the earth's inner core. Since the heat of impact is released near the surface of the growing planet, the liquid is stably stratified. We note that in this model, core formation is not a distinct event occurring subsequently to the accretionary process as envisioned by Urey (1952), Birch (1965), and Hanks and Anderson (1969). The core is produced by inhomogeneous accretion, and the processes of accumulation and core formation cannot physically, chronologically, or logically be separated (Turekian and Clark, 1969).

As the planetary nebula cools, condensation of olivine and pyroxenes takes place, and these abundant phases are added to nickel-iron alloy as the principal accumulates. No clean separation between the early stage (where metal is the primary accumulate) and the later stage (where silicates are the dominant infalling material) can be recognized. The fact that we are dealing after impact with metallic and silicate liquids which are immiscible ensures a sharp boundary between the core and the overlying mantle. During the main stage of mantle accumulation, temperatures near the surface of the earth may rise to the point where significant vaporization occurs. The vapors may be entrained by gases in the planetary nebula and swept away from the earth. This process removes energy from the earth and tends to cool it. A balance is reached between rate of infall and rate of evaporation, thus setting an upper limit to the temperature at the earth's surface. Near-surface temperatures of the accreting earth are shown schematically in Fig. 1.

During this high temperature stage, any of the volatile, late condensing substances that might have been incorporated in impacting material could not be retained within the body of the earth. They would be expelled into a primitive atmosphere, from which they would either be swept away by the nebular gases, or where they would remain suspended until accretion slowed and the surface layers of the earth cooled.

The rate of accretion may be expected to decline as the earth grows to a substantial fraction of its present mass. The bodies in its original ring that are in orbits favorable for capture will be swept up quickly, but the remaining bodies in less favorable orbits will be accreted at a significantly slower rate. The accompanying decrease in the rate of input of accretional energy will take place gradually, but it will eventually result in a reduction of the surface temperature, which in turn will produce a convective stage. The core and the part of the mantle that have accreted at this stage will each be internally homogenized by this process. With slower accretion, a veneer of material enriched in compounds that condensed at a relatively low temperature will grow on the earth. The veneer, which might comprise up to 20% of the mass of the earth, grows at such a rate that the earth is finally accreted, for all intents and purposes, in an additional period of 10^5 to 10^7 years. Actually, of course, the accretional process has never stopped, but our model supposes that we can neglect the

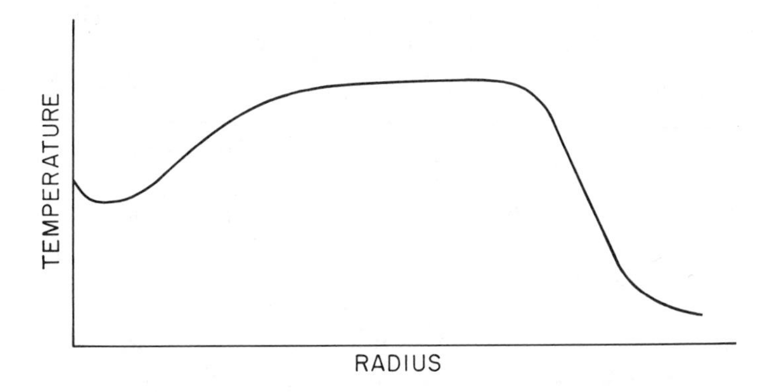

Fig. 1. Schematic temperatures at the surface of the accreting earth.

addition of mass after 10^7 years. The material of the veneer is a mixture of early and late condensates that have been exposed to the nebular gas at low temperatures. Any metallic iron-nickel will tend to be oxidized when the temperature falls below 400°K (Urey, 1952), and oxidation will proceed to the extent permitted by the state of aggregation of the metal and the kinetics of the process at these low temperatures.

An important feature of this model of the accumulation of the earth is that the late veneer that today comprises the crust, the upper mantle, and perhaps part of the transition zone has never been in contact with the core. It is a direct consequence of the short time scale of the main phase of accretion, which enables the core to settle to the center and be surrounded by the lower mantle before the outer layers are added. This feature of the model enables us to meet two objections noted by Ringwood (1959, 1966), that have raised difficulties with earlier theories of the origin of the earth. He pointed out that the Ni content of basalts and deep-seated ultramafic rocks was much higher than would be expected if these rocks had ever reached chemical equilibrium with an iron-nickel alloy. He also noted that the oxidation state of basalts, as measured by the Fe_2O_3/FeO ratio, is typically much higher than would be compatible with metallic Fe. He sought to avoid these difficulties by postulating chemical disequilibrium during the segregation of the core, which he presumed to take place later than the accumulation of the earth.

Although our model explicitly permits a veneer that is more oxidized than the deep interior, the high oxidation state of the upper mantle could be explained in an alternative way if two conditions are met early in the earth's history. First, water must be available at the surface to undergo photodissociation coupled with easy loss of hydrogen. Second, the resulting oxidized crustal material would have to be efficiently cycled into the upper mantle by mechanisms similar to that involved in present-day sea floor spreading. The dissociation and hydrogen loss from 10% of the present volume of the oceans would raise the ferric/ferrous ratio in the top 50 km of the mantle from essentially zero to the observed value.

The problem posed by the nickel content of the part of the earth that is sampled by basalts is much harder to escape. It is well established that the nickel content of basalts is about 200 ppm and that the nickel content of ultramafic material is in the range 2000 to 3000 ppm (Kuno and Aoki, 1970; Ross, Foster, and Myers, 1954; Turekian, 1963). In contrast, the nickel content of the olivines in pallasites is 50 to 100 ppm (Buseck and Goldstein, 1969), and this value is in keeping with what we know of the siderophile nature of nickel.

Some control on the rate of diffusion of nickel in silicates is provided by the results reported by Wood (1967). He found that the taenite grains in chondrites were characteristically enriched in nickel near their margins in a way that is closely similar to the nickel pattern in taenite grains which are in contact with kamacite in iron meteorites. Wood concluded from this result that nickel diffused from kamacite to taenite, in the appropriate temperature range, with about the same rate regardless of whether or not a silicate grain intervened between the two grains of alloy. This conclusion implies rapid diffusion of nickel through the silicates. It may have taken place on grain boundaries or through some liquid or gaseous interstitial phase, but it nevertheless must have been an efficient path of diffusion. A. M. Clark (1971) has made direct measurements of coefficients of diffusion of nickel in natural crystals of olivine. His results at temperatures of 1525°K and below show marked anisotropy of diffusion in different crystallographic directions, but when his measurements are extrapolated to the melting point of iron (1805°K), coefficients in different directions converge, leading to a nearly isotropic situation. The lowest diffusion coefficient at 1805°K is 2×10^{-11} cm^2/sec. Using this value and assuming isotropic diffusion in spherical grains, we find that the nickel content of grains 2 mm in diameter is reduced to 10% of its initial value in about 3 years and virtually to zero in about 10 years. Use of an average diffusion coefficient rather than a lowest one will reduce the calculated times.

The only previous earth model that explicitly faces the problem of the nickel content of the upper mantle is that of Ringwood (1960, 1970). His is an "inside out" model that assumes that material of the crust and upper mantle was sequestered in a cool primitive core during most of the period of accumulation of the earth. Formation of the present metallic core displaces this material, which then rises to the surface so rapidly that equilibration with the remainder of the earth cannot take place. Thus there is no equilibrium partitioning of nickel between metal and silicates. It is not clear to us, however, that Ringwood's model is physically possible. Our model escapes the nickel problem by providing a shield a good 2000 km thick between the core and the outer parts of the earth.

The history of accumulation of the moon must have been very different from the earth. Far less GPE is available to the moon, and in addition its process of accretion must be such as to account for two basic chemical contrasts with the earth. The moon's low mean density, 3.36 gm/cm^3 (Kaula, 1969), shows that the moon's iron core, if present at all, must make up a far smaller proportion of the whole body than does the earth's core. In addition, volatile elements at the moon's surface (Tranquility Base) are depleted by factors of 10 to 100 relative to the earth (Keays et al., 1970).

The GPE released by accumulation of the moon is not sufficient to melt it totally unless accretion took place while the nebula was still rather hot. Even in that case, allowances for radiative heat loss makes it improbable that complete melting ever occurred. Accretion on a time scale of 10^3 to 10^4 years leads to surface temperatures of 500 to 1000°K, high enough to prevent condensation of volatile elements and to expose them to entrainment by nebular gasses.

Two alternative sets of hypotheses have been advanced to account for the moon as a satellite of the earth. One set presumes that the moon was captured (Singer, 1968); the other supposes that the moon formed in orbit about the earth (Cameron, 1970; Ganapathy et al., 1970; Ringwood, 1970). Our model can accommodate either case provided that the moon appeared in orbit around the earth after the earth had swept up most of the nickel-iron in its orbit, but before the major accumulation of the low temperature, volatile fraction by the earth-moon system. The first constraint is occasioned by the absence of a massive metallic core in the moon and the last constraint is based on the depletion of volatile elements at its surface.

No really satisfactory explanation of the relatively low concentration of volatile elements at the lunar surface has thus far been offered. Ganapathy et al. (1970) suggested that during the time that the earth acquired its late veneer, it was competing with the moon for material. The capture cross section of the earth was calculated to be larger than that of the moon by a factor of 100 or more, with the result that the earth acquired a much thicker veneer than the moon. The model of Ganapathy et al. (1970) was reiterated by Anders (1970) and was further discussed by Singer and Bandermann (1970). The latter authors concluded that conditions were most favorable for preferential accumulation of the veneer on the earth rather than the moon if the two bodies were in separate heliocentric orbits at the time. The moon would subsequently have to be captured by the earth, an improbable event that reduces the attractiveness of the hypothesis of Ganapathy et al. (1970).

We cannot exclude the possibility that the moon formed from material volatilized from the earth during its high-temperature stage of rapid accretion, as suggested by Cameron (1970) and Ringwood (1970). If the low concentration of volatiles at the lunar surface can be shown to be a consequence of this mode of origin, acceptance of the hypothesis would be hard to deny.

ACKNOWLEDGMENT

A preliminary version of this paper was critically reviewed by E. Anders, A. G. W. Cameron, J. W. Larimer, A. E. Ringwood, and S. F. Singer. We are deeply indebted to them for their comments and suggestions. As is apparently inevitable in a field as speculative as this, many points of disagreement between us and our reviewers remain. Imperfections and impossibilities in our model that may be revealed in the future are solely our responsibility. They are not to be laid at the doors of our distinguished colleagues.

Financial support from the National Science Foundation and the National Aeronautics and Space Administration is also acknowledged.

REFERENCES

Anders, E., Water on the Moon? Science, **169**, 1309-1310, 1970.

Berlage, H. P., Jr., The Origin of the Solar System. Pergamon International Popular Science Series, London (Oxford), 1968.

Birch, F., Energetics of core formation, J. Geophys. Res., **70**, 6217-6221, 1965.

Buffon, G. L. L., De la formation des planètes. Paris, 1745.

Buseck, P. R., and Goldstein, J. I., Olivine compositions and cooling rates of pallasitic meteorites, Geol. Soc. Am. Bull., **80**, 2141-2158, 1969.

Cameron, A. G. W., The formation of the Sun and planets, Icarus, **1**, 13-69, 1962.

______, A new table of abundances of the elements in the solar system, *in* Origin and Distribution of the Elements, edited by L. H. Ahrens, Pergamon Press, London, 1178 pp., 125-143, 1967.

______, Formation of the earth-moon system (abstract), Trans. Am. Geophys. Union, **51**, 350, 1970.

Chamberlain, T. C., On a possible function of disruptive approach in the formation of meteorites, comets, and nebulae. Astrophys. J., **14**, 17-40, 1901.

Clark, A. M., Diffusion processes, *in* Proceedings of the Thomas Graham Memorial Symposium, edited by J. N. Sherwood, Gordon & Breach, London, in preparation, 1971.

Descartes, R., Principia Philosophiae, Amsterdam, 1644.

Eucken, A., Physikalisch-chemische Bertrachtungen über die Früheste Entwicklungsgeschichte der Erde, Nachr. Akad. Wiss. Göttingen, Math. Phys. Kl., **1**, 1-25, 1944.

Ezer, D., and Cameron, A. G. W., The early evolution of the Sun, Icarus, **1**, 422-441, 1963.

______, A study of solar evolution, Can. J. Phys., **43**, 1497-1517, 1965.

Ganapathy, R., Keays, R. R., Laul, J. C., and Anders, E., Trace elements in Apollo 11 lunar rocks: Implications for meteorite influx and origin of moon, *in* Proceedings of the Apollo 11 Lunar Science Conference, Geochim. Cosmochim. Acta Supplement 1 (A. A. Levinson, editor), Pergamon Press, New York, 1117-1142, 1970.

Gaustad, J. E., The opacity of diffuse cosmic matter and the early stages of star formation, Astrophys. J., **138**, 1050-1073, 1963.

Goody, R. M., Atmospheric radiation, 435 pp., Oxford Univ. Press, London, 1964.

Hanks, T. C., and Anderson, D. L., The early thermal history of the Earth, Phys. Earth Planet. Interiors, **2**, 19-29, 1969.

Hoyle, F., Note on the origin of the Solar System, Monthly Notices Roy. Astron. Soc., **105**, 175-178, 1945.

______, On the condensation of the planets, Monthly Notices Roy. Astron. Soc., **106**, 406-422, 1946.

Jeans, J. H., The motion of tidally-distorted masses, with special reference to theories of cosmogony, Memo. Roy. Astron. Soc., **62**, 1-48, 1917.

______, Problems of Cosmogony and Stellar Dynamics, 293 pp., Cambridge Univ. Press, London and New York, 1919.

______, Astronomy and Cosmogony, 420 pp., Cambridge Univ. Press, London and New York.

Jeffreys, H., On certain possible distributions of meteoric bodies in the Solar System, Monthly Notices Roy. Astron. Soc., **77**, 84-111, 1916.

______, On the early history of the Solar System, Monthly Notices Roy. Astron. Soc., **78**, 424-441, 1918.

______, Collision and origin of rotation in the Solar System, Monthly Notices Roy. Astron. Soc., **89**, 636-641, 1929.

Kant, I., Allgemeine Naturgeschichte und Theorie des Himmels. Königsberg und Leipzig, 1755.

Kaula, W. M., The gravitational field of the Moon, Science, **166**, 1581-1588, 1969.

Keays, R. R., Ganapathy, R., Laul, J. C., Anders, E., Herzog, G. F., and Jeffery, P. M., Trace elements and radioactivity in lunar rocks: Implications for meteorite infall, solar-wind flux, and formation conditions of the moon, Science, **167**, 490-493, 1970.

Kuno, H. and Aoki, K. I., Chemistry of ultramafic modules and their bearing on the origin of basaltic magmas, Phys. Earth Planet. Interiors, **3**, 273-301, 1970.

de Laplace, P. S., Exposition du système du monde, Paris, 1796.

Larimer, J. W., Chemical fractionations in meteorites–1. Condensation of the elements, Geochim. Cosmochim. Acta, **31**, 1215-1238, 1967.

Larimer, J. W., and Anders, E., Chemical fractionations in meteorites–II. Abundance patterns and their interpretation, Geochim. Cosmochim. Acta, **31**, 1239-1270, 1967.

Larson, R. B., Numerical calculations of the dynamics of a collapsing proto-star, Monthly Notices Roy. Astron. Soc., **145**, 271-295, 1969.

Latimer, W. M., Astrochemical problems in the formation of the Earth, Science, **112**, 101-104, 1950.

Lord, H. C., III, Molecular equilibria and condensation in a solar nebula and cool stellar atmospheres, Icarus, **4**, 279-288, 1965.

Lyttleton, R. A., On the origin of binary stars, Monthly Notices Roy. Astron. Soc., **98**, 646-650, 1938.

______, Note on the origin of planets and satellites, Monthly Notices Roy. Astron. Soc., **101**, 349-351, 1941.

Middleton, W. E. K., Vision Through the Atmosphere, 250 pp., University of Toronto Press, Toronto, 1952.

Moulton, F. R., On the evolution of the Solar System, Astrophys. J., **22**, 165-181, 1905.

Ringwood, A. E., On the chemical evolution and densities of the planets, Geochim. Cosmochim. Acta, **15**, 257-283, 1959.

______, Some aspects of the thermal evolution of the Earth, Geochim. Cosmochim. Acta, **20**, 241-259, 1960.

______, The chemical composition and origin of the Earth, *in* Advances in Earth Science, edited by P. M. Hurley, pp. 287-356, MIT Press, Boston, 1966.

______, Origin of the Moon: The precipitation hypothesis, Earth Planet. Sci. Letters, **8**, 131-140, 1970.

Ross, C. S., Foster, M. D., and Myers, A. T., Origin of dunites and olivine-rich inclusions in basaltic rocks, Am. Mineralogist, **39**, 693-737, 1954.

Singer, S. F., The origin of the moon and geophysical consequences, Geophys. J., **15**, 205-226, 1968.

Singer, S. F., and Bandermann, L. W., Where was the moon formed? Science, **170**, 438-439, 1970.

Spencer Jones, Sir Harold, The origin of the Solar System, Phys. Chem. Earth, **1**, 1-16, 1956.

ter Haar, D., and Cameron, A. G. W., Historical review of theories of the origin of the Solar System, *in* Origin of the Solar System, edited by R. Jastrow and A. G. W. Cameron, pp. 1-37, Academic Press, New York and London, 1963.

Turekian, K. K., The chromium and nickel distribution in basaltic rocks and eclogites, Geochim. Cosmochim. Acta, **27**, 835-846, 1963.

Turekian, K. K., and Clark, S. P., Jr., Inhomogeneous accumulation of the earth from the primitive solar nebula, Earth Planet. Sci. Letters, **6**, 346-348, 1969.

Urey, H. C., The Planets: Their Origin and Development, Yale Univ. Press, New Haven, Conn., 1952.

______, Evidence regarding the origin of the earth, Geochim. Cosmochim. Acta, **26**, 1-13, 1962.

Whipple, F. L., Kinetics of cosmic clouds, *in* Harvard Observatory Monograph No. 7, pp. 109-142, 1948.

Wood, J. A., Chondrites: Their metallic minerals, thermal histories and parent planets, Icarus, **6**, 1-49, 1967.

2

THE CHEMICAL COMPOSITION OF THE EARTH, THE MOON, AND CHONDRITIC METEORITES

by **Paul W. Gast**

Some restrictions on the chemical composition of the earth's upper mantle are derived from the isotopic composition of Sr, the composition of the earth's crust, and the composition of oceanic volcanic rocks; these restrictions suggest that the earth is depleted in elements more volatile than K and probably enriched in refractory elements like the alkaline and rare earths.

Analyses of lunar samples from two locations suggest that a similar situation occurs on the moon.

These conclusions lead to the inference that the chemical compositions of the smaller bodies that formed the earth and the moon varied from one terrestrial planet to another and may even have varied as individual planets increased in size.

INTRODUCTION

Our understanding of the evolution of the present chemical structure of the planet earth is probably in a rather primitive state when compared to the detailed modern theories of the evolution of a main sequence star like the sun. Neither our knowledge of the behavior of the solid matter nor our understanding of the internal composition of our own planet is comparable to that for our sun. Knowledge of chemical compositions based on direct measurements is limited to material that makes up less than 0.5% of the mass of the planet, i.e., the crust. In the last 10 years, some information on the composition of the upper mantle has been obtained from both direct measurements on possible mantle samples and inferences drawn from mantle-derived materials. Thus, one can with some confidence make some limiting statements regarding the composition, in particular, concerning trace elements. The usefulness of these limiting conditions in understanding the evolution of the earth would be much enhanced if the initial chemical condition of the earth were known. The similarity of the relative abundances of a very large number of elements in the sun and in carbonaceous chondrites suggests that these objects may be a relatively unfractionated sample of the nongaseous component (at $\sim 1000°C$ or less) of the primitive solar nebulae, suggesting that material of this composition may be the ultimate ancestor of the planet. Whether or not the whole planet or any part of the planet ever had this composition is a much more difficult question. Nevertheless, the concentrations and relative abundances of the elements found in chondritic meteorites have been found to be a very convenient geochemical frame of reference. This frame of reference will be frequently employed in this discussion; however, unless specifically stated, it is not intended to imply that it represents the primitive composition of the earth.

The primary objective of this study will be to review the data obtained from isotopic and trace element studies on crustal rocks that place significant constraints on the present composition and history of the upper mantle. As a corollary of these deductions, I will make several observations with regard to the formation of the earth as a planet and its subsequent chemical evolution. Finally, it will be of some interest to compare the earth with the moon. Recent data obtained on materials from the first two lunar landings will be briefly compared with data on materials from the earth.

The data used in this review will consist of (1) data on the isotopic composition of Sr in volcanic rocks, (2) trace element concentrations in the crust and atmosphere of the earth, (3) trace element abundances in mantle-derived rocks and the interpretation of these data, and (4) lead isotope compositions in oceanic volcanic rocks. I have deliberately neglected data from continental volcanic rocks and mantle-derived ultramafic inclusions and intrusions. Trace elements from continental volcanic rocks are often difficult to relate to the mantle because of interactions with the continental crust. Trace element concentrations of ultramafic rocks do not yield a consistent interpretation at the present time.

ISOTOPIC COMPOSITION OF Sr

The isotopic composition of Sr provided one of the first strong indications that the earth was chemically distinct from chondrites (Gast, 1960). In brief, the argument is as follows. The Sr^{87}/Sr^{86} ratio of the mantle can be inferred from that found in oceanic volcanic rocks; from this ratio and the presumed initial ratio for the whole earth, one can infer the mean Rb/Sr ratio of the upper mantle source. This inferred Rb/Sr ratio is at least 10 times less than the average ratio found in chondritic meteorites.

Data on the isotopic composition of Sr in a variety of oceanic volcanic rocks are summarized in Fig. 1. The recent very precise measurements of Papanastassiou and Wasserburg (1969) and Sanz and Wasserburg (1969) on a variety of meteorites and lunar samples have established the initial Sr^{87}/Sr^{86} ratio for the initial earth within very narrow limits. The results for volcanic rocks from more than 15 different localities, largely in oceanic regions (Fig. 1) show that there are real differences between some islands, e.g., Samoa and Ascension. These differences, discussed elsewhere (Gast, 1967), show a remarkably small range, 0.702 to 0.705, when compared to the range observed in meteorites (0.699 to 0.800) or in crustal rocks (0.701 to $>$ 0.9). The median value assumed for the present mantle, based on these analyses, is 0.703. From the initial value, the present value, and the 4.5-b.y. age for the mantle, we calculate an Rb/Sr ratio of 0.03 (+ 0.02, – 0.01) for the present mantle. This ratio compares to a range of ratios from 0.1 to 0.8 for all chondritic meteorites. From this comparison, we can readily infer that the Rb/Sr ratio of the earth's upper mantle is much less than that found in the primitive solar nebulae. The Sr^{87}/Sr^{86} data from the first two lunar landings (Papanastassiou et al., 1970) indicate that this is also true for the moon; in fact, the moon may have an even lower Rb/Sr ratio ($<$ 0.01). Some early enrichment of Rb relative to Sr in the crust of the earth may account for a small portion of the difference between the earth and chondrites. This question was discussed previously (Gast, 1968a) and does not appreciably alter the conclusions drawn here. In summary, one can conclude that the Rb/Sr ratio of the earth is about one-tenth that of the chondritic meteorites and the primitive solar nebula.

DIFFERENTIATION OF THE EARTH INFERRED FROM THE COMPOSITION OF THE CRUST AND THE ATMOSPHERE*

It has long been recognized that certain elements have been highly concentrated into the earth's crust. Birch (1958) showed that if the abundance of uranium in the earth is chondritic, the degree of concentration of this element in the crust approaches 100%. A similar situation exists for barium (Gast, 1960). It seems quite improbable that this situation can be entirely the result of a single early differentiation of the planet. I should like to explore this question further by comparing the crustal abundance of a number of additional elements with abundances derived from a hypothetical chondritic abundance. The mass of the earth's crust used in the calculations that follow is summarized in Table 1. Even though the oceans make up a large area, the oceanic crust accounts for less than one-third of the mass of the crust, and furthermore, the concentrations of U, Ba, K, and Rb in oceanic volcanic rocks are so low that this part of the crust can account for only a small portion of the crustal inventory of these elements.

*The first part of this section is taken from Gast (1968a).

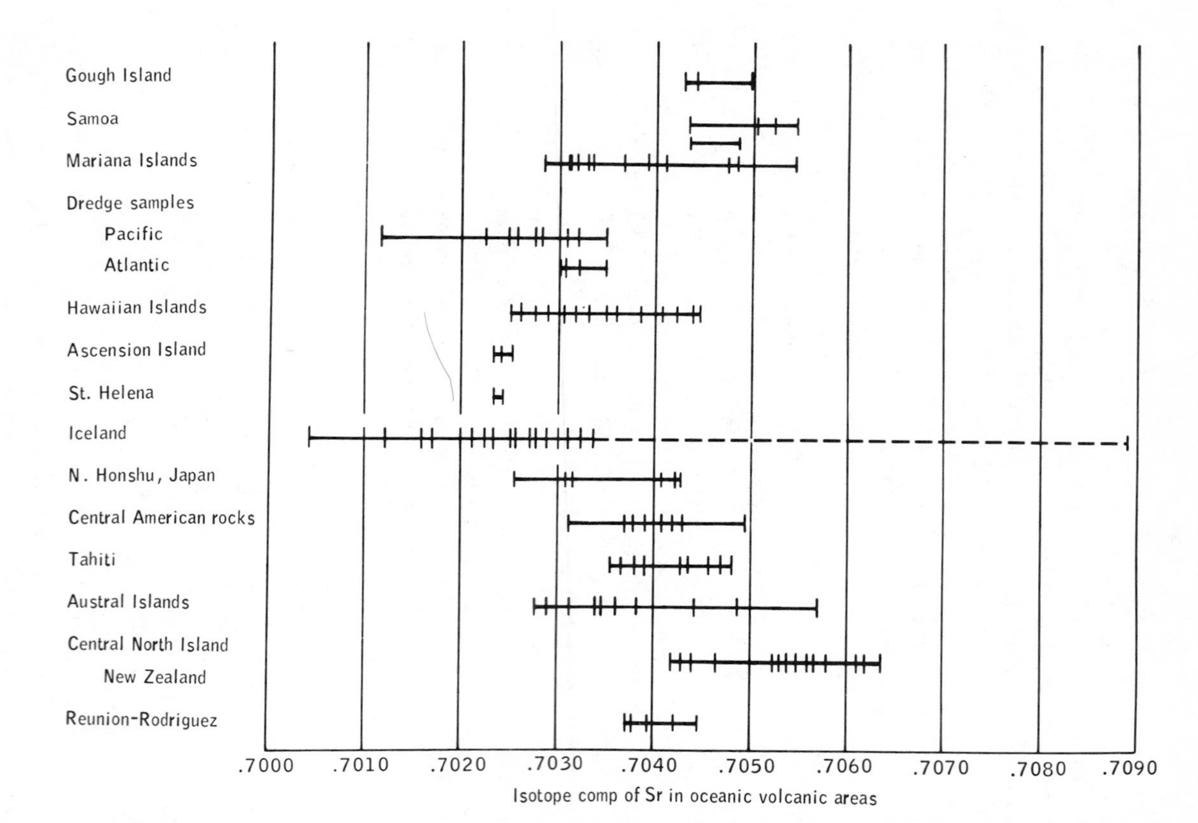

Fig. 1. Isotopic composition of Sr in oceanic regions and oceanic border regions. Data are taken from the literature and normalized to a value of $Sr^{87}/Sr^{86} = 0.708$ for the E&A standard.

Table 1. The mass of the earth's crust

Portion	Mass, gm
Shield areas	1.08×10^{25}
Continental shelves	0.48
Young folded belts	0.53
Oceanic crust	0.48
Total*	2.61

*Total = 0.44% of the mass of the entire earth; 0.65% of the mass of the present mantle.

Thus, our ignorance of this part of the crust is not a major limitation in determining a crustal abundance. A much more significant limitation arises from the inability to specify the composition of the lower continental crust. Poldervaart (1955), Gast (1960), and Taylor (1964) have used the concentrations determined for common surface or upper crustal rocks (basalts, diorites, etc.) in estimating the abundances of the elements in the entire crust. Heier (1965), Lambert and Heier (1966), and Zartman and Wasserburg (1969) have suggested that there may be very significant and systematic differences in the trace element complement of the lower and upper continental crusts. In particular, these authors find that pyroxene-granulite subfacies rocks are depleted in Th, U, and Rb relative to amphibolite facies rocks. Heier (1965) suggests that this may be the result of anatexis or partial melting of the lower crust. R. Lambert (oral communication, 1967) has made similar observations on the pyroxene-granulites of the Scottish Highlands. Thus, the previous estimates of crustal abundances may have overestimated the abundances of those elements that are enriched into the liquid during partial melting of the lower crust.

The concentrations of a number of elements in common rocks are summarized in Table 2. The crustal abundances, based on several different assumptions, are given in Table 3. Ignoring the Sr^{87} evidence for the moment, these crustal abundances are compared with the terrestrial abundances, based on a chondritic earth. (In the case of Tl, Zn, Cd, Pb, S, and Cl, the terrestrial abundance is based on carbonaceous chondrites.) The results shown here indicate that the refractory elements U and Ba are much more highly enriched than either the alkali metals or some highly volatile elements. Ringwood (1966) has suggested that the under-abundance of Tl, Zn, Cd, and Pb cannot be explained by their chalcophile properties, but that these elements, like Rb, must be depleted in the earth relative to their abundance in meteorites. This is not particularly surprising when one recalls that Tl, Bi, In, Pb, Zn, and Cd are highly depleted in normal chondrites, relative to carbonaceous chondrites (Anders, 1965).

One may now ask what would be the enrichment of Rb in the crust if the

Table 2. Trace element content of some common rocks

Sample	Concentration, ppm								
	U	Ba	Sr	Pb	Tl	Zn	Cd	K	Rb
Composite "granite"[a] > 70% SiO_2 (221)	4.2	–	132	28.7	–	–	–	[b]3.70	186
Composite "granodiorite"[a] 60 to 70% SiO_2 (191)	2.60	–	371	19.2	–	–	–	3.11	140
Composite "diorite"[a] < 60% SiO_2 (85)	1.63	–	635	14.6	–	–	–	2.41	93
Composite granite I[c] (27)	3.14	570	348	22.2	–	–	–	3.29	133
Composite granite II[d] (50)	2.3	685	283	–	–	–	–	3.04	125
Composite basalt[a] < 55% SiO_2 (282)	1.07	–	526	12.7	–	–	–	1.07	34
Composite basalt[d] (250)	1.07	334	461	6.59	–	–	–	0.95	29
Granulities[e]	0.5	–	–	–	–	–	–	2.5	(40)
High-calcium granite[f]	3.0	420	440	15	0.72	60	0.13	2.52	110
Oceanic tholeiites[g]	0.09	14	130	0.75	–	–	–	0.12	1.0
Basalt[f]	1.0	330	465	6	0.21	105	0.22	0.83	30
Granodiorite[h]	2.5	500	400	15	0.7	60	0.1	2.5	100
Diorite-granulite[i]	0.6	(100)	500	8	(0.1)	60	(0.1)	1.5	20
Abyssal basalt	0.1	15	130	0.75	(0.1)	100	0.2	0.12	1.0

[a]Composite of specimens analyzed in the Rock Analyses Laboratory, University of Minnesota.

[b]This value measured in percent.

[c]Composite of 27 Finnish Precambrian granites.

[d]Composite of basalts, analyzed by Turekian and Kulp (1956).

[e]Lambert and Heier (1966, 1967); rubidium content based on K/Rb ratios reported orally at American Geophysical Union meetings.

[f]Turekian and Wedepohl (1961).

[g]Engel et al. (1965); U and Pb content, Tatsumoto (1966).

[h]Parentheses indicate very uncertain or arbitrary estimates.

[i]Composite of 50 Precambrian granites from the western United States.

observed Sr isotope composition is explained entirely as the result of Rb depletion, as discussed earlier. In this case, the terrestrial abundance of Rb is reduced by a factor of 8. For this assumption, the concentration of Rb in the crust that is required by the various models given in Table 3 is nearly 100% in some cases; i.e., there is almost insufficient Rb in the earth to explain the "observed" crustal abundance.

POTASSIUM-URANIUM RATIO IN THE EARTH, THE MOON, AND METEORITES

Because of the chemical similarity of K and Rb, we may infer that the abundance of potassium in the earth may also differ from that inferred from

Table 3. Crustal abundance and degree of concentration into the crust of some trace elements

Element	Meteoritic abundance, ppm	Model A[a]		Model B[b]		Model C[c]	
		Abundance (10^{21} gm)	Fraction in crust	Abundance (10^{21} gm)	Fraction in crust	Abundance (10^{21} gm)	Fraction in crust
U	0.012[d]	0.056	0.77	0.036	0.50	0.033	0.46
Ba	3.6[d]	8.8	.41	8.7	.40	6.3	.30
Sr	11[e]	8.84	.13	9.0	.14	10	.15
Pb	2.0[d]	.26	.02	.22	.02	.24	.02
Zn	50[f]	–	–	2.15	.007	1.85	.005
Cd	1.0[g]	–	–	.004	.007	.003	.005
Tl	.10[d]	.009	.015	.009	.015	.008	.013
K	860	440	.085	330	.064	420	.0812
Rb	2.7[h]	19	.116	1.4	.088	1.35	.080
"Rb"	.32[i]	19	.98	1.4	.72	1.35	.67
Cl	300[j]	–	–	34	.02	3.34	.02

[a]Crustal abundances of Taylor (1964) in a crustal mass of 2.09×10^{25} gm.
[b]Continental crust (2.09×10^{25} gm) 1/2 basalt and 1/2 granodiorite; oceanic crust, abyssal basalt.
[c]Continental crust 1/2 granodiorite and 1/2 diorite-granulite; oceanic crust, abyssal basalt.
[d]Reed et al. (1960).
[e]Gast (1962); Pinson et al. (1965).
[f]Nishamura and Sandell (1964).
[g]Schmitt et al. (1963).
[h]Smales et al. (1964).
[i]Depleted by a factor of 8.
[j]Reed and Allen (1966).

chondritic meteorites. Birch (1965) and Wasserburg et al. (1964) have noted the apparent systematic differences in the K/U ratio of terrestrial materials and chondritic meteorites. These differences are summarized in Fig. 2. A number of recent determinations of these elements in meteorites, terrestrial rocks, and lunar samples are shown in this figure. It is indeed remarkable that two such

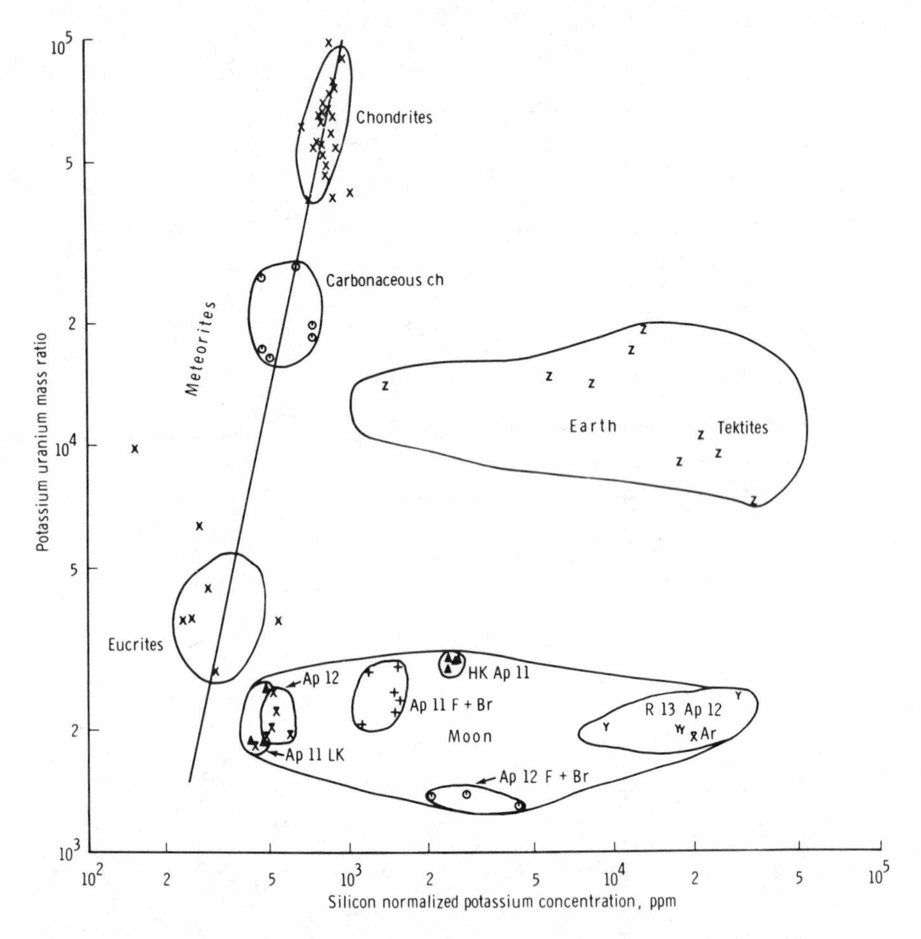

Fig. 2. Potassium-uranium mass ratios in the earth's crust, chondritic meteorites, calcium-rich achondrites, and lunar rocks and soil. The K concentrations have been normalized to a constant SiO_2 concentration, i.e., K plotted = K observed × Si observed/18.5. Terrestrial data are representative analyses of composites or averages taken from the literature. Chondrite analyses include 4 enstatite chondrites, 6 olivine bronzite chondrites, and 17 olivine hypersthene chondrites. Analyses are taken from the literature; K and U determinations are usually on different samples and by different analysts. Calcium-rich achondrites plotted are Nuevo Laredo, Stannern, Moore Co., Sioux Co., Pasamonte, Juvinas, and Macabini. Groups of lunar samples shown are Apollo 11 low-K (LK) rocks, fines (F) and breccia (Br), and high-K (HK) rocks; Apollo 12 igneous rocks; Apollo 12 fines and breccia; and Apollo 12 rock 12013 (R 13).

chemically dissimilar elements show so little variation in their relative abundances in rocks. The constancy of these ratios suggests that the difference between the crustal and perhaps upper mantle K/U ratio and the chondritic K/U ratio is a result of an earth-wide terrestrial depletion in potassium relative to uranium rather than an earth-wide chemical differentiation of these elements.

It is interesting to note that even for the first two sampling points on the moon, we see that the K/U ratio is approximately independent of the K content of the samples. The apparent between planet variations and between meteorite variations in this ratio, along with the lack of variation observed in igneous processes on the planets, suggest that the K/U ratio variations precede the formation of the planets.

ABUNDANCE OF POTASSIUM INFERRED FROM ATMOSPHERIC Ar^{40}

Hurley (1968) and Gast (1968a) have independently shown that the concentration of Ar^{40} in the atmosphere results in a significant restriction on the terrestrial K content and K/Rb ratio. If one makes the rather plausible assumption that Ar^{40} produced by decay of K^{40} is differentiated into the atmosphere at least as efficiently as Rb is differentiated into the crust of the earth, we can use the data we have just presented to derive the mean K content of the earth with a chondritic Sr content. By using the Sr^{87}/Sr^{86} corrected Rb enrichment in model C, we infer that 66% of all Ar^{40} produced in the earth during the last 4.6 b.y. is in the atmosphere, which gives a mean K content of 120 ppm, i.e., about one-seventh that of chondrites. In this case, the fraction of the earth's K that is today in the crust of the earth is about 0.63. Since well over one-half of the K and Rb in the earth is now in the crust according to this reasoning, the K/Rb ratio of the earth cannot differ very much from the crustal ratio, i.e., ~ 300. It is again of interest that the lunar soil from both the Apollo 11 and the Apollo 12 sites and possible igneous liquids from the lunar interior give K/Rb ratios in the range 300 to 500.

In summary, one can now conclude (1) that the earth (and the moon) are significantly depleted in K and Rb relative to U and Sr (and other refractory elements) and (2) that the earth is very efficiently differentiated if refractory elements are not enriched relative to chondrites. Almost all the K, Rb, Ba, U, and Th of the earth is found in the upper one-half percent of the mass of the earth (i.e., the crust).

UPPER MANTLE CHEMISTRY AS INFERRED FROM OCEANIC VOLCANIC ROCKS

The chemical and isotopic characteristics of igneous liquids brought to the surface of the earth from the upper mantle must carry some information about the compositions of their source. As a first approximation, we may assume that many liquids ascend from depths of 50 to 100 km without appreciable changes in the concentrations of many trace elements. Two different arguments in support of this assertion can be offered.

1. Gast (1968b) has suggested that certain elements, in particular Ni and Cr, are so sensitive to the removal of olivine and pyroxene, the most common primary liquid phases, that removal of more than ~ 20% by weight of these minerals will be seen in the abundance of Ni and Cr. For example, removing 30% by weight of olivine from a parent liquid enriches that liquid in Ba by less than a factor of 2, but reduces the Ni content by approximately a factor of 20.

2. Green (1970) has argued that many igneous liquids contain inclusions and xenocrysts that are stable only at high pressures. Their presence in their host liquids precludes extensive crystallization at low pressures. It is thus quite likely that both the absolute and the relative abundances of trace elements that do not enter early crystallizing phases (e.g., Ba, Sr, K, Rb, the rare earth elements, U, Th, Pb, etc.) in many basaltic rocks are essentially those of a liquid at the point where it was produced. In this case, one may use the abundances of these elements to infer something about their relative and absolute abundances in their mantle source.

Gast (1968b) and Shaw (1970) have proposed rather simple, but useful, schemes relating the phase composition, degree of melting, and bulk composition of the source to the trace element composition of a basaltic liquid. The distribution coefficients between the coexisting phases are essential parameters in these calculations. They are, however, in most cases only poorly known. Nevertheless, from these models, Gast (1968b) has suggested that many of the trace element characteristics of basaltic liquids can be related to the degree of partial melting. The abundances of K, Rb, Sr, Ba, and the rare earth elements have been most extensively interpreted in such calculations (Griffin and Murthy, 1969; Kay et al., 1970).

The chondrite-normalized abundances of these elements in a series of terrestrial and lunar basalts is shown in Fig. 3. Rocks that are grossly similar in major element composition (all having 9 to 13% CaO, 8 to 13% MgO, 12 to 16% Al_2O_3) have trace element concentrations that differ by as much as a factor of 1000 in some cases and by more than a factor of 100 in many cases.

In addition, some elements are extremely enriched with respect to chondrites. Gast (1968b) and Frey et al (1968) have pointed out that the concentrations of the rare earth elements are difficult to explain in a liquid derived from a mantle with chondritic rare earth element concentrations. This can be further illustrated by considering how such extreme variations may originate. If the degree of melting is the primary control of the trace element concentrations, a range of at least a factor of 20 to 40 in the degree of melting is required to produce the observed range of Ba and La concentrations. (This excludes the extremely low Ba concentrations in oceanic ridge basalts and the very high concentrations in the leucite basalts.) For this assumption, the range of partial melting implied for a chondritic mantle concentration is 7 to 10% for the oceanic ridge basalts and

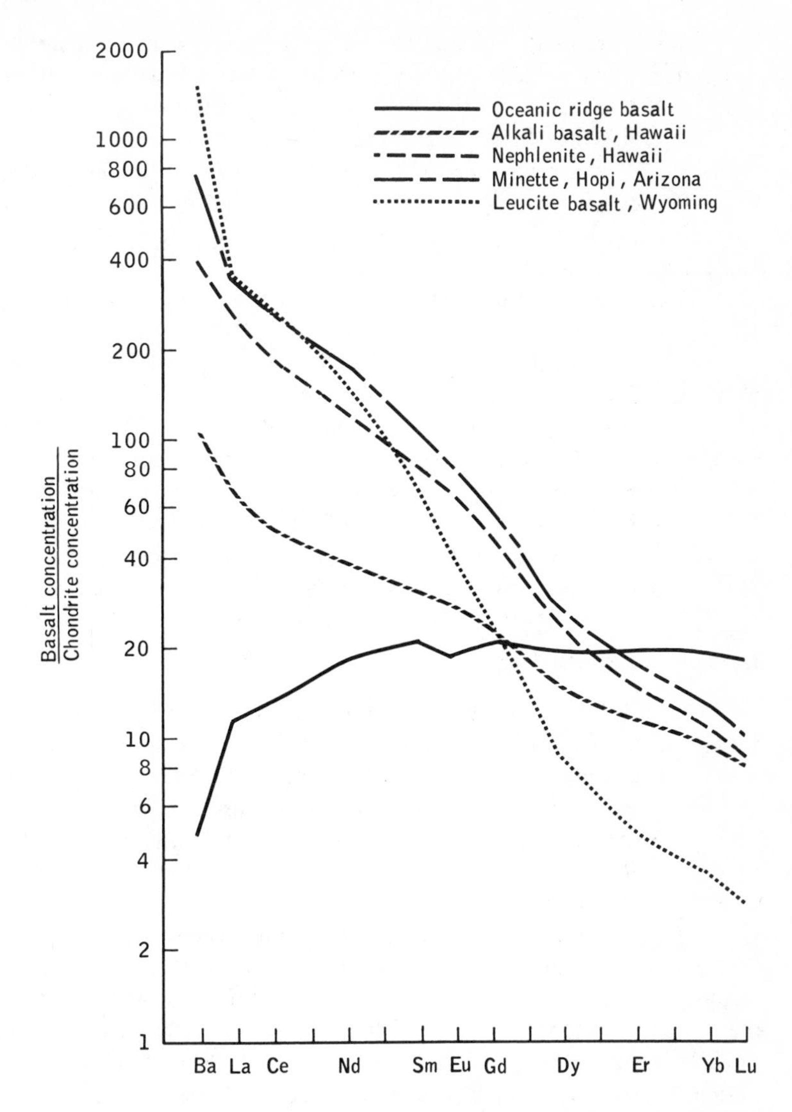

Fig. 3. Chondrite-normalized abundances of rare earth elements and Ba. Data are taken from Kay (written communication, 1970) and Kay et al. (1970). There is a regular progression from a horizontal (La and Ba depleted) pattern for the oceanic ridge basalts to highly fractionated, Ba-La enriched, and heavy rare earth element depleted patterns for undersaturated rocks. The latter rocks are generally rich in trace elements (e.g., U, Pb, Nb, and P_2O_5) when compared with the oceanic ridge basalts.

0.2 to 1% for nephelinites and alkali basalts. The latter values clearly present a serious problem; i.e., how does one segregate and extract such a small proportion of liquid and transport it to the surface. This situation is considerably alleviated if the concentrations of these elements in the present day mantle are 4 to 6 times those of chondrites. In this case, the observed range of concentrations requires ~1 to ~40% partial melting. Finally, in support of this observation, one notes that the abundances of these elements in many ultramafic materials are considerably greater than those found in chondrites (Sr content ranges from 2 to 100 ppm; see Fig. 4). One can, therefore, conclude that the earth's upper mantle is enriched in a number of large refractory ions when compared to carbonaceous chondrites.

A similar conclusion has been suggested for the lunar basalts from Apollo 11 and 12 (Gast and Hubbard, 1970; Gast et al., 1970). If the oceanic ridge basalts represent such extensive melting that they have removed most of the trace elements from the mantle source, their chemistry indicates that the mantle was already depleted in certain elements (in particular, Ba, La, Rb, and K) when they were formed. These are, of course, the very same elements that are strongly enriched in a liquid produced by small degrees of partial melting and also those that are most enriched in the crust (Table 2). Therefore, I suggest that the oceanic ridge basalts provide evidence for a much earlier removal of material from the mantle, perhaps the removal of the early crust.

ISOTOPE COMPOSITION OF Pb

I have already mentioned the isotopic composition of Sr in oceanic volcanic rocks and the limitations on the mantle Rb/Sr ratios that arise from these data. The isotopic composition of Pb can be similarly related to the U/Pb and Th/Pb ratios of the mantle. In this case, however, because two isotopes of Pb (Pb^{207} and Pb^{206}) are produced by two isotopes of uranium (U^{235} and U^{238}) with very different half-lives, it is possible in principle to infer from isotope variations when variations in the U/Pb ratio of the source of a given sample may have taken place. In practice, it has not been possible to determine whether volcanic leads have a complex history (i.e., whether the volcanic leads are the product of two or more systems with different U^{238}/Pb^{204} ratios) because it is not possible to specify uniquely the U^{238}/Pb^{204} ratios and times that were involved. It is possible, however, to argue from additional geochemical arguments that the observed isotope ratio variations require fractionations in the U^{238}/Pb^{204} ratio before 500 m.y. ago and in some instances prior to 1 b.y. ago (Gast, 1967, 1969).

The scope of these variations is illustrated in Fig. 5, which summarizes recent lead isotope data from oceanic volcanic rocks. If the deviations from the simple single stage evolution are ascribed to a second stage and if on the average the second stage has a U/Pb ratio some factor times that of the first stage, we can construct the locus of points that represent different times on the Pb^{206}/Pb^{204} and Pb^{207}/Pb^{204} plot. Such a curve is shown in Fig. 5 for enrichments of 50%. Very roughly, such curves fall in the range of the observed isotope ratios. I suggest that the Pb isotope compositions observed in oceanic volcanic rocks

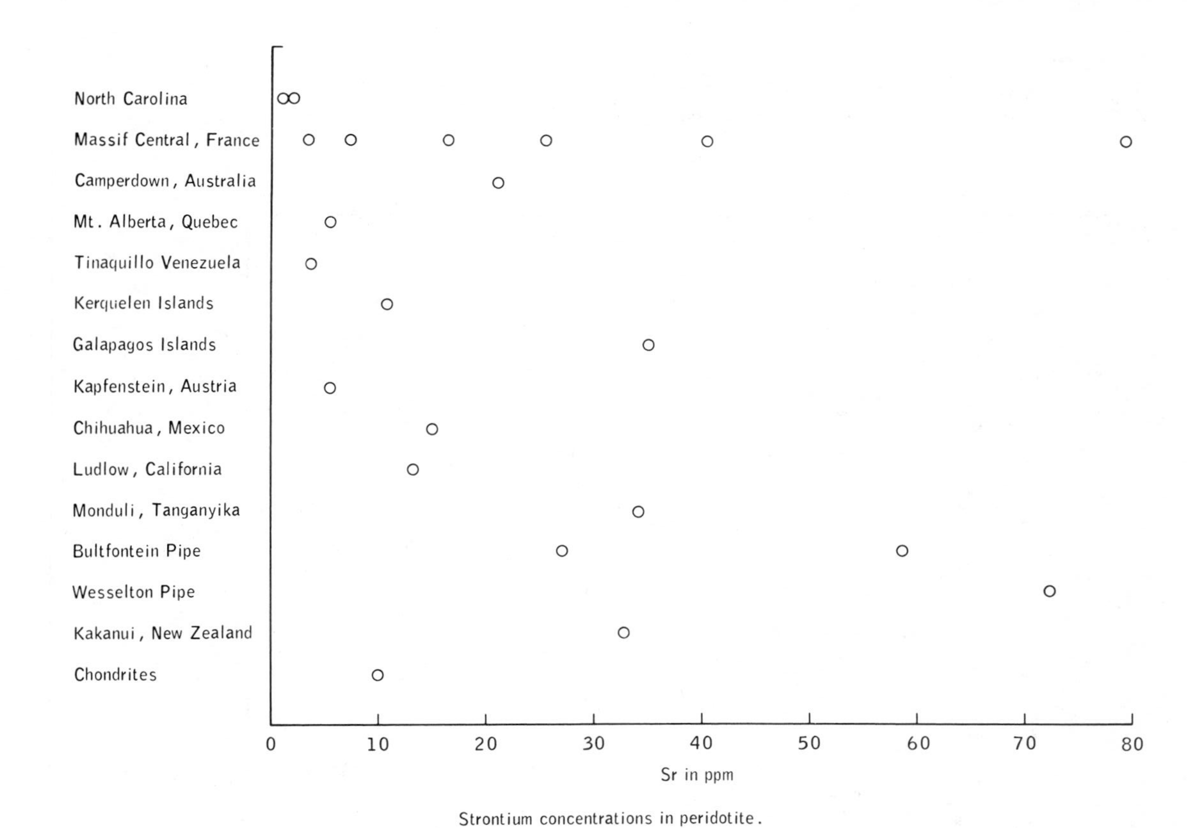

Fig. 4. Strontium concentrations in peridotite nodules and peridotite bodies. Data are taken from the literature. Most of the determinations shown here are made by stable isotope dilution methods.

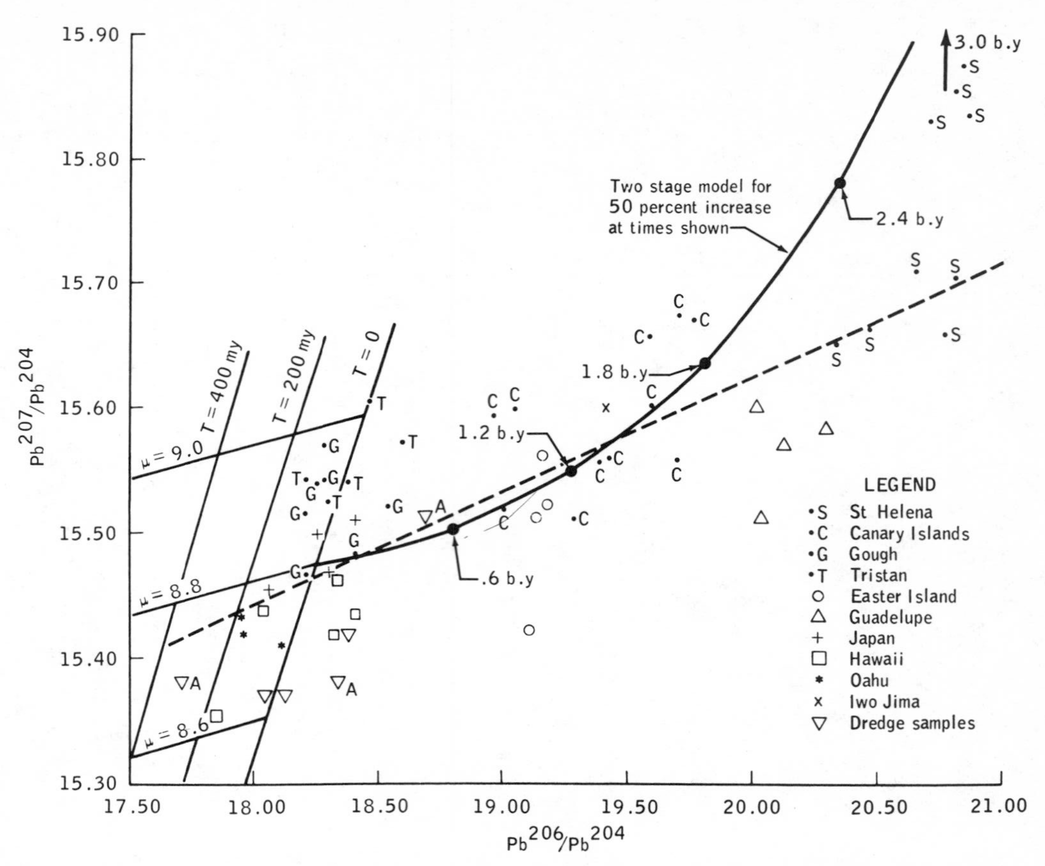

Fig. 5. Lead isotope compositions of young (< 20 m.y. old) oceanic volcanic rocks. Taken from Oversby and Gast (1970). The curved line shows the locus of isotope compositions that are generated by evolving lead in a two-stage system in which the U^{238}/Pb^{204} increased by 50% at various times in the past.

indicate that the suboceanic mantle has at various times in its history undergone fractionations of U and Pb by removal of a partial melt that is usually enriched in Pb relative to U. There is some evidence both from the trace elements and the isotopes in oceanic volcanic rocks that the mantle source of U and Pb has not remained chemically closed during the 4.5-b.y. history of the mantle.

One should also note that the isotopic composition of Pb fixes the U/Pb ratio of the earth as more than 10 times that of carbonaceous chondrites. Similar arguments for the moon indicate a fractionation of U and Pb by a factor of more than 100 relative to carbonaceous chondrites (Tatsumoto and Rosholt, 1970). This is quite consistent with the Rb/Sr fractionation inferred from the Sr^{87}/Sr^{86} ratios. That is, lead is more volatile than rubidium and should be less efficiently retained by the planet.

We can now summarize the deductions drawn from the previous observations.

1. Volatile elements (at temperatures of $\sim 1000°C$) are depleted both from the earth and the moon relative to chondrites, but degree of depletion varies and is not 100% even for quite volatile materials such as halogens and Bi, In, and Hg. The moon appears to be even more depleted in volatiles than the earth.
2. The upper mantle is enriched in many refractory elements relative to chondrites by a factor of as much as 3 to 5.
3. The present oceanic upper mantle has undergone chemical fractionation throughout its history.

DISCUSSION

The chemical characteristics inferred here for large portions of the earth and the moon constitute important restrictions on any scheme of origin proposed for these planets. The rather simple notion that volatile elements are depleted in some primitive materials that later accreted into these planets and that, conversely, refractory elements may be enriched in other primitive materials introduces significant restrictions on the temperatures at which solids and vapor were separated and on the time in the accretionary history of the planet when this separation took place.

A planet-wide depletion in volatile elements can be accounted for (1) by removal of these elements from the planet during or subsequent to its formation, i.e., the fractionation of volatile and nonvolatile material being localized in the solar nebula, or (2) by not retaining these volatile elements in some of the smaller objects that formed the planets (in the same way that the rare gases, e.g., Kr and Xe, were not retained), i.e., the fractionation is dispersed over large regions of the nebula.

The localized fractionation mechanisms can only account for planet-wide enrichments in refractory elements by removing a significant fraction of the mass of the planet and retaining a refractory residue. If, however, the fractionation between volatile and nonvolatile elements takes place throughout the nebula (before planets achieved their present mass), it is quite possible to produce material that is highly enriched in refractory elements, along with material that is depleted in volatile elements, by simply specifying the temperature where condensation from the gas phase takes place. Furthermore, by mixing materials that accumulate at different temperatures in different proportions, an even greater variety of chemical compositions is possible. Distinguishing between these two rather different situations remains as one of the significant unanswered questions concerning the origin of the planets.

The data for the earth clearly show that the extent of the depletion of the volatile elements varies. The depletion is never complete even for the most volatile elements. For example, at the temperatures where potassium and rubidium are fractionated from uranium and strontium, one would expect essentially complete removal of volatile elements like Cl, Tl, Bi, and Hg (Larimer, 1967). This is not observed. There is enough Pb, Tl, and Cl in the crust alone to account for approximately 1% of a carbonaceous chondrite earth. Therefore, the terrestrial depletion in these elements is certainly no more than 99% and probably no more than 90%.

The first two conclusions regarding the earth that are listed previously may be accounted for in two ways.

1. The earth accumulated from a mixture of volatile depleted material (~ 9/10) and volatile retaining material (~ 1/10) and subsequently differentiated to produce an upper mantle one-fifth to one-sixth the mass of the earth and very enriched in large refractory ions and some volatiles, e.g., K, Rb, Cs, Pb, etc. This upper mantle was subsequently differentiated to form the crust.

2. The earth accumulated from a mixture of volatile depleted and refractory enriched matter (~ 1/2 to 9/10) and volatile retaining material (~ 1/10) and underwent only moderate chemical differentiation, except for strong upward segregation of K, Rb, and H_2O.

Recent models and theories of the origin of the planets are for the most part variations of the first possibility. Ringwood (1966) has proposed a theory of the origin of the terrestrial planets which considers most of the observed chemical characteristics of these objects. He assumes, with previous investigators, that the chemical composition of the objects that formed the planets had the composition of carbonaceous chondrites. The chemical properties of the earth and the moon are explained by fractionation of this parent material between the earth, the moon, and an undetermined mass of escaped volatile substances. The loss of volatiles from the system is not sufficient, however, to enrich large volumes of residual material in refractory ions. Thus, the concentrations of elements like Ca, Al, U, Th, Ba, and the REE for the mean primitive earth and moon, in Ringwood's model, should be those of chondritic starting materials. In this case, the extreme differences in composition of the surface of the earth and the average earth must be ascribed to very efficient chemical differentiation of the planet.

The enrichment in refractory elements in the upper mantle of the earth and in the upper part of the moon is definitely limited by the crustal and whole planet inventory of refractory elements in Ringwood's model. In addition, this model postulates that elements like K, Rb, and Pb are totally lost from the earth-moon system at the time that a dense, hot atmosphere is dissipated. In order to retain these elements in such an atmosphere, its temperature must be maintained at 1000°C and hotter. The radiant energy loss from such an atmosphere must be very substantial. In fact, the equilibrium between radiative losses from the hot atmosphere and gravitational energy release resulting from accretion places significant time constraints on the way in which this atmosphere is dissipated. Thus, the temperature of the atmosphere that is established by the requirement that substantial fractions and K and Rb must be lost from the moon and the earth introduces a significant time constraint that must be investigated further before the plausibility of Ringwood's mechanism can be accepted.

Turekian and Clark (1969) have proposed a radically different model for the accretion of the earth. They suggest that the metallic core and most of the mantle accumulated at rather high temperatures while the upper mantle of the earth was always volatile-rich and partially oxidized due to late stage accumulation of low temperature condensates. From this model, one infers that the upper part of the earth should contain a significant proportion of material in which the relative abundance of the elements is essentially that of carbonaceous chondrites, in particular this part of the earth should have chondritic Rb/Sr ratios and very low U/Pb ratios, or one must accept the fact that the relative abundances of trace elements in material condensing at low temperatures can be arbitrarily specified to accommodate a particular model. According to the observations made here, the Turekian and Clark model requires that the late additions to the earth were simultaneously enriched in refractory elements, depleted in volatile elements like K and Rb, and enriched in other volatiles, particularly water. The validity of their model depends on the ability to produce such material in the solar nebula. If the carbonaceous chondrite composition is taken for the volatile-rich fraction added to the earth, large additions of U and Sr from the lower mantle of the earth are required to modify the chondritic Rb/Sr and U/Pb ratios of the volatile-rich outer layer of the earth in order to produce the presently observed ratios. If, for example, we assume that the upper 15 to 20% of the mass of the earth was initially made up of material with cosmic or chondritic Rb/Sr and U/Pb ratios, the Sr and U content of the mantle would have to be increased to approximately 80 and 0.1 ppm, respectively, to account for the observed element ratios. This would require a source with more than five times the mass of the outer shell of the U and Sr content if the inner part of the earth were chondritic.

When examined carefully, both the above models encounter some difficulty in explaining the apparent loss of volatile elements like K, Rb, and Pb from the earth and the moon and also require very extreme upward enrichment of refractory elements such as Ba, U, Th, and the REE in order to account for the present concentrations of these elements in the crust and upper mantle of the moon and the earth.

I suggest that much of the fractionation of volatile elements from the solids that formed the planet took place at an earlier stage (during condensation from the solar nebula); that is, that there may be significant separation of nebular gases from solids before significant accretion of the earth and the moon occurred. The separation of volatile elements in meteoritic materials (Larimer and Anders, 1967) is strong evidence that the temperatures necessary to produce this separation existed in the early solar system and were not a result of high temperatures associated with gravitational energy release. There is even some chemical evidence in meteorites that enrichment of refractory material may have taken place in some early condensates. Marvin et al. (1970) have found that the Ca-rich and Al-rich inclusions in the Allende meteorite are similar in composition to theoretical high-temperature condensates. Gast et al. (1970) report rare earth element abundances on such an inclusion. These data are in general accord with this hypothesis. Furthermore, this meteorite is depleted in both Na and K (Mason, written communication, 1970). It is, however, not as depleted in Pb and Ge as the normal chondrites. This suggests that this meteorite must also contain a significant fraction of volatile-retaining material. Ahrens and von Michaelis (1969) have shown that Ca and Al vary together in almost all meteorites. The observations both on the Allende meteorite and on meteorites in general seem to require the existence of both volatile-rich and refractory-rich components in the portion of the nebula where meteorites accreted.

When summarized, the evidence seems to be most plausibly explained by postulating that the most primitive solid bodies in the early solar system had a wide range of compositions ranging from volatile-rich materials like the carbonaceous chondrites to refractory-rich materials like the Ca-rich and Al-rich inclusions in the Allende chondrite. In this situation, the overall composition of the earth and the moon is largely determined by the relative proportions of the different compositions that were swept up by these planets.

THE IRON PROBLEM

One of the most difficult and also most intriguing questions regarding the origin of the planets is the origin of the earth's core and metallic iron in the early solar system. Two basic schools of thought prevail. First, the native iron is produced by carbon reduction within large bodies, and second, metallic iron is formed in the dispersed nebula. Both of these hypotheses encounter difficulties when we try to explain all the observations.

1. Much of the iron (i.e., $\sim 1/2$ of the Fe in the earth) is reduced and is in the earth's core, yet the nickel content and oxidation state of iron in the upper mantle suggest that this portion of the earth is not, and has not been, in equilibrium with native iron (Ringwood, 1966).

2. The mean density of the moon indicates that the moon can have very little free iron in its makeup (Urey, 1967), yet liquids produced from its interior are iron-rich, highly reduced, and appear to be near equilibrium with native iron (Reid et al., 1970). That portion of the moon sampled in the first two landings

does not appear to be significantly depleted in Fe relative to other major elements, e.g., Mg, Si, or Al. These observations suggest that the reducing capacity of the system in which the lunar material accumulated was insufficient to reduce a significant fraction of the lunar Fe.

3. Most meteorites are very reduced and show a very close coherence between Ni and Fe (Prior's law), where iron is only partially reduced; i.e., silicate phase and iron appear to be in equilibrium. Isotopic age determinations indicate that this state existed at the time when meteorites were formed. Nevertheless, it should also be noted that the Ni/Fe ratio in the earth's upper mantle is significantly different from that of chondritic meteorites (about one-third to one-half that of chondrites).

A completely consistent attempt to resolve these differences is beyond the scope of this review. It seems possible that kinetic factors that cannot be easily predicted may play a major role in the sequence of events in the formation of the planets (Blander and Katz, 1967). If, for example, the condensation of metallic iron is inhibited relative to silicate dust, iron may condense as oxidized silicate dust rather than metallic Fe. Also, if temperatures in the nebula from which large bodies are accumulating is changing very rapidly, large bodies (>1 km in diameter) may not equilibriate with the ambient temperature, resulting in material with a wide range of compositions.

If we accept the suggestion of Blander and Katz (1967) that the condensation of metallic iron may be kinetically inhibited and disregard some essential features of current astrophysical models, a sequence of events that explains most of the chemical observations can be constructed.

1. Matter that accumulated or accreted during the early stages of the formation of the earth and possibly the moon was carbon rich (similar to carbonaceous chondrites).

a. In the case of the earth, the release of gravitational potential energy heated this material at the surface of the body when the mass of the body reached one-fourth to one-third of its present mass (Ringwood, 1966). This resulted in reduction of iron and production of H_2O-rich and CO_2-rich silicate liquids.

b. In the case of the moon, temperatures at the surface of the accreting body would never be sufficient to reduce Fe and melt silicates during early stages of accumulation.

2. At some point during the accretion of the earth, i.e., when it reached about one-half of its present mass, the incoming materials became carbon free and refractory element enriched. Such a change could be the result of changing conditions in the nebula or accumulation of material from different parts of the nebula.

3. The upper 400 km of the moon also accumulated from carbon free, water free, and refractory element-enriched material.

Thus, in its earliest form, the outer one-half to one-third of the mass of the earth would contain Ni and Fe in their nebular proportions, i.e., aside from differences resulting from volatility differences. (The apparent difference in the abundance of Fe or in the Fe/Si ratio of the lunar and terrestrial mantle clearly requires further modification of this hypothesis.) The outer one-half to one-third of both the earth and the moon would then be strongly depleted in volatile elements and enriched to varying degrees in refractory elements. For the earth, this material would overlay a partially molten, iron-depleted, volatile-rich inner half. Water-rich and CO_2-rich liquids from this inner part would rise into the outer shell, resulting in some oxidation of iron and eventual crystallization of hydrous phases in the outer shell. Since the moon would be much cooler during its early growth, the outer shell of the moon would not be "contaminated" by addition of volatiles and would retain its primitive composition.

As our knowledge of the chemistry of the planets improves, any explanation of the origin of these planets from a dispersed nebula consisting of gas and dust will have to concern itself with a variety of factors not previously considered in detail, such as (1) how the temperature, pressure, and H/O ratio in the nebula varied with time; (2) how the distribution of mass between objects ranging in size from dust particles to planets changed as the temperature changed; and (3) how the ratio $\frac{H_2 + He + CH_4 + H_2O}{Fe + Al + Mg + Si}$ changed as temperature changed.

ACKNOWLEDGMENTS

Discussions with C. Meyer of the NASA Manned Spacecraft Center (MSC) have significantly improved several aspects of this paper. E. Schonfeld of MSC is largely responsible for Fig. 2. Donna Sanders of MSC cheerfully undertook the job of typing many undecipherable revisions of the manuscript. Their assistance is gratefully acknowledged. Finally, I would like to acknowledge that this paper would not have been completed without the patient, persistent encouragement of Eugene Robertson of the U.S. Geological Survey.

REFERENCES

Ahrens, L. H., and von Michaelis, H., The composition of stony meteorites–V, Some aspects of the composition of the basaltic achondrites, Earth Planet. Sci. Letters, **6**, 304-308, 1969.

Anders, E., Chemical fractionations in meteorites, Meteoritica, **26**, 17-25, 1965. (Also available as NASA CR-299.)

Birch, F., Differentiation of the mantle, Geol. Soc. Am. Bull., **69**, 483-486, 1958.

______, Speculations on the Earth's thermal history, Geol. Soc. Am. Bull., **76**, 133, 1965.

Blander, M., and Katz, J. L., Condensation of primordial dust, Geochim. Cosmochim. Acta, **31**, 1025-1034, 1967.

Engel, A. E. J., Engel, C. G., and Havens, R. G., Chemical characteristics of oceanic basalts and the Upper Mantle, Geol. Soc. Am. Bull., **76**, 719-734, 1965.

Frey, F. A., Haskin, M. A., Poetz, J. A., and Haskin, L. A., Rare earth abundances in some basic rocks, J. Geophys. Res., **73**, 18, 6085-6098, 1968.

Gast, P. W., Limitations on the composition of the Upper Mantle, J. Geophys. Res., **65**, 4, 1287-1297, 1960.

Gast, P. W., The isotopic composition of strontium and the age of stone meteorites–I, Geochim. Cosmochim. Acta, **26**, 927-943, 1962.

______, Isotopic geochemistry of volcanic rocks, *in* Basalts, The Poldervaart Treatise on Rocks of Basaltic Composition, edited by A. Poldervaart, Interscience, New York, 1967.

______, Upper mantle chemistry and evolution of the earth's crust, *in* The History of the Earth's Crust, edited by R. A. Phiney, pp. 15-27, Princeton University Press, 1968a.

______,Trace element fractionation and the origin of tholeiitic and alkaline magma types, Geochim. Cosmochim. Acta, **32**, 1057-1086, 1968b.

______, The isotopic composition of lead from St. Helena and Ascension Islands, Earth Planet. Sci. Letters, **5**, 353-359, 1969.

Gast, P. W., and Hubbard, N. J., Rare earth abundances in soil and rocks from the Ocean of Storms. Earth Planet. Sci. Letters, **10**, 94-100, 1970.

Gast, P. W., Hubbard, N. J., and Wiesmann, H., Chemical composition and petrogenesis of Basalts from Tranquillity Base, *in* Geochim. Cosmochim. Acta, Supplement 1, Proceedings of the Apollo 11 Lunar Science Conference, vol. 2, edited by A. A. Levinson, pp. 1143-1163, Pergamon Press, 1970.

Green, D. H., A review of experimental evidence on the origin of basaltic and nephelinitic magmas, Phys. Earth Planet. Interiors, **3**, 221-235, 1970.

Griffin, W. L., and Murthy, V. Rama, Distribution of K, Rb, Sr and Ba in some minerals relevant to basalt genesis, Geochim. Cosmochim. Acta, **33**, 1389-1414, 1969.

Heier, K. S., Metamorphism and the chemical differentiation of the crust, Geol. Foren. Stockholm, Forh.,**87**, 249-256, 1965.

Hurley, P. M., Absolute abundance and distribution of Rb, K and Sr in the Earth, Geochim. Cosmochim. Acta, **32**, 273-283, 1968.

Kay, R., Hubbard, N. J., and Gast, P. W., Chemical characteristics and origin of oceanic ridge volcanic rocks, J. Geophys. Res., **75**, 8, 1585-1613, 1970.

Lambert, I. B., and Heier, K. S., The vertical distribution of thorium and uranium in the continental crust, A. M. Geophy. Union Trans., **47**, 200, 1966.

______, The vertical distribution of uranium, thorium, and potassium in the continental crust, Geochim. Cosmochim. Acta, **31**, 377-390, 1967.

Larimer, J. W., Chemical fractionations in meteorites–I, Condensation of the elements, Geochim. Cosmochim. Acta, **31**, 1215-1238, 1967.

Larimer, J. W., and Anders, E., Chemical fractionations in meteorites–II, Abundance patterns and their interpretation, Geochim. Cosmochim. Acta, **31**, 1239-1270, 1967.

______, Chemical fractionations in meteorites–III, Major element fractionations in chondrites, Geochim. Cosmochim. Acta, **34**, 367-387, 1970.

Marvin, U. B., Wood, J. A., and Dickey, J. S., Jr., Ca-Al rich phases in the Allende meteorite, Earth Planet. Sci. Letters, **7**, 346-350, 1970.

Nishamura, M., and Sandell, E. B., Zinc in meteorites, Geochim. Cosmochim. Acta, **28**, 1055-1079, 1964.

Oversby, V. M., and Gast, P. W., Isotopic composition of lead from oceanic islands, J. Geophys. Res., **75**, 2097-2114, 1970.

Papanastassiou, D. A., and Wasserburg, G. J., Initial strontium isotopic abundances and the resolution of small time differences in the formation of planetary objects, Earth Planet. Sci. Letters, **5**, 361-376, 1969.

Papanastassiou, D. A., Wasserburg, G. J., and Burnett, D. S., Rb-Sr ages of lunar rocks from the Sea of Tranquillity, Earth Planet. Sci. Letters, **8**, 1-19, 1970.

Pinson, W. H., Schnetzler, C. C., Beiser, E., Fairbairn, H. W., and Hurley, P. M., Rb-Sr Age of stony meteorites, Geochim. Cosmochim. Acta, **29**, 455-466, 1965.

Poldervaart, A., Chemistry of the earth's crust, *in* Crust of the Earth, Geol. Soc. Am. Spec. Paper 62, pp. 119-144, 1955.

Reed, G. W., Kigoshi, K., and Turkevich, A., Determinations of concentrations of heavy elements in meteorites by activation analysis, Geochim. Cosmochim. Acta, **20**, 122-140, 1960.

Reed, G. W., Jr., and Allen, R. O., Jr., Halogens in chondrites, Geochim. Cosmochim. Acta, **30**, 779-800, 1966.

Reid, A. M., Meyer, C., Harmon, R. S., and Brett, P. R., Metal grains, *in* three Apollo 12 igneous rocks, Earth Planet. Sci. Letters, **9**, 1-5, 1970.

Ringwood, A. E., Chemical evolution of the terrestrial planets, Geochim. Cosmochim. Acta, **30**, 41-104, 1966.

Sanz, H. G., and Wasserburg, G. J., Determination of an internal ^{87}Rb-^{87}Sr isochron for the olivenza chondrite, Earth Planet. Sci. Letters, **6**, 335-345, 1969.

Schmitt, R. A., Smith, R. H., and Olehy, D. A., Cadmium abundances in meteoritic and terrestrial matter, Geochim. Cosmochim. Acta, **27**, 1077-1088, 1963.

Shaw, D. M., Trace element fractionation during anatexis, Geochim. Cosmochim. Acta. **34**, 237-243, 1970.

Smales, A. A., Hughes, T. C., Mapper, D., McInnes, C. A. J., and Webster, R. K., The determination of rubidium and caesium in stony meteorites by neutron activation analysis and by mass spectrometry, Geochim. Cosmochim. Acta, **28**, 209-233, 1964.

Tatsumoto, M., Genetic relations of oceanic basalts as indicated by lead isotopes, Science, **153**, 1094-1101, 1966.

Tatsumoto, M., and Rosholt, J. N., Age of the moon: An isotopic study of uranium-thorium-lead systematics of lunar samples, Science, **167**, 461-463, 1970.

Taylor, S. R., Trace element abundances and the chondritic earth model, Geochim. Cosmochim. Acta, **28**, 1989-1998, 1964.

Turekian, K. K., The terrestrial economy of helium and argon, Geochim. Cosmochim. Acta, **17**, 37-43, 1959.

Turekian, K. K., and Clark, S. P., Inhomogeneous accumulation of the Earth from the primitive solar nebula, Earth Planet. Sci. Letters, **6**, 346-348, 1969.

Turekian, K. K., and Kulp, J. L., The geochemistry of strontium, Geochim. Cosmochim. Acta, **10**, 245-296, 1956.

Turekian, K. K., and Wedepohl, K. H., Distribution of the elements in some major units of the Earth's crust, Geol. Soc. Am. Bull., **72**, 175-192, 1961.

Urey, H. C., Moon, Composition of, *in* International Dictionary of Geophysics, edited by S. K. Runcorn et al., pp. 1-3, Pergamon Press, Oxford, New York, 1967.

Wasserburg, G. J., MacDonald, G. J. F., Houle, F., and Fowler, W. A., Relative contributions of uranium, thorium and potassium to heat production in the Earth, Science, **143**, 465-467, 1964.

Wood, J. A., Chondrites–Their metallic minerals, thermal histories, and parent planets, Icarus, **6**, 1-49, 1967.

Zartman, R. E., and Wasserburg, G. J., The isotopic composition of lead in potassium feldspars from some 1.0-b.y. old North American igneous rocks, Geochim. Cosmochim. Acta, **33**, 901-942, 1969.

3
COMPOSITION OF THE MANTLE AND CORE

by **Don L. Anderson, Charles Sammis, and Tom Jordan**

Ultrasonic, shock wave, and seismic data are combined to estimate the composition of the upper and lower mantle and the outer core. The properties of the upper mantle are consistent with the pyrolite mineralogy or with a mineralogy containing slightly more pyroxene. The lower mantle contains more FeO than the upper mantle.

Finite strain theory is consistent with the seismic velocities in the lower mantle although slight departures from homogeneity cannot be ruled out. The zero-pressure properties of the lower mantle, including their pressure derivatives, are estimated with finite strain theory. There is some suggestion that the lower mantle has a pyroxene stoichiometry and a crystal structure that involves hexagonal close packing of oxygen ions.

Geophysical and geochemical arguments are advanced for sulfur in the core which are equally as compelling as the arguments advanced for silicon. Available laboratory data cannot distinguish between the two possibilities.

INTRODUCTION

In 1952 Francis Birch published a remarkable paper on the elasticity and constitution of the earth's interior. This publication laid the groundwork for all subsequent research in this field, much of which has served to verify Birch's conclusions.

A considerable body of seismic, ultrasonic, shock wave, static compression, and petrologic data has been accumulated since the publication of this pioneering work. The use of surface waves and large seismic arrays has led to refinements in the velocity structure of the earth's interior, particularly in the important transition region of the upper mantle. Free oscillations have made a direct determination of density possible and have supplied a new parameter, the seismic anelasticity of the various regions of the mantle. Shock wave data are now available for a number of metals, rocks, and minerals to pressures of the order of a megabar and more. Ultrasonic measurements have been made on a variety of important rocks and minerals and, in many cases, include the effect of temperature and pressure. High pressure experimental petrology techniques have provided information on the crystal structures and densities of high pressure phases of mantle minerals and have defined the stability fields and melting relations of many important geophysical systems. The use of diamond anvils and x-rays has considerably improved and extended the pressure range of static compression experiments.

The present paper uses many of the above data to interpret the properties of the various regions of the earth's interior but is far from being a complete synthesis of all available information.

STRUCTURE OF THE MANTLE AND CORE

The basic divisions of the earth, (crust, mantle, liquid outer core, and solid inner core) have been known for some time. Considerable refinement has been possible in the last 10 years with the advent of large aperture seimic arrays, the world-wide network of standardized seismic stations, large underground explosions, and the use of long-period surface waves. The compressional velocity and the shear velocity are now fairly well determined throughout the earth from body wave studies. The periods of the earth's free oscillations, which can be measured after large earthquakes, provide data relevant to the distribution of density in the earth although not with the resolving power available from body waves for the seismic velocities.

Figure 1 shows the compressional velocity V_p, the shear velocity V_s, and density ρ as a function of radius in the earth. The two models shown are consistent with the earth's mass, moment of inertia, periods of free oscillation, and travel times of body waves. Two parameter functional relationships between V_s, V_p, and ρ have been used to constrain the behavior of V_s and ρ in the regions between the vertical lines. The values of V_s and ρ have been constrained to be continuous across the regions separated by the light dashed vertical lines. First-order discontinuities are allowed between other regions. Without a smoothing process of some sort, the instability of the seismic inversion problem leads to physically unreasonable oscillatory solutions. Other smoothing techniques give very similar results. The above assumptions also considerably reduce the number of earth models that are consistent with available data.

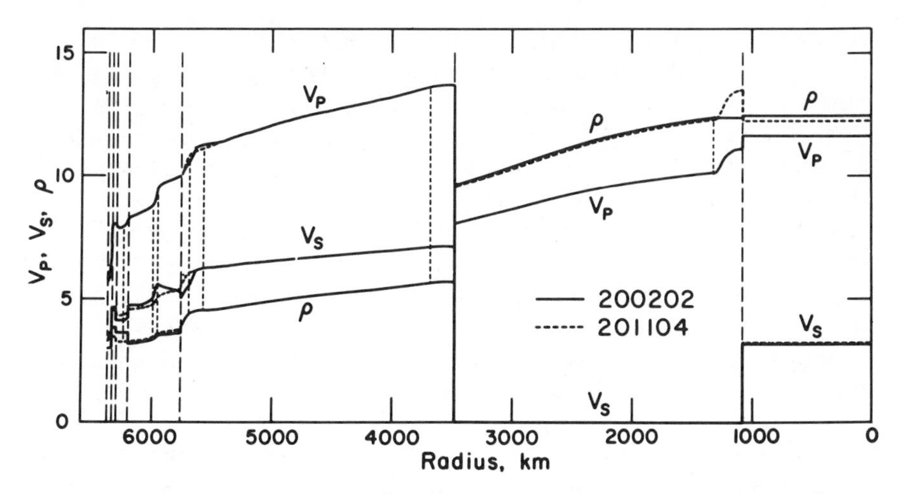

Fig. 1. Compressional velocity V_p, shear velocity V_s (km/sec), and density ρ (gm/cm^3) as a function of radius in the Earth, determined by the inversion of mass, moment of inertia, body wave travel time, and free oscillation data (Anderson and Smith, 1968).

The major features of the earth may be summarized as follows:

1. The upper mantle (Moho–400 km) is made up of a thin (0 to 50 km thick high-velocity layer) that together with the crust constitutes the lithosphere, the strong outer layer of the earth, overlying the low-velocity zone, a partially molten layer of the order of 100 km in thickness (the asthenosphere). The average density of the low-velocity layer and its lid appears, from free oscillation data, to be slightly greater than the 3.3 gm/cm^3 usually assumed from petrological arguments. This high density requires an upper mantle that contains more eclogite or more heavy elements such as Fe and Ti than usually assumed.

The important minerals of the upper mantle are olivine, pyroxene, and garnet in their familiar low-pressure phases. Spinel and amphibole are the dominant accessory minerals near the top of the upper mantle.

2. The transition region (400 to 800 km) is characterized by several abrupt increases in velocity that can be attributed to pressure induced changes in crystal structure resulting ultimately, in a mineral assemblage that is some 20% denser than the corresponding assemblage in the upper mantle. Below some 600 km, much of the silicon is in sixfold coordination with oxygen. Earthquakes do not occur below 700 km, and this is probably related to the major change of coordination that occurs in mantle minerals near this depth.

3. The lower mantle is relatively homogeneous, that is, the velocities and density increase monotonically because of self-compression with no major solid-solid phase changes. However, small discontinuities have been observed.

4. The lower mantle transition zone, just above the core, is a region where the velocities are relatively constant, but the interpretation is complicated by diffraction effects at the mantle-core boundary.

5. The density and compressional velocity change abruptly at the mantle-core boundary and no shear waves are able to penetrate this interface. Thus, the outer core is liquid; it has the properties of an iron-rich alloy. Motions in the fluid outer core are probably responsible for the earth's magnetic field although there is some evidence that these motions have not completely homogenized this region of the earth. The outer core may be chemically zoned.

6. The inner core, on the basis of indirect evidence and free oscillation data is solid although the low shear velocity, relative to the compressional velocity, suggests that it is near the melting point or is partially molten. The density appears to be continuous across the outer core-inner core boundary, but the compressional velocity increases substantially. The boundary between the outer and inner core may be a liquid-solid phase change boundary or a compositional boundary.

In this chapter, we will utilize finite strain theory and laboratory measurements of shock wave and ultrasonic properties to infer the chemical composition of the various regions of the earth.

Mantle

LOWER MANTLE

Finite Strain Theory and Seismic Velocities. Birch (1939), in the first attempt to calculate the variation of seismic velocities resulting from self-compression of a homogeneous spherical shell in a gravitational field, concluded that the density and velocities in the upper mantle could be adequately explained by finite strain theory but that the gradient of velocities in the lower mantle required a gradual change in composition.

Birch's equations for the compression velocity V_p and shear velocity V_s as a function of the hydrostatic strain ϵ and the Lamé parameters and density λ, μ, and ρ

$$\rho V_p^2 = (1 - 2\epsilon)^{5/2}[\lambda + 2\mu - \epsilon(11\lambda + 10\mu)] \tag{1}$$

and

$$\rho V_s^2 = (1 - 2\epsilon)^{5/2}[\mu - \epsilon(3\lambda + 4\mu)] \tag{2}$$

were derived by truncating the expansion of the strain energy function at the second-order term before differentiating with respect to the appropriate strain. The above expressions, in brackets, are therefore incomplete to first order in strain since the differentiation brings in the coefficients of the higher order terms in the strain energy expansion.

The complete first-order Eulerian expressions for velocity are (Sammis et al., 1970:

$$\rho V_p^2 = (1 - 2\epsilon)^{5/2}[\lambda + 2\mu - \epsilon(11\lambda + 10\mu - 18l - 4m)] \tag{3}$$

and

$$\rho V_s^2 = (1 - 2\epsilon)^{5/2}\left[\mu - \epsilon\left(3\lambda + 4\mu + \frac{3}{2}m + \frac{n}{2}\right)\right] \tag{4}$$

where l, m, and n are the third-order coefficients in the expansion of the elastic energy density in terms of the strain invariants, as given by Birch (1939).

For hydrostatic compression, it is not necessary to know the individual values of l, m, and n but only the combinations

$$\zeta = 18l + 4m \tag{5}$$

and

$$\eta = \frac{1}{2}(3m + n) \tag{6}$$

These parameters can be determined from hydrostatic ultrasonic measurements and can be expressed in terms of the pressure derivatives of the two seismic velocities

$$\frac{1}{V_p}\frac{\partial V_p}{\partial P} = \frac{1}{6K}\,\frac{13\lambda + 14\mu - \zeta}{\lambda + 2} \tag{7}$$

$$\frac{1}{V_s}\frac{\partial V_s}{\partial P} = \frac{1}{6K}\,\frac{3\lambda + 6\mu + \eta}{\mu} \tag{8}$$

evaluated at $P = 0$. In all the above expressions, K, λ, and μ refer to the zero-pressure bulk modulus and Lamé parameters.

Similar expressions can be derived using a Lagrangian coordinate system (Hughes and Kelly, 1953; Toupin and Bernstein, 1961):

$$\frac{1}{V_p}\frac{\partial V_p}{dP} = \frac{1}{6K(\lambda + 2\mu)}(7\lambda + 10\mu + 3\nu_1 + 10\nu_2 + 8\nu_3) \tag{9}$$

$$\frac{1}{V_s}\frac{\partial V_s}{\partial P} = \frac{1}{6K\mu}(3\lambda + 6\mu + 3\nu_2 + 4\nu_3) \tag{10}$$

where ν_1, ν_2, and ν_3 are the third-order Lamé constants as defined by Toupin and Bernstein (1961).

The elastic constants and velocity derivatives have been determined for a number of oxides and silicates by ultrasonic techniques. Hence, it is possible, from Eqs. (7) and (8) to calculate the third-order elastic constants ζ and η. These parameters are tabulated in Table 1 for 12 compounds. Note that these higher order parameters are of the same order as the ordinary elastic constants K_s and μ.

Birch (1939) attempted to fit Eqs. (1) and (2) to the seismic velocities in the lower mantle but was unable, by any adjustment of parameters, to satisfy the velocity gradients. He therefore speculated that the lower mantle was inhomogeneous and had a varying composition with depth.

The complete first-order expressions, Eqs. (3) and (4), have enough additional flexibility to satisfy the seismic data for the lower mantle very well, as shown by the middle curves in Fig. 2. The required second-order parameters are quite reasonable when compared with ultrasonic measurements of silicates and oxides of Table 1. Note that it is possible, within the framework of even first-order finite strain theory and with reasonable parameters to have the shear velocity *decrease* with pressure even though the corresponding compressional velocities are increasing with pressure. This is in agreement with lattice calculations of the properties of certain crystal structures (Anderson and Liebermann, 1970; Sammis, 1970). Certain materials, e.g., SiO_2, Ni_2FeO_4, ZnO, and $MgAl_2O_4$, do have shear velocities that decrease with pressure or increase very slowly. Anderson and Liebermann (1970) discuss the implications of a negative shear velocity pressure derivative with regard to crystal instability. It is noteworthy that the spinel structure ($MgO \cdot 2.6 \cdot Al_2O_3$, Ni_2FeO_4) has very small or negative shear velocity pressure derivatives (Table 1) and that certain seismic solutions for the spinel region of the mantle, between 400 and 600 km (Anderson and Julian, 1969) show negative shear velocity gradients. The effect of a temperature gradient would accentuate this tendency.

The equation of state consistent with Eqs. (3) and (4) is

$$P = -3K\epsilon(1 - 2\epsilon)^{5/2}(1 + 2\epsilon\xi) \tag{11}$$

where

$$\xi = \frac{3\zeta + 4n}{12K} \tag{12}$$

is the parameter introduced by Birch (1952). By simultaneously fitting Eqs. (3), (4), and (11) to seismic V_p, V_s, and ρ profiles, it is possible to solve for the zero-pressure parameters V_{p_0}, V_{s_0}, and ρ_0 and the second-order finite strain parameters ζ and η for the various regions of the mantle. Excellent fits can be obtained for most earth models as shown in Table 2, which contains fits to several seismic solutions for the lower mantle. The first two models are from Birch (1964). The rest of the models are the result of inverting free oscillation, surface wave and travel time data (Anderson and Jordan, 1970). The overall fit to the density and velocity profiles for a given lower mantle model is generally

Table 1. Velocity derivatives and finite strain parameters for oxides

		$\frac{1}{V_P}\frac{dV_P}{dP}$ $(10^{-12}\ cm^2/dy)$	$\frac{1}{V_S}\frac{dV_S}{dP}$ $(10^{-12}\ cm^2/dy)$	K_S $(10^{12}\ dy/cm^2)$	μ $(10^{12}\ dy/cm^2)$	ξ $(10^{12}\ dy/cm^2)$	$-\eta$ $(10^{12}\ dy/cm^2)$
Forsterite[a]	Mg_2SiO_4	1.249	.714	1.286	.811	- 1.8	2.6
Olivine[a]	$Fo_{0.93}Fa_{0.07}$	1.211	.737	1.294	.791	- 1.0	2.5
Periclase	MgO	.862	.665	1.622	1.308	- 0.2	1.6
Lime[b]	CaO	1.309	.603	1.059	.761	0.6	3.3
Bromelite[b]	BeO	.538	.0449	2.201	1.618	6.3	12.1
Zincite[b]	ZnO	.613	- 1.138	1.394	.442	10.3	10.2
Corundum	Al_2O_3	.478	.347	2.521	1.613	7.6	5.5
Hematite[b]	Fe_2O_3	.591	.151	2.066	.910	7.7	8.1
Spinel	$MgO \cdot 2.6\ Al_2O_3$	.494	.0762	2.020	1.153	11.1	9.6
Trevorite[c]	$NiFe_2O_4$	.610	- .0082	1.823	.713	9.0	8.4
Garnet	Al-Py	.919	.456	1.770	.943	- 1.5	4.5
Rutile[d]	TiO_2	.825	.101	2.155	1.124	- 3.9	9.3
Lower mantle[e]				1.94	1.30	6.9	.65
				2.19	1.43	15.9	7.9

[a] Polycrystalline.
[b] Kumazawa and Anderson (1969).
[c] Liebermann (1969).
[d] Manghnani (1969). All others from Anderson et al. (1968).
[e] Range of values from Table 2.

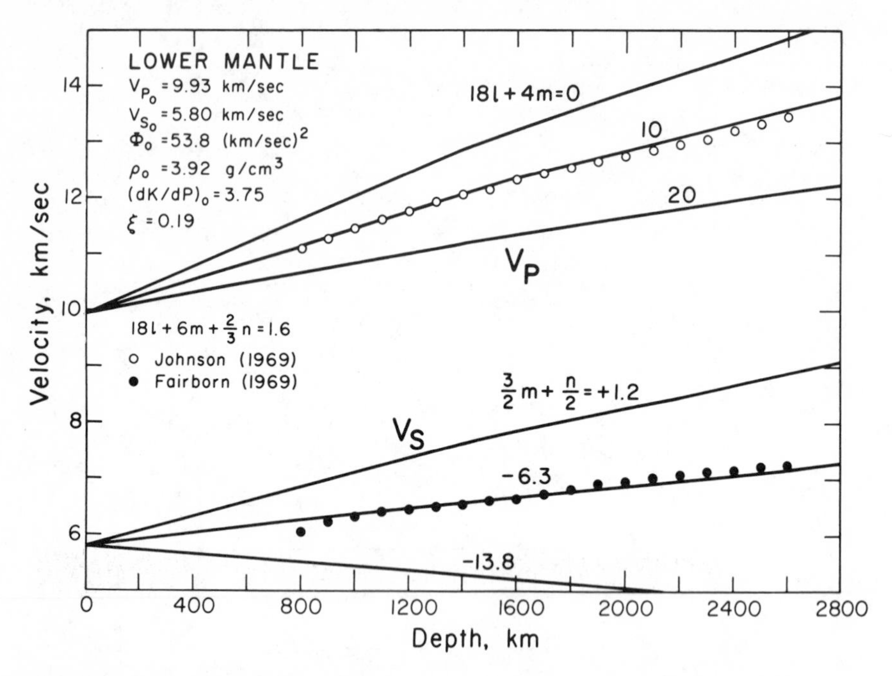

Fig. 2. Compressional velocity V_p and shear velocity V_s as a function of depth for various values of the parameters $18l + 4m$ and $(3m + n)/2$. For these calculations $18l + 6m + 2n/3$ is constrained to + 1.6, a value consistent with the variation of density with depth (Anderson and Jordan, 1970). Compression is calculated assuming homogeneity, but recent work indicates that this assumption is questionable.

better than 1%. We have ignored the distinction between the adiabatic and isothermal elastic constants. If the lower mantle has an adiabatic temperature gradient, the parameters given in Table 2 are appropriate for zero pressure and high temperature [~ 1600°C (Anderson and Jordan, 1970; Clark and Ringwood, 1964)], which represents the lower mantle adiabatic gradient extrapolated to $P = 0$.

The method outlined above and in Sammis et al (1970) provides a framework for using laboratory elastic constant data to calculate the seismic velocities at high pressure in a manner consistent with conventional methods of extrapolating density. Density computed at high pressures from ultrasonic data with the Murnaghan or Birch-Murnaghan finite strain equations of state compare favorably with those found by static or shock wave compression experiments. This gives some assurance that the ultrasonic data can be reliably extrapolated at least to moderate compressions with finite strain theory.

Likewise, a framework has been established to extrapolate mantle seismic data to zero pressure. In this application the finite strain formulation for the velocities is unique in that no assumption of composition is necessary. The more sophisticated finite strain formulations of Thomsen (1970) and the lattice models of Sammis (1970) and Anderson and Liebermann (1970) are all cast in terms of the elastic constants and thus require the specification of composition and crystal structure before they may be applied. Previous discussions of the composition of the deep mantle have been based on ρ_0 and $\Phi_0 = V_{p_0}^2 - (4/3)V_{s_0}^2$. The usefulness of the additional parameter now available is limited by the availability of ultrasonic data on appropriate crystal structures. The shear velocity and its pressure derivatives appear to be more sensitive to coordination effects and details of the crystal structure and may prove to be decisive in understanding the crystal structure of the lower mantle.

Using the modified form of the seismic equation of state (Anderson, 1967, 1969)

$$\frac{\rho_0}{\bar{M}} = 0.0492\ \Phi_0^{1/3}$$

where

$$\Phi_0 = V_{p_0}^2 - \left(\frac{4}{3}\right) V_{s_0}^2$$

the mean atomic weight, M, for the lower mantle models of Table 2, corrected to room temperature ranges from 21.6 to 22.6. Anderson and Jordan (1970) found values from 21.2 to 22.2 using a different set of earth models. The values for ζ and η are comparable to those for such close-packed oxides as Al_2O_3, Fe_2O_3, $MgAl_2O_4$, and $NiFe_2O_4$ but are significantly different from olivine, garnet, MgO, and TiO_2. The normalized velocity derivatives may be more meaningful parameters to compare with the ultrasonic data. For the lower mantle $d\ln V_p/dP$ is about 0.4 to 0.6 and $d\ln V_s/dP$ is about 0.2 to 0.3. Al_2O_3

Table 2. Finite strain parameters and fits for various lower mantle models, assuming homogeneity

Model	ρ_0 (g/cm^3)	V_{p_0}	V_{s_0}	ζ	$-\eta$	RMS error (%)		
		(km/sec)		(10^{12} dy/cm^2)		$[\rho]$	$[V_p]$	$[V_s]$
Birch 1	3.91	9.80	5.78	8.35	5.80	0.64	0.48	0.26
Birch 2	3.96	9.70	5.77	6.57	5.59	0.12	0.57	0.28
200202	4.10	9.89	5.84	9.80	6.92	0.31	0.66	0.27
201104	4.06	10.03	5.91	12.76	7.49	0.37	1.00	0.39
300711	3.96	10.04	6.01	14.30	7.79	0.45	1.71	0.67
301703	3.94	10.00	6.00	13.47	7.64	0.45	1.60	0.67
304702	3.86	10.11	6.06	15.92	7.89	0.47	1.93	0.77
402203	3.93	9.77	5.87	6.94	6.52	0.13	0.41	1.58
402803	3.93	9.85	5.93	8.44	6.95	0.16	0.39	1.48
402813	3.90	9.95	6.02	10.80	7.48	0.22	0.55	1.20

and Fe_2O_3, both having M_2O_3 stoichiometry and corundum structure, are the closest matches. Rutile, garnet, spinels, and olivines have quite different values. Whether this is evidence for a predominantly pyroxene stoichiometry and/or a crystal symmetry involving hexagonal close packing of oxygen ions remains to be seen. Ultimately, the second-order parameter ζ and η may provide useful information pertinent to the crystal structure of the lower mantle.

Composition of the Lower Mantle. Figure 3 illustrates how shock wave data, seismic data, finite strain theory, and the seismic equation of state can be used in discussions of the composition of the deep mantle. The dark circles are shock wave data of McQueen et al (1964) reduced by the method of Anderson and Kanamori (1968) and Ahrens, Anderson, and Ringwood (1969). The raw shock wave data are reduced to Birch-Murnaghan adiabats which are constrained, in the absence of ultrasonic data, to satisfy the seismic equation of state at zero pressure. This yields values for ρ_0 and $\Phi_0 = (\partial P/\partial \rho)_0$ for phases that are not available for measurement at standard conditions. The lower mantle point is a temperature corrected Birch-Murnaghan extrapolation of mantle ρ and Φ data

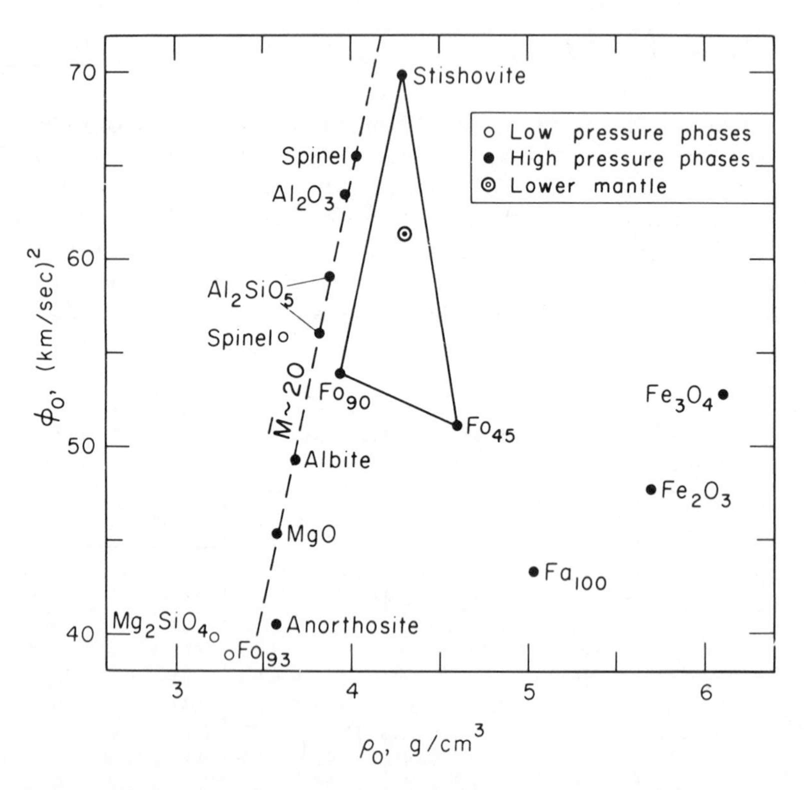

Fig. 3. Φ_0 versus ρ_0 from shock wave experiments for various rocks and minerals (Anderson, 1970) and one solution for the lower mantle (Anderson and Jordan, 1970).

(Anderson and Jordan, 1970), obtained from inversion of free oscillation data (model 200204). The mantle point falls in the field defined by Fo_{90} $[(Mg_{0.90}Fe_{0.10})_2SiO_4]$, Fo_{45} $[(Mg_{0.45}Fe_{0.55})_2SiO_4]$ and SiO_2 (stishovite). The inferred composition of the lower mantle for this model is 0.32 MgO + 0.18 FeO + 0.50 SiO_2 (molar) or $(Mg_{0.64}\ Fe_{0.36})SiO_3$. A slightly different technique (Anderson 1970) gave 0.38 MgO + 0.16 FeO + 0.46 SiO_2 (molar) for the same model and 0.49 MgO + 0.12 FeO + 0.39 SiO_2 (molar) for Birch 2. These compositions imply pyroxene contents of 80% and 43%, respectively. Other free oscillation models require less SiO_2 in the lower mantle but all models fall well above the high-pressure olivine line and require considerably more FeO than is typical of upper mantle petrologies. A conclusive study of the composition of the lower mantle will require shock wave data on garnets, relatively iron-rich pyroxenes, and such natural rocks as eclogites and pyroxenites.

Density-velocity Systematics. Birch (1961b) demonstrated a relationship between density, compressional velocity, and mean atomic weight for rocks and minerals. He speculated that there was a unique relation between density and compressional velocity at constant mean atomic weight regardless of whether the density change was a result of crystal structure effects, pressure, or temperature. In lieu of a theory that provided an estimate of the change of velocity with pressure, he proposed that the density in the earth could be found from the compressional velocity by applying the systematics found at low pressure. This conjecture can now be tested with available ultrasonic data.

Figures 4 and 5 are Birch diagrams showing the relation between velocity and density with mean atomic weight as a parameter. For compressional velocity, the linear relation between V_p and ρ for minerals of the same mean atomic weight is well substantiated. In addition, one can draw isostructural lines, for example, for the spinels and the olivines.

The light dashed lines in Figure 4 are $V_p - \rho$ trajectories calculated from the complete first-order finite strain expressions, Eqs. (3) and (4) and the ultrasonic data in Table 1. In most cases, the trend of the pressure trajectories is similar to the slope of the lines of constant mean atomic weight. The effect of a 1000°K rise in temperature is also shown.

An extrapolated value for the lower mantle is also shown in Figure 4. The hatched region is an estimate of the room temperature value of ρ_0 and V_{p_0} to provide direct comparison with the values for the oxides and silicates. The mean atomic weight consistent with the temperature corrected value for the lower mantle is 21.0 to 21.7, which can be compared with the estimates of 22.5 by Birch (1961b), 22.8 to 22.9 by Anderson (1970), and 21.0 to 23.7 by Anderson and Jordan (1970). Previous estimates of the zero-pressure density of the lower mantle are 3.93 ± 0.03 (Birch, 1961b), 4.15 ± 0.05 (Clark and Ringwood, 1964), and 4.0 ± 0.1 (Anderson and Jordan, 1970). The compressional velocity lies well above the spinel line, as might be expected for the closer packed phases of the lower mantle.

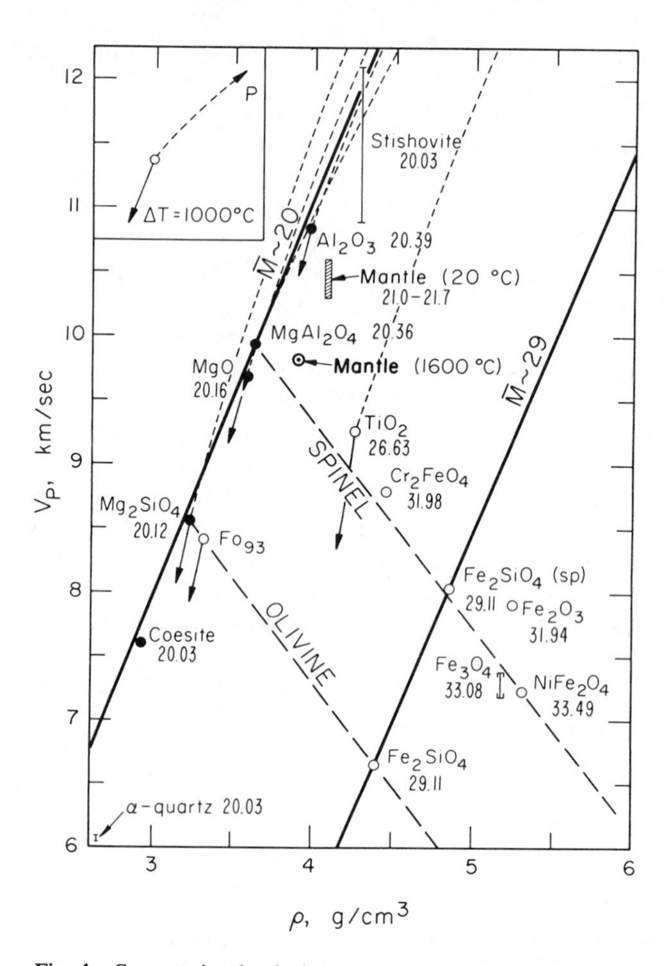

Fig. 4. Compressional velocity versus density for various oxides and silicates (data from Anderson et al, 1968; Liebermann, 1969; Manghnani, 1969; Mizutani et al, 1970). The dark circles are minerals with mean atomic weight near 20. The light dashed lines are pressure trajectories calculated from finite strain theory and the parameters of Table 1. The solid lines with arrows show the effect of a 1000°C rise in temperature.

The $V_s - \rho$ systematics, Fig. 5, are not as regular as the $V_p - \rho$ systematics and the $M \sim 20$ line is schematic. The dark circles are oxides with mean atomic weight near 20; the scatter indicates that structural effects may be more important for shear waves than for compressional waves.

The spinel and olivine lines are also schematic. The pressure trajectories show a much greater variability than the corresponding curves for the compressional velocity. This is again attributed to the crystal structure effects that apparently are more important for the pressure derivatives of the shear velocity. Some materials, e.g., rutile and spinel, have very low pressure derivatives of shear velocity or negative derivatives, e.g., ZnO, quartz, $NiFe_2O_4$. Between some 400 and 600 km in the mantle, olivine is in the spinel structure and the shear velocity gradient in this region is constant or slightly negative (Anderson and Julian, 1969).

An extrapolated lower mantle value is also shown on Fig. 5. Because of the scatter and the lack of a simple relation among density, V_s, and mean atomic weight, the only conclusion is that the mean atomic weight of the lower mantle is consistent with that found from the compressional velocity. When data on more close packed oxides is available it should be possible to speculate on the crystal structure of the lower mantle from these diagrams. For the moment, all that can be said is that the density and velocity are consistent with a denser ionic packing than occurs in such close packed oxides as MgO, $MgAl_2O_4$, and garnets. In these compounds the packing ratio, defined as the fraction of the cell volume that is occupied by ions, ranges from 0.63 to 0.66. The lower mantle has a

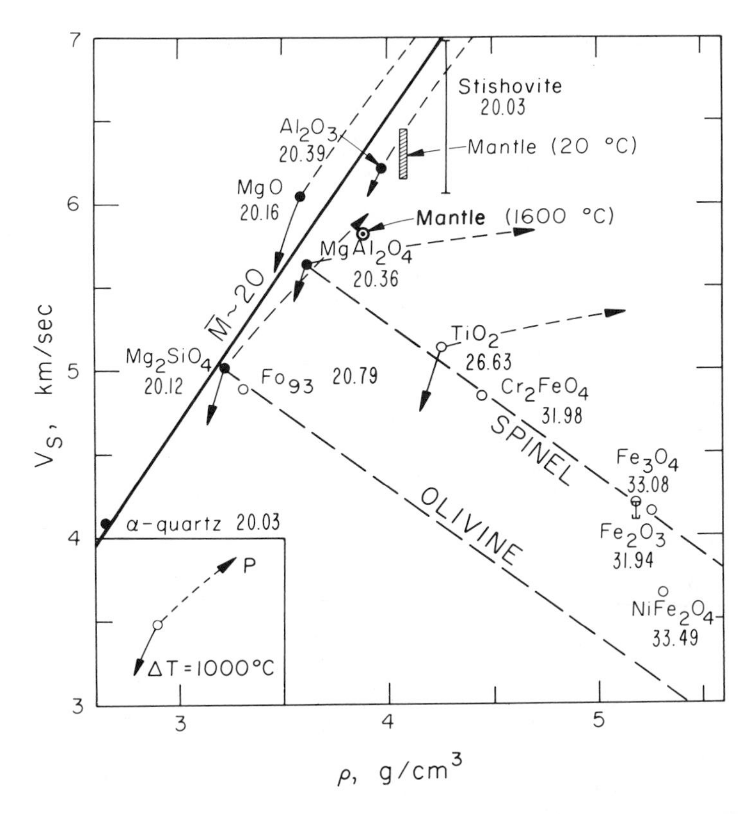

Fig. 5. Shear velocity versus density for various oxides and silicates, showing the effect of pressure (light dashed lines) and temperature (solid lines with arrows).

packing fraction close to that of Al_2O_3, 0.72. For comparison, stishovite, the high pressure form of SiO_2, with the Si ions in sixfold coordination with the oxygen ions, has a packing fraction of 0.84. High pressure structures of magnesium, aluminum silicates with Si in sixfold coordination would have packing fractions of approximately 0.70 (forsterite stoichiometry), 0.73 (pyrope stoichiometry), and 0.74 (enstatite stoichiometry).

UPPER MANTLE

Mineralogy of the Upper Mantle. Ultrasonic data is now available for many of the minerals that are thought to occur in the upper mantle (Anderson et al, 1968). Temperature and pressure derivatives are available for olivine, garnet, and spinel but not for amphibole, feldspars, or pyroxene. The interpretation of the seismic data is complicated by the lateral variability of the velocity in the upper 200 km of the mantle. Below 200 km, the various solutions tend to be fairly consistent although even at these depths there is some evidence for lateral heterogeneity.

For illustrative purposes we will take the recent solutions of Johnson (1967), Anderson and Julian (1969), and Anderson and Jordan (in preparation) which give, at 200 km, $V_p = 8.30$ km/sec, $V_s = 4.64$ km/sec, and $\rho = 3.5$ gm/cm^3. In order to compare these values with ultrasonic data, they must be corrected to room temperature and pressure. We will assume that the temperature and pressure at 200 km are 1200°C and 67 kb, respectively, and will adopt the following values for the derivatives:

$$\left(\frac{\partial V_p}{\partial P}\right)_T = 10 \times 10^{-3}\ \text{km/sec/kb}$$

$$\left(\frac{\partial V_s}{\partial P}\right)_T = 3 \times 10^{-3}\ \text{km/sec/kb}$$

$$\left(\frac{\partial V_p}{\partial T}\right)_P = -4 \times 10^{-4}\ \text{km/sec/°C}$$

$$\left(\frac{\partial V_s}{\partial T}\right)_P = -3 \times 10^{-4}\ \text{km/sec/°C}$$

$$\alpha = -\frac{1}{\rho}\frac{\partial \rho}{\partial T} = 20 \times 10^{-6}/°\text{C}$$

With these parameters, the properties of the mantle at standard conditions are as shown in the second column of Table 3. Also shown in this table are velocities and densities of olivine, pyroxene, and garnet, all natural single crystals, and computed properties for two mixtures of these minerals. If these minerals are representative of those found in the upper mantle, excellent agreement with all three properties can be obtained with a mineralogy containing 40% olivine, 50% pyroxene, and 10% garnet. The main discrepancy with the pyrolite mineralogy is with the compressional velocity. If pyroxene has a $(\partial V_p/\partial P)_T$ 20% less or a $(\partial V_p/\partial T)_P$ 40% greater than the values adopted here, the pyrolite mineralogy would be acceptable. Either possibility, or some combination, is acceptable considering the range of the parameters that have been measured. The definition of pyrolite is also flexible enough to cover the alternate mineralogy that would involve a larger ratio basalt: peridotite than the 1:3 ratio adopted by Ringwood (1970). Anderson (1970) using Birch's ultrasonic data (1961a) on ultrabasic rocks concluded that the composition of the mantle near 200 km could be explained with a mineralogy of 45 to 75% olivine, 25 to 50% pyroxene, and up to about 15% garnet. This range is probably a good estimate of the present uncertainty. Velocity and density depend, of course, on the composition of the olivine, pyroxene, and garnet as well as their relative proportions and these compositions may vary with depth and be quite different from those that have been measured. In addition, the pressure and temperature derivatives of density and velocity are not constant as assumed above. It is nevertheless encouraging that there is such good agreement among the seismic, ultrasonic, and petrologic data.

The Low-velocity Zone. The interpretation of seismic velocities in terms of composition is complicated by the presence of a region in the upper mantle of abnormally low velocities and high seismic attenuation. The velocities in the low velocity zone are too low to be explained by any reasonable mineralogy or temperature gradient (Anderson and Sammis, 1970) and, in addition, the boundaries of this zone are apparently quite sharp (Whitcomb and Anderson, 1970). By combining ultrasonic, petrologic, heat flow, and melting relations

Table 3. Velocities and densities of mantle minerals (after Anderson et al., 1968) and various mixes

	Mantle (200 km)		Olivine $Fo_{0.93}$	Pyroxene $En_{0.85}$	Garnet Al-Py	Mix 1[a]	Mix 2[b]
	(ambient)	(STP)					
V_P (km/sec)	8.30	8.11	8.42	7.85	8.53	8.22	8.14
V_S (km/sec)	4.52	4.80	4.89	4.76	4.76	4.83	4.81
ρ (gm/cm^3)	3.50	3.42	3.31	3.34	4.16	3.38	3.39

[a]Pyrolite; 56% olivine, 35% pyroxene, 9% garnet (Ringwood, 1970).
[b]40% olivine, 50% pyroxene, 10% garnet.

Anderson and Sammis (1970) concluded that this zone could be best explained by partial melting or dehydration of a mineral assemblage containing a small amount of water.

Walsh (1969) has used the theory of Eshelby (1957) to derive equations for the elastic wave velocities in a two-phase material composed of a solid isotropic matrix with randomly oriented penny-shaped melt zones. The velocities are a function of the elastic properties of the two phases, the melt concentration, and the aspect ratio of the melt zones. Birch (1969) calculated the velocities of an olivine matrix containing spherical inclusions of basaltic glass with zero rigidity as an approximation to partially molten material. He concluded that the low velocity zone could be explained by a melt concentration of approximately 6%. Penny-shaped cracks more nearly approximate the situation for grain boundary melting and adequately account for the only pertinent experimental data (Anderson and Spetzler, 1970; Spetzler and Anderson, 1968).

Figure 6 shows the compressional and shear velocity as a function of melt concentration using the Eshelby-Walsh theory and the parameters of Birch (1969). The upper curves are for aspect ratio $\alpha = 1$, which correspond to spherical melt zones, the case that Birch considered. Because of the importance of the geometry, the velocities in the low velocity zone cannot be unambiguously interpreted in terms of melt content. If the aspect ratio is as small as 10^{-2}, a melt content of only 1% is required to decrease the velocities to the lowest values measured in the low-velocity zone. Since the low-melting components will probably be concentrated at grain boundaries, this is not an unreasonable aspect ratio.

Core

The molten outer core is generally considered to be an Fe-Ni melt containing some lighter alloying element or elements. The zero-pressure density of molten iron at 1540°C is 7.02 gm/cm^3. At room temperature, the density would be 7.13 to 7.23 gm/cm^3, using the data discussed by Clark (1963). The zero-pressure, high temperature density for the outer core has been estimated to be 6.4 ± 0.5 gm/cm^3 (Birch, 1952), and 7.2 ± 0.3 gm/cm^3 (McQueen et al., 1964). These estimates are based on Adams-Williamson solutions for the core and assume that the entire outer core is homogeneous. Recent seismic solutions for the core make this assumption questionable (Kovach and Toksöz, personal communications, 1970). Attempts to fit, simultaneously, the density and compressional velocity of the whole outer core by finite strain theory have been unsuccessful, which suggests that this region, even though it is molten, is not homogeneous. The estimates of zero-pressure properties are therefore open to some question.

Fortunately, shock wave data exists for some Fe-Ni and Fe-Si alloys at core pressures. Some uncertainty exists in the temperature corrections but these are probably much less than errors involved in extrapolating core properties to standard conditions. McQueen and Marsh (1966), on the basis of shock wave experiments on iron-nickel alloys, concluded that pure iron is about 8% denser than the outer core. Knopoff and MacDonald (1960) and Birch (1961b) had

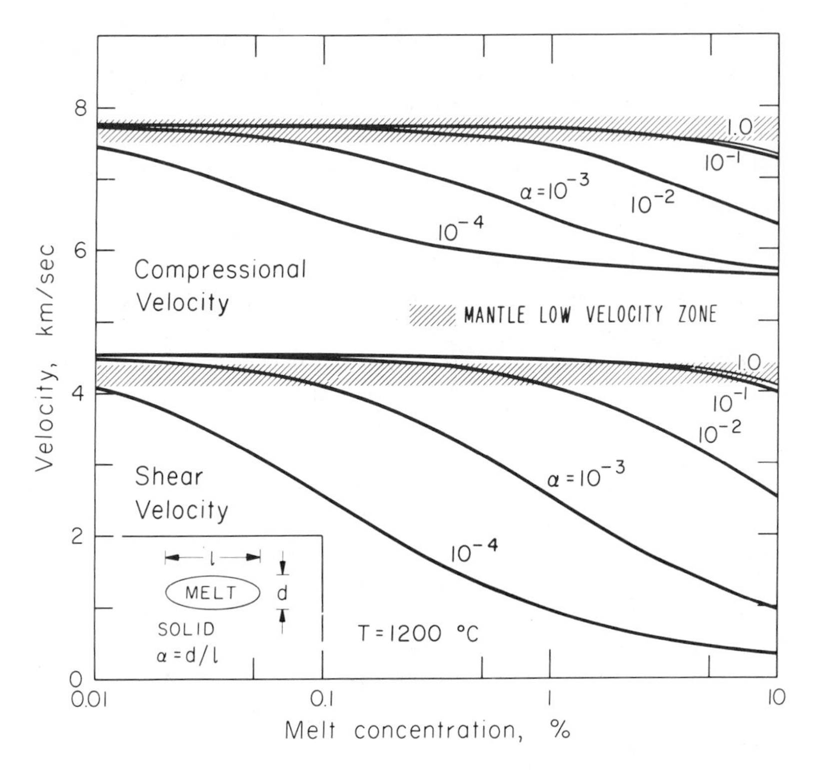

Fig. 6. Compressional and shear velocity as a function of melt concentration and melt zone aspect ratio for a solid olivine matrix containing ellipsoidal pockets of basalt melt. The stippled regions show the range of velocities in the low-velocity zone.

previously concluded that the outer core is roughly 15% less dense than iron at the same temperature and pressure. In a later paper, Birch (1964) reduced this discrepancy to 10%. Most of these authors have ignored the roughly 5% decrease in density upon melting, or have assumed implicitly that shocked metals behave as their melts.

The zero-pressure high temperature bulk sound speed, C_0, of the core has been estimated to be 5.05 km/sec (Birch, 1952) and 5.03 ± 0.15 km/sec (McQueen et al, 1964). The corresponding values for shocked Fe-Ni alloys range from 3.1 to 3.7 km/sec. The decrease in C_0 upon melting is roughly 7 to 15%. The shock wave values should probably be decreased further by a small amount in order to make them comparable to extrapolated core temperature values.

The above results suggest that a pure Fe-Ni core has too large a density and too small a bulk sound speed to be compatible with geophysical data. A lighter alloying element that would increase the bulk sound speed is required.

There are two major candidates for this element. One is silicon (Knopoff and

MacDonald, 1960; Ringwood, 1966). The other is sulfur. Both elements would have the desired properties of decreasing the density and increasing the seismic velocity of pure iron.

Ringwood (1966) has discussed at great length the geochemical arguments for silicon in the core. He argues for a hot origin for the earth and assumes that most of the sulfur escaped from the earth as volatiles.

Iron-sulfur Core. The possibility of sulfur in the core has been extensively discussed by Murthy and Hall (1970). They point out that the conditions under which sulfur and silicon go into the core are radically different. The presence of silicon in the core requires large-scale temperature reduction and the formation of a massive CO atmosphere that must be blown off after accretion, carrying with it other volatiles (Ringwood, 1966). On the other hand, even a relatively cold planet ($>990^\circ C$) can produce a sulfur-rich iron core. In the present earth, these temperatures are available below about 30 km.

Murthy and Hall (1970) compared the relative abundances of such volatile elements as the halogens, rare gases, sulfur, and water in the crust and mantle with chondritic abundances. Sulfur was shown to be depleted in the crust and mantle by several orders of magnitude relative to the other volatile elements. They concluded, on several geochemical grounds, that sulfur was more likely to be incorporated into the earth than not, and since it was obviously deficient in the crust and mantle, it had probably been segregated into the core as a sulfur-rich iron melt. They did not investigate the geophysical implications of their proposal.

Table 4 summarizes the major element compositions of three classes of meteorites and various models for the earth's mantle. No single class of meteorites has the proper overall composition to match the data for the earth. For example, the mean atomic weight of the earth is about 27.0. Meteorites have the following mean atomic weights; carbonaceous chondrites (H_2O free), 23.4 to 24.0; ordinary chondrites, 24.4; "high iron" chondrites, 25.1; enstatite chondrites, 25.6, and iron meteorites, 55. The earth's core comprises 32.5% of the earth's mass as compared with the FeS + Fe + Ni percentage of carbonaceous and ordinary chondrites of 24.0% and 19.2%, respectively. By mixing of 40% carbonaceous I, 46% ordinary, and 14% iron meteorites, we can obtain the proper core/mantle ratio. These are close to the proportions assumed by Murthy and Hall (1970). The mean atomic weight and the zero-pressure density of the resulting core are 50.5 and 6.34 gm/cm^3, respectively; the sulfur content is 14 wt %. The density will be reduced by up to 5% upon melting. The mantle, for this mix, contains 18.4 wt % FeO. There is good agreement between the mantle composition derived here and that determined in a previous section as can be seen by comparing columns 4 and 6. Note also the good agreement with Urey's (personal communication, 1970) estimate of the composition of the lunar interior (column 7). Note the large increase in FeO content of the mantle over that of pyrolite, column 8, which is presumably representative of the upper mantle (Ringwood, 1966).

Table 4. Composition of meteorites, earth, and moon (weight percent)

	1	2	3	4	5	6	7	8
	Carbonaceous chondrites I	Ordinary chondrites	Iron[a] meteorites	Mix[b]	Reduced carbonaceous chondrites I	Lower mantle	Lunar mantle	Upper mantle
SiO_2	32.5	38.8		49.6	47.7	51	49.1	49.6
MgO	21.9	24.3		32.0	42.0	28	33.8	41.1
FeO	14.5	12.1		18.4	10.3	21	17.1	9.3
	68.9	75.2		100.0	100.0	100	100.0	100.0
FeS	23.8	6.0	3.7	12.8	–			
Fe	0.2	11.8	89.1	18.0	25.9			
Ni	0.0	1.4	7.2	1.7	1.7			
Si	–	–	–	–	3.5			
	24.0	19.2	100.0	32.5	31.1			
$\overline{M}$ (core)				50.5	50.4			
ρ_0 (core)				6.34	6.24			

[a] Coarse octahedrites.
[b] 40% Type I carbonaceous chondrites, 46% ordinary chondrites, 14% iron meteorites. The composition for "mantle" for this mixture can be well approximated by mixing 75% "lower mantle" material with 25% "upper mantle" (pyrolite) material. These percentages would suggest that the lower mantle begins near 600 km.

The "mantle" of the meteorite mix, column 4, is more iron rich than solutions for the upper mantle, column 8, and slightly less iron rich than solutions for the lower mantle, column 6. If we assume that the upper 600 km of the mantle has the composition of pyrolite and the rest of the mantle has the composition determined for the lower mantle, then the average composition of the mantle is similar to column 4. Whitcomb and Anderson (1970) have previously concluded, on the basis of seismic wave reflection amplitudes, that the 600-km discontinuity probably involves a larger density contrast, or is sharper, than the other transition zones in the mantle. Either alternative is consistent with a compositional change, as well as a phase change, being involved at the 600-km discontinuity.

Iron-silicon Core. The presence of a light crust and a dense core indicate that the earth is a gravitationally differentiated body. It is usually assumed, however, that the great bulk of the earth, i.e., the mantle, is chemically homogeneous. The results of this paper suggest that the mantle itself may be gravitationally stratified and that pyrolite may be a differentiate of a more primitive material just as basalt is a differentiate of pyrolite.

The composition of column 5 was determined by reducing some of the FeO and SiO_2 of a C-, S-, and H_2O-free Type I carbonaceous chondrite in order to obtain a mantle composition consistent with pyrolite and to obtain the proper silicate metal or mantle/core ratio (Ringwood, 1966). The resulting core has 11 wt % silicon, a mean atomic weight of 50.4, and a zero-pressure density of 6.24 gm/cm^3; the latter two values are very close to those estimated above for the iron-sulfur core.

Balchan and Cowan (1966) determined the density of shocked Fe-Si alloys at conditions comparable to those in the core and concluded that their results were consistent with a core containing 14-20 wt % silicon in iron. The zero-pressure, room temperature densities of these compositions are 7.02 to 7.25 gm/cm^3. The density of α-iron, 7.87 gm/cm^3, is also much greater than the sulfides of iron; i.e., FeS (troilite), 4.83 gm/cm^3; FeS (sphalerite structure), 3.60 gm/cm^3; FeS (wurtzite structure), 3.54 gm/cm^3; FeS_2 (pyrite), 5.02 gm/cm^3; and FeS_2 (marcasite), 4.89 gm/cm^3.

The zero-pressure bulk sound speed C_0 of the Fe-Si alloys lies between 4.1 ± 0.1 (4% Si) and 5.4 ± 0.1 km/sec (20% Si), which bracket estimates for the core. Comparable shock wave data on iron sulfides is not available but zero-pressure ultrasonic data on pyrite (FeS_2) gives 5.23 km/sec (Simmons, 1965), which is much higher than the shock wave, 3.45 km/sec, for pure iron at zero pressure (McQueen and Marsh, 1966). The bulk sound speeds in such sulfides as CdS and ZnS are 40 to 45% greater than in the metal. The approximate zero-pressure bulk sound speed of the FeS-Fe core of Table 4 is 3.9 km/sec. For Ringwood's Fe-Si core, the corresponding value is about 4.2 km/sec. Thus it appears that both silicon and sulfur serve to decrease the density and increase the velocity of iron. These are very crude estimates and do not take into account phase changes or melting. The point is that the two core models have very similar physical properties (M, ρ_0, and C_0). Shock wave data on Fe-FeS mixtures may be able to resolve the possibilities.

Origin of the Mantle and Core

Accretion and Condensation. Larimer (1967) has outlined the condensation history of a cooling gas of cosmic composition. Compounds such as $CaTiO_3$, $MgAl_2O_4$, Al_2SiO_5, and $CaAl_2Si_2O_8$ will condense first at temperatures between 1740 and 1620°K. Iron will condense next at 1620°K. Magnesium-rich pyroxenes and olivines condense between 1470 and 1420°K. FeS condenses at 680°K and H_2O at 210°K. All the above temperatures were calculated assuming a total pressure of 6.6 x 10^{-3} atm. Larimer and Anders (1967) conclude that the fractionation patterns in meteorites occurred in the solar nebula as it cooled from high temperatures and could not be produced in the meteorite parent bodies. Specifically, they infer the following accretion temperatures from the abundance patterns: carbonaceous chondrites, $\leqslant$400°K, enstatite and ordinary chondrites 400 to 650°K, and the major fraction of iron meteorites $\geqslant$1100°K.

Accretion of the planets presumably involved planetesimals that condensed over the entire temperature range although the different planets may have incorporated different proportions of the various condensates. As a planet grows, the accretional energy increases and the temperature at the surface of the body is controlled by a balance between the available gravitational or kinetic energy, the heat capacity and thermal conductivity of the surface layer, the heats of reaction involved in chemical reactions occurring at the surface, and the reradiation of energy to the dust-gas cloud. As the planet grows, it will be less and less capable of retaining the volatiles brought in by the accreting particles and the fractionation, in time, will be the reverse of the condensation procedure. When the surface temperature reaches 680°K, the reaction $FeS + H_2 \rightarrow Fe + H_2S$ will occur if the planet is accreting in a H_2-rich environment and the planet will thereafter not be able to incorporate much FeS into its interior. With the previously discussed meteorite mix model for the earth, this temperature was reached when the planet had assembled some 50% of its present mass. This would require that the time scale for accretion be much longer than the 10^5 to 10^6 years discussed by Hanks and Anderson (1969). Alternatively, the FeS-H_2 reaction could start earlier in the accretional history of the planet if it were not 100% efficient, or it would not occur at all if the H_2 had already left the environment (Ringwood, 1966).

Reactions such as the following:

$$Mg_3Si_4(OH)_2O_{10} \rightleftharpoons 3MgSiO_3 + SiO_2 + H_2O$$

$$M(OH)_2 \rightleftharpoons MO + H_2O \quad (M = Mg, Ca, Fe)$$

$$CH_4 \rightleftharpoons C + 2H_2$$

$$FeS_2 + MgSiO_3 + H_2O \rightleftharpoons FeS + FeMgSiO_4 + H_2S$$

between the atmosphere and the surface of the accreting planet could tend to buffer the surface temperature and keep it below 500°K for long periods of time during the accretion process. Substantial amounts of heat could also be buried in the planet by the impacting bodies and be unavailable for reradiation if the rate of accretion is faster than the rate of heat conduction to the surface. The above considerations would prolong the period available for trapping FeS in the interior.

Implicit in the above discussion is the assumption that the planets did not start to accrete until the condensation process was complete and the solar nebula was relatively cold. A growing planet would incorporate the solid portion of cosmic abundances in its interior in its initial stages of growth but could retain only the more refractory compounds as it grew. A large-scale redistribution of material must occur later in its history in order to form a core and to transport the retained volatiles and the less refractory compounds from the center of the body to the surface to form the crust and atmosphere. Thus the earth, for example, must turn itself inside out to obtain its present configuration.

If the planets accreted while condensation was taking place, a chemically zoned planet would result with such compounds as $CaTiO_3$, $MgAl_2O_4$, Al_2SiO_5, $CaAl_2Si_2O_8$, and Fe forming the central part of the body, these compounds condensing above 1620°K. In the next 160°K range of cooling, compounds such as Ca_2SiO_4, $CaSiO_3$, $CaMgSi_2O_6$, $KAlSi_3O_8$, $MgSiO_3$, SiO_2, and Ni would condense. Between 1420 and 1120°K, the rest of the important mantle minerals, Mg_2SiO_4, $NaAlSi_3O_8$, and $(Na, K)_2SiO_3$ would condense. Further cooling would bring in such metals as Cu, Ge, Au, Ga, Sn, Ag, Pb, the sulfides of Zn, Fe, and Cd, and finally, Fe_3O_4 (400°K), the hydrated silicates (~ 300°K), H_2O (210°K), and the rare gases. Turekian and Clark (1969) have discussed some of the implications of this model of planetary formation. The scarcity of sulfur, particularly relative to water, in the crust and upper mantle, is a major problem since neither the core nor volatility can be invoked nor can it be argued that carbonaceous chondrites are enriched relative to solar abundances.

Formation of an Fe-S Core. In both the above models of accretion, the composition of the planet changes as it grows and there is no compelling reason to assume that the composition of the various planets should be the same or the same as any single class of meteorites. The work of Larimer (1967) and Larimer and Anders (1967) suggests that carbonaceous chondrites simply represent the last material to condense from a cooling nebula rather than representing primordial material. In the cold accretion model, this material would be mixed with material that condensed at higher temperatures, such as ordinary chondritic and iron meteorites, to form the nucleus of the accreting planet which would contain most of the volatiles such as H_2O and FeS. As the planet grows the material that can be retained on the surface and incorporated into the interior becomes progressively less representative of the material available. In the terminal stages of accretion the surface temperatures will decrease with time (Hanks and Anderson, 1969) and eventually will be controlled by solar heating. The amount of volatiles and low-temperature minerals retained by a planet will depend on the fraction of the planet that accretes during the initial and terminal stages. In the accretion-during-condensation model, the full complement of volatiles and sulfides must be brought in during the terminal stage. In both these models, much of the core-forming material, Fe + FeS in the former and Fe in the latter, is already near the center of the earth. In alternate proposals, the core-forming material is either distributed evenly throughout the mantle (Birch, 1965) or at the surface of the accreting planet (Ringwood, 1966).

One important boundary condition for the formation of the earth is that the outer core at present, and probably 3×10^9 years ago, be molten. If we take the temperature in the mantle to be 1880°C at 620 km (Anderson, 1967) and assume an adiabatic temperature gradient of 0.5°C/km from that depth to the core-mantle boundary, the temperature at the top of the core will be about 3000°C, which is of the order of the melting temperature of iron at core pressures. For the condensation model, the temperature of the primitive iron core is 1150 to 1350°C and will increase to 2300 to 2500°C because of adiabatic compression when the planet is earth sized. This can be compared with Kennedy's (1966) estimate of 3725°C for the melting temperature of pure iron at the boundary of the inner core. These temperatures are all subject to great uncertainty, but it appears difficult with the condensation model to have a molten iron core. It is even more difficult to raise the central part of the earth above the melting temperature of pure iron in the cold accretion model (Hanks and Anderson, 1969). It appears that the extra component in the core must also serve to decrease its melting temperature. This is easily accomplished with sulfur.

The eutectic temperature for the system Fe-FeS is 990°C and is remarkably insensitive to pressure up to at least 30 kb (Brett and Bell, 1969). The eutectic composition is 31 wt % sulfur at 1 atm and 27 wt % sulfur at 30 kb. In an earth of meteoritic composition, a sulfur-rich iron liquid would be the first melt to be formed. Core formation could proceed under these conditions at a temperature some 600°C lower than would be required to initiate melting in pure iron. In the vicinity of the core, the melting temperature is some 1600°C lower than that for pure iron. If the earth accreted from cold particles under conditions of radiative equilibrium, the temperatures will be highest in the upper mantle, which is where melting will commence. In most plausible thermal history calculations, the melting point of iron increases with depth much more rapidly than the actual temperatures and any sinking molten iron will refreeze. Unless the earth was very hot during most of the accretional process, it will be difficult to get the iron to the center of the planet. This difficulty does not occur for the Fe-S model of the core since gravitational accretion energy, adiabatic compression, and radioactive heating bring the temperatures throughout most of the earth above 1000°C early in its history and probably, during most of the accretional process (Hanks and Anderson, 1969). Since the eutectic temperature seems to be independent of pressure, core formation will be self-substaining. The increase of gravitational energy resulting from core formation leads to a temperature rise of about 1600 to 2000°C throughout most of the earth (Birch, 1965; Hanks and Anderson, 1969). This would be adequate to melt the rest of the iron in the mantle which would drain into the core and to cause extensive melting of silicates and differentiation of the crust and upper mantle. The short time scale of accretion required by Hanks and Anderson (1969) to form a molten iron core within the first 10^9 years of earth history is no longer required but, of course, cannot be ruled out.

Possible Core of Mars. The ease of manufacturing a core in the earth might suggest that Mars should have a substantial core. Current estimates of the initial temperature for Mars, assuming cold accretion, allow it to retain all infalling FeS. The average temperature in Mars below some 600 km is about 1500°C if the planet accreted in 3×10^5 years (Hanks and Anderson, 1969), which is well above the Fe-FeS eutectic temperature. The liquidus for an FeS-rich FeS-Fe system is 1100°C at atmospheric pressure, 1600°C at 60 kb, and about 3400°C at the center of Mars if the slope of the melting curve of FeS (Sharp, 1969) remains constant to hundreds of kilobars. Thus, most of the deep interior of

Mars is much closer to the eutectic temperature than to the liquidus temperature, and only a fraction of the planet's complement of Fe and FeS will be molten. For a carbonaceous chondrite I composition, the total FeS plus Fe content is 24 wt %. Binder (1969) estimated that the mass of a Martian core would be 2.7 to 4.9% of the mass of the planet, or about 11 to 20% of the metallic and sulfide content available. The core can be much larger if it is less dense than pure iron. With this model, the core of Mars would be richer in sulfur than the core of the earth, and much Fe would be left in the mantle of Mars. The small size of the Martian core and the small size of the planet suppress the importance of gravitational heating resulting from core formation.

ACKNOWLEDGMENTS

This work was supported by National Science Foundation Grant GA-12703, and NASA Grant NGL 05-002-069. The authors would like to thank Rama Murthy and Hall for a preprint of their paper.

REFERENCES

Ahrens, T. J., Anderson, D. L., and Ringwood, A. E., Equations of state and crystal structures of high-pressure phases of shocked silicates and oxides, Rev. Geophys., **7**, 667-707, 1969.

Anderson, D. L., A seismic equation of state, Geophys. J. Roy. Astron. Soc., **13**, 9, 1967.

______, Chemical inhomogeneity of the mantle, Earth Planet. Sci. Letters, **5**, 89-94, 1968.

______, Bulk modulus-density systematics, J. Geophys. Res., **74**, 3857-3864, 1969.

______, Petrology of the mantle, *in* The Mineralogy and Petrology of the Upper Mantle, edited by B. A. Morgan, Min. Soc. Am. Special Paper 3, 1970.

Anderson, D. L., and Jordan, T., The composition of the lower mantle, Phys. Earth Planet. Interiors, **3**, 23-35, 1970.

Anderson, D. L., and Julian, B. R., Shear velocities and elastic parameters of the mantle, J. Geophys. Res., **74**, 3281-3286, 1969.

Anderson, D. L., and Kanamori, H., Shock-wave equations of state for rocks and minerals, J. Geophys. Res., **73**, 6477-6501, 1968.

Anderson, D. L., and Sammis, C., Partial melting in the upper mantle, Phys. Earth Planet. Interiors, **3**, 41-50, 1970.

Anderson, D. L., and Smith, M., Mathematical and physical inversion of gross Earth geophysical data, Amer. Geophys. Union Trans., **49**, 283, 1968.

Anderson, D. L., and Spetzler, H., Partial melting and the low-velocity zone, Phys. Earth Planet. Interiors, **4**, 62-64, 1970.

Anderson, O. L., Schreiber, E., Liebermann, R. C., and Soga, N., Some elastic constant data on minerals relevant to geophysics, Rev. Geophys., **6**, 491-524, 1968.

Anderson, O. L., and Liebermann, R. C., Equations for the elastic constants and their pressure derivatives for three cubic lattices and some geophysical applications, Phys. Earth Planet. Interiors, **3**, 61-85, 1970.

Balchan, A. S., and Cowan, G. R., Shock compression of two iron-silicon alloys to 2.7 megabars, J. Geophys. Res., **71**, 3577-3588, 1966.

Binder, A. B., Internal structure of Mars, J. Geophys. Res., **74**, 3110-3118, 1969.

Birch, F., The variational seismic velocities within a simplified Earth model in accordance with the theory of finite strain, Bull. Seism. Soc. Am., **29**, 463-479, 1939.

______, Elasticity and constitution of the Earth's interior, J. Geophys. Res., **57**, 227-286, 1952.
______, The velocity of compressional waves in rocks to 10 kilobars–II, J. Geophys. Res., **66**, 2199-2224, 1961a.
______, Composition of the Earth's mantle, Geophys. J., **4**, 295-311, 1961b.
______, Density and composition of the mantle and core, J. Geophys. Res., **69**, 4377-4388, 1964.
______, Energetics of core formation, J. Geophys. Res., **70**, 6217-6222, 1965.
______, Density and composition in the upper mantle: First approximation as an olivine layer, *in* The Earth's Crust and Upper Mantle, edited by P. J. Hart, pp. 18-36, Geophysical Monograph 13, American Geophysical Union, Washington, D.C., 1969.
Brett, R., and Bell, P. M., Melting relations in the Fe-Rich portion of the system Fe-FeS at 30 kb pressure, Earth Planet. Sci. Letters, **6**, 479-482, 1969.
Clark, S. P., Jr., Variation of density in the Earth and the melting curve in the mantle, *in* The Earth Sciences; Problems and Progress in Current Research, edited by T. W. Donnelly, pp. 5-42, University of Chicago Press, Chicago, 1963.
Clark, S. P., Jr., and Ringwood, A. E., Density distribution and constitution of the mantle, Rev. Geophys., **2**, 35, 1964.
Eshelby, V. D., The determination of the elastic field of an ellipsoidal inclusion, and related problems, Proc. Roy. Soc. (London), Ser. A, 241, 376, 1957.
Fairborn, J. W., Shear wave velocities in the lower mantle, Bull. Seism. Soc. Am., **59**, 1983-1999, 1969.
Hanks, T., and Anderson, D. L., The early thermal history of the Earth, Phys. Earth Planet. Interiors, **2**, 19-29, 1969.
Hughes, D. S., and Kelly, J. L., Second-order elastic deformation of solids, Phys. Rev., **92**, 1145-1149, 1953.
Johnson, L. R., Array measurements of *P* velocities in the upper mantle, J. Geophys. Res., **72**, 6309-6325, 1967.
Kennedy, G. C., cited by E. A. Kraut and G. C. Kennedy, New melting law at high pressures, Phys. Rev. Letters, **16**, 608, 1966.
Knopoff, L., and MacDonald, G. J. F., An equation of state of the core of the Earth, Geophys. J., **3**, 68-77, 1960.
Larimer, J. W., Chemical fractionations in meteorites–I, Condensation of the elements, Geochim. Cosmochim. Acta, **31**, 1215-1238, 1967.
Larimer, J. W., and Anders, E., Chemical fractionations in meteorites–II, Abundance patterns and their interpretation, Geochim. Cosmochim. Acta, **31**, 1239-1270, 1967.
Liebermann, R. C., Effect of iron content upon the elastic properties of oxides and some applications to geophysics, Ph.D. thesis, Columbia University, 1969.
Manghnani, M., Elastic constants of single-crystal rutile, J. Geophys. Res., **74**, 4317-4328, 1969.
McQueen, R. G., Marsh, S. P., and Fritz, J. N., On the composition of the Earth's interior, J. Geophys. Res., **69**, 2047-2965, 1964.
McQueen, R. G., and Marsh, S. P., Shock wave compression of iron-nickel alloys and the Earth's core, J. Geophys. Res., **71**, 1751-1756, 1966.
Mizutani, H., Hamano, Y., Ida, Y., and Akimoto, S., Compressional wave velocities of fayalite, Fe_2SiO_4 spinel and coesite, J. Geophys. Res., **75**, 2741-2797, 1970.
Murthy, V. R., and Hall, H. T., The chemical composition of the Earth's core: Possibility of sulfur in the core, Phys. Earth Planet. Interiors, **2**, 276-282, 1970.
Ringwood, A. E., The chemical composition and origin of the Earth, *in* Advances in Earth Science, edited by P. M. Hurley, pp. 287-356, MIT Press, Cambridge, Mass., 1966.
______, Phase transformations and constitution of the mantle, Phys. Earth Planet. Interiors, **3**, 109-155, 1970.
Sammis, C. G., The pressure dependence of the elastic constants of cubic crystals in the

NaCl and spinel structures from a lattice model, Geophys. J. Roy. Astron. Soc., **19**, 285-297, 1970.

Sammis, C. G., Anderson, D. L., and Jordan, T., A note on the application of isotropic finite strain theory to ultrasonic and seismological data, J. Geophys. Res., **75**, 4478-4480, 1970.

Sharp, W. E., Melting curves of sphalerite, galena, and pyrrhotite and the decomposition curve of pyrite between 30 and 65 kilobars, J. Geophys. Res., **74**, 1645-1652, 1969.

Simmons, G., Single crystal elastic constants and calculated aggregate properties, J. Grad. Res. Center, SMU, **34**, 1-269, 1965.

Spetzler, H., and Anderson, D. L., The effect of temperature and partial melting on the velocity and attenuation in a simple binary system, J. Geophys. Res., **73**, 6051-6060, 1968.

Thomsen, L., On the fourth-order anharmonic equation of state of solids, J. Chem. Phys. Solids, **31**, 2003-2016, 1970.

Toupin, R. A., and Bernstein, B., Sound waves in deformed perfectly elastic materials: Acoustoelastic effects, J. Acoust. Soc. Am., **33**, 216-225, 1961.

Turekian, K., and Clark, S. P., Jr., Inhomogeneous accumulation of the Earth from the primitive solar nebulae, Earth Planet. Sci. Letters, **6**, 346-348, 1969.

Walsh, J. B., New analysis of attenuation in partially melted rock, J. Geophys. Res., **74**, 4333-4337, 1969.

Whitcomb, J. H., and Anderson, D. L., Reflection of P′P′ (PKPPKP) seismic waves from discontinuities in the mantle, J. Geophys. Res., **75**, 5713-5728, 1970.

4

MINERALOGY OF THE DEEP MANTLE: CURRENT STATUS AND FUTURE DEVELOPMENTS

by **A. E. Ringwood**

The hypothesis that a series of major phase transformations occurs at depths of 300 to 1000 km in the mantle was proposed by Birch in 1952 on the basis of his studies of mantle elasticity. Experimental methods by which this hypothesis was subsequently tested are described and their limitations discussed. These include direct static high pressure experiments, indirect germanate-silicate analogue methods, and shock wave techniques. The combined application of these methods has verified Birch's hypothesis in all essential aspects. The probable nature of the mineral phases that occur with increasing depth in the mantle is discussed in the light of the experimental results. It is now possible to provide a widely self-consistent explanation of the depth variation of major physical properties of the mantle in terms of the probable natures, stability fields, and properties of the mineral phases present. Among the unsolved problems, the constancy (or otherwise) of the iron-magnesium ratio with depth is currently debated. Resolution of this problem will require more precise experimental and seismic data than is now available.

The broader significance of our knowledge of the present constitution of the mantle is discussed and possible future developments are explored. Major advances are possible in knowledge of the detailed physical properties of the deep mantle (electrical conductivity, elastic and anelastic properties, thermodynamic properties) by means of investigations on appropriate analogue compounds (germanates) under comparatively low pressures.

INTRODUCTION

It is a particular pleasure to contribute this chapter to a volume honoring Francis Birch. The publication of his paper *Elasticity and Constitution of the Earth's Interior* in 1952 was a landmark in solid earth geophysics. His interpretation of the elastic properties and density of the deep mantle and of the rapid increases in seismic velocities between 400 and 900 km in terms of phase changes presented an immediate challenge to experimentalists to attempt to verify these predictions. Much of my own experimental career has been devoted to this problem, including one and a half stimulating and rewarding early years working in Birch's Dunbar Laboratory. All those who were fortunate enough to work in the ideal research environment provided by this laboratory owe a great deal to Birch's influence, characterized by a unique blend of scientific and human qualities. This chapter will be principally concerned with a review of the experimental developments that finally verified Birch's 1952 hypothesis on the constitution of the mantle and a consideration of possible future developments in this field.

The principal source of physical information on the mantle is seismology, particularly the variation of P and S wave velocities with depth, which provide the basis for subdivision of the mantle into three regions (Fig. 1). In the upper mantle, extending down to about 350 km, velocity gradients are generally small, except possibly in the vicinity of the low-velocity zone. In the transition zone between about 350 and 1000 km, velocity gradients are high on the average. Recent observations by Anderson and others indicate that most of the velocity increase in this region is concentrated in two narrow zones around 400 and 650 km. Below 1000 km and extending to the core at 2900 km lies the lower mantle, where velocity gradients are relatively small and uniform. This distribution of velocities clearly reflects a corresponding variation of other important physical properties and must ultimately be explained in terms of the nature and properties of the mineral phases that are stable in the various regions of the mantle.

The first clue to the nature of the transition zone arose from the classical studies on the density of the mantle by Bullen (1936) who demonstrated that this region is physically inhomogeneous. Bernal (1936) and Jeffreys (1937) suggested that the inferred inhomogeneity was located near a depth of 400 km and might be caused by a phase transformation of olivine to the spinel structure. This was based upon Goldschmidt's observation (1931) that the compound Mg_2GeO_4 was dimorphous, displaying both olivine and spinel structures. Although highly speculative at the time, this suggestion was to prove prescient.

A systematic investigation of the nature of the inhomogeneity was conducted by Birch (1939, 1952) using an equation of state based upon finite strain theory. Birch concluded that the properties of the upper mantle were consistent with

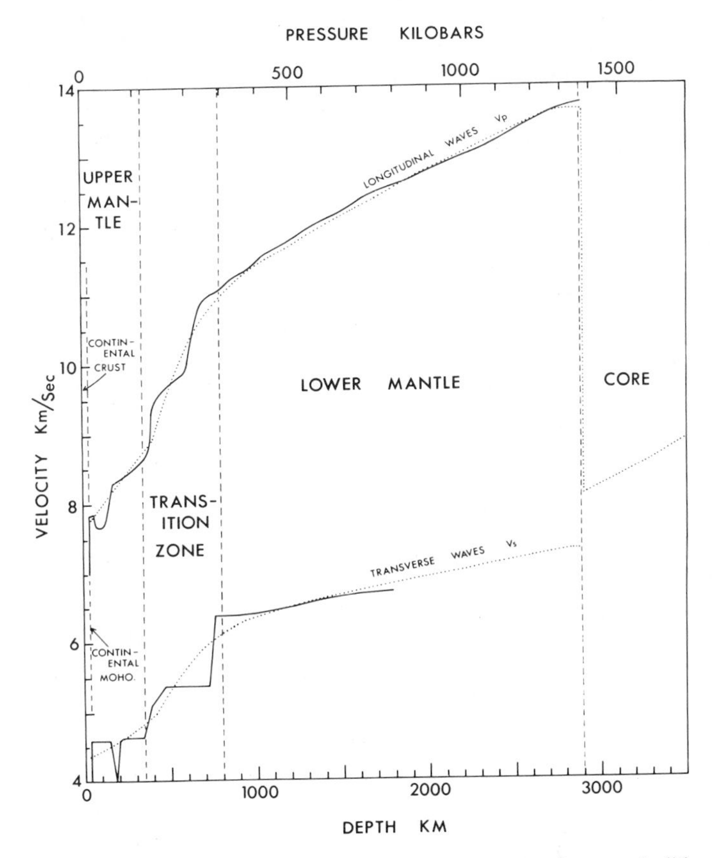

Fig. 1. Seismic velocity distributions in the mantle: *P* waves (solid line), Johnson (1967, 1969); *S* waves (solid line), Nuttli (1969). Broken lines, *P* and *S* waves, Jeffreys (1937, 1939).

this region being composed of familiar minerals such as olivines, pyroxenes, and garnets. The elastic properties of the lower mantle were very different, however, and resembled those possessed by relatively close-packed oxides such as corundum, periclase, rutile, and spinel. Birch proposed that the transition zone was characterized by a series of major transformations resulting from the instability of olivines, pyroxenes, and garnets at high pressure. He argued that these minerals transformed into a new assemblage of close-packed polymorphs, that the transformations were complete by about 1000 km, and that between 1000 and 2900 km, no further transformations occurred, this region being essentially homogeneous.

EXPERIMENTAL METHODS

Birch's hypothesis required that under P-T conditions equivalent to those existing at depths of 400 to 1000 km, the major minerals of the upper mantle (olivines, pyroxenes, and garnets) should become unstable and transform to a new assemblage of close-packed "oxide" phases that possessed the elastic properties and densities required to explain the properties of the lower mantle. Although Birch's hypothesis provided an attractive explanation of the elastic properties of the lower mantle, its basic conclusions were nevertheless dependent upon the assumption of a rather simple equation of state. This equation described the compressions of alkali metals remarkably well, but had not been demonstrated to be applicable to structurally more complex compounds such as oxides and silicates. Sceptics who questioned the equation of state accordingly preferred other interpretations of the constitution of the lower mantle. The most popular alternative hypothesis was that the lower mantle consisted of a mixture of metallic iron and olivine. Other interpretations held that the abnormal elastic properties of the transition zone were caused by a change from ionic to covalent bonding, by nonhydrostatic conditions, or exclusively by changes in chemical composition.

Clearly, it would be necessary to verify Birch's hypothesis by experimental methods before it could be regarded as finally established. Experimentally, this presented a difficult challenge at the time. The pressures in the 400- to 1000-km depth interval vary from 130 to 380 kb and the temperatures are probably in the vicinity of 1500 to 3000°C. Experimental methods capable of attaining these conditions did not exist, and pressure transformations in mafic and ultramafic silicate minerals and in oxides were all but unknown. The first encouragement came with the pioneering experiments by Coes (1953) and his successful synthesis of a new high pressure polymorph of quartz. Nevertheless, many sceptics remained and diligent efforts were made by some to prove that certain phase transformations that had been previously suggested (e.g., the olivine-spinel transition) were impossible on the grounds of lattice dynamics (Shimazu, 1958; Wada, 1960).

It was clear that during the 1950's, evidence bearing on Birch's hypothesis could be obtained only by indirect methods, using information obtained at relatively low pressures. Such methods included comparative crystal chemistry and studies of model systems by thermodynamic methods and by experiments within the limited pressure range then available. We will review the ways in which these methods were used. Although in more recent years important advances in static and dynamic methods of generating high pressures have been made, the use of these indirect techniques continues to be essential to an understanding of the probable mineralogy of the deeper regions of the mantle.

Indirect Methods

The indirect methods make use of comparative crystal chemistry, thermodynamics, and particularly, of studies of germanate isotypes of silicates. Both silicon and germanium readily form tetravalent ions that possess similar outer electronic structures and radii. Accordingly, the crystal chemistry of silicates is very closely related to that of germanates. Corresponding silicates and germanates are usually isostructural and capable of forming continuous solid solutions. In general terms, it appears that if a germanate with a new structure should be synthesized, there would be a reasonable probability that under some appropriate P-T conditions a corresponding isostructural silicate would be stable.

A further important relationship between silicates and germanates also emerged: germanates often behave as high pressure models for the corresponding silicates. If a germanate is found to display a given phase transformation at a particular pressure, the corresponding silicate often displays the same transformation but at a much higher pressure (Table 1). The reverse of this relationship has not been observed.

Alternatively, a germanate may crystallize at atmospheric pressure in a structure that is only attained by the corresponding silicate at very high pressure. This behavior is readily explicable in terms of crystal chemical principles (Bernal, 1936; Ringwood, 1970) and ultimately is a consequence of the slightly larger radius of the germanium ion, as compared to silicon.

For these reasons, the study of germanates as high pressure models of silicates offers the possibility of obtaining useful information about phase transformations that may occur in silicates at pressures beyond the range of currently available experimental techniques. These relationships may be taken advantage of in several ways. The first is purely qualitative. A systematic study at high pressure is made of the stabilities of germanate isotypes of the principal mantle minerals. In nearly all cases investigated in the author's laboratory, the germanate has been found to transform to denser structures. From the considerations advanced above, there is a fairly high probability that silicates will

Table 1. Germanates as high pressure models for silicates

In each case, the germanate is stable at a lower pressure than the silicate at the same temperature

Structure type	Germanate	Silicate
Rutile	GeO_2	SiO_2
Garnet	$Ca_3Al_2Ge_3O_{12}$	$Ca_3Al_2Si_3O_{12}$
	$Na_2Ca\ Ti_2Ge_3O_{12}$	$Na_2Ca\ Ti_2Si_3O_{12}$
	$Ca_3Mg\ Ti\ Ge_3O_{12}$	$Ca_3Mg\ Ti\ Si_3O_{12}$
Spinel	Ni_2GeO_4	Ni_2SiO_4
	Co_2GeO_4	Co_2SiO_4
	Fe_2GeO_4	Fe_2SiO_4
	$(Mg_{0.8}Fe_{0.2})_2GeO_4$	$(Mg_{0.8}Fe_{0.2})SiO_4$
	$LiAlGeO_4$	$LiAlSiO_4$
Kyanite	Al_2GeO_5	Al_2SiO_5
Scheelite	$ZrGeO_4$	$ZrSiO_4$
	$HfGeO_4$	$HfSiO_4$
Hollandite	$KAlGe_3O_8$	$KAlSi_3O_8$
Jadeite	$NaAlGe_2O_6$	$NaAlSi_2O_6$

transform to these structures at much higher pressures. This is a most valuable aid to exploration.

It is also possible to take advantage of solid solubility relationships between silicates and germanates in systems in which the germanate forms a high pressure phase but not the silicate. Thus, by measuring the solid solubility of a silicate in a high pressure germanate structure over an experimentally accessible range of pressures, the pressure required for the transition of the silicate to the new structure may be estimated by extrapolation. Using this method applied to the system Ni_2GeO_4-Mg_2SiO_4, the pressure for the olivine-spinel transformation in Mg_2SiO_4 was estimated as 130 kb at 600°C (Ringwood, 1958a; Ringwood and Seabrook, 1962).

Finally the solid solubility relationships between germanates and silicates provide data on which thermodynamic calculations of the pressure required for a given transformation in a silicate can be made. Where a low pressure silicate phase (A) with a relatively open structure displays substantial solid solubility in a closer packed germanate phase (B), the pressure P required to transform the silicate into the structure possessed by the germanate is approximately

$$P = -\frac{RT}{\Delta v} \ln \frac{a_2}{a_1}$$

where a_1 represents the activity of the silicate component in phase B, a_2 represents the activity of the silicate component of phase A in equilibrium with B and Δv is the difference in molar volumes between the pure silicate in its low pressure structure A and its hypothetical high pressure structure B. Because of the similarity between the germanium and silicon ions, it appears probable that germanate-silicate solid solutions will behave approximately ideally at elevated temperatures, so that the activities may be obtained from the compositions of the equilibrium solid solutions that are measured directly. More generally, the observation of a substantial solid solubility of a silicate in a closer packed germanate under high pressure conditions is an extremely valuable indicator as to the probable transformation mode and to the approximate pressure of transformation of the silicate. Examples of the use of this prediction method are given by Ringwood (1956, 1970).

As experimental observations on high pressure transformations have expanded during recent years, it has become possible to make some wider crystal-chemical generalizations than those reached previously. An important result is the strong tendency of high pressure phases to crystallize in structures that are already known. Of about 70 high pressure transformations discovered in the author's laboratory, only 8 have been to structures that were unknown or not closely related to known structures. Bearing in mind the large number and complexity of existing silicate and germanate structures, this is a fortunate circumstance that could not have been confidently predicted before the advent of high pressure experimentation. The probability that transformations deep in the mantle will involve presently known structure types greatly simplifies the interpretation of the mineralogy of this region.

The indirect methods described above provided the first solid experimental evidence supporting Birch's hypothesis. All the germanate olivines and pyroxenes that were studied in the pressure range 0 to 90 kb were found to be unstable at high pressures and transformed to denser phases, suggesting strongly that the corresponding silicates would transform similarly at higher pressures. Whenever quantitative estimates of the pressures required for transformation of upper mantle silicates were made by the methods described above, the transformation pressures were found to be within the pressure range in the transition zone (Ringwood, 1956, 1958a, 1962, 1966b). These results enabled a model for the mineralogical constitution of the mantle to be constructed which provided a reasonable explanation of the inferred density distribution and elastic properties.

Indirect methods nevertheless have certain limitations. They do not predict with certainty, since a significant number of cases is known where a silicate does not have a germanate analogue (e.g., coesite) or vice versa. Also it sometimes happens that the indirect methods indicate more than one possible transformation for a silicate at high pressure. Which of the possible transformations will actually be displayed can only be determined by direct experiment. Accordingly it is also desirable to utilize direct static or dynamic methods to study the predicted phase changes in mantle silicates under the P-T conditions actually existing in the transition zone and lower mantle. These have recently been employed with considerable success and will be discussed. These techniques nevertheless also possess important limitations, and present knowledge of the probable nature of the mineral phases that exist in the mantle below 600 km is still based largely upon the use of the indirect methods already discussed. These methods have an important future potential in studying a wide range of properties in the deep mantle and are further discussed in the section entitled Future Developments.

Direct Static Methods of Investigating Mantle Phase Transformations

In order to synthesize directly the minerals occurring in the transition zone, apparatus capable of generating pressures substantially exceeding 150 kb simultaneously with high temperatures is required. Apparatus of this type have been developed only comparatively recently (Akimoto and Ida, 1966; Bundy, 1963; Kawai, 1966; Ringwood and Major, 1966, 1968). These have made it possible to synthesize most of the major mineral phases that probably occur in the mantle down to a depth of 600 km or so. Most of the transformations in olivines, pyroxenes, and garnets that are relevant to an interpretation of the constitution of the transition zone have been discovered with the apparatus described by Ringwood and Major (1966, 68) and shown in Figs. 2 and 3. This consists essentially of a pair of heavily supported Bridgman anvils that compress a cell containing an internal heater. The cell is of novel design and is constructed

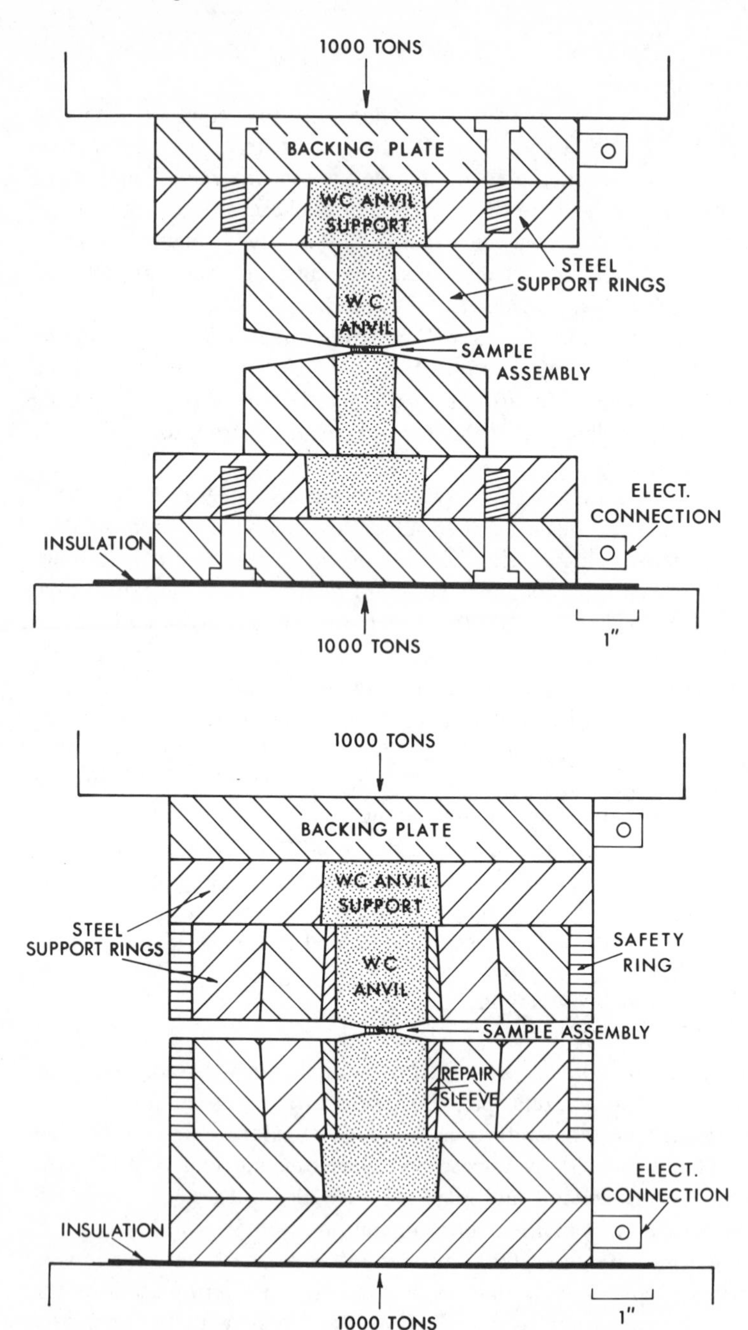

Fig. 2. Bridgman anvils used in high P-T investigations. (*a*) single support ring, (*b*) large anvil with compound support ring. After Ringwood and Major (1968).

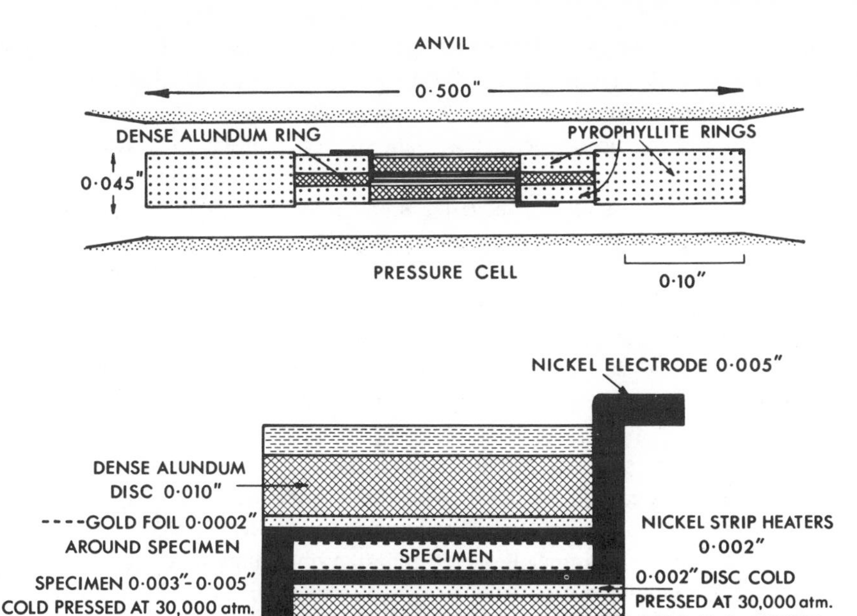

Fig. 3. Internally heated pressure cell used in conjunction with Bridgman anvils. After Ringwood and Major (1968).

so that it acts as a pressure intensifier, with the pressure on the specimen in the center being 60 to 70 kb higher than the mean pressure across the piston faces. This apparatus is capable of developing pressures substantially exceeding 200 kb. Between 80 and 150 kb, the pressure on the sample in individual runs is believed to be correct to ± 10%, assuming a pressure of 98 kb for the coesite-stishovite transition at 1000°C which is the principal calibration point. With care and with frequent repetition of runs, the uncertainty in determining a given phase boundary can be reduced to ± 5%. Outside the 80- to 150-kb interval, errors in pressure become much larger. While under pressure, the sample is heated by passing a current through the nickel strip furnace. The temperature of the center of the specimen is maintained at about 1000°C which is sufficiently high to enable reactions to proceed within reasonable time intervals. Temperature is not employed as an independent variable in phase equilibria determined with this apparatus. After completion of an experimental run, temperature is swiftly lowered by terminating the power supply; pressure is then lowered and the sample extracted to be examined by optical and x-ray methods.

This apparatus and those of most other workers in this field are only capable of discovering new phases that can be thermally quenched and that are retained metastably when removed from the apparatus. Major reconstructive phase transformations in silicates, germanates, and oxides are usually relatively sluggish permitting quenching and retention of metastable phases. This technique has therefore been successful in discovering many transformations below about 150 kb. With increasing pressure required for transformation (assuming a substantial volume change), the high pressure phase becomes increasingly metastable at atmospheric pressure and increasingly liable to revert spontaneously to its stable low pressure state during quenching and release of pressure.

This situation appears likely to provide a serious obstacle to extensions of the quenching technique. Recently the author has developed further the device shown in Figs. 2 and 3 and has attained pressures in the vicinity of 300 kb. Results so far have been somewhat disappointing. Clear evidence that the sample has transformed to another structure while under pressure has been obtained in a number of cases, but the new phases have not been retained when pressure was released. It appears possible therefore that the quenching method may not be successful in resolving the mineralogical nature of the mantle below about 600 km, unless improved methods of quenching (e.g., cooling to liquid nitrogen temperature) are introduced. The problems of removing the sample from the pressure cell under these conditions are formidable.

Alternatively, one might attempt to use techniques that make in situ measurements of properties of the phase while it is under pressure. The most promising technique here is the x-ray diffraction by a sample compressed between two diamond anvils. Basset et al. (1967) have generated pressures above 300 kb at room temperature by this method. In order to activate the typically sluggish major silicate transformations anticipated above 200 kb, it would probably be necessary to heat the sample and anvils to about 1000°C. Whether the diamonds will be able to maintain pressures above 200 kb at these temperatures remains to be seen. A further difficulty with this technique is the long time required for each run (up to several weeks) particularly when the new phase has a complicated x-ray diffraction pattern and does not produce intense low angle lines. If the structure of the new phase can be anticipated, as for example, by making use of the model relationships between germanates and silicates, the interpretation becomes much easier. Thus the indirect methods will probably continue to be of value, even if the problems associated with the use of diamond anvils are solved.

The first direct transformation discovered in a silicate relevant to the constitution of the transition zone was the olivine-spinel transformation in fayalite (Ringwood, 1958b). This was followed in 1961 by Stishov and Popova's important discovery of the rutile-type polymorph of SiO_2. This mineral, "stishovite," was later discovered in a meteoritic impactite (Chao et al., 1962). It possesses a density of 4.28 gm/cc and has the silicon ions in octahedral coordination with respect to oxygen. This transformation incidentally is modeled by an analogous quartz-rutile transformation in GeO_2. During 1966-1967, a number of important new transformations in natural mantle minerals were discovered. These included the transformation of olivine into spinel and beta Mg_2SiO_4 structures and the transformation of pyroxenes into new garnet structures containing some octahedrally coordinated silicon. These are reviewed in detail by Ringwood and Major (1970) and Ringwood (1970) and more briefly in the section entitled Phase Transformations in the Mantle.

Shock-wave Methods

Shocks produced directly or indirectly by high explosives enable the generation of pressures up to several million bars in many metals, oxides, and silicates. Although these pressures are maintained only for very short intervals of time, usually less than a microsecond, they are sufficient to permit the measurement of some geophysically important properties–principally pressure and density. Differentiation of this data provides (with lesser accuracy) the hydrodynamical sound velocity $(\partial P/\partial \rho)_s^{1/2}$. The corresponding temperatures, however, are poorly known, particularly if the sample possessed a relatively low initial density and this introduces difficulties in the use of shock wave data to interpret geophysical phenomena.

The first application of shock wave methods to mantle minerals was by Hughes and McQueen (1958). This failed to demonstrate the occurrence of the phase transformation in olivine predicted by indirect methods. These experiments, however, were only the prelude to a much more comprehensive investigation by McQueen et al. (1964, 1967), who determined the Hugoniots of a number of rocks and minerals of geophysical significance. These important experiments have provided invaluable raw material for many interpretive papers dealing with shock wave data reduction and its application to the constitution of the lower mantle (Ahrens et al., 1969; Birch, 1964; Takeuchi and Kanamori, 1966; Wang, 1968, 1970). McQueen and coworkers demonstrated that many rocks and minerals including olivines and pyroxenes displayed major phase transformations at pressures of a few hundred kilobars. The estimated "zero-pressure densities" of these high pressure phases were generally similar to the densities of isochemical mixtures of oxides (SiO_2 as stishovite). This implied that the olivines and pyroxenes had been shocked to new structures denser than the spinels and garnets obtained by static experiments at pressures up to 200 kb. As Birch (1964) has shown, the pressure-density relationship for these shocked phases above about 400 kb agrees well with the pressure-density relationship for the lower mantle determined by other methods.

Shock wave methods have made a most important contribution toward the problem of the constitution of the lower mantle. Nevertheless, the application of these methods encounters some difficult problems. The first is the estimation of temperature along the Hugoniot, which requires dubious assumptions about the form of the equation of state (Knopoff and Shapiro, 1969; Shapiro and Knopoff, 1969). The uncertainties in temperature cause corresponding uncertainties in the estimation of densities and elastic ratios [$\phi = (\partial P)/(\partial \rho)_s$] at temperatures appropriate for the lower mantle and increase the difficulty of resolving some important problems, for example, the iron content of the lower mantle (Anderson, 1967a; Anderson and Jordan, 1970; Ringwood, 1970; Wang, 1968, 1970). A second problem is concerned with the short period (< 1 μsec) over which the specimen is subjected to peak pressure. In many cases, this is probably insufficient for the attainment of chemical equilibrium, and the high pressure phases for which data is being obtained may not be identical with the phases that actually occur in the deep mantle (Ringwood, 1970). Finally, although shock wave data are of great value in demonstrating the occurrence of phase transformations in silicates and oxides at pressures well beyond the range of static pressure apparatus, they do not serve to identify the nature of the high pressure phases. The indirect methods based upon crystal chemistry and germanate analogue studies described earlier go far toward complementing the shock wave data in this respect. In many cases where shock transformations occur, the probable nature of the high pressure structure may be revealed by these indirect methods (Ahrens et al., 1969). Although they may sometimes suggest several alternative modes of transformation, the requirement that the proposed structure be consistent with the inferred zero-pressure density of the high pressure phase together with certain kinetic considerations constitute strong constraints.

PHASE TRANSFORMATIONS IN THE MANTLE

A detailed review of experimental data bearing on this subject has been published elsewhere (Ringwood, 1970; Ringwood and Major, 1970). To minimize repetition, we will discuss only the principal results. For further detail the reader is referred to the papers referenced.

It has been argued elsewhere that the primary chemical composition of the upper mantle corresponds to a mixture of about 3 parts of peridotite to 1 part basalt (Green and Ringwood, 1967a, b; Ringwood, 1966a, 1969). This composition termed "pyrolite" is given in Table 2. The relative abundances of the principal metallic (except Fe) components of pyrolite, Mg, Si, Al, Ca, are similar to those in chondritic meteorites and in the solar photosphere. Accordingly, it is reasonable to assume that to a first approximation, these abundances are applicable to the entire mantle. The Fe/Fe + Mg ratio in pyrolite is 0.11 molecular. Initially, it is assumed that this ratio is also applicable to the entire mantle. Subsequently, the effects of varying this ratio can be considered. It is convenient for purposes of discussion to further subdivide the mantle into zones distinguished by characteristic seismic properties (Fig. 1).

Upper Mantle (Extending to a Depth of About 350 Km)

A large number of investigations (Green and Ringwood, 1967b; Ito and Kennedy, 1968; O'Hara, 1968; Ringwood, 1966a, 1969) have shown that

Table 2. Model pyrolite composition (after Ringwood, 1966a)

SiO_2	45.16
MgO	37.49
FeO	8.04
Fe_2O_3	0.46
Al_2O_3	3.54
CaO	3.08
Na_2O	0.57
K_2O	0.13
Cr_2O_3	0.43
NiO	0.20
CoO	0.01
TiO_2	0.71
MnO	0.14
P_2O_5	0.06
	100.00

throughout most of this region, material of pyrolite composition would crystallize to a mineral assemblage composed of olivine, pyroxenes, and garnet. The seismic velocity structure within this region is complex and varies widely upon a regional scale. A low-velocity zone possibly caused by a very small degree of partial melting (Anderson, 1962, 1967b) is widespread. Other variations in the physical properties of this region are probably caused by chemical zoning and by mineralogical transformations between the (olivine + pyroxene + garnet) assemblage and the (olivine + aluminous pyroxenes) assemblage. Below 150 km and extending to 350 km, the stable mineral assemblage for pyrolite composition (Table 2) is as follows:

	wt %
Olivine $(MgFe)_2SiO_4$	57
Orthopyroxene $(MgFe)SiO_3$	17
Omphacitic clinopyroxene $(CaMgFe)_2Si_2O_6$-$NaAlSi_2O_6$ s.sn.	12
Pyrope garnet $(MgFeCa)_3$ (Al, Cr) Si_3O_{12}	14

The zero-pressure density (ρ_0) of this assemblage is 3.38 gm/cm^3. This mineral assemblage is characterized by fourfold coordinated Si and by sixfold and eightfold coordinated Mg, Fe, Ca. This region appears to be homogeneous and despite intensive search, phase transformations of the above minerals to denser structures have not been discovered in the relevant P-T range.

The "20-Degree Discontinuity" (Approximately 350 to 420 Km)

The seismic velocity distributions of Jeffreys and Gutenberg were based upon deliberately smoothed travel time data, and hence the derived velocity distributions between 400 and 1000 km were also smooth. Jeffreys' velocity distribution showed a discontinuity in gradient near 400 km (Fig. 1). Subsequent refinements in velocity structures (Johnson, 1967; Niazi and Anderson, 1965) have shown that major increases in seismic velocities occur in a limited depth interval around 400 km. For P waves this increase amounts to 0.6 to 1.0 km/sec and is regarded as constituting a major seismic discontinuity (Fig. 1). Experimental investigations show that this discontinuity is almost certainly caused by two major phase transformations.

The Olivine-Spinel-Beta Mg_2SiO_4 *Transformations.* Olivine is probably the most abundant mineral in the upper mantle, and, accordingly, it is of crucial importance to establish its high pressure behavior. Indirect evidence suggesting that magnesium olivines should transform to the spinel structure around 400 to 500 km has been reviewed. A large measure of confirmation of this prediction was obtained when Ringwood and Major (1966) succeeded in synthesizing a continuous series of spinel solid solutions from pure Fe_2SiO_4 to $(Mg_{0.8}Fe_{0.2})_2SiO_4$ at 170 kb and 1000°C. The spinels were found to be about 10% denser than the olivines. Timely complementation of this experimental work was provided by Binns, Davis, and Reed's discovery (1969) of a silicate spinel in the Tenham meteorite. Textural evidence showed that the spinel, $(Mg_{0.74}Fe_{0.26})_2SiO_4$, had been isochemically transformed from olivine in situ by the action of shock waves probably generated by extraterrestrial collisions.

In compositions between $(Mg_{0.8}Fe_{0.2})_2SiO_4$ and pure Mg_2SiO_4, Ringwood and Major (1966) were unable to synthesize true spinels. These olivines invariably transformed above 150 kb into a phase of lower symmetry that was about 8% denser than olivine. This phase was called beta-Mg_2SiO_4. Relationships between the olivine (α), spinel (γ), and the new β phase have recently been determined by Ringwood and Major (1970) in a comprehensive investigation of the system Mg_2SiO_4-Fe_2SiO_4 over the pressure range 50 to 200 kb at 1000°C (Fig. 4). A detailed study of the iron-rich portion of this system at pressures below 100 kb and at 800, 100, and 1200°C has also been carried out by Akimoto and Fujisawa (1966, 1968).

New high pressure phases of Co_2SiO_4 and Mn_2GeO_4 which are isostructural with beta-Mg_2SiO_4 have been discovered by Akimoto and Sato (1968) and by Morimoto, Akimoto, Koto, and Tokonami (1969). The latter authors, and independently, Moore and Smith (1969), have solved the crystal structure of the beta phase. The oxygen anions are approximately in cubic close packing as in the spinel structure. The A and B cations possess similar oxygen coordinations to the spinel structure; however, their distributions are different. Whereas, in an A_2BO_4 normal spinel, the BO_4 tetrahedra are isolated, in the beta phase, the BO_4 tetrahedra share one of their oxygen atoms leading to the formation of B_2O_7 groups. As a result, some oxygen atoms are not bonded to any B atoms.

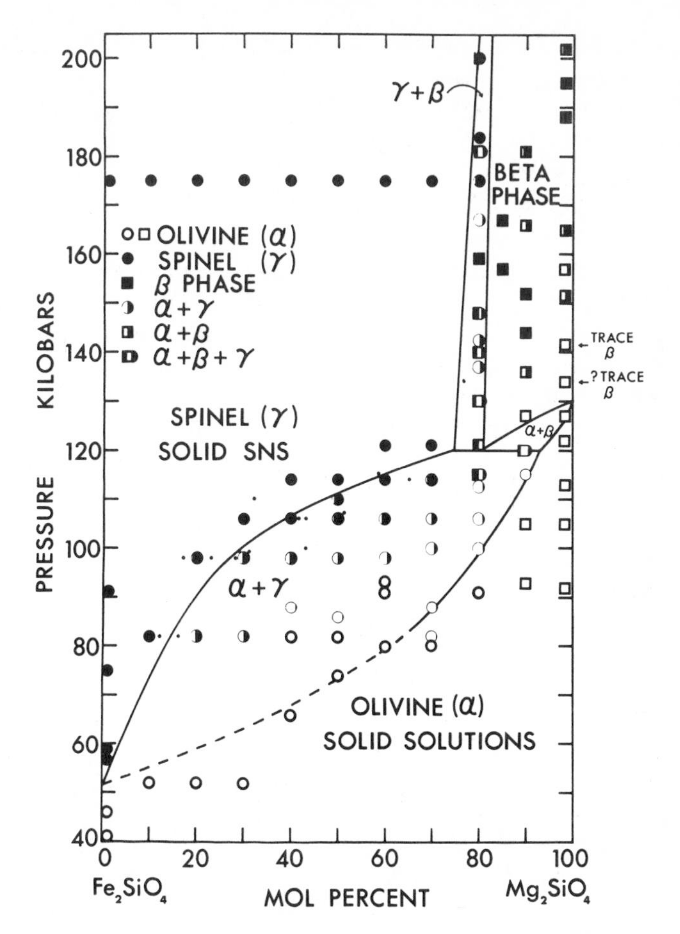

Fig. 4. Phase relationships in the system Mg_2SiO_4-Fe_2SiO_4 at 50 to 200 kb and 1000°C. After Ringwood and Major (1966, 1970).

The formula of this phase may thus be expressed as $A_4O \cdot B_2O_7$ where A = Mg, Mn, Co and B = Si, Ge.

Applying the phase diagram in Fig. 4 to the mantle, olivine of composition $Mg_{89}Fe_{11}$ (pyrolite) would first begin to transform with increasing pressure at 109 kb to a spinel of composition $(Mg_{0.46}Fe_{0.54})_2SiO_4$. As pressure increases, the amount of spinel increases, while the spinel becomes richer in magnesia, reaching $(Mg_{0.75}Fe_{0.25})_2SiO_4$ at 116 kb. At this point, the spinel phase completely reacts to form a beta phase of composition $(Mg_{0.8}Fe_{0.2})_2SiO_4$. Above this pressure, spinel is absent and the field of olivine + beta phase is entered. With further slight increase of pressure, olivine finally transforms to the beta phase by 118 kb. The total transformation is thus spread over an interval of 9 kb with a median value of 114 kb.

Now let us assume that the temperature at a depth of 400 km in the mantle is about 1600°C (Clark and Ringwood, 1964) and that the gradients dP/dT of the phase boundaries in Fig. 4 are 30 bars/°C as indicated by the results of Akimoto and Fujisawa (1968). This would have the effect of increasing the pressures of Fig. 1 by 18 kb, and the median pressure for the transition would be 132 kb. Thus, the transformation of olivine through spinel into the beta phase would occur at a median depth of 397 km with a width of about 27 km. This is in excellent agreement with the depth of the seismic discontinuity (Fig. 1) considering the uncertainties. Although the phase transformations extend over an interval of 27 km, the existence of the spinel-beta phase reaction point at 116 kb (134 kb or 403 km, at 1600°C) causes a discontinuous change in mineralogy and density at this depth and a further rapid increase of density between 403 and 410 km. These effects will produce a first-order seismic discontinuity, rather than a second-order discontinuity, as was suggested in earlier investigations. The reflections observed by Bolt et al. (1968) may well have been produced by this feature. It is most satisfactory that when we take an Mg/Mg + Fe ratio for the upper mantle which is derived directly from petrological and geochemical considerations and a temperature at 400 km which is consistent with geothermal considerations, the laboratory investigations are so closely consistent with seismic investigations.

Pyroxene-Garnet Transformation. The pyrolite upper mantle model contains about 60% of olivine together with 40% of pyroxenes and garnets. Clearly, the transformations displayed by these latter minerals will have an important influence on the properties of the mantle. Several different kinds of high pressure transformations have been discovered, the most important being the transformation of pyroxene and pyroxenoids into a new kind of garnet (Ringwood, 1967, 1970). Examples are $CaGeO_3$ and $MnSiO_3$, the latter transforming to a garnet $Mn_3(MnSi)Si_3O_{12}$ at about 100 kb. In this structure, the octahedral sites are shared by Mn^2 and Si^4 ions, which are presumably ordered. Thus, one-quarter of the silicon atoms change from 4 to 6 coordination in this transformation. Of greater relevance to the mantle are the compositions $MgSiO_3 \cdot 10\%\ Al_2O_3$, $FeSiO_3 \cdot 10\%\ Al_2O_3$, $CaSiO_3 \cdot 10\%\ Al_2O_3$, and $CaMgSi_2O_5 \cdot 10\%\ Al_2O_3$, which transform to garnet structures at about 100 kb and 1000°C. The experimental data demonstrate the existence of a continuous series of solid solutions at high pressure extending from ordinary pyrope garnet $Mg_3Al_2Si_3O_{12}$ at least 70% of the way towards a garnet, $Mg_3(MgSi)Si_3O_{12}$. Ca-rich and Fe-rich pyroxenes behave similarly. The ideal end-member $Mg_3(MgSi)Si_3O_{12}$ has not yet been synthesized, but has been discovered in heavily shocked chondritic meteorites (Binns et al., 1969; Smith and Mason, 1970).*

The solid solution of pyroxene in the garnet structure sets in rather suddenly at about 100 kb, 1000°C (Fig. 5) and is probably of considerable importance in

*A garnet phase of the composition discovered in chondritic meteorites has recently been synthesized at pressures of 250 to 300 kb (Ringwood and Major, unpublished results, 1971).

the mantle. The gradient dP/dT of the transformation is not known, but if it is in the vicinity of 20 to 30 bars/°C (a common range for many solid-solid reconstructive transformations), the transformation may be expected to occur in the mantle at a slightly shallower depth than the olivine transformations and perhaps in the vicinity of 350 to 400 km. The transformation is accompanied by an effective increase in density of the pyroxene component by about 10%, and may contribute to the existence of a complex fine structure in the mantle around 350 to 420 km, indications of which have recently been found by seismologists (Bolt et al., 1968).

Directly observed phase transformations in olivine and pyroxene of a model pyrolite composition thus provide a satisfactory explanation of the depth of the 400-km discontinuity which is one of the major seismic features of the mantle. Moreover, they also provide a quantitative explanation of the magnitude of the velocity increase across the discontinuity (Ringwood, 1970). This is calculated to be 0.7 km/sec for the mineral assemblages appropriate to the pyrolite model, compared with the range 0.65 to 0.9 km/sec inferred by seismologists (Fig. 1).

Depth Interval 450 to 650 Km

The P velocity in this region increases at a constant and relatively small rate with depth (Fig. 1) which suggests that this region is homogeneous, following the arguments of Birch (1952). Extensive experimental studies of the stabilities of silicate garnets and beta phase up to pressures equivalent to depths of 600 km

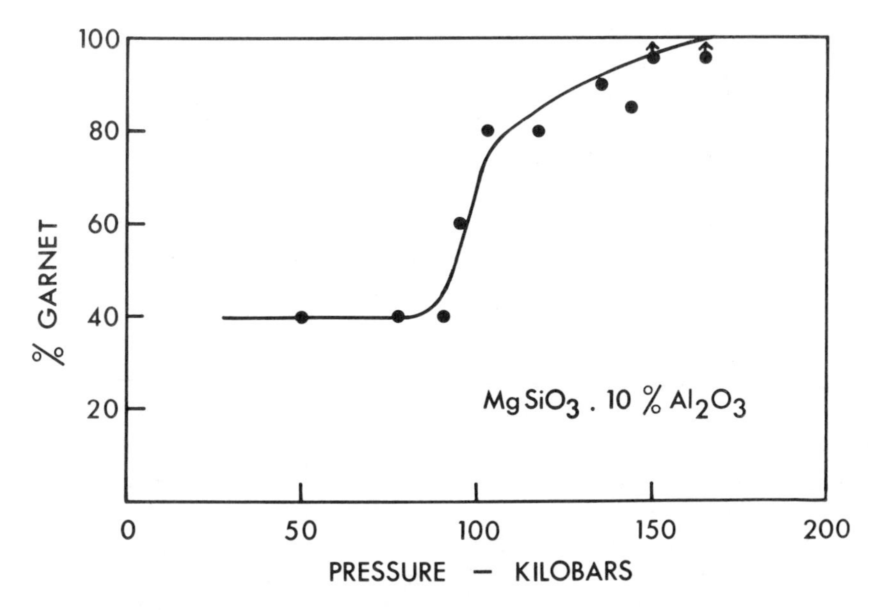

Fig. 5. Proportions of garnet crystallizing from an $MgSiO_3 \cdot 10\%\ Al_2O_3$ glass as a function of pressure at approximately 1000°C. After Ringwood (1967). Garnet crystallizing at pressures above 100 kb consists of a solid solution between ideal pyrope $Mg_3Al_2Si_3O_{12}$ and an end member $Mg_3(MgSi)Si_3O_{12}$.

have failed to reveal any further transformations to denser phases. The experimental observations therefore suggest that a mantle of pyrolite composition would be homogeneous between 450 and 600 km in agreement with the seismic observations. The mineralogy of this region would be as follows:

	wt %
Beta-$(MgFe)_2SiO_4$	57
Complex garnet solid solution	39
Jadeite $NaAlSi_2O_6$	4

The density of this mineral assemblage reduced to atmospheric pressure and temperature would be 3.62 gm/cm^3.

The 650-Km Discontinuity

Figure 1 shows a major seismic discontinuity occurring close to 650 km. Direct static high pressure methods have so far failed to synthesize the phases occurring below this discontinuity and accordingly their nature must be studied by indirect methods or by shock wave techniques.

Germanate model studies suggest that the following transformations may occur (Ringwood, 1970).

1. Beta-$(MgFe)_2SiO_4$ may transform to the strontium plumbate structure. The transformation is displayed by Mn_2GeO_4 (Wadsley et al., 1968) and by $FeMnGeO_4$. A slightly less probable alternative is for beta-Mg_2SiO_4 to disproportionate into an $(MgFe)SiO_3$ ilmenite plus free (MgFe)O. This transformation is displayed by a number of titanates and stannates (Akimoto and Syono, 1968; Reid and Ringwood, in preparation).

2. The pyrope-rich component is likely to transform to the ilmenite structure, as displayed by the high pressure modifications of $MgGeO_3$ and $Mg_3Al_2Ge_3O_{12}$ (Ringwood and Major, 1967a; Ringwood and Seabrook, 1963).

3. The calcium-rich component of garnet transforms to the perovskite structure. This transformation is exhibited by $CaGeO_3$ (Ringwood and Major, 1967b).

4. Jadeite is likely to disproportionate to yield the calcium ferrite structure (Reid et al., 1967), e.g., $NaAlGe_2O_6$ (jadeite) $\rightarrow$ $NaAlGeO_4$ (calcium ferrite) + GeO_2 (rutile).

Thermodynamic studies of the solid solubility of silicates in high pressure germanate structures (Mg_2SiO_4 in Mn_2GeO_4 strontium plumbate, $MgSiO_3$ in $MgGeO_3$ ilmenite, $Mg_3Al_2Si_3O_{12}$ in $Mg_3Al_2Ge_3O_{12}$ ilmenite, $CaSiO_3$ in $CaGeO_3$ perovskite) have been made using the indirect methods. In each case, the calculated pressures required for the silicate to transform to the high pressure germanate structure are in the range 200 to 300 kb (Ringwood, 1970). These estimates, although admittedly imprecise, are in good agreement with the pressure at a depth of 650 km (230 kb), strongly suggesting that these transformations may be responsible for the discontinuity.

Shock wave investigations on several olivines and pyroxenes (Birch, 1964; McQueen et al., 1967) have revealed the occurrence of major transformations into phases that are substantially denser than spinels and garnets between pressures of 400 to 800 kb. The estimated zero-pressure densities of the high pressure phases formed under shock pressures exceeding 600 kb corresponds to those of isochemical mixtures of oxides: $MgO + FeO + SiO_2$ (as stishovite). It is very significant that the density of the mineral assemblage proposed in items 1 through 4 is also identical with that of its isochemical oxide mixture. The germanate model and thermodynamic studies and the shock wave results are thus self-consistent and complementary. Furthermore, calculations (Ringwood, 1970) of the magnitude of the velocity change at 650 km which would be caused by the sequence of phase transformations proposed (1.0 km/sec) are in good agreement with the seismically observed value (0.9 to 1.1 km/sec; Fig. 1).

It is unlikely that these transformations would occur at exactly the same depth. By far the most important would be of beta-$(MgFe)_2SiO_4$ to the strontium plumbate structure. This transformation may well occur within a relatively narrow depth range, and it may be primarily responsible for the occurrence of the 650-km seismic discontinuity. The transformations of garnets to ilmenite and perovskite structures involve the formation of very complex solid solutions. They will probably be smeared out more than the strontium plumbate transformation, and it would be coincidental if they were to occur at exactly the same pressure. It appears that they may contribute in this region to a rather complex fine structure which has yet to be resolved by seismology.

The Region Between 700 and 1000 Km

Assuming that the above transformations are complete by about 700 km, the mineralogy below this depth would consist of the following:

	wt %
$(MgFe)_2SiO_4$ (strontium plumbate str.) alternatively $(MgFe)SiO_3$ ilmenite + $(MgFe)O$	55
Ilmenite-type solid solution $(Mg, Fe)SiO_3$-$(Al, Cr, Fe)AlO_3$	36
Perovskite $CaSiO_3$	6.5
Calcium ferrite $NaAlSiO_4$	2.5

The zero-pressure density of this mineral assemblage is 3.99 gm/cm^3. It is probable that there would be some solid solubility of $FeSiO_3$ and minor components such as $MnSiO_3$ in the $CaSiO_3$ perovskite. Allowance for this effect would bring the zero-pressure density of this assemblage up to about 4.03 gm/cm^3. This mineral assemblage is characterized by sixfold coordination of Mg, Fe, and Si and is almost identical with the mean density of an isochemical mixture of component oxides (SiO_2 as stishovite).

The Lower Mantle (1000 to 2900 Km)

The general nature of the phase transformations responsible for increasing the zero-pressure density of pyrolite up to that of the equivalent isochemical mixed oxides appears to be fairly well understood (Fig. 6). If we accept the above interpretation that these transformations are complete by about 700 km, then the possibility of further transformations to denser states must be considered. The seismic velocity distribution of Johnson (1969) in Fig. 1 shows regions about 50 km wide of anomalously high gradients near 830, 1000, 1230, 1540, and 1910 km. These regions account for about 17% of the total velocity increase between 800 and 2300 km. Johnson suggests that phase changes may be responsible for these features. Further phase transformations in the deep mantle over the pressure range 300 to 1300 kb are not unlikely when considered in relation to the number of transformations that occur at pressures less than 300 kb. Perhaps the most surprising aspect is the small total magnitude of the density changes that is permitted for possible further transformations by constraints arising from the density distribution and elasticity of the deep mantle.

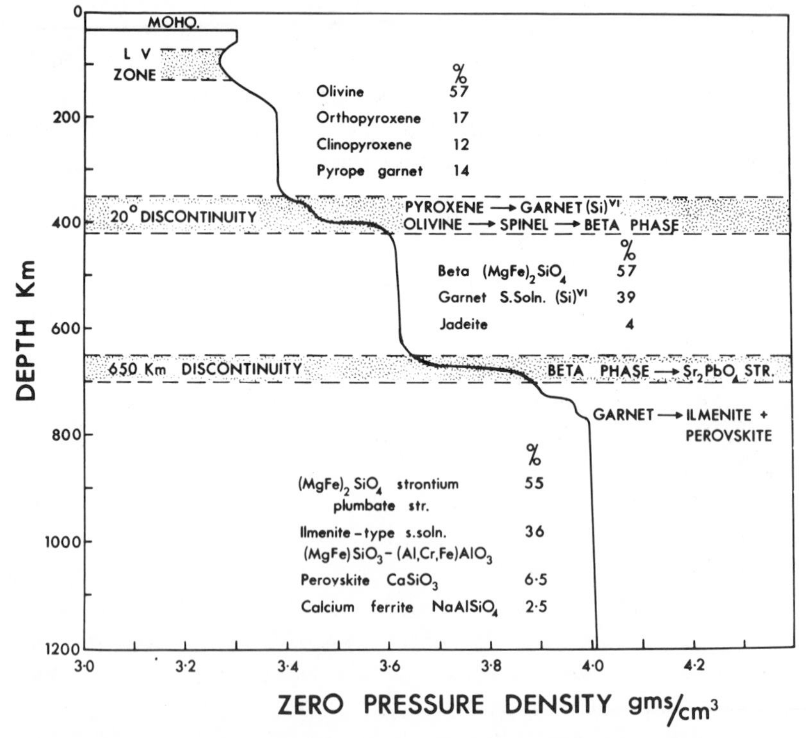

Fig. 6. Probable mineral assemblages and corresponding zero-pressure densities for a model mantle of pyrolite composition. Mineralogy to depth of 600 km is based upon direct static experiments. Mineralogy at greater depths is inferred on the basis of indirect methods and shock wave methods.

The nature of the possible transformations is much more speculative than those discussed previously. Nevertheless, some suggestions can be made on the bases of general crystal chemical relationships and phase transformations in analogue compounds. A possible mineral assemblage for pyrolite composition which is suggested by analogue studies is as follows:

$(CaMgFe)SiO_3$	perovskite structure
$NaAlSiO_4$	calcium ferrite structure
$(MgFe)(Al, Cr, Fe)_2O_4$	calcium ferrite structure
$(MgFe)_2SiO_4$	calcium ferrite structure

Extensive or perhaps complete solid solution would probably occur between the calcium ferrite type components. The zero-pressure density of this assemblage would be about 5% higher than that of the mixed oxides, and the assemblage is characterized by Mg and Fe coordinations higher than 6.

Studies of the density distribution of the lower mantle combined with densities of silicates at comparable pressures obtained from shock wave data (Anderson, 1968; Anderson and Jordan, 1970; Birch, 1964; Clark and Ringwood, 1964; Ringwood, 1970; Wang, 1968, 1970) have led to the generally accepted conclusion that the lower mantle is significantly denser (by about 5%) than an oxide mixture (SiO_2 as stishovite) isochemical with pyrolite. A possible explanation is that further transformations in the lower mantle have resulted in the formation of a mineral assemblage intrinsically denser than the isochemical mixed oxides. An alternative explanation is that the lower mantle is characterized by a higher Fe/Mg ratio than the upper mantle and that transformations to phases denser than the mixed oxides do not occur.

In principle, a choice between these alternatives could be made through a comparison of elastic ratios ϕ ($\phi = k/\rho$, where k = bulk modulus, ρ = density) which are obtained directly from the observed seismic velocities with the ϕ values estimated for the lower mantle composition given above largely using shock wave data. Since increasing Fe/Mg values and phase transformations to denser structures affect ϕ in opposite directions, this comparison should provide a constraint upon alternative choices.

The use of this method has so far led to conflicting interpretations. Anderson (1967a, 1968), Anderson and Jordan (1970), and Press (1968) interpret the data to imply an increase in Fe/Mg ratio with depth, whereas Ringwood (1970) and Wang (1970) argue that this increase is not required. The conflict arises from fundamental uncertainties in the treatment of the shock wave data, particularly in estimating the temperature correction needed to compare shock wave ϕ values with seismically inferred ϕ values for the lower mantle. Future refinements in shock wave equations of state for mantle materials combined with increased accuracy in determining ϕ for the lower mantle from seismic data may be expected to resolve this problem.

FUTURE DEVELOPMENTS

Experimental investigations carried out during the last 15 years have decisively confirmed Birch's hypothesis on the role of phase transformations in the mantle. Directly discovered or indirectly inferred phase transformations in a mantle of pyrolite composition explain quantitatively the occurrences, depths, and velocity changes of the two major seismic discontinuities near 400 and 650 km. These studies also show that the regions between 150 and 350 km and between 450 and 650 km probably do not contain important phase changes. This is consistent with recent determinations of seismic velocity gradients in these regions which are inferred to be homogeneous.

With the achievement of a broad explanation of some of the major physical properties of the mantle in terms of the probable minerals present and their stability fields, attention is being directed toward more detailed problems, particularly those connected with the fine structure of the mantle. Resolution of the fine structure, including the problem of variation of iron content with depth will require intensive collaborative efforts in the fields of seismology, experimental mineralogy, shock wave experimentation, crystal chemistry, and solid state physics. It appears likely that as our knowledge of the constitution of the mantle becomes more detailed, important clues to the origin and subsequent evolution of the earth will emerge.

The importance of determining the detailed mineralogy of the mantle as a function of depth should be emphasized. Many important mantle properties are strongly dependent upon the specific mineral structures present and cannot be interpreted in terms of intrinsic densities (as obtained, for example, from shock wave and seismic data) combined with appropriate equations of state. For example the shear modulus and its pressure and temperature derivatives are sensitively related to crystal structure (Anderson et al., 1968) and detailed understanding of these relationships is essential if the variation of shear wave velocity with depth in the mantle is to be interpreted quantitatively. This is a fundamental geophysical property that has not yet been adequately exploited. Consider the region between 450 and 650 km where the *S* wave velocity gradient is very low (Fig. 1) and may even be negative (Julian and Anderson, 1968), while, over this same interval, *P* wave velocity increases at a substantial rate with depth. D. Anderson (personal communication) has suggested that this behavior might be characteristic of the spinel lattices and related lattices such as beta-Mg_2SiO_4 in which the oxygen sublattice is similar to that of spinel. This suggestion is based upon the work of Anderson et al, (1968) which showed that for $MgAl_2O_4$ spinel, the pressure derivative of the shear modulus (0.75 kb), was much smaller than the corresponding derivative of the bulk modulus, (4.2 kb). The critical temperature gradient $(\partial T/\partial P)_{Vp}$, $(\partial T/\partial P)_{Vs}$, at which the opposing effects of increased temperature and pressure upon seismic velocities cancel each other out are 2°C per kb for *S* waves and 16°C per kb for *P* waves (Anderson et al., 1968). Clearly, a more thorough understanding of the pressure and temperature derivatives of the elastic moduli of spinel-related minerals present in this region should ultimately lead to determinations of temperature gradients. The same applies also to other homogeneous regions of the mantle. This would be of tremendous importance. The uncertainties surrounding the temperature distribution in the mantle provide the greatest obstacle to more detailed investigations of the fine structure and affect in a major way the interpretation of a wide range of mantle properties.

Other geophysically important properties that are sensitive to crystal structure include mechanical properties (deformation, slip and creep processes, attenuation of seismic waves, and electrical conductivity). The latter appears likely to provide an increasingly important clue to the nature of the earth's interior, particularly the temperature distribution. Some thermodynamic properties are also strongly influenced by structure. The entropy of stishovite was first estimated by extrapolation of the entropies of the rutile-type compounds GeO_2, SnO_2, TiO_2, and PbO_2 (Stishov, 1963) and was found to be in good agreement with direct experimental measurements subsequently obtained.

Clearly, the investigation of these properties under the P-T conditions actually existing within the deep mantle is likely to be a formidable experimental undertaking. Nevertheless, because they are structure sensitive, it should be possible to obtain a vast amount of relevant information by investigating analogue compounds, especially germanates. By studying properties of isostructural sequences (plumbates, stannates, titanates and germanates), the corresponding property of the ideal silicate isotype should be obtainable in many cases simply by extrapolation against mean atomic weight or against metal-oxygen bond length.

One other use of these methods in connection with shock wave experimentation may be mentioned. Interpretation of shock wave Hugoniot data on silicates has hitherto been complicated by the high and uncertain temperatures along Hugoniots because of high initial compressibilities and phase changes. This problem could be simplified, however, by using germanate high pressure phases synthesized under moderate static pressures as the starting materials for shock wave experiments. Because of lower compressibilities, these phases would not reach very high temperatures when shocked to pressures on the order of a megabar, and reduction of the primary Hugoniot data to isothermal conditions could be achieved without introducing substantial further uncertainty. Another kind of experiment that would be of great assistance in the interpretation of shock wave data on silicates would be to compare the Hugoniots of both high and low pressure germanate phases. For example, shock wave data on GeO_2 (quartz) and GeO_2 (rutile) starting materials would doubtless aid in the interpretation of existing shock data on SiO_2.

It seems likely that the next decade will see a considerable expansion in the use of analogue methods to study the physical and chemical properties of the deep mantle. These methods, combined with improved shock wave and static high pressure techniques, are capable of leading to a detailed understanding of the constitution of the mantle. This has always been a major objective of the solid earth sciences. It has, in addition, a direct bearing on some other major geophysical problems, particularly those concerning the dynamics of the mantle. While the occurrence of large-scale horizontal motions in the crust driven by motions in the mantle is widely accepted, the nature of motions within the mantle and the engine that drives them are poorly understood. Advances in understanding of the mechanical and thermodynamic properties of the mantle and of the relationships between phase transformations and mantle motions are likely to play a decisive role in obtaining a solution to this problem.

REFERENCES

Ahrens, T. J., Anderson, D. L., and Ringwood, A. E., Equations of state and crystal structures of high pressure phases of shocked silicates and oxides, Rev. Geophys., **4**, 667-707, 1969.

Akimoto, S., and Fujisawa, H., Olivine-spinel transition in system Mg_2SiO_4– Fe_2SiO_4 at 800°C, Earth Planet. Sci. Letters, **1**, 237-240, 1966.

______, Olivine-spinel solid solution equilibria in the system Mg_2SiO_4-Fe_2SiO_4, J. Geophys. Res., **73**, 1467-1479, 1968.

Akimoto, S., and Ida, Y., High pressure synthesis of Mg_2SiO_4 spinel, Earth Planet. Sci. Letters, **1**, 358-359, 1966.

Akimoto, S., and Sato, Y., High pressure transformation in Co_2SiO_4 olivine and some geophysical implications, Technical Rept. Inst. Solid State Physics, University of Tokyo, Series A, No. 328, 1968.

Akimoto, S., and Syono, Y., The coesite-stishovite transition, Technical Rept. Inst. Solid State Physics, University of Tokyo, Series A., No. 327, 1968.

Anderson, D. L., The plastic layer of the earth's mantle, Sci. Am., **205**, 2–9, 1962.

______, Phase changes in the upper mantle, Science, **157**, 1165-1173, 1967a.

______, A review of upper mantle seismic data (abstract), Trans. Am. Geophys. Union, **48**, 254, 1967b.

______, Chemical inhomogeneity of the mantle, Earth Planet. Sci. Letters, **5**, 89-94, 1968.

Anderson, D. L. and Jordan, T., The composition of the lower mantle, Phys. Earth Planet. Interiors, **3**, 23-35, 1970.

Anderson, O. L., Schreiber, E., and Liebermann, R. C., Some elastic constant data on minerals relevant to geophysics, Rev. Geophys., **6**, 491–524, 1968.

Bassett, W. A., Takahashi, T., and Stook, P. W., X–ray diffraction and optical observations or crystalline solids up to 300 kb, Rev. Sci. Instr., **38**, 37-42, 1967.

Bernal, J. D., 1936, Discussion, Observatory, **59**, 268, 1936.

Binns, R. A., Davis, R. J., and Reed, S. B. J., Ringwoodite, natural (MgFe) SiO_4 spinel in the Tenham meteorite, Nature, **221**, 943-944, 1969.

Birch, F., The variation of seismic velocities within a simplified earth model in accordance with the theory of finite strain, Bull. Seism. Soc. Am., **29**, 463-479, 1939.

______, Elasticity and constitution of the Earth's interior, J. Geophys. Res., **57**, 227-286, 1952.

______, Density and composition of mantle and core, J. Geophys. Res., **69**, 4377-4388, 1964.

Bolt, B. A., O'Neill, M., and Qamar, A., Seismic waves near 110°: is structure in core or upper mantle responsible? Geophys. J. Roy. Astron. Soc., **16**, 475-487, 1968.

Bullen, K. E., The variation of density and the ellipticities of strata of equal density within the earth, Monthly Notices Roy. Astron. Soc., Geophys. Suppl., **3**, 395-401, 1936.

Bundy, F. P., Direct conversion of graphite to diamond in static pressure apparatus, J. Chem. Physics, **38**, 631-643, 1963.

Chao, E. C. T., Fahey, J. J., Littler, J., and Milton, D. J., Stishovite, SiO_2, a very high pressure new mineral from meteor crater, Arizona, J. Geophys. Res., **67**, 419, 1962.

Clark, S. P., and Ringwood, A. E., Density distribution and constitution of the mantle. Rev. Geophys., **2**, 35-88, 1964.

Coes, L., A new dense crystalline silica, Science, **118**, 131-132, 1953.

Goldschmidt, V. M., Zur Kristallchemie des Germaniums, Nachr. Akad. Wiss. Goettingen, Nach., Math. Phys. Kl., 184-190, 1931.

Green, D. H., and Ringwood, A. E., The genesis of basaltic magmas, Contr. Mineral. Petrol., **15**, 103-190, 1967a.

______, The stability fields of aluminous pyroxene peridotite and garnet peridotite and their relevance in upper mantle structure, Earth Planet. Sci. Letters, **3**, 151-160, 1967b.

Hughes, D. S., and McQueen, R. G., Density of basic rocks at very high pressures, Trans. Am. Geophys. Union, **39**, 959-965, 1958.

Ito, K., and Kennedy, G. C., Melting and phase relations in the plane tholeiite-lherzolite-nepheline basanite to 40 kilobars with petrological implications, Contr. Mineral. Petrol., **19**, 177-211, 1968.

Jeffreys, H., On the materials and density of the earth's crust. Monthly Notices Roy. Astron. Soc., Geophys. Suppl., **4**, 50-61, 1937.

______, The times of P, S, and SKS and the velocities of P and S, Monthly Notices Roy. Astron. Soc., Geophys. Suppl., **4**, 498-533, 1939.

Johnson, L., Array measurements of P velocities in the upper mantle, J. Geophys. Res., **72**, 6309-6325, 1967.

______, Array measurements of P velocities in the lower mantle, Bull. Seism. Soc. Am., **59**, 973-1008, 1969.

Johnson, Q., Keeler, R. N., and Lyle, J. W., X-ray diffraction experiments in nanosecond time intervals, Nature, **213**, 1114-1115, 1967.

Julian, B., and Anderson, D. L., Travel times, apparent velocities and amplitudes of body waves, Bull. Seism. Soc. Am., **58**, 339-366, 1968.

Kawai, N., A static high pressure apparatus with tapering multipistons forming a sphere–I, Proc. Japan Acad., **42**, 385-388, 1966.

Knopoff, L., and Shapiro, J. N., Comments on the inter-relationships between Gruneisen's parameter and shock and isothermal equations of state, J. Geophys. Res., **74**, 1439-1450, 1969.

McQueen, R. G., Fritz, J. N., and Marsh, S. P., On the composition of the earth's interior, J. Geophys. Res., **69**, 2947-2978, 1964.

McQueen, R. G., Marsh, S. P., and Fritz, J. N., Hugoniot equation of state of twelve rocks, J. Geophys. Res., **72**, 4999-5036, 1967.

Moore, P. B., and Smith, J. V., High pressure modification of Mg_2SiO_4: Crystal structure and crystallochemical and geophysical implications, Nature, **221**, 653-655, 1969.

Morimoto, N., Akimoto, S., Koto, K., and Tokonami, M., Modified spinel, beta manganous orthogermanate: Stability and crystal structure, Science, **165**, 586-588, 1969.

Niazi, M., and Anderson, D. L., Upper mantle structure of western North America from apparent velocities of P waves, J. Geophys. Res., **70**, 4633-4640, 1965.

Nuttli, O. W., Travel times and amplitudes of S waves from nuclear explosions in Nevada, Bull. Seism. Soc. Am., **59**, 385-398, 1969.

O'Hara, M. J., The bearing of phase equilibria studies in synthetic and natural systems on the origin and evolution of basic and ultrabasic rocks, Earth Sci. Rev., **4**, 69–133, 1968.

Press, F., Earth models obtained by Monte Carlo inversion, J. Geophys. Res., **73**, 5223-5234, 1968.

Reid, A. F., Wadsley, A. D., and Ringwood, A. E., High pressure $NaAlGeO_4$, a calcium ferrite isomorph and model structure for silicates at depth in the earth's mantle, Acta Cryst., **23**, 736-739, 1967.

Ringwood, A. E., The olivine-spinel transition in the earth's mantle, Nature, **178**, 1303-1304, 1956.

______, Constitution of the mantle–I, Thermodynamics of the olivine-spinel transition, Geochim. Cosmochim. Acta, **13**, 303-321, 1958a.

______, Constitution of the mantle–II, Further data on the olivine-spinel transition, Geochim. Cosmochim. Acta, **15**, 18-29, 1958b.

______, Mineralogical constitution of the deep mantle, J. Geophys. Res., **67**, 4005-4010, 1962.

______, The chemical composition and origin of the earth, *in* Advances in Earth Science, edited by P. M. Hurley, pp. 287-356, MIT Press, Boston, 1966a.

______, Mineralogy of the mantle, *in* Advances in Earth Science, edited by P. M. Hurley, pp. 357-399, MIT Press, Boston, 1966b.

______, The pyroxene-garnet transformation in the earth's mantle, Earth Planet. Sci. Letters, **2**, 255-263, 1967.

______, Composition and evolution of the upper mantle, *in* The Earth's Crust and Upper Mantle, edited by P. J. Hart, Geophysical Monograph No. 13, pp. 1-17, Am. Geophys. Union, Washington, D.C., 1969.

______, Phase transformations and the constitution of the mantle, Phys. Earth Planet. Interiors, **3**, 109-155, 1970.

Ringwood, A. E., and Major, A., Synthesis of Mg_2SiO_4–Fe_2SiO_4 solid solutions, Earth Planet. Sci. Letters, **1**, 241-245, 1966.

______, The garnet-ilmenite transformation in Ge-Si pyrope solid solutions, Earth Planet. Sci. Letters, **2**, 331-334, 1967a.

______, Some high pressure transformations of geophysical interest, Earth Planet. Sci. Letters, **2**, 106-110, 1967b.

______, Apparatus for phase transformation studies at high pressures and temperatures, Phys. Earth Planet. Interiors, **1**, 164-168, 1968.

______, The system Mg_2SiO_4-Fe_2SiO_4 at high pressures and temperatures, Phys. Earth Planet. Interiors, **3**, 89-108, 1970.

Ringwood, A. E., and Seabrook, M., Olivine-spinel equilibria at high pressure in the system Ni_2GeO_4–Mg_2SiO_4, J. Geophys. Res., **67**, 1975-1985, 1962.

______, High pressure phase transformations in germanate pyroxenes and related compounds, J. Geophys. Res., **68**, 4601-4609, 1963.

Shapiro, J. N., and Knopoff, L., Reduction of shock-wave equations of state to isothermal equations of state, J. Geophys. Res., **74**, 1435-1438, 1969.

Shimazu, Y., A chemical phase transition hypothesis on the origin of the C–layer within the mantle of the earth, J. Earth Sci., Nagoya University, **6**, 12-30, 1958.

Smith, J. V., and Mason, B., Pyroxene-garnet transformation in Coorara meteorite, Science, **168**, 832-833, 1970.

Stishov, S. M., Equilibrium between coesite and the rutile-like modification of silica (in Russian), Dokl. Akad. Nauk. SSR, **148**, 5, 1186-1188, 1963.

Stishov, S. M., and Popova, S. V., New dense polymorph modification of silica, Geochimiya, No. 10, 837–839, 1961.

Takeuchi, H., and Kanamori, H., Equations of state of matter from shock wave experiments, J. Geophys. Res., **71**, 3985-3994, 1966.

Wada, T., On the physical properties within the B-layer deduced from olivine model and on the possibility of polymorphic transition from olivine to spinel at the 20° discontinuity, Bull. 37, pp. 1–20, Disaster Prevention Institute, University of Kyoto, 1960.

Wadsley, A. D., Reid, A. F., and Ringwood, A. E., The high pressure form of Mn_2GeO_4, a member of the olivine group, Acta Cryst., B24, 740–742, 1968.

Wang, C., Constitution of the lower mantle as evidenced from shock wave data for some rocks, J. Geophys. Res., **73**, 6459-6476, 1968.

______, Density and constitution of the mantle, J. Geophys. Res., **75**, 3264-3284, 1970.

5

SEDIMENTARY CYCLING IN RELATION TO THE HISTORY OF THE CONTINENTS AND OCEANS

by **Robert M. Garrels, Fred T. Mackenzie, and Raymond Siever**

Examination of the chemical and mineralogical composition of sedimentary rocks as a function of their ages reveals the following general relations. (1) Variation in composition from system to system, or any similar length stratigraphic interval, is so great that it is necessary to integrate over time intervals of the order of 200 m.y. to see systematic time trends; these time intervals are of the same order of magnitude as basinal or geosynclinal cycles. (2) Mesozoic-Cenozoic shales are more diverse in their mineralogy and have larger (metal oxide)/Al_2O_3 ratios, except for those of K_2O and FeO, than older rocks. (3) Metamorphism of sedimentary rocks tends either to be isochemical or because of addition and subtraction of materials produces chemical compositions similar to those of igneous rocks. There are relatively few metamorphic rocks with chemical compositions intermediate between typical sedimentary and typical igneous rock types. (4) Age comparison of similar rock type groups (e.g., shales, mudstones, and slates; sandstones and quartzites), shows smaller compositional differences between Paleozoic and older rocks than between Paleozoic and Mesozoic-Cenozoic rocks. Furthermore, differences between Precambrian muddy metasediments (slates, schists, and phyllites) and igneous rock compositions are generally greater than compositional differences between Paleozoic and younger shales and mudstones.

The ratios of rock types of sedimentary origin show striking changes over the past 2 b.y. Carbonate rocks make up about 25% of the sedimentary rock mass of Phanerozoic age, but only about 5% of rocks 600 m.y. to 2 b.y. old. Evaporites are roughly 5% of the Phanerozoic mass, but much less than 1% of older rocks.

The preceding gross secular relations can be explained by a model in which: (1) the composition of the sedimentary rock mass at any given instant in geologic time has been the same for approximately the last 2 b.y.; (2) the total sedimentary mass has been deficient in carbonate and evaporite minerals relative to freshly deposited or young sediments; and (3) long-term trends in mineralogic and chemical composition of individual rock types, as well as variations in the ratios of contemporaneous rock types among the rocks that now remain, have been caused by differential erosion rates of the chemical components of the sedimentary mass and the effects of long-term diagenesis. Alternative models would have to invoke a combination of gradual secular changes with sudden changes in compositions and sedimentary regimes at era boundaries, changes for which we have no independent evidence.

Consequently, the hypothesis is advanced that, for at least the past 2 b.y., there may have been no marked continuous changes in the weathering and erosion of the continents, nor in the composition of oceans. Short-term variations (a few millions or tens of millions of years) may be regarded as more important than long-term trends (hundreds of millions of years).

INTRODUCTION

Sufficient data have now been accumulated to make preliminary estimates, for the whole earth, of the mass of the crust, the mass and composition of the oceans, the total mass of sedimentary rocks, present-day erosion rates, approximate percentages and masses of various sedimentary rock types as functions of geologic time, as well as chemical and mineralogical compositions of some rock types as functions of time.

Despite the uncertainties inherent in these estimates, it may not be too soon to apply them to some of the basic questions about earth history, such as: Are gross differences among the chemical and mineralogical compositions of existing sedimentary rocks related to their ages? If so, are these differences indicative of the earth's surface environment at the time of deposition? For example, has the composition of the ocean changed significantly through time? To put it another way, can the compositions and rock-type ratios of contemporaneous sedimentary rocks, as measured today, be roughly equated to their compositions and ratios at the time of deposition? What is the total mass of sedimentary rock that has been deposited over the past several billion years? Can the rate of deposition during a specified time interval be determined?

In our attempts to investigate these and similar questions, we will look first at the compositional variations of sedimentary rock types as functions of their ages.

CHEMISTRY, MINERALOGY, AND AGE OF SEDIMENTARY ROCKS

The average of chemical compositions of any group of sediments, as a function of geologic age, depends in an important way on the time interval over which the compositions are averaged and on the part of the geological time scale in which the interval is located. The limitation on the smallest stratigraphic interval that can be chosen for averaging is the ability of the stratigrapher to map time-equivalent units over the map area specified. For most intervals of the stratigraphic column in local areas of the order of 10^4 square miles, the practical limit of resolution is about 1 m.y. This is the best that can be done; the resolution is ordinarily much weaker.

Stratigraphic resolution diminishes as larger and larger areas are included. When the entire earth is sampled for averaging, resolution is at a minimum. The dependence of stratigraphic resolution on geologic age is shown in Fig. 1. The numbers are broad estimates and there are many exceptions. We can resolve without much difficulty to about 5 m.y. on a world-wide basis in the Tertiary or Mesozoic but cannot do much better than 100 m.y. in the Precambrian, even with the best radioactive age dating methods available.

Consequently, we have ordinarily not tried to obtain compositional or mineralogical data for rocks representing time intervals of less than a geologic period. To discern continuous compositional trends it is necessary to average over time spans of era proportions. This point is illustrated by data from the Russian Platform (Vinogradov and Ronov, 1956a, b).

In Fig. 2, data are plotted for the weight ratios of Na_2O/Al_2O_3 in shaly rocks versus that of the weight ratios of Mg/Ca in contemporaneous limestones and dolomites. It can be seen that the values for the ratios have a wide range during the Paleozoic era, that the Permian period has unique ratios, and that there is great variability within the Mesozoic and Cenozoic eras. All the points for the Paleozoic can be enclosed in an area completely distinct from the area enclosing Mesozoic and Cenozoic data, however. Therefore, although there is no obvious

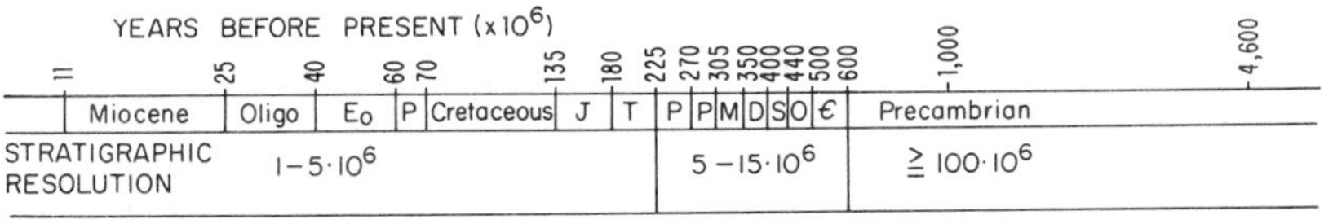

YEARS BEFORE PRESENT ($\times 10^6$)

11 | 25 | 40 | 60 | 70 | 135 | 180 | 225 | 270 | 305 | 350 | 400 | 440 | 500 | 600 | 1,000 | 4,600

Miocene | Oligo | Eo | P | Cretaceous | J | T | P | P | M | D | S | O | € | Precambrian

STRATIGRAPHIC RESOLUTION: $1-5 \cdot 10^6$ | $5-15 \cdot 10^6$ | $\geq 100 \cdot 10^6$

Fig. 1. Global stratigraphic resolution of the geologic time scale for purposes of reliable geochemical averages. The minimum number represents the best possible equivalent to stratigraphic stages; the maximum number is the poorest resolution likely with good stratigraphic mapping.

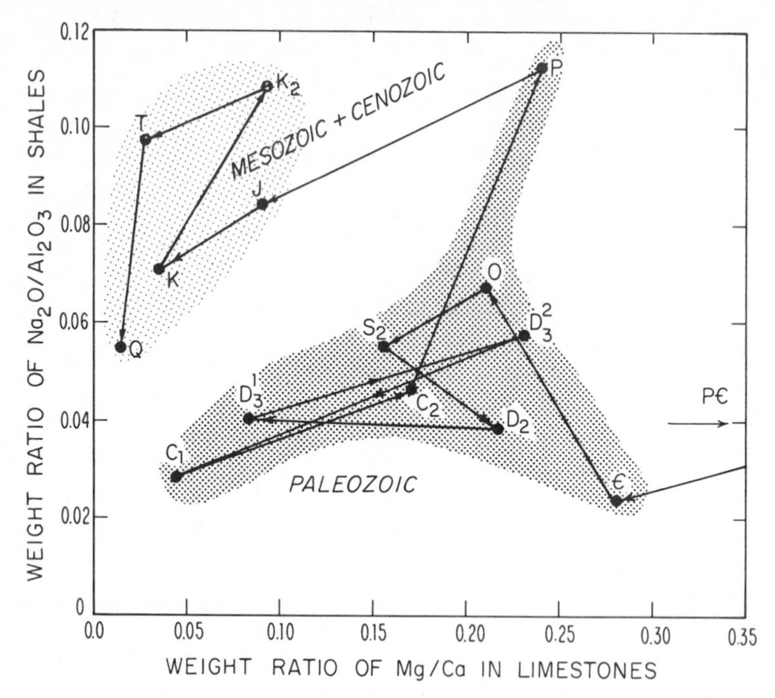

Fig. 2. Chemical composition of shales and limestones of different ages from the Russian Platform. The average weight ratio of Na_2O/Al_2O_3 in the shaly rocks of each geologic system is plotted versus the average weight ratio of Mg/Ca in the limestones of that system. Data from Vinogradov and Ronov (1956a, b).

time trend for the points within the Paleozoic, or within the Mesozoic and Cenozoic, it can be stated that the ratio of Na_2O/Al_2O_3 in shales for the most recent 200 m.y. is about twice that for the preceding 400 m.y., whereas the ratio of Mg/Ca in younger carbonate rocks is only about one-third or one-quarter that of Paleozoic carbonates. It is also apparent that Paleozoic carbonates show a much greater range of Mg/Ca ratios than Mesozoic and Cenozoic carbonates.

This particular plot is similar to plots of other chemical parameters from the Russian Platform data.

Shaly Rocks

In Fig. 3, we show four sets of composite analyses of shaly rocks; an estimate of the average composition of modern muddy sediments (authors), that of Mesozoic and Cenozoic shales (Clarke, 1924), of Paleozoic shales (Clarke, 1924), and of Precambrian slates, schists, and phyllites (Nanz, 1953). The ratios of the various metal oxides to Al_2O_3 were chosen as the ordinate because of the relative immobility of alumina.

Lack of precise absolute age data forced us to average, as a group, all analyses of Precambrian shaly rocks. Our only justification for assigning an age of 2000 m.y. to these older rocks is that many of the analyses used were from "middle" Precambrian.

The most striking aspect of the diagram is that there is but one reversal in trend, namely, the decrease in Na_2O/Al_2O_3 between Mesozoic-Cenozoic and Recent, and it is a change so minor that it may not be real. Second, all oxides except K_2O and FeO diminish with respect to Al_2O_3 with increasing rock age. Water, CO_2, and CaO behave similarly, going from high to very low values. MgO, Na_2O, and SiO_2 form a second behavioral group, but change much less than water, CO_2, and CaO. There is a reciprocal relation between FeO and Fe_2O_3; young rocks are high in oxidized iron, old rocks are low, but total iron oxide remains almost constant.

The chemical trends are reflected in mineralogical differences. An age effect in the relative abundance of major clay mineral groups has been discussed by Grim (1968, pp. 551-554). Figure 4 shows the relative percentages of clay mineral types from Cambrian to Recent using data from Weaver (1967). The more restricted chemistry of Paleozoic shales is a reflection of the dominance of only two clay minerals, illite and chlorite, with minor kaolinite and expandable (smectite or montmorillonite group) clays. In younger rocks, kaolinite and expandable clays add up to about 50% of total clay mineralogy. The increase in illite percentage in older rocks correlates with their relative enrichment in K_2O.

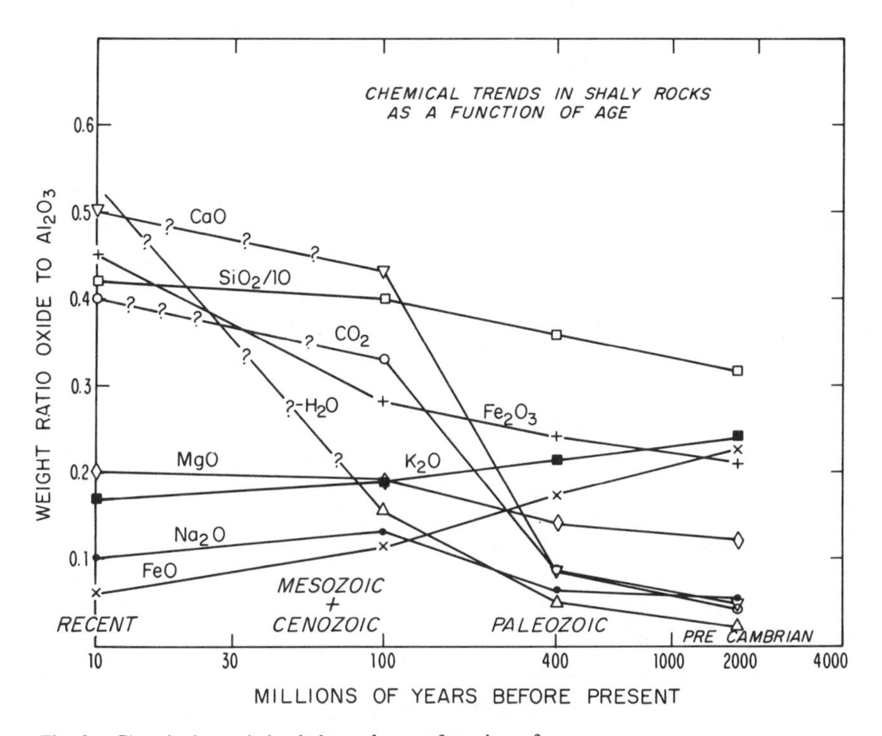

Fig. 3. Chemical trends in shaly rocks as a function of age.

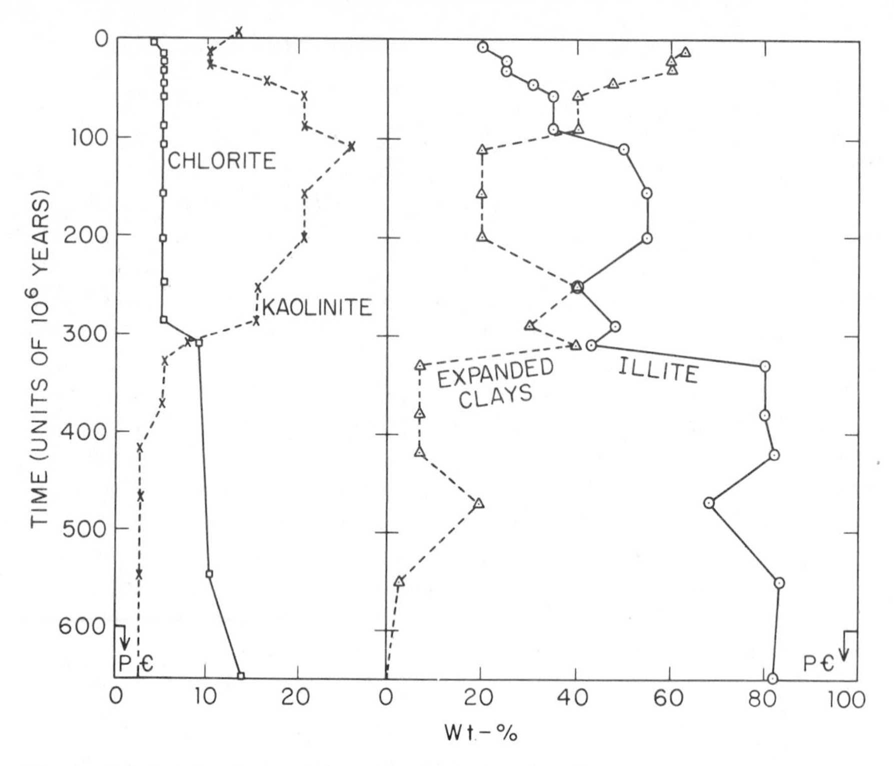

Fig. 4. Relative abundance of clay minerals as a function of age. Data from Weaver (1967).

Increased percentages of expandable clays in younger rocks account for increases in SiO_2, MgO, and Na_2O, the three oxides that behave similarly on the plot of shaly rock chemical relations (Fig. 3). The marked mineralogical change at about the end of the Paleozoic is readily apparent from Fig. 4, as are the large fluctuations in clay mineralogy over short time intervals in the Mesozoic and Cenozoic. The low percentages of kaolinite and expandable clays in Paleozoic shales are characteristic of Precambrian shales as well.

Whatever the reasons for variations of the compositions of shales as a function of their ages, the differences are sufficient to cause difficulties in obtaining a chemical analysis for the average shale. One widely used value (Clarke, 1924) has been obtained by taking a mean of the analyses of Paleozoic and Mesozoic-Cenozoic shales, on the assumption that roughly equal masses are involved, and indirectly implying that Precambrian shales or their metamorphosed compositional equivalents can be neglected. As we shall see later, the mass of Precambrian sedimentary rocks may be nearly equivalent to the Paleozoic and Mesozoic-Cenozoic total. If so, reference to Fig. 2 shows that the average composition of Paleozoic shales (Table 1) may be closer to a true composite than a 1/1 mixture of Paleozoic and Mesozoic-Cenozoic. The errors involved in

Table 1. Chemical analyses of average rocks (wt %)

	1	2	3	4	5
	Average igneous rock (after Brotzen, 1966)	Average limestone (after Clarke, 1924)	Average Paleozoic shale (after Clarke, 1924)	Average sandstone (after Pettijohn, 1963)	Average sedimentary rock (authors)
SiO_2	63.5	5.2	61.9	78.0	59.7
Al_2O_3	15.9	0.8	16.9	7.2	14.6
Fe_2O_3	2.9	{0.5	4.2	1.7	3.5
FeO	3.3		3.0	1.6	2.6
MgO	2.9	8.0	2.4	1.2	2.6
CaO	4.9	43.0	1.5	3.2	4.8
Na_2O	3.3	0.1	1.1	1.2	0.9
K_2O	3.3	0.3	3.7	1.3	3.2
CO_2	–	41.9	1.5	2.6	4.7
H_2O (+ 110°C)	–	0.2	3.9	2.0	3.4

using the average Paleozoic are small, whether we mix Mesozoic-Cenozoic, Paleozoic, and Precambrian on a 1/1/1 mass basis, or even in a 1/1/2 ratio. The major chemical effect is a change in the percentages of CaO, CO_2, and H_2O. The greater the mass of older rocks mixed with the post-Precambrian in obtaining an average, the lower the content of these three oxides.

Carbonate Rocks

The composition of the average carbonate rock is given in Table 1. Again, there is a difficult problem in weighting the analytical data because of compositional change with age. Recent carbonate rocks are nearly pure calcium carbonate, with Mg/Ca weight ratios of about 1/50 whereas Precambrian carbonates approach the composition of the mineral dolomite, with a Mg/Ca ratio of 1/1.7. Figure 5 shows the variation in the Mg/Ca ratio as a function of time for North American (Chilingar, 1956) and Russian Platform (Vinogradov and Ronov, 1956a, b) carbonate rocks. The similarity of the general trends, as well as the period to period "noise," is well illustrated. Also plotted is the MgO/Al_2O_3 ratio in shales from Fig. 3, to show the more or less reciprocal relation of Mg in shales to that in carbonate rocks.

Carbonate rocks less than 100 m.y. old have the low average magnesium content expected of carbonate deposits formed by accumulation of the skeletal debris of organisms. With increasing age, the Mg content rises irregularly and the trend apparently extends into the older Precambrian, as suggested by the extrapolated dashed lines on Fig. 5. Voluminous research on the "dolomite problem" has shown that the reasons for high Mg content of carbonates are diverse and complex. Some dolomitic rocks are primary precipitates; others were deposited as $CaCO_3$ and then converted entirely or partially to dolomite before

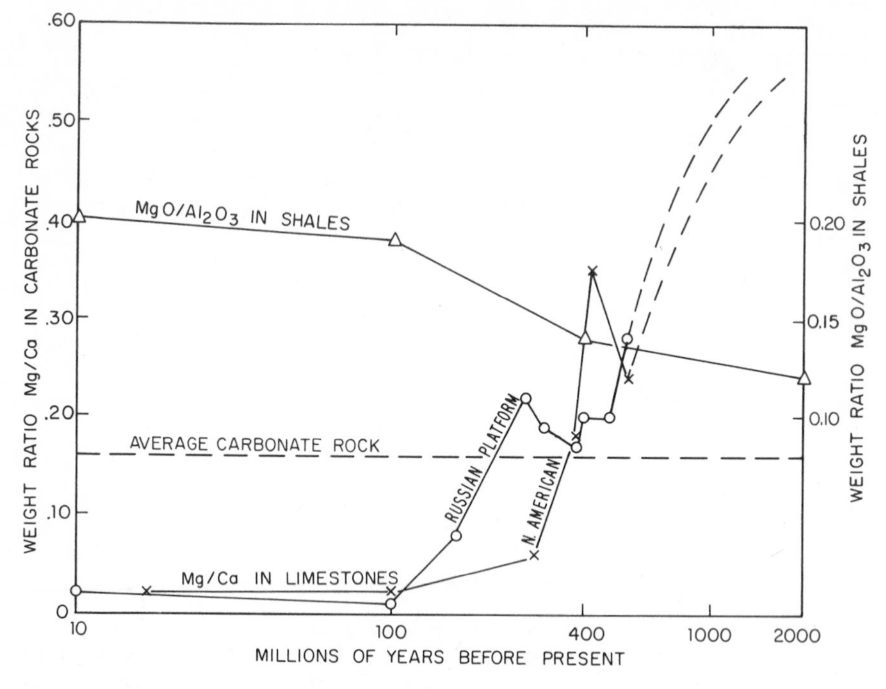

Fig. 5. Abundance of Mg in shales and limestones as a function of age. Weight ratio of Mg/Ca is used for carbonate rocks and weight ratio of MgO/Al_2O_3 is used for shales.

deposition of a succeeding layer; still others were dolomitized by migrating underground waters tens or hundreds of millions of years after deposition.

The mass of Precambrian carbonate rocks is small compared to the mass of younger ones; the average limestone is a composite of largely Phanerozoic limestones and thus may be a little low in its MgO content. However, an adjustment for the poorly known mass and composition of Precambrian carbonates does not seem justified at this time.

Sandstones

There seem to be few marked changes in the major element composition of sandstones with time. Composite analyses of Paleozoic, Mesozoic, and Cenozoic sandstones of the Russian Platform show no particular trends; their average composition looks much like that of Pettijohn's estimate of the world average (Table 1). The Russian samples are lower in SiO_2 (70% versus Pettijohn's 78%) and lower in the ratio of Na_2O/K_2O but similar in their contents of other oxides. The difference in SiO_2 percentages comes from the inclusion of analyses of quartzites that contain 95% or more quartz in the Pettijohn estimates. The composition of quartzites evolves by the virtual elimination of constituents other than SiO_2. The lower Na_2O/K_2O ratio of the Russian Platform sandstone is traceable to the relative deficiency of graywacke sandstones, which as a group have high Na_2O/K_2O ratios regardless of age.

Because sandstones are more heterogeneous in chemical composition than other sedimentary rocks, a reflection of the range of diverse petrographic types, secular changes in the chemical composition of the average sandstone are easily obscured by poor averaging among rock types. Though data are scarce, a few highly tentative conclusions may be made. Mesozoic and Cenozoic arkoses are as a group slightly higher in Na_2O/K_2O than older arkoses, dividing at a ratio of about 0.6. The graywackes, in contrast, show no significant age effects on compositions even when we include samples of the sandy, volcanic-rich sediments of the Columbia River, shown by Hawkins and Whetten (1969) to be graywacke precursors. If all non-graywacke sandstones are grouped, an age effect appears similar to that for arkoses, that is, Mesozoic and Cenozoic sands as a group tend to have slightly higher Na_2O/K_2O ratios than Paleozoic and Precambrian ones.

Despite the lack of documentation of time differences in sandstones by chemical analyses, the possiblity of major differences in the proportions of the four sandstone types with time must be considered. Very old geologic terrains, especially those in the plus 2500 m.y. range, are now being studied intensively, and according to several investigators, the percentages of graywackes and arkoses are increased relative to quartzose sandstones (Engel, 1963, Ronov, 1964). If so, the average sandstone composition might be shifted from the analyses given in Table 1 to a composition somewhat less siliceous and higher in the other oxides.

CONVERSION OF AVERAGE IGNEOUS ROCK INTO AVERAGE SEDIMENTARY ROCK (EXCLUDING EVAPORITES)

The preceding pages should have sufficed to illustrate the difficulties of choosing a chemical composition for each major rock type that represents a properly composited sample, weighted in terms of the relative masses of different compositions. We also recognize that more useful (but difficult to accumulate) averages would be those of the chemical composition of all carbonate minerals and all clay minerals and other silicates rather than the rock types that contain these minerals in large amounts, for the presence of many mixed rock types, such as calcareous shales and argillaceous limestones blurs differences and obscures true total masses. A good test of the analyses finally chosen, however, can be made by attempting to convert the average igneous rock into limestone, shale, and sandstone. This type of calculation is by no means new and has been made by many investigators. Goldschmidt's estimate (1933) is perhaps the best known.

If a mixture of these three rock types can be made that will yield the proportions of oxides found in the average igneous rock, the sedimentary rock analyses chosen to represent each type can be validated. Additionally, calculated proportions of the three types can be compared with the observed ones. The test is quite sensitive to the chemical analyses chosen for the individual rock types, because the sum of *each* oxide component in the three sedimentary rock types should add up to that in the igneous rock. If a mixture is chosen so that total K_2O, for example, adds up to that in the igneous rock, the proportions of all other oxides in the individual rock type analyses must be such that they also sum correctly.

The procedure that we used to get a balance can be followed by reference to Table 2. We started with a kilogram of average igneous rock (column 1). Then 70% of the CaO, or 34 grams of the original 49 in the igneous rock, was assigned to limestones. The percentage of CaO transferred was determined by trial and error as that producing the best final balance, and controls the proportion of limestone in the sedimentary mass. Then, from the chemical analysis of the average limestone, we calculated the number of grams of each other oxide component required by 34 grams of CaO. These requirements (column 2) were subtracted from the igneous rock to yield a remainder to be used to make shale and sandstone. Carbon dioxide is required to make limestone; because it is not present, except in traces, in the igneous rock analysis, the 34 grams of CO_2 were carried forward into the igneous rock remainder as a deficit.

In the next step (column 4), 97% of the K_2O in the igneous rock remainder was assumed to be present in shales. Again this particular amount is equivalent to selecting the proportion of shale in the sedimentary rock mixture. As before, the chemical analysis of the average shale was employed to calculate the number of grams of other oxides required by the 32 grams of K_2O taken from the igneous rock remainder, and these amounts subtracted from the igneous remainder to provide a second remainder. This second remainder (column 5), if the assumed proportions of limestone and shale were correct, should correspond to the composition of the average sandstone. In column 6, the composition of the average sandstone was calculated, adjusting the amounts of the various oxides to the SiO_2 content of the remainder in column 5.

Finally, column 7 is the test of the validity of the average analyses chosen and of the proportions of rock types mixed. Of the oxides in the original average igneous rock, all except Na_2O and iron oxides balance out to within a few percent of their original amounts, a result as close as could be expected when the difficulties of choosing averages are considered. A deficit in Fe_2O_3 and an excess of FeO in the final remainder are to be expected, because ferrous iron should be oxidized during weathering of the igneous source rock. The balance of *total iron* is comparable to that of the other oxides. The oxygen required is about 0.5 gm/kg of igneous rock converted to sediments.

The excess in Na_2O is certainly significant, and if the balance of the other oxides is used as a measure, is somewhere between 21 and 25 gm/kg of rock. This excess Na_2O, which disappears in the conversion of igneous to sedimentary rocks, has long been the basis for a variety of geochemical studies. It is found today converted to NaCl dissolved in the oceans and in the pore waters of sediments, or in deposits of solid NaCl in evaporites. Just as CO_2 was required to

Table 2. Conversion of igneous to sedimentary rocks (analyses in grams)

	1	2	3	4	5	6	7
	Average igneous rock (after Brotzen, 1966) (gm/kg)	Average limestone (assuming 70% of CaO of igneous rock)	Remainder	Average shale (assuming 97% of K_2O in igneous remainder)	Remainder	Average sandstone	Remainder, found in ocean, pore waters, evaporites
SiO_2	635	4	631	535	96 → (100%)	96	0
Al_2O_3	159	1	158	146	12	9	3
Fe_2O_3	29	–	29	36	-7	2	-9
FeO	33	–	33	26	7	2	5
MgO	29	6	23	21	2	1	1
CaO	49 → (70%)	34	15	13	2	4	-2
Na_2O	33	–	33	9	24	1	23
K_2O	33	–	33 → (97%)	32	1	2	-1
CO_2	–	34	-34	13	-47	3	-50
H_2O	–	–	–	34	-34	2	-36
Total	1,000	79		865		122	

Note.–HCl required to convert oxides to chlorides, 27 grams; CO_2 required for carbonate minerals, 50 grams.

convert CaO and some of the MgO of the average igneous rock to carbonate minerals, so HCl must be invoked to convert the excess Na_2O to NaCl. The reaction is

$$Na_2O + 2HCl = 2NaCl + H_2O$$

23 gm 27 gm 43 gm 6.7 gm

A "perfect" balance would show a residue of oxides in ratios similar to those in the combination of sea water, sediment pore waters, and evaporite and other miscellaneous rock types, but the amounts of oxides other than Na_2O in these residual sites are so small that their ratios cannot be resolved within the errors of the calculation.

The mass balance calculation permits us to calculate weight ratios of the three major sedimentary rock types. Note that the sum of the masses of the sediments is a little greater than that of the original igneous rock mass, 1066 versus 1000 grams, because added H_2O and CO_2 more than balanced the Na_2O loss to the oceans. According to the calculations, limestones, sandstones, and shales should occur in the ratio of 8/11/81. This ratio is in fair accord with the geochemical calculation made by Mead in 1907: 6/12/82; revised by Wickman (1954) to: 9/8/83; but checks less well with the most recent revaluation of earlier stratigraphic estimates by Horn and Adams (1966), used for an elaborate computer program of element balances, that gives a ratio of 9/17/74. The chemical composition of the 8/11/81 mixture is given in Table 1 (column 5).

Our results are comparable to those of other investigators who have made similar calculations, but they emphasize two important points. First, the calculated proportions of rock types do not agree with those obtained either by integration of field measurements of rock types found on the continents (Kuenen, 1941; Schuchert, 1931) or by stratigraphic estimates that include adequate representation of deep sea sediment (Horn and Adams, 1966). The obvious difference is the smaller percentage of limestone obtained by calculation. Second, the balance calculation is a demanding test of the relative proportions of the three rock types, so we can conclude that the average igneous rock would indeed give rise to limestones, sandstones, and shales in the relative quantities calculated.

ALTERATION OF MAFIC VOLCANIC ROCKS AS A SOLUTION TO THE GEOCHEMICAL BALANCE BETWEEN SEDIMENTARY AND IGNEOUS ROCKS

Recent studies of the submarine alteration of volcanic rocks (chiefly mafic) may provide the answers to the deficiency of carbonate rocks that typically result from chemical mass balance calculations. The excellence of the balance suggests that subaerial alteration of continental rocks into sedimentary rocks does indeed produce limestones, sandstones, and shales in the proportions calculated, and that what is required to bring calculated and measured ratios into agreement is an independent source of carbonate rocks.

In the oceans, volcanic rocks that have never been exposed to the atmosphere are commonly observed to be altered. For mafic rocks, the chemical change seems to be a loss of FeO, MgO, and CaO, accompanied by retention or even increase, of other constituents. Hart (1969) and Hart and Nalwalk (1970) have shown that K, Rb, and Cs are enhanced 2, 5, and 20 times, respectively, in the altered margins of submarine basalt fragments. Table 3 shows a comparison of analyses of a typical unaltered basaltic glass, altered glass from the same specimen, an altered volcanic Pacific pelagic sediment, and the Kitchi schist of the Archean basement of Northern Michigan. The brown altered glass shows marked loss of Fe, MgO, and CaO and apparently some increase of K_2O. If Al_2O_3 is assumed constant during alteration, there has been little change in Na_2O or SiO_2. The volcanic Pacific pelagic sediment shows low MgO and CaO compared to the basalt, much higher K_2O, but an iron content comparable to the basalt. The Kitchi schist, derived by alteration of a basaltic agglomerate, shows the same pattern of alkali enrichment, a low content of CaO, MgO, and Fe_2O_3, and some apparent enrichment in SiO_2.

It appears that the pattern of alteration in these three cases of basaltic material of vastly different ages is the same. It is likely that these chemical changes are a result of a mild attack on the basaltic materials by sea water. The combination of the affinity of basaltic rocks for K and the high Na content of the sea water inhibits loss of alkali metals while permitting the alkaline earths to go into solution. Rain water weathering, on the other hand, strips the alkalis as well as the alkaline earths.

Thus it appears that submarine alteration of mafic rocks can provide Ca and

Table 3. Comparison of unaltered basaltic glass with altered glass and sediment[a]

	Unaltered green basaltic glass[b]	Brown glass, altered from green[b]	Volcanic Pacific pelagic sediment[c]	Kitchi schist older Precambrian altered mafic agglomerate[d]
SiO_2	49.0	58.0	53.4	65.0
Al_2O_3	17.0	20.0	17.7	17.5
Fe_2O_3	11.0	6.3	14.1	6.0
MgO	7.0	3.7	3.9	3.3
CaO	11.1	4.3	3.7	1.0
Na_2O	2.5	3.2	2.2	5.6
K_2O	0.5	2.8	3.3	1.1
TiO_2	1.8	2.1	1.7	0.3

[a] Analyses adjusted to 100% for oxides selected, iron reported as Fe_2O_3.
[b] Muffler et al., 1969.
[c] El Wakeel and Riley, 1961.
[d] Van Hise, 1897.

Mg (and perhaps ferrous iron under reducing conditions) to ocean waters. The released Ca and Mg migrate elsewhere and eventually precipitate as carbonate rocks. Today oceanic bottom waters are almost everywhere oxygenated and released iron is precipitated as ferric oxide within or upon the basaltic host. This may account for the high iron content in the analyses of the volcanic Pacific pelagic sediment. On the other hand, several billion years ago when atmospheric oxygen was probably lower or nearly absent, there would have been a tendency for the Fe to travel in solution with the Ca and Mg or to precipitate as a ferrous sulfide.

In summary, we can look upon submarine alteration of volcanic rocks as a process in which unaltered mafic rock plus CO_2 give rise to an altered rock plus limestones or dolomites. There are not enough chemical analyses to permit a good quantitative estimate of the amount of carbonate rock that could be produced by a given mass of basaltic material, but comparing unaltered green volcanic glass with its altered brown counterpart, or comparing the composition of the Kitchi schist with the green basaltic glass, it appears that production of as much as 200 grams of carbonate rock from a kilogram of unaltered basaltic material is a reasonable estimate. Figure 6 shows an adjusted balance in which the CO_2 requirement for the submarine alteration of volcanic rocks and the addition to limestones from them are added to the classical limestone/sandstone/shale balance previously discussed. This adjustment increases the total CO_2 demand for the formation of sediments plus altered volcanics from 50

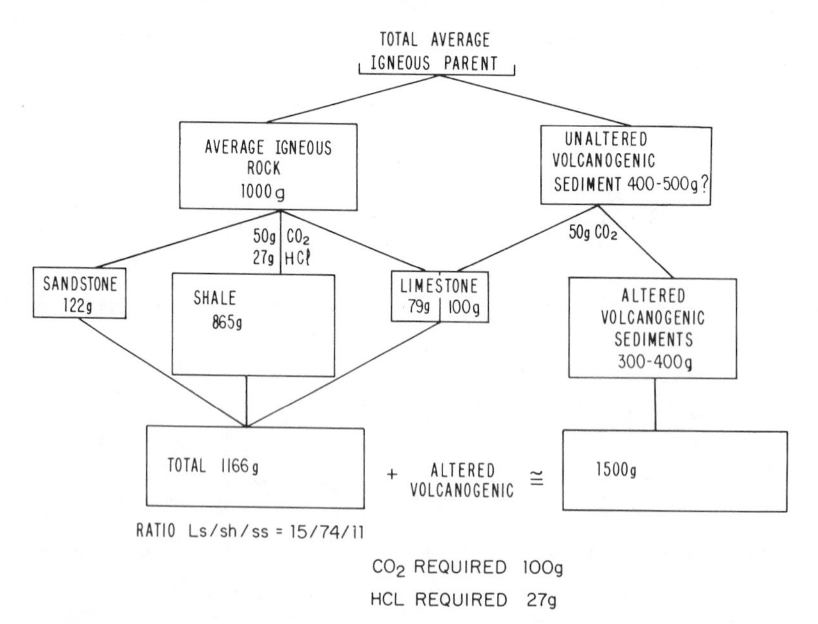

Fig. 6. Model calculation of the total sediment produced by the weathering of 1000 grams of igneous rock.

grams for the production of about 1200 grams of sediment from 1000 grams of rock to 100 grams CO_2 for the production of 1500 grams of total sediment plus altered submarine volcanics. On the other hand, the HCl requirement for the production of the 1500 grams total remains the same. In the new balance, limestone is 15% of the total of limestone plus sandstone plus shale, as compared with the 8% obtained without consideration of submarine alteration.

It is possible that submarine alteration of mafic rocks may have been of greater relative importance in the early history of the earth. It may have been the major process by which the primary acid volatiles were neutralized. The important world-wide deposits of cherty iron formations have characteristics that fit well into a general picture of oceans with many anaerobic areas from which ferrous iron, Mg, and Ca could move in solution, leaving the rest of the rock behind, and with fractional precipitation yield the classical sequence we find today of quartzite, slate, dolomite in one area and quartzite, slate, iron formation in another.

CALCULATION OF MASS OF SEDIMENTARY ROCKS

We can now estimate the relative masses and the total mass of sedimentary rocks including recognizable metasediments, using the numbers from Fig. 6. First we use an adaptation of the "chloride method," which depends upon an estimate of the total Cl in the crust that represents HCl neutralized by reaction with igneous rocks. The most recent estimated total Cl in the crust, exclusive of that in igneous rocks, is given by Horn and Adams (1966) as 560×10^{20} grams. As we have shown, 27 grams of HCl, equivalent to 26 grams of Cl, correspond to the production of 1500 grams of sedimentary rock. The calculation for total sedimentary mass is:

$$\frac{560 \times 10^{20} \times 1500}{26} = 32{,}000 \times 10^{20}\ \text{gm}$$

A similar calculation can be done for Na but the Cl method is preferable, because Cl is a direct measure of the acid neutralized, whereas Na can be bound or freed without acid consumption by cation exchange reactions.

The value of $32{,}000 \times 10^{20}$ grams is substantially larger than that of Ronov (1968). If we accept Gregor's (1968) value of about $18{,}000 \times 10^{20}$ grams for the Phanerozoic mass, however, $14{,}000 \times 10^{20}$ remain for the body of Precambrian sediments. Ronov's estimate of about $24{,}000 \times 10^{20}$ grams total sediment leaves only $6{,}000 \times 10^{20}$ grams for rocks older than Cambrian. Such a relatively small mass implies more destruction of old rocks than most geologists are willing to admit.

The figure $32{,}000 \times 10^{20}$ also leads to a consistency in carbonate rock relations. If we accept the Phanerozoic ratios of limestone/shale/sandstone of about 20/65/15 proposed by several investigators, we can inquire, from the overall ratios of 15/74/11 we have calculated, what the ratios would be for Precambrian rocks, if we know the relative total rock masses of Precambrian and post-Precambrian rocks. The calculation is given in tabular form here:

	Limestone	Shale	Sandstone	Volcanogenic	Total
			(in units of 10^{20} grams)		
Total	3,500	17,300	2,600	8,600	32,000
Post-Precambrian	2,600	8,600	2,000	4,800	18,000
Precambrian (difference)	900	8,700	600	3,800	14,000
Precambrian percent (ratios)	6	62	4	27	

The calculations show that, whatever the actual mass numbers, the mass of Precambrian sediments relative to post-Precambrian is substantial, of the order of 1/1 as opposed to ratios like 1/3 or 1/4. A large mass of low carbonate Precambrian sediments is consistent with the 20 to 25% of carbonate rocks of Phanerozoic age and the 15% of carbonate rocks calculated for the total carbonate-shale-sandstone sedimentary mass. The total mass of $32{,}000 \times 10^{20}$ grams is, of course, based on Gregor's value of $18{,}000 \times 10^{20}$ grams for the post-Precambrian, and his estimate is subject, in our opinion, to a probable error of ± 15%.

SEDIMENTARY ROCK MASS AGE RELATIONS AND SEDIMENTARY CYCLING

The data on secular variation of chemical composition with age already have been presented. Now we need the information available on the distribution of sedimentary mass as a function of the ages of the rocks, as well as the same information for individual rock types. Then simple models of erosion, deposition, and sediment distribution can be devised and tested against what we know of the actual age distribution of sediment types, and the chemical variations of each type.

Figure 7 is an estimate of the relative distribution of sedimentary mass as a function of age, based on the maximum thickness data of Kay (1955) and the volume estimates by Ronov (1959) for individual periods. We divided the mass by the duration of the period to find the average mass remaining for each period per unit of time. The numbers obtained were used to plot the histogram. The area of each block is proportional to the mass of rock of that period that remains today.

The following generalizations are probably justified, despite the uncertainties involved in making the estimates and the fact that the precise shape of a histogram is affected by class intervals, in this case the choice of system boundaries. The late Precambrian (600 to 700 m.y. old), the Cambrian, the

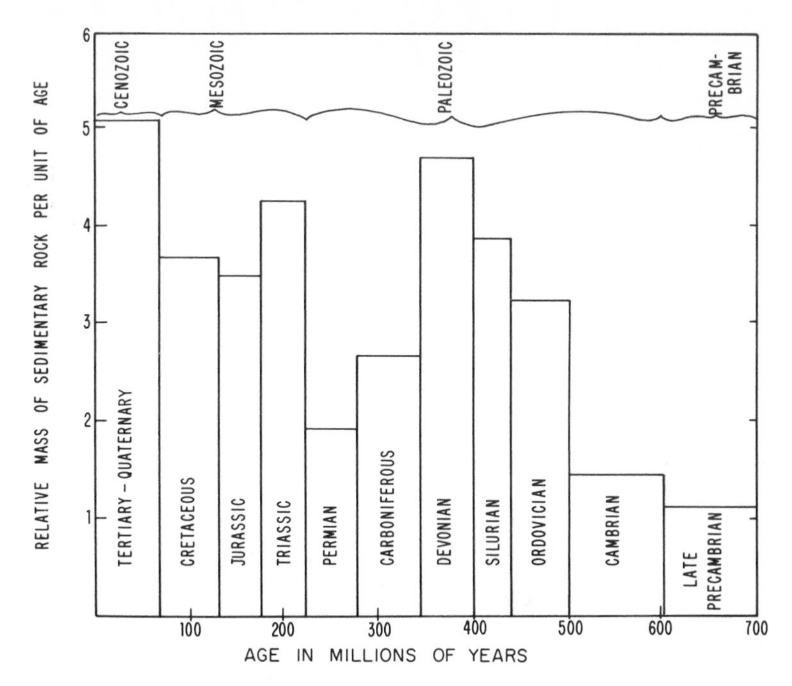

Fig. 7. Relative mass of observed sedimentary rocks as a function of age. The area of each histogram block is proportional to the mass of rock of that period that is found today divided by the duration of the period.

Carboniferous, and the Permian have much less rock preserved per year of their respective time spans than do other periods.

There is a clear-cut minimum of mass per unit time between 200 and 300 m.y. ago, and a low, which may or may not be a minimum, during the interval between 500 and 700 m.y. ago. The center of gravity of the mass distribution shown is at about 300 m.y.; inasmuch as the total time portrayed is 700 m.y., the gross mass distribution is almost uniform with age, but with somewhat more mass remaining per unit time for the most recent 350 m.y. than for the preceding 350 m.y. The Mesozoic-Cenozoic mass is somewhat larger than the Paleozoic mass, but this difference is attributable almost entirely to the large mass of Tertiary and Quaternary sedimentary rocks.

According to our preceding estimate, the mass of Precambrian sedimentary rocks is somewhat less than the total of Phanerozoic ($14{,}000 \times 10^{20}$ versus $18{,}000 \times 10^{20}$ grams), but we know next to nothing about its distribution with age. We know the mass is spread over a total of at least 3 b.y., and that the amount remaining per unit time diminishes with increasing age. We are also fairly certain that there are maxima and minima in the spread of $14{,}000 \times 10^{20}$ grams of rock over the 3 b.y. time span. Sedimentary rocks with ages of about 1 b.y. are found in many areas. There are relatively few occurrences of rocks in the

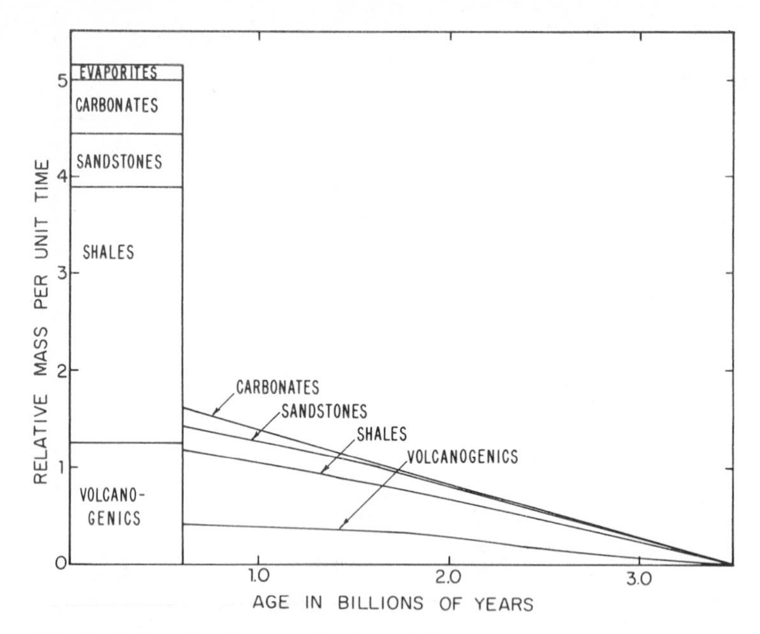

Fig. 8. Schematic diagram showing relative masses of major sediment lithologic types as a function of age. Data from Ronov (1968).

600 to 800 m.y. range, so the low value for the late Precambrian shown on Fig. 7 may indeed be a minimum.

Figure 8 has been derived from Ronov's (1964) estimates of the relative percentages of rock types and from our estimate of the relative masses of Phanerozoic and Precambrian sediments. It shows schematically that the Precambrian mass is distributed over a tremendous time span and the marked concentration of mass is toward the present. It also illustrates the much greater relative importance of limestones and evaporites in the Phanerozoic than in the Precambrian and the constancy of the ratio of shale to sandstone. Though limestones of the 2 to 3 b.y. old range may be of great importance in early historical geology they are a vanishingly small proportion of the total.

Sedimentary Cycling

Now that we have rough estimates of the total mass of preserved sedimentary rocks, as well as some notion of the distribution of that mass with age and rock type, we can construct models of the distribution, making assumptions concerning rates of deposition, destruction, and sediment accumulation throughout geologic time and see which of the models gives the best fit to the little we know.

There are two chief types of models: one in which a mass of sediments comparable to that preserved today was formed early in earth history, and has been continuously destroyed and redeposited, and a second in which the preserved mass increased steadily with time.

Constant Mass Models. The underlying basis for constant mass models is an early degassing of the earth, with the emission of all the water to the hydrosphere and all the CO_2, HCl, and other acid gases that reacted with primary igneous rocks to form sedimentary rocks. This model implies that since that time there has been no significant increase in the total mass of sediments because there have been no great amounts of new acid gases released from the deep interior to create them, although the sediments have been turned over by erosion and destroyed by metamorphism and crustal plate resorption into the mantle, with concomitant recycling of CO_2 and HCl. This view, of course, is an extreme one, but it should provide us with one set of limiting conditions.

The assumptions of constant mass models are the following.

1. A mass of sediments equal to the existing mass was formed very early in earth history from primary igneous rock.
2. This mass has been redeposited and destroyed (by erosion, metamorphism, or crustal plate resorption) at a constant rate through time.
3. The probability of a given mass of sediments being destroyed at any time is proportional to the ratio of the given mass to the total mass.

Figure 9 shows the calculated fraction remaining today of the sediments of the past as a function of various rates of deposition relative to the total mass (M).

Linear Accumulation Models. The major alternative to the constant mass model is one in which water, HCl, and CO_2 are continuously being degassed from the interior of the earth, and as a result, the mass of sediments existing at any

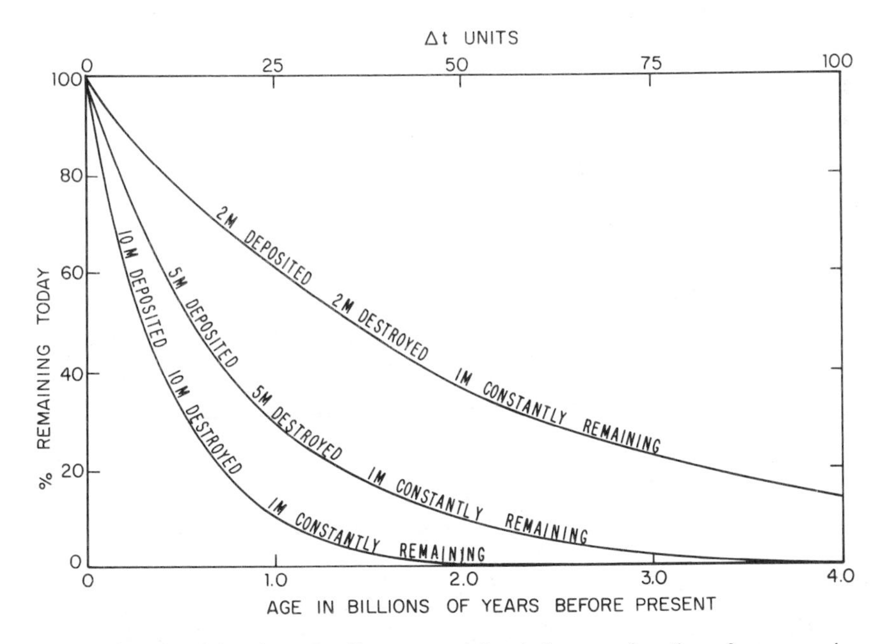

Fig. 9. Calculated fraction of sediment remaining today as a function of age assuming several constant rates of deposition and destruction for a constant sedimentary mass model.

given time continually increases with time. A constant rate of degassing giving rise to a linear accumulation of mass would appear to represent the extreme alternative condition to a constant mass model, because most hypotheses support a model with a high early degassing rate, followed by a continuously decreasing rate that probably is somewhat irregular. The assumptions of the linear accumulation models are as follows.

1. The mass of sediments has grown linearly through time from zero to the currently existing mass.

2. Deposition, destruction, and accumulation have all gone on at constant rates, so that deposition rate minus destruction rate equals accumulation rate.

3. As before, the probability of a given mass of sediments being destroyed at any time is proportional to the ratio of the given mass to the total mass.

In this model, the oceans also have grown linearly with time. Note, however, the model assumes that the ratios of $H_2O/CO_2/HCl$ in the released gases are constant and that the oceans and the sediment mass grow together. Implied by such a model is the linear accumulation of total continental masses.

Figure 10 shows curves of percent remaining sediment versus geologic age for models in which the ratios of total mass deposited to mass accumulated are 2/1, 5/1, and 10/1. Details of the calculations required for plotting Figs. 9 and 10 are given by Garrels and Mackenzie (1971).

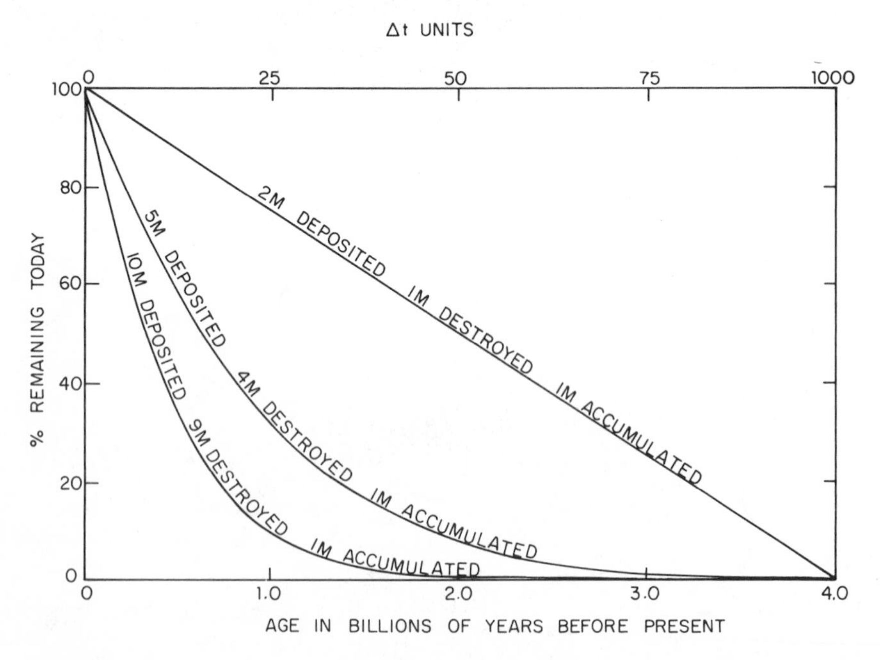

Fig. 10. Calculated fraction of sediment remaining today as a function of age assuming various rates of deposition and destruction for a linear accumulation sedimentary mass model.

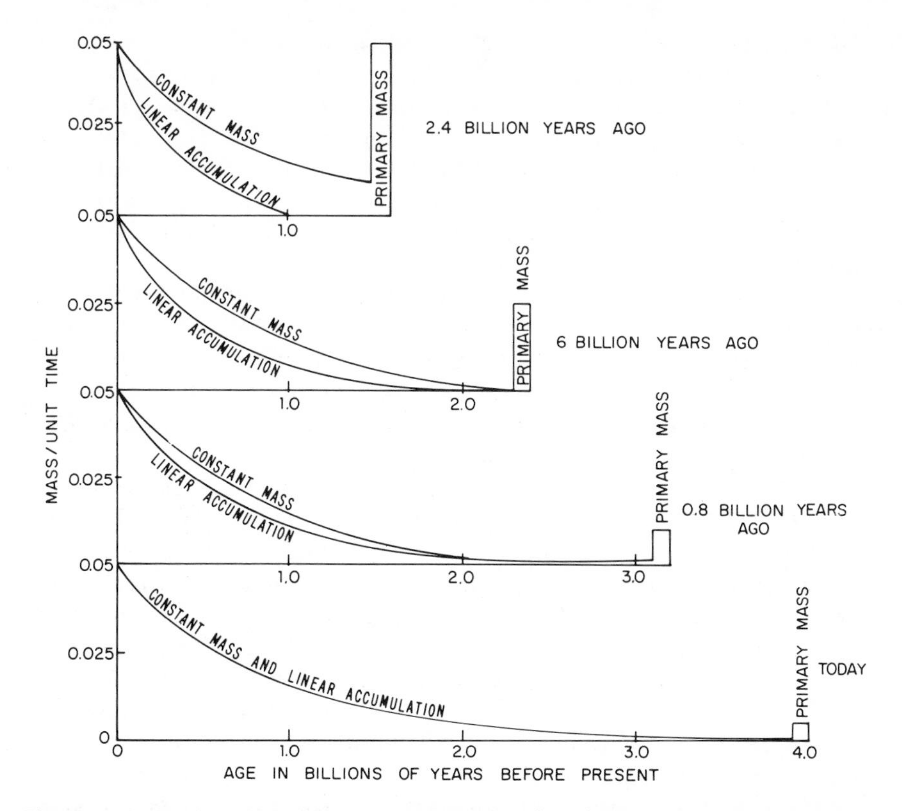

Fig. 11. Distribution of a sedimentary mass as a function of age as it would have appeared at various times in earth history for a constant mass as opposed to a linear accumulation model.

Discussion of Models. Figures 9 and 10 show that the predicted age distributions of sedimentary rocks are almost independent of the model used if the total mass that has been deposited through time is much larger than that preserved today. The effect of the presence of an initial mass of sediment is wiped out by the continuous processes of deposition and destruction, and the curves for the two models converge. Figure 11 shows the distribution of mass as a function of age as it would have been at various times in earth history according to a constant mass model as opposed to a linear accumulation model. In both cases, the depositional rate chosen is sufficient to have deposited five times the mass of sedimentary rocks existing today. In the uppermost diagram, which is drawn to show the distribution of mass 2.4 b.y. ago, the influence of the initial nearly instantaneous deposition of a sediment mass M is still strongly apparent, and more than 30% of the sedimentary rocks at that time would still be "primeval." The lower diagrams show successively how, with reworking, the constant mass and linear accumulation models converge, until today they are apparently indistinguishable.

It is an interesting and somewhat ironical conclusion that man apparently evolved just at the time of coincidence of the two models, and too late to have much chance of finding evidence of a primeval sediment mass, if such existed. On the other hand, we have arrived in time to find at least fragments of very old sedimentary masses. Let us hope, for the sake of future geologists, that the present-day rate of destruction of the sedimentary record does not last too long; another 100 m.y. at the present rate would destroy half the currently existing deposits.

Figure 12 (Garrels and Mackenzie, 1969) shows the relation of the curves for depositional rates sufficient to have deposited 2, 5, and 10 times the existing sediment mass and fitted to the estimates of the mass-age distribution of the existing mass. Despite the obvious important departures of the actual mass distribution from the smooth curves of the simple models, a fivefold turnover of mass clearly gives a far better fit than a twofold or a tenfold turnover.

The departures of the actual mass-age relation from the models can be explained by changing or modifying the original assumptions. The mid-Paleozoic maximum, followed by the Permo-Carboniferous minimum, for example, could be explained either by varying the depositional rate or by changing the assumption concerning equal probabilities of destruction. The former explanation, however, does not seem to be valid, for the Paleozoic mass maximum comes at a time when we are assured, from other criteria, that depositional rates were at a minimum, whereas the Permo-Carboniferous minimum comes at a time of great changes of many types and is regarded as a probable interval of high depositional rates. It seems more likely, then, that the minima in the mass-age distribution reflect deposition of sediments under conditions that increased the subsequent chances of their destruction relative to those of older rocks. Perhaps parts of the Permian deposits were formed at the beginning of a time of rifting of the continents and were swept beneath the continental blocks at a greater rate than before or since. Conversely, it seems as though the mid-Paleozoic maximum might be the result of better than average protection of those deposits, as compared to older and younger rocks. We have not yet followed up many of the relations suggested by the mass-age relations; for example, the apparent protection of much of Paleozoic deposition leads to the possibility that the chief sources of Mesozoic-Cenozoic rocks are from Precambrian rocks or from cannibalism of the Mesozoic-Cenozoic sediments and igneous rocks themselves. Conversely, the relative absence of late Precambrian and early Cambrian sediments implies that they were a major source for the later Paleozoic sedimentary mass.

In passing, we should note that the curves drawn from the models are independent of the proportion of sedimentary rock that is destroyed by erosion, as contrasted with destruction by metamorphic or igneous processes. Consequently, the calculations permit no direct predictions about this important aspect of earth history.

Differential Cycling Rates. The mass half-age of the total sedimentary mass has

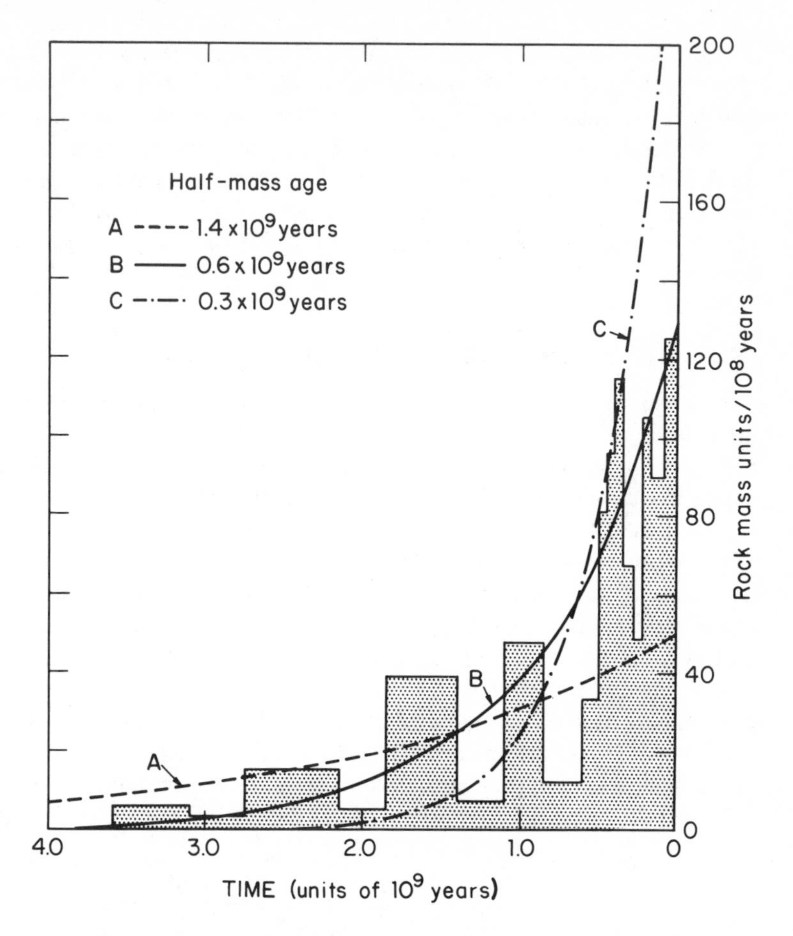

Fig. 12. Curves for depositional rates sufficient to have deposited 2(A), 5 (B), and 10 (C) times the existing sedimentary mass fitted to estimated mass-age distribution of existing mass.

been estimated at about 500 m.y., i.e., half the rocks are younger and half are older than 500 m.y. On the other hand, if we use Ronov's estimate (1964) of the distribution of rock types as a function of time, we find that the mass half-age of evaporites is about 200 m.y., and that of carbonate rocks is about 300 m.y. Shales and sandstones have mass half-ages of about 600 m.y. Therefore, if we consider only evaporites, using the same kinds of cycling models as before, we would have to turn over a total evaporite mass 15 times to account for the present relations. Similarly, the total limestone cycled would be about 10 times the existing mass to account for its mass half-age. These conclusions are consistent with our general information regarding the ease of erosion of these rocks. Modern dissolved stream loads are dominated by the constitutents of limestones and evaporites.

Figure 13*a* shows the result of a calculation of the relative percentage of evaporites, carbonate rocks, and shales and sandstones, based on the assumption that evaporites have been cycled 15 times; carbonate rocks 10 times; and shales and sandstones 5 times. For comparison we have redrawn the part of Ronov's diagram (Figure 13*b*) that pertains chiefly to the observed distribution of these rocks. It is evident, if we remove the kinds of perturbations that result from short-time variations in geologic processes, that the predicted distributions are in general agreement with the observed ones.

SELECTIVE DESTRUCTION VERSUS PRIMARY FEATURES

It is necessary to try to present the current situation with respect to the distribution of rock types with time, as well as to look at the chemical trends of individual rock types. To what extent do these distributions reflect initial depositional conditions? To what extent do the ratios of types of a given age represent the ratios that existed at the time of deposition?

It is a consequence of the differential cycling model that the composition of the sediment mass deposited at any given time would always be the same, if the model chosen is the constant mass model. The initial mass would be formed quickly, and the proportions of rock types would be immutable; the total mass at any time, regardless of the age distribution of types, would contain the same proportions. As soon as the initial mass began to be eroded and new sediments deposited, the proportions of rock types in the new deposits would reflect the differential rates of cycling of the components, so that the deposits would have a different ratio of rock types from those of the original mass. The deposits would

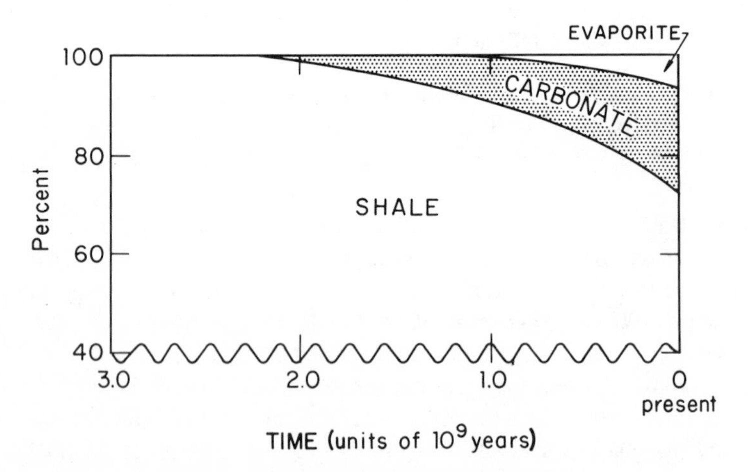

Fig. 13*a*. Calculation of relative abundance of shale, carbonate, and evaporite as a function of age, assuming total evaporite deposited was 15 times the existing mass, total carbonate rock was 10 times, and total shale and sandstone was 5 times.

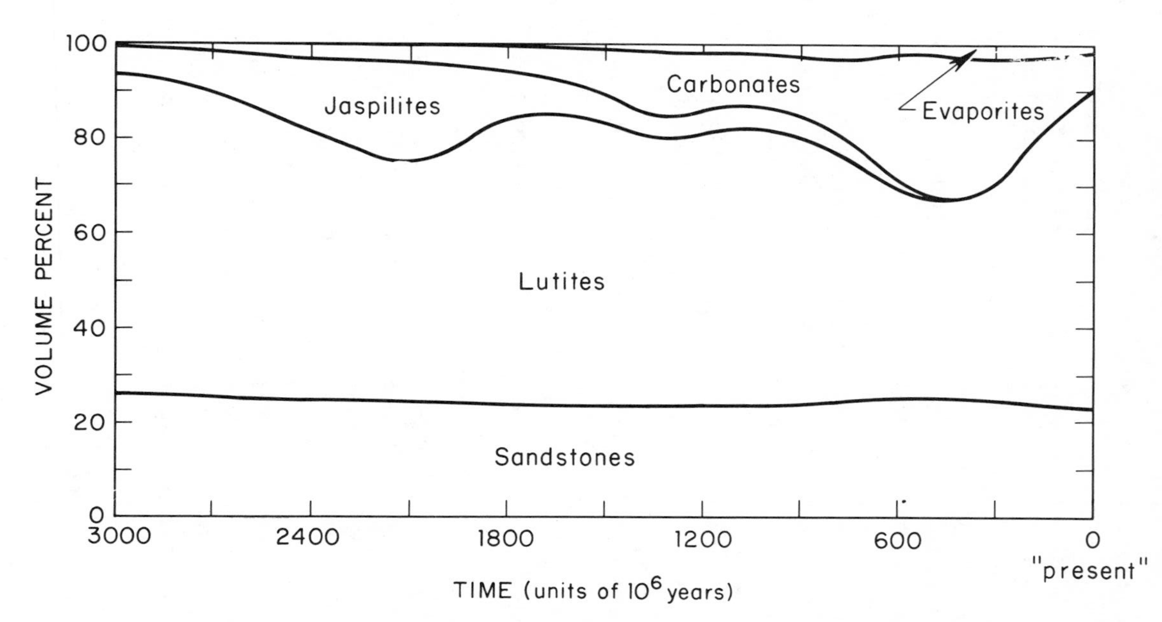

Fig. 13*b*. Relative abundance of sediment lithologies as a function of age, after Ronov (1968).

be enriched, relative to the original mass, in evaporites and carbonate rocks, and impoverished in shales and sandstones. Because of the assumption of equal probability of exposure to erosion of all equal masses, the same source materials of the total mass would always be available to erosion, so that the result of deposition of a constant ratio of rock types, derived from a different ratio of source rock types, results in changes in the proportions of rock types as a function of age.

The present is probably not a good test of this concept, but because of the immensity of geologic time, we can probably average the Cenozoic and Mesozoic deposits to derive the proportions of rock types that would be expected as a result of the depositional ratios being constant through time.

Such a model leads to a concept of chemical uniformitarianism; that is, the sediments deposited have always been the same in their proportions of types, and the differences we see in sediments and metasediments as a function of age are post-depositional in their origin, engendered by the continuous differential selection of contemporary materials from the total sedimentary mass.

There are a number of secular trends of individual rock types that are in harmony with this general concept. Mesozoic and Cenozoic rocks are diverse in mineralogy; they contain higher proportions of carbonate rocks and of evaporites than the average for all sedimentary rocks. Mesozoic and Cenozoic muddy sediments contain many minerals that disappear or diminish with time, such as aragonite, magnesian calcites, sepiolite, kaolinite, and abundant mixed-layer clays. The chemical differences between Paleozoic shales and the younger ones are consistent with a gentle water leach, loss of Ca, CO_2, Mg, Na, and Si (all relative to Al), and a gain of K only. The number of minerals is decreased, and the number of incompatible phases also is decreased. It looks very much as if the Mg, once perhaps in Paleozoic shales, has been transferred irregularly through a few hundred million years, to the contemporaneous limestones, and has made them dolomitic. The decrease in the O^{18}/O^{16} ratios of carbonate sediments and in cherts from Cambrian to Recent has been interpreted by some investigators as the result of extensive post-depositional exchange with fresh waters, another suggestion of important post-depositional modification.

The transfer of chemical constituents from one rock type to another has a time scale of the order of magnitude of 100 m.y. that corresponds to a time lag of the same magnitude between the primary production of detritus from tectonic events that generate weathering and erosion and the eventual arrival in relatively stable marine basins of most of the reworked and recycled debris. This time lag comes about because much of the weathered and unweathered detritus produced by a mountain building episode is at first tectonically trapped in intermontane basins within the mobile belt, where it has a high probability of being quickly recycled. The rapid cycling results in continual loss of detritus, and the chemical constituents leached during weathering, to areas of the seas and continents not immediately involved in the mobile belt, where the rate of cycling drops drastically. So the post-depositional modification that gives rise to chemical differences between average Paleozoic and Mesozoic-Cenozoic rocks may be related to the long times it takes for most of the original source material to be "weathered" by a complex series of sedimentary cycles, each of which contributes its statistical share of leaching and transfer of chemical constituents.

We can now look at the other side of the coin. Let us interpret some of the same changes as primary deposition and add some examples of deposits that surely represent different initial conditions. The outstanding classic example is the cherty iron deposits. As shown in Fig. 13*b*, the jaspilites, as Ronov terms them, occur only in the Precambrian. Even though their relative abundance may have been enhanced since their deposition by differential removal of other rocks, they definitely composed at that time a significant part of the sedimentary materials deposited. The occurrence of jaspilites has most frequently been related to low levels of oxygen in the atmosphere, high frequency of anaerobic environments, and chemical differences resulting from the primitive stage of the development of life.

The occurrence of the jaspilites may be related to another aspect of the early Precambrian scene that is emphasized by Ronov–a high proportion of volcanogenic sediments and of graywackes. There is no particular reason why these two rock types would have relatively slow cycling rates, which would be necessary to increase their percentages at the expense of lutites, for example, and thus their high early percentages are undoubtedly primary. The record of the early Precambrian is so fragmental that most of what has been said about cycling and rock types may break down entirely when we get to ages greater than 3 b.y. At any rate, the early high percentages of mafic lavas and graywackes may indicate that the average igneous source rock for sediments 3 b.y. ago was considerably higher in Ca, Fe, and Mg than the one we chose for today, and the sedimentary mass would be expected to be enriched in carbonate rocks, chert, and in iron compounds, corresponding to the alteration of volcanogenic rocks discussed before. Again, however, we would expect to have most of the carbonate rocks that might have been an important fraction of the early deposits to have been cycled forward in time, leaving the volcanogenic sediments behind.

We must remember that the large differences between the simple cycling models and the actual distribution of rocks, as shown in Fig. 12, represent in part primary sedimentary mass differences. As pointed out, the high percentages of limestones in the middle Paleozoic probably mean a slowing at that time of the cycling rate of the clastic materials, and thus an original depositional difference in rock type ratios of that time.

The noise within eras, that is to say, the large periodic fluctuations in ratios and compositions of sedimentary types, also must be world-wide primary sedimentary features. They are modified, to be sure, by post-depositional differential cycling, but the initial variations of conditions still control what we see. It is in the Precambrian rocks that selective loss of rock types has had its most important effects. If our estimates of rates of sedimentation are roughly valid, about 90% of the Precambrian once deposited is gone, and the residual record must be interpreted with great care to avoid mistaking selective loss for a primary record.

REFERENCES

Brotzen, O., The average igneous rock and the geochemical balance, Geochim. Cosmochim. Acta, **30**, 863-868, 1966.

Chilingar, G. V., Relationship between Ca/Mg ratio and geologic age, Bull. Am. Assoc. Petrol. Geologists, **40**, 2256-2266, 1956.

Clarke, F. W., Data of geochemistry, 841 pp., U.S. Geol. Surv. Bull., **770**, 1924.

El Wakeel, S. K., and Riley, J. P., Chemical and mineralogical studies of deep-sea sediments, Geochim. Cosmochim. Acta, **25**, 110-146, 1961.

Engel, A. E. J., Geologic evolution of North America, Science, **140**, 143-152, 1963.

Garrels, R. M., and Mackenzie, F. T., Sedimentary rock types: Relative proportions as a function of geological time, Science, **163**, 570-571, 1969.

______, Evolution of Sedimentary Rocks, W. W. Norton & Co., New York, 1971.

Goldschmidt, V. M., Grundlagen der quantitativen Geochemie, Fortschr. Mineral. Krist. Petrog., **17**, 112-156, 1933.

Gregor, C. B., The rate of denudation in Post-Algonkian time, Kon. Ned. Akad. Wetensch. Proc., **71**, 22-30, 1968.

Grim, R. C., Clay Mineralogy, 2d ed., 596 pp., McGraw-Hill, New York, 1968.
Petrog., **17**, 112-156, 1933.

Hart, S. R., K, Rb, Cs contents and K/Rb, K/Cs ratios of fresh and altered submarine basalts, Earth Planet. Sci. Letters, **6**, 295-303, 1969.

Hart, S. R., and Nalwalk, A. J., K, Rb, Cs and Sr relationships in submarine basalts from the Puerto Rico trench. Geochim. Cosmochim. Acta, **34**, 145-156, 1970.

Hawkins, J. W., Jr., and Whetten, J. T., Graywacke matrix minerals: hydrothermal reactions with Columbia River sediments, Science, **166**, 868-870, 1969.

Horn, M. K., and Adams, J. A. S., Computer-derived geochemical balances and element abundances, Geochim. Cosmochim. Acta, **30**, 279-290, 1966.

Kay, M., Sediments and subsidence through time, *in* Crust of the Earth, edited by A. Poldervaart, pp. 665-684, G.S.A. Special Paper **62**, 1955.

Kuenen, Ph. H., Geochemical calculations concerning the total mass of sediments in the earth, Am. J. Sci., **239**, 161-190, 1941.

Mead, W. J., Redistribution of elements in the formation of sedimentary rocks, J. Geol., **15**, 238-256, 1907.

Muffler, L. J., Short, M., Keith, E. C., and Smith, V. C., Chemistry of fresh and altered basaltic glass from the upper Triassic Hound Island volcanics, Southeastern Alaska, Am. J. Sci., **267**, 196-209, 1969.

Nanz, R. H., Jr., Chemical composition of pre-Cambrian slates with notes on the geochemical evolution of lutites, J. Geol., **61**, 51-64, 1953.

Pettijohn, F. J., Chemical composition of sandstones—Excluding carbonate and volcanic sands, *in* Data of Geochemistry, 6th ed., edited by M. Fleischer, U.G.S.G. Prof. Paper **440-S**, 21 pp., 1963.

Ronov, A. B., On the post-Precambrian geochemical history of the atmosphere and hydrosphere, Geochemistry, **5**, 493-506, 1959.

______, Common tendencies in the chemical evolution of the earth's crust, ocean, and atmosphere, Geochemistry, **8**, 715-743, 1964.

______, Probable changes in the composition of sea water during the course of geological time, Sedimentology, **10**, 25-44, 1968.

Schuchert, C., Geochronology of the age of the earth on the basis of sediments and life, *in* The Age of the Earth, pp. 10-64, Bull. Nat'l. Res. Council, 1931.

Van Hise, C. R., The Marquette iron-bearing district of Michigan, 608 pp., U.S. Geol. Surv. Monographs, **28**, 1897.

Vinogradov, A. P., and Ronov, A. B., Evolution of the chemical composition of clays of the Russian Platform, Geochemistry, **2**, 123-139, 1956a.

______, Composition of the sedimentary rocks of the Russian Platform in relation to the history of its tectonic movements, Geochemistry, **6**, 533-559, 1956b.

Weaver, C. E., Potassium, illite, and the ocean, Geochim. Cosmochim. Acta, **31**, 2181-2196, 1967.

Wickman, F. E., The "total" amount of sediment and the composition of the "average igneous rock," Geochim. Cosmochim. Acta, **5**, 97-110, 1954.

PHYSICAL MODELS OF THE EARTH

6

INVERSE PROBLEMS FOR THE EARTH'S NORMAL MODES

by **Freeman Gilbert**

Recent progress in the normal mode inverse problem is summarized. The nonuniqueness in the inverse problem is exploited to construct earth models whose computed gross earth data are in agreement with observations. We show how to determine whether a given finite set of gross earth data can be used to specify an earth structure uniquely except for fine-scale detail and how to determine the shortest length scale that the given data can resolve at any particular radius. Finally, when there are errors in the data, the variance of the error in computing a local average for a given length scale can be found. The variance decreases as the length scale increases.

INTRODUCTION

Whenever there is a large earthquake, the earth vibrates for days afterwards. The vibrations consist of the superposition of the elastic-gravitational normal modes of the earth that are excited by the earthquake. Dissipation causes the modes to decay; typical values of Q falling in the range 100 to 400. The lowest observed frequencies belong to a quintet of modes at nearly 1.1 cycles/hour (c/hr). The Q for this quintet is close to 350, so, after 10 days, the amplitude is diminished by a factor of 10. By far the highest Q belongs to the fundamental radial mode. With a period of 20.46 min, this mode has a motion where every point moves in or out along a radius, all points at the same radius having the same amplitude.

There is a small amount of shear in this mode, but it is nearly purely dilatational and has a Q of about 6000 so that its amplitude is diminished by a factor of 10 in 2 months (Slichter, 1967).

Until recently instruments were not sufficiently sensitive and quiet to permit the observation of the lower frequency normal modes for earthquakes with Richter magnitude below 7 3/4. In the past 20 years there have been nine earthquakes with Richter magnitude equal to or greater than 7 3/4, the latest being the Peruvian shock of May 31, 1970.

In the past year, newly developed accelerometers have allowed geophysicists to lower the magnitude threshold for low frequency modes from 7 3/4 to 6 1/2. Since October 1969, when the first of the new accelerometers became operational (Block and Moore, 1970), there have been more than 20 earthquakes of magnitude 6 1/2 or greater. In principle, then, we could have accumulated more normal mode data in the past 10 months with the new instruments than was accumulated in the past 2 decades with more traditional instruments. In practice, however, operational difficulties with the systems for data acquisition and recording have prevented our achieving such a goal. Nevertheless, the deployment and operation of the new instruments promises to give a great improvement in both the quality and the amount of seismological data at our disposal.

As an example of what can be achieved (Block, Dratler, and Moore, 1970) we refer to Fig. 1. This is a recording of the New Hebrides earthquake of October 13, 1969 made in San Diego by a Block-Moore vertical accelerometer. In the frequency band 1 to 30 c/hr the signal is amplified 100 times with respect to other frequency bands. We see the earthquake signal superimposed on the bodily tide, whose amplitude is about 10^{-7} gm. A standard numerical Fourier analysis of the earthquake signal is shown in Fig. 2. For comparison an averaged Fourier spectrum of background noise, recorded prior to the earthquake, is shown as the ambient spectrum. Clearly there are several peaks, such as the one labeled ${}_0S_{30}$, that stand significantly above the ambient spectrum. The continuation of the spectrum to higher frequencies is shown in Fig. 3. We interpret the positions of spectral peaks in Figs. 2 and 3 as the eigenfrequencies of the normal modes of the earth.

If we knew the mechanical properties of the earth, we could calculate its eigenfrequencies and the observations would merely serve to corroborate our theory. We do believe that our knowledge, although incomplete, is good enough to make such a calculation meaningful. Seismologists have become convinced that the interpretation of travel time data has led to models of the earth that are grossly correct, which differ from the real earth only in small, but important, details. Consequently, the calculated eigenfrequencies may not agree exactly with the observed ones, but the differences could reasonably be expected to be small. One of the inverse problems for the earth's normal modes is to find perturbations to a standard earth model so that the differences between calculated and observed eigenfrequencies are minimized (Backus and Gilbert,

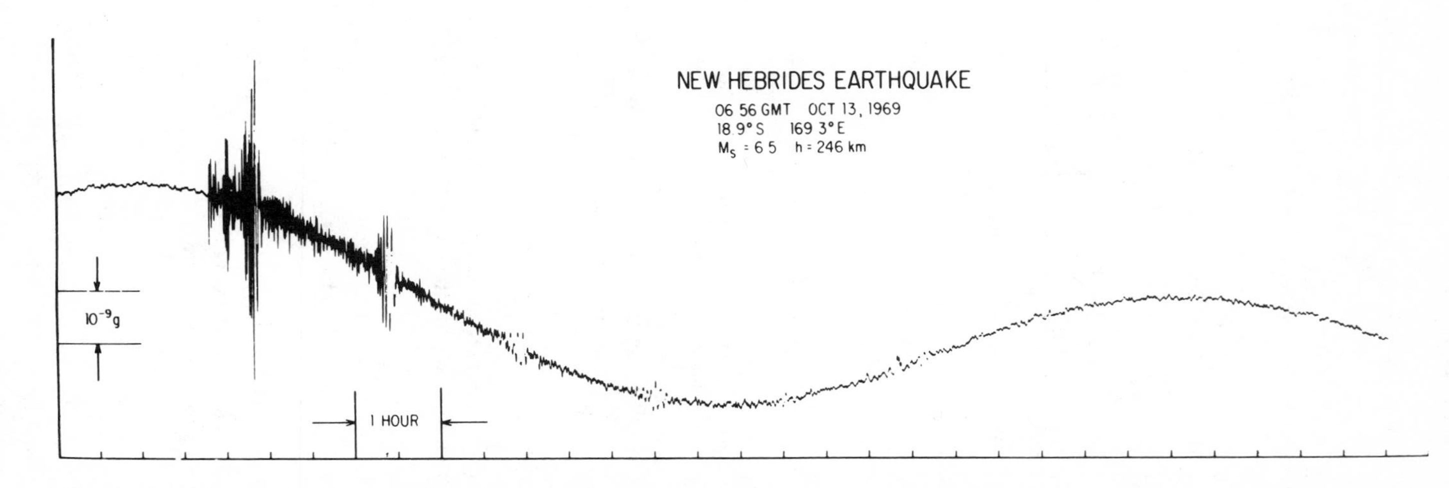

Fig. 1. Accelerogram for New Hebrides earthquake. The frequency band 1 to 30 c/hr is amplified 40 db with respect to the bodily tidal frequencies. The vertical scale, 10^{-9} gm, is to be used for the band 1 to 30 c/hr.

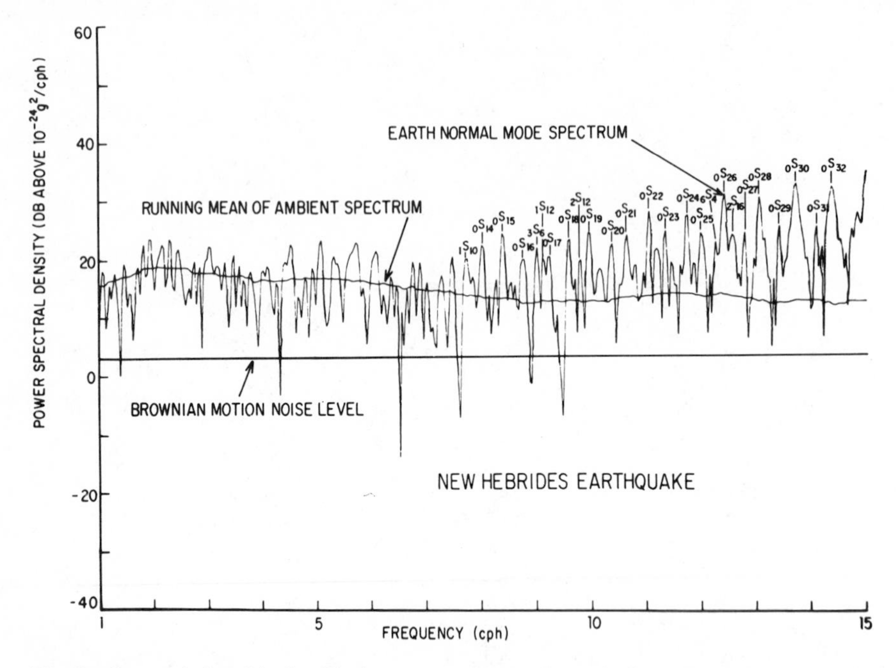

Fig. 2. Spectra below 14 c/hr. The instrumental Brownian noise is shown as the heavy solid line. The 0-db point refers to a spectral density of 7.4×10^{-25} gm^2/c/hr.

1967). We shall use the nomenclature and symbolism of the classical theory of small oscillations in our exposition.

INVERSION OF LINEAR SYSTEMS

Consider a conservative system of N particles in small oscillation. Let the linearized equation for the conservation of linear momentum for the αth particle be

$$m_\alpha \frac{d^2}{dt^2} \mathbf{u}_\alpha + \sum_{\beta=1}^{N} \mathbf{V}_{\alpha\beta} \cdot \mathbf{u}_\beta = \mathbf{f}_\alpha \tag{1}$$

where m is the mass, $\mathbf{u}$ the displacement, $\mathbf{V}$ the symmetric positive definite potential energy matrix, and $\mathbf{f}$ the applied force. The solution to (1) can be represented as a superposition of the normal modes $\mathbf{s}_{\alpha,n}$, $n = 1, \ldots, 3N$ that satisfy the equation

$$-m_\alpha \omega_n^2 \mathbf{s}_{\alpha,n} + \sum_\beta \mathbf{V}_{\alpha\beta} \cdot \mathbf{s}_{\beta,n} = 0 \tag{2}$$

where the eigenfrequencies are ω_n.

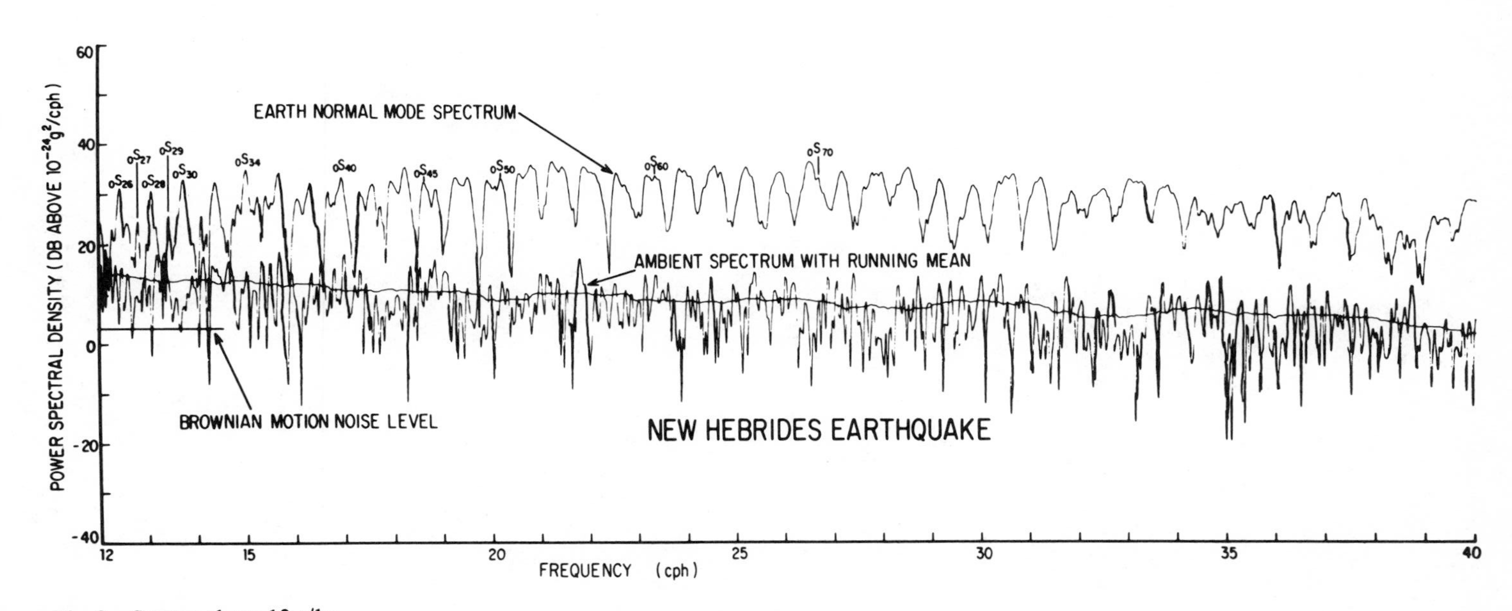

Fig. 3. Spectra above 12 c/hr.

The Laplace transform of (1) is

$$m_\alpha p^2 \bar{\mathbf{u}}_\alpha + \sum_\beta \mathbf{V}_{\alpha\beta} \cdot \bar{\mathbf{u}}_\beta = \bar{\mathbf{f}}_\alpha \tag{3}$$

where

$$\bar{\mathbf{u}}_\alpha(p) = \int_0^\infty \mathbf{u}_\alpha(t)\, e^{-pt}\, dt \;\; ; \;\; \mathrm{Re}\, p > 0$$

and

$$\mathbf{u}_\alpha(0) = \left[\frac{d}{dt} \mathbf{u}_\alpha(t)\right]_{t=0} = 0$$

The normal modes in (2) are orthonormal (* denotes complex conjugate)

$$\sum_\alpha m_\alpha \mathbf{s}^*_{\alpha,n} \cdot \mathbf{s}_{\alpha,l} = \delta_{nl} \tag{4}$$

and form a complete basis for the finite-dimensional vector space of the $3N$ degrees of freedom. Consequently, $\bar{\mathbf{u}}_\alpha$ has the representation

$$\bar{\mathbf{u}}_\alpha = \sum_n a_n \mathbf{s}_{\alpha,n} \tag{5}$$

where

$$a_n = \sum_\alpha m_\alpha \mathbf{s}^*_{\alpha,n} \cdot \bar{\mathbf{u}}_\alpha \tag{6}$$

If we take the scalar product of (3) with $\mathbf{s}^*_{\alpha,n}$ and sum over α, we get

$$p^2 a_n + \sum_{\alpha\beta} \mathbf{s}^*_{\alpha,n} \cdot \mathbf{V}_{\alpha\beta} \cdot \left(\sum_l a_l \mathbf{s}_{\beta,l}\right) = \sum_\alpha \mathbf{s}^*_{\alpha,n} \cdot \bar{\mathbf{f}}_\alpha \tag{7}$$

If we take the scalar product of (2) with $\mathbf{s}^*_{\alpha,l}$, sum over α, and use (4), we get

$$\sum_{\alpha\beta} \mathbf{s}^*_{\alpha,l} \cdot \mathbf{V}_{\alpha\beta} \cdot \mathbf{s}_{\beta,n} = \omega_n^2 \delta_{nl} \tag{8}$$

which can be used to reduce (7) to

$$a_n(p^2 + \omega_n^2) = \sum_\alpha \mathbf{s}^*_{\alpha,n} \cdot \bar{\mathbf{f}}_\alpha \tag{9}$$

Thus, the representation (5) becomes

$$\bar{\mathbf{u}}_\alpha = \sum_n \left(\sum_\beta \mathbf{s}^*_{\beta,n} \cdot \bar{\mathbf{f}}_\beta\right) \frac{\mathbf{s}_{\alpha,n}}{(p^2 + \omega_n^2)} \tag{10}$$

Most earthquakes are modeled as step functions so that $\bar{\mathbf{f}}_\beta = \mathbf{f}_\beta p^{-1}$. Then the Laplace inversion of (10) is, for $t > 0$,

$$\mathbf{u}_\alpha(t) = \sum_n \mathbf{s}_{\alpha,n} \left(\sum_\beta \mathbf{s}^*_{\beta,n} \cdot \mathbf{f}_\beta\right) \frac{1 - \cos\omega_n t}{\omega_n^2} \tag{11}$$

When there is a small amount of dissipation ($Q \gg 1$) we have

$$\mathbf{u}_\alpha(t) = \sum_n \mathbf{s}_{\alpha,n} \left(\sum_\beta \mathbf{s}^*_{\beta,n} \cdot \mathbf{f}_\beta\right) \frac{1 - \cos\omega_n t\, e^{-\omega_n t/2Q_n}}{\omega_n^2} \tag{12}$$

so that, after a long time, all that remains is the static displacement

$$\lim_{t\to\infty} \mathbf{u}_\alpha(t) = \sum_n \mathbf{s}_{\alpha,n} \left(\sum_\beta \mathbf{s}^*_{\beta,n} \cdot \mathbf{f}_\beta\right) \frac{1}{\omega_n^2} \tag{13}$$

The study of static displacements (strain offsets) following large earthquakes has become an important research subject.

For the earth, which we regard as a classical continuum, a sum such as

$$\sum_\beta \mathbf{s}^*_{\beta,n} \cdot \mathbf{f}_\beta \text{ becomes } \int dV\, \mathbf{s}^*_n(\mathbf{r}) \cdot \mathbf{f}(\mathbf{r})$$

and $\mathbf{f(r)}$ is the body force per unit volume. Almost all earth models that we use are radially stratified spheres. Spherical symmetry, plus the properties of perfect elasticity and isotropy, permit a normal mode eigenfunction to be written in spherical coordinates as $\mathbf{s(r)} = \mathbf{s}(r, \theta, \phi) = {}_n\boldsymbol{\sigma}_l^m$ or ${}_n\boldsymbol{\tau}_l^m$ where

$$
\begin{aligned}
{}_n\boldsymbol{\sigma}_l^m &= \hat{\mathbf{r}}\,{}_nU_l(r)\,Y_l^m(\theta, \phi) + r\,{}_nV_l(r)\,\nabla Y_l^m(\theta, \phi) \\
{}_n\boldsymbol{\tau}_l^m &= -{}_nW_l(r)\,\mathbf{r} \times \nabla Y_l^m(\theta, \phi)
\end{aligned}
\tag{14}
$$

The modes denoted by $\boldsymbol{\sigma}$ are spheroidal or poloidal modes. The curl of $\boldsymbol{\sigma}$ is nonradial. The modes denoted by $\boldsymbol{\tau}$ are torsional or toroidal modes. The curl of $\boldsymbol{\tau}$ is radial. The scalars U, V, and W are solutions to ordinary differential equations with prescribed boundary conditions (Alterman, Jarosch, and Pekeris, 1959). The eigenvalues are the squared eigenfrequencies.

For a particular l and m, the smallest eigenfrequency is labeled $n = 0$, the next $n = 1$, etc. We call l the total angular order, m the azimuthal order $(-l \leqslant m \leqslant l)$, and n the radial order. The azimuthal order m does not appear in the differential equations or boundary conditions so that, for each l and n, there are $2l + 1$ normal mode eigenfunctions, all of which belong to the same eigenfrequency. This is the familiar phenomenon of degeneracy. Small perturbation resulting from the rotation of the earth and from the ellipticity of the earth's figure completely remove the degeneracy and lead to a split multiplet (Dahlen, 1968, 1969); the splitting formula having the form

$$
{}_n\omega_l^m = {}_n\omega_l^0 (1 + {}_n\alpha_l + m\,{}_n\beta_l + m^2\,{}_n\gamma_l), -l \leq m \leq l \tag{15}
$$

The splitting effect is on the order of 1 in 1000 and can interfere with measurements of dissipation. Symmetrical splitting, resulting from the Coriolis force and represented by β in (15), has been observed for $n = 0$, $l = 2$, 3, but there have been no reports of the asymmetries resulting from γ. For the toroidal modes, ${}_n\beta_l = 1/l(l + 1)$, which provides a mechanical analogue of the Zeeman effect. For the spheroidal modes, all the splitting parameters depend on the earth's mechanical structure so that their observation, with the exception of α, would provide new data for the inverse problem.

Because of the near degeneracy in m, the azimuthal order is customarily not used in identifying a normal mode. Thus, a spheroidal spectral line is denoted ${}_nS_l$ and a toroidal line ${}_nT_l$. The line labeled ${}_0S_{30}$ in Fig. 2 really consists of all 61 members of the spheroidal multiplet that has the smallest unperturbed eigenfrequency $(n = 0)$ and total angular order $(l = 30)$. The amplitude, but of course not the frequency, of each member of the multiplet varies over the earth's surface according to its spherical hamonic pattern (14); consequently the whole multiplet's shape will also vary with position. That is, the apparent peak of the line ${}_0S_{30}$ will not be the same in San Diego as in, say, New York. If we knew the splitting formula and splitting parameters we could conceivably "super resolve" the multiplet, but, so far, that has not been attempted.

Naturally there is ambiguity in the identification of spectral peaks, but, for many observed spectral lines, there is a very close correlation, both in frequency and amplitude, with theoretical calculations, and the best fitting theoretical normal mode is assigned to the observed one. In fact, that is how the line in Fig. 2 near 13.7 c/hr is identified as ${}_0S_{30}$.

We assume that it is possible to identify spectral lines, to measure their eigenfrequencies and Q's, and to assign experimental errors to the observations. Such data are raw material for the inverse problem.

Let us return to Rayleigh's principle (8) in the form

$$\omega_l^2 \sum_{\alpha} m_\alpha \mathbf{s}^*_{\alpha,l} \cdot \mathbf{s}_{\alpha,l} = \sum_{\alpha\beta} \mathbf{s}^*_{\alpha,l} \cdot \mathbf{V}_{\alpha\beta} \cdot \mathbf{s}_{\beta,l} \tag{16}$$

We suppose, for the purposes of demonstration, that the mass of each particle is changed by a small amount δm_α but that $\mathbf{V}_{\alpha\beta}$ is not changed. We seek the perturbation in ω^2 resulting from the perturbation in mass. A first variation of (16) gives

$$\begin{aligned} &\delta\omega_l^2 \sum_{\alpha} m_\alpha \mathbf{s}^*_{\alpha,l} \cdot \mathbf{s}_{\alpha,l} + \omega_l^2 \sum_{\alpha} \delta m_\alpha \mathbf{s}^*_{\alpha,l} \cdot \mathbf{s}_{\alpha,l} + \\ &+ \omega_l^2 \sum_{\alpha} \left(m_\alpha \mathbf{s}^*_{\alpha,l} - \sum_{\beta} \mathbf{V}_{\alpha\beta} \cdot \mathbf{s}^*_{\beta,l} \right) \cdot \delta\mathbf{s}_{\alpha,l} + \\ &+ \omega_l^2 \sum_{\alpha} \left(m_\alpha \mathbf{s}_{\alpha,l} - \sum_{\beta} \mathbf{V}_{\alpha\beta} \cdot \mathbf{s}_{\beta,l} \right) \cdot \delta\mathbf{s}^*_{\alpha,l} = 0 \end{aligned} \tag{17}$$

The use of (2) and its complex conjugate allows us to simplify (17)

$$\delta\omega_l^2 \sum_{\alpha} m_\alpha \mathbf{s}^*_{\alpha,l} \cdot \mathbf{s}_{\alpha,l} = -\omega_l^2 \sum_{\alpha} \delta m_\alpha \mathbf{s}^*_{\alpha,l} \cdot \mathbf{s}_{\alpha,l} \tag{18}$$

If we knew the small perturbations δm_α, we could calculate the small perturbation $\delta\omega_l^2$. Using the normalization (4) and the notation

$$M_{\alpha,l} = -\omega_l^2 \mathbf{s}^*_{\alpha,l} \cdot \mathbf{s}_{\alpha,l}$$

we simplify (18)

$$\delta\omega_l^2 = \sum_{\alpha} \delta m_\alpha M_{\alpha,l} \tag{19}$$

and we suppose now that $\delta\omega_l^2$ is known and that we want to find δm_α. That is

$$\delta\omega_l^2 = (\omega_{\text{obs}}^2 - \omega_{\text{comp}}^2)_l$$

where ω_{obs} is the observed value of ω_l and ω_{comp} is the computed value. Now there are N values of δm_α and $3N$ values of $\delta\omega_l^2$ so the inverse problem appears to be overdetermined. In reality N is very large, infinite for continuum mechanics, and it is usual that the number of observations is much less than N. Thus, we assume that we know $\delta\omega_l^2$, $l = 1, \ldots, L \ll N$ and we want to find δm_α. Formulated in this way, the inverse problem appears to be hopelessly nonunique, and it is customary to give up and pursue some other line of work. If we are satisfied to have any δm_α satisfying (19), however, we can exploit the nonuniqueness to our advantage. For example, suppose we demand $\sum_\alpha (\delta m_\alpha)^2$ be a minimum (Backus and Gilbert, 1967). We regard the central problem to be the minimization of $\sum_\alpha (\delta m_\alpha)^2$ with constraints (19). This is the classical isoperimetric problem in the calculus of variations, and we introduce Lagrange multipliers ν_l to find the solution

$$\delta m_\alpha = \sum_l \nu_l M_{\alpha,l} \tag{20}$$

Substituting (20) into (19) gives

$$\sum_l \left(\sum_\alpha M_{\alpha,l} M_{\alpha,n} \right) \nu_l = \delta\omega_n^2 \tag{21}$$

The matrix $\sum_\alpha M_{\alpha,l} M_{\alpha,n}$ is an inner product matrix so it is symmetric and positive definite. Consequently, there is a unique solution ν_l to (21) such that (20) gives a perturbation δm_α that satisfies (19). Now we have a new mass distribution, $m_\alpha + \delta m_\alpha$ with which we can return to (2) and calculate new eigenfrequencies and normal modes. The process (19) through (21) can be repeated. Like any modification of Newton's method, the process we have outlined should converge provided the initial values of $\delta\omega_n^2$ are not too large. (It would prove embarrassing if $\delta m_\alpha < - m_\alpha$ in one of the iterations.) This is exactly the process that has been used in finding solutions to the inverse normal mode problem for the free oscillations of the earth (Gilbert and Backus, 1968). The only difference in computational detail is that $\sum_\alpha \ldots$ is replaced by $\int dV \ldots$.

One obvious modification of this method is to include observational errors in solving (21). For example, if σ_l is the standard error in $(\omega^2_{\mathrm{obs}})_l$ we replace (21) by

$$\delta\omega_n^2 - \sigma_n \leq \sum_l \left(\sum_\alpha M_{\alpha,l} M_{\alpha,n} \right) \nu_l \leq \delta\omega_n^2 + \sigma_n \tag{22}$$

and we find that solution to (22) that minimizes $\sum\limits_\alpha (\delta m_\alpha)^2$.

As an example of the application of the method, we use the set of data listed in Table 1. Our initial model, used to start the iterative process, is shown in Fig. 4. Its computed eigenfrequencies are compared to the observed ones in Table 2, and we can see that it is not a bad fit to the data, although a better fit is desirable. After two iterations, we have the model shown in Fig. 5. It is only slightly different from the initial model, but is in better agreement with the data (Table 3).

THE AVERAGING KERNEL

Now that we have a model that fits a chosen set of data, we must face the problem of uniqueness. Since the number of data we have is much smaller than the number of eigenfrequencies of the earth, we cannot hope that our model is a unique one. Our difficulty is further compounded when we admit observational errors in the data.

To begin, we shall suppose that there is a *linear* relation between the data, γ_n, ($\delta\omega^2$ in our example) and the model, $m(r)$, (δm_α, $\delta\rho$, δv_p, δv_s in our example)

$$\gamma_n = \int_0^1 dr\, G_n(r)\, m(r) \tag{23}$$

and that G does not depend on m. For the moment all observed data are assumed to be exact. Thus we can interpret (23) by saying that each datum can be regarded as a linear average of the model. Consequently, any linear average that we can calculate must necessarily be a linear combination of the data that we have.

Table 1. Data used to obtain Fig. 5 from Fig. 4 in two iterations

mass	$\bar{\rho} = 5.517$ gm/cm^3
moment	$z = 0.33089$
modes	${}_0S_0$, ${}_1S_0$, ${}_2S_0$, ${}_3S_0$, ${}_0S_2$, ${}_1S_2$, ${}_2S_2$, ${}_1S_4$, ${}_0S_{25}$, ${}_0S_{49}$, ${}_0S_{73}$, ${}_0S_{97}$, ${}_0T_{27}$, ${}_0T_{53}$, ${}_0T_{105}$

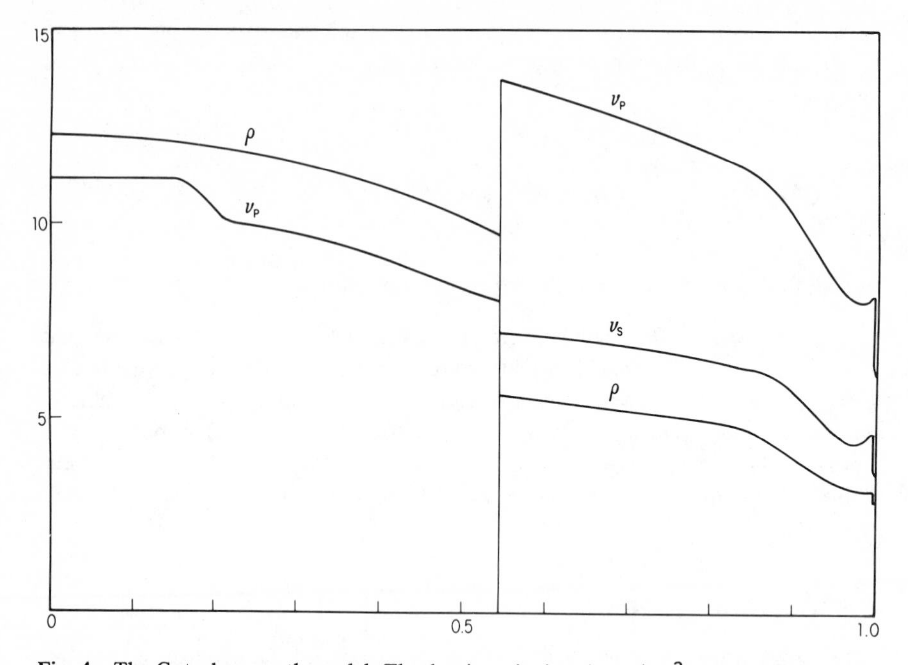

Fig. 4. The Gutenberg earth model. The density ρ is given in gm/cm^3. The S-wave velocity v_s and the P-wave velocity v_P are given in km/sec. The radius of the earth is normalized to unity.

Table 2. Comparison of observed eigenfrequencies with ones computed for Fig. 4 (values are given in rad/sec)

Mode	ω (observed) $\times 10^2$	ω (computed) $\times 10^2$
${}_0S_0$	0.51162	0.50734
${}_1S_0$	1.0403	1.0358
${}_1S_0$	1.5745	1.5876
${}_3S_0$	2.0735	2.0710
${}_0S_2$	0.19411	0.19610
${}_1S_2$	0.42705	0.43328
${}_2S_2$	0.68771	0.69039
${}_1S_4$	0.73494	0.74727
${}_0S_{25}$	2.1050	2.1466
${}_0S_{49}$	3.4614	3.5246
${}_0S_{73}$	4.8715	4.9331
${}_0S_{97}$	6.2641	6.3861
${}_0T_{27}$	2.2208	2.2516
${}_0T_{53}$	4.0005	4.0455
${}_0T_{105}$	7.5445	7.6443

rms relative error = 0.013.

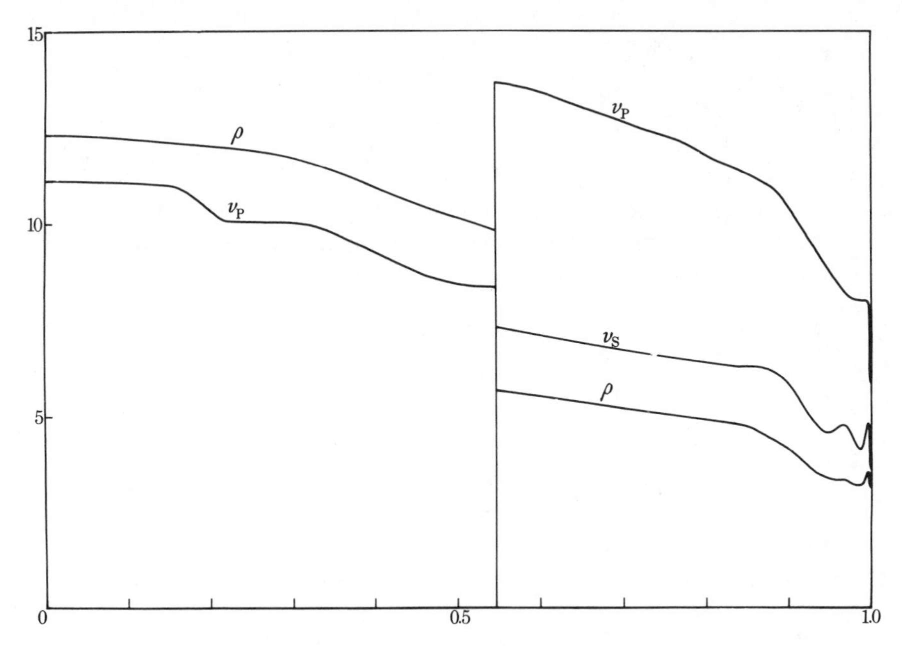

Fig. 5. Model 8734 (computer job number) derived in two iterations from the Gutenberg model by using the data in Tables 1 and 2.

Table 3. Comparison of observed eigenfrequencies with ones computed for Fig. 5 (values are given in rad/sec)

Mode	ω (observed) × 10^2	ω (computed) × 10^2
$_0S_0$	0.51162	0.51165
$_1S_0$	1.0403	1.0400
$_2S_0$	1.5745	1.5744
$_3S_0$	2.0735	2.0727
$_0S_2$	0.19411	0.19424
$_1S_2$	0.42705	0.42713
$_2S_2$	0.68771	0.68812
$_1S_4$	0.73494	0.73508
$_0S_{25}$	2.1050	2.1048
$_0S_{49}$	3.4614	3.4607
$_0S_{73}$	4.8715	4.8700
$_0S_{97}$	6.2641	6.2617
$_0T_{27}$	2.2208	2.2208
$_0T_{53}$	4.0005	3.9997
$_0T_{105}$	7.5445	7.5432

rms relative error = 0.00031.

$$\sum_n a_n \gamma_n = \int_0^1 dr \left[\sum_n a_n G_n(r) \right] m(r) \tag{24}$$

Let $A = \sum_n a_n G_n(r)$. Then the averaging kernel A is a linear combination of the data kernels $G_n(r)$. For the sake of argument, suppose $G_n(r) = \sin n\pi r$, $n = 1, 2, \ldots, N$. Without necessarily being aware of this fortuitous circumstance, suppose we decide to choose $a_n = 2 \sin n\pi r_0$. Then

$$\sum_{n=1}^{N} a_n G_n(r) = 2 \sum_{n=1}^{N} \sin n\pi r_0 \sin n\pi r \tag{25}$$

$$\sum_{n=1}^{N} a_n G_n(r) = \frac{\sin[(2N+1)(r-r_0)\pi/2]}{2\sin[(r-r_0)\pi/2]} - \frac{\sin[2N+1)(r+r_0)\pi/2]}{2\sin[(r+r_0)\pi/2]} \tag{26}$$

Thus, when r_0 is not close to 0 or 1 and when r is close to r_0, A is a function with a tall central spike at $r = r_0$ and is small elsewhere; that is, A looks like a traditional approximation to the δ function. In this special case, (24) would become

$$\sum_n a_n \gamma_n = \int_0^1 dr\, Am(r) \simeq m(r_0) \tag{27}$$

It is conceivable, then, that a particular linear combination of the observed data can yield a good estimate of the model at a particular place, r_0 (Backus and Gilbert, 1968). Our principal task is to determine whether A can be made to resemble a δ function.

We could try to minimize the quadratic form

$$J = \int_0^1 dr[A(r) - \delta(r - r_0)]^2 \tag{28}$$

Let us write $a = \sum_n a_n G_n(r)$ as $A = \mathbf{a} \cdot \mathbf{G}(r)$. Also, let $\mathbf{S} = \int_0^1 dr\ \mathbf{G}(r)\mathbf{G}(r)$ and denote the elements of **S** by S_{ij}. Then (28) becomes

$$J = \mathbf{a}\cdot\mathbf{S}\cdot\mathbf{a} - 2\mathbf{a}\cdot\mathbf{G}(r_0) + J_\delta \tag{29}$$

where $J_\delta = \int_0^1 dr\,[\delta(r - r_0)]^2$. Although J_δ is infinite it does not depend on **a**, so the minimum of J is the solution to

$$\mathbf{S}\cdot\mathbf{a} = \mathbf{G}(r_0) \tag{30}$$

and

$$A = \mathbf{G}(r_0)\cdot\mathbf{S}^{-1}\cdot\mathbf{G}(r) \tag{31}$$

For $G_n(r) = \sin n\pi r$, the solution to (30) yields (25) for A. More generally, if we regard $G_n(r)$, $n = 1, \ldots, N$ as a set of basis functions for the finite dimensional space of observed data, then $\sum_{l=1}^{N} S_{nl}^{-1} G_l(r)$; $n = 1, \ldots, N$, is the dual basis and the averaging kernel is the scalar product of the two.

For those who dislike the presence of J_δ in (29), another approach is to minimize

$$K = \int_0^1 dr\,(r - r_0)^2 [A(r) - \delta(r - r_0)]^2 \tag{32}$$

Obviously, the minimum of (32) is $\mathbf{a} = 0$, so some constraint is needed. A simple and reasonable one is to demand unit area for A

$$\int_0^1 A(r)\,dr = 1$$

If $\mathbf{G} = \int_0^1 \mathbf{G}(r)dr$, then we want to minimize (32) with the constraint $\mathbf{a}\cdot\mathbf{G} = 1$. We note that K has the same dimensions as r. Since $K = 0$ for $A(r) = \delta(r - r_0)$, and $K = 1/12$ for $A(r) = w^{-1}$ for $r_0 - w/2 \leqslant r \leqslant r_0 + w/2$, and $A = 0$ for other values of r, we take $12K$ as a measure of the width of A when A is centered on r_0. Let $s = 12K$. We call s the spread of A. When A is centered on r_0, s is the width of A. We could add another constraint $\int A(r) r dr = r_0$ to center A but we do not want to become too fancy too soon. Let

$$\mathbf{S} = 12\int_0^1 dr\,(r - r_0)^2\,\mathbf{G}(r)\,\mathbf{G}(r) \tag{33}$$

Then we want to minimize $\mathbf{a} \cdot \mathbf{S} \cdot \mathbf{a}$ with the side condition $\mathbf{a} \cdot \mathbf{G} = 1$. Clearly

$$\mathrm{a} = \frac{\mathrm{S}^{-1} \cdot \mathrm{G}}{\mathrm{G} \cdot \mathrm{S}^{-1} \cdot \mathrm{G}} \tag{34}$$

To compare minimizing J in (28) with K in (32), we refer to Fig. 6.

We subjectively prefer the K criterion to the J criterion because the K criterion gives lower sidebands for A. The slight increase in the peak width of A is the price we have to pay for lower sidebands.

To summarize briefly, we minimize $s = \mathbf{a} \cdot \mathbf{S} \cdot \mathbf{a}$ with $\mathbf{a} \cdot \mathbf{G} = 1$. If $\mathbf{a} \cdot \mathbf{G}(r)$

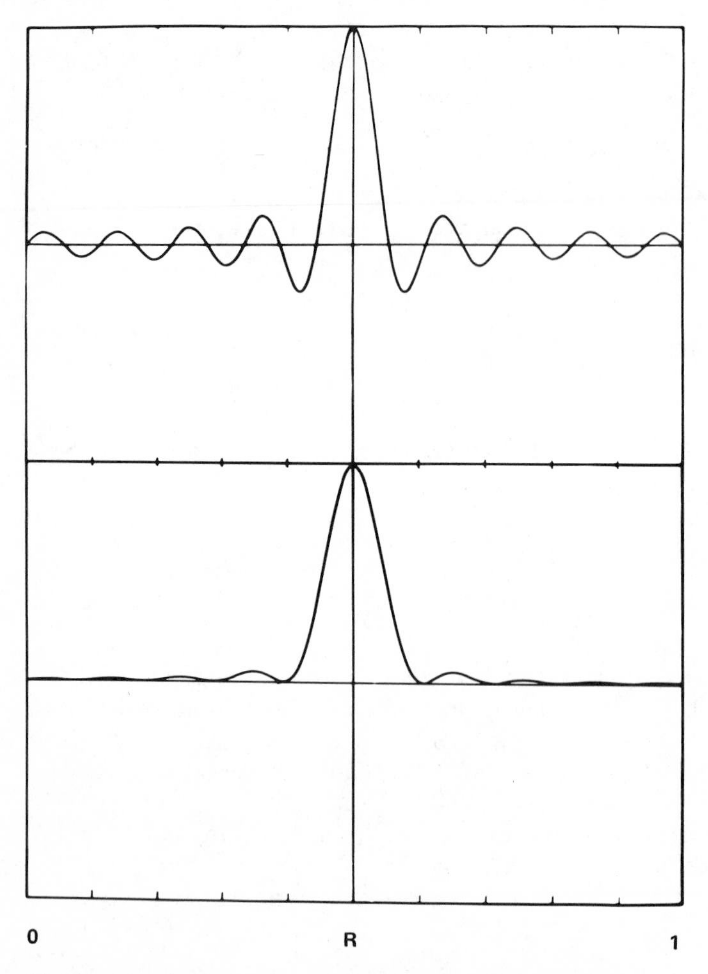

Fig. 6. Averaging kernels A computed at $r_0 = 1/2$ from $\sin n\pi r$, $n = 1, \ldots, 17$. Above is the J kernel (28) and below is the K kernel (32).

resembles $\delta(r - r_0)$, then $\mathbf{a} \cdot \boldsymbol{\gamma} \cong m(r_0)$ and $s = \mathbf{a} \cdot \mathbf{S} \cdot \mathbf{a}$ is the spread of A, or the resolving length of A.

In the problems where the data are nonlinear functionals of the model, we restrict attention to models that are "close." Suppose that m_1 and m_2 are "close." We mean

$$\delta\gamma = \int_0^1 dr[m_1(r) - m_2(r)]\,\mathbf{G}(r) + 0[(m_1 - m_2)^2] \tag{35}$$

and $0[(m_1 - m_2)^2]$ may be neglected. Then

$$\mathbf{a} \cdot \delta\gamma = \int_0^1 dr[m_1(r) - m_2(r)]A(r)\,dr \tag{36}$$

If both models agree with the data, $\delta\boldsymbol{\gamma} = 0$ and

$$\int_0^1 drm_1(r)\,A(r) = \int_0^1 drm_2(r)\,A(r) \tag{37}$$

both models have the *same* linear averages (Backus and Gilbert, 1968). If one model is the real earth and another model is close to it, then certain linear averages of the structure of the real earth can be found by (37). Such linear averages could be of interest if s is not too big.

As an illustration of this technique, we use the set of data listed in Table 4 and pretend that both v_p and v_s are known exactly, and that ρ is "close" to the density of the real earth. For various r_0, the averaging kernels A are shown in Fig. 7. The density model ρ is shown as the piecewise continuous curve in Fig. 8. Each dot represents the linear average of ρ (at a value of r_0) made with the appropriate A in Fig. 7.

ERRORS IN THE DATA

It is time to admit that if we use the K criterion (32) to find A, then the averaged m will not satisfy (23). In fact, only the J criterion leads to an averaged m that satisfies (23). An obvious response is to be satisfied with an averaged m if

Table 4. Data used in computations for Figs. 7 to 9 (Mass and moment are given in Table 1)

modes	${}_0S_0$, ${}_1S_0$, ${}_2S_0$, ${}_3S_0$, ${}_1S_1$, ${}_2S_1$, ${}_0S_2$, ${}_2S_2$, ${}_1S_3$, ${}_0S_4$, ${}_1S_4$, ${}_2S_4$, ${}_4S_4$, ${}_0S_7$, ${}_1S_8$, ${}_0S_{25}$, ${}_0S_{49}$, ${}_0S_{73}$, ${}_0S_{97}$, ${}_0T_7$, ${}_0T_{14}$, ${}_0T_{27}$, ${}_0T_{53}$, ${}_0T_{105}$

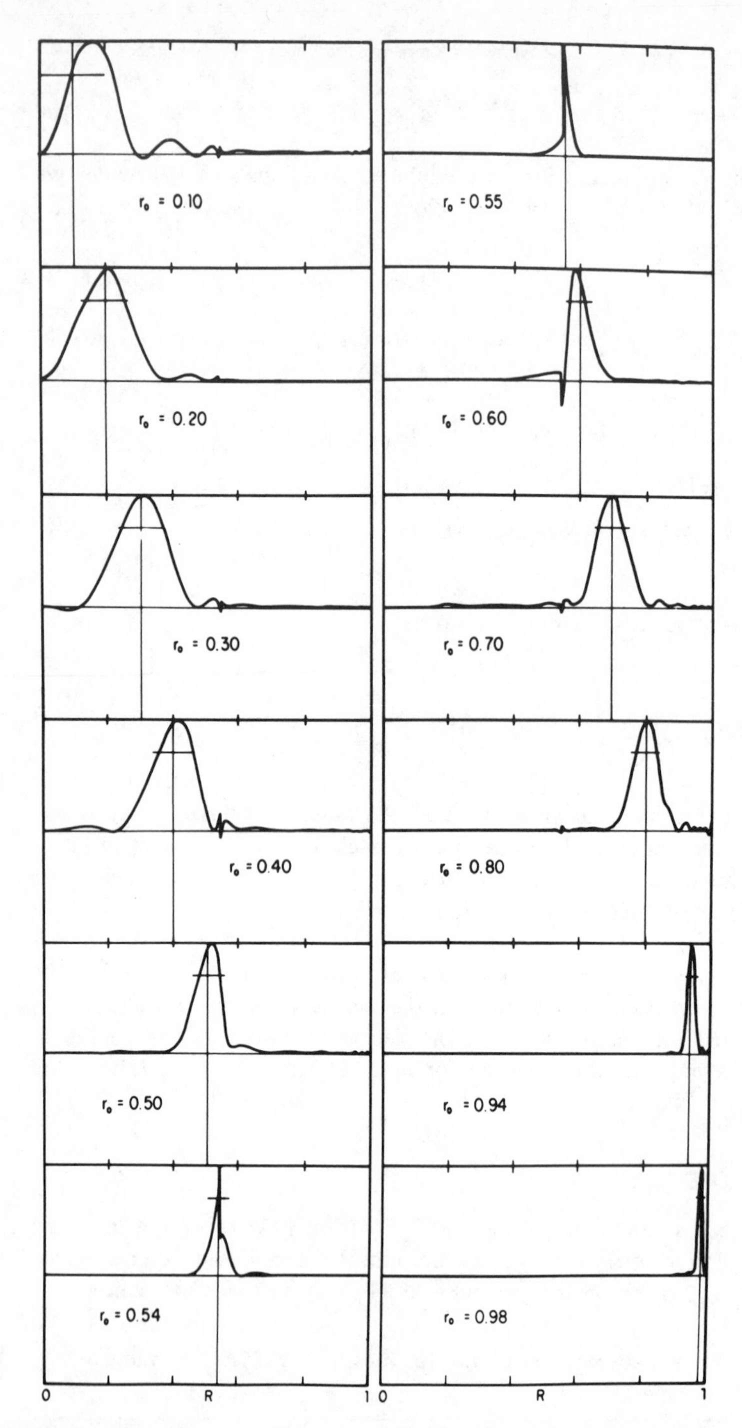

Fig. 7. The averaging kernels A for ρ at selected r_0 (vertical lines). The K criterion (32) is used and the data are given in Table 4. The spread s is represented by the horizontal bars.

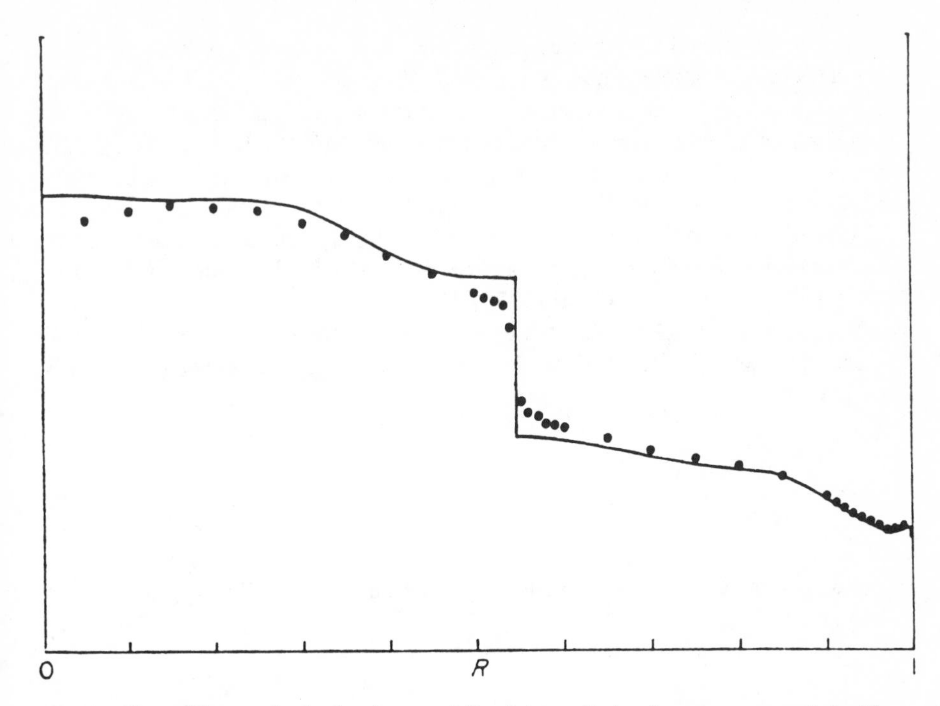

Fig. 8. The solid curve is the density ρ and the dots are its local averages computed with the kernels in Fig. 7.

it satisfies (23) to within one standard error in the data. If there are errors, $\delta\boldsymbol{\gamma}$, in the data, however, there will be associated with them uncertainties or errors in the averaged model

$$\delta\gamma = \int_0^1 dr\,\mathbf{G}(r)\,\delta m(r)$$

$$\mathbf{a}\cdot\delta\gamma = \int_0^1 dr A(r)\,\delta m(r) = \delta m(r_0) \tag{38}$$

Experimentally, we should be able to find the variance of the data

$$\mathbf{a}\cdot\overline{\delta\gamma\delta\gamma}\cdot\mathbf{a} = \overline{\delta m(r_0)\,\delta m(r_0)} = \epsilon^2 \tag{39}$$

where $\mathbf{V} = \overline{\delta\boldsymbol{\gamma}\delta\boldsymbol{\gamma}}$ is the experimentally estimated variance matrix. Given $\mathbf{V}$, we find $\mathbf{a}$ and A and determine $\mathbf{a}\cdot\boldsymbol{\gamma}$ to find the linear average of m and we then calculate ϵ, the uncertainty or error in the average.

Clearly, we should like to minimize ϵ. That is, we want not only to minimize $s = \mathbf{a} \cdot \mathbf{S} \cdot \mathbf{a}$, but also to minimize $\epsilon^2 = \mathbf{a} \cdot \mathbf{V} \cdot \mathbf{a}$, still satisfying the constraint $\mathbf{a} \cdot \mathbf{G} = 1$. Except under the most unusual and unrealistic circumstances, it is obvious that ϵ^2 and s cannot be minimized simultaneously. It is equally obvious that some linear combination of ϵ^2 and s can be minimized (Backus and Gilbert, 1970).

A geometrical sketch of the basic idea may help one develop an intuitive understanding. For a proof of the assertions made here, the reader is referred to the paper by Backus and Gilbert (1970).

The constraint $\mathbf{a} \cdot \mathbf{G} = 1$ can be regarded as a hyperplane in the N-dimensional space of observed data. Both $\mathbf{a} \cdot \mathbf{S} \cdot \mathbf{a}$ and $\mathbf{a} \cdot \mathbf{V} \cdot \mathbf{a}$ are hyperspheres. When s is minimized

$$\mathbf{a} = \frac{\mathbf{S}^{-1} \cdot \mathbf{G}}{\mathbf{G} \cdot \mathbf{S}^{-1} \cdot \mathbf{G}} = \mathbf{a}_s \tag{40}$$

and $\mathbf{a}_s$ is in $\mathbf{a} \cdot \mathbf{G} = 1$ and represents a point (hyperpoint). Similarly when ϵ^2 is minimized

$$\mathbf{a} = \frac{\mathbf{V}^{-1} \cdot \mathbf{G}}{\mathbf{G} \cdot \mathbf{V}^{-1} \cdot \mathbf{G}} = \mathbf{a}_\epsilon$$

and $\mathbf{a}_\epsilon$ is another point on the hyperplane. It has been proved that there is a line in $\mathbf{a} \cdot \mathbf{G} = 1$ joining $\mathbf{a}_s$ and $\mathbf{a}_\epsilon$ such that to each point on the line there corresponds an $\mathbf{a}$, and, therefore, an s and ϵ^2, and that the hypercircle S_G, representing the intersection of $\mathbf{a} \cdot G = 1$ with $\mathbf{a} \cdot \mathbf{S} \cdot \mathbf{a}$, and the hypercircle $\epsilon_G{}^2$, representing the intersection of $\mathbf{a} \cdot \mathbf{G} = 1$ with $\mathbf{a} \cdot \mathbf{V} \cdot \mathbf{a}$, are tangent at the point $\mathbf{a}$. In other words, for each value of s on the line joining $\mathbf{a}_s$ and $\mathbf{a}_\epsilon$, ϵ^2 is minimized. Furthermore, $\epsilon^2(s)$ is a monotonically decreasing function of s.

Let

$$\mathbf{W} = \mathbf{S}(1 - \alpha) + V\alpha, \quad 0 \leq \alpha \leq 1 \tag{42}$$

Then when $\alpha = 0$, we minimize s, and when $\alpha = 1$, we minimize ϵ^2. For some α, $0 \leqslant \alpha \leqslant 1$, let

$$\mathbf{a}_\alpha = \frac{\mathbf{W}^{-1} \cdot \mathbf{G}}{\mathbf{G} \cdot \mathbf{W}^{-1} \cdot \mathbf{G}} \tag{43}$$

Then

$$s_\alpha = \mathbf{a}_\alpha \cdot \mathbf{S} \cdot \mathbf{a}_\alpha$$

and

$$\epsilon_\alpha^2 = \mathbf{a}_\alpha \cdot \mathbf{V} \cdot \mathbf{a}_\alpha$$

and α parameterizes the curve $\epsilon(s)$. This curve is called the tradeoff curve of absolute error versus spread.

For nonlinear problems we appeal to (35) through (37). When $\delta\gamma$ is less than one standard error, both models have effectively the same tradeoff curves.

In most cases, one is more interested in the relative error than the absolute

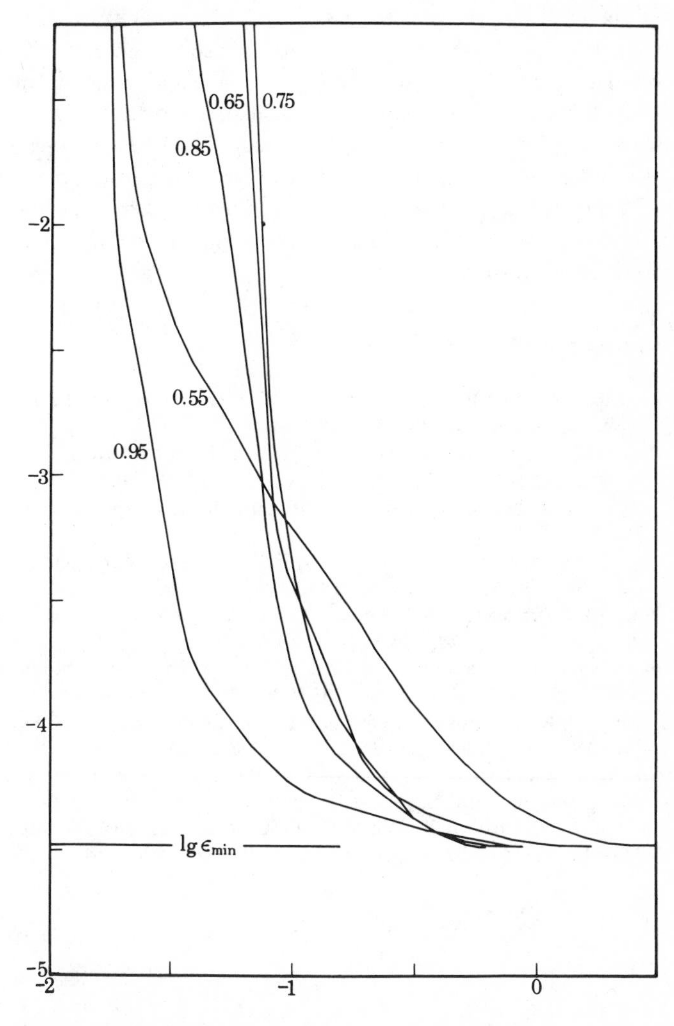

Fig. 9. Tradeoff curves of absolute error ϵ as a function of spread s at selected values of r_0.

error. The analysis is more involved, and not even a sketch will be given here (see Backus and Gilbert, 1970).

For an illustration of the application of tradeoff curves to geophysical data, the only example we give is an artificial one. A simple, even though physically unrealizable, model for dissipation (see (12)) is

$$Q_n^{-1} = \int_0^1 dr G_n(r) Q^{-1}(r) \tag{44}$$

In the earth, $G_n(r)$ is 0 for $r < r_c = 0.545$, where r_c is the radius of the fluid core. For $r > r_c$, we take $Q^{-1}(r) = 0.004r$ so that $Q(1) = 250$. For the 24 modes listed in Table 4, we calculate Q_n^{-1} in (44) and assume standard errors of 5%. Further, we assume that the errors are uncorrelated so $\mathbf{V}$ is diagonal. For selected values of r_0, $r_c < r_0 < 1$, tradeoff curves are presented in Fig. 9. In every case, a very slight increase in s above s_{min} leads to an enormous decrease of ϵ below ϵ_{max}. If we had used a linear, rather than logarithmic, scale the tradeoff curves would have closely resembled the shape L. The place to be is down at the corner.

REFERENCES

Alterman, Z, Jarosch, H., and Pekeris, C. L., Oscillations of the Earth, Proc. Roy. Soc., (London), Ser. A, **252** 80–95, 1959.

Backus, G., and Gilbert, F., Numerical applications of a formalism for geophysical inverse problems, Geophys. J. Roy. Astron. Soc. **13**, 247-276, 1967.

______, The resolving power of gross Earth data, Geophys. J. Roy. Astron. Soc., **16**, 169-205, 1968.

______, Uniqueness in the inversion of inaccurate gross Earth data, Phil. Trans. Roy. Soc. (London), Ser. A, **266**, 123-192, 1970.

Block, B., Dratler, J., and Moore, R. D., Earth normal modes from a 6.5 magnitude earthquake, Nature, **226**, 343-344, 1970.

Block, B., and Moore, R. D., Tidal to seismic frequency investigations with a quartz accelerometer of new geometry, J. Geophys. Res., **75**, 1493-1505, 1970.

Dahlen, F. A., The normal modes of a rotating, elliptical Earth, Geophys. J. Roy. Astron. Soc., **16**, 329-367, 1968 and **18**, 397-436, 1969.

Gilbert, F. and Backus, G., Approximate solutions to the inverse normal mode problem, Bull. Seism. Soc. Am., **58**, 103-131, 1968.

Slichter, L. B., Free oscillations of the Earth, *in* International Dictionary of Geophysics, pp. 331-343, edited by S. Runcorn, Pergamon, London, 1967.

7

THE EARTH'S INTERIOR AS INFERRED FROM A FAMILY OF MODELS

by **Frank Press**

The latest geophysical data, including travel times, eigenperiods, and surface waves for oceanic, shield, and tectonic regions have been used to obtain a large number of earth models that fit the data equally well. The family of solutions are used to appraise how well currently available data constrain earth models and lead to the suggestion that the mantle is divided into three zones.

1. A dense lithosphere, laterally heterogeneous, 70 km thick under oceans and twice as thick under continental shields, formed from the densest fraction of the underlying asthenosphere.

2. The asthenosphere and transition zone, a mostly chemically homogeneous region characterized by partial melting at the top and by solid-solid phase transitions.

3. The lower mantle, initially enriched in iron with respect to zone 2, either constant in composition or diminishing in Fe/Mg ratio with depth.

INTRODUCTION

The first-order structure of the earth's interior is reasonably well known. The thickness of the crust, the velocity of compressional and shear waves in the crust and at the top of the mantle is known for ocean basins, continental shields, and tectonic regions as a result of seismic refraction studies. The precise analysis of body wave travel times, using multiple P arrivals and the spatial gradient of the travel time curves ($dt/d\Delta$) as observed with arrays of seismographs, has led to the discovery of two regions of rapid velocity increase in the transition zone at depths near 400 and 700 km. Travel time data constrain the compressional velocities in most of the mantle to about 1% and depth to the core to within about 10 km. A major goal of geophysics is to elucidate the second-order details of earth structure, using these results together with surface wave and eigenvibration data, travel times of body waves, mass, and moment of inertia of the earth. Hopefully, if the elastic velocities and densities in the interior can be specified within narrow ranges, they could be interpreted in terms of the composition and physical state of the interior.

There are many problems to be overcome before achieving these goals. Because of formal mathematical difficulties and our use of imprecise and incomplete data, it is not possible to determine uniquely the elastic velocity and density distributions (Backus and Gilbert, 1968; Dziewonski, 1970). Our difficulty is further compounded by the lack of uniqueness in interpreting velocity and density distributions in terms of variations in composition and state.

Despite these impediments, it may still be possible to make meaningful statements about the interior. Geophysical data may not lead to unique solutions but could constrain possible solutions to a narrow band. Auxiliary information such as abundance of elements, petrological studies of crust and mantle, and laboratory data relating elastic properties to composition and state could be used to delimit compositional models and prescribe certain physical states at specific depths.

This paper represents an extension of earlier efforts to attack the problem using Monte Carlo inversion procedures. The numerical methods have been improved and the data set is more extensive than in the earlier work. For these reasons the results supersede those presented earlier (Press, 1968, 1969, 1970).

METHOD

Monte Carlo methods have been used widely in many fields. Their exploitation for the inversion of geophysical data was introduced in the USSR (Keilis-Borok and Yanovskaya, 1967). The use of Monte Carlo procedures, in which large numbers of randomly generated models are tested against the data offers several advantages. Successful models are found without bias or preconceived notions. Assumptions regarding relationships between composition, velocity, and density are not needed in the inversion process since what is sought is a family of successful solutions rather than the elusive unique model. Monte Carlo methods are most powerful when a sufficiently large number of models are examined so that the retained solutions are representative of the class of successful models that fit the data. Unfortunately, this is not demonstrable with the inversion of geophysical data because of the large number of parameters required to specify an earth model and the very large number of possible models that this implies. We will use simplifications, numerical experiments, and heuristic arguments, however, and suggest that solutions based on the same data set and the same parameterization of the earth which differ significantly from the family of models presented probably would not be found.

The flow diagram of our Monte Carlo program is shown in Fig. 1 in the form of a diagnostic report printed with each run of the program. In our most recent improvement a "coarse-fine" search is used to find models, which roughly fit the data, and which are subsequently refined to fit within the prescribed tolerances. The flow diagram shows branching according as several tests are passed or failed. ALPHA, BETA, and RHO-SLTMOD randomly generate compressional velocity, shear velocity, and density distributions. ALPHA and BETA TTT test these models against observed travel times for *P* and *S* waves. MASMOM tests the density distribution for compatibility with the observed mass and moment of inertia of the earth. VRPR is the final test of the complete earth model against the surface wave and free oscillation data. In the particular run shown, 41 successful models were found in 257 sec on an IBM 360-65 computer for a cost of 10 dollars. This high speed results from the use of a table of variational parameters (Wiggins, 1968) $\delta T/\delta\rho_i$, $\delta T/\delta\beta_i$, $\delta T/\delta\alpha_i$, and the relationship

$$\Delta T = \sum_i \left(\frac{\delta T}{\delta\rho_i} \Delta\rho_i + \frac{\delta T}{\delta\beta_i} \Delta\beta_i + \frac{\delta T}{\delta\alpha_i} \Delta\alpha_i \right)$$

to compute the change in period T for a particular mode resulting from perturbations $\Delta\rho_i$, $\Delta\beta_i$, $\Delta\alpha_i$ in density, shear velocity, and compressional velocity, respectively, of the ith layer. Numerical tests show that eigenperiods calculated by this approximation are sufficiently accurate for the precision of the observational data. The data against which the randomly generated models are tested and the constraints placed on the models are as follows.

1. Compressional velocity distribution in the mantle is fixed to the values determined by Johnson (1969) for shield or tectonic belts (shield data are used for the oceanic distribution with a suitable modification for the oceanic crustal structure). The distributions are based on a fit of travel times and $dt/d\Delta$ data determined from the apparent velocity of compressional waves across large seismic arrays. Although our procedure can include random distributions of compressional velocity, no purpose would be served since the compressional velocity is already highly constrained by travel time and $dt/d\Delta$ data and slight variations consistent with these data would have minimal effect on surface wave velocities and on free oscillation periods.

2. Compressional velocities in the core are fixed to the values determined by Toksöz (personal communication, 1969) using travel time and $dt/d\Delta$ data. (See Table 1.) Shear velocity in the inner core can have any value in the range 3 to 6 km/sec without affecting our results.

3. Shear velocity distributions in the mantle are forced to lie within the bounds shown in Figs. 14 to 20. These bounds contain most recent models

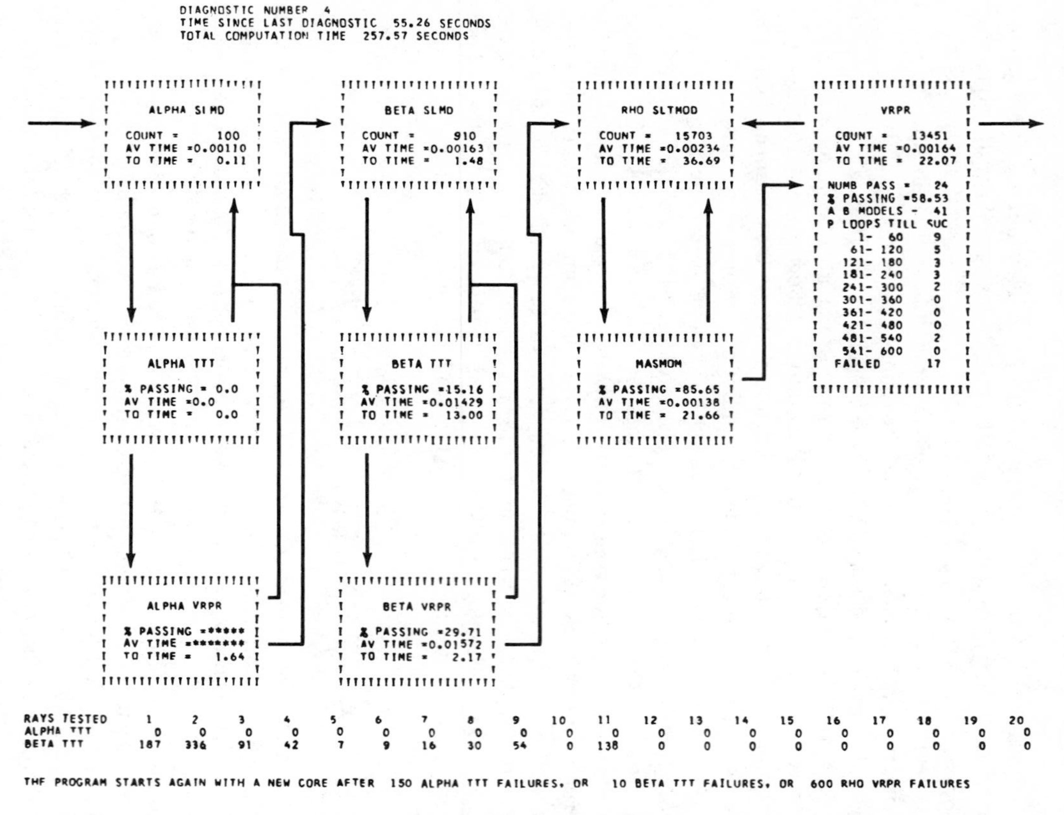

Fig. 1. Sample flow diagram showing random model generator, various tests of models against geophysical data, and efficiency of each test.

Table 1. Assumed compressional velocity distribution in the core (linear connection between points)

Depth (km)	Compressional velocity (km/sec)
2898	8.18
3471	9.00
3671	9.30
5118	10.40
5118	11.03
6371	11.32

satisfying travel time and $dt/d\Delta$ data, particularly the latest models of Hales and Roberts (1970). High-velocity gradients are required in two zones in the transition zone, corresponding to those in Johnson's compressional velocity model (1969). Wide bounds are used in the upper mantle where the distribution is uncertain, whereas narrow bounds are used below 350 km where travel time data are more constraining. Random models generated within these bounds are required to fit travel time data for S and ScS as follows: $JB \leqslant t \leqslant JB + 10$ sec for ocean and shield paths and $JB + 2 \leqslant t \leqslant JB + 12$ for tectonic paths, in the distance range 21 to 99° for S and 0° for ScS.

4. Crustal thickness was fixed at 33 km for shields and tectonic regions and 10 km for oceans. The initial shear velocity β (km/sec) and density ρ (gm/cm^3) in the mantle at the M-discontinuity was fixed as follows: $4.6 \leqslant \beta \leqslant 4.7$, $3.28 \leqslant \rho \leqslant 3.45$ for oceans and shields and $4.2 \leqslant \beta \leqslant 4.5$, $3.2 \leqslant \rho \leqslant 3.3$ for tectonic regions. These densities were obtained by relating densities to compressional velocities using Birch's law and allowing some "jitter" in the mean atomic weight.

5. Successful models were required to fit Kanamori's phase velocities (1970) for surface waves traversing shield, tectonic, and oceanic paths within the standard errors quoted in his paper. These data are shown in Figs. 2 and 3. For modes graver than ${}_0S_{26}$ and ${}_0T_{23}$, eigenperiod data were used. This procedure is an approximate method to allow for lateral variations in upper mantle and crustal structure by using regionalized data for those modes that are most affected by lateral variation. Eigenperiod data for overtones were used as shown in Fig. 4, which summarizes all of the surface wave and eigenperiod data used to find models topped by oceanic structure. The standard errors adopted for the data are shown in the figure. We follow current practice and fit our models to eigenperiod data to ± 0.2% (except for ${}_0S_2$ where a fit to ± 4 sec was required). A higher resolution is unwarranted until the mode splitting resulting from earth rotation and flattening have been resolved. Larger errors were assumed for some overtones. No attempt was made to fit ${}_0S_0$ to better than 0.2% since this mode

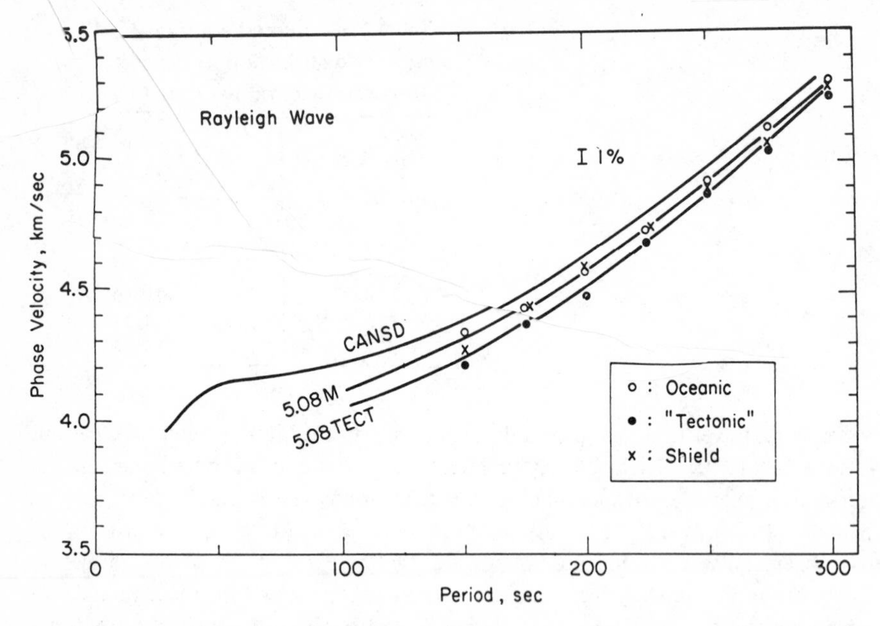

Fig. 2. Kanamori's Rayleigh wave phase velocities for oceanic, shield, and tectonic paths, used to find earth models appropriate for these regions.

provides minimal constraint on the model and slight changes in the velocity in the core produce period changes of this amount.

6. Density in the mantle was required to fall within the bounds shown in Figs. 6 to 12. Rapid density increases were required, corresponding to the high velocity gradients in the transition zone. Below 1000 km, a limit was placed on

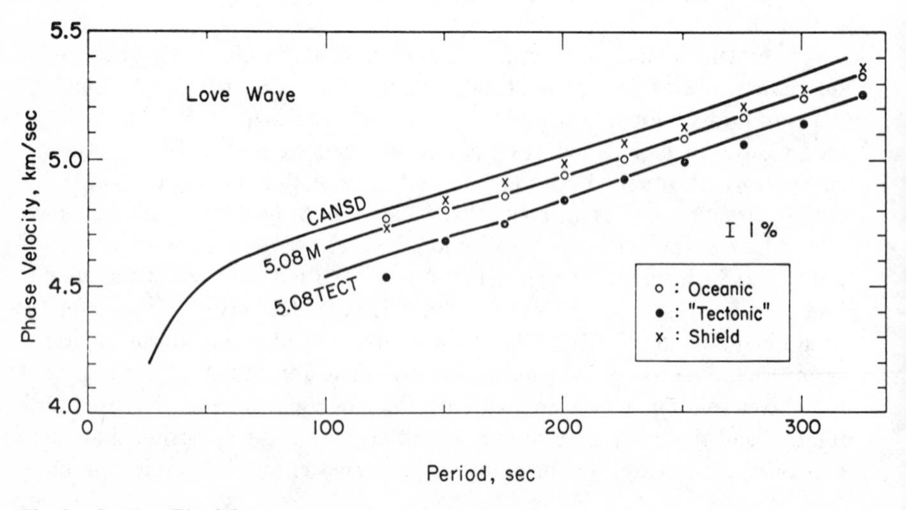

Fig. 3. Same as Fig. 2 for Love waves.

```
S TRAVEL TIME TEST WITH  58 LAYERS USED
INITIAL, FINAL AND INCREMENTAL RAY PARAMETERS   16.60  7.50  1.10
UPPER TOLERANCE 10.0 AND LOWER TOLERANCE   0.0

THE FOLLOWING OSCILLATION MODES ARE TESTED ALLOWING VARIATIONS OF 2.00 STANDARD DEVIATIONS
0 S *  2 3229.0  4:0  *  4 1547.0  3.1  *  6  963.9  2.0  * 14  448.4  0.9  * 34  239.3  0.3  * 50  177.8  0.3  * 60  152.7  0.9
       3 2134.0  4.3     5 1189.0  2.4     7  811.7  1.6     8  707.6  1.4     9  634.0  1.3    10  580.0  1.3    11  537.5  1.1
      12  502.0  1.0    13  473.1  0.9    15  426.2  0.8    16  406.5  0.8    17  389.4  0.8    18  373.4  0.7    20  347.4  0.7
      21  335.8  0.7    22  324.0  0.6    24  305.7  0.6    26  288.9  0.4    28  274.1  0.4    32  249.9  0.3    30  261.4  0.4
      36  229.6  0.3    39  216.3  0.3    42  204.2  0.3  . 45  193.5  0.3    55  164.4  0.4  / 19  360.6  0.7
0 T *  4 1304.0  2.6  *  8  737.0  1.5  * 11  575.7  1.1  * 50  164.2  0.3  * 60  138.8  0.3  * 17  408.3  0.8  * 33  237.5  0.4
       5 1076.0  2.1     6  925.4  1.8     7  817.9  1.6     9  672.0  1.3    10  619.4  1.2    15  452.3  0.9    12  536.9  1.1
      13  504.7  1.1    14  477.0  1.0    16  429.6  1.0     2 2640.0 30.0    18  390.2  0.8    19  374.5  0.8    20  360.6  0.7
      21  347.1  0.7    22  334.3  0.7    23  321.0  0.6    25  299.9  0.5    27  281.2  0.5     3 1708.0  6.0    30  257.4  0.4
      37  215.0  0.4    45  180.5  0.3    40  200.5  0.4    55  150.5  0.3    66  126.6  0.3
0 R    1 1227.6  1.0
1 S    2 1472.0  5.0     3 1064.0  4.0     5  728.7  2.5     6  660.0  2.0     8  555.4  2.0    12  396.8  1.5
1 T
1 R
2 S /  2  905.5  3.0     4  724.9  2.5     6  594.6  2.0    10  415.2  1.5
2 T
2 R
```

Fig. 4. Free oscillation and surface wave data (converted to equivalent eigenperiods) and standard deviations assumed in testing earth models. For example, the mode ${}_0S_2$ has the value 3229.0 sec with $\sigma = 4.0$ sec.

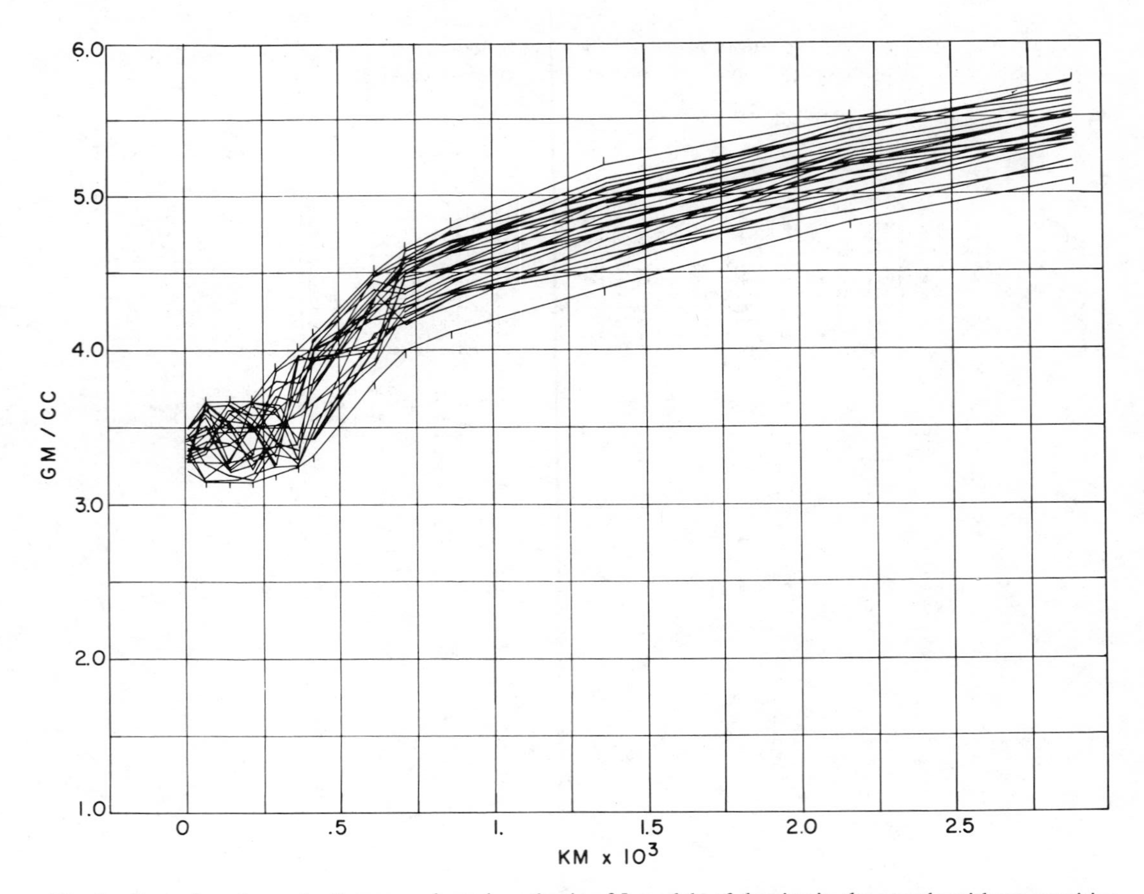

Fig. 5. Test of random selection procedures by selecting 25 models of density in the mantle without requiring fit to geophysical data.

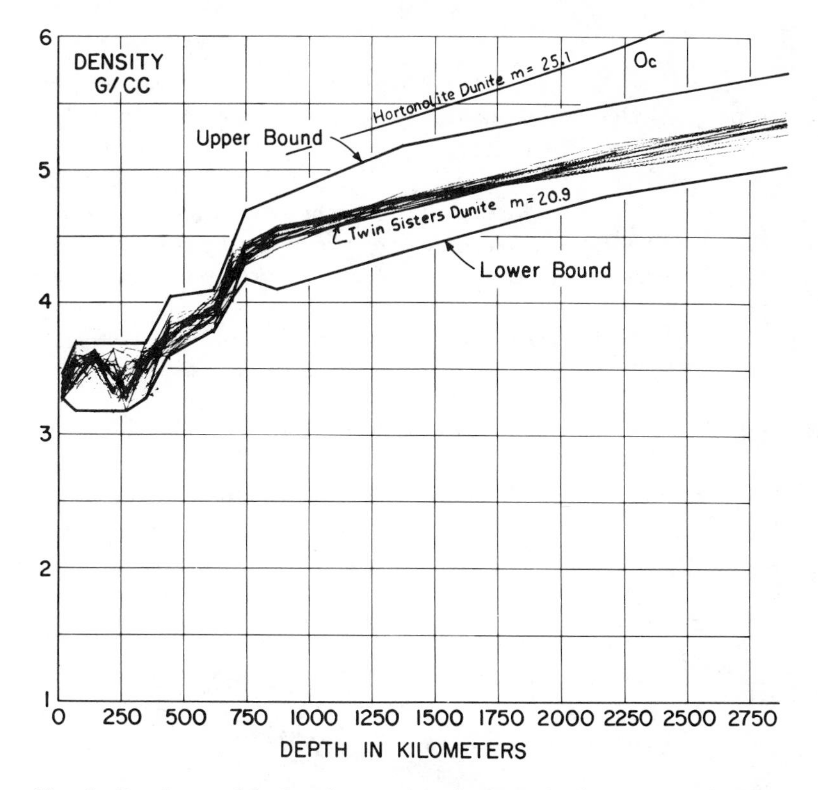

Fig. 6. Density models for the mantle using oceanic data compared with densities of two dunites (inferred from shock wave data) with mean atomic weights m = 20.9 and 25.1.

the amount that the density gradient could exceed the gradient for the homogeneous adiabatic case. Most models that allow for reasonable composition changes or sub- or superadiabatic temperature gradients could pass these strictures. The bounds and restraints on gradients are sufficiently broad to allow most models that have been proposed. Density in the fluid core was required to have an adiabatic gradient.

7. All density models were required to fit the mass and moment of inertia of the earth (5.976×10^{27} grams and 0.3308 for I/Ma^2, respectively).

8. The earth was parameterized by selecting specific depth points (shown by the ticks on the bounds in Figs. 6 to 20) at which shear velocity and density were randomly selected. Linear connection between these points was assumed in the travel time and eigenperiod calculations. In this parameterization, we were guided by the P velocity distribution, the assumption that the olivine-spinel transition initiates the transition zone, and refraction data for the different regions.

9. A maximum of one low-velocity and low-density channel was allowed even though models with two such channels could be found.

RESULTS

Preliminary to finding models that fit the data, a test was made to see that randomly selected models uniformly filled the space between the permissible bounds. A run in which 25 models were randomly selected but not tested against the data is shown in Fig. 5. No bias is apparent in the selection procedure.

The results of a search for models that fit the data are presented as follows.

Figures 6 to 8: Density models for a mantle topped by oceanic structure.

Figures 9 and 10: Density models for mantle topped by continental shield structures.

Figures 11 and 12: Density models for a mantle topped by tectonic structure.

Figure 13: Results of a special search for upper mantle with pyrolite density distribution.

Figures 14 to 16: Shear velocity models for a mantle topped by oceanic structure.

Figures 17 and 18: Shear velocity models for a mantle topped by continental shield structure.

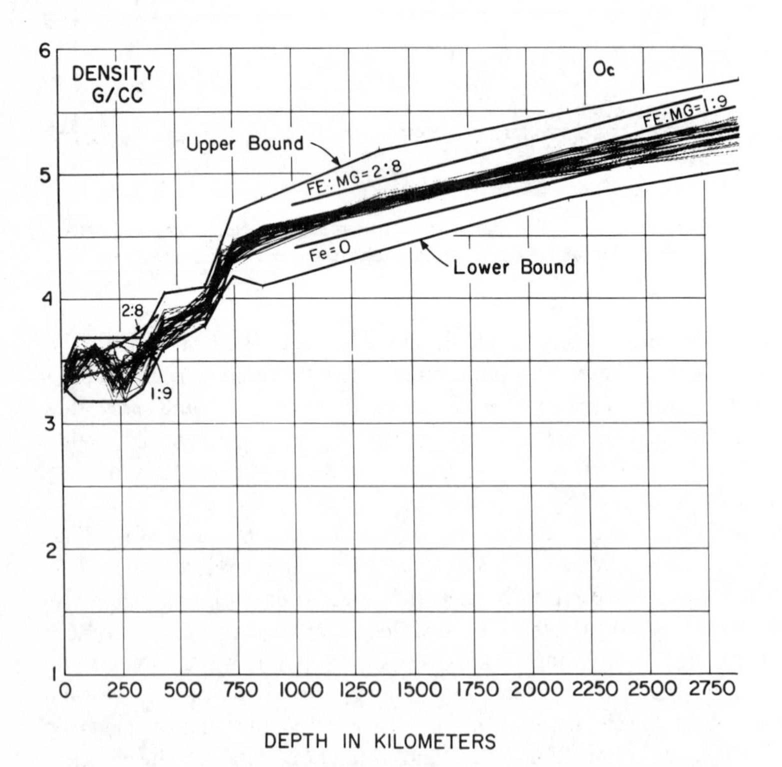

Fig. 7. Density models for the mantle using oceanic data compared with Fujisawa's calculated densities for a mantle of ferromagnesian silicates with varying Fe/Mg ratios.

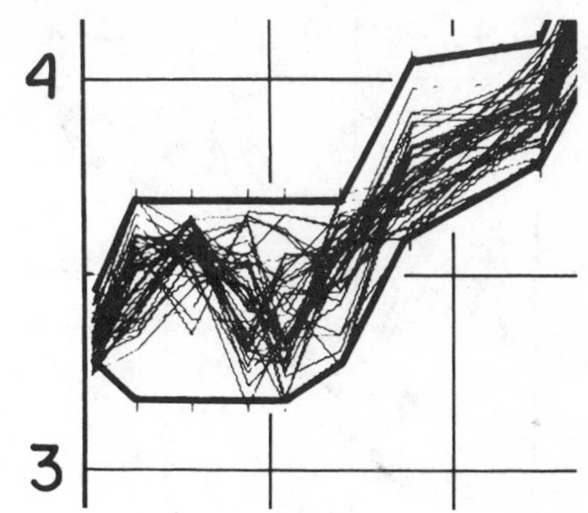

Fig. 8. Enlargement of oceanic upper mantle density distributions of Fig. 6.

Figures 19 and 20: Shear velocity models for a mantle topped by tectonic structure.

Figure 21: Density models for the core.

In each group of models, the upper and lower bounds within which models were selected for testing are shown by heavy lines.

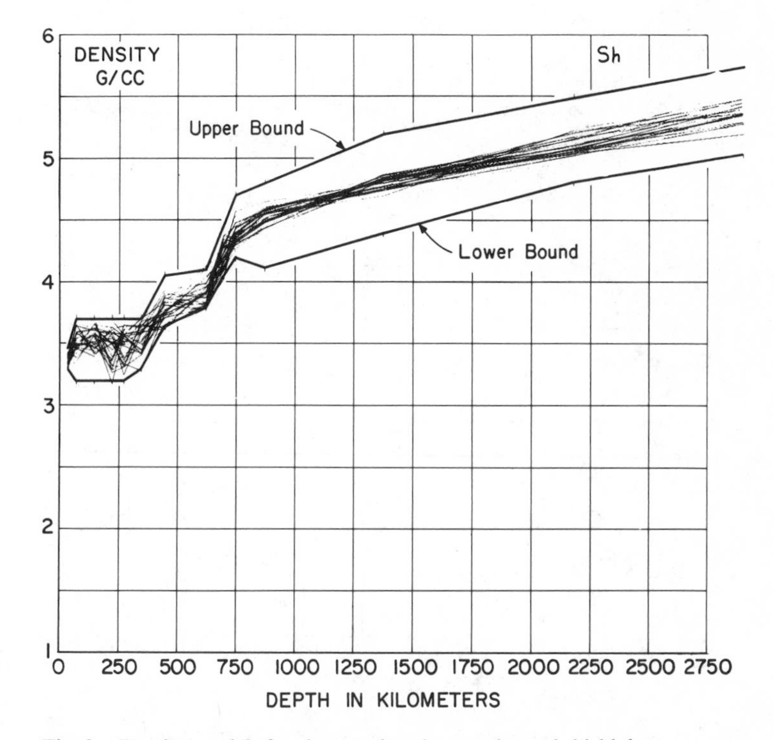

Fig. 9. Density models for the mantle using continental shield data.

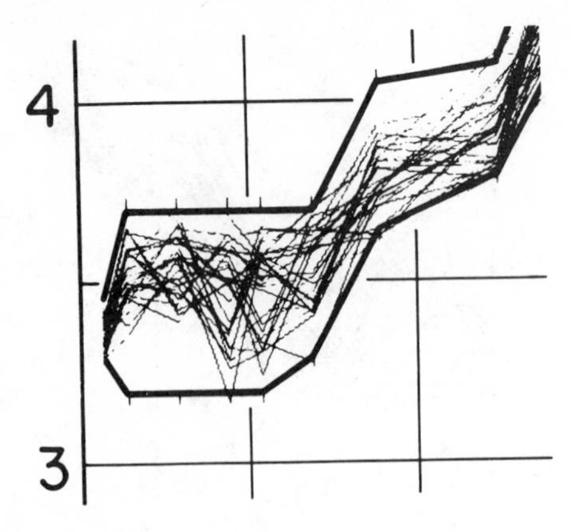

Fig. 10. Enlargement of upper mantle shield density distributions of Fig. 6.

INTERPRETATION OF RESULTS

The density models for the mantle below 800 km are similar for all three upper mantle structures. This is to be expected since only the lower order modes affect this region and a common suite of data were used for modes graver than ${}_0S_{26}$ and ${}_0T_{23}$. Over most of the range below 800 km, the density solutions fall

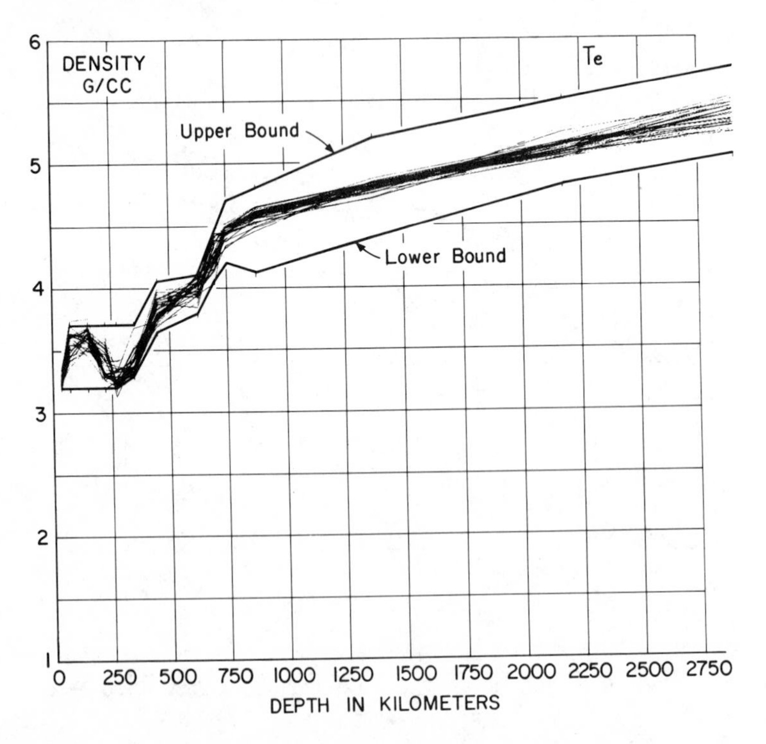

Fig. 11. Density models for the mantle using tectonic regional data.

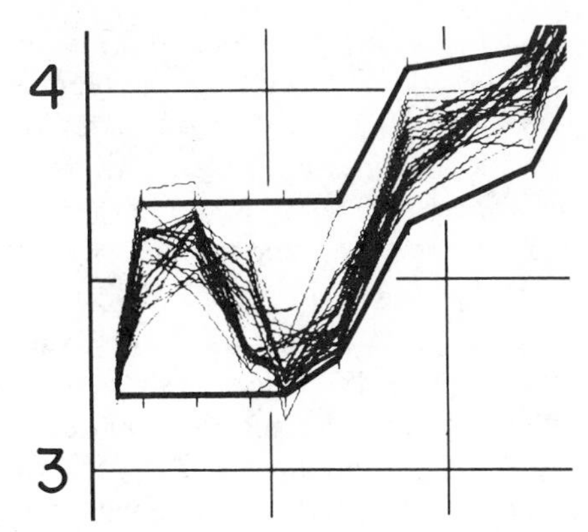

Fig. 12. Enlargement of upper mantle tectonic density distributions of Fig. 11.

within a band of width 0.25 gm/cm^3. Superimposed on the density models in Fig. 6 are the expected density, based on shock wave data, for Twin Sister dunite with mean atomic weight m = 20.9 and hortonolite dunite with m = 25.1 (Anderson and Jordan, 1970). The lower mantle solutions are consistent with constant composition ($m \sim 21$) or a changing composition in the direction of a reduction in m, which is primarily an indicator of the Fe/Mg ratio. This can also be seen in Fig. 7 where the same solutions are presented with Fujisawa's estimated densities (1968) for varying Fe/Mg values. In computing density, Fujisawa made the reasonable assumption that the Fe/Mg ratio in olivine and pyroxene is approximately the same and that both species break down into a mixture of oxides MgO, FeO, and SiO_2 under lower mantle conditions. Figure 7 implies an Fe/Mg ratio of 1/9 for the mantle below 800 km or a gradual reduction in this ratio from a beginning value of 1/9 at about 800 km. Although a superadiabatic temperature gradient would show a density variation similar to that for a reduction in Fe/Mg this possibility is minimized since the corresponding decrease in the gradients of compressional and shear velocities are not present.

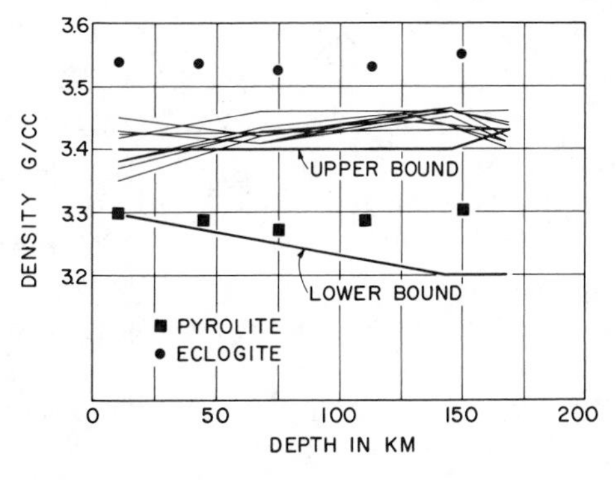

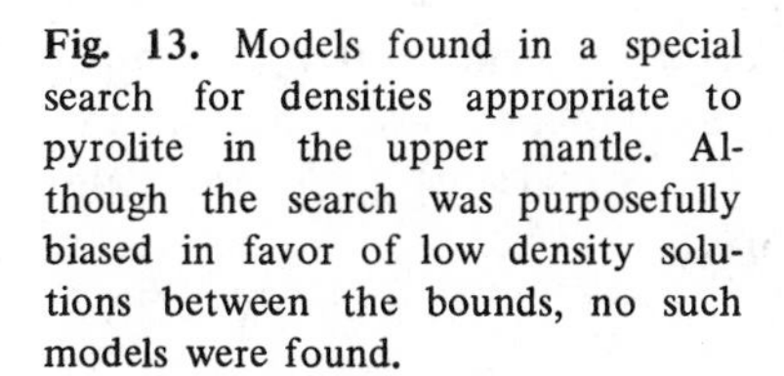
Fig. 13. Models found in a special search for densities appropriate to pyrolite in the upper mantle. Although the search was purposefully biased in favor of low density solutions between the bounds, no such models were found.

Similar results are obtained by a comparison of density ρ and bulk sound velocity c, as obtained from the Monte Carlo solutions with values for dunite of varying Fe/Mg content, as calculated from shock wave data. Unfortunately the reduction of the shock wave data by two different groups lead to somewhat different $c - \rho$ plots. Wang's calculation (1968) for Twin Sisters dunite is a straight line connecting the two crosses in Fig. 24. The slope of this line differs from that found by Ahrens, Anderson and Ringwood (1968), using the same shock wave data. The difference persists for the more iron-enriched materials. Despite these differences the results indicate a lower mantle with $m \sim 21$-22, either chemically homogeneous or decreasing in iron content with depth.

Clark and Ringwood (1964) and Wang (1970) recognized that a $c - \rho$ comparison of earth models with laboratory data offers the advantage of minimizing the effect of temperature to first order, since the proportional changes in c and ρ are the same for a given temperature variation.

The offset in the band of solutions on both sides of the depth range 620 to 870 km in Fig. 24 suggests an increase in iron content with depth through the lower part of the transition zone corresponding to $\Delta m \sim 1$. The region 350 to 400 km where olivine transforms to the spinel structure shows no offset, implying unchanging composition.

The shear velocities in the lower mantle fill the permissible range indicating that the travel time and $dt/d\Delta$ data that are used to establish the bounds are more constraining than the eigenperiod data.

The density and elastic velocities in the upper mantle are constrained by seismic refraction, surface wave and eigenperiod data. The refraction experiments place bounds on the initial velocity (hence density just below the *M*-discontinuity) if Birch's law is assumed together with an educated guess of mean atomic weight for the top of the mantle. The surface wave data provide whatever resolving power is available for upper mantle structures and the eigenperiod data make the upper mantle solutions consistent with the whole earth. For example, absolute densities for the upper mantle cannot be assigned without fitting with high precision the graver eigenmodes. Another example is that the resolution of the low-velocity zone of the suboceanic mantle deteriorates if the initial shear velocity at the *M*-discontinuity is allowed to take values below that indicated by refraction measurements.

The upper mantle density solutions are indicated in Figs. 6 to 12 and Table 2. The solutions are all characterized by positive initial density gradients and by an absence of solutions with $\rho \sim 3.3$ gm/cm^3 between 70 and 150 km. The mean density over the depth range 70 to 170 km exceeds 3.4, implying material closer to eclogite in composition than pyrolite or peridotite. The scatter of density solutions below 150 km shows a lack of resolving power at these depths for the oceanic and shield models. A reversal of density is indicated in some models, whereas others increase monotonically, implying an inability to resolve a low-density zone with these data. Surprisingly, all tectonic solutions show a low density channel beneath the high-density cap (Figs. 11 and 12). Apparently, the low starting density taken for the top of the mantle in tectonic regions and the low surface wave phase velocities constrain the solutions in this case. The absence of low densities between 70 and 150 km and of high densities between 250 and 300 km is striking.

The indication of a high-density ($\rho > 3.4$ gm/cm^3) cap to the mantle is an important result and deserves further verification. Anderson (1970) reports an upper mantle density of at least 3.5 gm/cm^3. A numerical search for lower density models was attempted in the selection procedure by placing the upper bound at 3.4 ± .07 gm/cm^3 for depths above 140 km. The results are shown in Fig. 13. No models with $\rho < 3.4$ gm/cm^3 were found at depths below 50 km, the successful models all falling above this value. The solutions in Fig. 13 all show higher densities than the pyrolite models despite their origin in a search for lower density models. With the 3-point parameterization used in this search, there are about 200 significantly different density models for the region above 140 km. Since several hundred thousand density models were generated in this numerical test we feel that the density space between the bounds shown in Fig. 13 was adequately examined in the search for lower density models. These results are difficult to reconcile with those of Wang (1970) who reports finding a model with densities in the upper mantle appropriate for pyrolite. Perhaps our requirement for a precise fit to both surface wave and eigenperiod data and our use of crustal structure and initial velocities appropriate for oceanic, shield and tectonic regions with the corresponding surface wave data, accounts for this difference.

Dziewonski (1970) questions the resolving power for density at depths between 33 and 200 km because of the compensatory effect of shear velocity changes between 250 and 800 km. Since we vary shear velocity and density, the compensation is automatically taken into account. Our shear velocity distribution below 350 km is highly constrained by body wave and $dt/d\Delta$ data, so that Dziewonski's uncompensated density perturbation (which would have been rejected by our tests) is more pertinent. Because of questions like this, we undertook the special searches for low-density models as described earlier. The failure of these searches, though not a proof of the absence of such solutions,

Table 2. Average density 70 to 170 km

Ocean	$3.42 \leqslant \bar{\rho} \leqslant 3.64$
Shield	$3.43 \leqslant \bar{\rho} \leqslant 3.62$
Tectonic	$3.47 \leqslant \bar{\rho} \leqslant 3.72$
Eclogite mantle	$\bar{\rho} \sim 3.55$
Pyrolite mantle	$\bar{\rho} \sim 3.30$
Spinel mantle	$\bar{\rho} \sim 3.70$

casts doubt on their existence for the data set and parameterizations we have adopted.

Every model of shear velocity distribution in the mantle beneath oceans and continents (Figs. 14 to 18) requires a low velocity channel. This is a consequence of the assumed initial shear velocity and the low surface wave velocities. Since our surface wave data are drawn from world-circling paths, we believe this to be a world-wide feature. The suboceanic channel tends toward lower velocities than occurs under shields. If the thickness of the lithosphere is defined arbitrarily by the depth at which the shear velocity drops below 4.5 km/sec, then the subshield lithosphere is twice as thick as the suboceanic lithosphere as can be seen from the pattern of solutions in Figs. 15 and 18. Kanamori and Press (1970) subjected Monte Carlo solutions to an additional test against a new and highly precise group velocity curve. The successful solutions fall in a remarkably narrow band (Fig. 16) which indicates a sharp drop in shear velocity below about 70 km. They associated this abrupt reduction in rigidity with the solidus, which marks the lower boundary of the lithosphere.

The mantle shear velocity under tectonic regions (Figs. 19 and 20) is forced to begin with values in the range 4.2 to 4.5 km/sec, based on seismic refraction data. Solutions with and without a low-velocity channel are found. In the latter

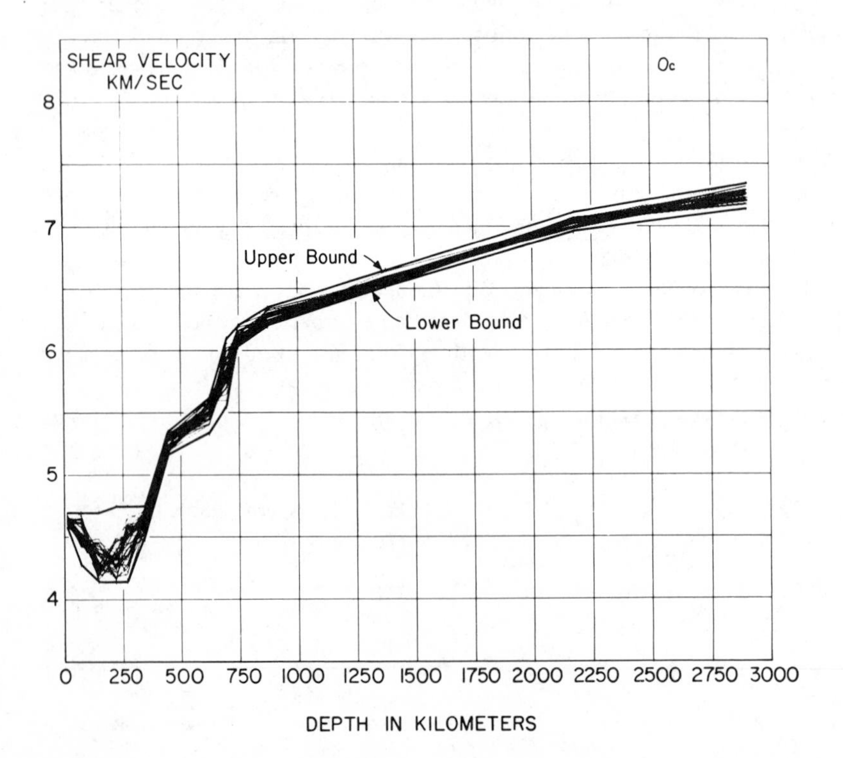

Fig. 14. Shear velocity models for the mantle using oceanic data.

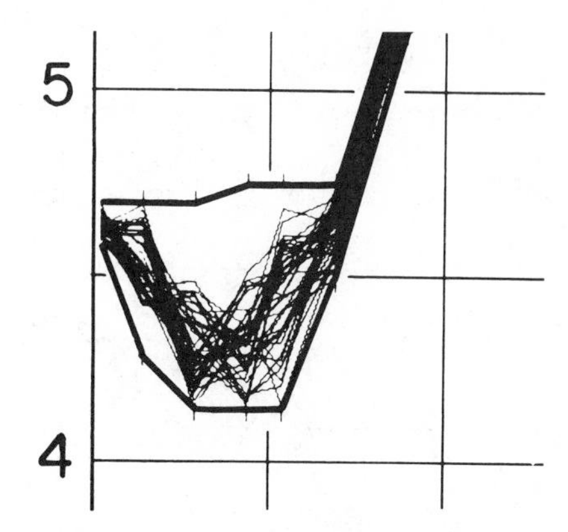

Fig. 15. Enlargment of upper mantle oceanic shear velocity distributions of Fig. 14.

case, the initial velocity falls below 4.5 km/sec as if the center of the low velocity zone migrated to the M-discontinuity in these regions.

The density distribution in the fluid, outer core falls within a surprisingly narrow band of about 0.3 gm/cm^3. Apparently, the combination of eigenperiod data, mass, moment of inertia, and the requirement of adiabaticity constrain the densities in the outer core. The wide divergence of solutions for the inner core imply that this region is uncontrolled by the data presently available.

The distribution of the seismic parameters ϕ and k/μ for 42 models are given in Figs. 22 and 23.

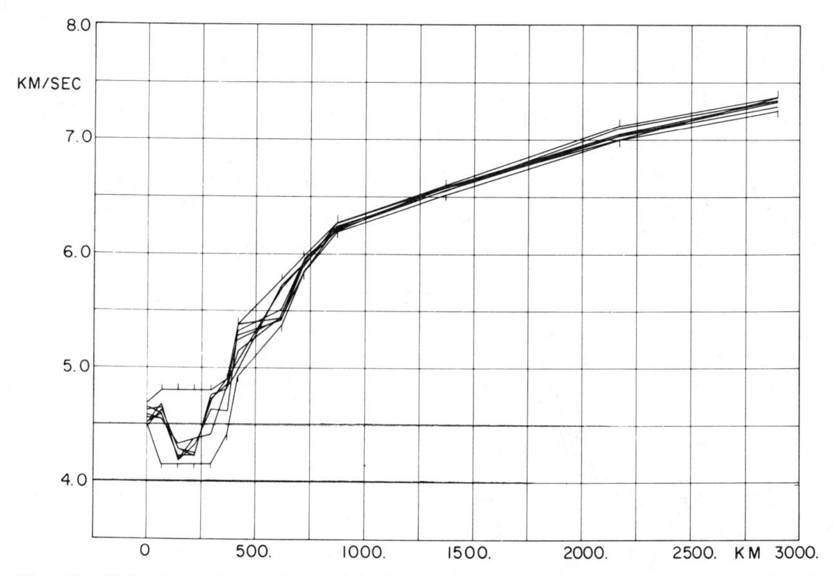

Fig. 16. Selection of oceanic models fitting high precision group velocity data. All successful models show sharp drop in shear velocity at 70 km, which marks the boundary between lithosphere and asthenosphere according to Kanamori and Press (1970).

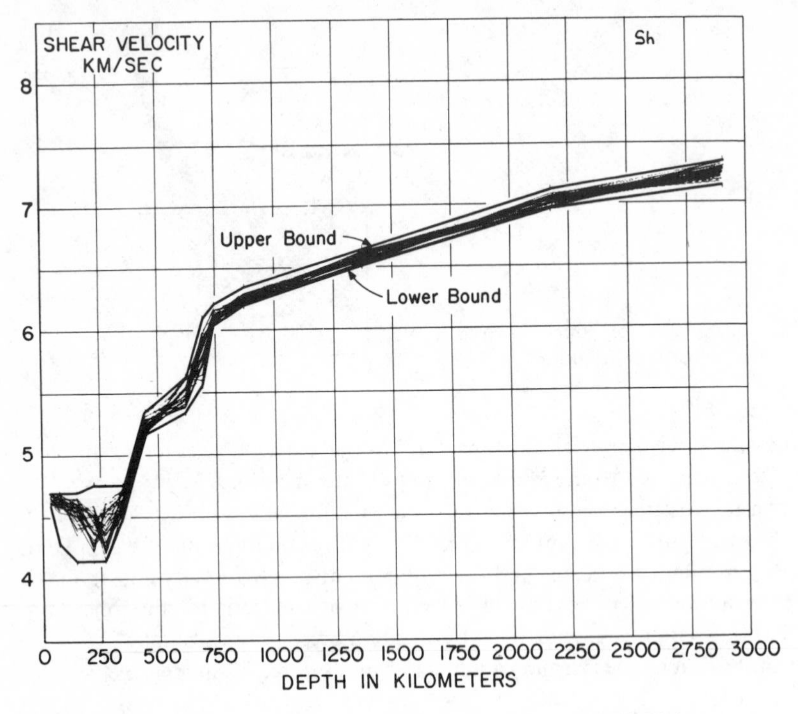

Fig. 17. Shear velocity models for the mantle using continental shield data.

CONCLUSIONS

One hesitates to draw overly specific conclusions about the earth's interior in view of the unresolved question of uniqueness inherent in the inversion of currently available geophysical data and the uncertain correlation of physical properties with mineral assemblages and state. Nevertheless, we have used the best available data in this study, and in certain regions of the earth, the solutions

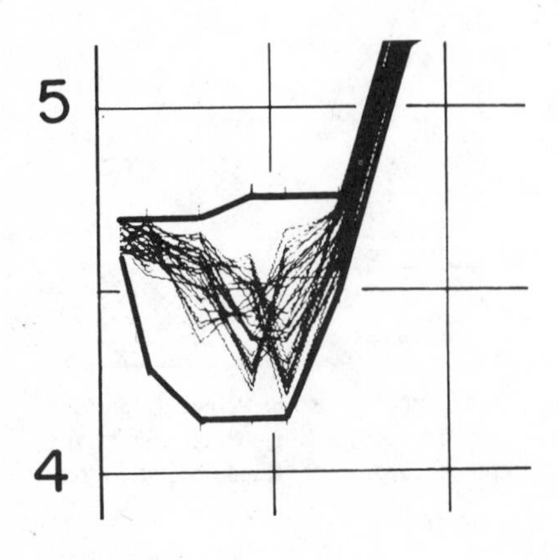

Fig. 18. Enlargment of upper mantle continental shear velocity distributions of Fig. 17.

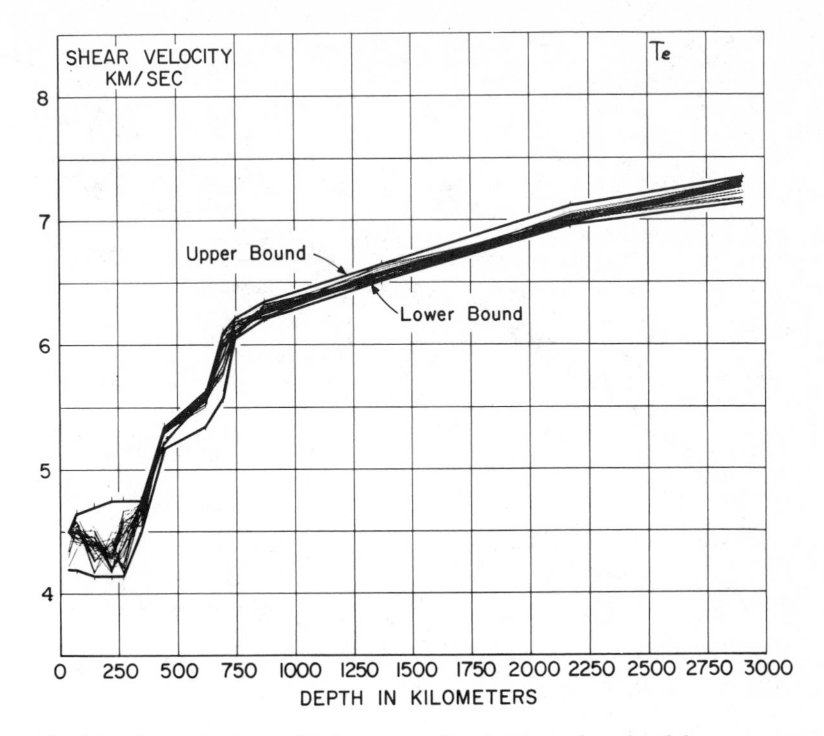

Fig. 19. Shear velocity models for the mantle using tectonic regional data.

seem sufficiently constrained to warrant advancing serious hypotheses if not firm conclusions.

It is currently widely held that the mantle can be divided into three zones: (1) an upper mantle dominated by the minerals olivine, pyroxene, and garnet, chemically zoned and physically divided into the lithosphere and asthenosphere; (2) the transition zone characterized by major phase changes; (3) a chemically

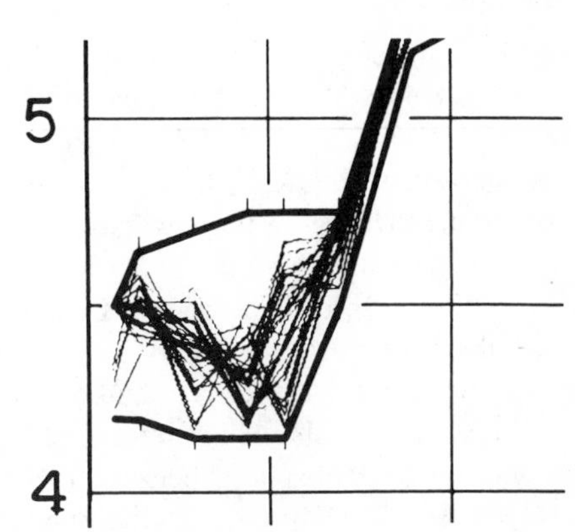

Fig. 20. Enlargment of upper mantle tectonic shear velocity distributions of Fig. 19.

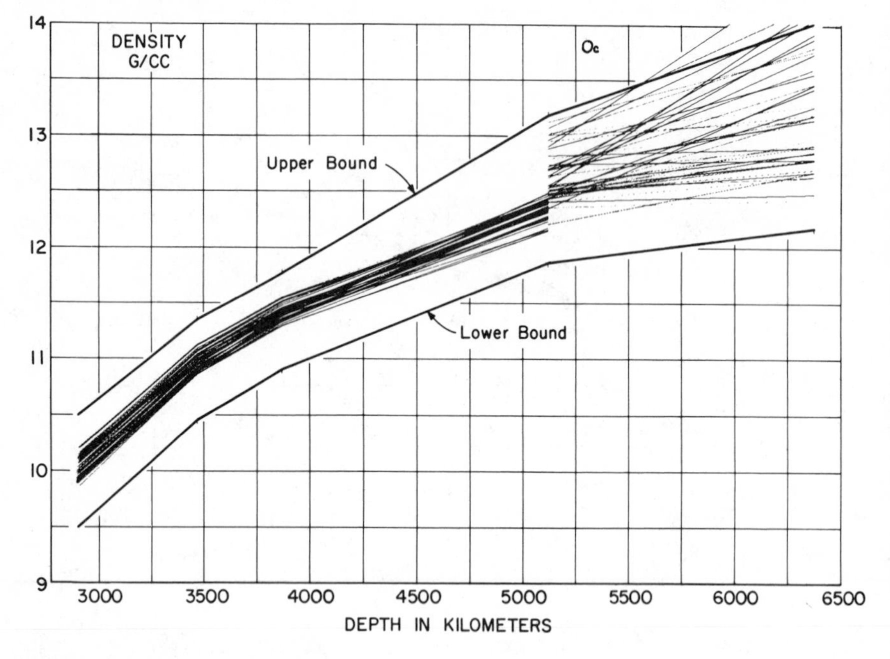

Fig. 21. Density models for the core.

homogeneous lower mantle composed primarily of a mixture of the oxides FeO, MgO, SiO_2.

Guided by the Monte Carlo solutions (Figs. 6, 7, 24), we propose a division into three zones based on mean atomic weight m or Fe/Mg ratio, which may be more pertinent to the evolution of the mantle. The three zones are clearly indicated by three sections of the $c - \rho$ plot of the models in Fig. 24: (1) 10 to 70 km, (2) 150 to 870 km, (3) 870 to 2900 km.

Lithosphere

The lithosphere is a zone about 70 km thick under ocean basins, twice as thick under shields. The Fe/Mg values for the lithosphere fall in the range 1/9 to 2/8 and $21 < m < 23$. The lithospheric plate is more dense than the underlying asthenosphere because the basalts that are a major component originate in the fractional melting of the densest components of the asthenosphere. The basalts rise as low density fluids along mid-ocean ridges cool and transform to denser rocks in the spreading process. Below about 20 km, the lithosphere is laterally zoned with gabbro near the ridge, transforming into garnet granulite and eclogite with increasing distance (Ito and Kennedy, 1970; Press, 1969). The lithospheric plates so formed show high velocity, high Q, are strong and effectively decoupled from the asthenosphere because the boundary between the two zones is the solidus where a marked reduction in strength occurs. The gravitational instability

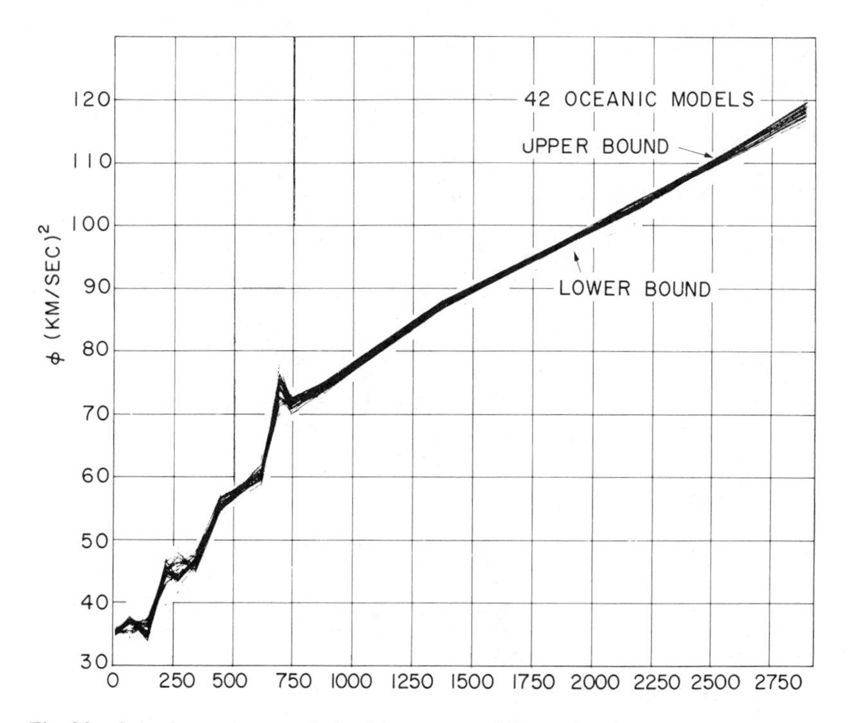

Fig. 22. Seismic parameter ϕ derived from successful mantle solutions.

of "heavy" lithosphere over "light" asthenosphere is an element of a complex upper mantle convection process involving rising fluids, transformation to strong, dense plates by lateral spreading, sinking, and remelting of these plates. The unusual density maximum and minimum found under tectonic regions may

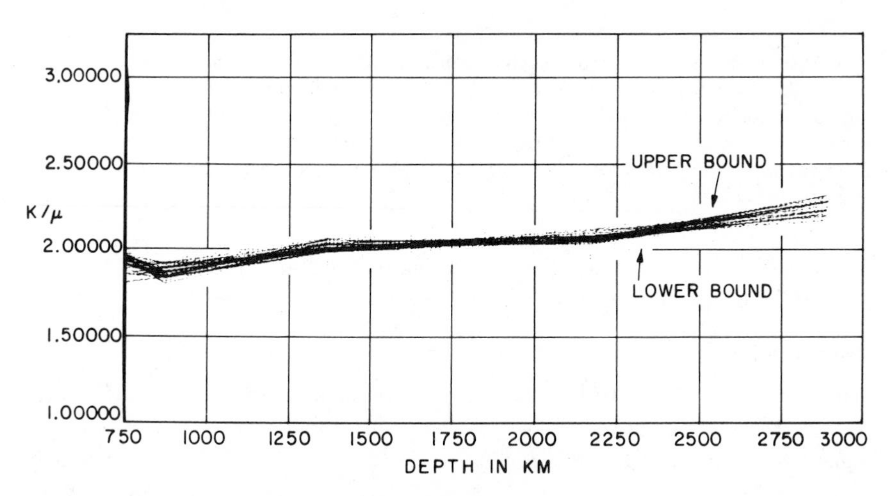

Fig. 23. Ratio K/μ derived from successful mantle solutions.

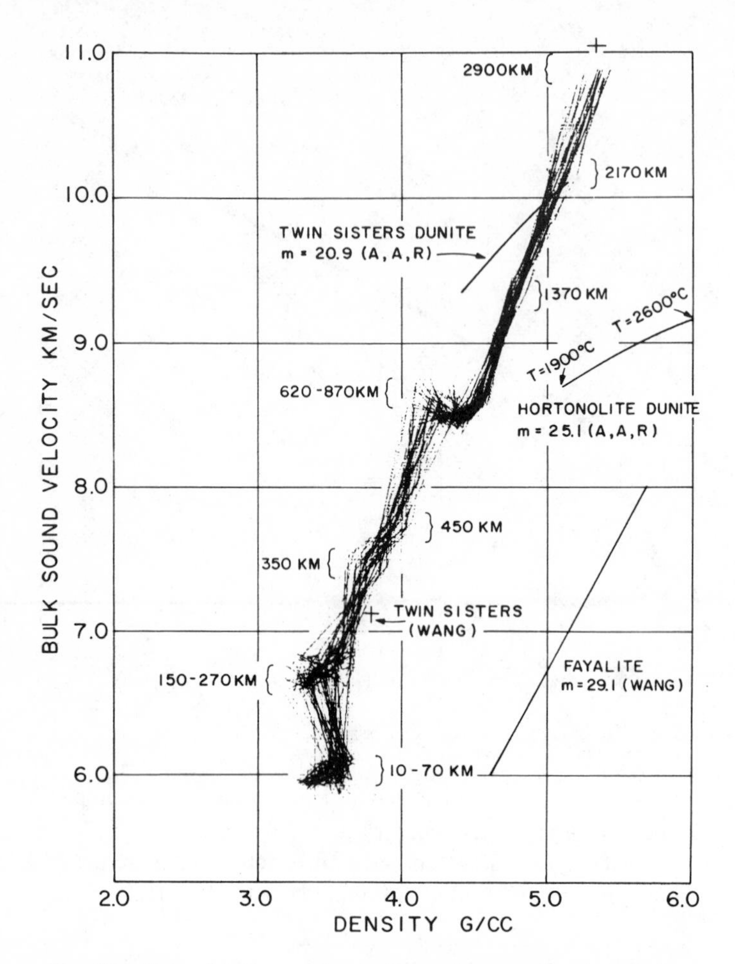

Fig. 24. Plot of bulk sound velocity against density derived from successful solutions. Comparison with shock wave derived curves for ultrabasic rocks according to Ahrens, Anderson, and Ringwood (1969) and Wang (1968). Wang's results for Twin Sisters dunite is a straight line connecting the two crosses. Ahrens', Anderson's, and Ringwood's results are indicated by *A, A, R*.

be indicative of an asthenosphere extending to the *M*-discontinuity, into which the dense lithosphere is thrust.

Asthenosphere and Transition Zone

This is a region, 150 to 870 km, of pyrolitic or peridotitic composition with vertical variation in physical state. The Fe/Mg ratio for this zone falls in the range from less than 1/9 to 2/8, but closer to the lower value, and the mean atomic weight is bracketed by $20 < m < 21$. The asthenosphere is defined by the low-velocity, low-Q zone. A state of incipient or partial melting not only accounts for the low velocity and low Q but also is predicted by the intersection of most proposed geotherms with the solidus of ultrabasic rocks in the presence of small amounts of water (Anderson, 1970; Birch, 1970; Lambert and Wyllie, 1970; Press, 1959). The asthenosphere extends to a depth of about 250 km. Lambert and Wyllie propose that the low-velocity zone terminates downward where the water content diminishes, reducing the quantity of melt. The transition zone 350 to 870 km is initiated by the olivine-spinel phase transformation at 350 to 400 km (Ringwood and Major, 1970). There is no offset in the $c - \rho$ plot through the olivine-spinel transition as would be expected under chemically homogeneous conditions. Other phase changes are also required to account for the rapid density and velocity increase (Anderson, 1967; Ringwood, 1970). Birch (1952) proposed a final transition to a mixture of close-packed oxides. The offset in trend on the $c - \rho$ plot (Fig. 24) between 620 and 870 km carries the suggestion that the deepest phase transition is accompanied by an increase $\Delta m \sim 1$, a value close to that proposed by Press (1970) and Anderson and Jordan (1970).

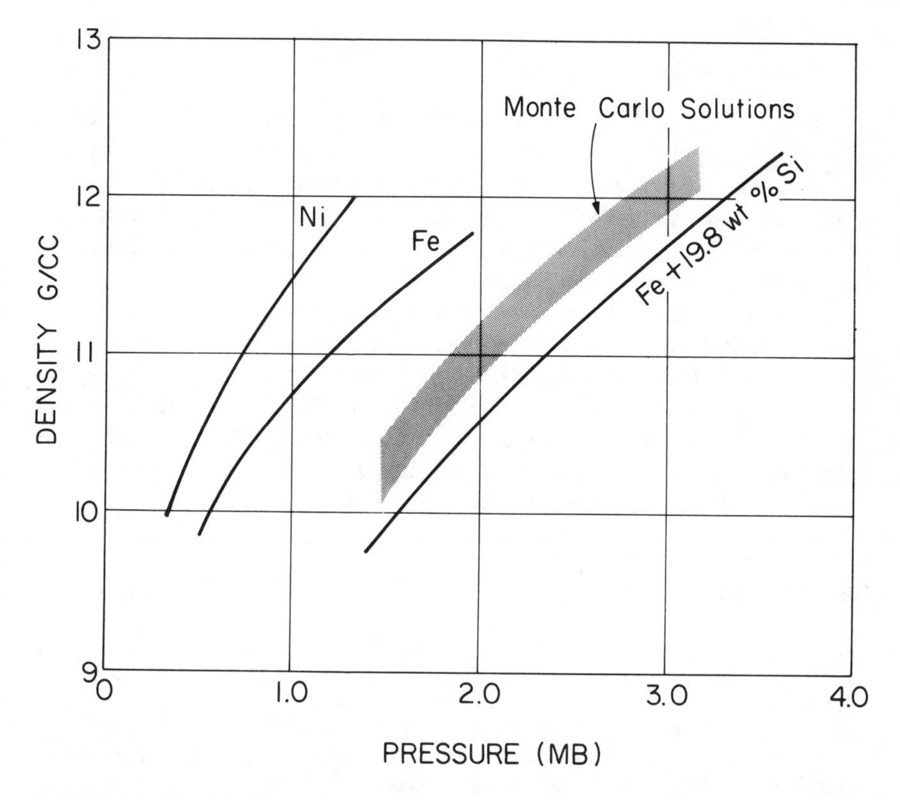

Fig. 25. Successful density models for the core compared with densities predicted from shock wave data on Ni, Fe, and Fe + Si mixture.

Lower Mantle

Characterized by $21 < m < 22$, from a comparison of density and bulk sound velocity with shock data on ultrabasic rocks. Other workers who report m values in this range are McQueen et al. (1967), Wang (1968), Anderson and Jordan (1970). Using Fujisawa's densities for a mixture of oxides with varying Fe/Mg ratios, the density near 1000 km indicates a value of Fe/Mg near 1/9. There is a definite suggestion of a decreasing Fe content with depth in the lower mantle, some models showing near constant composition, others indicating a total depletion of iron at the bottom of the mantle, most models falling between these extremes. This is an important point that should be resolved in future studies since it bears on fractionation processes in the lower mantle and possibly on the evolution of the core.

As many recent investigators have proposed, the core densities imply a mixture of iron and a lighter element. On geochemical grounds, the candidate elements are silicon and sulfur. Shock wave data on mixtures of Fe and Si are compared with our density solutions in Fig. 25. About 10 to 15% Si content is implied if this is the "lightening" element.

ACKNOWLEDGMENTS

I benefited greatly from discussion with Drs. R. Wiggins, H. Kanamori, A. Hales, J. Derr, and G. Kennedy. These gentlemen along with Drs. C. Wang, D. Anderson, M. N. Toksöz, A. E. Ringwood, P. Wyllie, and H. Fujisawa were kind enough to send me preprints in advance of publication. This research was sponsored by the Advanced Research Projects Agency and monitored by the Air Force Office of Scientific Research.

Note added in proof: I am indebted to Professors Francis Birch and C. Wang for pointing out to me that a recent revision of the paper by Ahrens, Anderson, and Ringwood (1969) carried out by Davies and Anderson (in press) brings the reduced shock wave data on Twin Sisters dunite into closer agreement with Wang's results. Removal of this large discrepancy brings the earth models into closer agreement with the shock-data and reinforces the gross conclusions about mean atomic weight reached in the discussion of Fig. 24.

REFERENCES

Ahrens, T. J., Anderson, D. L., and Ringwood, A. E., Equations of state and crystal structure of high pressure phases of shocked silicates and oxides, Rev. Geophys., **7**, 667, 1969.

Anderson, D. L., Phase changes in the upper mantle, Science, **157**, 1165, 1967.

______, Petrology of the mantle, *in* The Minerology and Petrology of the Upper Mantle, edited by B. A. Morgan, Min. Soc. Am. Spec. Paper **3**, 85, 1970.

Anderson, D. L., and Jordan, T., The composition of the lower mantle, Phys. Earth Planet. Interiors, **3**, 23, 1970.

Backus, G. E., and Gilbert, F., The resolving power of gross earth data, Geophys. J., **16**, 169, 1968.

Birch, F., Elasticity and constitution of the earth's interior, J. Geophys. Res., **57**, 227, 1952.

______, Interpretations of the low velocity zone, Phys. Earth Planet. Interiors, **3**, 178, 1970.

Clark, S. P., and Ringwood, A. E., Density distribution and constitution of the mantle, Rev. Geophys., **2**, 35, 1964.

Dziewonski, A. M., Correlation properties of free period partial derivatives and their relation to the resolution of gross earth data, Bull. Seism. Soc. Am., **60**, 741, 1970.

Fujisawa, H., Temperature and discontinuities in the transition layer within the earth's mantle: geophysical application of the olivine-spinel transition in the Mg_2SiO_4 - Fe_2SiO_4 system, J. Geophys. Res., **73**, 3281, 1968.

Hales, A. L., and Roberts, J. L., Shear velocities in the lower mantle and the radius of the core, Seism. Soc. Am. Bull., **60**, 1427, 1970.

Ito, K., and Kennedy, G. C., The fine structure of the basalt-eclogite transformation, *in* Minerology and Petrology of the Upper Mantle, edited by B. A. Morgan, Min. Soc. Am. Spec. Paper **3**, 77, 1970.

Johnson, L. R., Array measurements of *P* velocities in the lower mantle, Bull. Seism. Soc. Am., **59**, 973, 1969.

Kanamori, H., Velocity and *Q* of mantle waves, Phys. Earth Planet. Interiors, **2**, 259, 1970.

Kanamori, H., and Press, F., How thick is the lithosphere? Nature, **226**, 330, 1970.

Keilis-Borok, V. I., and Yanovskaya, T. B., Inverse problems of seismology, Geophys. J., **13**, 223, 1967.

Lambert, I. B., and Wyllie, P. J., Low velocity zone of the earth's mantle: Incipient melting caused by water, Science, **169**, 764, 1970.

McQueen, R. G., Marsh, S. P., and Fritz, J. N., Hugoniot equation of state of twelve rocks, J. Geophys. Res., **72**, 4999, 1967.

Press, F., Some implications on mantle and crustal structure from *G* waves and Love waves, J. Geophys. Res., **64**, 565, 1959.

______, Earth models obtained by Monte Carlo inversion, J. Geophys. Res., **73**, 5223, 1968.

______, The suboceanic mantle, Science, **165**, 174, 1969.

______, Earth models consistent with geophysical data, Phys. Earth Planet. Interiors, **3**, 3, 1970.

Ringwood, A. E., and Major, A., The system Mg_2SiO_4 - Fe_2SiO_4 at high pressures and temperatures, Phys. Earth Planet. Interiors, **3**, 89, 1970.

Ringwood, A. E., Phase transformations and the constitution of the mantle, Phys. Earth Planet. Interiors, **3**, 109, 1970.

Wang, C., Constitution of the lower mantle as evidenced from shock wave data, J. Geophys. Res., **73**, 6459, 1968.

______, Density and constitution of the mantle, in press, J. Geophys. Res., **75**, 3264, 1970.

Wiggins, R. A., Terrestrial variation tables for the periods and attenuation of the free oscillations, Phys. Earth Planet. Interiors, **1**, 201, 1968.

8

TRAVEL TIMES OF SEISMIC WAVES

by **Anton L. Hales and Eugene Herrin**

Studies of the travel times of P and S waves fall naturally into two sections: the first covering distances up to 30° and the second distances from 30 to 100°. It is clear that the uppermost two or three hundred kilometers of the upper mantle vary a great deal laterally. These variations are probably related to variations of the S low-velocity layer (the existence of which has been established by surface wave dispersion studies) and to lateral variations of temperature in the upper mantle. The resulting variations in travel times at distances less than 30° are such that discussions of times at these shorter distances must be restricted to well defined geographic regions of the earth. On the other hand, it is possible to discuss travel times for distances greater than 30° with reasonable assurance that these average values are generally applicable to the entire earth.

The travel times of both P and S waves beyond 30° have been shown to be made up of three components: a standard travel time, and source and station anomalies, which are dependent on upper mantle structure at the source and station, respectively. The standard travel times of P and S from 30 to 96° can be fitted to within 0.7 and 2.0 sec, respectively, by quadratics. Thus the discontinuities below 800 km, if any, must be minor as compared to those at 350 and 650 km.

Since the times are so well fitted by quadratics, the values of $dT/d\Delta$ should be approximately linear functions of Δ. The values of $dT/d\Delta$

found from studies on array stations are in fact approximately linear in Δ from 30 to 95°. The deviations from linearity found in studies of different array systems, however, show certain similarities, especially near 50°, which may well be significant in respect to the structure and properties of the lower mantle.

The station anomalies reflect upper mantle structure. The arrivals are consistently late in areas of large-scale tectonic uplift and volcanic activity within the past 50 m.y., but they are consistently early in the shield and relatively old platform areas. The ratio of S to P station anomalies is high, being about 4. This ratio can only be explained in terms of the late arrivals being in regions of high upper mantle temperatures and by some degree of partial melting.

TRAVEL TIMES FOR DISTANCES GREATER THAN 30°

The Jeffreys-Bullen travel times for arc distances greater than 30° differ very little from those of Gutenberg, and thus the velocity distributions of Jeffreys and Gutenberg are very similar below 800 km, at which depth the seismic waves bottom for an arc distance of 30°. This is illustrated in Fig. 1, reproduced from Birch (1952). The quantity plotted is $(1 - \Delta\phi/g\Delta r)$, which was introduced by Bullen and may be regarded in some sense as a measure of homogeneity. It is clear from this figure that the major departures from homogeneity occur in the region designated C by Bullen. This region extends from a depth of 413 to 1000 km. It is now known both from surface wave studies and body wave studies that there are two zones of rapid velocity change centered around 380 and 650 km. It will be seen later that there is still some uncertainty about the behavior of the velocities between depths of 800 and 1200 km, but the effect on the travel times is very small (less than 1 sec). Discontinuities in the lower mantle were proposed by Dahm (1936) and Repetti (1930). Once again, however, the effect on the travel times was relatively small.

In spite of the close agreement of the Jeffreys-Bullen and Gutenberg travel times beyond 30°, travel times for this distance range have been studied extensively since 1962. The interest in more precise determination of seismic travel times arose for four reasons.

1. Study of the travel times of the nuclear explosions in the Marshall Islands showed significant differences from the Jeffreys–Bullen travel times (Carder, 1964; Jeffreys, 1954, 1966).

2. Comparison of the observed periods of the free oscillations of the earth with those calculated from current earth models showed that there were significant differences for some of the graver modes. It was suggested that the shear velocity distribution required modification (MacDonald and Ness, 1961), that the density distribution should be changed (Anderson, 1965, 1967; Landisman, Satô, and Nafe, 1965), and that the core radius and the shear velocity close to the core-mantle boundary should be modified (Bullen and

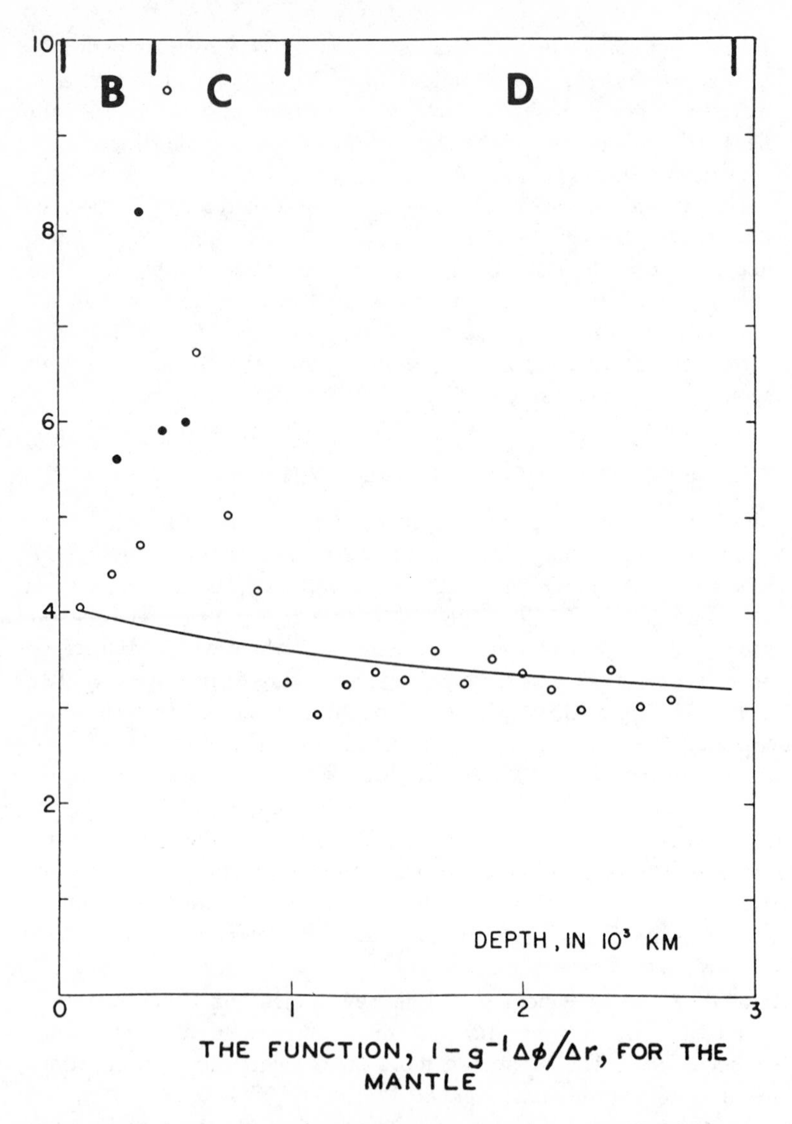

Fig. 1. The function, $1 - \Delta\phi/g\Delta r$, for the mantle. This function is a measure of the homogeneity of the mantle. The figure shows clearly the relatively inhomogeneous region *C*. From Birch (1952).

Haddon, 1967; Cleary, 1969; Dorman, Ewing, and Alsop, 1965). The problem of the inversion of travel time and free oscillation data subject to the constraints imposed by the known mass and moment of inertia of the earth has been studied extensively (Derr, 1969; Dziewonski, 1970; Gilbert and Backus, 1968; Press, 1968; Wiggins, 1968). It is clear from these studies that there is considerable interaction between the effects of parameters such as the shear velocity, the density, and the radius of the core and that further progress demands that the body wave travel times should be determined as precisely as possible. With this information in hand, it is possible to redetermine the core radius. Then, hopefully, the free oscillation data can be used to control the density distribution.

3. It became evident from studies of the dispersion of surface waves that there were considerable regional variations in the properties of the upper mantle. Furthermore, the precise determination of *P* travel times and apparent velocities at distances of less than 20° from the Gnome explosion (Herrin, 1969; Herrin and Taggart, 1962; Romney et al., 1962) showed considerable regional variation (see Fig. 2). It was apparent that significant regional variation in teleseismic travel times was probable but that determination of the regional variation of teleseismic travel times necessarily required that the travel times themselves should be examined at the same time.

4. It was found that the locations of the sites of nuclear explosions determined by the standard methods used for the location of earthquake epicenters often deviated considerably from the known sites. It was thought that improvements in the travel times, including their regional variations, would lead to more precise determination of earthquake epicenters. It has been found, however, that anomalous structures at the source contribute significantly to the errors in location, and these errors are difficult to determine for single events.

P Wave Travel Times

The studies by Carder (1964) and Gogna (1967) of the travel times of *P* waves from nuclear explosions showed that there were significant deviations from the Jeffreys-Bullen tables. Herrin and Taggart (1968a) also considered the travel times from nuclear explosions, but later extended their study to the bulletin readings from a large number of earthquakes. Cleary and Hales (1966) relied on their own readings from the records of earthquakes at North American stations and did not make use of nuclear explosion travel times except for the purpose of determining a baseline. The study by Arnold (1965) was based on travel times of Japanese earthquakes.

Methods of analysis differed. Herrin and Taggart determined the means of the deviations from the J. B. tables for 1° cells and then determined station corrections from the residual deviations. The earthquakes were then relocated using the corrected travel times and the station corrections. This process was repeated 12 times. In the final stages of analysis, these authors determined azimuthally dependent station corrections (Bolt and Nuttli, 1966; Cleary and Hales, 1966; Herrin and Taggart, 1968a). Herrin and Taggart used a method of smoothing proposed by Jeffreys and modified by Arnold (1968). The Herrin and Taggart times were reported as standard tables in a special issue of the Seismological Society of America Bulletin (Herrin, 1968). Cleary and Hales restricted their study to the observations at North American stations of the arrivals from a relatively small set of earthquakes chosen in order to be

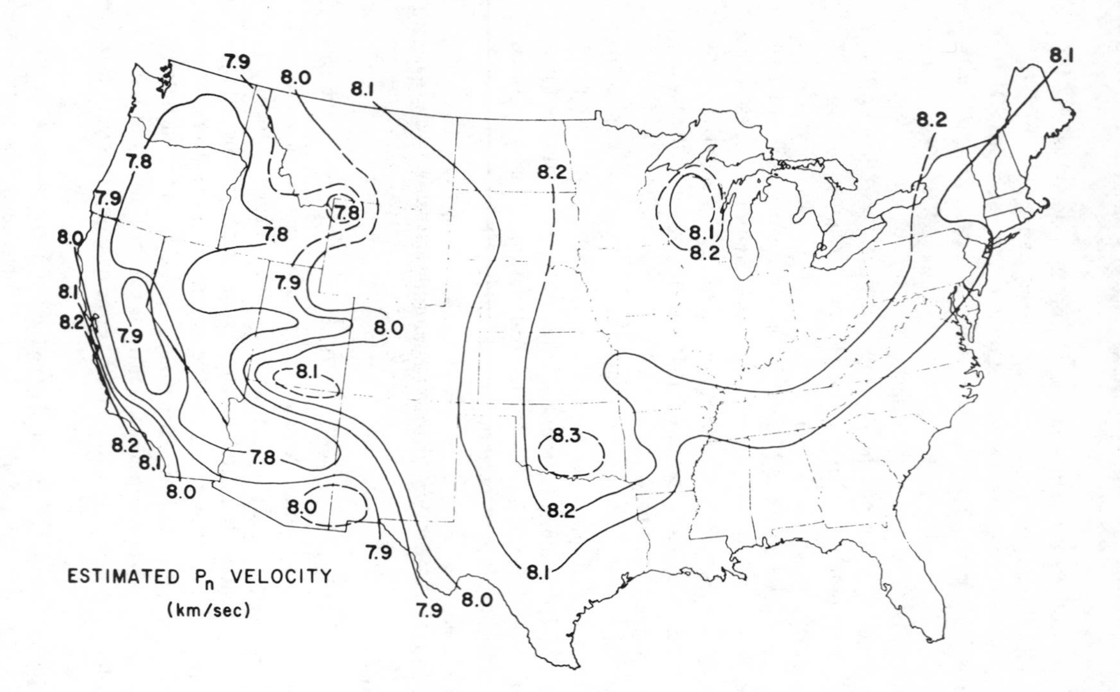

Fig. 2. The regional variation of P_n velocities from Herrin (1969). Based on Herrin and Taggart (1962).

reasonably uniformly distributed in azimuth with respect to the stations. (The restriction to North American stations was adopted in order to minimize the effect of azimuthal variation of the source effect.) They separated the deviations from the J.B. tables into three components: (1) one associated with the source, (2) one with the station, and (3) one representing a correction to the J.B. tables. The deviations from the J.B. tables were smoothed using two polynomials.

All these studies showed earlier arrivals than the J.B. tables centered around 60° arc distance. They differed systematically, however, between 30 and 90°, the Cleary and Hales times, for example, being later than those of Herrin and Taggart by about 1.5 sec at 30° (see Fig. 7 of Herrin et al., 1968). Herrin and Taggart (1968a) suggested that bias might have arisen in the Cleary and Hales analysis as a result of the neglect by these authors of the azimuthal component of the station correction combined with an irregular distribution of events to the east of North America. Hales, Cleary, and Roberts (1968) analyzed the Herrin and Taggart data for North American stations using the Cleary-Hales method. They found deviations from the J.B. tables like those of Herrin and Taggart for a set which, similarly to that of Herrin and Taggart, was dominated by western events and deviations like those of Cleary and Hales for a set reasonably uniformly distributed in azimuth with respect to North America. Cleary and his colleagues (Cleary, 1967a; Cleary and Hales, 1968; Cleary and Muirhead, 1969; Muirhead and Cleary, 1969) have analyzed nuclear explosion times for the Marshall Island sources, the Sahara sources, a Nevada source, and Longshot. In all cases they corrected for the Herrin and Taggart station residuals and found deviations from the Herrin and Taggart times similar to those of Cleary and Hales from the same tables. The differences are illustrated in Fig. 3. It is possible that the explanation for the differences lies in the effect on the travel times of the structure along the Alaskan-Aleutian arc. Studies of the travel times from Longshot (Cleary, 1967a; Herrin and Taggart, 1968b) show considerable anomalies. These anomalies have been explained by Herrin and Sorrells (1969) and Sorrells, Crowley, and Veith (1971) by three-dimensional ray tracing through a structure based on the down thrust plate concept of Isacks, Oliver, and Sykes (1968).

Similar regional variations of the *P* travel times have been found by Kondorskaya (1957) for stations in the USSR. The practice of the USSR observational network appears to be to use special travel time tables for each region.

Since anomalies occur both at the source and the station, there is a problem of baseline determination associated with any modified time term method such as that of Cleary and Hales. In their first paper, Cleary and Hales determined the baseline using travel times from the nuclear explosions at Bikini and Eniwetok. Thus, their times are uncertain to the extent of the source anomaly for those islands. Hales, Cleary, and Roberts (1968) preferred to use travel times from a Nevada explosion and assumed that the source anomaly was the same as the station anomaly for that region. These travel times are keyed to a zero station

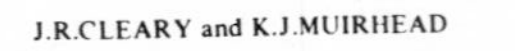

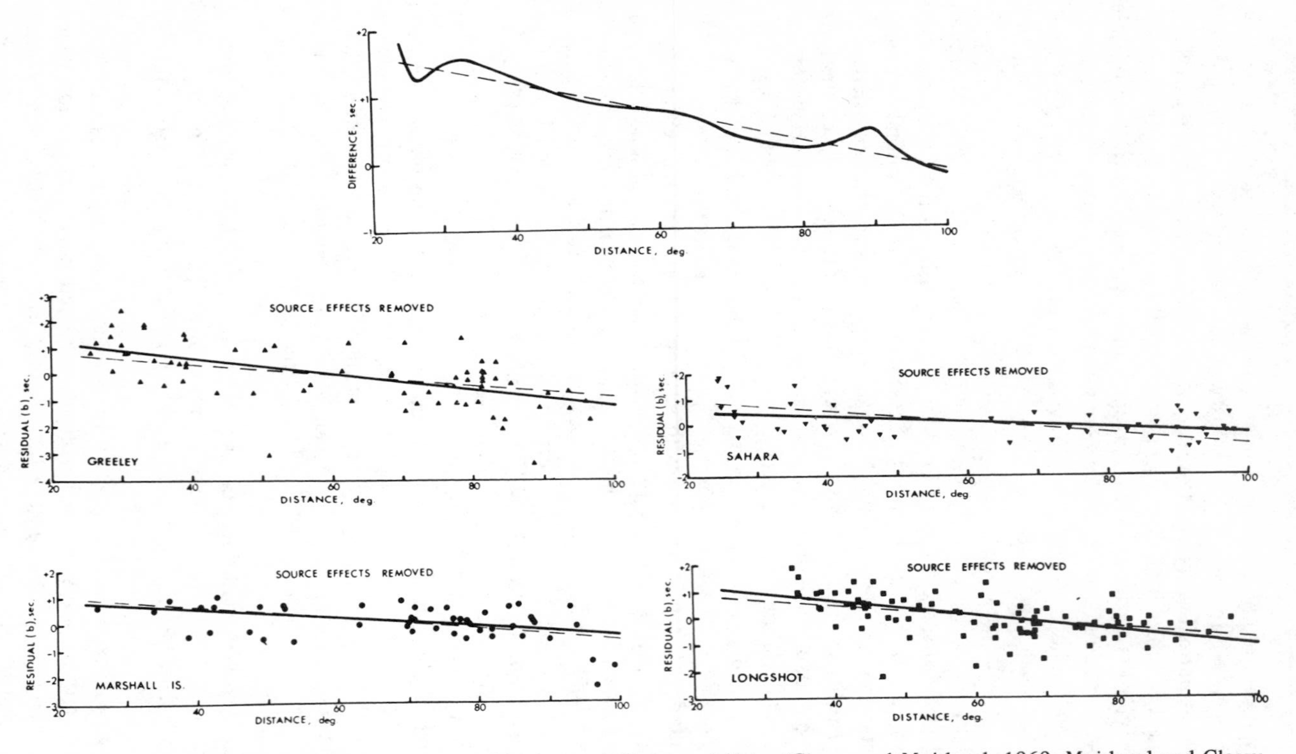

Fig. 3. Comparison of the 1968 travel times with those of Cleary and Hales (Cleary and Muirhead, 1969; Muirhead and Cleary, 1969). The top figure shows the difference between the Cleary and Hales times and the 1968 tables. The four lower figures show the analysis by Cleary and his colleagues of nuclear explosion travel times.

anomaly in the Great Plains region of the United States and are such that *P* station anomalies world-wide range from about + 1 to − 1 sec. (The Herrin and Taggart station anomalies are offset by about 0.3 sec relative to those of Cleary and Hales, as shown in Fig. 4.) A case could be made for keying the standard travel time to zero station anomalies in the shield areas where arrivals are earliest. It was thought that the advantage of having the station anomalies numerically small was significant enough to warrant the choice made by Hales, Cleary, and Roberts (1968).

S Wave Travel Times

A study of *S* travel times by Jeffreys (1966) covered distances up to 95°, whereas Ibrahim and Nuttli (1967) restricted their study to distances less than 65°. Doyle and Hales (1967) determined *S* travel times from 30 to 82° using the procedure applied by Cleary and Hales (1966) for *P* waves. The baseline was determined from the travel times of *S* waves from nuclear explosions reported

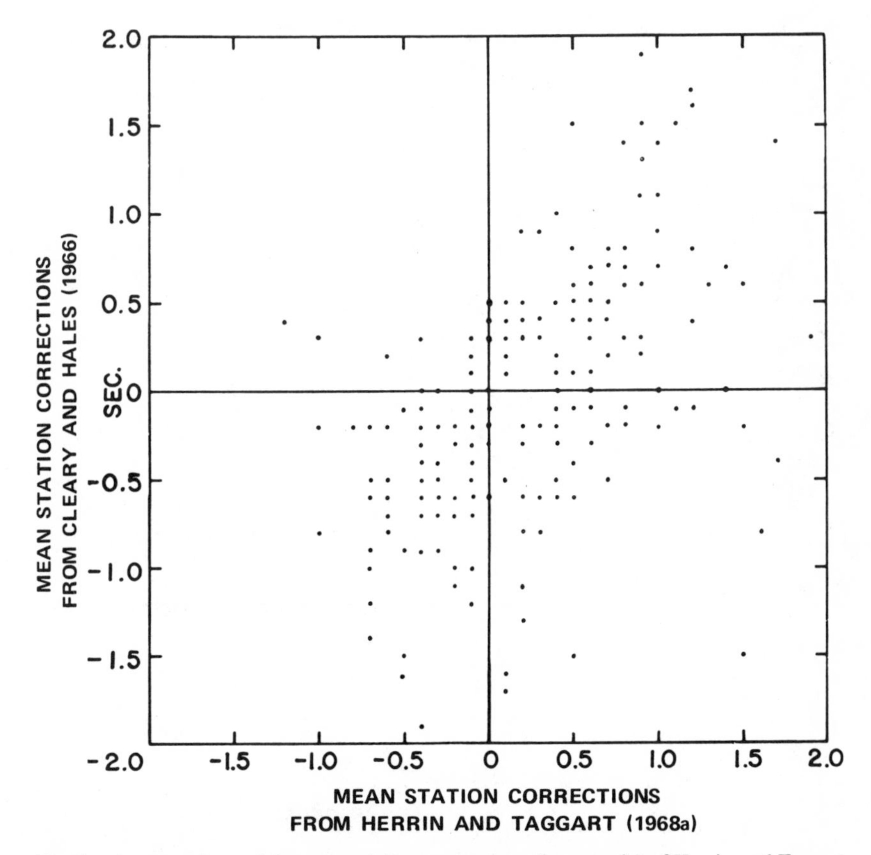

Fig. 4. A comparison of the mean station corrections (in seconds) of Herrin and Taggart (1968) and Cleary and Hales (1966). From Herrin and Taggart (1968a).

by Kogan (1960). Later Hales and Roberts (1970a) extended the Doyle and Hales study of S to arc distances from 30 to 100° and SKS from 83 to 126°. In this case, the baseline was determined from Nuttli's observations (1969) of S travel times from Nevada nuclear explosions.

Cleary, Porra, and Read (1967) reported travel times of diffracted S (or diffracted ScS in Bolt's notation) from 99 to 130°. The apparent velocity was constant and equal to 1/8.9 deg/sec corresponding to a velocity of 6.81 km/sec round the core boundary, assuming the core radius to be 3473 km. This apparent velocity for diffracted S was confirmed by Hales and Roberts (1970) and Bolt et al. (1970).

The travel times found are compared in Fig. 5. The differences from the Jeffreys-Bullen times are small everywhere except between 85 and 100° where they increase to between 4 and 5 sec. Jeffreys (1966) finds small negative corrections to the Jeffreys-Bullen (1940) tables between 30 and 95°.

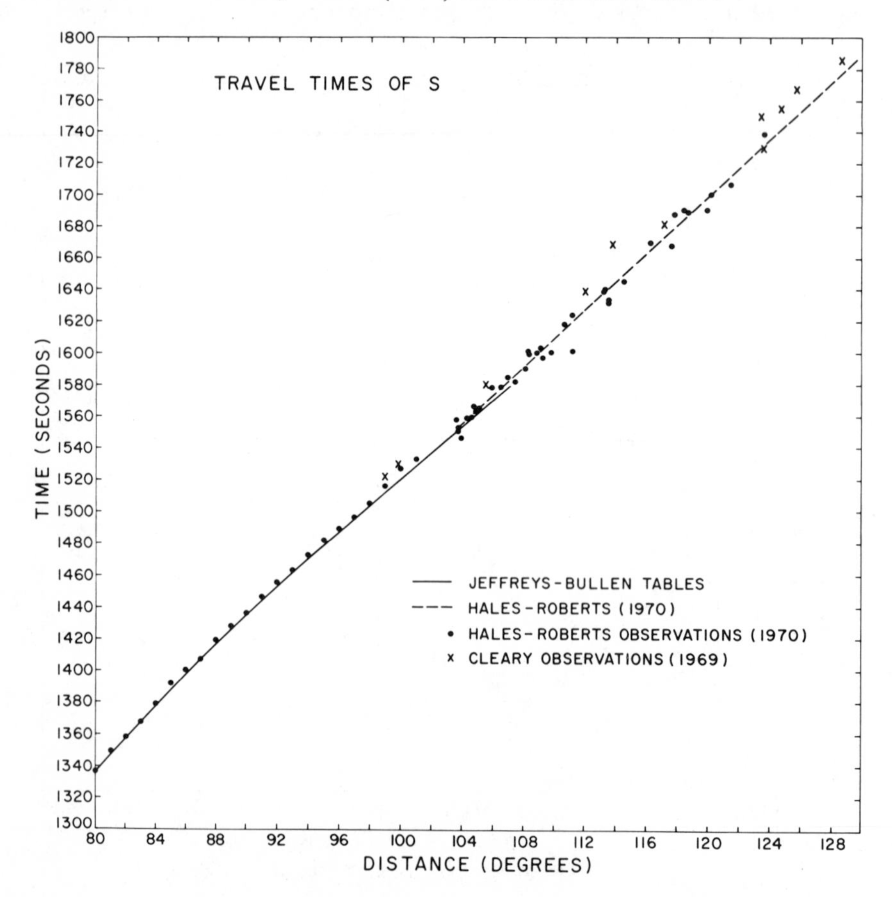

Fig. 5. A comparison of the travel times of S phases from Hales and Roberts (1970) and Cleary (1969) with the Jeffreys-Bullen 1940 table times.

Determinations of $dT/d\Delta$

During the 1960's, a number of seismic array stations were built. Observations at these stations make possible the direct determination of $dT/d\Delta$. Niazi and Anderson (1965) used the records of the Tonto Forest array for the determination of $dT/d\Delta$ for P waves at distances of less than 30°. Johnson (1967) redetermined $dT/d\Delta$ for distances less than 30° from the same array and later (Johnson, 1969) extended the study to the distance range of 30 to 100°. Similar studies using the LASA Montana array were made by Chinnery and Toksöz (1967), Toksöz, Chinnery, and Anderson (1967), Chinnery (1969); and Greenfield and Sheppard (1969). Other determinations of $dT/d\Delta$ have been made by Kanamori (1967) from the records of the Wakayama Micro-Earthquake Observatory, Wright (1970) from the records of the Warramunga array, and Corbishley (1970) from four UKAEA array stations.

The advantage of the use of the array lies in having a fairly large number of stations with accurate relative timing in a limited area. Thus, it is possible to determine $dT/d\Delta$ over a limited range of Δ with considerable accuracy. This is of special importance for the determination of the minor irregularities in the $dT/d\Delta$ curve. Although minor, these irregularities may have considerable significance in questions relating to the physical properties and composition of the lower mantle and in the interpretation of amplitude variations. Chinnery (1969) suggested that it is hard to achieve high resolution in the observation of the absolute times of body waves for the reason that it is extremely difficult to separate the contributions to the travel time of deviations arising in the crust and upper mantle from those resulting from velocity structure deep in the earth. Although often neglected, the same difficulty exists for $dT/d\Delta$ measurements from arrays as has been pointed out by Corbishley (1970). Roughly speaking, P station anomalies have a range of 2 sec. The North American LRSM and WWSSN stations used by Cleary and Hales (1966) for P and Doyle and Hales (1967) for S have linear array dimensions of the order of 3000 or more km. Thus, to achieve the same accuracy as Cleary and Hales from a 300-km array such as LASA requires that the crust-upper mantle contributions should have a range of less than 0.2 sec over the array. For a 20- or 30-km array, these contributions must be less than 0.02 sec for comparable accuracy. Assuming that on the average the Herrin and Taggart (1968a) or Cleary and Hales station corrections were determined with 10% accuracy there is the further requirement that the crust and upper mantle contributions should be known at the arrays to 0.02 sec for a 300-km array or 0.002 sec for a 30-km array.

There is a further complication. The station corrections shown in plots such as that of Fig. 6 are based on a relatively small number of stations and therefore can represent satisfactorily only the longer wavelength anomalies in the upper mantle. Shorter wavelength anomalies undoubtedly exist as is easily seen by comparing vectors showing the azimuthal components (Herrin and Taggart, 1968a, Fig. 3) with normals to the contours of delay times (Fig. 6). Chinnery (1969), in fact, remarked that the travel time anomalies at LASA indicated a

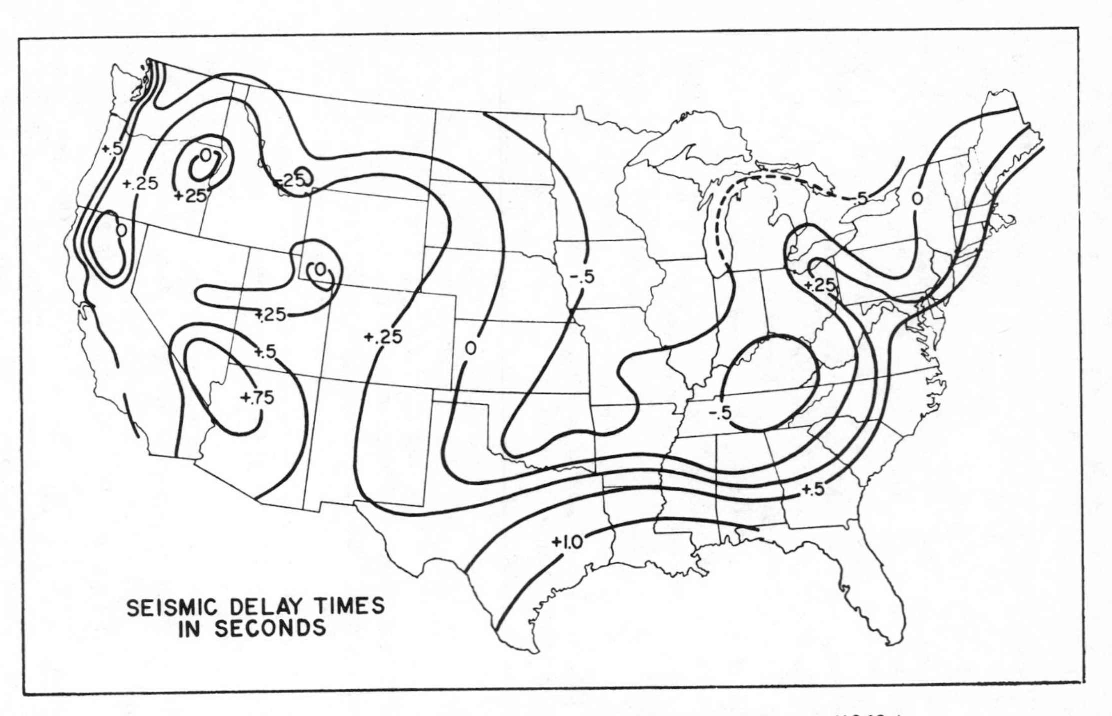

Fig. 6. *P* seismic delay times in seconds for North America. From Herrin and Taggart (1968a).

high degree of complexity varying both as functions of azimuth and distance from the array. Thus, as is clear from Otsuka's analysis (1966), it is as necessary to allow for the effect of crust and upper mantle structure in array determinations of $dT/d\Delta$ as it is in standard travel time studies. The array studies, however, are more satisfactory for resolving the minor perturbations in the $dT/d\Delta$ curve. Hopefully, it will be possible in time to analyze the later arrivals in the upper mantle transition regions in the manner of Johnson (1969) and solve the difficult questions of whether some of the transitions are multiple and how rapid the velocity changes actually are.

In the Hales, Cleary, and Roberts (1968) analysis it was found that P arrival times from 30 to 96° could be fitted to within 0.33 sec by a cubic in Δ as follows:

$$T(\Delta) = 72.77 + 10.9210\Delta - 3.2087 \times 10^{-2}\Delta^2 - 2.003 \times 10^{-5}\Delta^3$$

Raw values of $dT/d\Delta$ over 2° intervals from 30 to 96° are presented in Fig. 7,

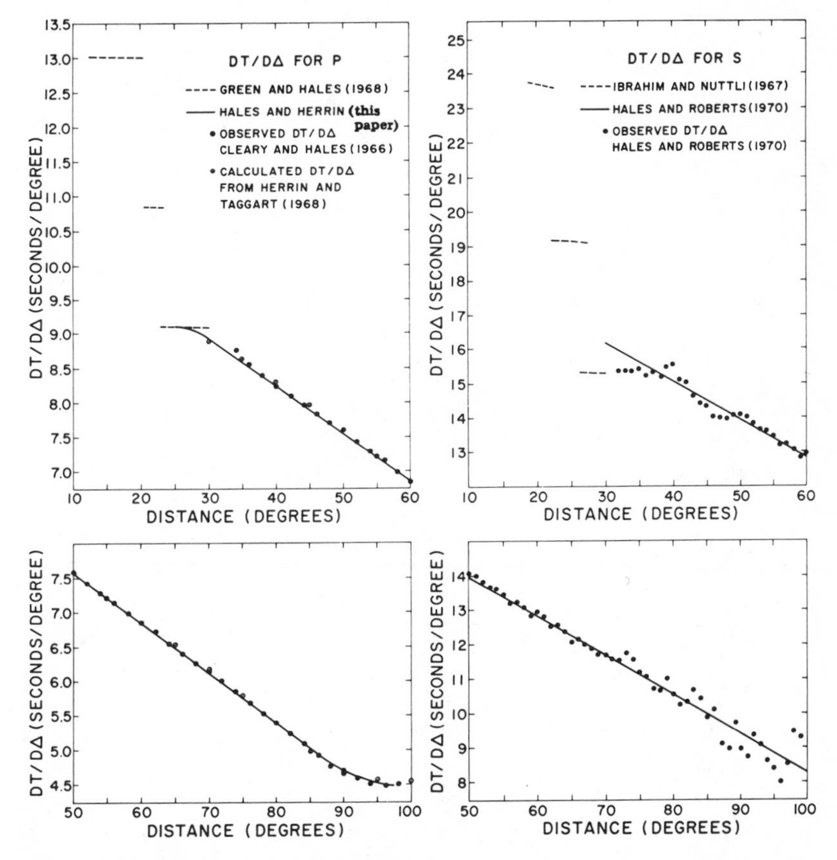

Fig. 7. $dT/d\Delta$ for P and S from 10 to 100°.

(a)

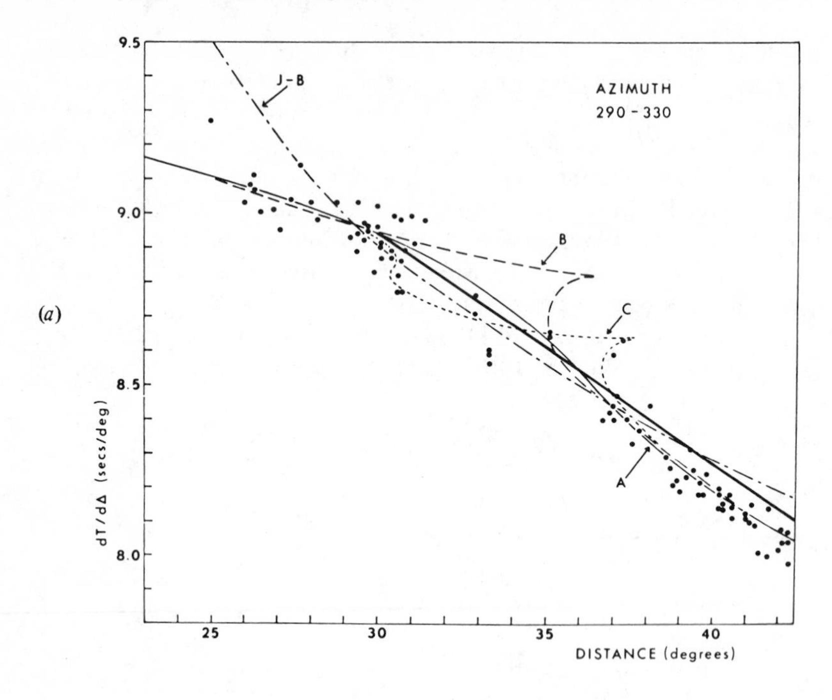

(b)

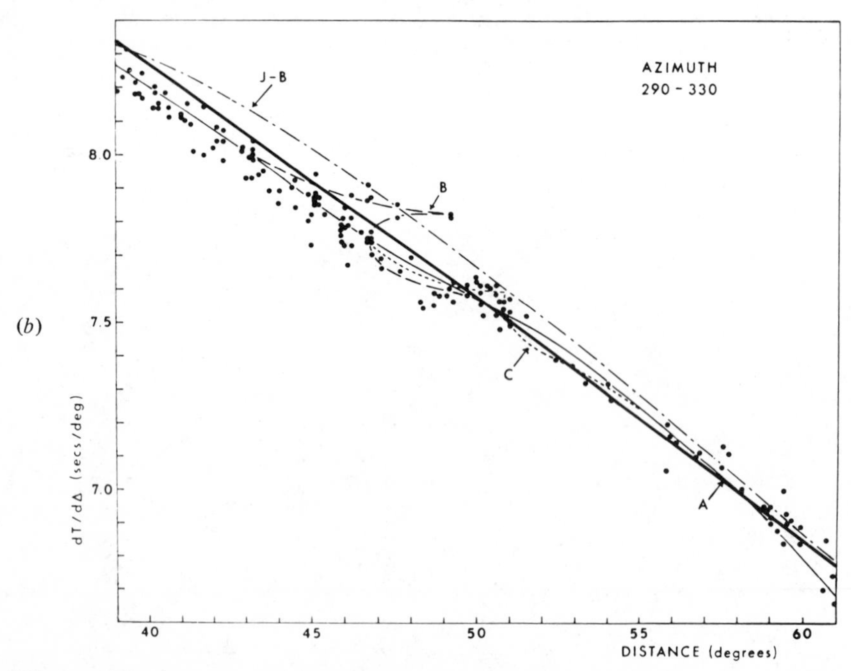

Fig. 8. $dT/d\Delta$ determinations from Chinnery (1969), (a) 9.5 to 8 sec/deg, and (b) 8 to 7 sec/deg.

(*c*)

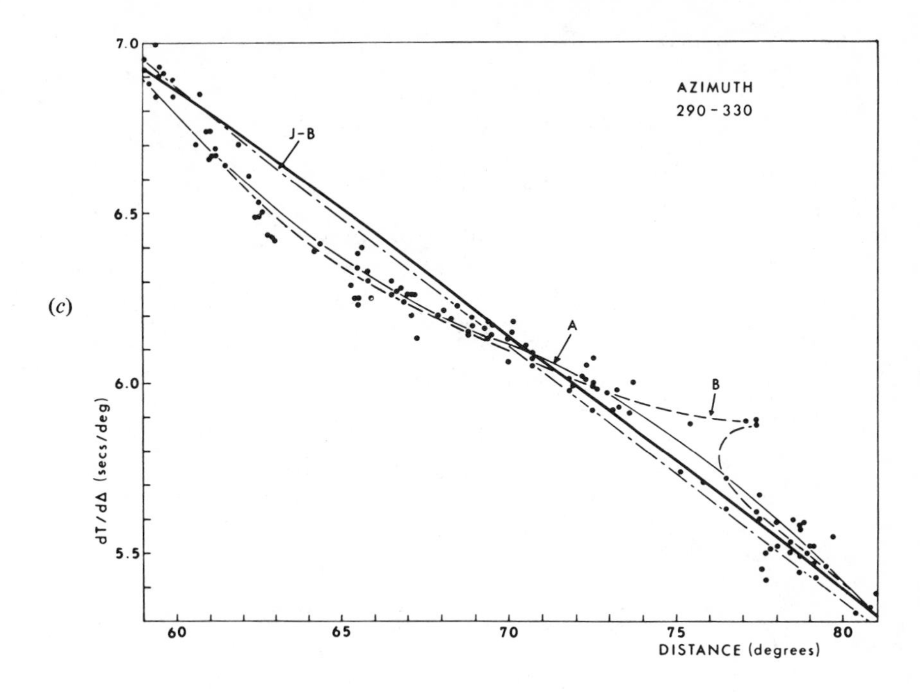

(*d*)

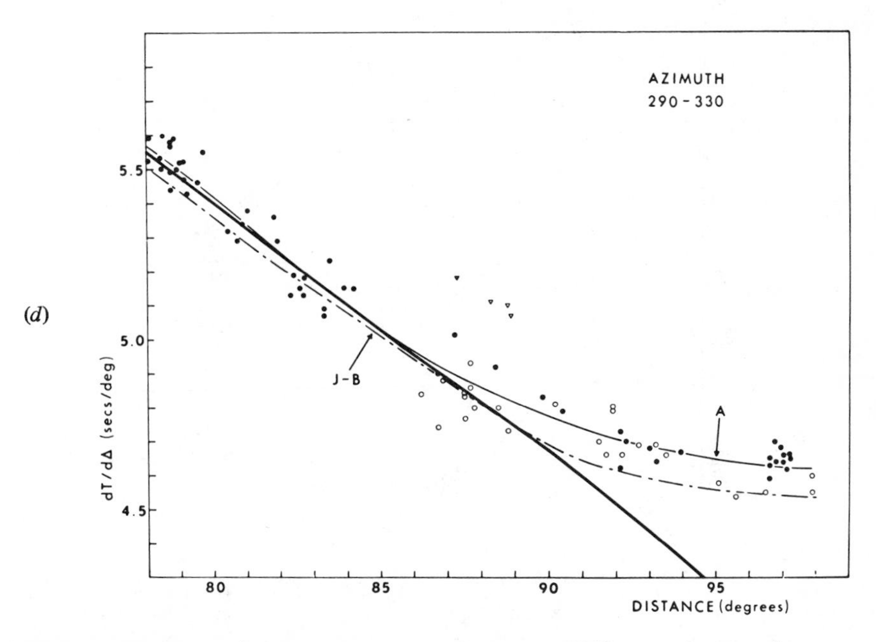

Fig. 8 (continued). $dT/d\Delta$ determinations from Chinnery (1969), (*c*) 7 to 5.5 sec/deg, and (*d*) 5.5 to 4.5 sec/deg.

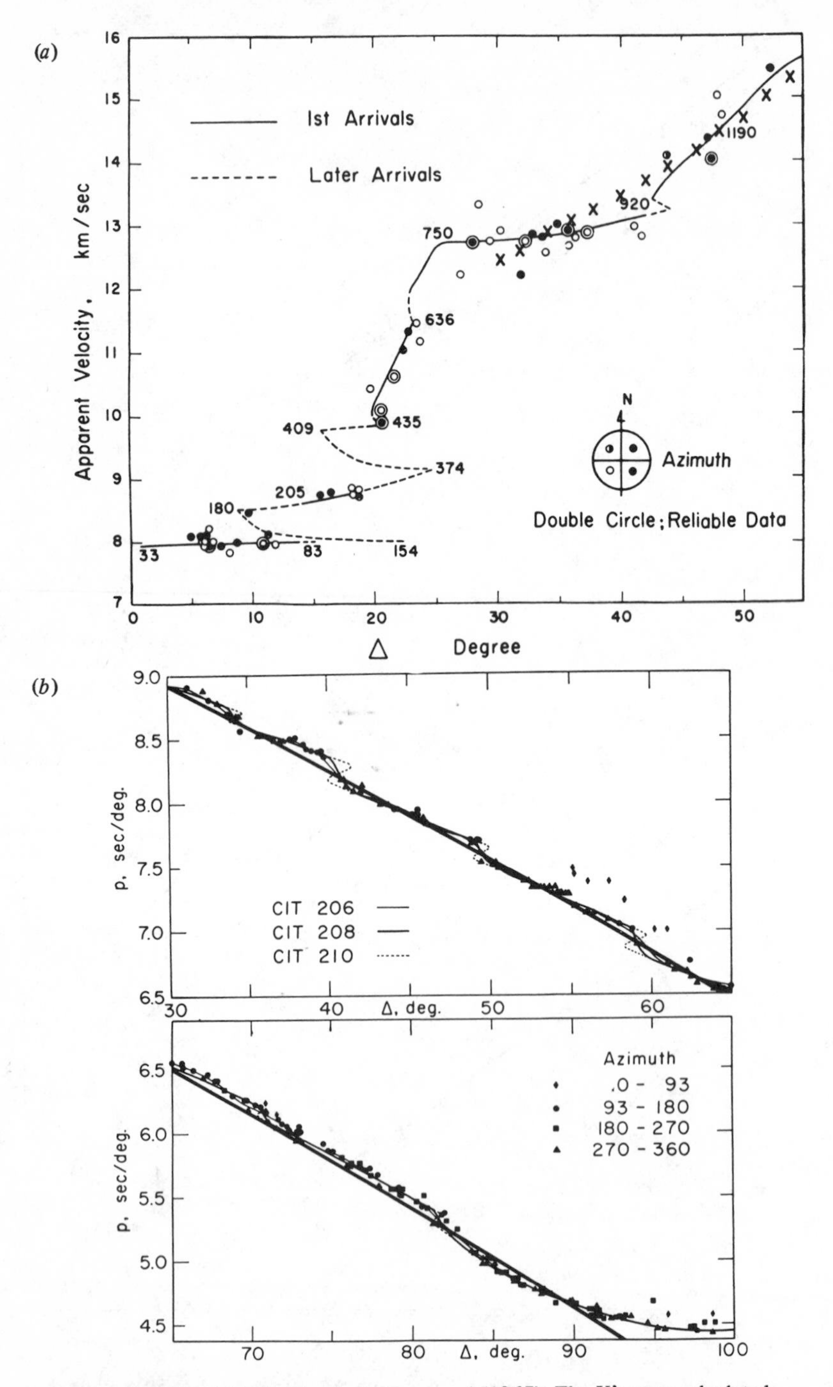

Fig. 9. (*a*) Apparent velocities from Kanamori (1967). The X's were calculated from the cubic. (*b*) $dt/d\Delta$ from Johnson (1969).

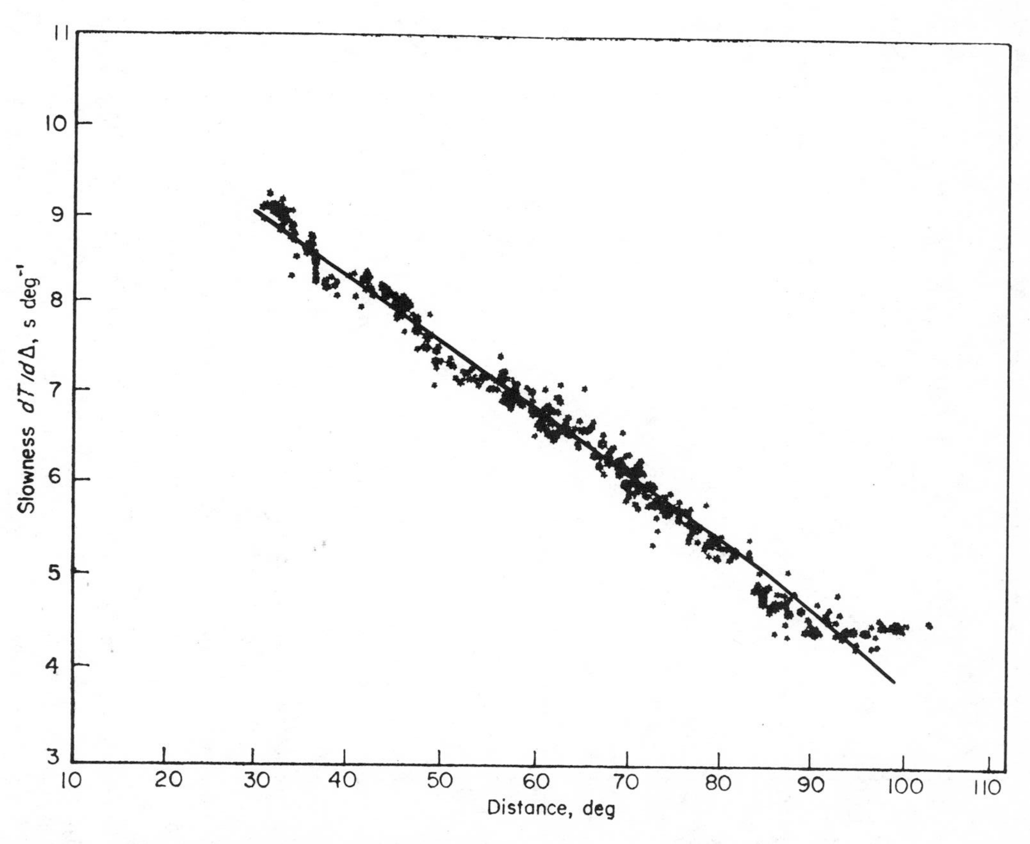

Fig. 10. $dT/d\Delta$ determinations from Corbishley (1970).

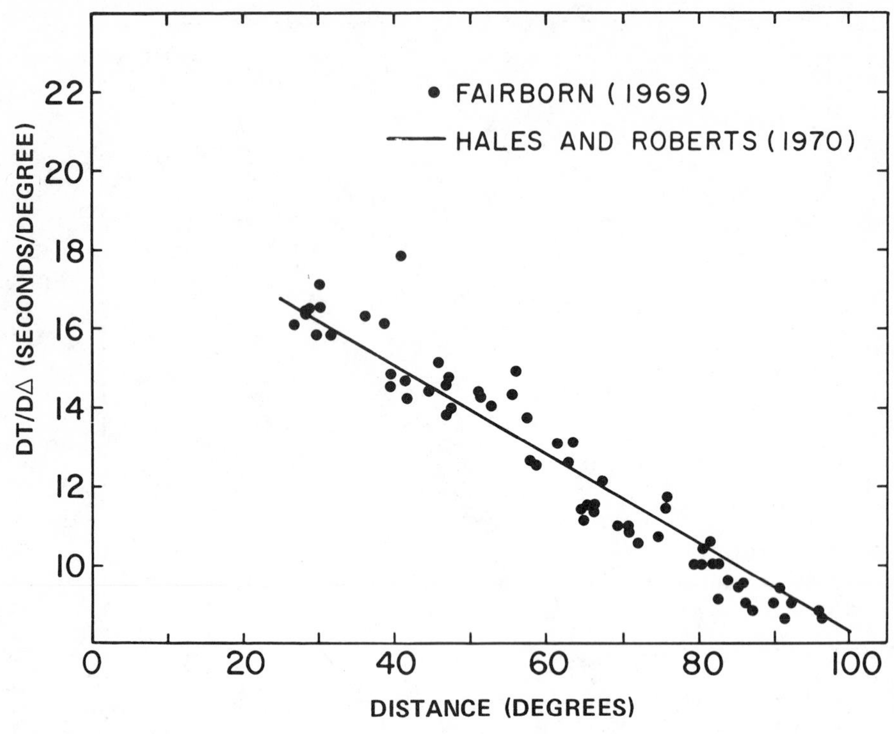

Fig. 11. $(dT/d\Delta)_S$ from Fairborn (1969).

together with calculated $dT/d\Delta$'s from Herrin and Taggart (1968a). Lines showing the values of $dT/d\Delta$ for P from Green and Hales (1968) for the distance ranges to 20, 20 to 24 and 25 to 30° are also shown in Fig. 7. There are two major features in the distribution of $dT/d\Delta$ for both P and S: (1) there are no discontinuous changes of $dT/d\Delta$ beyond 30° arc distance comparable in magnitude with the step changes at 20 and 24° and (2) beyond 30° the slope of $dT/d\Delta$ versus Δ curve, i.e., $d^2T/d\Delta^2$, is sensibly constant and much greater than that of the segments between 20 and 30°.

Comparison of the published values of $dT/d\Delta$ from the arrays is difficult because almost all the data is available only in the form of plots. To facilitate comparison, $dT/d\Delta$ from the cubic equation above has been included on all the plots of $dT/d\Delta$ reproduced as Fig. 8 (from Chinnery, 1969), Fig. 9 (Johnson, 1969; Kanamori, 1967), and Fig. 10 (Corbishley, 1970). Similar information for S based on the travel times of Hales and Roberts (1970a) and $dT/d\Delta$ study of Fairborn (1969) is presented in Figs. 7 and 11.

Comparison of the observations of $(dT/d\Delta)_P$ shown in Figs. 8, 9, and 10 with the reference curve on the figures, shows that the various $(dT/d\Delta)_P$'s have a number of features in common. All show a decrease in the slope of the $dT/d\Delta$ curve between 85 to 90 and 98°. This implies that the P velocity increases less rapidly in the 500 km above the core-mantle boundary than the values found by Hales, Cleary, and Roberts (1968) by inversion of the travel times represented by the cubic equation above. Most show changes in character at arc distances of 80, 59, 48 to 50, 40 and about 34°. Some of the differences appear to depend on the density of the observations. The interpretations in terms of smooth curves vary considerably. The Johnson interpretation (1969) involves changes such as those in Fig. 12(*a*) corresponding to small rapid or discontinuous increases in velocity. Chinnery and Toksöz (1967); Toksöz et al. (1967), Chinnery (1969), and Wright (1970) show breaks such as those shown in Fig. 12(*b*), which correspond to smoothing of the $dT/d\Delta$'s for thin low-velocity zones.

Values of $d^2T/d\Delta^2$ derived from the $dT/d\Delta$ curves for the arrays do not show as good agreement with the amplitude-distance curve as might be expected (Corbishley, 1970). Amplitudes should be proportional to $|d^2T/d\Delta^2|$. The experimental data (Carpenter et al., 1967 or Cleary, 1967b) do not vary as much as $|d^2T/d\Delta^2|$ (see Fig. 7 of Corbishley, 1970). The notable exception is that $d^2T/d\Delta^2$ tends to zero between 95 and 100°, which agrees with the reduced amplitudes noted by Sacks (1967) at 96°. The magnitude calibration functions Q, Gutenberg and Richter (1956) and β, Vanek and Stelzner (1960) and Vanek et

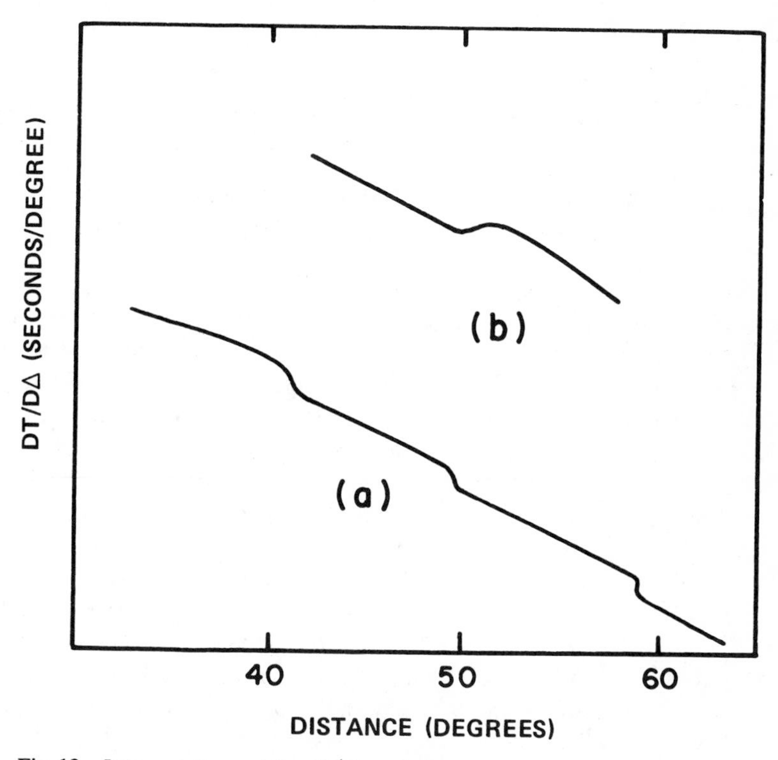

Fig. 12. Interpretations of the $dT/d\Delta$ variations.

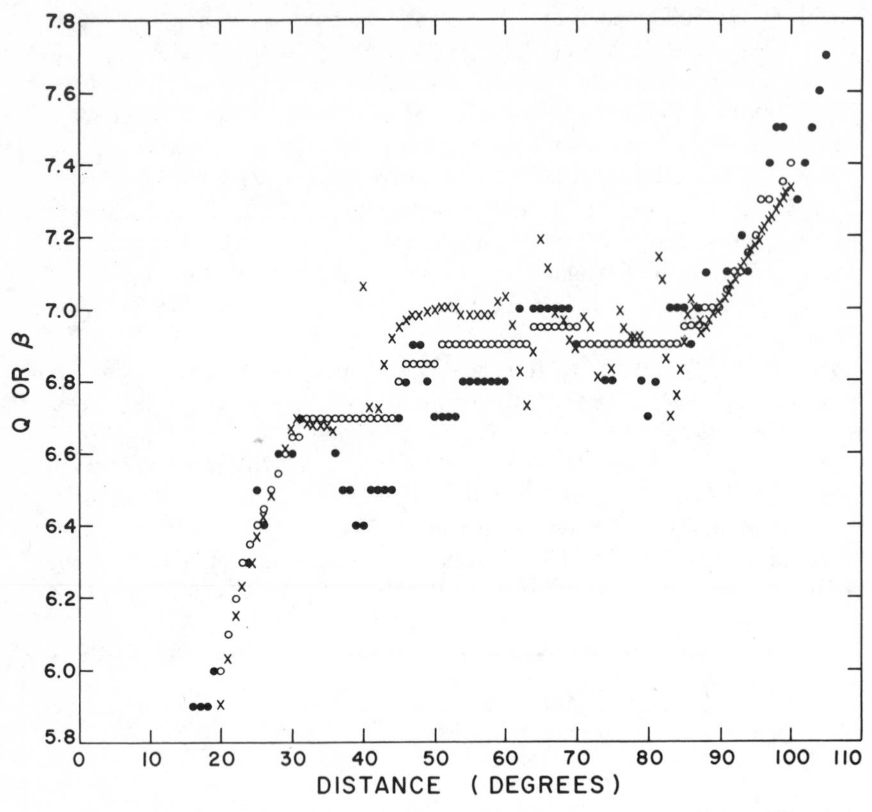

Fig. 13. The functions Q of Gutenberg and Richter, 1956 (solid circles) and β of Vanek et al., 1962 (open circles), Vanek and Stelzner, 1960 (crosses).

al. (1962) increase systematically from 85 to 100° as shown in Fig. 13. These calibrating functions are effectively correction functions for the variation of log A/T with distance. Increases in the functions correspond therefore to decreases in the observed amplitude with distance. Thus, these curves provide independent evidence of a decrease in $|d^2T/d\Delta^2|$ from 85 to 100°. Additional support for the decrease in amplitude between 85 and 100° is provided by the analysis of Willey, Cleary, and Marshall (1970) and for S by the data of Bolt et al. (1970).

In general, the experimentally determined $dT/d\Delta$ curves at distances of less than about 30° show less slope than the average slope from 30 to 85°. This is true also of the slopes determined from the travel time studies for $\Delta < 30°$. It is not clear how the slopes blend into one another between 30 and 35°. Kanamori's interpretation (1967) is that the slope of the $dT/d\Delta$ curve increases rapidly beyond 30° and blends smoothly into the average slope at about 40°.

The interpretation of the relatively flat slopes of the $dT/d\Delta$ versus Δ segments at distances less than 30° is not clear. It is possible that it is a reflection of higher temperature gradients in the upper mantle above 700 km as compared with those below 700 km. Further study of the behavior of travel times and $dT/d\Delta$ between 20 and 32° is desirable for the differences in travel time involved are relatively small, less than 0.5 sec, and it is not certain that the effects of source and station anomalies have been entirely eliminated at these distances.

The foregoing remarks apply to the shape of the observed $dT/d\Delta$ curves. In so far as absolute values are concerned, it is obvious that the array $dT/d\Delta$'s scatter about the values calculated from the cubic in T so much that it is unlikely that the velocities calculated from the cubic will be much in error except at the base of the mantle. It has been suggested by Toksöz et al. (1967) that there are differences between the values of $dT/d\Delta$ at LASA for different azimuths especially for distances between 65 and 83°. Their estimates of the slope of the $dT/d\Delta$ curve are – 0.065 sec/deg and – 0.093 sec/deg for the azimuths 300 to 320° and 140 to 160°, respectively. (Note that the decimal point is misplaced in Toksöz et al., 1967; the misplacement has been corrected here.) The difference in slopes quoted would amount to relative differences in travel time of 5 sec at some point of the distance range. Such differences in travel time are improbable. It is probable that the variation in $dT/d\Delta$ for different azimuths is a result of the complexity of structure at LASA mentioned earlier coupled perhaps with effects arising at the source of western events. Johnson's analysis (1969) of the Tonto Forest observations does not show a dependence of $dT/d\Delta$ on azimuth, but does show discrepant values for some mid-Atlantic ridge events.

The Velocities

The P velocities found by Hales, Cleary, and Roberts are compared with other determinations of the P velocity in the lower mantle, namely those of Jeffreys, Gutenberg (Press, 1966), Herrin (1968) and Johnson (1969), in Fig. 14.

It has been noted that the $dT/d\Delta$ studies are consistent in showing a decrease in $|d^2T/d\Delta^2|$ from about 85° onward, and their decrease is consistent with what is known of the observed amplitudes. We have redetermined the velocities for the lower mantle to correspond to a model in which (1) $dT/d\Delta$ changes slope smoothly between 27 and 33° as suggested by Kanamori; (2) between 34 and 85°, $dT/d\Delta$ is fitted by a quadratic in Δ; and (3) beyond 85°, $dT/d\Delta = (dT/d\Delta)_{85°} + k(\Delta - 85°) + l(\Delta - 85°)^2 + m(\Delta - 85°)^3$, where k, l, and m were determined so that $(d^2T/d\Delta^2)$ is continuous at 85°, $(d^2T/d\Delta^2)_{97°} = 0$, and $(dT/d\Delta)_{97°} = 45$ sec/deg. The values of k, l, and m were 0.07547, 0.001693, and 0.00008065, respectively. The corresponding values of $dT/d\Delta$ are shown by the solid line in Fig. 7. The travel time curve so modified has been inverted using the Herglotz-Wiechert procedure.

The revised velocities are shown in Figs. 14 and 15 and are tabulated in Table 1. These velocities are to be preferred to those of Hales, Cleary, and Roberts (1968). Figure 15 shows also the relation between bottoming depth and distance. On this solution, the velocity reaches a maximum value at a depth of

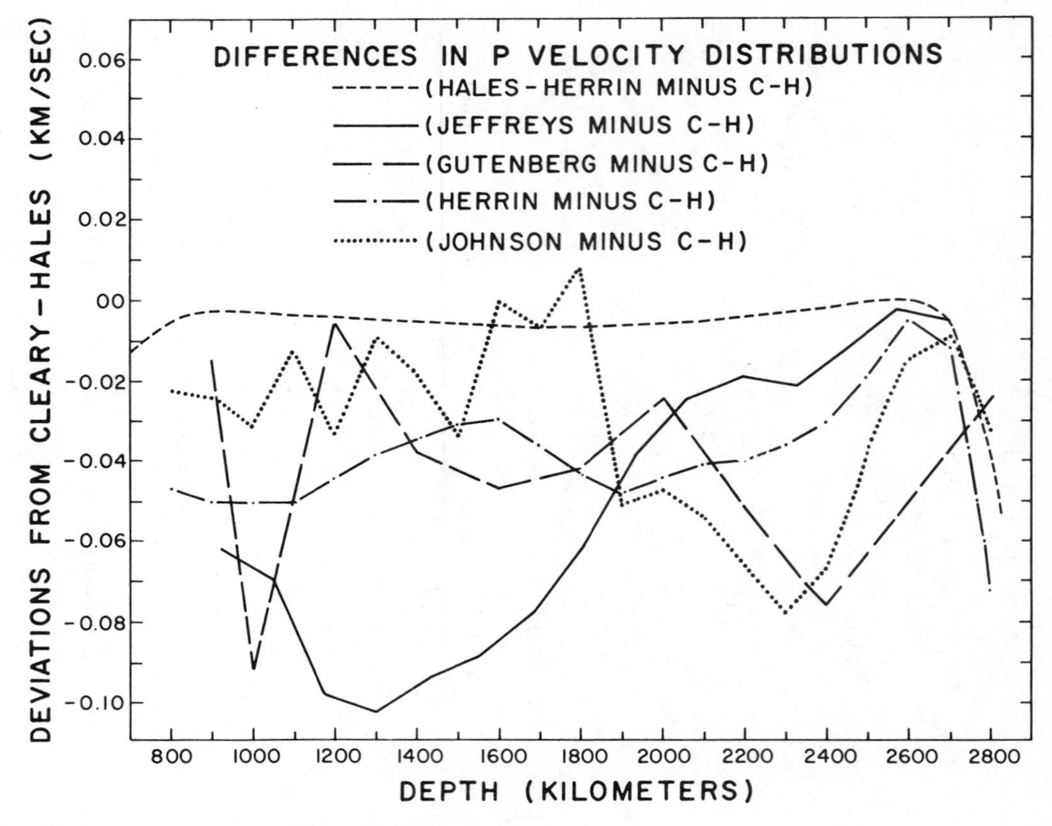

Fig. 14. Comparison of P velocity distributions.

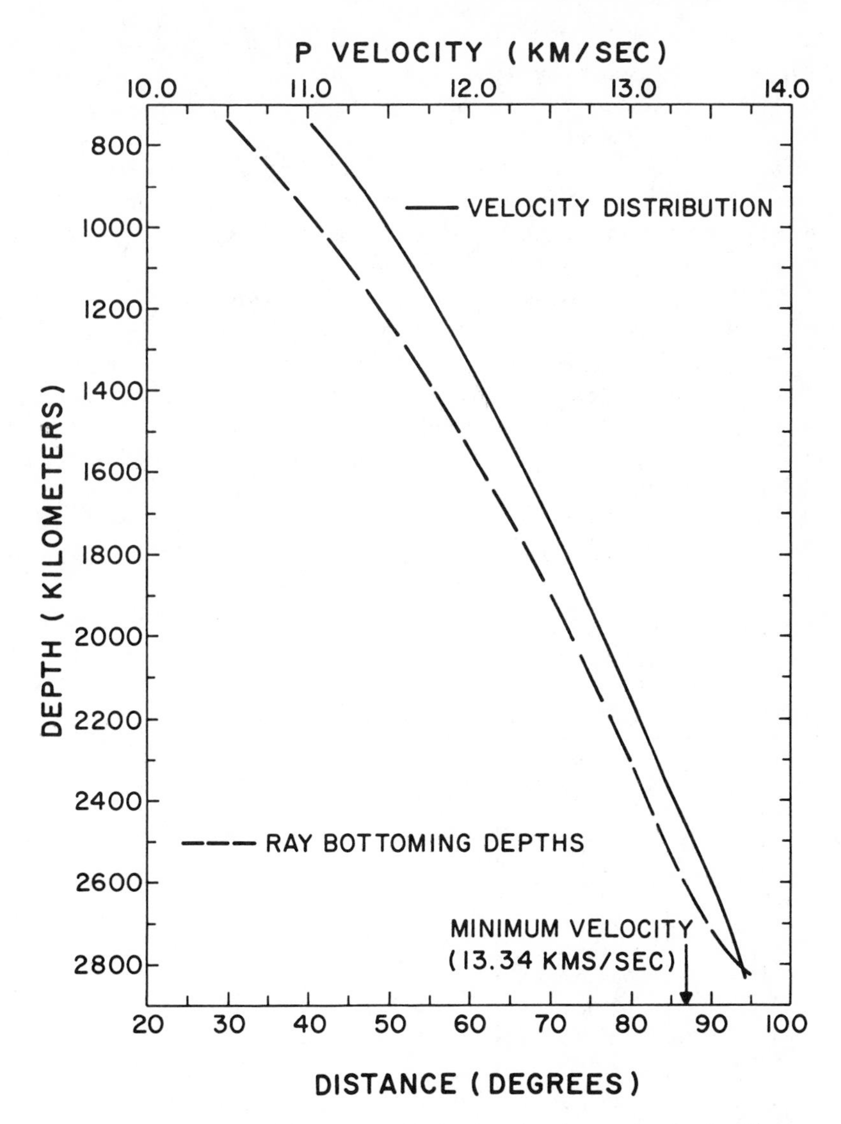

Fig. 15. The variation of *P* velocity with depth and of bottoming depth with arc distance.

2825 km, and the velocities below that depth are not determined by the inversion.

The *S* velocities found by Hales and Roberts are shown in Fig. 16 and compared with those of Jeffreys, Gutenberg (Press, 1966), Fairborn (1969), and a second model of Hales and Roberts (1970b) in Fig. 17. The travel times used in the Hales and Roberts analysis were expressed as a quadratic in *T*. There is again a suggestion in the $dT/d\Delta$ plots in Fig. 7 and 11 that $(d^2T/d\Delta^2)_S$ decreases beyond some distance between 85 and 100°. Decreasing $(d^2T/d\Delta^2)_S$ in this distance range is supported by the amplitude studies of Gutenberg and Richter (1956) and Bolt et al. (1970).

The lower mantle velocities found in all recent studies differ very little from those of Jeffreys or Gutenberg. Some minor modifications of the velocity may be expected when agreement has been reached on the shape of the $dT/d\Delta$ curve in the "peculiar" regions at 40, 50, 59, and 80°.

The behavior of the velocities near the core-mantle boundary requires further examination. This is especially true of the shear velocities. MacDonald and Ness (1961) pointed out that there was a discrepancy between the calculated and observed periods of the free oscillations and suggested a modification of the shear velocity distribution that would remove the discrepancy. It is clear now that this modification was more drastic than would be permitted by the travel times. Other modifications were proposed by Landisman et al. (1965), Dorman, Ewing, and Alsop (1965), Bullen and Haddon (1967), Cleary (1969), and Press (1968). These modifications include changes in the radius of the core, the shear

Table 1. Tabulation of revised velocities with depth in the lower mantle

Radius (km)	Depth (km)	Velocity (km/sec)
5671.2	700.0	10.9297
5471.2	900.0	11.3116
5271.2	1100.0	11.6505
5071.2	1300.0	11.9457
4871.2	1500.0	12.2133
4671.2	1700.0	12.4633
4471.2	1900.0	12.7019
4271.2	2100.0	12.9332
4071.2	2300.0	13.1605
3871.2	2500.0	13.3867
3851.2	2520.0	13.4093
3831.2	2540.0	13.4319
3811.2	2560.0	13.4544
3791.2	2580.0	13.4764
3771.2	2600.0	13.4984
3751.2	2620.0	13.5202
3731.2	2640.0	13.5415
3711.2	2660.0	13.5624
3691.2	2680.0	13.5827
3671.2	2700.0	13.6024
3651.2	2720.0	13.6213
3631.2	2740.0	13.6319
3611.2	2760.0	13.6558
3591.2	2780.0	13.6709
3571.2	2800.0	13.6838

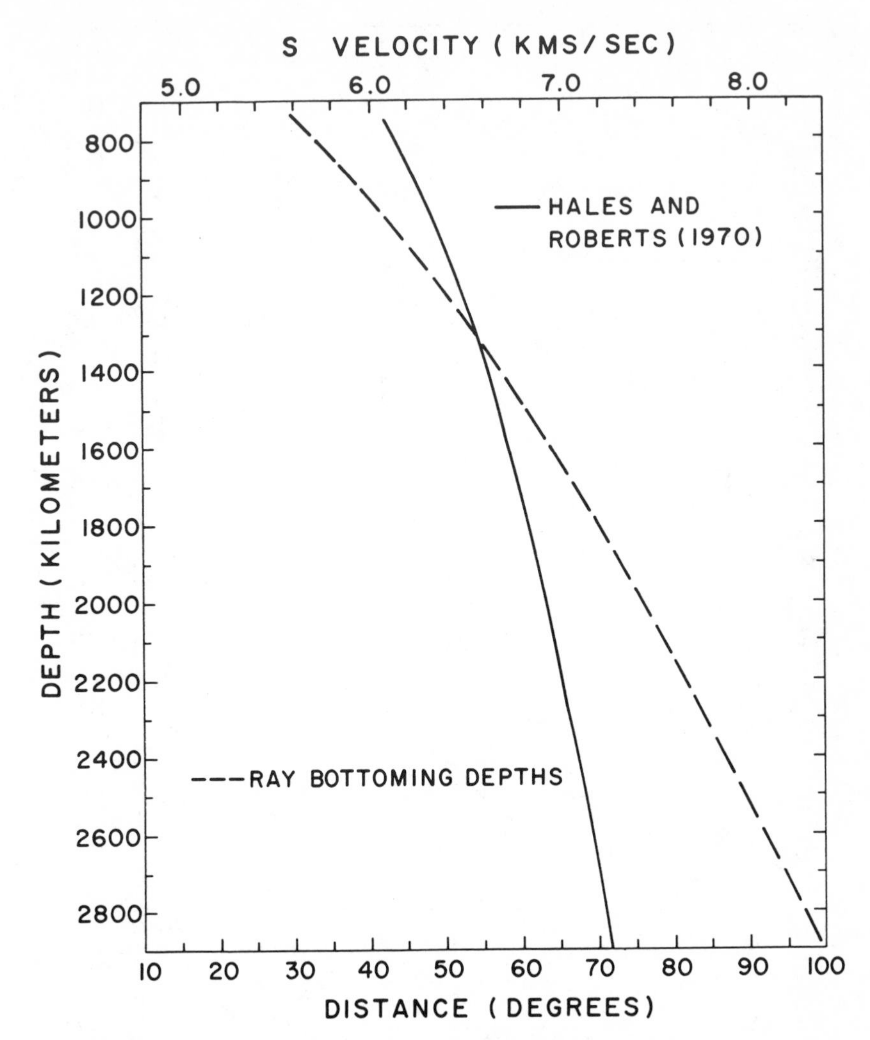

Fig. 16. The variation of *S* velocity with depth and of bottoming depth with arc distance.

velocity distribution, and the density distribution. Hopefully, it will be possible to use body wave information to set narrower limits on the radius of the core and on the velocity distribution near the core-mantle boundary so that the free oscillation information can then be used to control the estimates of the density distribution.

Observations of short period *PcP* such as those of Taggart and Engdahl (1968) provide the best current estimate of the core radius as 3477 km. The Hales and Roberts (1971) estimates of 3489.9 and 3486.1 km based on *ScS* observations are greater, but do not differ significantly from those of Taggart and Engdahl at the

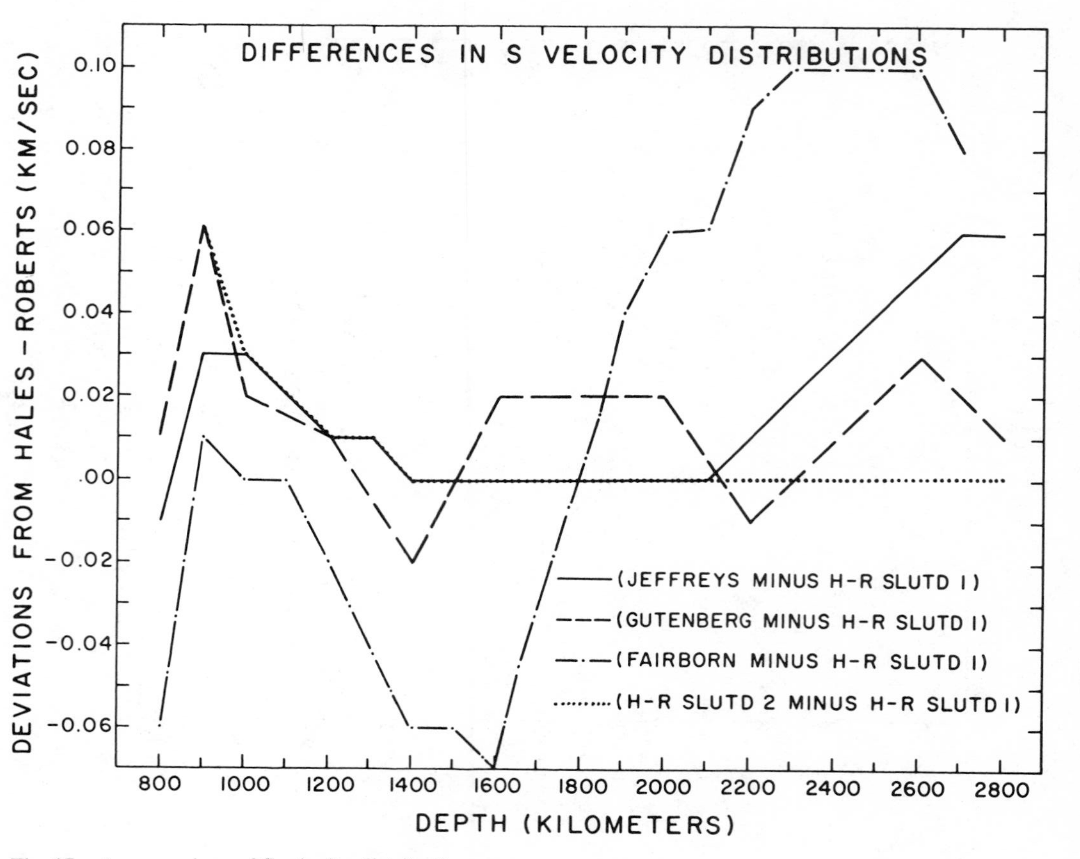

Fig. 17. A comparison of *S* velocity distributions.

99% confidence level. Followill and Nuttli's estimate (1970) agrees with that of Taggart and Engdahl to within 2 km.

The principal uncertainty concerns the possible existence of a low-velocity zone near the base of the mantle. Observations of the *P* phase diffracted around the core boundary give velocities of 13.62 km/sec (Sacks, 1967). Similarly, observations of the diffracted *S* phase [Cleary, Porra, and Read (1967), Cleary (1969), Hales and Roberts (1970a), and Bolt et al. (1970)] result in a velocity of 6.8 to 6.9 km/sec. Both diffracted phase velocities are lower than the maximum velocities found by inversion of body wave travel times. Intuitively, it would be expected that the velocity of a long period phase of period 15 to 30 sec would be related to the average velocity for a shell of thickness 15 to 30 times the velocity (some 200 to 400 km for *P* and 120 to 240 km for *S*). For the short period *P* phase, the shell would be only 10 to 30 km thick and might therefore have a higher velocity. Sacks (1966, 1967) and Alexander and Phinney (1966) show, however, that the diffracted *P* phase is not dispersed. Theoretical studies by Phinney and Alexander (1966) and Phinney and Cathles (1969) suggest that the diffracted phases are nondispersive; however, Richards (1970) suggests that the velocities should vary with period.

It is possible that the phases interpreted as diffracted *P* should be interpreted as direct phases through a layer of decreasing velocity between a depth of 2800 km and the core. In that case, there is no shadow zone, so that $\eta = r/v$ continues to decrease even though v decreases. Subject to this condition, the minimum value of v at the core boundary is 13.34 km/sec, taking r_c=3477 km after Taggart and Engdahl (1968). Bolt (1970) has calculated model velocities for this zone using a power law for the variation of the velocities. The velocities at the core-mantle boundary for two of the Bolt models are 13.20 and 13.34 km/sec. Ergin (1967) proposed a low-velocity region at the base of the mantle in order that the PKP_2 phase should extend to at least 180° as required by the observations. Ergin's velocity at the core-mantle boundary was 13.37 km/sec.

THE STATION ANOMALIES OR DELAY TIMES

P Station Anomalies

There have been a number of studies of the regional variation of travel times in recent years (Bolt and Nuttli, 1966; Carder et al., 1966; Cleary and Hales, 1963, 1966; Douglas and Lilwall, 1970; Herrin and Taggart, 1968a; Husebye, 1965; Press and Biehler, 1964). The residuals found by the different authors using different material all correlate well as is shown by Fig. 4 reproduced from Herrin and Taggart (1968a) and Fig. 18 from Hales et al. (1968). Thus, there is no doubt that the effect is a real one. Douglas and Lilwall have claimed that there is bias introduced into the world-wide station anomalies of Herrin-Taggart and Cleary-Hales by reason of the azimuthal variation of the source anomaly. A study by Cleary and Hales (1970) of *PKIKP* station anomalies suggests that this is not the case.

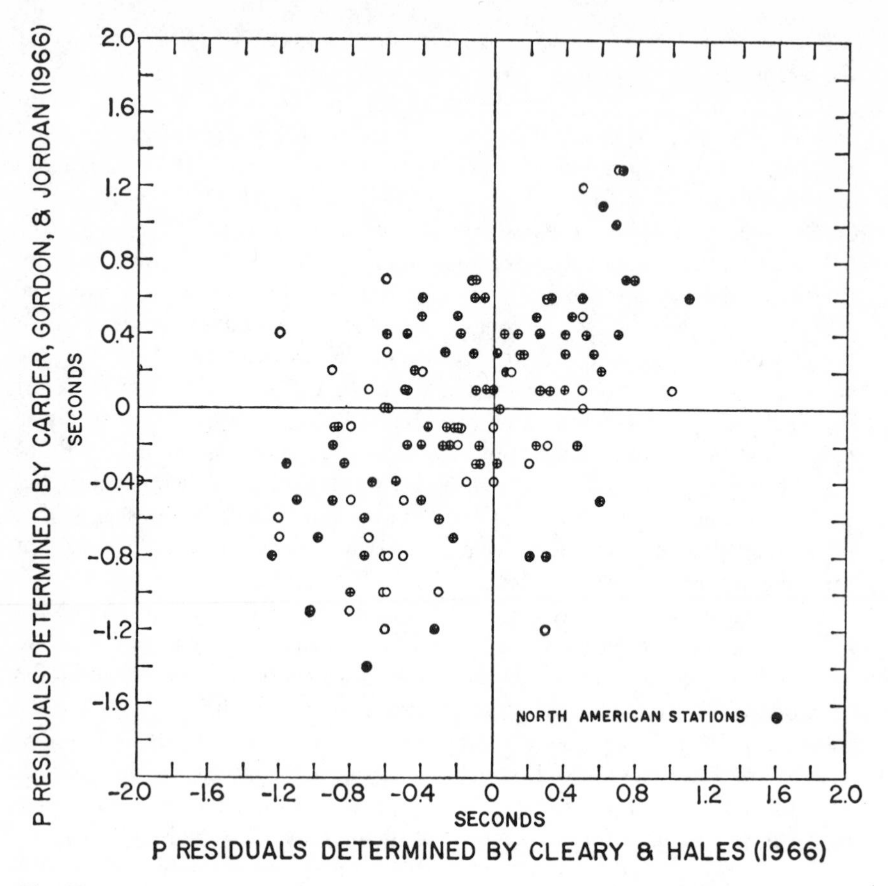

Fig. 18. A comparison of the Cleary and Hales (1966) *P* station anomalies with those of Carder et al. (1966).

Press and Biehler associated variations in *P* travel time to the stations of the Pasadena network with variations in the velocity and thickness of the crustal layers in that region. Nuttli and Bolt (1969) have found it possible to explain *P* station anomalies relative to Berkeley in terms of undulations of the low-velocity layer. Cleary and Hales (1966) pointed out that late *P* arrivals, both in America and elsewhere in the world, were associated with regions of high heat flow that were tectonically active, and they regarded the anomalies as originating in the upper mantle. In North America, for example, as is shown by Fig. 6 reproduced from Herrin and Taggart (1968a), it is clear that *P* arrivals in the western United States are systematically late with respect to those in the central and eastern United States. The range of anomalies shown in this figure is 2 sec. It can be shown by model calculations that differences as large as 2 sec cannot be accounted for in terms of differences in crustal structure. The close similarity between this figure and Herrin and Taggart's figure (1962) showing the variation in apparent P_n velocity in the United States (or the more recent version reproduced as Fig. 2 of this paper) lends additional support to the hypothesis that the regional variations in travel time arise in the upper mantle.

S Station Residuals or Anomalies

Doyle and Hales (1967) determined *S* station anomalies for North American stations and found that these were strongly correlated with the *P* station anomalies of Cleary and Hales. The relation is shown in Fig. 19 from Hales and Roberts (1970a) and Fig. 20. There is a strong similarity between the distribution of station anomalies shown in Fig. 20 and the Rayleigh wave phase velocities for periods of 51 sec found for North America by Pilant (1967), the regions of late arrivals corresponding to regions of low phase velocity. The most important result of the Doyle-Hales *S* wave study was that the ratio of the *S* station anomalies to the corresponding *P* station anomalies was of order 4. This surprising result was confirmed by Hales and Roberts (1970a) and by Followill and Nuttli (1970) and Followill (personal communication, 1970). The Followill and Nuttli *S* residual distribution closely resembles that of Herrin and Taggart for *P*.

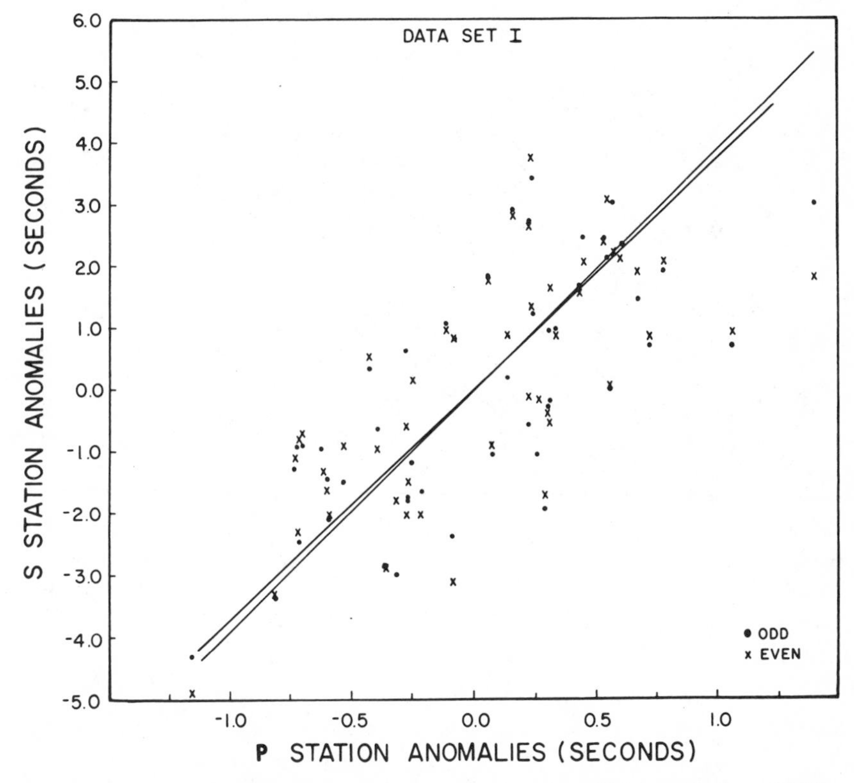

Fig. 19. The correlation of *P* and *S* station anomalies. From Hales and Roberts (1970).

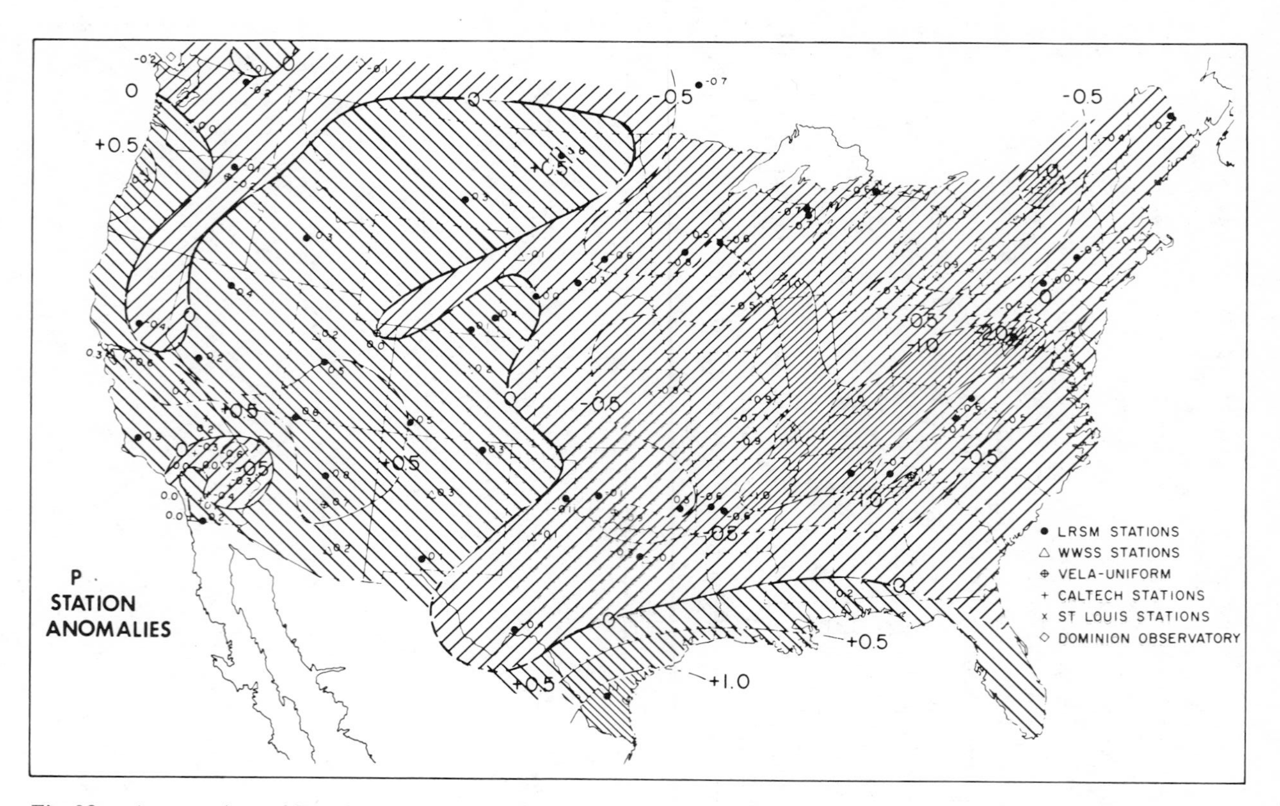

Fig. 20*a*. A comparison of *P* station anomalies in the U.S.A.

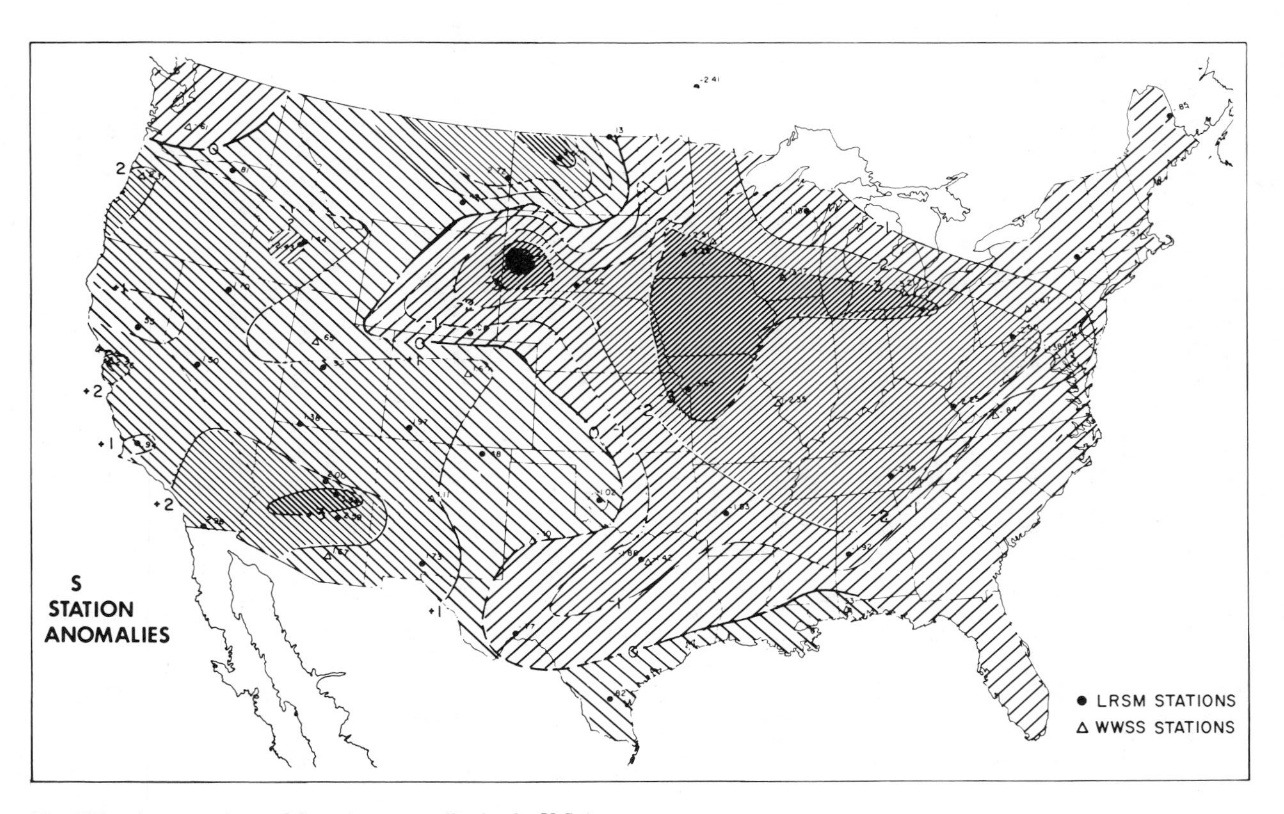

Fig. 20*b*. A comparison of *S* station anomalies in the U.S.A.

Kaila, Reddy, and Narain (1968) have determined P and S travel times for earthquakes and nuclear explosions in central Asia. They find that the ratio of the observed S residuals to the observed P residuals is 1.78 and remark that this ratio "agrees with the observed longitudinal to transverse velocity ratio over most parts of the Earth." The low value of the $\delta t_S/\delta t_P$ ratio found by these authors is not unexpected for the majority of their observations were at distances of less than 30° and about half at distances of less than 20° (Fig. 2 of Kaila et al., 1968). It is reasonable to suggest that for their observations the P and S station anomalies arose from differences in the crust and upper mantle structure above the low-velocity layer.

The Interpretation of the Ratio $\delta t_S/\delta t_P$

The meaning to be attached to the high ratios found for the ratio $\delta t_S/\delta t_P$ was discussed by Hales and Doyle (1967).

If it be assumed that the "standard" values of the velocities of P and S waves are α_0 and β_0, respectively, and that in some other region the velocities are $\alpha_0 - \delta\alpha$ and $\beta_0 - \delta\beta$, then the differences in travel time over a path length D are $\delta t_P = D\delta\alpha/\alpha_0^2$ and $\delta t_S = D\delta\beta/\beta_0^2$, provided that the effects of ray geometry are neglected. (In some cases this assumption may not be justified.) Thus, $\delta t_S/\delta t_P = \delta\beta/\delta\alpha \cdot \alpha_0^2/\beta_0^2$.

It follows from laboratory measurements of the elastic constants (or seismic velocities) by Birch and his colleagues that for the rocks of the crust and upper mantle α_0/β_0 varies from 1.60 for quartz rich rocks to 1.80 for basic rocks (Hales, 1951). It appears that the relation $\alpha/\beta = \sqrt{3}$, which corresponds to $\lambda = \mu$ and Poissons ratio equal to 1/4, is a fairly representative value. It will be assumed in this section that $\alpha_0/\beta_0 = \sqrt{3}$.

With this value of α_0/β_0, it is found that $\delta t_S/\delta t_P = 4.2$ implies that $\delta\beta/\delta\alpha = 1.4$ and $\delta\beta/\beta_0 = 2.4\,(\delta\alpha/\alpha_0)$.

Let us now consider some possible models. In the first place, if the regional differences are differences between different rocks of the standard type then it would be expected that $\delta\beta/\delta\alpha = 1/\sqrt{3} = 0.58$ or $\delta\beta/\beta_0 = \delta\alpha/\alpha_0$. It would in fact be expected that travel time anomalies arising from differences in the crust would satisfy this relation. Clearly, however, this model is incompatible with the observed value of $\delta t_S/\delta t_P$ of 4.2.

The observed value of $\delta t_S/\delta t_P$ implies that the deviations in β are relatively larger than would be expected from standard rocks. It is well known that as materials approach their melting points the changes in μ, the rigidity modulus, are larger than those in the other elastic constants. In the extreme case when melting occurs, μ approaches zero. Thus, it is natural to consider a model in which the differences in travel time arise from differences in μ, the other elastic constants being unchanged. For λ constant, it follows that $\delta\beta/\beta_0 = 1.5\,(\delta\alpha/\alpha_0)$ and, for k constant, that $\delta\beta/\beta_0 = 2.25\,(\delta\alpha/\alpha_0)$. Thus, the model in which the differences arise because of changes in μ, k remaining constant, is closely in accord with observation. Since k decreases by about 10% on the average at liquefaction it would be expected that $\delta\beta/\beta_0$ should be somewhat less than 2.25 $(\delta\alpha/\alpha_0)$ instead of greater than 2.25 $(\delta\alpha/\alpha_0)$ as observed. This suggests that the observed value of $\delta t_S/\delta t_P$ is somewhat too high.

It is interesting to note that studies of the attenuation of surface waves by Anderson and Archambeau (1964) and Anderson, Ben Menahem, and Archambeau (1965) have shown that the attenuation of Rayleigh and Love waves in the upper mantle can be explained satisfactorily by models in which the elastic constant μ is complex but k remains real. Thus, in the case of the attenuation of surface waves also the major effect is in μ, the incompressibility k being unaffected.

It was noticed by Cleary and Hales (1966) that the P arrivals were late in regions from which high heat flow values had been reported. The correlation between heat flow and P station anomalies has been studied by Toksöz and Arkani-Hamed (1967), Simmons and Horai (1968), Horai and Simmons (1969), and Horai (1969). The correlation is significant as is shown by Fig. 3 of Horai (1969).

Since late arrivals (positive station anomalies) are associated with high heat flow and, therefore, presumably with high temperatures in the upper mantle, it is necessary to consider whether the regional differences in travel time can be accounted for wholly in terms of temperature differences in the upper mantle.

Birch (1969) has reviewed the variation of the seismic velocities as functions of pressure and temperature. Soga et al. (1966) have measured the temperature effect on the seismic velocities for synthetic aggregates. Table 2 summarizes the results.

The values of $(\delta\beta/\delta\alpha)\,\alpha/\beta$ in the third column of each set are to be compared with the value of 2.4 derived from the observed value of $\delta t_S/\delta t_P$. The values found for olivine from the Soga et al. (1966) measurements fall far short of the value derived from $\delta t_S/\delta t_P$. The values found from the laboratory measurements on corundum are closer, but this mineral is not likely to be a major constituent of upper mantle rocks. It is unfortunate that similar measurements on the pyroxenes and garnets have not yet been made for these minerals are with olivine the most probable upper mantle constituents.

Table 2. Temperature coefficients of velocity (parts per million per degree) for synthetic aggregates (at zero pressure)

$T°$K	Corundum			Periclase			Forsterite (6% porosity)		
	$\frac{1}{\alpha}\left(\frac{\partial\alpha}{\partial T}\right)$	$\frac{1}{\beta}\left(\frac{\partial\beta}{\partial T}\right)$	$\frac{\delta\beta}{\delta\alpha}\frac{\alpha}{\beta}$	$\frac{1}{\alpha}\left(\frac{\partial\alpha}{\partial T}\right)$	$\frac{1}{\beta}\left(\frac{\partial\beta}{\partial T}\right)$	$\frac{\delta\beta}{\delta\alpha}\frac{\alpha}{\beta}$	$\frac{1}{\alpha}\left(\frac{\partial\alpha}{\partial T}\right)$	$\frac{1}{\beta}\left(\frac{\partial\beta}{\partial T}\right)$	$\frac{\delta\beta}{\delta\alpha}\frac{\alpha}{\beta}$
300	42	57	1.36	59	76	1.29	53	65	1.23
1000	49	67	1.37	71	94	1.32	70	79	1.13
1500	55	93	1.69	80	108	1.35	83	89	1.07
2000	61	107	1.75	90	125	1.39	98	102	1.04

A similar conclusion that the regional differences in travel time cannot be explained in terms of regional differences in upper mantle temperature alone can be reached on the basis of a discussion of the ranges of the P and S station anomalies separately. The range of the P station anomalies is roughly 2 sec (from Table 2 at 1500°K, $|1/\alpha \; \partial\alpha/\partial T| = 83 \times 10^{-6}$ for forsterite). A difference of 500°C in temperature over 400 km is required to produce a difference in δt_P of 2 sec.

Similarly, the range of S station anomalies is 8 sec and at 1500°K $1/\beta \; \partial\beta/\partial T = 89 \times 10^{-6}$ for forsterite. In this case, a difference in temperature of 1000°C is required to produce a difference in δt_S of 8 sec. The required temperature differences are improbably large.

For the other minerals measured by Soga et al. (1966), the required temperature differences are even larger. It is not inconceivable that temperature differences large enough to account for the P station anomalies might occur without partial melting taking place. If temperature differences of the magnitude required to account for the S station anomalies existed, however, then melting would occur. Considerations such as these led Hales and Doyle (1967) to remark cautiously, probably too cautiously, that an explanation had to be sought in terms of "an approach of one constituent" to melting. Birch (1969) observed that "actual partial melting appears to provide an explanation." It should be emphasized that if partial melting occurs beneath such regions as the western United States, then the temperatures in the upper mantle will be higher in such regions and, therefore, some part of the travel time anomaly will be contributed by the decreased velocities arising from the increased temperature.

It has long been recognized that basaltic magmas originate in the upper mantle, probably as a result of partial fusion. Shimozuru (1963a, b) and Beloussov (1966) have discussed the effect of partial fusion upon the properties of the upper mantle. The hypothesis is that in the low-velocity zone liquid basalt exists in the form of films between the crystals or in droplets. Birch (1969) has examined the hypothesis quantitatively, making use of a theory of the elastic properties of two component systems developed by Mackenzie (1950) for a material with holes and extended by Oldroyd (1956) and Hashin (1959) to the case in which the holes are filled by another material, in this case a liquid.

For the case of holes

$$\frac{\delta k}{k_0} = \frac{9C}{4} \quad \text{and} \quad \frac{\delta\mu}{\mu_0} = 2C$$

where C is the fractional volume of the holes and $\mu_0 = 3/5 \; k_0$. From these equations, it follows that $\delta\beta/\beta_0 = 0.9 \; \delta\alpha/\alpha_0$, approximately.

When the holes are filled with liquid the rigidity is not changed but $\delta k/k_0$ is reduced by an amount that depends on the incompressibility of the liquid. For the case in which the incompressibility of the liquid is half that of the solid, Birch finds that $\delta k/k_0 = 0.7C$, assuming that the liquid density is 5/6 that of the density of the solid.

For this case, $\delta\beta/\beta_0 = 1.65\ \delta\alpha/\alpha_0$. This value is closer to that derived from the observed ratio $\delta t_S/\delta t_P$, but it still falls short of it. It is probable that in partial melting the intercrystalline constraints are relaxed and drops of melt formed so that the effect on μ will be greater than that found for the model with spherical cavities filled with liquid.

Walsh (1969) has studied a model in which effectively the intercrystalline or intergranular constraints are reduced as a result of partial melting. Walsh remarks that a large decrease in rigidity but a relatively small decrease in bulk modulus are the features of this model.

For this model, $\delta\alpha/\delta\rho = 10\alpha_0/3\rho_0 = 8$ approximately and $\delta\beta/\delta\rho = 11\beta_0/2\rho_0 = 8$ approximately. These values, as Birch (1969) points out, are quite unlike those for isothermal compression.

A related possibility is that there exists in some regions a layer of liquid accumulated as a result of the partial fusion process. At first sight, it might appear that the existence of a liquid layer is incompatible with tidal deformation and the transmission of long period shear waves to distances greater than 30°. Since any layer of this kind would be much less than one wavelength thick for waves of 20-sec period and the viscosity high, it is probable that there would be coupling across the layer by the viscous stresses. In addition, SV waves could be transmitted across any such layer as P waves, reconverting to S at the lower boundary. This conversion would lead to early arrivals. It is, in fact, fairly common to find forerunners for the S phases at great distances. The effect of viscous liquid layers requires further study.

Finally, it should be noted that Bolt (personal communication, 1969) has suggested that the hypothesis of an undulating low-velocity layer different for P and S is a possible explanation for the large ratio $\delta t_S/\delta t_P$.

For the present, we lean toward the conclusion that the regional differences in travel time arise in part from regional variations of upper mantle temperature, but that in the regions of markedly late arrivals or positive station anomalies, the major contribution to the increased travel time arises from a layer in which there has been extensive partial melting.

The Core Phases

The early interpretations of the core phases were relatively simple and are illustrated by the schematic travel times and ray paths illustrated in Fig. 21. The rays that just penetrate into the core were refracted as shown by the ray marked (2) and reached the point called A, which was close to 180°. There was an angle of minimum deviation close to 143° at which there was a large amplitude caustic B. The observations at distances of less than 143° were called P'' and interpreted as diffracted phases from the caustic.

Lehmann (1936) proposed instead that these early arrivals arose from refraction at an inner core. Lehmann's interpretation was adopted by Gutenberg and Richter (1938) and Jeffreys (1939). The Jeffreys interpretation is illustrated in Fig. 22 from Ergin (1967) and the corresponding velocity

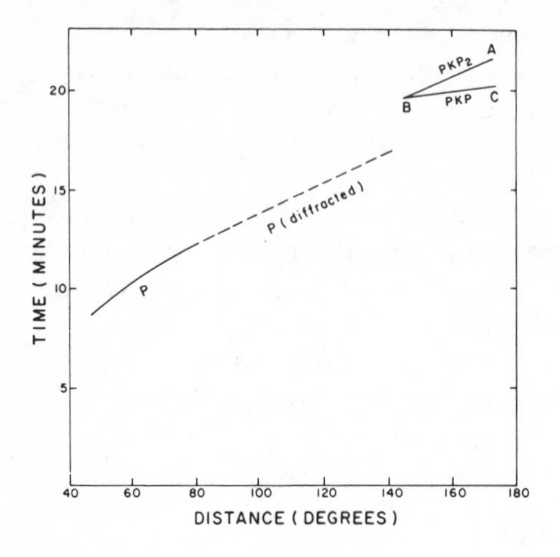

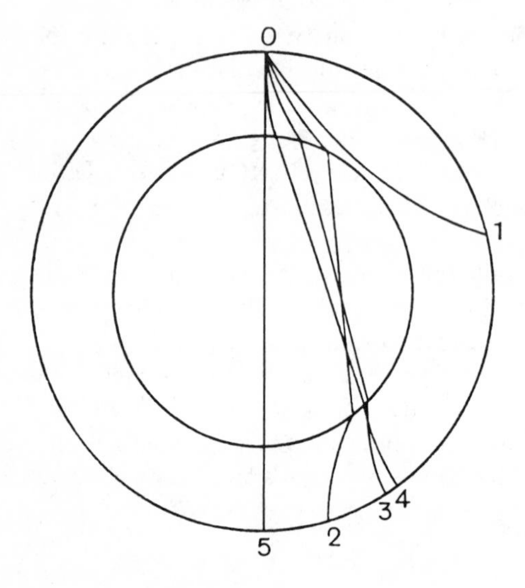

Fig. 21. Early interpretations of the core phases illustrated by schematic travel times and ray paths (without the inner core).

distribution in Fig. 23 from Hannon and Kovach (1966). The sharp dip in velocities outside the inner core was necessary in order that *PKIKP* should reach back as far as 110°. Gutenberg and Richter's analysis did not indicate this low-velocity zone.

It was later found by Bullen and Burke-Gaffney (1958) that the records of the 1954 Marshall Islands stations showed arrivals earlier than *PKIKP*. These were interpreted by Bullen and Burke-Gaffney as diffracted phases. Bolt (1959) found additional evidence of arrivals earlier than *PKIKP*. Later, Bolt (1962, 1964, 1968) interpreted these arrivals as arising from a refraction at a new interface some hundreds of kilometers outside the inner core. In his most recent paper, Bolt calls this phase *PKHKP*.

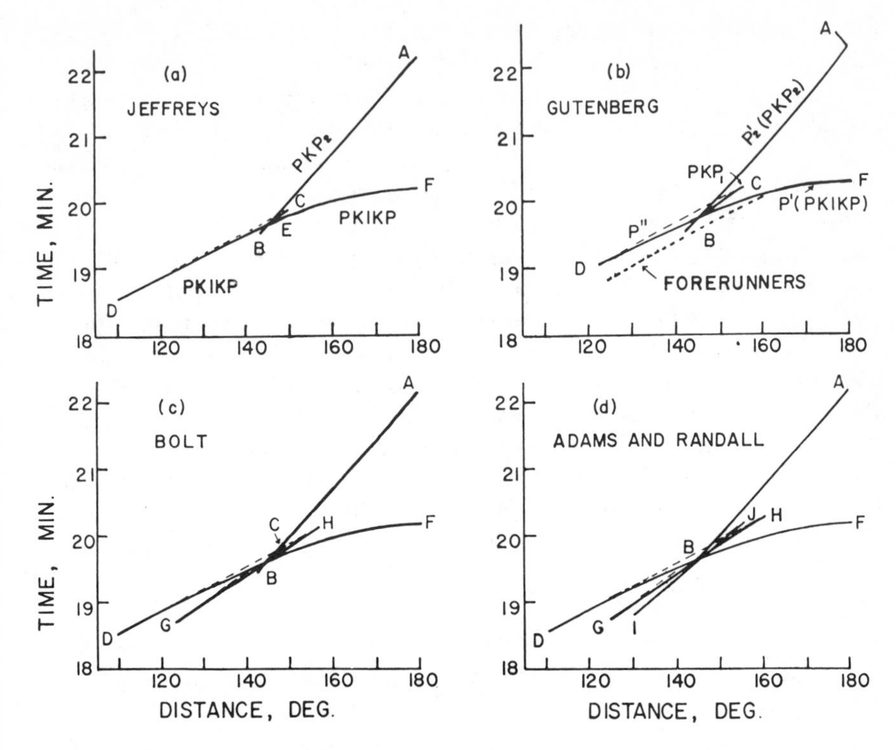

Fig. 22. Travel time interpretations of the core phases of Jeffreys, Gutenberg, Bolt, and Adams and Randall. From Ergin (1967).

The core phases were also studied by Nguyen-Hai (1961), Adams and Randall (1964), Ergin (1967), Gogna (1968), Kovach and Glover (1968), and Husebye and Madariaga (1970). All find arrivals ahead of the DF or *PKIKP* branch. These arrivals continue through the confused zone near the "caustic" *B* to a point *H* at about 156°. The GH branch begins at about 125°. Adams and Randall found arrivals ahead of the GH branch on a branch called the IJ and introduced still another discontinuity in the outer core. The arrivals ahead of DF occurred on the Marshall Islands records also (Bullen and Burke-Gaffney, 1958). Adams and Randall's interpretation (1964) with two branches earlier than DEF was confirmed by Hannon and Kovach (1966) for distances less than 143°, but Hannon and Kovach could not identify the IJ branch beyond the caustic with certainty.

On the other hand, Gogna (1968) concluded that his analysis "supplied no evidence in support of the *two* intermediate branches suggested by Adams and Randall." A similar conclusion was reached by Husebye and Madariaga (1970).

Kovach and Glover (1968) found considerable scatter in their observations of the precursors of the DEF branch. Although their times for *PKIKP* from 115 to 140° agreed with those of Bolt to within 0.4 sec and were usually much closer, their times for the GH branch differed from Bolt's by as much as 3.9 sec. The slopes of the GH branch found by different authors vary widely. Kovach and Glover give 2.2 ± 0.1, Bolt, 2.74, Gogna, 3.2 from 130 to 140°, and Husebye and Madariaga, 2.85; for all observations, 3.05 sec/deg from 132 to 140°. Clearly, the slope of the travel time curve for the GH branch is not well determined. It is possible, as Kovach and Glover suggest, that two precursors are involved but that these cannot be separated with certainty.

Ergin proposed an even more complex pattern of phases in the core and raised some interesting points with regard to the travel times and amplitudes. In the first place, Ergin remarks that the arrivals on the DF branch at distances of less than 124° might be interpreted as reflections from the inner core boundary (*PKiKP*) rather than as refractions. This suggestion had been made earlier by Gutenberg, who assumed that the DF branch (P'') began at 124°. Second, he suggested that the refracted branch *PKIKP* probably begins at a cusp at a distance of 125 to 132°. Third, Ergin pointed out that it was necessary to lower the mantle *P* velocities for depths of 2800 to 2900 km in order to have the $(PKP)_2$ branch extend beyond 180° as required by the observations. Ergin noted

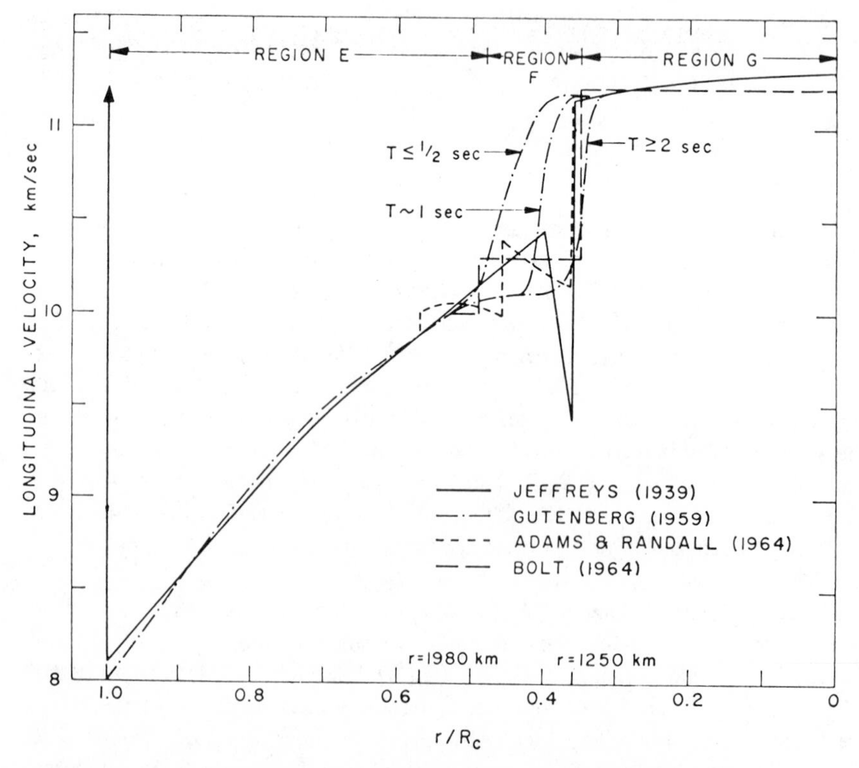

Fig. 23. Velocity distributions in the core. From Hannon and Kovach (1966).

that with this modification of the velocities at the base of the mantle, the direct P phase then extended with small amplitude as far as 130°.

Another question concerns the caustic at B. The rise in amplitudes near 143° is relatively sharp (Fig. 14 of Ergin, 1967), but Bolt (1968) points out that these large amplitudes near B may not be the result of a caustic but instead of the interaction of several phases.

There is some question whether the rays for the minimum deviation phase intersect the interfaces in the outer core corresponding to the GH phase or not, but little doubt that the rays would intersect the interface corresponding to Adams and Randall's IJ phase. The apparent velocity curve for the Adams and Randall model (Fig. 24) shows rapid variation in the apparent velocity, and

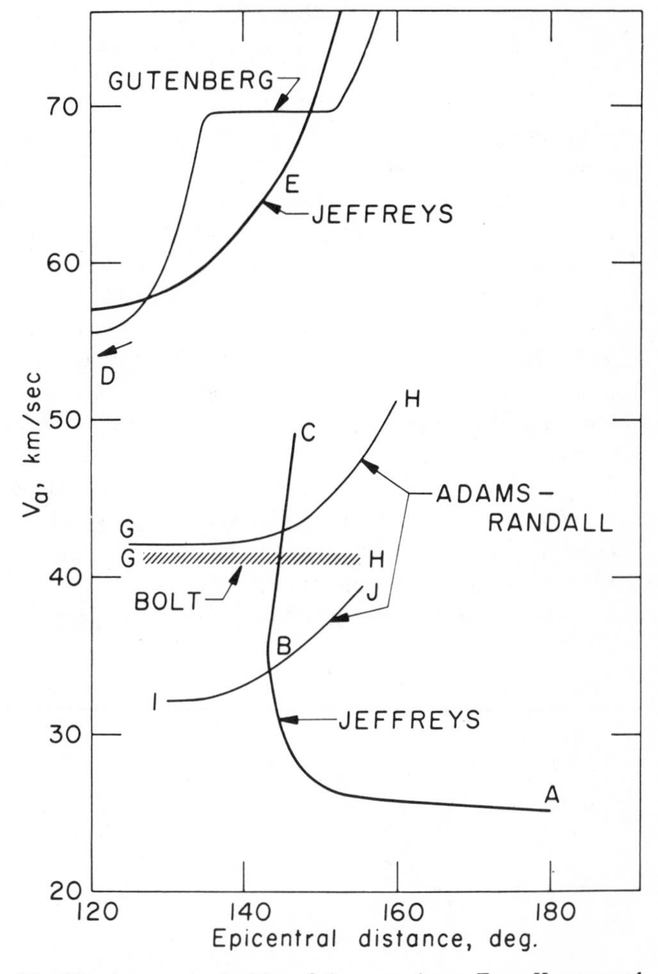

Fig. 24. Apparent velocities of the core phases. From Hannon and Kovach (1966).

therefore, large $d^2T/d\Delta^2$ for the BA branch near 145°. On the Adams and Randall model, therefore, large amplitudes would be expected near 145°. This may be the most satisfactory explanation for the observed amplitudes.

There is a still further complication. Bolt (1968) remarks on a series of arrivals between 156 and 162°, which are about 1 sec earlier than his DF branch. Bolt suggests that these arrivals may arise from a discontinuity in the inner core. Cleary and Hales (1970) prefer to regard these early arrivals as defining the DF branch.

Recent studies of the multiply reflected phases *PKKP, PKKKP,* etc., by Engdahl (1968) and of reflections from the inner core by Engdahl, Flinn, and Romney (1970) provide powerful constraints on any proposed variation in the velocity and other properties of the core.

PKIKP station residuals are strongly correlated with those for *P* as is shown by the comparison in Fig. 25 based on data of Cleary and Hales (1970). The slope of the regression line is closer to 1 than had been expected from the comparison of *P* and *PcP* station anomalies for model structures made by Hales et al. (1968). The implication is that a major part of the regional variation effect arises in a comparatively thin layer.

The steeper ray paths in the upper mantle of the core phases suggest that the azimuthal part of the station anomaly should be less for the core phases than for *P*. This has been found to be the case in general, but there is a notable exception at Uinta Basin Observatory as was mentioned earlier.

The *P* core phases are refracted strongly down when they enter the core, and in consequence, they do not provide information on the velocities in the outer few hundred kilometers of the outer core. For information on the velocities for the outermost core region, it is necessary to consider the *SKS* travel times. Even so, travel times in the outermost part of the core are somewhat uncertain for *SKS* is not easily observable for $\Delta < 84°$, which corresponds to a Δ at the core boundary of about 30°. A preliminary study by Hales and Roberts (1971) suggests that the velocity at the core boundary is about 2% less than that of the Jeffreys model. In this connection, it should be noted that Buchbinder (1968) has suggested that the velocity inside the core mantle boundary should be reduced.

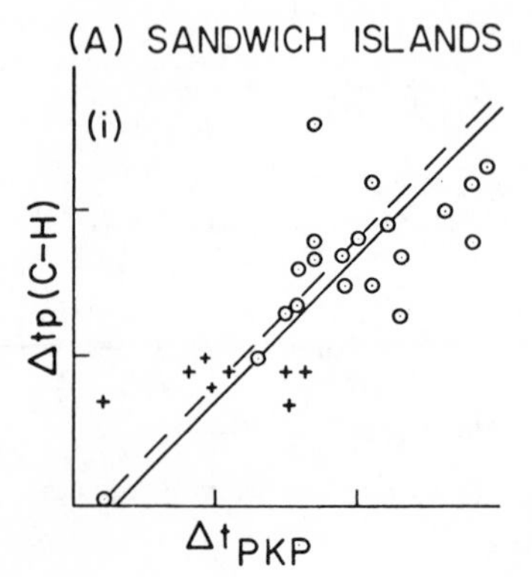

Fig. 25. A comparison of *PKP* and *P* station anomalies from Cleary and Hales (1970).

ACKNOWLEDGMENTS

The authors are grateful to Drs. D. L. Anderson, Bruce Bolt, R. Kovach, and Mr. C. Wright for preprints of papers and to Mr. F. Followill and Mr. R. Robinson for information on work in progress. We thank John Cleary for helpful discussion, Jeanne Roberts for assistance in the preparation of this paper, and Bruce Bolt and Eugene Robertson for critical reviews of the manuscript.

Studies of the travel times of *P* and *S* waves by Hales has been supported by the Air Force Cambridge Research Laboratories, O.A.R., under Contract No. AF19(628)-2936 and F19(628)-67-C-0419 for the Advanced Research Projects Agencies' Project VELA-Uniform.

The work reported in this paper was supported under NASA general grant NGL-44-004-001 to the Southwest Center for Advanced Studies and The University of Texas at Dallas.

REFERENCES

Adams, R. D., and Randall, M. J., The fine structure of the Earth's core, Bull. Seism. Soc. Am., **54**, 1299-1313, 1964.

Alexander, S. S., and Phinney, R. A., A study of the core-mantle boundary using *P* waves diffracted by the Earth's core, J. Geophys. Res., **71**, 5943-5958, 1966.

Anderson, D. L., An equation of state for the Earth (abstract), Trans. Am. Geophys. Union, **46**, 43-44, 1965.

______, Latest information from seismic observations, *in* The Earth's Mantle, edited by T. F. Gaskell, pp. 355-420, Academic Press, New York, 1967.

Anderson, D.L., and Archambeau, C.B., The anelasticity of the Earth, J. Geophys. Res., **69**, 2071-2084, 1964.

Anderson, D. L., Ben-Menahem, A., and Archambeau, C. B., Attenuation of seismic energy in the upper mantle, J. Geophys. Res., **70**, 1441-1448, 1965.

Arnold, E. P., The revision of the seismological tables, Ph.D. thesis, Cambridge University, England, unpublished, 1965.

_____, Smoothing travel-time tables, Bull. Seism. Soc. Am. **58**, 1345-1351, 1968.

Beloussov, V.V., Modern concepts of the structure and development of the Earth's crust and the upper mantle of continents, Quart. J. Geol. Soc., London, **122**, 293-314, 1966.

Birch, F., Elasticity and constitution of the Earth's interior, J. Geophys. Res., **57**, 227-286, 1952.

_____, Density and composition of the upper mantle: First approximation as an olivine layer, *in* The Earth's Crust and Upper Mantle, Geophysical Monograph **13**, edited by P. J. Hart, pp. 18-36, Am. Geophys. Union, Washington, D.C., 1969.

Bolt, B. A., Travel-times of *PKP* up to 145°, Geophys. J., **2**, 190-198, 1959.

_____, Gutenberg's early *PKP* observations, Nature, **196**, 122-124, 1962.

_____, The velocity of seismic waves near the earth's center, Bull. Seism. Soc. Am., **54**, 191-208, 1964.

______, Estimation of *PKP* travel times, Bull. Seism. Soc. Am., **58**, 1305-1324, 1968.

______, *PdP* and *PKiKP* waves and diffracted *PcP* waves, Geophys. J. Roy. Astron. Soc., **20**, 367-382, 1970.

Bolt, B. A., Niazi, M., and Somerville, M. R., Diffracted *ScS* and the shear velocity at the core boundary, Geophys. J. Roy. Astron. Soc., **19**, 299-305, 1970.

Bolt, B.A., and Nuttli, O.W., *P* wave residuals as a function of azimuth, J. Geophys. Res., **71** 5977-5985, 1966.

Buchbinder, G., Properties of the core-mantle boundary and observations of *PcP*, J. Geophys. Res., **73**, 5901-5923, 1968.

Bullen, K.E., and Burke-Gaffney, T.N., Diffracted seismic waves near the *PKP* caustic, Geophys. J. Roy. Astron. Soc., **1**, 9-17, 1958.

Bullen, K.E., and Haddon, R.A.W., Derivation of an Earth model from free oscillation data, Proc. Nat'l Acad. Sci., **58**, 846-852, 1967.

Carder, D.S., Travel times from central Pacific nuclear explosions and inferred mantle structure, Bull. Seism. Soc. Am., **54**, 2271-2294, 1964.

Carder, D.S., Gordon, D.W., and Jordan, J.N., Analysis of surface-foci travel times, Bull. Seism. Soc. Am., **56**, 815-840, 1966.

Carpenter, E.W., Marshall, P.D., and Douglas, A., The amplitude distance curve for short period teleseismic *P*-waves, Geophys. J. Roy. Astron. Soc., **13**, 61-70, 1967.

Chinnery, M. A., Velocity anomalies in the lower mantle, Phys. Earth Planet. Interiors, **2**, 1-10, 1969.

Chinnery, M. A., and Toksöz, M. N., *P*-wave velocities in the mantle below 700 km., Bull. Seism. Soc. Am., **57**, 199-226, 1967.

Cleary, J. R., Azimuthal variation to the Longshot source term, Earth Planet. Sci. Letters, **3**, 29-37, 1967a.

______, Analysis of the amplitudes of short-period *P* waves recorded by Long-Range Seismic Measurements stations in the distance range 30° to 102°, J. Geophys. Res., **72**, 4705-4712, 1967b.

______, The *S* velocity at the core-mantle boundary, from observations of diffracted *S*, Bull. Seism. Soc. Am., **59**, 1399-1405, 1969.

Cleary, J. R., and Hales, A. L., Travel times of *P* waves in the range 30° to 100° across the United States (abstract), Trans. Am. Geophys. Union, **44**, 888, 1963.

______, An analysis of the travel times of *P* waves to North American stations, in the distance range 32° to 100°, Bull. Seism. Soc. Am., **56**, 467-489, 1966.

______, *P* times from the nuclear explosion 'Greeley', Earth Planet. Sci. Letters, **4**, 305-309, 1968.

______, *PKIKP* residuals at stations in North America and Europe, Earth Planet. Sci. Letters, **8**, 279-282, 1970.

Cleary, J. R., and Muirhead, K. S., Comparison of the 1968 *P* tables with times from nuclear explosions (I) Longshot and Greeley, Earth Planet. Sci. Letters, **7**, 119-124, 1969.

Cleary, J. R., Porra, K., and Read, L., Diffracted *S*, Nature, **216**, 905-906, 1967.

Corbishley, D.J., Multiple array measurements of the *P*-wave travel-time derivative, Geophys. J. Roy. Astron. Soc., **19**, 1-14, 1970.

Dahm, C.G., Velocities of *P* and *S* waves calculated from the observed travel times of the Long Beach earthquake, Bull. Seism. Soc. Am., **26**, 159-171, 1936.

Derr, J.S., Internal structure of the Earth inferred from free oscillations, J. Geophys. Res., **74**, 5202-5220, 1969.

Dorman, J., Ewing, J., and Alsop, L., Oscillations of the Earth: New core-mantle boundary model based on low order free vibrations, Proc. Nat'l. Acad. Sci., **54**, 364-368, 1965.

Douglas, A., and Lilwall, R., Estimation of *P* wave travel times using the Joint Epicentre method, Geophys. J., **19**, 165-181, 1970.

Doyle, H.A., and Hales, A.L., An analysis of the travel times of *S* waves to North American stations, in the distance range 28° to 82°, Bull. Seism. Soc. Am., **57**, 761-771, 1967.

Dziewonski, A.M., Correlation properties of free period partial derivatives and their relation to the resolution of gross Earth data, Bull. Seism. Soc. Am., **60**, 741-768, 1970.

Engdahl, E.R., Seismic waves within the Earth's outer core: multiple reflection, Science, **161**, 263-264, 1968.

Engdahl, E. R., Flinn, E. A., and Romney, C. F., Inner core reflections (abstract), Geological Society of America, pp. 89-90, 66th Annual Meeting, Hayward, California, 1970.

Ergin, K., Seismic evidence for a new layered structure of the Earth's core, J. Geophys. Res., **72**, 3669-3687, 1967.

Fairborn, J.W., Shear wave velocities in the lower mantle, Bull. Seism. Soc. Am., **59**, 1983-1999, 1969.

Followill; F.E., and Nuttli, O.W., Shear body wave travel times and velocity distribution in the lower mantle (abstracts), p. 92, vol. 2, The Geological Society of Am., 66th Annual Meeting, 1970.

Gilbert, F., and Backus, G.E., Approximate solutions to the inverse normal mode problem, Bull. Seism. Soc. Am., **58**, 103-131, 1968.

Gogna, M.L., Travel times from central Pacific nuclear explosions, Geophys. J. Roy. Astron. Soc., **13**, 503-527, 1967.

_____, Travel times of *PKP* from Pacific earthquakes, Geophys. J. Roy. Astron. Soc., **16**, 489-514, 1968.

Green, R.W.E., and Hales, A.L., The travel times of *P* waves to 30° in the central United States and upper mantle structure, Bull. Seism. Soc. Am., **58**, 267-289, 1968.

Greenfield, R.J., and Sheppard, R.M., The Moho depth variations under the LASA and their effect on $dT/d\Delta$ measurements, Bull. Seism. Soc. Am., **59**, 409-420, 1969.

Gutenberg, B., Physics of the Earth's Interior, Academic Press, New York, 1959.

Gutenberg, B., and Richter, C.F., *P*' and the Earth's core, Monthly Notices Roy. Astron. Soc., Geophys. Suppl., **4**, 363-372, 1938.

_____, Magnitude and energy of earthquakes, Annali di Geofisica, **9**, 1-15, 1956.

Hales, A. L., Crustal layers of the earth, Nature, **168**, 163, 1951.

Hales, A.L., and Doyle, H.A., *P* and *S* travel time anomalies and their interpretation, Geophys. J. Roy. Astron. Soc., **13**, 403-415, 1967.

Hales, A.L., Cleary, J., and Roberts, J., Velocity distributions in the lower mantle, Bull. Seism. Soc. Am., **58**, 1975-1989, 1968.

Hales, A. L., Cleary, J. R., Doyle, H. A., Green, R., and Roberts, J., *P*-wave station anomalies and the structure of the upper mantle, J. Geophys. Res., **73**, 3885-3896, 1968.

Hales, A. L., and Roberts, J. L., The travel times of *S* and *SKS*,, Bull. Seism. Soc. Am., **60**, 461-489, 1970a.

_____, Shear velocities in the lower mantle and the radius of the core, Bull. Seism. Soc. Am., **60**, 1427-1436, 1970b.

_____, The velocities in the outer core, Bull. Seism. Soc. Am., **61**, in press, 1971.

Hannon, W.J., and Kovach, R.L., Velocity filtering on seismic core phases, Bull. Seism. Soc. Am., **56**, 441-454, 1966.

Hashin, Z., On non-homogeneity in elasticity and plasticity, Proceedings of the IUTAM Symposium, Pergamon Press, London, pp. 463-478, 1959.

Herrin, E., 1968 seismological tables for *P* phases, Bull. Seism. Soc. Am., **58**, 1193, 1968.

_____, Regional variations of *P*-wave velocity in the upper mantle beneath North America, *in* The Earth's Crust and Upper Mantle, Geophysical Monograph, **13**, p. 242-246, 1969.

Herrin, E., and Sorrells, G.G., Travel time models (abstract), AGU Transactions, 243, 1969.

Herrin, E., and Taggart, J., Regional variations in P_n velocity and their effect on the location of epicenters, Bull. Seism. Soc. Am., **52**, 1037-1046, 1962.

_____, Regional variations in *P* travel times, Bull. Seism. Soc. Am., **58**, 1325-1337, 1968a.

_____, Source bias in epicenter determinations, Bull. Seism. Soc. Am., **58**, 1791-1796, 1968b.

Herrin, E., Tucker, W., Taggart, J., Gordon, D., and Lobdell, J., Estimation of surface focus *P* travel times, Bull. Seism. Soc. Am., **58**, 1273-1291, 1968.

Horai, K., Cross-covariance analysis of heat flow and seismic delay times for the Earth, Earth Planet. Sci. Letters, **7**, 213-220, 1969.

Horai, K., and Simmons, G., Spherical harmonic analysis terrestrial heat flow, Earth Planet. Sci. Letters, **6**, 386-394, 1969.

Husebye, E. S., Correction analysis of Jeffreys-Bullen travel time tables, Bull. Seism. Soc. Am., **55,** 1023-1038, 1965.

Husebye, E., and Madariaga, R., The origin of precursors to core waves, Bull. Seism. Soc. Am., **60,** 939-952, 1970.

Ibrahim, A.K., and Nuttli, O., Travel time curves and upper mantle structure from long period *S* waves, Bull. Seism. Soc. Am., **57,** 1063-1092, 1967.

Isacks, B., Oliver, J., and Sykes, L.R., Seismology and the new global tectonics, J. Geophys. Res., **73,** 5855-5899, 1968.

Jeffreys, H. The times of core waves, Monthly Notices Roy. Astron. Soc., Geophys. Suppl., **4,** 548-561, 1939.

______, The times of *P* in Japanese and European earthquakes, Monthly Notices Roy. Astron. Soc., Geophys. Suppl., **6,** 557-565, 1954.

______, Revision of travel times, Geophys. J. Roy. Astron. Soc., **11,** 5-12, 1966.

Jeffreys, H., and Bullen, K.E., Seismological Tables, 55 pp., Brit. Assoc. Advancement of Sci., Gray-Milne Trust, London, 1940.

Johnson, L. R., Array measurements of *P* velocities in the upper mantle, J. Geophys. Res., **72,** 6309-6325, 1967.

______, Array measurements of *P* velocities in the lower mantle, Bull. Seism. Soc. Am., **59,** 973-1008, 1969.

Kaila, K. L., Reddy, P. R., and Narain, H., *P*-wave travel times from shallow earthquakes recorded in India and inferred upper mantle structure, Bull. Seism. Soc. Am., **58,** 1879-1897, 1968.

Kanamori, H., Upper mantle structure from apparent velocities of *P* waves recorded at Wakayama micro-earthquake observatory, Bull. Earthquake Res. Inst., **45,** 657-678, 1967.

Kogan, S.D., Travel times of longitudinal and transverse waves, calculated from data on nuclear explosions made in the region of the Marshall Islands, Bull. Acad. Sci. USSR (Izvest.), Geophys. Series, **3,** 246-253, 1960.

Kondorskaya, N.V., Regional peculiarities in the travel times of seismic waves, Bull. (Izvest.) Acad. Sci. USSR, Geoph. Ser. (English translation), No. 7, 67-86, 1957.

Kovach, R. L., and Glover, P., Travel times of *PKP* in the range $115° \leqslant \Delta \leqslant 140°$, Geophys. J. Roy. Astron. Soc., **15,** 367-376, 1968.

Landisman, M., Satô, Y., and Nafe, J., Free vibrations of the Earth and the properties of the deep interior regions, Geophys. J. Roy. Astron. Soc., **9,** 439-468, 1965.

Lehmann, I., *P*', Publication Bur. Central Seism. Intern., A, **14,** 3-31, 1936.

MacDonald, G.J.F., and Ness, N., A study of the free oscillations of the Earth, J. Geophys. Res., **66,** 1865-1911, 1961.

Mackenzie, J. K., The elastic constants of a solid containing spherical holes, Proc. Phys. Soc. (London), Ser. B, **63,** 2-11, 1950.

Muirhead, K.J., and Cleary, J. R., Comparison of the 1968 *P* tables with times from nuclear explosions (II), the Marshall Islands and Sahara series, Earth Planet. Sci. Letters, **7,** 132-136, 1969.

Nguyen-Hai, Propagation des ondes longitudinales dans le noyeau terrestre d'après les séismes profonds de Fidji, Ann. Geophys., **17,** 60-66, 1961.

Niazi, M., and Anderson, D., Upper mantle structure of western North America from apparent velocities of *P* waves, J. Geophys. Res., **70,** 4633-4640, 1965.

Nuttli, O.W., Travel times and amplitudes of *S* waves from nuclear explosions in Nevada, Bull. Seism. Soc. Am., **59,** 385-398, 1969.

Nuttli, O.W., and Bolt, B.A., *P*-wave residuals as a function of azimuth, 2, undulations of the mantle low-velocity layer as an explanation, J. Geophys. Res., **74,** 6594-6602, 1969.

Oldroyd, J.G., The effect of small viscous inclusions on the mechanical properties of an elastic solid, Deformation and Flow of Solids, edited by R. Grammel, Springer Verlag, Berlin, 304-313, 1956.

Otsuka, M., Azimuth and slowness anomalies of seismic waves measured on the central California seismographic array. Part I. Observations, Bull. Seism. Soc. Am., **56,** 223-239, 1966.

Phinney, R.A., and Alexander, S.S., *P* wave diffraction theory and the structure of the core mantle boundary, J. Geophys. Res., **71**, 5959-5975, 1966.

Phinney, R.A., and Cathles, L.M., Diffraction of *P* by the core: A study of long-period amplitudes near the edge of the shadow, J. Geophys. Res., **74**, 1556-1574, 1969.

Pilant, W., Tectonic features of the Earth's crust and upper mantle, Final Technical Report, AFOSR 67-1797, Air Force Office of Scientific Research, 43 pp., 1967.

Press, F., Handbook of Physical Constants, edited by S.P. Clark, Jr., Geol. Soc. Am. Mem., **97**, pp. 195-218, 1966.

______, Earth models obtained by Monte-Carlo inversion, J. Geophys. Res., **73**, 5223-5234, 1968.

Press, F., and Biehler, S., Inferences on crustal velocities and densities from *P* wave delays and gravity anomalies, J. Geophys. Res., **69**, 2979-2995, 1964.

Repetti, W.C., New values for some of the discontinuities in the Earth, cited in Theoretical Seismology, edited by J.B. Macelwane and F.W. Sohon, pp. 181 and 221, John Wiley & Sons, Inc., 1936.

Richards, P.G., A contribution to the theory of high frequency elastic waves, with applications to the shadow boundary of the Earth's core, Ph.D. thesis, California Institute of Technology, 1970.

Romney, C., Brooks, B.G., Mansfield, R.H., Carder, D.S., Jordan, J.N., and Gordon, D.W., Travel times and amplitudes of principal body phases recorded from Gnome, Bull. Seism. Soc. Am., **52**, 1057-1074, 1962.

Sacks, I.S., Diffracted wave studies of the Earth's core: 1. Amplitude, core size, and rigidity, J. Geophys. Res., **71**, 1173-1181, 1966.

______, Diffracted *P*-wave studies of the Earth's core: 2. Lower mantle velocity, core size, lower mantle structure, J. Geophys. Res., **72**, 2589-2594, 1967.

Shimozuru, D., The low velocity zone and temperature distribution in the upper mantle, J. Phys. Earth, **11**, 19-24, 1963a.

______, On the possibility of the existence of the molten portion in the upper mantle of the Earth, J. Phys. Earth, **11**, 49-55, 1963b.

Simmons, G., and Horai, K., Heat flow data 2, J. Geophys. Res., **73**, 6608-6629, 1968.

Soga, N., Schreiber, E. and Anderson, O.L., Estimation of bulk modulus and sound velocities of oxides at very high temperatures, J. Geophys. Res., **71**, 5315-5320, 1966.

Sorrells, G.G., Crowley, J.B., and Veith, K., Methods for computing ray paths in complex geologic structures, Bull. Seism. Soc. Am., **61**, 27-53, 1971.

Taggart, J.N., and Engdahl, E.R., Estimation of *PcP* travel times and the depth to the core, Bull. Seism. Soc. Am., **58**, 1293-1303, 1968.

Toksöz, M.N., and Arkani-Hamed, J., Seismic delay times: correlation with other data, Science, **158**, 783-788, 1967.

Toksöz, M.N., Chinnery, M.A., and Anderson, D.L., Inhomogeneities in the Earth's mantle, Geophys. J. Roy. Astron. Soc., **13**, 31-59, 1967.

Vanek, J., and Stelzner, J., The problem of magnitude calibrating functions for body waves, Annali di Geofisica, **13**, 393-407, 1960.

Vanek, J., Zatopek, A., Karnik, V., Kondorskaya, N. V., Riznichenko, Yu. V., Savarensky, E. F., Solov'ev, S. L., and Shebalin, N. V., Standardization of magnitude scales, Bulletin (Izvest.) Acad. Sci. USSR, Geoph. Ser., (English translation), No. 2, 108-111, 1962.

Walsh, J.B., New analysis of attenuation in partially melted rock, J. Geophys. Res., **74**, 4333-4337, 1969.

Wiggins, R.A., Terrestrial variational tables for the periods and attenuation of the free oscillations, Phys. Earth Planet. Interiors, **1**, 201-266, 1968.

Willey, G., Cleary, J. R., and Marshall, P. D., Comparison of least squares analyses of long and short period *P* wave amplitudes, Geophys. J. Roy. Astron. Soc., **19**, 439-445, 1970.

Wright, C., *P* wave travel time gradient measurements and lower mantle structure, Earth Planet. Sci. Letters, **8**, 41-44, 1970.

9

A COMPARATIVE STUDY OF UPPER MANTLE MODELS: CANADIAN SHIELD AND BASIN AND RANGE PROVINCES

by **Eugene Herrin**

Physical models are constructed for the upper mantle beneath the Canadian Shield and Basin and Range provinces of North America in order to test the hypothesis that the inferred differences in upper mantle structure beneath these two regions can be caused by lateral variations in temperature alone. Temperature, seismic velocity, and density profiles are constructed subject to the constraints imposed by measured values of heat flow, mean radioactivity of exposed basement rocks, gravity, elevation, and seismic delay times and by inferred information such as crustal thickness, seismic velocity as a function of depth, and general rules concerning the distribution of radioactivity within the crust and upper mantle. The models are shown to be consistent with the data presently available and with the measured properties of materials thought to exist in the crust and upper mantle.

Within the limits imposed by the accuracy of available data, it is concluded that the physical differences in the upper mantle beneath the Canadian Shield and the Basin and Range provinces can be explained satisfactorily by lateral differences in temperatures.

NOTATION AND UNITS

ρ density (gm/cm^3)
T temperature (°C)

α	volume thermal expansion (1/deg)
P	pressure (dynes/cm^2)
Bt	isothermal compressibility (cm^2/dyne)
Bs	adiabatic compressibility (cm^2/dyne)
Vp	velocity of compressional waves (km/sec)
Vs	velocity of shear waves (km/sec)
ϕ	$= (\rho \text{ Bs})^{-1} = \text{Vp}^2 - 4/3 \text{ Vs}^2$ (cm^2/sec^2)

Heat flow unit (hfu) 10^{-6} cal/cm^2 sec
Heat generation unit (hgu) 10^{-13} cal/cm^3 sec
Thermal conductivity unit (tcu) 10^{-3} cal/cm sec °C

INTRODUCTION

Whereas travel times of seismic body waves at distances greater than about 3000 km can be discussed in terms of world averages, travel times at shorter distances must be considered on a regional basis. In many cases because of low-velocity layers and lateral variations in velocity, it may be practically impossible to observe first arrivals of seismic energy over wide intervals of distance. In this case, the observed travel times become ambiguous and thus cannot be inverted to obtain estimates of seismic velocity as a function of depth.

An alternative approach to the determination of velocities from travel times alone involves the construction of physical models of the upper mantle beneath particular regions of the earth. Implicit in these models is the distribution of seismic velocities with depth. In this way, all the available measurements pertinent to the determination of upper mantle properties can be brought to bear on the problem, and the data can be weighted according to their reliabilities.

The approach taken in this paper involves the comparison of two tectonic regions in North America. By working with two regions, it is possible to concentrate on the differences in physical properties between the mantles beneath the regions rather than on the absolute value of such properties as density and seismic velocity. Of course, in constructing the models every attempt should be made to meet all constraints imposed by observed and implied relationships among the physical properties of rocks thought to exist in the upper mantle at the pertinent temperatures and pressures.

Models of density, composition, and seismic velocity of the upper mantle have been constructed by a number of workers. For example, Birch (1969) considered models in which a low-velocity zone for compressional and shear waves results from (1) the temperature-pressure effect on a homogeneous layer, (2) partial fusion, and (3) variation in iron content with depth. These studies have considered an "average" upper mantle where parameters such as seismic velocity are inferred from data obtained in the western United States or in oceanic regions. It is now well established that temperatures, seismic velocities, and densities in the upper mantle show significant lateral variations that are often correlated with regional differences in tectonic history. In this paper, an attempt is made to construct models for the upper mantle beneath two greatly different tectonic provinces in North America: The Canadian Shield and Basin and Range provinces. In particular, we wish to test the hypothesis that inferred differences in upper mantle structure beneath these two regions can be caused by lateral variations in temperature alone.

Upper mantle models for the Canadian Shield (CS) and Basin and Range (BR) must fit the constraints imposed by measured values of heat flow, mean radioactivity of exposed basement rocks, gravity, elevation, and seismic delay times. In addition, inferred constraints such as crustal thickness, seismic velocity as a function of depth, and general rules concerning the distribution of radioactivity within the crust and mantle must be considered.

GEOTHERMAL MODELS

The surface heat flux assigned to the Basin and Range (BR) and Canadian Shield (CS) models are those values one would expect to measure in each province if the measurements were made in basement rocks of average radioactivity for the region and away from known temperature anomalies evidenced by such features as geysers and hot springs. The values chosen (Roy, Blackwell, and Decker, 1971) are 1.2 heat flow units in the CS and 1.9 hfu in the BR province.

A further constraint on the geothermal model is imposed by seismic evidence of a discontinuous increase in Vp at a depth of about 360 ± 10 km beneath both the CS and BR provinces (Archambeau, Flinn, and Lambert, 1969; Barr, 1967; Hales, 1969; Johnson, 1967). This discontinuity is generally attributed to the transformation of olivine to a more closely packed phase. Ringwood (1968) estimated the slope of this transition, for an olivine composition pertinent to the upper mantle, to be about 100°C for 10 km of depth. Thus, from the seismic evidence, it is inferred that the temperatures at 350 to 400 km beneath the BR and CS provinces should not differ by more than 200°C. Because of the higher surface heat flow in the BR province, the temperatures at 350 km there, if different from those in the CS, might be expected to be higher. In this case, the gradient should be lower so that the temperatures for the two regions would converge at a greater depth. These considerations become constraints to the geothermal model.

Crustal thicknesses (Table 1) for the two regions are based on studies reported by Berry and West (1966), Hill and Pakiser (1966), Green and Hales (1968), Mereu and Hunter (1969), and Prodehl (1970). Values of thermal conductivity in the CS are 6.5 thermal conductivity units (tcu) in layer 1 and 5.0 tcu in layer 2. In the BR, they are 6.0 tcu in layer 1 and 4.8 tcu in layer 2. These are average values of conductivity for granites and gabbros with a slight adjustment for the difference in crustal temperature between the two regions (Clark, 1966). In the mantle, radiative transfer of heat should become a significant factor, along with thermal

Table 1. Crustal thickness

Province	Layer 1 (km)	Layer 2 (km)
Canadian Shield	15	20
Basin and Range	19	11

diffusion, in determining the effective thermal conductivity. Clark and Ringwood (1964), in their thermal model of the mantle, took the radiative component of the conductivity to be proportional to the cube of the absolute temperature; however, they noted factors that may cause a departure from the T^3 law at lower temperatures. Shankland (1970) agrees with the results of Fukao (1969) that the effect of radiative heat transport is to maintain the thermal conductivity of minerals thought to exist in the upper mantle at roughly a constant value at temperatures below 1000°C. At higher temperatures, following the conclusion of Clark and Ringwood (1964, p. 52), we expect the conductivity to rise with temperature at a rate less than or equal to that implied by the T^3 law.

The thermal conductivity model for the upper mantle used in these calculations assumes a constant value of 10.0 tcu for temperatures below 1000°C. This number is based on the conductivity of forsterite powder (Fukao, 1969) and magnesium-rich olivine (Kanamori et al., 1968). Above 1000°C, the conductivity rises linearly to a value of 25 tcu at 2000°C, a rate of increase smaller than would be implied by the T^3 law at the higher temperatures. Thus, the conductivity model between 1000° and 2000°C is rather arbitrary, being designed to insure a convergence of temperatures for the two regions near a depth of 400 km.

Heat production at the top of layer 1 is 5.0 heat generation units (hgu) for both models. In the CS, the exponential decrement (the depth interval in which the heat production decreases by a factor of $1/e$) is 7.5 km, and in the BR, it is 9.5 km. At the top of layer 2 in both models, the heat production is 0.68 hgu and the exponential decrement is 80 km throughout layer 2 and into the mantle. The heat production in layer 1 is consistent with the observed relationship between surface heat flow and measured radioactivity in the two regions (Birch, Roy, and Decker, 1968; Lachenbruch, 1968; Lachenbruch, 1970; Roy et al., 1968; Roy, Blackwell, and Decker, 1971). In the BR province, two layers of constant heat generation are inserted in the model in order to provide the required higher heat flow. From 30 to 37 km, the heat production is 4.5 hgu, and from 37 to 50, it is 3.2 hgu. This model is in no sense unique nor does it imply any particular physical mechanism for the heat generation. No very different arrangement of heat sources was found which would meet the constraints of the geothermal model without producing excessively high temperatures in the lower crust and upper mantle of the BR model. The

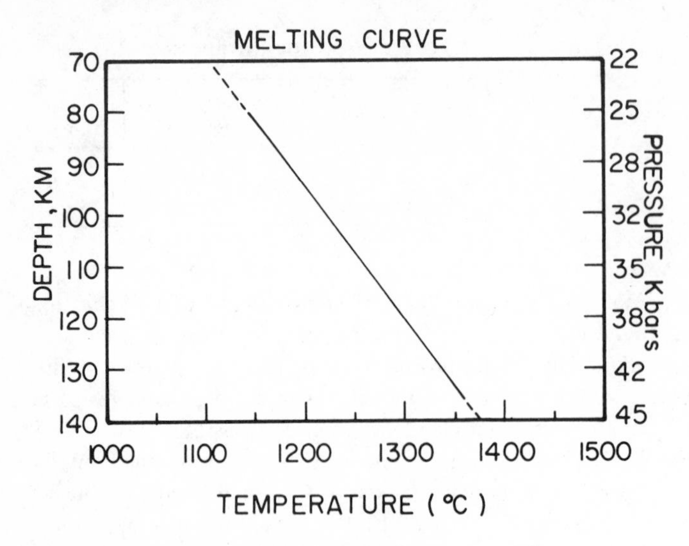

Fig. 1. The assumed melting curve for the upper mantle.

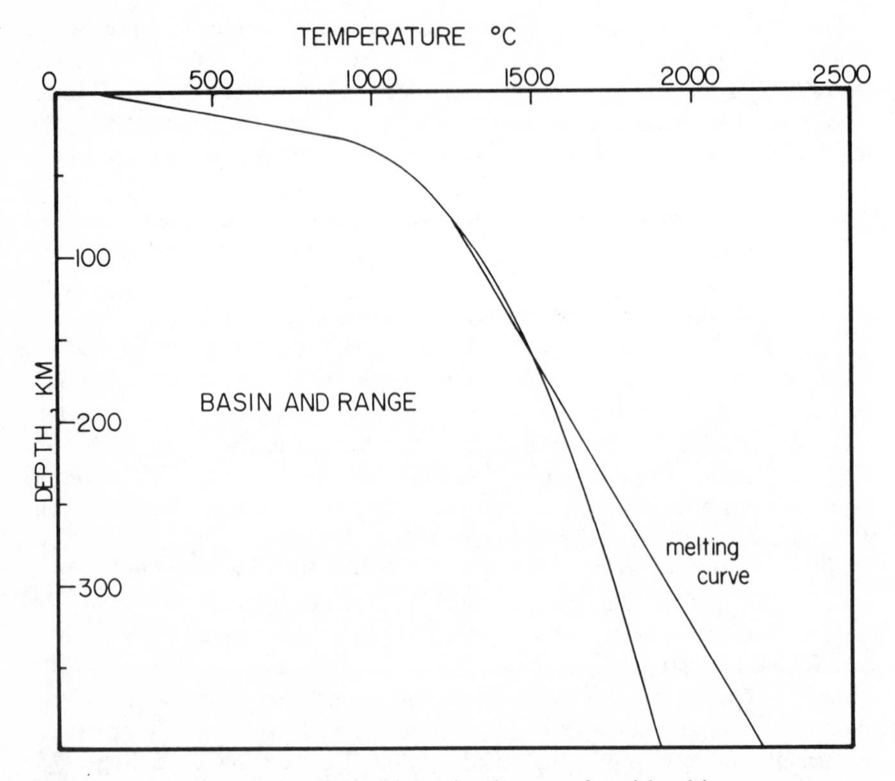

Fig. 2. Temperatures for the BR model showing the zone of partial melting.

constants used in these geothermal models are the same as those used by Roy, Blackwell, and Decker (1971) in the CS and BR provinces.

The assumed melting curve for the upper mantle is shown in Fig. 1. This curve, which has an intercept at 1015°C and a slope of 3°/km, is consistent with the melting of a slightly "damp" pyrolite (Ringwood, 1969). Figure 2 gives temperature as a function of depth below sea level for the BR model. The zero-depth temperature is 63°C because the average elevation in the central part of this province is about 1 km. The temperature profile intersects the melting curve at 80 and 159 km. Between these depths, partial melting occurs and the temperature profile follows the melting curve. Basaltic magma produced by partial melting at these depths would have, according to the model, a temperature of 1300 to 1400°C. The resulting BR and CS temperature profiles are shown in Fig. 3 and the temperature gradients are shown in Fig. 4. The temperatures at a depth of 350 km are 1823°C in the BR and 1630°C in the CS. The heat flow at this depth is slightly greater than 0.3 hfu in both models, but the temperature gradient is a bit higher in the CS because of the temperature effect on conductivity. All the imposed restraints are thus met by these geothermal models, and they can now be used to infer the physical differences in

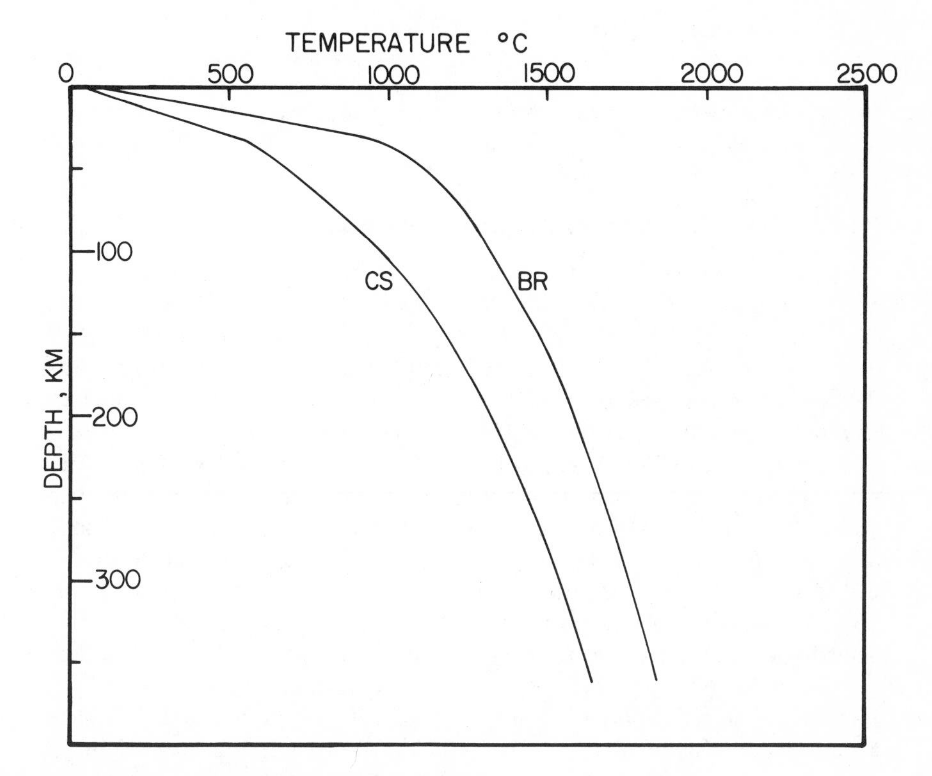

Fig. 3. Temperatures for the BR and CS models.

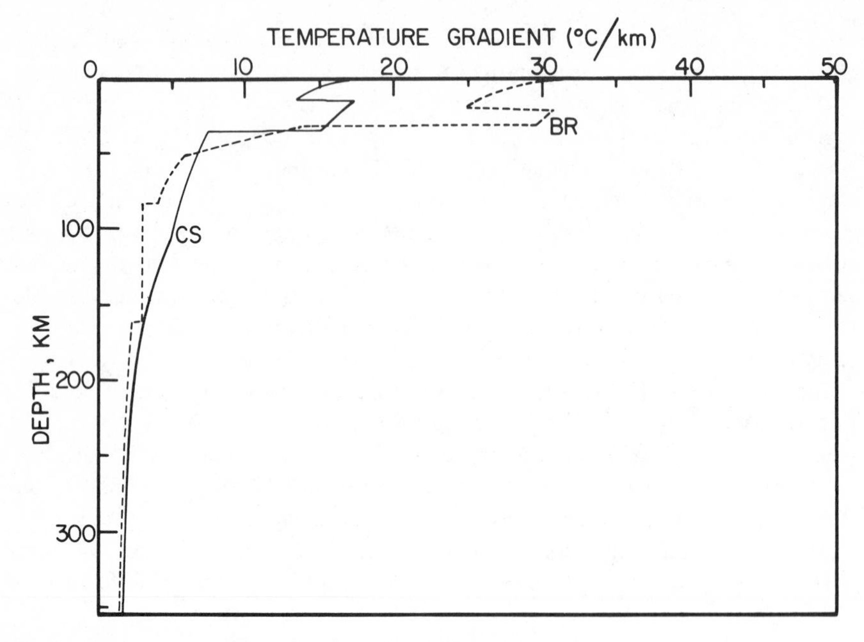

Fig. 4. Temperature gradients for the BR and CS models.

the upper mantle of the two provinces, which would result from differences in temperature alone.

PHYSICAL MODEL–CANADIAN SHIELD PROVINCE

The profile of Vp as a function of depth in this province (Fig. 5) is a composite from the work of Barr (1967), Green and Hales (1968), Hales (1969), and Mereu and Hunter (1969). The discontinuity at a depth of 90 km was described by Hales (1969) and attributed to a phase change from spinel peridotite to garnet peridotite. In this profile, there is a slight, but continuous, increase in Vp with depth throughout the upper mantle. The profile of Vs as a function of depth (Figs. 5 and 6) is a modification of the CANSD shear wave model (Brune and Dorman, 1963) which allows an increase in Vs at the Hales 90-km discontinuity. The profiles show a jump in Vp of 0.30 km/sec and in Vs of 0.12 km/sec at 90 km. The Vs in the CS model shows a very slight decrease with depth in the upper mantle. In the mantle, the temperature gradients in the CS geothermal model are about 3 to 4°/km. According to Liebermann and Schreiber (1969), a gradient in this range would be expected to produce a slight increase in Vp and a slight decrease in Vs with depth in upper mantle rocks.

The physical parameters used in the CS model are shown in Table 2. These parameters are for granite in layer 1, gabbro in layer 2, and magnesium-rich olivine in layer 3. The values of the physical constants used in the models,

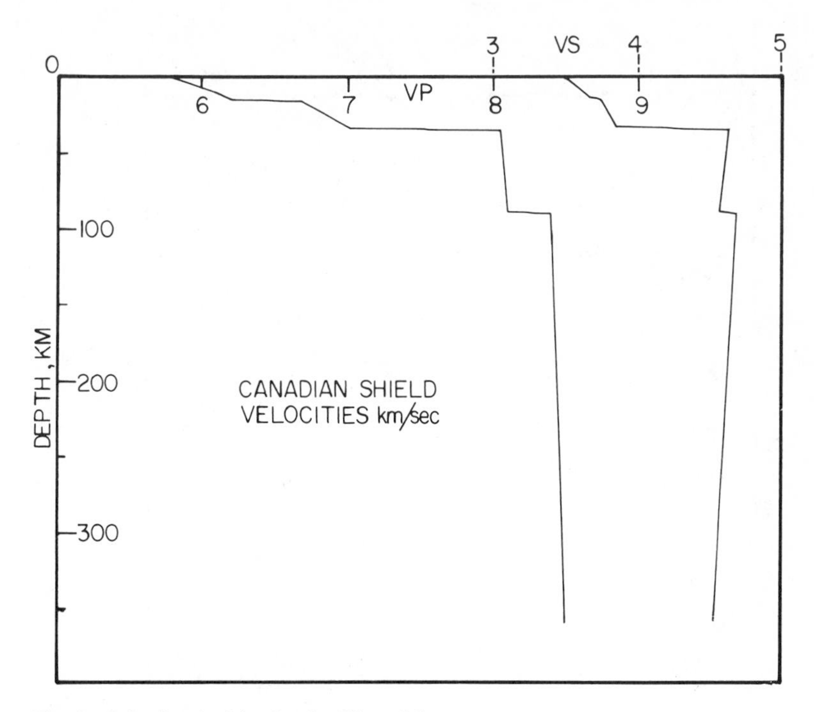

Fig. 5. Seismic velocities for the CS model.

including the heat capacity required to make the small correction from Bs to Bt, are from Birch, Schairer, and Spicer (1942), Clark (1966), Birch (1969), and Mao et al. (1970). The density change at 90 km is inferred from the change in velocity and is supported by the work of Press (1969). The change in Bs at 90 km is calculated from the values of ρ, Vp, and Vs in the model.

Pressure and density as a function of depth for the CS model were calculated using the parameters in Table 2 (Bt was calculated from Bs). The seismic parameter ϕ was then computed from

$$\phi = (\rho \text{ Bs})^{-1}$$

and is shown as $\phi 2$ in Fig. 7. A second estimate, $\phi 1$, was calculated from the seismic velocities. The physical model and velocity model were judged to be consistent if the two estimates of ϕ agreed within 10%. The divergence of $\phi 1$ and $\phi 2$ within each layer (Fig. 7) occurs because no attempt was made to model the change in Bs with temperature and pressure.

PHYSICAL MODEL—BASIN AND RANGE PROVINCE

Using the physical constants from the CS model, it is now possible to determine whether the seismic data from the central BR is consistent with an

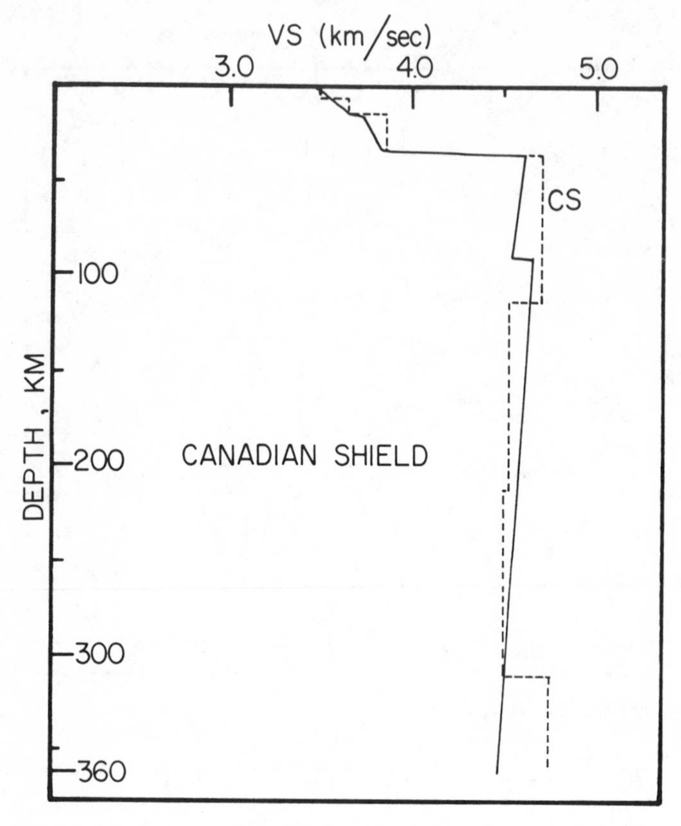

Fig. 6. Shear velocity for the CS model. The dashed curve (CANSD) is from Brune and Dorman (1963).

upper mantle of the CS composition at the higher temperatures implied by the BR geothermal model. The profile of Vp as a function of depth in the BR province, from Hill and Pakiser (1966) and Archambeau, Flinn, and Lambert (1969), is shown in Fig. 8. The low-velocity layer for *P* waves in this profile coincides with the zone of partial melting implied by the BR geothermal model.

The physical model for the BR province, excluding the melting zone, uses the

Table 2. Canadian Shield model

Layer	Depth (km)	ρ (at top of layer)	Bs	α
1	16	2.64	2.0×10^{-12}	1.7×10^{-5}
2	35	2.90	1.2	2.0
3	90	3.40	0.8	3.6
4	360	3.50	0.6	4.0

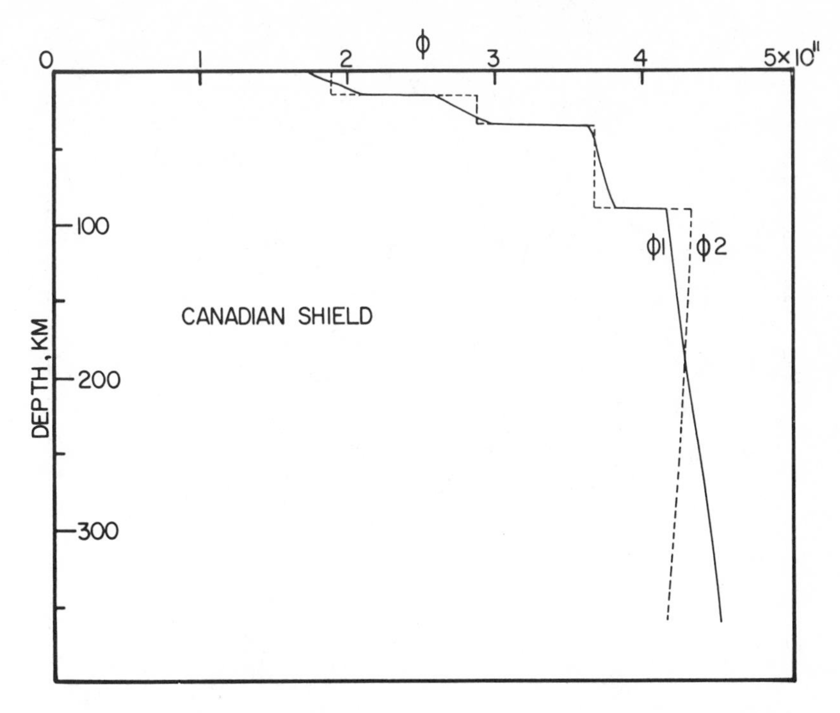

Fig. 7. The parameter ϕ for the CS model. $\phi 1$ is calculated from the seismic velocities. $\phi 2$ is calculated from the density and compressibility.

same physical constants as used in the CS model. In the melting zone, the model used by Birch (1969) for olivine plus basaltic glass was used to determine the change in the physical constants. The percent melt required to produce the decrease in Vp was determined, then ρ, Bs, and the decrease in Vs were calculated in the zone of partial melting. These values are shown in Table 3.

The increase in Vp at the base of the low-velocity layer (Fig. 8 in this paper and Fig. 34 in Archambeau, Flinn, and Lambert, 1969) is probably too sharp to be explained by the termination of melting. In the BR model the Hales discontinuity, found at a depth of 90 km in the CS province, is postulated to occur near the base of the melting zone at a depth of 160 km. Lewis and Meyer (1968) report a jump in Vp of about 0.3 at a depth of 126 km from data obtained in Montana where the upper mantle temperatures are probably higher than those in the CS province. Thus, the phase change assumed to give rise to the Hales discontinuity would occur at about 28 kb and 1000°C (90 km) in the CS and at about 50 kb and 1500°C (160 km) in the BR province. The slope of the phase boundary would be about 23°C/kb. Ringwood (1970, p. 6464) shows a phase diagram for felspathic pyroxenite, with a composition determined from Apollo 11 samples, in the range 0 to 2000°C and 0 to 50 kb. The position and slope of the boundary between the pyroxenite and garnet pyroxenite fields for a

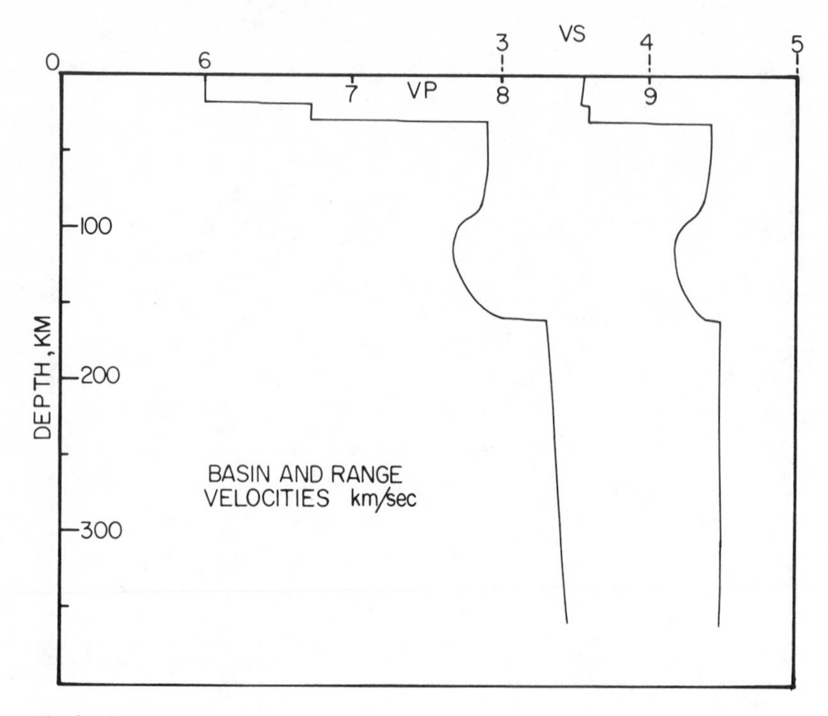

Fig. 8. Seismic velocities for the BR model.

composition containing 2 to 3% alumina (ρ = 3.48 gm/cm^3) nearly coincides with the presumed phase boundary for the Hales discontinuity. Similar data for compositions more likely to be present in the earth's upper mantle were not found by the writer; however, Ringwood's diagram suggests that the assumption that the same phase transition occurs at 90 km in the CS and 160 km in BR province and gives rise to the Hales discontinuity at these two depths is not unreasonable.

The profile of Vs as a function of depth shown in Fig. 8 has been constructed

Table 3. Basin and Range model

Layer	Depth (km)	ρ (at top of layer	Bs·	α	% Melt
1	19	2.640	2.00×10^{-12}	1.7×10^{-5}	–
2	30	2.900	1.20	2.0	–
3	80	3.400	0.80	3.6	–
4	95	3.392	0.81	3.7	1.5
5	135	3.368	0.82	3.9	4.0
6	160	3.389	0.81	4.1	2.0
7	360	3.500	0.66	4.3	–

to follow the P wave profile. In the low-velocity layer, the drop in Vs is determined by the percent of partial melting. In Fig. 9, the seismic parameter ϕ, estimated from the physical model ($\phi 2$) and from the seismic velocities ($\phi 1$), is plotted as a function of depth. The distribution of Vs outside of the partial melt zone was adjusted so that the two estimates of ϕ would agree within 10%. In Fig. 10, the Vs profile for the BR model is compared with the shear velocity determinations made by Wickens and Pec (1968) from an analysis of surface waves north of Tucson, Arizona. The fit is poor, and there is no evidence in the surface wave data for a low-velocity zone for shear waves similar to that implied by the P wave data.

A shear velocity model for the BR province was proposed by Pilant (1967) based on an extensive study of the phase and group velocities of Rayleigh waves crossing the North American continent. This model is in much better agreement with the implied shear velocity profile shown in Fig. 10 than is the Wickens and Pec model. Pilant shows a low-velocity zone in the mantle between about 100 and 200 km in which the shear velocity is 0.2 km/sec less than the velocity beneath the M-discontinuity. At the base of this zone, the velocity rises to 4.5 km/sec and remains at this value to 300 km. Although the low-velocity zone shown in Pilant's model is somewhat thicker and perhaps 20 km deeper than the

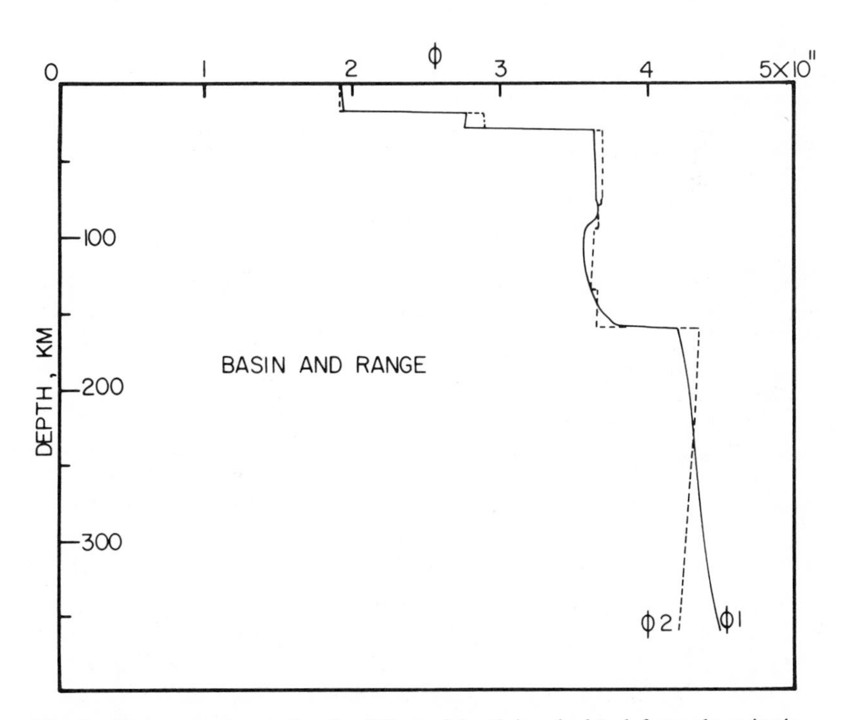

Fig. 9. The parameter ϕ for the BR model. $\phi 1$ is calculated from the seismic velocities. $\phi 2$ is calculated from the density and compressibility.

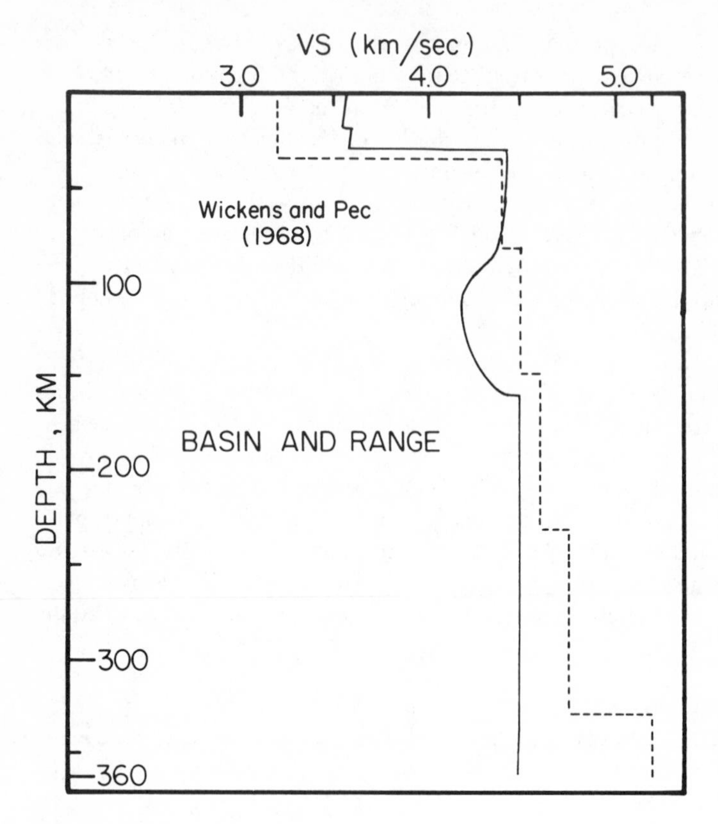

Fig. 10. Shear velocity for the BR model. The dashed curve is from Wickens and Pec (1968).

zone in the implied, shear velocity profile in Fig. 10, the differences may not exceed the resolution of the surface wave technique.

CONCLUSIONS

Density as a function of depth for the BR and CS physical models are shown in Fig. 11. From the density and velocity profiles, the difference in mass per unit area, P wave transit time, and S wave transit time from a depth of 360 km to sea level were computed for both provinces. Allowing for an average elevation of 1 km, the P wave delay in the BR province is 1.3 sec and the S wave delay is 4.0 sec relative to the CS transit times. These delay times generally agree with station corrections published by Cleary and Hales (1966), Herrin and Taggart (1968), and Doyle and Hales (1967). Hales (personal communication, 1970) suggests, however, that, based on more recent studies of shear wave travel times in North America, the S delays should be about four times the P delay in the BR province.

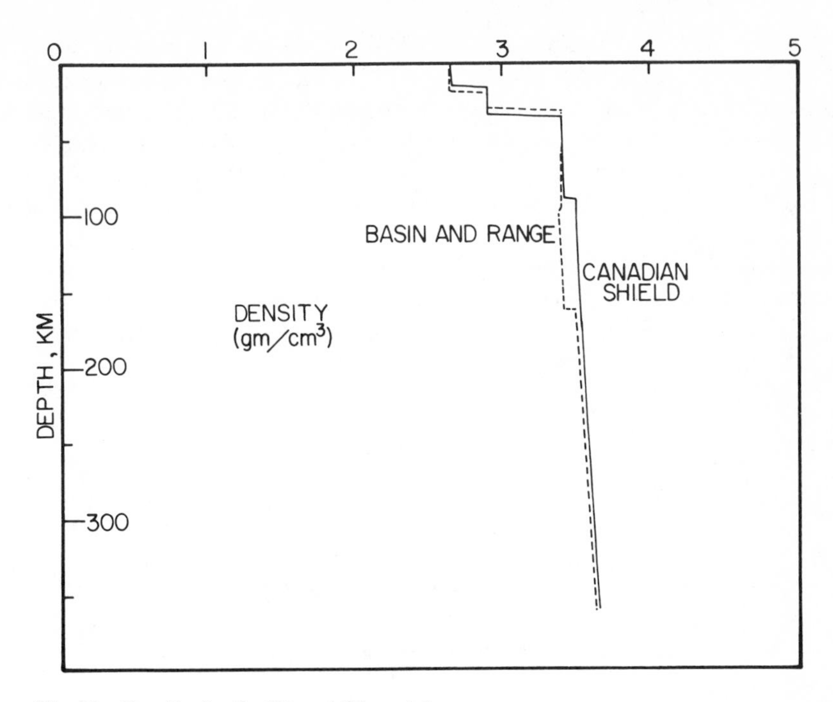

Fig. 11. Density for the BR and CS models.

The mass per unit area to a depth of 360 km in the BR province is less than that in the CS province by about 10^6 gm/cm^2. Assuming isostatic equilibrium, this mass difference is sufficient to support an average elevation of about 4 km, but the average elevation in the BR province is nowhere near that great. In discussing the melting model used in this study, Birch (1969) stated: "The numbers used in these calculations are such that the estimates of the amount of melt (required to produce the *P* wave low velocity zone) are probably upper limits, since the solid matrix has been assumed to be olivine." The melting model also assumes that the melted fraction exists in spherical cavities within the solid matrix. It is probable that a melting model in which the molten fraction fills nonspherical cavities and the solid matrix has a more realistic composition would predict a smaller fraction of melt in the low-velocity zone. In this case, the density between 80 and 160 km in the BR province would be greater than shown in Fig. 11. In addition, such a melting model would imply still lower values for Vs in the melt zone.

The increase in average density from the surface to 360 km beneath the BR province needed to make the density models consistent with the observed gravity data is about 0.02 gm/cm^3. Uncertainties in the density of crustal rocks, crustal thickness, thermal expansion, and, of course, the possibility of slight lateral variation in iron content in the mantle all tend to make the comparison of density models for the two regions rather difficult at this level of precision.

The geothermal and physical models for the CS and BR provinces presented in this paper are generally consistent with the available heat flow and seismic data. The BR shear velocity model is in fairly good agreement with the work of Pilant (1967) but disagrees with the Wickens and Pec (1968) model. Considering the difference between these two surface wave models, it is not possible with available seismic data to test adequately the BR shear velocity model shown in Fig. 10.

Within the limits imposed by the accuracy of available data, this comparative study supports the hypothesis that the physical differences in the upper mantle beneath the CS and the BR provinces can be explained by lateral differences in temperature.

ACKNOWLEDGMENTS

The author wishes to thank David Blackwell for the many valuable discussions that preceded the writing of this paper and Nancy Cunningham who wrote the computer programs used to make the model calculations. This work was partly supported by the Advanced Research Projects Agency under Air Force Grant No. AF-AFOSR 414-70 monitored by the Air Force Office of Scientific Research of the Office of Aerospace Research.

REFERENCES

Archambeau, C. B., Flinn, E. A., and Lambert, D. G., Fine structure of the Upper Mantle, J. Geophys. Res., **74**, 25, 5825-5865, 1969.

Barr, K. G., Upper Mantle structure in Canada from seismic observations using chemical explosions, Can. J. Earth Sci., **4**, 961-975, 1967.

Berry, M. J., and West, G. F., An interpretation of the first-arrival data of the Lake Superior experiment by the Time-Term method, Bull. Seism. Soc. Am., **56**, 141-171, 1966.

Birch, F., Density and composition of the upper mantle: First approximation as an olivine layer, *in* The Earth's Crust and Upper Mantle, Geophysical Monograph 13, edited by P. J. Hart, p. 18-36, American Geophysical Union, Washington, D.C., 1969.

Birch, F., Schairer, J. F., and Spicer, H. C., (Eds.), Handbook of Physical Constants, Geol. Soc. Am., Special Paper No. 36, 1942.

Birch, F., Roy, R. F., and Decker, E. R., Heat flow and thermal history in New York and New England, Studies in Appalachian Geology: Northern and Maritime, edited by E-an Zen, W. S. White, J. B. Hadley, and J. B. Thompson, Jr., pp. 437-451, Interscience Publishers, New York, 475 pp., 1968.

Brune, J., and Dorman, J., Surface waves and earth structure in the Canadian Shield, Bull. Seism. Soc. Am., **53**, 1, 167-210, 1963.

Clark, S. P., Jr. (Ed.), Handbook of Physical Constants, Rev. ed., Geological Society of America, Memoir 97, New York, 587 pp., 1966.

Clark, S. P., Jr., and Ringwood, A. E., Density distribution and constitution of the mantle, Rev. Geophys., **2**, 35-88, 1964.

Cleary, J., and Hales, A. L., An analysis of the travel times of *P* waves to North American stations, in the distance range 32° to 100°, Bull. Seism. Soc. Am., **56**, 2, 467-489, 1966.

Doyle, H. A., and Hales, A. L., An analysis of the travel times of *S* waves to North American stations, in the distance range 28° to 82°, Bull. Seism. Soc. Am., **57**, 4, 761-771, 1967.

Fukao, Y., On the radiative heat transfer and the thermal conductivity in the upper mantle, Bull. Earthquake Res. Inst., **47**, 549-569, 1969.
Green, R. W. E., and Hales, A. L., The travel times of *P* waves to 30° in the central United States and Upper mantle structure, Bull. Seism. Soc. Am., 58, 267-289, 1968.
Hales, A. L., A seismic discontinuity in the lithosphere, Earth Planet. Sci. Letters, **7**, 44-46, 1969.
Herrin, E., and Taggart, J., Regional variations in *P* travel times, Bull. Seism. Soc. Am., **58**, 1325-1337, 1968.
Hill, D. P., and Pakiser, L. C., Crustal structure between Nevada test site and Boise, Idaho, from seismic-refraction measurements, *in* The Earth Beneath the Continents, edited by J. S. Steinhart and T. J. Smith, pp. 391-419. Geophy. Monograph No. 10, American Geophysical Union, Washington, D.C., 1966.
Johnson, L. R., Array measurements of velocities in the upper mantle, J. Geophys. Res., **72**, 25, 6309-6325, 1967.
Kanamori, H., Fujii, N., and Mizutani, H., Thermal diffusivity measurement of rock-forming minerals from 400°K to 1100°K, J. Geophys. Res., **73,** 595-605, 1968.
Lachenbruch, A. H., Preliminary geothermal model for the Sierra Nevada, J. Geophys. Res., **73**, 6977-6989, 1968.
______, Crustal temperature and heat production: Implications of the linear heat-flow relation, J. Geophys. Res., **75**, 17, 3291-3300, 1970.
Lewis, B., and Meyer, R. P., A seismic investigation of the upper mantle to the west of Lake Superior, Bull. Seism. Soc. Am., **58**, 2, 565-596, 1968.
Liebermann, R. C., and Schreiber, E., Critical thermal gradients in the mantle, Earth Planet. Sci. Letters, **7**, 77-81, 1969.
Mao, N., Ito, J., Hays, J. F., Drake, J., and Birch, F., Composition and elastic constants of Hortonolite dunite, J. Geophys. Res., **75**, 20, 4071-4076, 1970.
Mereu, R. F., and Hunter, J. A., Crustal and upper mantle structure under the Canadian Shield from project Early Rise data, Bull. Seism. Soc. Am., **59,** 147-165, 1969.
Pilant, W. L., Tectonic features of the earth's crust and mantle, Final Report, Contract No: AF 49 (638)-1534, Research, Office of Aerospace Research, U.S. Air Force, August, 1967.
Press, F., The suboceanic mantle, Science, **165**, 174-176, 1969.
Prodehl, C., Seismic refraction study of crustal structure in the western United States, Bull. Geol. Soc. Am., **81**, 2629-2646, 1970.
Ringwood, A. E., Phase Transformations in the mantle, Publ. No. 666, Department of Geophy. and Geochem., Australian National University, Canberra, 31 pp. with appendix, 1968.
______, Composition and evolution of the upper mantle, *in* The Earth's Crust and Upper Mantle, Geophysical Monograph 13, edited by P. J. Hart, pp. 1-17, American Geophysical Union, Washington, D.C., 1969.
______, Petrogenesis of Apollo 11 basalts and implications for lunar origin, J. Geophys. Res., **75**, 32, 6453-6479, 1970.
Roy, R. F., Decker, E. R., Blackwell, D., and Birch, F., Heat flow in the United States, J. Geophys. Res., **73**, 16, 5207-5221, 1968.
Roy, R. F., Blackwell, D. D., and Decker, E. R. (in this volume) Continental heat flow, 1971.
Shankland, T. J., Pressure shift of infrared absorption bands in minerals and the effect on radiative heat transport, J. Geophys. Res., **72**, 2, 409-413, 1970.
Wickens, A. J., and Pec, K., A crust-mantle profile from Mould Bay, Canada, to Tucson, Arizona, Bull. Seism. Soc. Am., **58**, 6, 1821-1831, 1968.

10

GEOMAGNETIC DYNAMOS

by **Edward Bullard**

It is widely believed that the magnetic fields of the earth and the sun are produced by dynamos. In the earth, the dynamo action is supposed to be produced by motions in the fluid electrically conducting core. There is no direct demonstration of the existence of these dynamos; the reasons for believing in them are the absence of any other satisfactory theory and rather vague qualitative arguments about their possible properties.

There now exist powerful theorems about the circumstances in which dynamos can and cannot function including a proof by G. Roberts that almost all three-dimensional spatially periodic motions in infinite fluids are dynamos. There is a conspicuous lack of detail about possible terrestrial or solar dynamos. This is perhaps a result of the use of too highly symmetric models that do not satisfy Braginskii's criterion and also of the difficulty of including any realistic account of the dynamics in the theory.

INTRODUCTION

The dynamo theory of the origin of the earth's magnetic field was first suggested by Larmor and, after long neglect, was revived by Elsasser. Today it is fashionable for lack of anything better rather than from any great achievements of its own. Qualitatively, it is the right sort of theory; one feels that it does have a hope of accounting for the main facts; the trouble is that there is no detailed theory.

The magnetic field for which the theory attempts to account is a somewhat idealized version of the actual field. First, it is a theory of the part of internal origin; this removes most of the variations with periods of less than 5 years, the rest of these rapid variations are ascribed to induction in the mantle and ignored for the purposes of the theory. Then the spatial variations of short wavelength are ascribed to the magnetization of rocks near the surface and are also ignored.

The dynamo theory looks at the field through a spatio-temporal low pass filter and sees only the parts with wavelengths longer than 1000 km and periods above 5 years. The division is quite natural from the point of view of theory and is certainly justified but, in spite of statements to the contrary by the writer and others (Alldredge et al., 1963; Bullard, 1967; Bullard et al., 1962), there is no gap in either the wave number or the frequency spectra of the field to show where the division should be made.

The main features of the filtered field are that it is approximately (± 20%) a dipole field, that it changes by a large fraction of itself in a few hundred years, that it has repeatedly and rather accurately reversed its dipole component, and that the nondipole part and its rate of change have length scales of a few thousand kilometers, are quite complex, have a tendency to drift westward, and are not correlated with geology or geography (except for a systematic tendency to small rates of change over the Pacific).

Some of these features have been known since the 17th century, and Halley (1692) realized that the rapid changes imply an origin deep within the earth, in a place where relatively rapid motions can take place without catastrophic consequences. The argument is still valid; the place for a geomagnetic theory is the earth's core where, since the material is fluid, motions can take place with a time scale of hundreds rather than millions of years. There are not many ways in which a complicated changing, reversing field can be produced in a sphere of molten iron shut in a rigid container. The only plausible one is to assume that there are motions in the material and that these act as a self-exciting dynamo. The object of the dynamo theory is to develop the consequences of this idea. The main problems are the following.

1. Can such dynamos exist?
2. What sort of motions are required and what sort of fields are produced?
3. What forces are needed to drive such dynamos and what is their origin?
4. Will the field reverse and, if so, what is the pattern of the reversals in time?
5. Can the theory be extended to the sun and magnetic stars?

The first question is the only one to which we have an unambiguous answer; to the second we are rapidly approaching an answer; for the rest we have speculations of varying degrees of plausibility. In a sense these are meager results from 30 years of effort, some of it by very clever and sophisticated mathematicians. The inadequacy of the results is some measure of the difficulty of the problems.

EXISTENCE THEOREMS

In a bounded, stationary, electrically conducting body, any system of electric currents will decay. The field or the current may be analyzed into normal modes each of which decays exponentially with its own time constant. The time constant is proportional to kl^2, where k is the electrical conductivity and l is a characteristic length representing the distance in which the field changes by an appreciable part of itself. For a sphere of radius a, the most slowly decaying mode is reduced to $1/e$ in a time $\mu_0 ka^2/\pi^2$, where μ_0 is the permeability ($4\pi \times 10^{-7}$ Henry/m for a nonferromagnetic material). The core of the earth has a radius of 3500 km and a conductivity of about 3×10^5 ohm^{-1} m^{-1}. These give a time constant of 15,000 years, which is very short from a geological point of view. It is therefore essential to maintain the field; in a self-exciting dynamo, this is done by the currents produced by electromagnetic induction resulting from the movement of the conducting material through the field.

It is easy to show from Maxwell's equations that the field **B** must satisfy

$$\dot{\mathbf{B}} = \left(\frac{1}{\mu_0 k}\right)\nabla^2 \mathbf{B} + \text{curl}\,(\mathbf{v} \times \mathbf{B}) \tag{1}$$

$$\nabla \mathbf{B} = 0 \tag{2}$$

where **v** is the velocity of the fluid relative to the rotating earth. The only other condition is that at the boundary of the sphere all components of the field should be continuous with an external field that vanishes at infinity at least as rapidly as $1/r^3$, where r is the distance from the center of the sphere.

The $\nabla^2\mathbf{B}$ term causes the decay already discussed, and for a dynamo, it is essential that this term should be balanced and the decay prevented by the last term, which represents the interaction of the velocity and the field. This term may be written thus:

$$(\mathbf{v} \cdot \nabla)\mathbf{B} - (\mathbf{B} \cdot \nabla)\mathbf{v}.$$

The first part does not help much; it is the rate of change of the field in the direction of the velocity and merely represents the tendency of the motion to drag the field with it and therefore to make it increase at a point where the upstream field is greater and decrease where it is less; this term does not feed energy from the motion into the field. The second term is the rate of change of the velocity in the direction of the field and can be thought of as a stretching of the lines of force by the motion; it is this term that maintains the energy of the field.

It is far from obvious whether a given motion can maintain a field or, indeed, whether any motion can. There are a number of "non-dynamo theorems" that prohibit particular types of motion or field in a sphere from acting as dynamos, these exclude most kinds of symmetry. The theorems are of three kinds.

1. *Theorems that restrict the types of motion possible.* For example: the motion must have a radial component, and it must not be confined to planes perpendicular to an axis.

2. *Theorems that restrict the field (Cowling's theorem).* The field must not be symmetric about an axis and it must not have a closed line of force or of zero field around which the integral of the curl of the field does not vanish.

3. *Asymptotic theorems (Braginskii's theorem).* If the motion is nearly axially symmetric, the departure from symmetry must contain terms in both the sine and the cosine of some multiple of the longitude relative to the axis of symmetry. If this is so, the stream lines will have no plane of symmetry including the axis.

The methods of proof used for these theorems are somewhat *ad hoc,* and a more unified treatment is much to be desired. Some care is needed to remember just what has been proved; for example, an axially symmetric field cannot be produced by a dynamo, and it might seem natural to conjecture that an axially symmetric motion cannot act as a dynamo. Unpublished work by G. Roberts has shown this conjecture to be false; a symmetric motion can maintain an asymmetric field.

Proofs that particular motions can act as dynamos are few. Herzenberg (1958) has shown that a particular motion, the rigid rotation of two spheres within a spherical body of fluid, can act as a steady dynamo provided various kinds of symmetry are avoided (for example, the axes of the two spheres must not lie in the same plane). The proof involves expanding the velocity and the field in spherical harmonics and establishing convergence by a limiting process in which the rotating spheres become small. This is a matter of some delicacy, and it is still not known whether Herzenberg's dynamo would work in an infinite body of fluid. The corresponding dynamo with three rotating spheres will work even in an infinite fluid (Gibson, 1968).

Another existence proof has been provided by Backus (1958) who obtains convergence by stopping the motion at intervals to allow the higher harmonics to decay.

Recently, a remarkable series of investigations has been made by Childress (1969) and by G. Roberts (1970, 1971) into dynamos in infinite bodies of fluid. Roberts has shown that "almost all" motions spatially periodic in three dimensions act as dynamos. Of three-dimensional motions spatially periodic in two dimensions, he has shown that "about half" (i.e., not almost all and not only a set measure zero) act as dynamos. In unpublished work, he has further shown that these spatially periodic dynamos can be wrapped around an axis and enclosed in a sphere. Lortz (1968) has given an example of a dynamo with a motion periodic in two dimensions in an infinite fluid for which the equations have a simple solution.

G. Roberts' dynamos and those of Childress are remarkable in that, although the motion extends to infinity, the field does not. Also the field, unlike the motion, is not spatially periodic. The periodic motion maintains a field on a larger scale than itself; this large-scale field has a ripple with the wavelength of the motion superposed on it. Another remarkable feature of these dynamos is that they will work for indefinitely slow motions. As the motion gets slower the scale of the field must increase, otherwise the curl $(\mathbf{v} \times \mathbf{B})$ could not continue to balance the dissipative term. Such an increase in scale is only possible in an infinite body of fluid, and it is not to be expected that this feature will persist in a sphere.

Moffat (1970) has considered dynamos with small-scale random motions supporting a large-scale field and has shown that the motion must have no center of symmetry. The results are in some ways similar to those of Roberts.

FIELDS AND MOTIONS

It is not difficult to develop formal methods for computing the field produced by any specified motion. The direct solution of (1) by finite difference methods on a grid of points is not convenient, since it is awkward to fit the solution to an external field that vanishes at infinity and to satisfy (2). Expansion in spherical harmonics removes both these difficulties and reduces (1) to an infinite set of differential equations containing the radial functions for each harmonic, their first and second radial derivatives, and their first time derivatives. For a given set of initial conditions, these could be integrated in time steps and the nature of the solutions exhibited. Little has been done along these lines and most workers have preferred to look for steady solutions by setting the time derivative to zero. If this is done, an infinite set of ordinary differential equations is left and the problem requires choice of a velocity field and finding an eigenvalue that gives the magnitude of the velocity that will support a steady field. A velocity below the lowest critical one will give a field that decays to zero. One above the highest critical velocity (if there is one) will give a field that increases without limit. It is not known whether there is a highest critical velocity or what happens if the velocity lies between two critical values. The answers may well depend on the form of the velocity field assumed.

The earlier calculations were strikingly unsuccessful because of the difficulty of including enough spherical harmonics to give a reasonable approximation to the true solution or, indeed, to establish that such a solution exists. An attempt by Bullard and Gellman (1954) including harmonics up to degree and order four looked promising, but it was later shown by Gibson and P. Roberts (1969) that the solution including harmonics up to degree and order five differed greatly from the earlier one. The trouble may be that the velocity field chosen was approximately symmetric about an axis and had planes of symmetry. It therefore violated Braginskii's criterion and could not act as a dynamo in the limit of vanishing asymmetry and might well fail for small asymmetry. Lilley (1970) introduced asymmetry and obtained a smaller critical velocity and better convergence. Lilley's motion does possess a center of symmetry; perhaps, in view of Moffat's result, it might be wise to remove this. The methods used by Lilley are capable of improvement, and it seems that it should be possible to get good approximations to the solutions using existing computers.

Perhaps better convergence would be obtained by using a velocity field with a shorter wavelength. The magnetic field would then be expected to divide into two parts, a large-scale field analogous to that in G. Roberts' calculations (1970) for an unbounded fluid and a part with the wavelength of the velocity field.

An alternative approach to the theory of dynamos is to consider the phenomena involved in a pictorial way. The results do not prove anything in the strict sense, but they do give a "feel" for the subject and have been a guide to the kind of systems that are worth analytical or numerical investigation.

The basis of such studies is the idea that lines of force tend to move with the fluid. In a perfect conductor, this is precisely true in the sense that a closed curve that moves with the fluid always embraces the same magnetic flux no matter how it is deformed by the motion. In an imperfect conductor, the lines of force may be thought of as slipping through the fluid and behaving somewhat like elastic strings in a very viscous liquid (P. Roberts, 1967). In particular a closed loop of field will tend to shrink to a point and disappear. It is easy to imagine motions that will result in an increase in magnetic energy in a perfectly

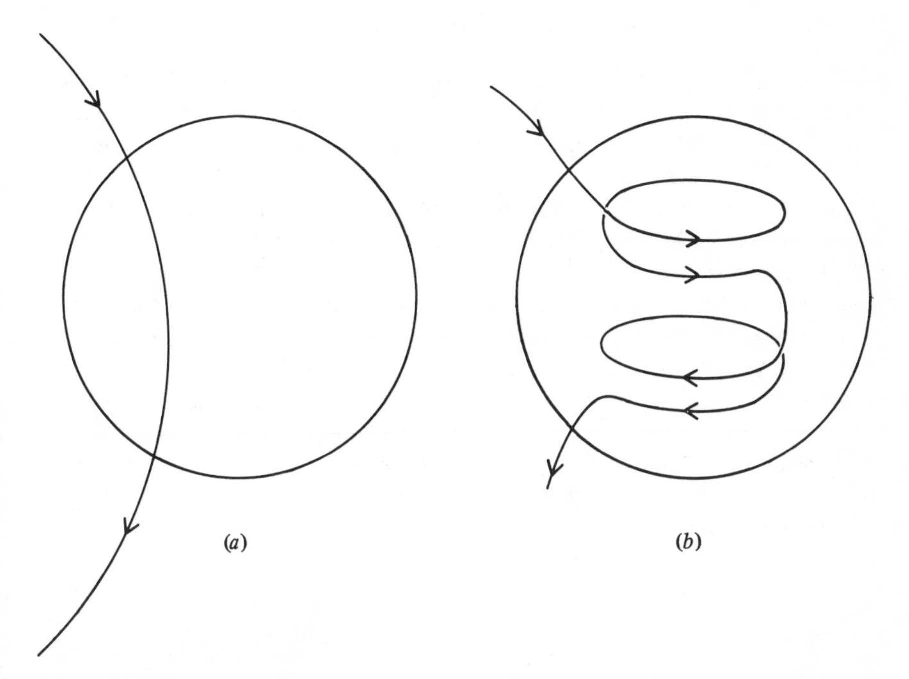

Fig. 1. Distortion of a line of force by rotation of the inner part of a sphere relative to the outer part: (*a*) undistorted, and (*b*) distorted.

conducting fluid. For example, suppose a line of force runs through a sphere (Fig. 1*a*) in which the inner part is rotating more rapidly than the outer part. The line of force will be drawn out along lines of latitude (Fig. 1*b*) and may encircle the axis many times, the field being in opposite directions in the two hemispheres. Such a field without a radial component is called a toroidal field. It may be shown that in an imperfect conductor the field along circles of latitude is $\pi^3 t_1/5t_2$ times the original field, where t_1 is the natural decay time and t_2 is the time needed for the inner part of the sphere to gain a complete rotation on the outer part. This is not a self-maintained dynamo, since it requires the original line of force to be provided and maintained against decay by some external agency. To produce a dynamo, we need a way of distorting the toroidal field to maintain the original line of force. One might, for example, suppose that a rising and twisting motion in the fluid pushed out a loop and rotated it into the meridian plane. It is easy to be deceived by such arguments. For example, one might suppose that a current rising all round the equatorial plane and sinking at the poles would be able to keep an equatorial ring of current suspended against its own tendency to collapse. In fact, this violates Cowling's theorem and is impossible. Perhaps it is fair to say that if we know that a particular dynamo works, such arguments will give a physical picture of what is happening. If we do not know whether it works, however, such arguments are not to be trusted.

THE FORCES DRIVING THE DYNAMO

Solutions in which Maxwell's equations are solved for specified velocities are of limited geophysical interest since there is no guarantee that there are forces in the earth's core that can sustain them. In a dynamical theory, the velocities would be calculated from assumed forces. Almost nothing has been done on this problem, largely because the failure to solve Maxwell's equations for plausible assumed velocities has discouraged attempts at the more difficult problem.

It is important to decide what terms in the hydrodynamic equations are important, and for this we need orders of magnitude of the relevant quantities. The angular velocity of the earth is 7×10^{-5} rad/sec, the corresponding peripheral velocity of the core is 200 m/sec, which is very large compared with the velocity of probable internal motions. The latter can be estimated from the westward drift of the nondipole field and from the condition that the motion must be capable of changing the field substantially in 100 years; 0.1 to 1 mm/sec appears reasonable. The equations of motion for the velocity **u** in inertial coordinates for an incompressible fluid are as follows:

$$\frac{\rho \partial \mathbf{u}}{\partial t} + \rho(\mathbf{u} \cdot \nabla)\mathbf{u} = \nabla p_1 + \mathbf{I} \times \mathbf{B} + \mathbf{F}_1 \tag{3}$$

$$\nabla \cdot \mathbf{u} = 0$$

where p_1 is the pressure, ρ the density, **I** the electric current (which equals curl $\mathbf{B}/\mu_0$), and $\mathbf{F}_1$ the applied force per unit volume. Let **v** be the velocity in coordinates rotating with the earth with angular velocity Ω, then

$$\mathbf{u} = \mathbf{v} + \Omega \times \mathbf{r}$$

since $|\Omega \times \mathbf{r}|$ is five orders of magnitude greater than $|\mathbf{v}|$, it is plausible to ignore terms in v^2 in comparison to those in $|\mathbf{v}| \cdot |\Omega \times \mathbf{r}|$. The acceleration, $\partial \mathbf{v}/\partial t$, may also plausibly be ignored; it is of the order of magnitude of the velocity divided by the time in which it changes appreciably, say 100 years, whereas $\Omega \cdot \mathbf{v}$ is of the order of the velocity divided by 4 hours. The term in Ω^2 may be incorporated with the pressure as also may a part, 1/2 grad B^2, of the electromagnetic force and any part of the applied force derivable from a potential. Let p be the pressure and **F** the force after these adjustments. Then, after a little rearrangement and transformation to rotating axes, (3) becomes

$$2\rho(\Omega \times \mathbf{v}) = \nabla p + \mu_0^{-1}(\mathbf{B} \cdot \nabla)\mathbf{B} + \mathbf{F} \tag{4}$$

This equation expresses a balance between the Coriolis forces, the pressure gradient, the electromagnetic forces, and the applied force. The pressure can be removed by taking the curl of (4) leaving an expression from which the curl of the force can be found for any dynamo for which the velocity and the field are known. If we had a solution to Maxwell's equations for a specified velocity field, we could calculate the forces needed to drive the system. To solve (1), (2), and (4) with specified forces is a harder problem, many schemes can be suggested but only experience can show which will work within the restrictions of available computing power. An iterative scheme, in which a trial velocity field is used to get the magnetic field from (1) and then (4) is used with the assumed forces and the computed field to give the next approximation to the velocity, might be attractive.

In the above discussion, the fluid has been assumed incompressible and viscous forces have been ignored. The reason for this is not so much the belief that they really are of no importance in the earth's core as that they are, in a sense, unnecessary complications. If we are to see our way in a complicated problem, it is necessary to study, not the real problem, but a skeleton of it that includes the essential structure. The object of computation is insight not numbers. If accurate numbers are needed, the inessential features can easily be added later.

In fact the range of density in the core is only about 20%, and the viscosity probably is negligible except in a boundary layer at the surface of the core. The force per unit volume, $\mathbf{I} \times \mathbf{B}$ in (3) or $\mu_0^{-1}(\mathbf{B} \cdot \nabla)\mathbf{B}$ in (4), is of order $\mathbf{B}^2/\mu_0 l$, where l is the scale length. If **B** is $10^{-2} T$ (100 oe) and l is 1000 km, this gives $8 \times 10^{-5}\ N/m^3$. The viscous force per unit volume is of order $\eta \mathbf{v}/l^3$, where η is the viscosity; the viscosity of the core is unkown but is unlikely to exceed 10^3 kg/m sec (a million times that of water). With this value and v = 1 mm/sec, the force is

10^{-12} N/m^3 for $l = 1000$ km and only becomes equal to the electromagnetic force for $l = 1$ m. The viscosity, the field, and the velocity are all only vaguely known, but it does seem that in neglecting the viscous forces we are not omitting an essential feature.

It is almost certain that there is motion in the core. Without it, there can be no dynamo, and it seems impossible to account for the short time scale of the magnetic variations. If there is motion, there must be forces to maintain it. In the absence of the term **F** in (4), the electromagnetic force cannot be balanced by the other two terms, since neither can supply energy, as can be seen by scalar multiplication by **v** and integrating over the whole fluid. Thus the energy dissipated by the electromagnetic forces would be supplied from the neglected term $\rho\partial\mathbf{v}/\partial t$, and the motion would be stopped in a time of order $\rho\mu_0 l\mathbf{v}/B^2$ which, with the values used above, is about 1 day.

The core is well protected from external influences, no great variety of forces can plausibly be supposed to produce the motion. The earlier workers suggested thermal convection resulting from radioactive heating. Whether, in fact, the radioactivity of the core is high enough and whether the processes for removing heat from the outside of the core are sufficiently effective to produce the temperature gradient needed to initiate convection is unknown. The main requirement is that the temperature gradient exceed the adiabatic; the heat flow required for this greatly exceeds the energy absorbed by the dynamo. To investigate this matter in detail, the equations of thermodynamics would need to be considered along with the electromagnetic and hydrodynamic equations (1) and (4). Nothing but order of magnitude estimates are available, but there seems no reason to suppose the process impossible.

Malkus (1968) has suggested that the precession of the earth may produce turbulent motion in the core and that this is the mechanism that drives the dynamo. It was previously thought (Lamb, 1932) that the core would precess with the mantle like a rigid body, but experiments by Malkus have shown that this is not so; presumably the state considered by Lamb (although a solution of the equations) is unstable. A detailed theory of a dynamo in a precessing turbulent core would be difficult, but the orders of magnitude seem reasonable, and it is possible that it is this mechanism that drives the terrestrial dynamo. It may be difficult to decide between the two sets of forces. Perhaps a study of planetary fields may be helpful. It would be interesting to know if Malkus's precessional turbulence can occur when the temperature gradient is below the adiabatic gradient.

REVERSALS OF THE FIELD

The rocks of the earth act as fossil compasses showing the direction of the field at the time they were formed. If we study a suite of rocks whose formation occupied a few thousand years at any time during the last 50 m.y., then the variations with periods of a hundred years or so are averaged out and we find that the mean direction of the horizontal component of magnetization points either to the north or to the south. For earlier periods, the results are complicated by continental drift and polar wandering but still show two opposite directions of magnetization. For many years there was controversy about the meaning of this result. Has the field repeatedly reversed, or have the rocks become magnetized in the opposite direction to the field? It is outside the scope of this study to discuss the evidence and arguments on this matter. A review with numerous references has been given by Bullard (1968). There now seems no doubt that the field has reversed in an apparently random way all through geological time. The interval between reversals is typically a few hundred thousand years, but intervals between reversals as long as 30 m.y. and as short as a few thousand years are known.

This remarkable phenomenon is a challenge to any theory of the origin of the earth's field and is sufficient to eliminate theories that ascribe the field to the magnetization of ferromagnetic materials within the earth. It is easy to see that dynamos can produce a field in either direction. Equation (1) is linear and homogeneous in the field and equation (4) is inhomogeneous and quadratic. Thus, if a given velocity distribution will support either a steady or a varying magnetic field, then it will also support the reversed field and the same forces will drive it. As far as it goes, this result is encouraging, but we require more. We need to show that not only does the reversed field satisfy the equations, but also that reversal will take place. We do not necessarily need a complete reversal of the field; a reversal of the dipole part would be sufficient.

No calculations have been made that show whether fluid dynamos will or will not reverse; in view of the difficulties experienced with calculations on steady dynamos, this is not surprising. A substantial amount of work has, however, been done on very simple systems that may give some hint of what to expect. The simplest dynamo is shown in Fig. 2; it consists of a disk that rotates in a magnetic field parallel to the axis of rotation. An emf is produced between the axle and the periphery of the disk; this drives a current through a coil and produces the magnetic field. If the disk is driven by a constant couple the

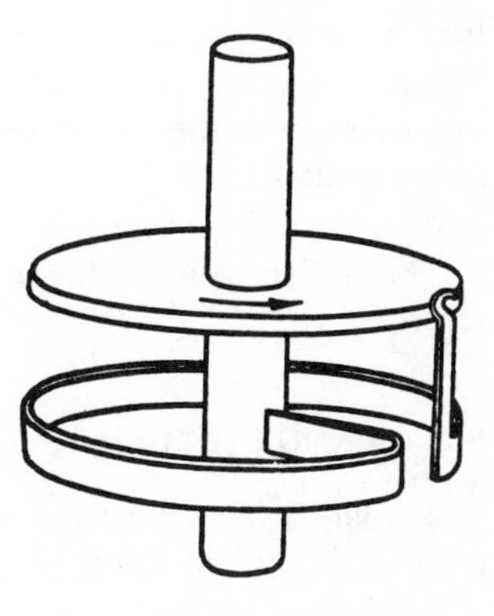

Fig. 2. A disk dynamo.

behavior of the system is easily investigated; it has two steady states, with fields in opposite directions. If the steady state is disturbed, a very curious aperiodic behavior may ensue, but the field does not reverse (Bullard, 1955). If two of these dynamos are connected together so that each disk feeds current to the coil of the other, then reversals do occur which have a surprising resemblance to the reversals of the earth's magnetic field (Allan, 1962). The period between two reversals is very irregular and shows no signs of periodicity.

The disk dynamo is far from anything that can be imagined to exist in the earth's core; there is, however, a certain similarity in the structure of the equations that control it to our Eqs. (1) and (3). In particular the equations do possess a nonlinear term containing the product of the velocity and the field, which corresponds to the $(\mathbf{v} \times \mathbf{B})$ term in (1) and a term in the product of the current and the field corresponding to $(\mathbf{I} \times \mathbf{B})$ in (3). The principal difference is that the disk dynamo does not contain anything corresponding to the Coriolis term, $(\Omega \times \mathbf{v})$ in (4). Also, of course, the disk, with its finite number of electrical and mechanical degrees of freedom is controlled by ordinary differential equations whereas the fluid dynamo has an infinite number of degrees of freedom and requires space as well as time derivatives and partial differential equations, such as (1) and (3). The prospect of getting reversals seems good and to demonstrate their occurrence and properties would be one of the great prizes of the subject.

If the equations give unstable solutions that flip from one direction to the opposite one, then a reversal has no specific "cause"; it is simply a consequence of the unfolding in time of the solution of the equations. It is not certain that this is the real state of affairs, it might be that the reversals were a result of disturbance of the motion by some catastrophic event, such as the fall of a very large meteorite. It is also possible that the cause is statistical and is associated with random fluctuations in the electric currents and the motions or the forces. It is known that random emf's can cause reversals in a disk dynamo (Bullard, 1955). Cox (1968) has suggested that random fluctuations in the nondipole part of the field may precipitate a reversal. He has developed a particular model in some detail and has shown that it leads to a Poisson distribution of time intervals between reversals. Comparison with the observed distribution then leads to an estimation of the parameters of the model. Inevitably the model is based on intuitions about the dynamo process rather than on any fundamental theory; it is nonetheless suggestive, and it seems possible, that the reversals of the field are associated with changes in the complex nondipole field. Parker (1969) has suggested that random fluctuations in the distribution of eddies produce the reversals.

SOLAR AND STELLAR MAGNETIC FIELDS

The sun has a complicated magnetic field that has been studied in some detail particularly by Babcock (1961). Fields of 1000 gauss or more appear in sunspots and frequently emerge from one spot of a pair and arch over above the sun's surface into the companion spot (Bray and Loughhead, 1964). It is tempting to believe that we see here a loop of toroidal field forced out from the interior of the sun. This view avoids the difficulties of time scale that are otherwise so perplexing. The field of a spot grows from almost nothing in a few days or weeks, whereas the electromagnetic time constant $\mu_0 k l^2$ must be many years. If the field is not created, but merely convected from a toroidal field in the interior of the sun, then the paradox disappears. Since the toroidal field of §3 runs along circles of latitude in opposite directions in the two hemispheres, it is natural that the two spots of a pair should be on a circle of latitude and that the polarity of pairs in the two hemispheres should be opposite, as they are observed to be.

In addition to the stray field associated with sunspots, the sun also possesses a field of a few gauss that bears some resemblance to a dipole field, or at any rate, has lines of force emerging near one pole and re-entering near the other. Both this field and the sunspot field reverse with the 11 year sunspot cycle, but not simultaneously. Babcock (1961) has sketched a theory similar to the pictorial theory in §3 to account for these facts, but no detailed calculations have been made.

Some stars are known to have magnetic fields, and some show reversals often associated with spectroscopic changes. In particular, the star HD125248 shows a waxing and waning of the spectral lines of chromium and europium synchronous with the reversals of the magnetic field. Almost nothing has been done to explain these very curious phenomena.

It is an advantage of the dynamo theory that it depends on a mechanism that will work both in the core of the earth and in the sun and stars. It is conceivable that the field of the sun might have an entirely different origin from that of the earth; it is, however, more satisfactory if the two phenomena can be regarded as manifestations of a single process.

Hope has been long deferred, but it does seem likely that, in the next few years, we shall have a version of the dynamo theory that carries conviction.

A more detailed account of many of the matters discussed in this paper can be found in a book by Rikitake (1966).

Note added in proof. Higgins and Kennedy (1971) have shown that the density in the earth's core may increase rapidly enough with depth for it to be in stable equilibrium. If this is so there will be no large scale steady motions and we must consider dynamos with wave-like motions. It seems that such oscillatory motions can maintain a steady field.

The emergence of this new type of dynamo is welcome since there has been unexpected difficulty in finding dynamos with large-scale steady motions. Further computations suggest that Lilley's dynamo, mentioned above, does not work.

REFERENCES

Allan, D. W., On the behavior of systems of coupled dynamos, Proc. Cambridge Phil. Soc., 58, 671, 1962.

Alldredge, L. R., Voorhis, G. D., and Davis, T. M., A magnetic profile round the World, J. Geophys. Res., **68**, 3679, 1963.

Babcock, H. W., The topology of the sun's magnetic field and the 22-year cycle, Astrophys. J., **133**, 572, 1961.

Backus, G., A class of self-sustaining dissipative spherical dynamos, Ann. Phys., **4**, 372-447, 1958.

Bray, R. J., and Loughhead, R. E., Sunspots, 303 pp., Chapman & Hall, London, 1964.

Bullard, E. C., The stability of a homopolar dynamo, Proc. Cambridge Phil. Soc., **51**, 744, 1955.

______, The removal of trend from magnetic surveys, Earth Planet. Sci. Letters, **2**, 293, 1967.

______, Reversals of the earth's magnetic field, Phil. Trans. Roy. Soc. (London), Ser. A, **263**, 481, 1968.

Bullard, E. C., and Gellman, H., Homogeneous dynamos and terrestrial magnetism, Phil. Trans. Roy. Soc. (London), Ser. A, **247**, 213, 1954.

Bullard, E. C., Hill, M. N., and Mason, C. S., Chart of the total force of the earth's magnetic field for the north-eastern Atlantic Ocean, Geomagnetica, **185**, 1962.

Childress, S., A class of solutions of the magnetohydrodynamic dynamo problem, *in* The Application of Modern Physics to the Earth and Planetary Interiors, edited by S. K. Runcorn, p. 629, Wiley & Sons, New York, 1969.

Cox, A., Lengths of geomagnetic polarity intervals, J. Geophys. Res., **73**, 3247, 1968.

Gibson, R. D., The Herzenberg dynamo, I and II, Quart. J. Mech. Appl. Math., **21**, 243, 1968.

Gibson, R. D., and Roberts, P. H., The Bullard-Gellman dynamo, *in* The Application of Modern Physics to the Earth and Planetary Interiors, edited by S. K. Runcorn, p. 577, Wiley & Sons, New York, 1969.

Halley, E., An account of the cause of the change of the variation of the magnetical needle; with an hypothesis of the structure of the internal parts of the earth, Phil. Trans. Roy. Soc., **16**, 563, 1692.

Herzenberg, A., Geomagnetic dynamos, Phil. Trans. Roy. Soc. (London), Ser. A, **250**, 543, 1958.

Higgins, G., and Kennedy, G. C., The adiabatic gradient and the melting point gradient in the core of the earth, J. Geophys. Res., **76**, 1870-1878, 1971.

Lamb, H., Hydrodynamics, 738 pp., Cambridge University Press, London, 1932.

Lilley, F. E. M., On kinematic dynamos, Proc. Roy. Soc. (London), Ser. A, **316**, 153, 1970.

Lortz, D., Exact solutions of the hydromagnetic dynamo problem, Plasma Phys., **10**, 967, 1968.

Malkus, W. V. R., Precession of the earth as the cause of geomagnetism, Science, **160**, 259, 1968.

Moffat, H. K., Turbulent dynamo action at low magnetic Reynolds number, J. Fluid Mech., **41**, 435, 1970.

Parker, E. N., The occasional reversal of the geomagnetic field, Astrophys. J., **158**, 815, 1969.

Rikitake, T., Electromagnetism and the Earth's Interior, 308 pp., Elsevier, Amsterdam, 1966.

Roberts, G. O., Spatially periodic dynamos, Phil. Trans. Roy. Soc. (London), Ser. A, **266**, 1179, 555-558, 1970.

______, Dynamos periodic in two dimensions, Phil. Trans. Roy. Soc. (London), Ser. A, (in press).

Roberts, P. H., An Introduction to Magnetohydrodynamics, 264 pp., Longmans, New York, 1967.

11

THE PACIFIC GEOMAGNETIC SECULAR VARIATION ANOMALY AND THE QUESTION OF LATERAL UNIFORMITY IN THE LOWER MANTLE *

by **Richard R. Doell and Allan Cox**

The excitation of geomagnetic signals in the earth's core may be partially controlled by lateral variations in the physical properties of the lower mantle or by undulations in the core-mantle interface. If this is true, variations in lower mantle properties will produce differences in the geomagnetic secular variation and strength of the nondipole field observed at different localities. From observatory records and paleomagnetic data, we show that these geomagnetic properties have been markedly attenuated in the central Pacific region for an extended period of time.

INTRODUCTION

Among the more important recent advances in geophysics has been the discovery through geophysical inversion techniques that the earth's upper mantle is surprisingly heterogeneous. There have been several attempts to determine whether the lower mantle is also heterogeneous and to discover something about the shape of the core-mantle interface. Geophysical information of two types has been used in this search: teleseismic signals and geomagnetic secular variation as recorded in observatories and paleomagnetically in rocks and archeological objects.

The main difficulty in the seismic approach is that any ray passing through the lower mantle also passes through two different parts of the upper mantle.

*Publication authorized by the Director, U.S. Geological Survey.

Therefore, it is difficult to ascribe differences, in travel times, for example, to heterogeneity along any particular part of the ray path. One may assume, as Vogel (1960) did, that the mantle is homogeneous and that observed differences in the travel times of waves reflected from the core are a result of variations in the radius of the core-mantle interface; the amplitude of these undulations then turns out to be about 100 km. This analysis does not rule out other models equally consistent with the data in which the core-mantle interface is a perfect ellipsoid and the mantle is heterogeneous.

In an attempt to reduce this inherent ambiguity of analyzing travel times, Alexander and Phinney (1966) studied *P* waves that had been diffracted by the core and restricted their attention to a parameter that depends mainly on the diffraction process itself. This parameter is the attenuation coefficient along paths of constant azimuth from an epicenter. Along ray paths diffracted by patches of the core beneath the Pacific Ocean, they found that at certain frequencies the attenuation coefficient has sharp absorption peaks which, on the basis of theoretical considerations, they attribute to the existence of a 30- to 160-km-thick layer at the base of the mantle that has reduced shear velocity and increased density (Phinney and Alexander, 1966). For rays diffracted by the core beneath the Atlantic, the absorption peaks are absent and a low shear velocity is not required. In a subsequent analysis, Phinney and Alexander (1969) interpreted the regional difference between the Atlantic and Pacific paths as being a result of variations in the radial gradient of the *P* wave velocity in the lower mantle. For the Pacific path an average gradient of 0.2 km/sec per 100 km is needed, whereas for the Atlantic path no gradient is required. For present purposes the importance of this line of research lies less in the details of the models than in the observational evidence that seismic parameters sensitive to physical properties in the lower 200 km of the mantle are different beneath the Pacific and the Atlantic regions.

The use of geomagnetic secular variation to determine mantle properties began with the classic determination of Lahiri and Price (1939) on the electrical conductivity of the upper mantle using short-period magnetic variations that originate in the ionosphere. The mantle is thicker than the electrical skin depth of the lowest frequency waves excited in the ionosphere, whose periods are only of annual and semiannual duration (Currie, 1966); thus this approach does not yield very much information about the lower mantle. More information about the conductivity of the lower mantle has come from analyzing magnetic signals generated in the earth's core because electrical induction in the lower mantle serves as a low pass filter to attenuate higher frequency signals during the propagation of these waves through the mantle (Elsasser, 1950; McDonald, 1957). From spectral analysis of observatory records over the period range of 40 days to 22 years, Currie (1967) determined that the cutoff period was about 4 years, and he used this to place constraints on the conductivity structure of the lower mantle. He noted, however, that models of conductivity are not unique because of uncertainty about the input spectrum of geomagnetic signals originating in the core. As he did not find a significant difference between the spectra of records from different parts of the world, there is no experimental evidence for lateral variations in the electrical conductivity of the lower mantle.

A different approach to interpreting the longer period part of the geomagnetic spectrum is based on the hypothesis that it is not attenuation in the mantle that varies laterally, but rather the spectrum of the geomagnetic source function (Cox, 1962; Cox and Doell, 1964). The nature of the magnetic signals originating in the core may be controlled by lateral variations in mantle properties or by undulations of the core-mantle interface in the following way. An essential feature of the geomagnetic dynamo theory of Larmor, Elsasser, and Bullard is that the fluid motion in the earth's core is partially controlled by Corioli's forces and therefore is similar to the planetary movement of the earth's atmosphere, both motions possessing first-order symmetry about the earth's rotation axis. Moreover, fluid motions in the earth's core and atmosphere both undergo transient departures from axial symmetry because of turbulence and other random processes. The short-term asymmetry disappears when averages are taken over time intervals that are long compared with the time constants of the turbulent motions. In addition, permanent departures from axial symmetry can be expected if the fluid is coupled to some condition at the fluid-solid boundary that is axially asymmetric. Both for the geomagnetic dynamo and for the earth's atmosphere, the boundary conditions that may be coupled with the fluid motion are (1) the temperature of the solid-fluid interface and (2) the topography of the interface. Both exert partial control on atmospheric motions, and either might be coupled to the flow of fluid in the earth's core (Cox and Doell, 1964; Hide, 1967).

The type of asymmetry we are concerned with in this study is lateral variation in the amplitude of the secular variation. Essentially, we wish to determine whether the earth's core is magnetically "stormier" beneath certain parts of the mantle than beneath others. Secular variation in the geomagnetic field arises primarily from changes in the irregular, or nondipolar, part of the geomagnetic field (Gaibar-Puertas, 1953; Vestine et al., 1947). Therefore, if it can be shown that there are significant lateral differences in the strength of the nondipole field, then this fact might also indicate that lateral differences exist in the lower mantle. In view of the present rather sketchy state of the theory of the geomagnetic dynamo, information of this type cannot be inverted to yield a model of the lower mantle. This approach, however, is able to give a yes or no answer to the question of whether lateral inhomogeneities or topographic irregularities exist at all. As we shall show, it appears that there is such a lateral inhomogeneity under the Pacific Ocean.

PRESENT GEOMAGNETIC FIELD

At the present time, the earth's field is far from axially symmetric, and it is likely that at any instant in the past, asymmetrical irregularities have been present comparable to those that exist today. The potential V of the magnetic field may be expressed as a series of spherical harmonic functions:

$$V = a \sum_{n=1}^{\infty} \sum_{m=0}^{n} \left(\frac{a}{r}\right)^{n+1} P_n^m(\theta)(g_n^m \cos m\phi + h_n^m \sin m\phi) \tag{1}$$

where a is the radius of the earth, r the radial distance of the point of observation, θ is the colatitude, φ the longitude, $P_n^m(\theta)$ are associated Lengendre functions, and g_n^m and h_n^m are the gauss coefficients. For a field with axial symmetry, the only terms that appear are those for which $m = 0$. Values of the gauss coefficients for the 1965 International Geomagnetic Reference Field (IGRF) (IAGA, 1969) are given in Table 1. Strong departure from axial symmetry may be seen in the relatively large value of coefficients with $m > 0$, which are comparable in magnitude to many of the zonal coefficients ($m = 0$). The largest departure from axial symmetry is produced by the equatorial dipole terms g_1^1 and h_1^1, which together with the g_1^0 term define the geomagnetic dipole. This dipole is inclined 11.4° from the rotation axis, lies in the meridional plane with longitude 69.8°W, and is situated at the earth's center. The magnetic field it produces is essentially the "best-fit" geocentric dipole field to the observed field over the entire earth. The geomagnetic dipole has not changed its position significantly during the time (about 200 years) that sufficient measurements have been available for spherical harmonic analyses; however, its strength has been decreasing at a rate of about 7% per century. Because of its geocentric location, changes in the geomagnetic dipole will not yield secular variation evidence for lateral inhomogeneities in the mantle.

The largest departures from the field of the geomagnetic dipole, including those with which the present study is primarily concerned, are best seen on contour maps showing the nondipole field. The following equations may be used to define the elements of the nondipole field. The nondipole potential is

$$V_N = a \sum_{n=2}^{\infty} \sum_{m=0}^{n} \left(\frac{a}{r}\right)^{n+1} P_n^m(\theta)(g_n^m \cos m\phi + h_n^m \sin m\phi) \tag{2}$$

The north, east, and vertical components are

$$X_N = \frac{1}{r}\left(\frac{\partial V_N}{\partial \theta}\right)_{r=a} \tag{3}$$

$$Y_N = \left(-\frac{1}{r}\sin\theta \frac{\partial V_N}{\partial \phi}\right)_{r=a} \tag{4}$$

Table 1. International Geomagnetic Reference Field 1965.0 coefficients

n	m	Main field (gammas)		Secular change (gammas/year)	
		g_n^m	h_n^m	$\dot{g}_n^m$	$\dot{h}_n^m$
1	0	- 30339		15.3	
1	1	- 2123	5758	8.7	- 2.3
2	0	- 1654		- 24.4	
2	1	2994	- 2006	0.3	- 11.8
2	2	1567	130	- 1.6	- 16.7
3	0	1297		0.2	
3	1	- 2036	- 403	- 10.8	4.2
3	2	1289	242	0.7	0.7
3	3	843	- 176	- 3.8	- 7.7
4	0	958		- 0.7	
4	1	805	149	0.2	- 0.1
4	2	492	- 280	- 3.0	1.6
4	3	- 392	8	- 0.1	2.9
4	4	256	- 265	- 2.1	- 4.2
5	0	- 223		1.9	
5	1	357	16	1.1	2.3
5	2	246	125	2.9	1.7
5	3	- 26	- 123	0.6	- 2.4
5	4	- 161	- 107	0.0	0.8
5	5	- 51	77	1.3	- 0.3
6	0	47		- 0.1	
6	1	60	- 14	- 0.3	- 0.9
6	2	4	106	1.1	- 0.4
6	3	- 229	68	1.9	2.0
6	4	3	- 32	- 0.4	- 1.1
6	5	- 4	- 10	- 0.4	0.1
6	6	- 112	- 13	- 0.2	0.9
7	0	71		- 0.5	
7	1	- 54	- 57	- 0.3	- 1.1
7	2	0	- 27	- 0.7	0.3
7	3	12	- 8	- 0.5	0.4
7	4	- 25	9	0.3	0.2
7	5	- 9	23	0.0	0.4
7	6	13	- 19	- 0.2	0.2
7	7	- 2	- 17	- 0.6	0.3
8	0	10		0.1	
8	1	9	3	0.4	0.1
8	2	- 3	- 13	0.6	- 0.2
8	3	- 12	5	0.0	- 0.3
8	4	- 4	- 17	0.0	- 0.2
8	5	7	4	- 0.1	- 0.3
8	6	- 5	22	0.3	- 0.4
8	7	12	- 3	- 0.3	- 0.3
8	8	6	- 16	- 0.5	- 0.3

$$Z_N = \left(\frac{\partial V_N}{\partial r}\right)_{r=a} \tag{5}$$

and the nondipole field intensity is

$$F_N = (X_N^2 + Y_N^2 + Z_N^2)^{1/2} \tag{6}$$

Values of the nondipole intensity, F_N, over the earth's surface are given in Fig. 1, and individual features may be seen to be characterized by wavelengths of about 60° and amplitudes of 0.12 to 0.16 oe. Of particular interest in the present study is the fact that the intensity of the nondipole field is unusually small over a region centered in the Pacific Ocean (Cox, 1962; Cox and Doell, 1964). Whether this low in nondipole intensity is a persistent characteristic of the earth's field or is a transient feature resulting from the random distribution of a small number of highs and lows over the entire earth is the main question we will attempt to answer.

The nondipole field intensity as plotted in Fig. 1 can be obtained only after spherical harmonic analysis of essentially instantaneous world-wide data have been made to yield the coefficients with $n \geqslant 2$. It is not directly comparable to paleomagnetic results that measure the total field. The nondipole field adds vectorially to the dipole field to change the direction and the intensity of the latter, and it is these two effects that may be seen in suitable paleomagnetic data. If $\mathbf{F}_D$ is the dipole field at some point on the earth's surface attributable to the dipole coefficients g_1^0, g_1^1, and h_1^1 and $\mathbf{F}_N$ the nondipole field vector as defined above, then their vector sum is $\mathbf{F}_T = \mathbf{F}_D + \mathbf{F}_N$ and the change in intensity ΔF resulting from addition of the nondipole field is given simply by the scalar difference in lengths:

$$\Delta F = F_T - F_D \tag{7}$$

where either large negative or large positive values indicate a large nondipole component. Values of ΔF given on Fig. 2 show that the Pacific region is characterized by small values of ΔF and by the absence of the well-defined highs and lows flanked by steep gradients that exist elsewhere. The ΔF diagram displays the component of the nondipole field $\mathbf{F}_N$ that is parallel to the dipole field $\mathbf{F}_D$.

The components of $\mathbf{F}_N$ perpendicular to $\mathbf{F}_D$ produce changes δ in the direction of the field, and it is these changes that can be most easily and reliably determined from the paleomagnetic record. The change in direction δ resulting from the nondipole field is given by

$$\delta = \cos^{-1} \frac{(\mathbf{F}_T \cdot \mathbf{F}_D)}{F_T F_D} \tag{8}$$

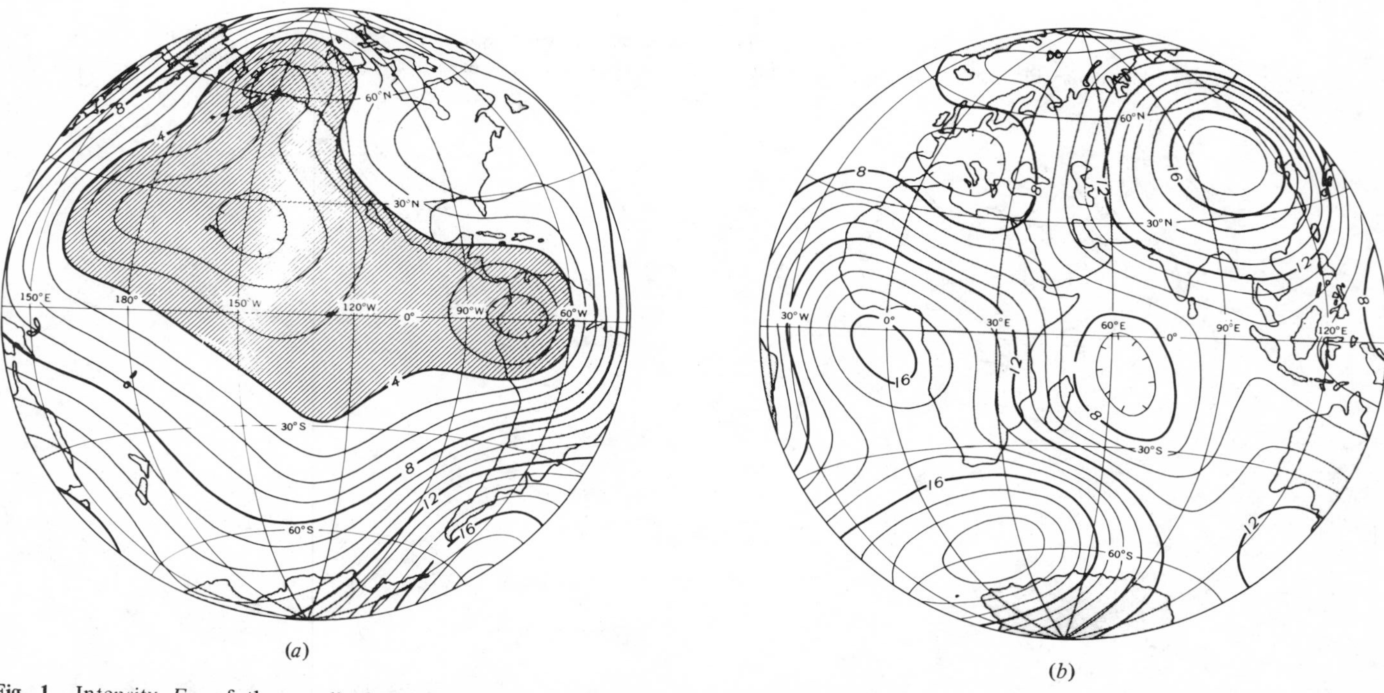

Fig. 1. Intensity F_N of the nondipole 1965 International Geomagnetic Reference Field (IGRF). Contour interval, 1000 gammas (100,000 gammas = 1 oe). Shaded areas, $F_N < 4000$ gammas. Maps are Lambert equal area with poles of projection on the equator at 130°W (*a*) and 50°E (*b*) longitude. The contours are based on values at grid intervals of 10°.

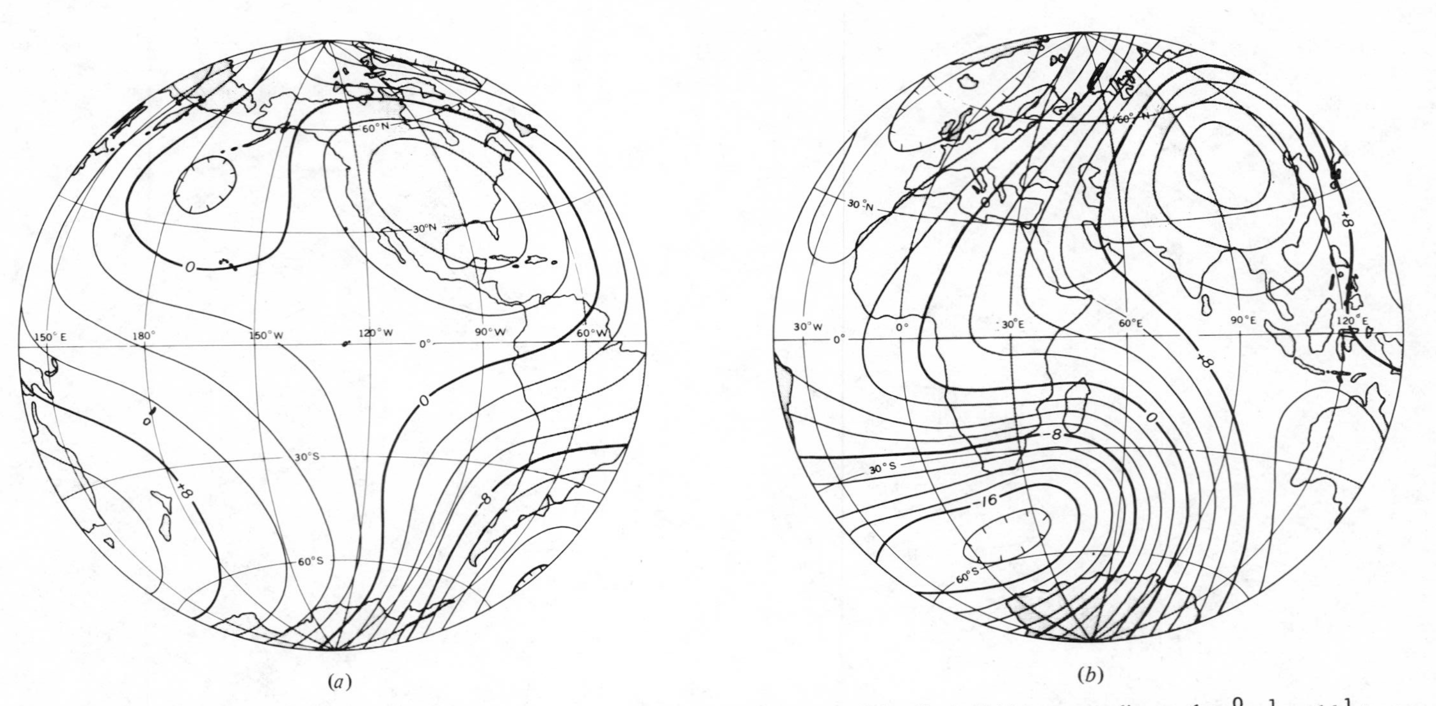

Fig. 2. Scalar difference ΔF between the intensity of the 1965 IGRF and the intensity of the dipole field corresponding to the g_1^0, g_1^1, and h_1^1 terms alone. Contour interval, 1000 gammas. Projection information and contour control as in Fig. 1.

The global distribution of values of δ may be seen in Fig. 3 to be generally small toward the equator and especially small in the central Pacific.

HISTORICALLY OBSERVED SECULAR VARIATION

During the several centuries spanned by observatory measurements, the nondipole field has been observed to slowly change in shape and to drift westward at a rate of about 0.2° of longitude per year (Bullard et al., 1950; Yukutake, 1962). A substantial part of the time derivative of the field at a fixed locality is equal to $\nu \, \partial F_N/\partial \varphi$, where ν is the velocity of westward drift and φ is the longitude. Hence, where the relief of the nondipole field is small, the secular variation should also be small. Several decades ago Fisk (1931) noted that the rates of secular variation were anomalously low in the Pacific. The rate of change calculated from the time derivatives of the gauss coefficients (Table 1) for the 1965 IGRF is also somewhat lower in the Pacific than elsewhere (Fig. 4), but it should be noted that experimentally the time derivatives are less well determined than the nondipole field itself.

The obvious way to determine whether the Pacific low is permanent or transient would be to make maps of the nondipole field for earlier epochs. Maps of the nondipole field for the years 1650, 1700, 1780, 1829, 1845, 1885, and 1922 have been constructed by Yukutake and Tachinaka (1968) from spherical harmonic coefficients for these epochs. Their maps for 1650 and 1700 show large values for the nondipole field in the central Pacific, but it is difficult to ascertain whether these are the result of using spherical harmonic functions to extrapolate from distant observational sites or whether they are real. Even for the 1780 field, there is considerable uncertainty about the nondipole field, as judged by large inconsistencies between two sets of coefficients calculated independently for this epoch (Yukutake and Tachinaka, 1968). By the 19th century, more measurements had been made and the results of the analyses are more consistent. The earliest reliable analysis is for the year 1829, for which values of F_N are shown in Fig. 5; a more accurate analysis for 1895 is shown in Fig. 6. The nondipole field intensity has been about as low in the central Pacific since 1829 as it is now.

These maps point to the main obstacles to the determination of whether the Pacific low is a persistent or transient phenomenon. The first is that the rate of westward drift of the nondipole field is only 0.2° of longitude per year. During the interval of 140 years for which adequate records are available, the total drift is only 28°. Thus, there has not been time for a major nondipole center to move into the central Pacific. The second problem is that the rate at which old features of the nondipole field decay and new ones form is so slow that there has not been time for the growth of a major new nondipole feature. The phenomena of greatest interest for determining the persistence of the Pacific anomaly appear to occur in a part of the geomagnetic spectrum just beyond the reach of direct observatory measurement.

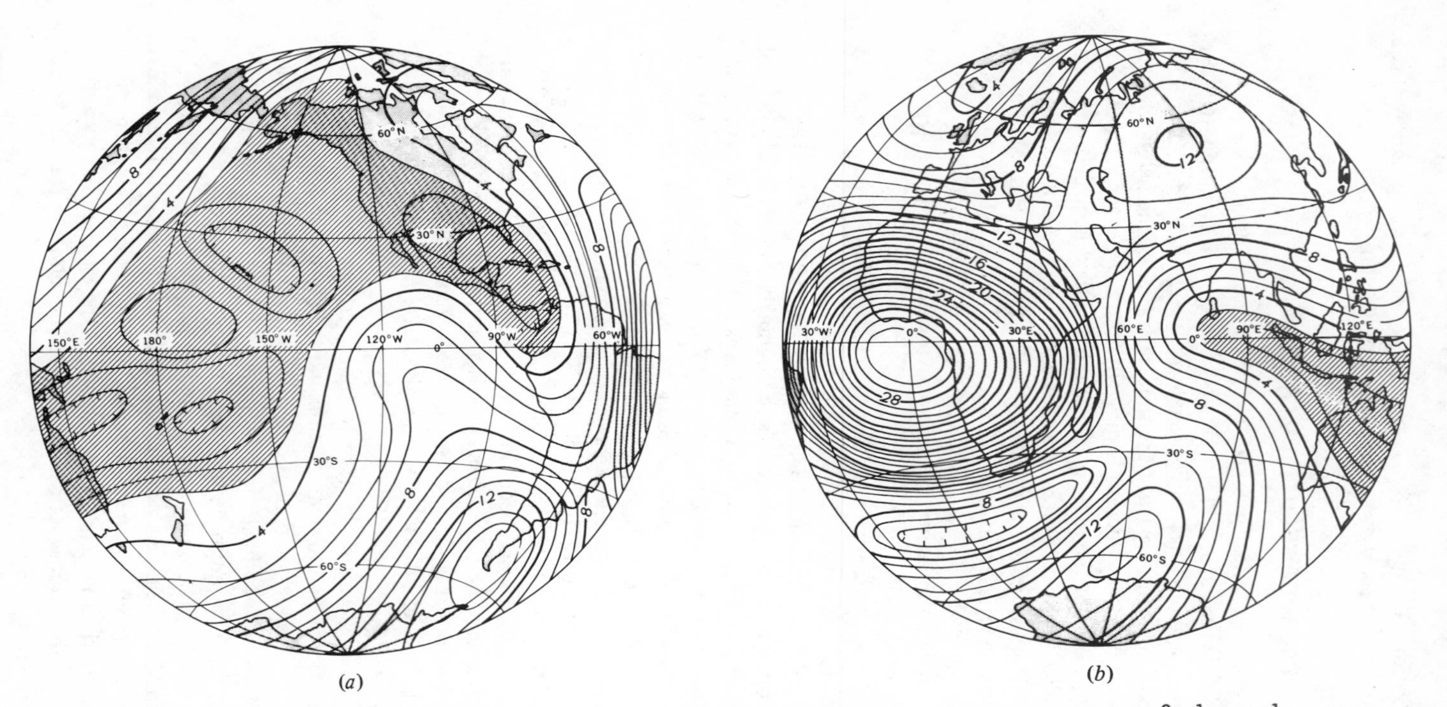

Fig. 3. The angle δ between the 1965 field and the best dipole approximation to the 1965 field, as described by the g_1^0, g_1^1, and h_1^1 coefficients of the 1965 IGRF. Contour interval, 1°. Shaded areas, $\delta < 3°$. Projection information and contour control as in Fig. 1.

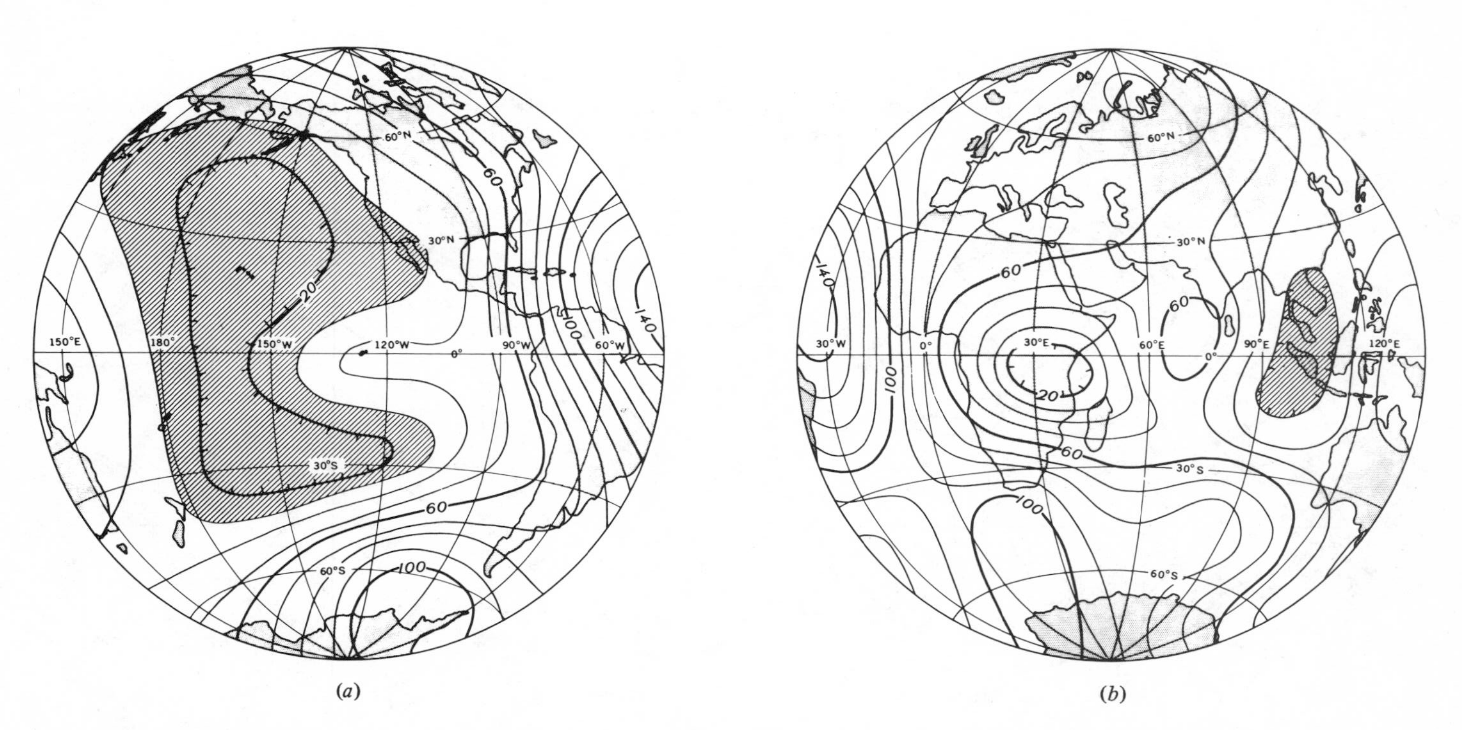

Fig. 4. Time derivative dF_N/dt of the nondipole field intensity. Contour interval, 10 gammas/year. Shaded area, rate of change < 30 gammas/year. In calculating dF_N/dt, Eqs. (2) to (6) were used, substituting $\dot{g}_n^m$ and $\dot{h}_n^m$ from Table 1 for g_n^m and h_n^m. Projection information and contour control as in Fig. 1.

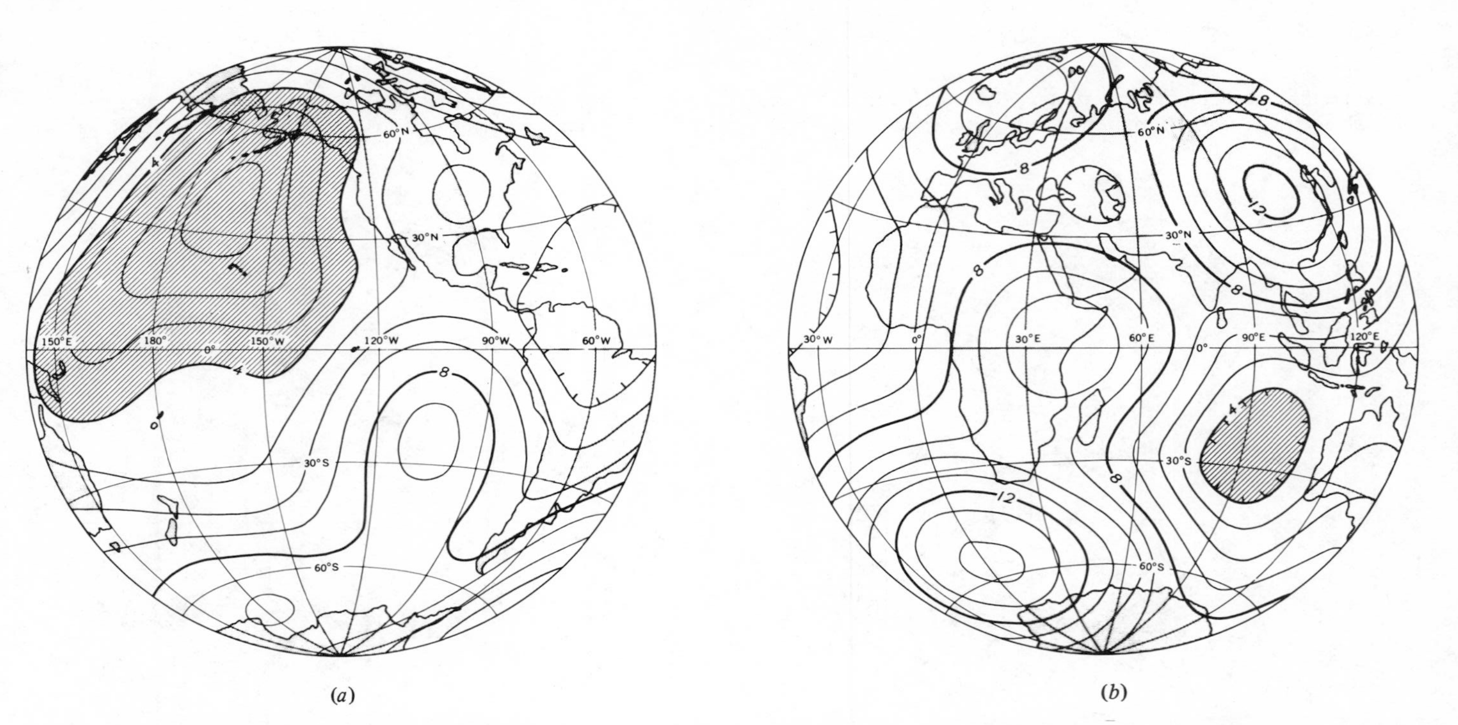

Fig. 5. Intensity F_N of the 1829 nondipole field. Contour interval, 1000 gammas. Shaded area, $F_N < 4000$ gammas. Reconstructed from Erman and Petersen's coefficients through $n = 4$ and $m = 4$ as given by Mauersberger (1959). Projection information and contour control as in Fig. 1.

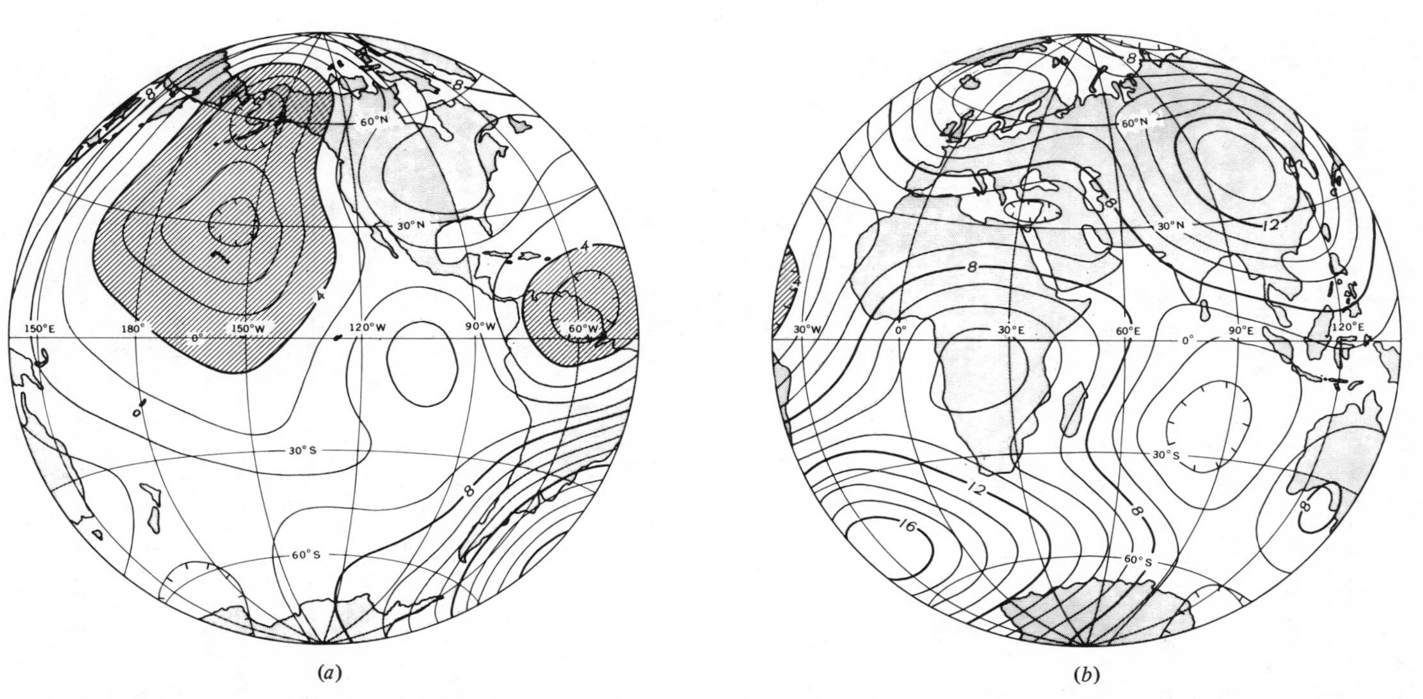

Fig. 6. Intensity of the 1895 nondipole field. Contour interval, 1000 gammas. Shaded area, $F_N < 4000$ gammas. Reconstructed from Schmidt's coefficients through $n = 6$ and $m = 4$ as given by Mauersberger (1959). Projection information and contour control as in Fig. 1.

GEOMAGNETIC SPECTRUM

The high-frequency end of the spectrum of geomagnetic field changes is well determined from observatory records. Two records are given in Fig. 7: Honolulu, Hawaii, where the nondipole field is small, and Hermanus, South Africa, where it is large. The step discontinuities are a result of changes in calibration or of site relocations. The continuous fluctuations are magnetic signals coming from the earth's core.

Spectral analysis of such records indicates that the power level $P(\tau)$ increases sharply with increasing period τ (Fig. 8), $P(\tau)$ being proportional to $\tau^{3.2}$ up to the longest period observed, which was 22 years (Currie, 1967). The steep rise in spectral power with period suggests that most of the power is present in periods much greater than 22 years. The power should continue to increase to periods of at least 700 years if, as appears to be true, a substantial part of the world-wide secular variation is produced by the westward drift of a nondipole field similar to the one that exists today. At the longest period end of the spectrum, substantial secular variation power is to be expected from wobble of the main dipole and from changes in its magnetic moment. A period of around 10^4 years may characterize both dipole wobble (Cox and Doell, 1964) and changes in magnetic moment (Cox, 1968; Smith, 1968) although neither period is yet well defined. These contributions to the power are shown schematically in Fig. 9. Also shown is a schematic spectrum for the central Pacific assuming that larger features of the nondipole field generally do not drift across nor originate in this area. The high-frequency parts of the two spectra are known from direct observation to be the same, and the lowest frequency end corresponding to dipole changes must also be the same, as these effects are world-wide. The difference between the spectra for the Pacific and other regions, if it exists, must be in the band between about 200 and 2000 years.

VIRTUAL POLE ANALYSIS

Before turning to paleomagnetic data in order to learn something about longer period secular variation, it will be useful to introduce the analytical techniques of calculating virtual geomagnetic poles. Spherical harmonic analyses and power spectra are not useful for analyzing most paleomagnetic data for two reasons. The first is that spherical harmonic analysis requires simultaneous measurements of the earth's field at numerous sites well distributed over the entire earth, and rocks suitable for paleomagnetic analysis of a given age are generally not at all well distributed over the earth's surface. Second, except for rocks and archeological objects 10^4 years old or less, the ages of rocks are not known accurately enough either to establish simultaneity for spherical harmonic analysis or to permit spectral analysis. Although it is possible in a few instances to establish sequence in a set of paleomagnetic data, the errors in age

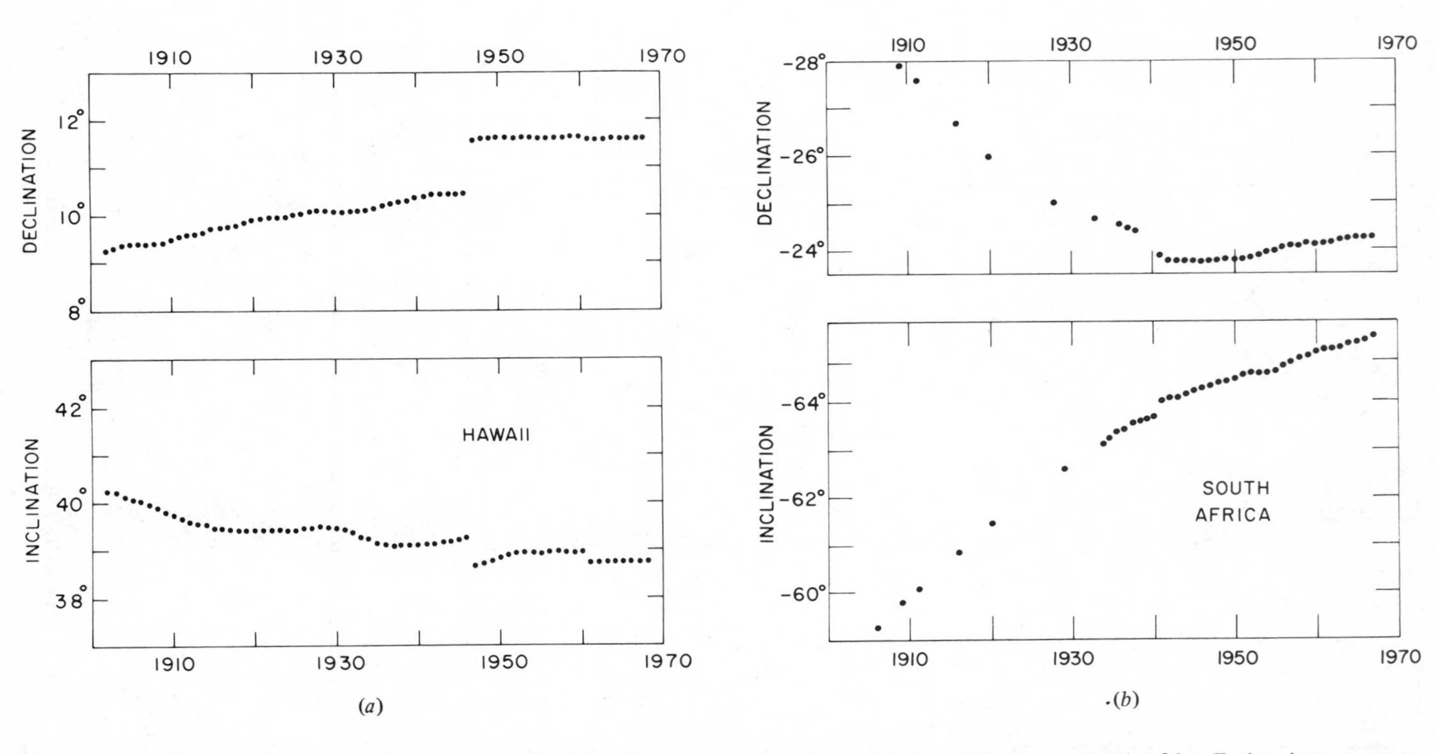

Fig. 7. Observatory measurements of magnetic inclination and declination at Honolulu, Hawaii, and Hermanus, South Africa. Each point represents an annual mean. Data through 1945 are from Vestine et al. (1947), after 1945 from the archives of the U.S. Coast and Geodetic Survey, Rockville, Maryland. Discontinuous offsets are a result of station relocation or recalibration.

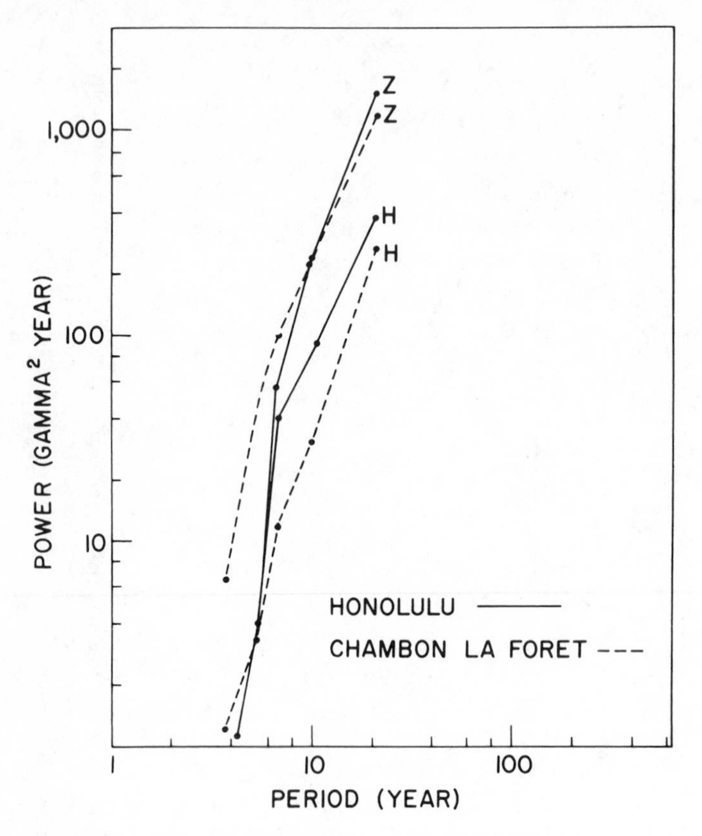

Fig. 8. Power spectra of observatory measurements of the vertical component Z and horizontal component H of the earth's field at Honolulu, Hawaii, and Chambon la Foret, France (Currie, 1967).

determinations are still generally larger than the longest periods of the secular variation. Other analytical tools are therefore needed.

Virtual geomagnetic poles provide a convenient method for comparing magnetic field directions at widely separated localities and also for assessing the relative intensities of the dipole and nondipole fields. Given a geocentric dipole field defined by the three gauss coefficients g_1^0, g_1^1, and h_1^1 the corresponding values of the inclination I, the declination D, and the intensity F are uniquely determined at any point on the globe. Conversely, given D, I, and F at any point on the globe, g_1^0, g_1^1 and h_1^1 are uniquely determined if it is assumed that the field is dipolar. If only D and I are known, then the ratio $g_1^0 : g_1^1 : h_1^1$ can be found, which determines the colatitude θ and longitude φ of the magnetic pole defined by these three coefficients. The geographic coordinates found in this way by inverting a pair of D and I values using the coordinates of the sampling site is termed a virtual geomagnetic pole. For a pure central dipole field ($g_n^m = h_n^m = 0$

for $n > 1$), the virtual geomagnetic poles calculated for points distributed over the surface of a sphere all coincide at one point determined by g_1^0, g_1^1, h_1^1. When there is a nondipole field containing terms with $n > 1$ added to the dipole field, then the virtual poles are dispersed, and a measure of the nondipole field is provided by the variance

$$S^2 = \frac{1}{N} \sum_{i=1}^{N} \Delta_i^2 \tag{9}$$

where N is the number of observations distributed over the sphere and Δ_i is the displacement of individual poles from the mean position of all poles.

Returning now to the present actual field, a set of values of D and I along a circle of latitude may be transformed to a set of virtual geomagnetic pole coordinates, which define an irregular path (Fig. 10). The pole paths for most latitude circles possess several loops, each corresponding to one of the main

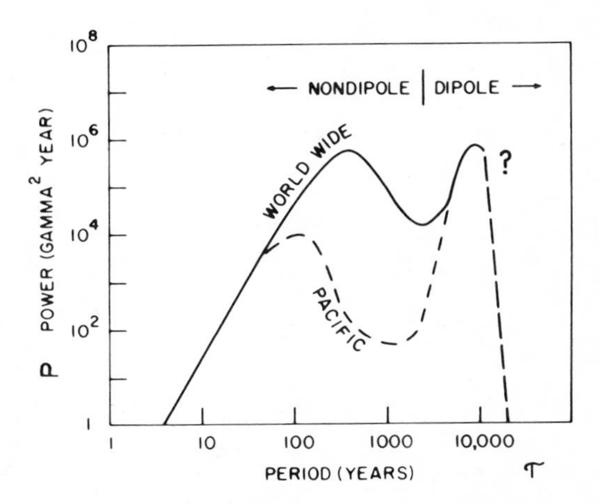

Fig. 9. Schematic power spectrum for the vertical component of the field. The initial slope fits the $P(\tau) \sim \tau^{3.2}$ relation found by Currie (1967) for short periods. The nondipole peak shows schematically the spectrum to be expected from the westward drift of the present nondipole field at a rate of 0.2° of longitude per year. The dipole peak has a period corresponding to the apparent period of changes in the intensity of the field during the past 4000 years, as determined from archeomagnetic results (Cox, 1968). Part of the dipole spectrum is a result of wobble of the main geomagnetic dipole, the frequency characteristics of which are not well determined. The breadth of the dipole peak is uncertain.

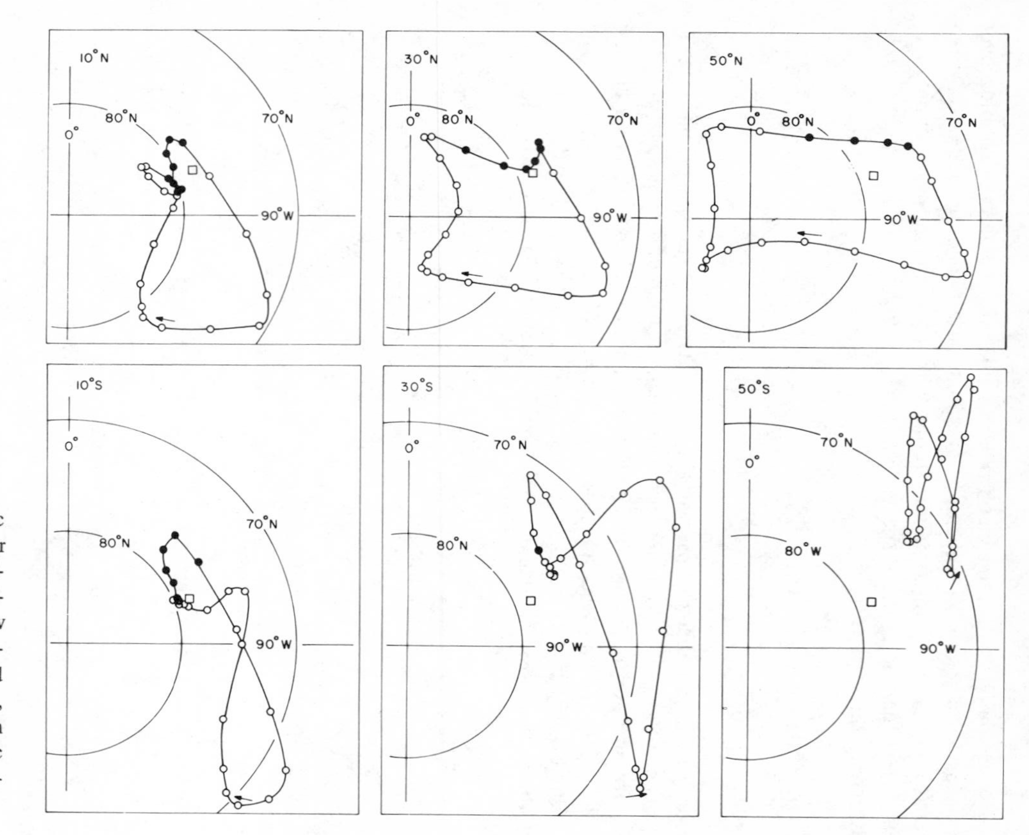

Fig. 10. Paths of virtual geomagnetic poles calculated for values of D and I for the 1965 IGRF along the circles of latitude indicated. Open circles mark longitude crossings at 15° intervals; the arrow is at 0° longitude pointing in the direction of increasing east longitude. Filled circles are in the central Pacific region, throughout most of which Δ is less than 3°. The square marks the geomagnetic pole. The map projection is polar Lambert equal area.

features of the nondipole field. The angular scatter of virtual poles is smallest at the equator and somewhat smaller in the northern hemisphere than in the southern.

Virtual poles calculated from values of D and I in the central Pacific (filled circles in Fig. 10) generally lie much closer to the average magnetic pole than those from other parts of the world. This is clearly seen in Fig. 11, where values of Δ appropriate to D and I at localities on a 10° grid have been contoured. Fig. 11 is therefore a one-to-one mapping of the angular departure of the observed field from the geomagnetic dipole field as contoured in Fig. 3 and thus exhibits similar features. It is worth noting, in Figs. 3 and 11, that low angular departures do not necessarily indicate the absence of a nondipole field. For example, the area of low δ and Δ values west of the Pacific Ocean near the equator contains considerable nondipole field, as may be seen in Figs. 1 or 2. As this portion of the nondipole field drifts westward, however, it will undoubtedly soon produce angular departures in at least some parts of this region. It is the long-term absence of angular departures in a given area that indicates the absence of the nondipole field.

PALEOMAGNETIC RESULTS FROM THE HAWAIIAN ISLANDS

The Hawaiian Islands, situated in the central Pacific Ocean and built up primarily of successive lava flows, provide a rather fortunate situation for paleomagnetic studies of the Pacific secular variation anomaly. There are two features that make this so. First, extrusive volcanic rocks of the Hawaiian types are ideal for paleomagnetic studies because they acquire a relatively strong and stable remanent magnetization after they have solidified during cooling. Moreover, it has been established that this magnetization very closely parallels the magnetic field in which the lavas cool (Doell and Cox, 1963), and under certain conditions, it has an intensity proportional to the intensity of the ambient field during cooling (Doell and Smith, 1969). This magnetization is therefore a "spot" measurement of the ancient field, as few Hawaiian lavas require more than a year in which to cool. The second favorable feature is that there has been more or less continuous volcanic activity in the Hawaiian Islands during the past several millions of years.

Historic Lavas

The existence on Hawaii of lava flows with ages that are well determined from historical records provides an opportunity to extend the magnetic record back beyond the beginning of observatory measurements in this area to determine whether the nondipole field was small in earlier times. With modern paleomagnetic techniques, it is possible to determine an ancient field direction from a lava flow with an accuracy of about 4° at the 95% confidence level (Doell and Cox, 1963). This is much less accurate than observatory measurements, yet accurate enough to detect changes in direction resulting from the presence of a

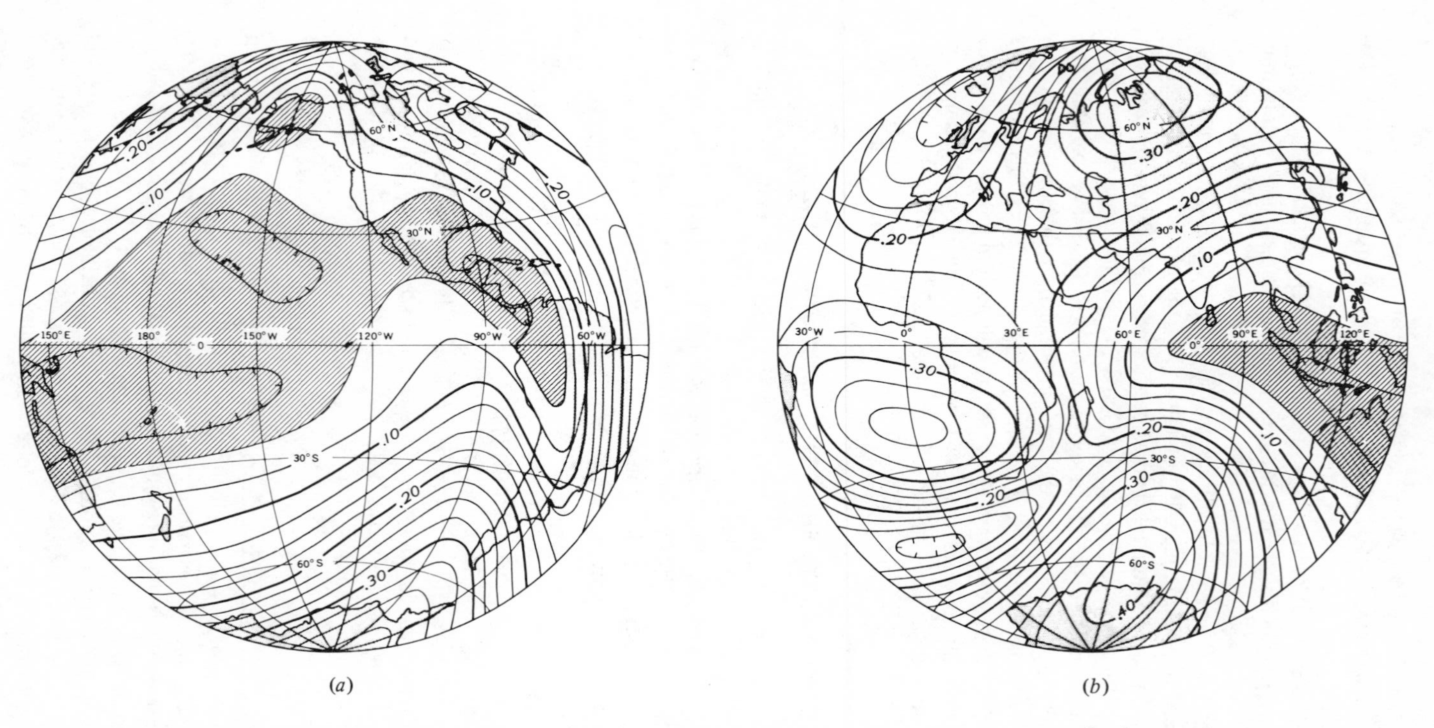

Fig. 11. Displacement Δ of virtual poles from the 1965 IGRF geomagnetic pole corresponding to the terms g_1^0, g_1^1, and h_1^1. At any locality the contours show the value of Δ corresponding to D and I at that locality. Field values were generated from the coefficients of the 1965 IGRF. Contour interval, 0.025 rad (1.43°). Shaded area, $\Delta < 0.05$ rad (2.89°). Projection information and contour control as in Fig. 1.

large nondipole field. As an example of the amplitude of such fluctuations, between 1750 and 1945 the direction of the earth's field changed by 45° on Ascension Island and by 22° at Capetown, Africa. Both would have been easily detectable paleomagnetically.

On Hawaii, paleomagnetic measurements were made on nine flows with known ages back to 1750 (Doell and Cox, 1963). Most of the values of inclination and declination shown in Fig. 12 are within 4° (shaded bands) of the magnetic field direction in Hawaii corresponding to the terms g_1^0, g_1^1, and h_1^1 of the 1965 IGRF. The only exceptions are the values of declination for 1750 and 1840. The angular difference of 3.5° between the paleomagnetic directions for the years 1840 and 1843 also suggests that much of the observed scatter is a result of experimental error. These results cannot preclude a 5% possibility of angular changes up to about 4° at individual data points, but they are also consistent at a

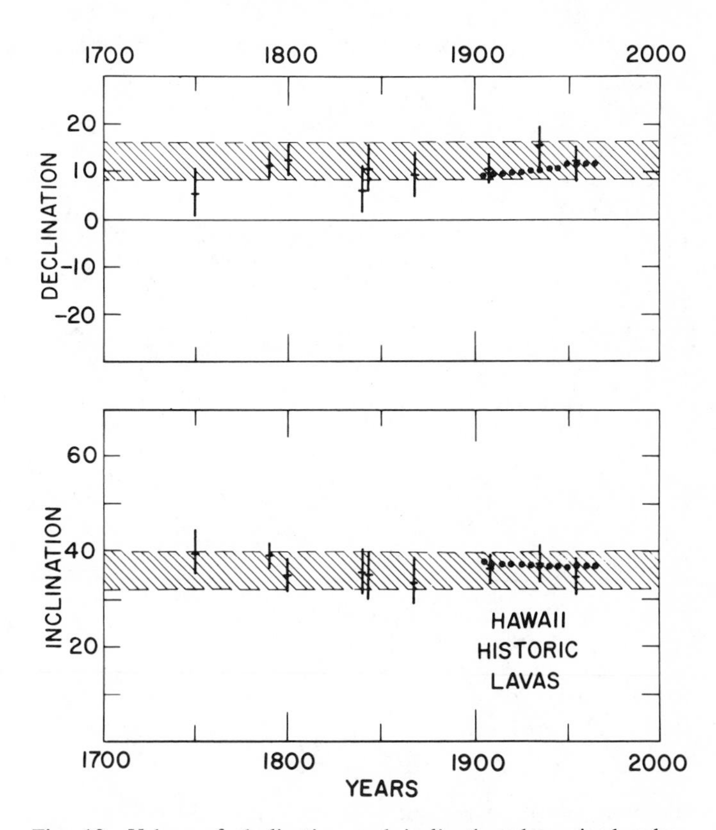

Fig. 12. Values of declination and inclination determined paleomagnetically from lava flows in Hawaii of known historic age (Doell and Cox, 1963). Points: observatory measurements at Honolulu extrapolated to the island of Hawaii. Shaded area: zone of ± 4° centered on the field in Hawaii attributable to the dipole terms (g_1^0, g_1^1, and h_1^1) of the 1965 IGRF. Vertical bars are 95% confidence limits.

much higher probability level, with an unchanging field having essentially the direction of the dipole component of the present field. These same data are shown in Fig. 13 by virtual pole representation along with VGP's calculated from other locations along the latitude circle of 20°N. The historic Hawaiian VGP's are much closer to the geomagnetic pole than is common at present for this latitude elsewhere. As an estimate of an upper limit on the nondipole field since 1750, an angular change of 4° would correspond to a nondipole field component of 2400 gammas normal to the dipole field component. A nondipole field component of 3400 gammas at an angle of 45° to the dipole field direction would also produce a change of 4°. Such values are within the 4000-gamma level that was shaded on Fig. 1 and are therefore consistent with the idea that the nondipole field in the central Pacific has been anomalously low throughout this extended period of time.

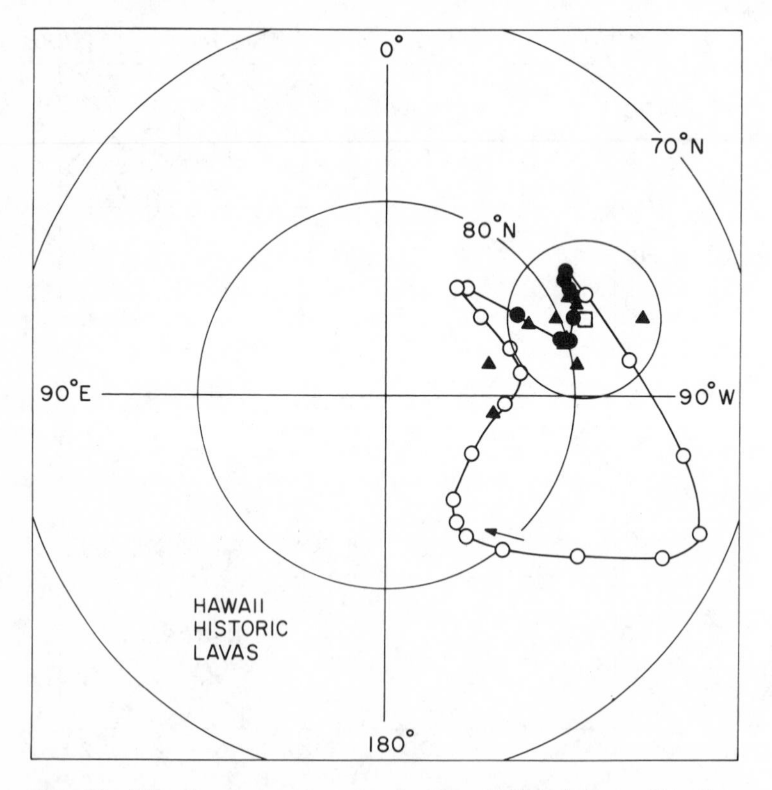

Fig. 13. Virtual geomagnetic poles for Hawaii historic lavas. Triangles, virtual poles corresponding to values of D and I from each flow. Open circles, virtual poles from values of D and I of the 1965 IGRF along 20°N latitude exclusive of central Pacific. Solid circles, virtual poles from values of D and I along 20°N latitude within central Pacific. Square, present geomagnetic pole. Large circle has a radius of 4° centered on the geomagnetic pole.

Kau Volcanic Series

Thousands of lava flows suitable for paleomagnetic measurements are exposed on the Hawaiian Islands. The geophysical usefulness of paleomagnetic data from these flows, however, is severely limited by lack of accurate information about the ages of the lavas. Radiometric dating techniques suitable for dating basalt, such as the potassium-argon method, lack the precision needed to resolve age differences much smaller than 10^5 years (Cox and Dalrymple, 1967), ruling out the possibility of formal spectral analysis. The situation is not completely hopeless, however. Where conditions are favorable, as they are for the Kau Volcanic Series, enough is known about the ages of the lavas from geologic evidence to permit some useful conclusions to be drawn about the ultra-low frequency characteristics of the geomagnetic field.

The Kau Volcanic Series is exposed near the summit of Mauna Loa, Hawaii, in a caldera wall 140 meters high. Accurate paleomagnetic measurements were made on 54 superimposed lava flows to produce the set of directions given in Fig. 14 (Doell, 1969). Each point shows the direction of the field at the time one of the flows cooled. The sequence of measurements in this "time series" is known from the superposition of the lava flows in the section, but the length of time between flows is unknown and can only be estimated indirectly.

The most useful piece of geologic information about the age of the Kau Volcanic Series is that flow number 1, at the bottom of the section, can be no older than 10,000 years. This follows from the observation that the section in the caldera is not deep enough to expose the locally ubiquitous Pahala Ash layer, whose age is 10,000 years as determined by the C^{14} method (Rubin and Berthold, 1961). The minimum age of the youngest flow in the series is determined by the date of collapse of the northwest wall of the caldera, which could have been as recent as several hundred years ago. The maximum duration of the entire series is therefore no more than about 10,000 years. An independent estimate of the total time represented has been made by calculating the time required to add a layer 145 meters thick to the summit of the volcano, assuming that the rate and style of volcanic extrusion was the same as it has been since 1832. The result of this calculation was that 9000 years would be required (Doell, 1969), which is consistent with the previous estimate. The average time between successive measurements of the field is about 200 years, although the actual time between successive eruptions of the field was probably irregular.

The amplitude of the changes in declination and inclination is surprisingly small for a record this long. To provide a basis of comparison, the present variation of declination and inclination with longitude is given in Fig. 15 for the 20° latitude circles in the northern and southern hemispheres, 20° being the latitude of Hawaii. The amplitudes are those expected at a fixed station as the result of (1) the westward drift of the nondipole field and (2) changes in the longitude of the equatorial component of the geomagnetic dipole shown by the continuous curves in Fig. 15. The rate of westward drift is 0.2° of longitude per year, or 1800 years for a complete circuit around a circle of latitude. The

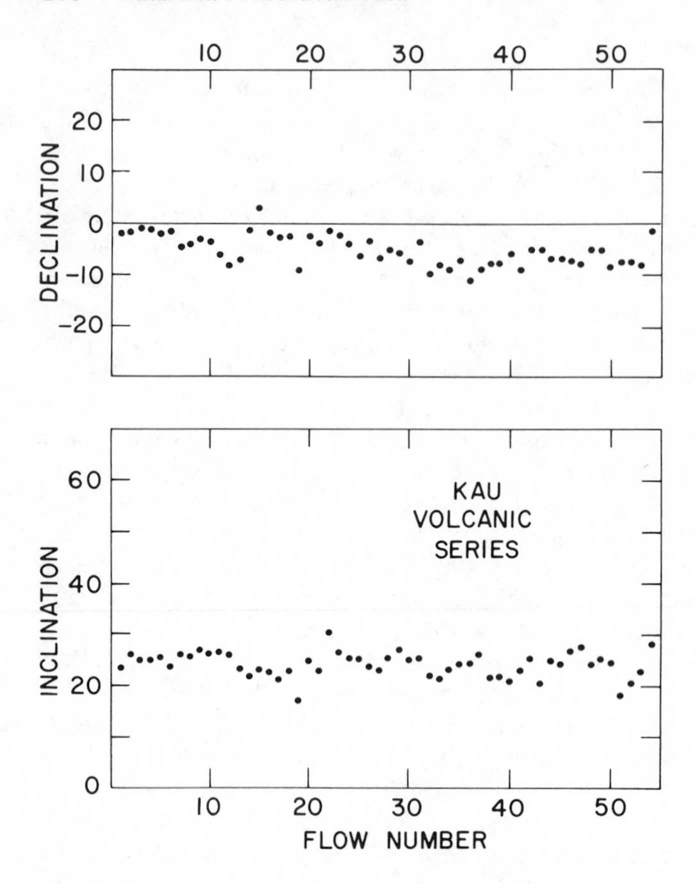

Fig. 14. Values of the declination and inclination of the Kau Volcanic Series plotted against flow number (Doell, 1969). Flow numbers increase with decreasing age. Each point represents the direction determined from paleomagnetic measurements on from five to eight samples from one lava flow. The mean 95% confidence limit for these data is about 3°, with an upper limit of 6°.

magnetic record of the Kau Volcanic Series appears to be long enough to provide a good sample of angular variations resulting from westward drift of a nondipole field similar to the present one and might have been expected to show variations at least as large as the differences between points and curves in Fig. 15.

The important question of whether this record is long enough to have recorded changes in the orientation of the geocentric dipole is more difficult to answer because little is known about the rate or nature of the changes in dipole orientation. During the past few centuries, the change in geomagnetic dipole orientation has been too small to detect using spherical harmonic analysis. From paleomagnetic research (Cox and Doell, 1960; Hospers, 1955; Irving, 1964), it is known that during the past few million years, the average orientation of the dipole has been parallel to the earth's rotation axis. There can be no question therefore that the dipole's present 11.5° displacement from the rotation axis is a transient phenomenon, but the nature of dipole movement about the geographic axis is uncertain. The main experimental difficulty in tracing changes of dipole orientation paleomagnetically is the presence of the nondipole field. The virtual pole paths in Fig. 10 for values of D and I of the 1965 IGRF along circles of latitude are typically displaced about 10° from the geomagnetic dipole, reflecting the effect of the nondipole field; however, the amplitude of the dipole wobble itself is also about 10° (Cox and Doell, 1964). Therefore, the ratio of dipole signal to nondipole noise is about one to one. Even in localities such as Great Britain where high-quality archeomagnetic measurements have been made

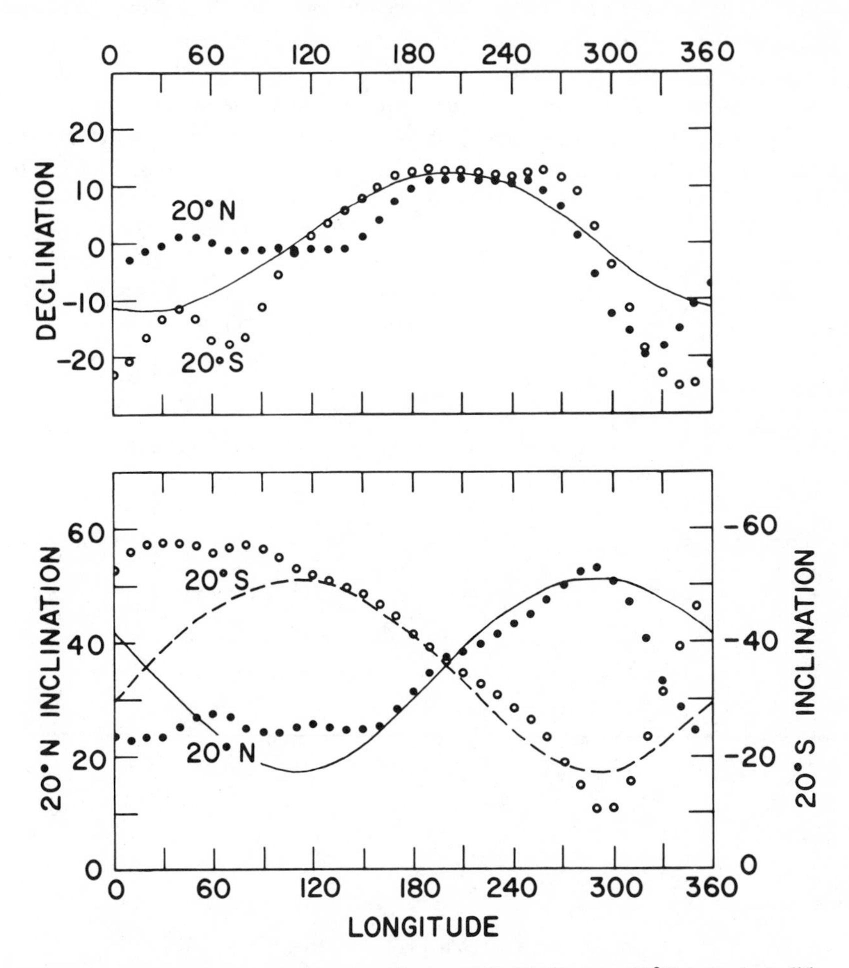

Fig. 15. Declination and inclination of the 1965 IGRF along 20°N latitude (solid circles) and 20°S latitude (open circles). The continuous curves represent values for the dipole portion of the field only.

at numerous sites with well-determined ages (Aitken and Weaver, 1965), the history of changes in the orientation of the dipole are uncertain even though the virtual pole path (Fig. 16) is well determined for the past 970 years.

Returning to the Kau Volcanic Series, the virtual geomagnetic poles (Fig. 17) are all clustered in one quadrant. The scatter is much less than that for the 970-year-long record from Britain, the angular standard deviation of all poles of the Kau Volcanic Series from their mean position being only 3.1°. This is remarkably small for a sequence of magnetic measurements that probably spans about 10^4 years. By comparison, the virtual poles along the 20°N latitude circle, shown in Fig. 17, have an angular standard deviation from the geomagnetic pole of 6.5°. Even if the geologic estimates of age are too large by a factor of 5, the time interval of the Kau Volcanic Series is still long enough to adequately sample the nondipole field. The amplitude of nondipole field variations is seen to be only half that of the expected world-wide average at 20°N, even if all of the observed angular dispersion is attributed to the nondipole field.

The weakest point in this interpretation of the Kau paleomagnetic data lies in the estimate of the time required for extrusion of the series. Imagine that,

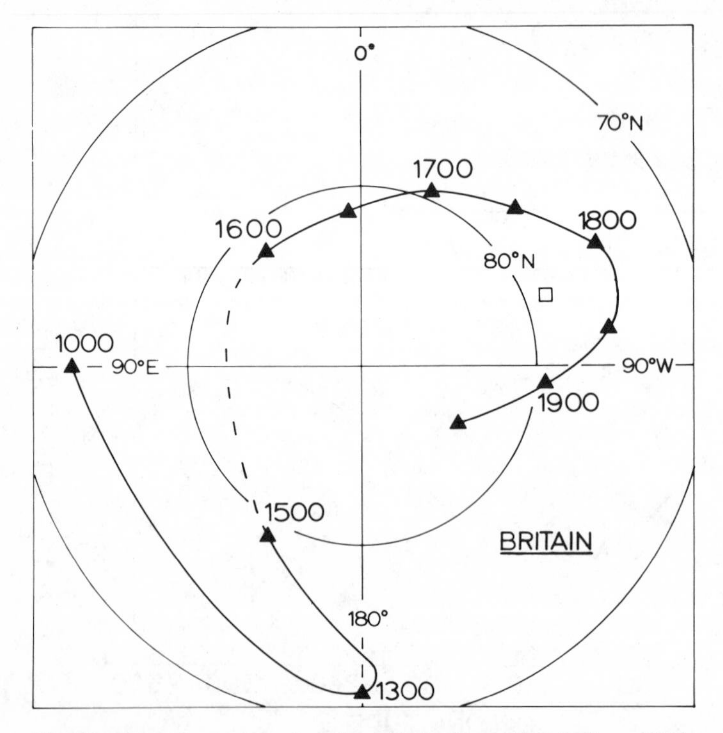

Fig. 16. Virtual geomagnetic pole path for Great Britain from 1000 A.D. to the present. The path is a smoothed average curve redrawn from the summary curve of Aitken and Weaver (1965). Square, present geomagnetic pole. Compare with 50°N curve in Fig. 10.

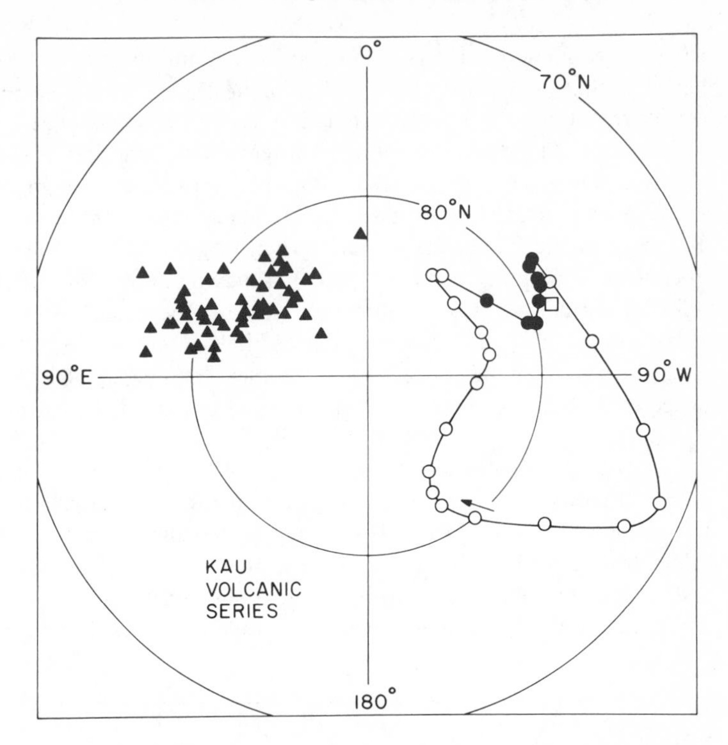

Fig. 17. Virtual geomagnetic poles for the Kau Volcanic Series. Triangles, virtual poles corresponding to values of D and I from each flow. Other conventions as in Fig. 13.

contrary to the geological evidence, the entire sequence formed during a few centuries or less. It might then be argued that secular variation in the Pacific happened to be unusually low then, as it is today, and that the Kau paleomagnetic data do not therefore imply that it was characteristically low in the Pacific or that the nondipole field was generally absent. The argument is a strong one. No matter how many individual lava flows were sampled, if they all formed during a short time, then they cannot supply very much information about the long-period characteristics of the field.

It has proven possible to obtain additional information about the time interval represented by the Kau Volcanic Series by determining paleomagnetically the intensity of the magnetic field during the time the lavas formed. World-wide archeomagnetic determinations of the ancient intensity of the earth's field during the past 8000 years have been averaged in 500-year intervals by Cox (1968) in order to obtain an estimate of changes in the earth's dipole moment. The contribution of the nondipole field should be small in these averages. In Fig. 18, these have been converted to show the changes in the intensity of the earth's field at the latitude of Hawaii. The most recent fluctuation in the amplitude of the dipole field appears to have a period of about 8000 years and an amplitude of

± 0.2 oe (± 20,000 gammas) about a mean value of 0.4 oe. The connection between time and magnetic intensity in the Kau data can be appreciated by considering the field intensity variation to be expected under two different assumptions: (1) the lavas formed during a few centuries when the secular variation was small, or, alternatively, (2) the lavas formed over an interval of 10^4 years. In case (1), the intensity variation will be small for the following reason. To produce the maximum observed changes of about 10° in the Kau field directions, nondipole field vectors with a maximum intensity of 5000 gammas oriented normal to the mean direction would be needed. Thus, ± 5000 gammas (± 0.05 oe) is also the expected range of field intensities expected in the Kau data. For the second case, which assumes an interval 10^4 years long, the expected intensity change is at least as large as that of the earth's dipole field (0.4 oe).

For 10 of the flows from the Kau Volcanic Series, determinations were made of the intensity of the earth's field at the time the lavas originally cooled, using the method of Smith (1967). The data (Fig. 19) show a range of paleointensity values from a high of about 0.6 oe (flow 18) to a low of about 0.3 oe (flows 5 and 45). The overall accuracy of these limited data is not well determined; however, the 0.07-oe average spread in several duplicate determinations is well

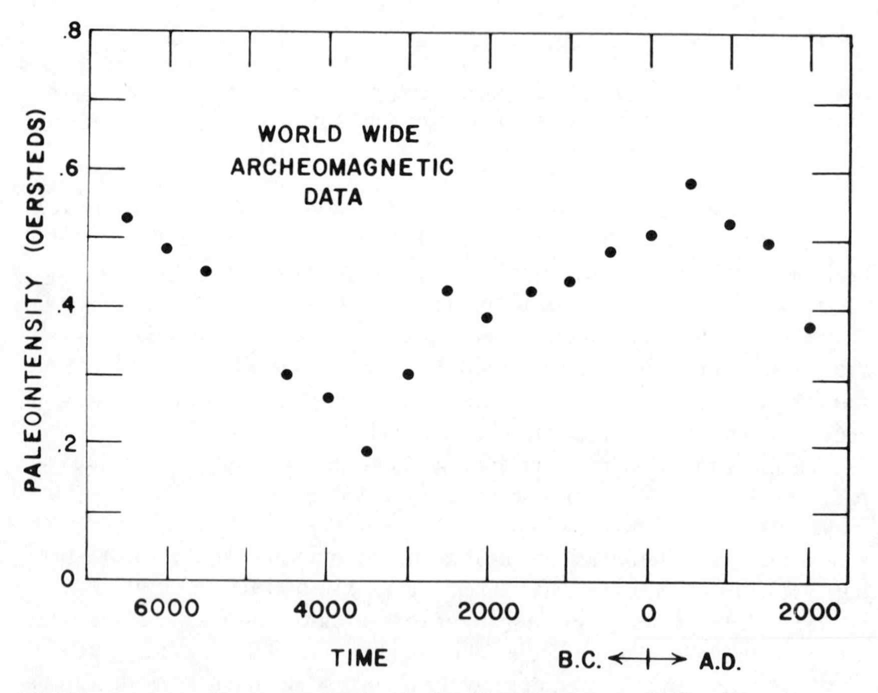

Fig. 18. Intensity of the earth's field in Hawaii resulting from changes in the earth's dipole moment during the past 8000 years. Each point represents an average of world-wide data for an interval of 500 years (Cox, 1968; Smith, 1968).

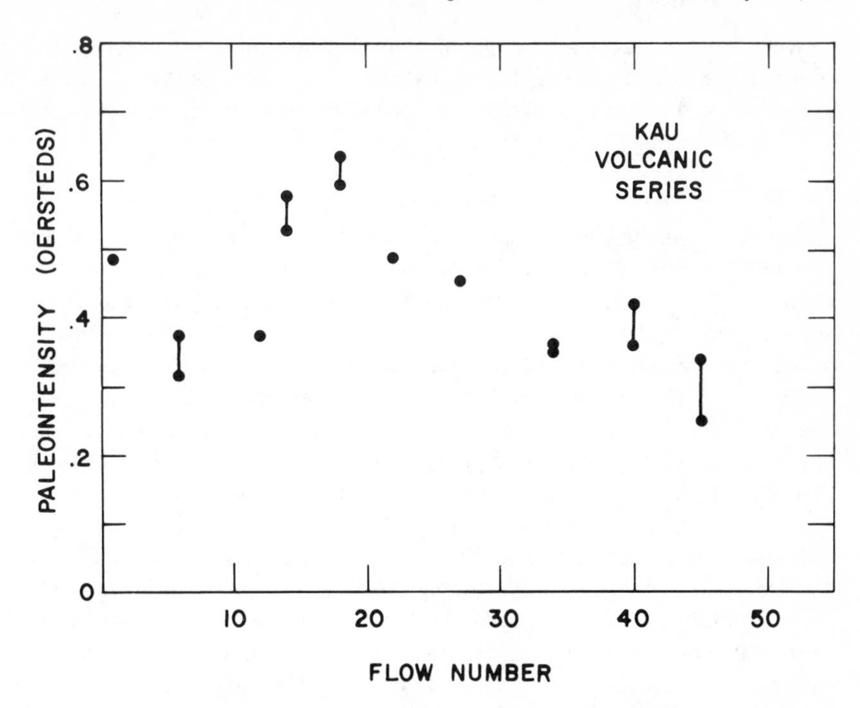

Fig. 19. Paleointensity of the earth's field at the time of formation of 10 flows from the Kau Volcanic Series. Duplicate determinations on different samples from the same flow are joined by bars.

below the observed total change; it should also be noted that the data show clearly defined serial changes, further suggesting that real paleointensity variations have been observed. The 0.61-oe peak in the intensity of the earth's field at the time of flow 18 may correspond to the peak in the world-wide data at 500 A.D., the peak at 7000 B.C., or possibly to an earlier peak. The importance of the Kau Volcanic Series intensity results, however, does not lie in making a point-to-point correlation with the world-wide data but rather with the demonstration that the intensity of the earth's field varied by a factor of 2 during the time flows of the Kau Volcanic Series were extruded. This is not consistent with the flows having formed during a few centuries but requires, at the very least, a few thousand years.

In summary, the geologic evidence suggests that the time represented by the Kau Volcanic Series is about 10^4 years and the paleointensity results are roughly consistent with this, because they show at least one major fluctuation of the dipole field within the series. If these time estimates are correct within a factor of 5, then the Kau Volcanic Series contains an adequate sample of the nondipole field, which appears to have been much smaller than in non-Pacific areas at the same latitude. In short, large intensity variations combined with small angular changes in a sequence such as the Kau Volcanic Series is incompatible with the presence of an appreciable nondipole field. Of additional interest to geomagnetism is the indication that the orientation of the main dipole remained fixed, or nearly so, for 3000 to 10,000 years while its intensity changed by a factor of 2.

Puna Volcanic Series

A similar study has been made of 30 lava flows of the Puna Volcanic Series exposed along the northern wall of Kilauea caldera (Doell and Cox, 1965). The age of the youngest flow is > 2500 years on the basis of a C^{14} age from charcoal under a pumice layer that covers the flow (Powers, 1955). The range of permissible geological ages for the Puna and the Kau Volcanic Series thus overlaps. The paleomagnetic evidence shows, however, that their ages cannot be the same because their directions of magnetization are considerably different. The evidence for the total age of this series is similar to the evidence for the Kau Volcanic Series; several thousand years is likelier than several hundred.

For flows 10 to 30, the results (Fig. 20) are similar to those for the Kau Volcanic Series. Changes in the direction of the field are considerably smaller than the changes that exist today along 20°N. Virtual poles for flows 10 to 30, however, although tightly grouped like those of the Kau Volcanic Series, are all in the quadrant 90°W to 180°W longitude (Fig. 21), which is considerably different from the position of the Kau virtual poles. Changes in the intensity of the field (Fig. 22) at the time the Puna Volcanic Series formed are remarkably similar to those for the Kau Volcanic Series, both series having an intensity maximum and minimum during a time when the dipole orientation did not change significantly. Again, small angular and large intensity changes preclude the presence of a significant nondipole field.

The older flows (1 to 9) indicate much larger changes in the direction of the field. These may reflect the presence of a transient nondipole component, or, alternatively, they may record movement of the main dipole toward the stable orientation recorded by flows 10 to 30. Whichever explanation is correct, the average intensity of the field during the change in direction was between 0.5 and 0.6 oe; the present intensity is 0.36 oe.

Even without precise age data for the times and duration of extrusion of the Kau and Puna Volcanic Series, the paleomagnetic data from these series reasonably can be interpreted only in terms of an absence of geomagnetic nondipole features in the central Pacific for at least the past several thousand years–a considerable extension beyond the record of magnetic observatories. Paleomagnetic studies have also been made on several other older volcanic series exposed on Hawaii (Doell and Cox, 1965) that give similar results, that is, remarkably small changes in direction throughout thick sequences of superposed lava flows. These data will be considered in the next section, where mean angular variances in paleomagnetic data for rocks with ages spanning the past several hundreds of thousands of years are compared for different geographical locations.

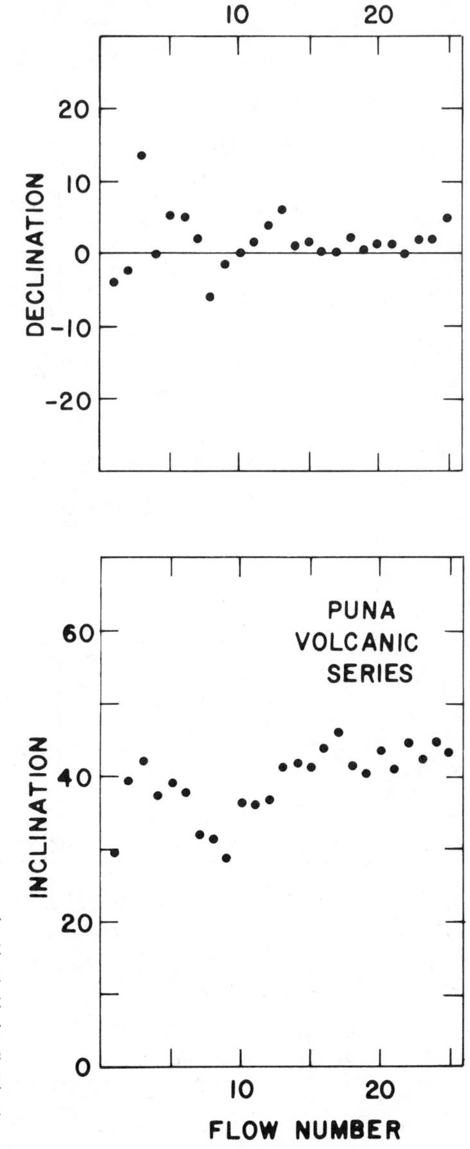

Fig. 20. Declination and inclination for the Puna Volcanic Series plotted against flow number. Age decreases with increasing flow number. Each point represents the average declination or inclination of multiple samples from one lava flow. Part of the data are from Doell and Cox (1965) and part are new with this report.

VARIATION OF ANGULAR DISPERSION WITH LATITUDE

Because there is little information about the lengths of time between successive data points in most paleomagnetic studies, all the information about ancient geomagnetic secular variation resulting from nondipole field changes is contained in a single parameter, the angular standard deviation of ancient field directions from their mean. Where the average value of the nondipole field has

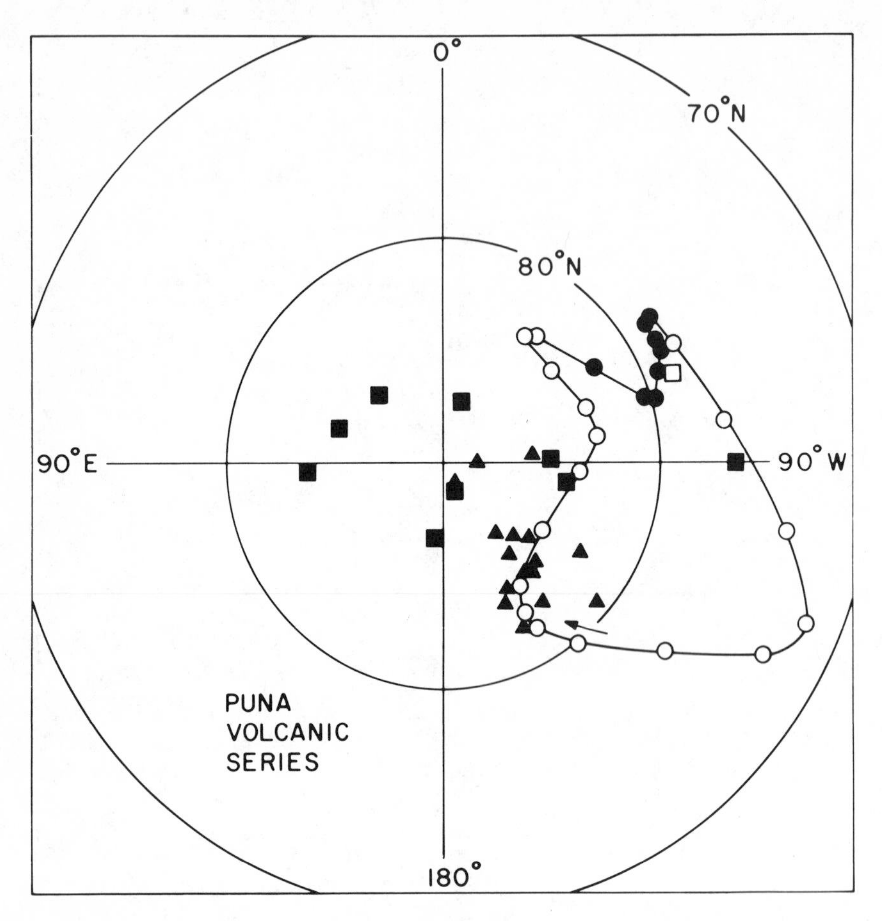

Fig. 21. Virtual geomagnetic poles for the Puna Volcanic Series. Solid squares, flows 1 to 9. Triangles, flows 10 to 30. Other conventions are as in Fig. 13.

been small, this parameter will be small, and the converse is true. Thus, the use of angular dispersion to test for the longer term permanency of the Pacific low might appear to be a simple matter of comparing the angular deviation parameter from Hawaiian paleomagnetic data with those from other localities over the globe. In practice, however, several complications arise which need to be considered before discussing the data available.

The first problem is that the wobble of the geomagnetic dipole is so large that it tends to mask changes in direction resulting from changes in the nondipole field. Our approach for handling this has been to transform ancient field directions to equivalent virtual geomagnetic poles. The advantage of analyzing paleomagnetic data in this way is that the angular dispersion resulting from dipole wobble is the same at all latitudes, whereas it is strongly latitude dependent if field directions are analyzed, the field resulting from the dipole alone being twice as large at the poles as it is along the equator. Using virtual poles, the effects of the nondipole field are seen as an increase in angular dispersion rising above a base level common to all sampling sites, regardless of their geographic location. If S_D is the angular standard deviation of the dipole wobble and S_N that of virtual poles produced by the nondipole field, then by simple analysis of variance arguments, the angular standard deviation S_T resulting both from the dipole and from nondipole components of the field is

$$S_T = (S_D^2 + S_N^2)^{1/2} \tag{10}$$

The lowest value of S_T determined paleomagnetically at any site with adequate sampling can be taken as an upper bound for S_D, the dipole wobble. Our best estimate of the value of S_D is about 11° (Cox and Doell, 1964; Doell, 1969, 1970).

A second problem in interpreting angular dispersion data is that paleomagnetic sampling sites are not well distributed for testing the permanency of the Pacific low. For the interval from 0.7 m.y. to the present, which constitutes the Brunhes normal polarity epoch, there are several paleomagnetic studies suitable for determinations of angular dispersion. This parameter is well determined for the island of Hawaii (Doell, 1969; Doell and Cox, 1965); however, there are no comparable paleomagnetic results from sites at the same latitude in other parts of the world where the nondipole field is large at present. On the other hand,

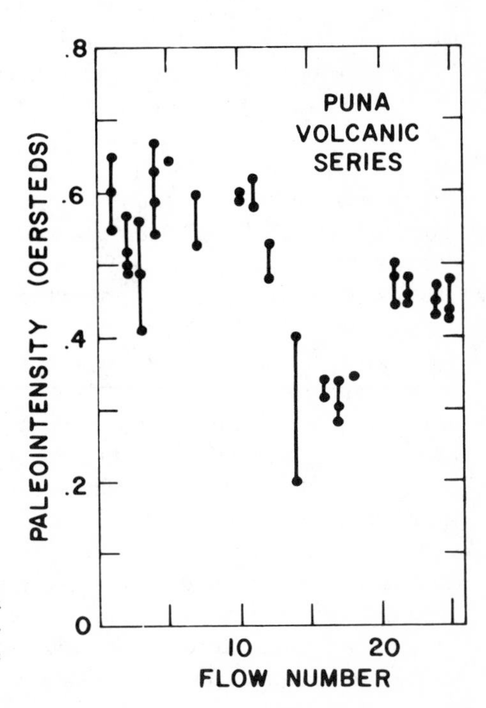

Fig. 22. Paleointensities for the Puna Volcanic Series. Replicate determinations on different samples from the same flow are joined by bars.

suitable results are available from other latitudes. What is needed is a model describing the variation of angular dispersion with latitude.

This latitude variation depends on three variables: variations in the direction and intensity of the nondipole field and the relative strength of the nondipole and dipole fields, which in turn depends on the wobble of the geomagnetic dipole and changes in its intensity. Cox (1970) has extended earlier model studies by Creer et al. (1959), Creer (1962), Cox (1962), and Irving and Ward (1964) to show how these variables may be combined to produce various models of secular variation. The solid line in Fig. 26 shows the result of a model for which the dipole wobble is 11° and the angular dispersion of the nondipole field is the same as that of the present nondipole field along circles of latitude, averaging results for the northern and southern hemispheres. If the angular standard deviation of virtual poles from any sampling site falls below this curve, then the average amplitude of the nondipole field at that site was anomalously low relative to the present world-wide field.

The final analytical problem is to establish confidence limits for the determinations of angular standard deviation. This is straightforward from a formal statistical point of view (Cox, 1969a). There is a practical problem of determining the relevant number of degrees of freedom in the sampling, however, which in turn depends on whether a set of ancient field determinations constitute a truly random sample. A necessary condition for randomness is that the time between successive paleomagnetic data points must be at least as long as the longest periods present in the geomagnetic spectrum. Our experience has been that this condition is commonly not met (Doell, 1969; Doell and Cox, 1965). Consequently, in many investigations, including some cited in the present study, the indicated confidence limits on S_T are probably too low.

A second sampling problem arises because the earth's field appears to undergo large, rapid excursions in direction at widely spaced intervals. Three such excursions recorded in pre-Brunhes age lavas extruded on the island of Oahu are shown in Fig. 23. These excursions appear to represent a phenomenon that is different from that responsible for ordinary secular variation for two reasons. The first is that the excursions are so large that they do not appear to belong to the same population as angular changes resulting from ordinary secular variation. This may be seen in Fig. 24, where we have plotted a histogram of colatitudes of virtual geomagnetic poles for 1035 lavas with ages less than about 5 to 10 m.y. A very small proportion of the virtual geomagnetic poles lie more than 40 to 50° away from the geographic axis, and several of these poles are parts of excursions of the type shown in Fig. 23. The second is that during the excursions, the virtual poles move along simple great circle paths, whereas ordinary secular variation produces much more complex paths (Figs. 10 and 16). Possible explanations are that these are abortive polarity reversals, or that they are produced by rapid tilts of the main dipole. Whether or not either of these is the correct explanation, there can be little doubt that the excursions exist as rare, widely spaced, rather rapid events. Because their amplitudes are large, they make

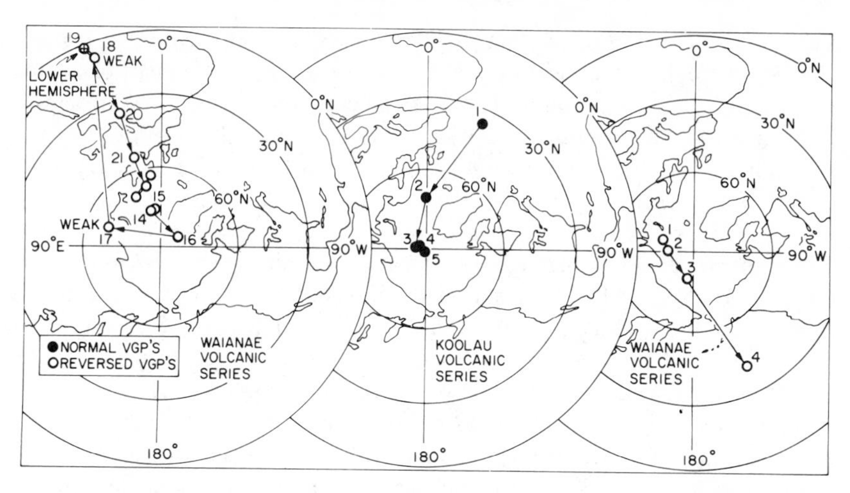

Fig. 23. Large excursions of virtual poles recorded in successive lava flows on Oahu. Each circle is an average pole based on paleomagnetic measurements of multiple samples from one flow. Arrows show sequence of pole positions, the poles with larger numbers corresponding to younger flows. Note change of scale from earlier plots to show the entire northern hemisphere.

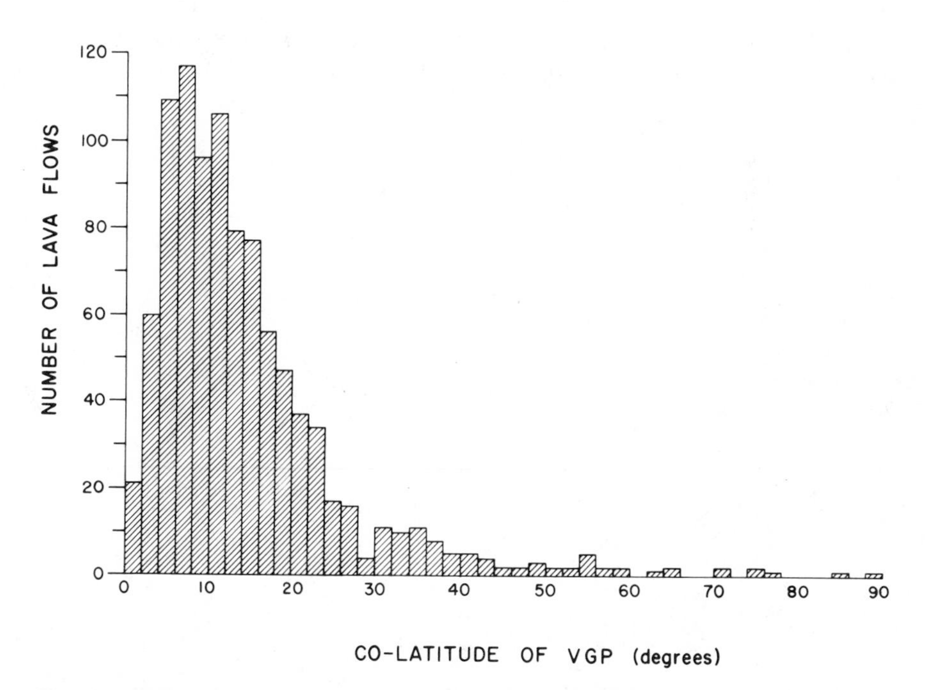

Fig. 24. Histogram of the co-latitude of virtual geomagnetic poles calculated from paleomagnetic studies of 1035 lava flows from Galapagos, Hawaii, Western United States, New Zealand, France, Alaska, Iceland, and Antarctica with ages less than about 5 to 10 m.y.

a sizable contribution to the total angular dispersion when they occur in a set of paleomagnetic data; and because they occur only during a small part of the total time, they are almost impossible to sample adequately. To circumvent the resultant problem of unstable statistics, we have excluded from our analysis virtual poles displaced more than 44° from the earth's rotation axis. Truncation of the data in this way regardless of site latitude is appropriate if, as we believe, the excursions reflect changes in the main dipole.

In a further effort to maintain statistical conformity, we have removed from the analysis all individual flow data with 95% confidence limits greater than 9° from their mean direction. As can be seen from Fig. 25, most of the paleomagnetic data are known (at the 95% confidence level) to within 5 or 6° and few are discarded by the above restriction. Experimental errors of this magnitude (3° SD) contribute very little to the observed angular variances of 11° or more. For example, if $S_E = 3°$ is the experimental error, $S_O = 11°$ the observed variance, and S_T the variance resulting from geomagnetic field changes, then $S_T = \sqrt{S_O^2 - S_E^2} = 10.6°$, which for present purposes is not significantly different from the observed value of 11°.

One further question should be considered and that is whether the angular standard deviations for each locality should be calculated about their mean virtual geomagnetic pole position or some other location (Cox, 1969b; Doell, 1970). The analyses should, of course, be made about the true average virtual geomagnetic pole position for the period of time in question, but this is not

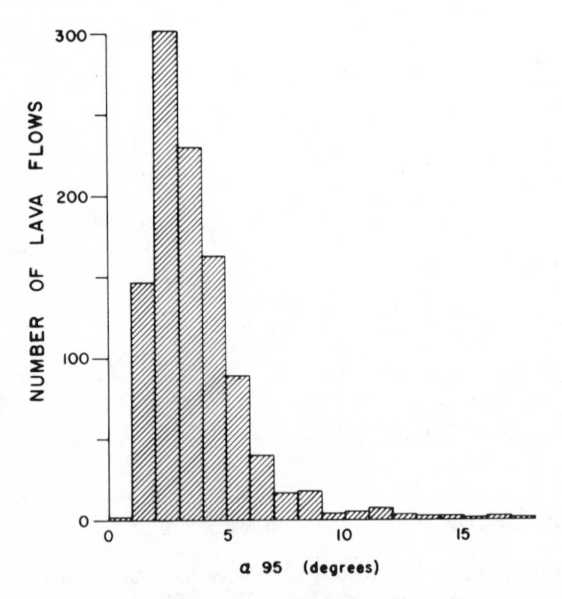

Fig. 25. Histogram of the 95% confidence limits for the 1035 lavas referred to in Fig. 24.

known precisely from the paleomagnetic data. The mean virtual geomagnetic pole position for the 555 lava flows from the eight different localities we wish to compare is displaced 1.3° (with a longitude of 59.4°W) from the rotation axis and has a formal 95% confidence limit of 1.1°. It is very difficult to establish, however, the proper number of degrees of freedom to use in calculating confidence limits when sequences of data such as those from the Puna and Kau Volcanic Series are included. Thus, the 1.1° confidence limit for the mean Brunhes age virtual geomagnetic pole is certainly too small; the true limit may be large enough to include the rotation axis. For this reason, and because the earth's rotation plays such a strong part in geomagnetic theory, we have

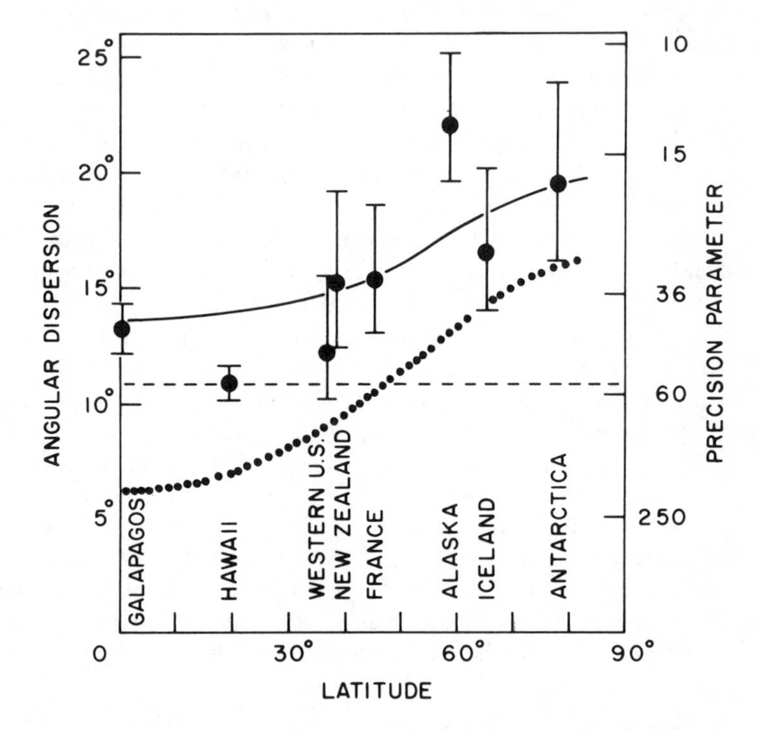

Fig. 26. Angular standard deviation S_T of virtual geomagnetic poles calculated with respect to the geographic axis. Dotted curve: angular dispersion of virtual geomagnetic poles expected solely from a rotation of the nondipole component of the 1965 IGRF, averaging between hemispheres. Dashed curve: angular dispersion resulting solely from a dipole wobble of 11°. Solid curve: total angular dispersion resulting from a combination of both of the above. Circles: angular standard deviation of virtual poles from paleomagnetic data of rocks less than 0.7 m.y. old collected at the sampling sites indicated, error bars being 95% confidence limits (Cox, 1969a). Numbers of independent data points at each site are as follows: Galapagos, 147; Hawaii, 219; Western United States, 21; New Zealand, 21; France, 31; Alaska, 60; Iceland, 28; Antarctica, 28.

calculated the angular variances for each locality with respect to the geographic axis.

The results of our studies of the angular variance of Brunhes age lavas from different parts of the world are given in Fig. 26, where the solid curve is the model previously described. All the experimental results fit the model except that for Hawaii, which has less angular dispersion than the model, and that for Alaska, which has more. We believe that the confidence limits on the Alaskan result may be too small because of sampling problems of a statistical nature and that the indicated deviation from the curve may not be statistically significant. As the sampling on Hawaii was more extensive, the angular dispersion indicated is on a firmer experimental basis. Although the indicated confidence limit is probably also too small, it appears likely that the indicated value is significantly below the curve. For example, if the true confidence limits for the calculated Hawaiian angular variance of 10.8° were to include the solid curve in Fig. 26, then, using Cox's methods (1969a), this would imply that only 18 independent determinations of the field direction were made in the paleomagnetic studies of the 219 lava flows used in the calculations. Thus, when compared with similar Brunhes age paleomagnetic data from other localities, the angular variance of the Hawaiian results are again anomalously low. The most straightforward conclusion is that the nondipole field has been greatly subdued (or missing) throughout the past 0.7 m.y.

We have made paleomagnetic studies on 19 lava flows of the Honolulu Volcanic Series on the island of Oahu. These lavas are also in the Brunhes epoch and were issued from many widely scattered vents throughout southeast Oahu. Moreover, they appear also to cover a wide range of ages (Stearns and Vaksvik, 1935); it is therefore probable that the statistical problems associated with the Hawaiian data are not present here. The angular dispersion with respect to the geographic axis is 9.8° with upper and lower limits of 12.6° and 8.0°, respectively, which exclude the expected world-wide curve at the 95% confidence level. The angular dispersion of the paleomagnetic data from the Honolulu Volcanic Series is thus also consistent with anomalously low values of the nondipole field in the central Pacific during the past 0.7 m.y.

SUMMARY AND CONCLUSION

We have considered geomagnetic observatory data, paleomagnetic data from lava flows with known historical ages, paleomagnetic data from undated but sequentially related lava flows, and paleomagnetic data from more than a thousand lava flows less than 0.7 m.y. old. In all of these data, we find strong evidence for a subdued secular variation and a subdued nondipole field in the central Pacific. Both are consistent with a pronounced attenuation of the geomagnetic spectrum in the period range 200 to 2000 years. This in turn implies that a lateral inhomogeneity exists in the lower mantle coupled to the core in such a way as to partially prevent the generation, beneath the central

Pacific Ocean, of nondipole magnetic fluctuations with periods in this range. With the present state of knowledge concerning the processes that give rise to the earth's magnetic field, these findings cannot be inverted directly to indicate the exact nature of this inhomogeneity.

REFERENCES

Aitken, M. J., and Weaver, G. H., Recent archaeomagnetic results in England, *in* The symposium on magnetism of the Earth's interior, J. Geomagnetism Geoelectricity, **17**, 391, 1965.

Alexander, S. S., and Phinney, R. A., A study of the core-mantle boundary using *P* waves diffracted by the earth's core, J. Geophys. Res., **71**, 5943, 1966.

Bullard, E. C., Freedman, C., Gellman, H., and Nixon, J., The westward drift of the earth's magnetic field, Phil. Trans. Roy. Soc. London, **243**, 67, 1950.

Cox, A., Analysis of present geomagnetic field for comparison with paleomagnetic results, J. Geomagnetism Geoelectricity, **13**, 101, 1962.

______, Lengths of geomagnetic polarity intervals, J. Geophys. Res., **73**, 3247, 1968.

______, Confidence limits for the precision parameter κ, Geophys. J., **17**, 545, 1969a.

______, A paleomagnetic study of secular variation in New Zealand, Earth Planet. Sci. Letters, **6**, 257, 1969b.

______, Latitude dependence of the angular dispersion of the geomagnetic field, Geophys. J., **20**, 253, 1970.

Cox, A., and Doell, R. R., Review of paleomagnetism, Bull. Geol. Soc. Am., **71**, 645, 1960.

______, Long period variations of the geomagnetic field, Bull. Seism. Soc. Am., **54**, 2243, 1964.

Cox, A., and Dalrymple, G. B., Statistical analysis of geomagnetic reversal data and the precision of potassium-argon dating, J. Geophys. Res., **72**, 2603, 1967.

Creer, K. M., The dispersion of the geomagnetic field due to secular variation and its determination for remote times from paleomagnetic data, J. Geophys. Res., **67**, 3461, 1962.

Creer, K. M., Irving, E., and Nairn, A. E. M., Palaeomagnetism of the Great Whin Sill, Geophys. J., **2**, 306, 1959.

Currie, R. G., The geomagnetic spectrum–40 days to 5.5 years, J. Geophys. Res., **71**, 4579, 1966.

______, Magnetic shielding properties of the earth's mantle, J. Geophys. Res., **72**, 2623, 1967.

Doell, R. R., Paleomagnetism of the Kau volcanic series, Hawaii, J. Geophys. Res., **74**, 4857, 1969.

______, Paleomagnetic secular variation study of lavas from the Massif Central, France, Earth Planet. Sci. Letters, **8**, 352, 1970.

Doell, R. R., and Cox, A., The accuracy of the paleomagnetic method as evaluated from historic Hawaiian lava flows, J. Geophys. Res., **68**, 1997, 1963.

______, Paleomagnetism of Hawaiian lava flows, J. Geophys. Res., **70**, 3377, 1965.

Doell, R. R., and Smith, P. J., On the use of magnetic cleaning in paleointensity studies, J. Geomagnetism Geoelectricity, **21**, 579, 1969.

Elsasser, W. M., The earth's interior and geomagnetism, Rev. Mod. Phys., **22**, 1, 1950.

Fisk, H. W., Isopors and isoporic movement, Bull. Int'l. Geodetic Geophys. Union, no. 8, Stockholm Assembly 1930, 280, 1931.

Gaibar-Puertas, C., Varacion secular del campo geomagnetico, 475 pp., Tarragona, Tortosa Obser. del Ebro Mem. no. 11, 1953.

Hide, R., Motions of the earth's core and mantle, and variations of the main geomagnetic field, Science, **157**, 55, 1967.

Hospers, J., Rock magnetism and polar wandering, J. Geol., **63**, 59, 1955.
IAGA Commission 2 Working Group No. 4, Analysis of the geomagnetic field, The international geomagnetic reference field 1965.0, J. Geomagnetism Geoelectricity, **21**, 569, 1969.
Irving, E., Paleomagnetism and its Application to Geological and Geophysical Problems, 399 pp., John Wiley & Sons, New York, 1964.
Irving, E., and Ward, M. A., A statistical model of the geomagnetic field, Pure Appl. Geophys., **57**, 47, 1964.
Lahiri, B. N., and Price, A. T., Electromagnetic induction in nonuniform conductors, and the determination of the conductivity of the earth from terrestrial magnetic variations, Phil. Trans. Roy. Soc. London, Ser. A, **237**, 509, 1939.
Mauersberger, P., Mathematische Beschreibung und Statistische Untersuchung des Hauptfeldes und der Säkularvariation, Geomagnetismus Aeronomie, **3**, 95, 1959.
McDonald, K. L., Penetration of the geomagnetic secular field through a mantle with variable conductivity, J. Geophys. Res., **62**, 113, 1957.
Phinney, R. A., and Alexander, S. S., *P* wave diffraction theory and the structure of the core-mantle boundary, J. Geophys. Res., **71**, 5959, 1966.
______, The effect of a velocity gradient at the base of the mantle on diffracted *P* waves in the shadow, J. Geophys. Res., **74**, 4967, 1969.
Powers, H. A., A new date in Kilauea's history, U.S. Geol. Surv. Volcano Letter, no. 527, 3, 1955.
Rubin, M., and Berthold, S. M., U.S. Geological Survey radiocarbon dates VI, Am. J. Sci. Radiocarbon Suppl., **3**, 86, 1961.
Smith, P. J., The intensity of the Tertiary geomagnetic field, Geophys. J., **12**, 239, 1967.
______, Ancient geomagnetic field intensities–3, Historic and archeological data H13X and H14-15: Geological data, G22-29, Geophys. J., **16**, 457, 1968.
Stearns, H. T., and Vaksvik, K. N., Geology and ground-water resources of the Island of Oahu, Hawaii, Bull. Hawaii Div. Hydrog., **1**, 478 pp., 1935.
Vestine, E. H., LaPorte, L., Cooper, C., Lange, I., and Hendrix, W. C., Description of the Earth's Main Magnetic Field and its Secular Change, 1905-1945, Carnegie Inst. Washington Pub. 578, Washington, D.C., 532 pp., 1947.
Vogel, A., Über unregelmassigkeiten der äusseren Begrenzung des Erdkerns auf Grund von am Erdkern reflektierten Erdbebenwellen, Gerlands Beitr. Geophysik, **69**, 150, 1960.
Yukutake, T., The westward drift of the magnetic field of the earth, Bull. Earthquake Res. Inst. Tokyo Univ., **40**, 1, 1962.
Yukutake, T., and Tachinaka, H., The non-dipole part of the earth's magnetic field, Bull. Earthquake Res. Inst. Tokyo Univ., **46**, 1027, 1968.

12

EARTH TIDES

by **Louis B. Slichter**

The interaction of ocean and land tides and instrumentation suitable for the observation of both earth tides and slower deformations of tectonic origin are central topics in modern earth tide studies. Current work on the interaction of ocean and land tides is illustrated with four of many potential examples. The ocean loads of known periodicity and less perfectly known geometric pattern, the response thereto recorded by tiltmeters, gravity tide meters, strainmeters, and the structure of the earth being deformed comprise three coupled elements of a system whose understanding is promoted by improved knowledge of each of its three parts and of their interrelationship. This subject has long been of interest, but modern instrumentation and computational facilities have only recently stimulated comprehensive studies of the relationships.

Each of three major types of earth tide meters is finding applications beyond its original purpose, in studies of the more rapid secular tectonic changes.

Observational problems, not unique to that station, are briefly discussed in the context of the South Pole station. The ocean tide correction for tidal gravity at South Pole is very small—about 0.14 μgals on a tidal signal of 14.91 μgals (of which 2.06 is geophysically pertinent). Methods for interpreting short records and for removing instrumental drift are being developed.

An interval of 39 days of the better data obtained in 1969 at the South Pole provides the value 0.121 (+ 0.027, – 0.028) for $G - 1$. In this interval, that of the most constant instrumental drift, gave $G - 1 = 0.178$, or 0.166 after introducing a 7% correction for ocean tides. The new observations for 1970 with two gravimeters, one of which has a very low drift rate, are expected to make 1969 results obsolete.

INTRODUCTION

An early tidal experiment that remains significant and precise even by present-day standards is the series of tilt measurements performed at Lake Geneva, Wisconsin, made by Michelson and Gale (1919), using two buried steel level tubes 150 meters long, half filled with water. The changing water level resulting from the daily tidal tilts was monitored with a Michelson interferometer. The experiment demonstrated the following three points: (1) that the mean rigidity of the earth is 8.6×10^{11} dynes/cm^2, which is within a few percent of today's findings; (2) that the rigidity does not differ in the east-west and north-south directions; and (3) that the value of the earth's tidal tilt parameter is 0.690 ± 0.004, which is the same as the mean value preferred today.

A dramatic announcement in the newspapers about 1935 described how the varying gravitational attraction of the moon at a point on the earth, as the earth revolved under the moon, had been clearly observed for the first time by a new gravimeter developed by Hoyt at Gulf Laboratories. About the same time, the early paper of Lucien LaCoste (1935) appeared concerning his "zero length" spring. The long-period gravimeters so attained presaged the development of today's highly sensitive earth tide gravimeters. Soon thereafter many earth tide stations with a variety of instruments were operating over the earth, coordinated by a world center under the direction of Prof. Paul Melchior in Brussels. In 1969, he listed 307 different earth tide stations employing about 400 earth tide meters of the three main types: tiltmeters, gravimeters, and extensometers.

It is not possible in this brief account to review the complex mathematical and theoretical aspects of earth tides, but some of the present activity and areas of probable improvements and progress can be described. To this end, we may examine the sources of error in a typical earth tide measurement made with a gravimeter, although a tilt measurement would serve as well.

Errors in Earth Tide Measurements

A measurement of gravitational acceleration, g_s, is the resultant of four components, g_1, g_2, g_3, and g_4, where g_1, represents the direct acceleration that would be observed on a rigid, undeformable earth; g_2 represents the contribution from a symmetrical, oceanless, but otherwise realistic earth model; g_3 is the contribution from the asymmetrical distribution of the ocean tides; and g_4 is a geologic contribution, to take account of anomalous mechanical properties and response of the specific local and regional geologic structures. In places where the local crustal and mantle mechanics are "normal," g_4 is zero by definition. The larger terms g_1 and g_2 have essentially the same time phase, which is taken as the zero of phase-reference. The terms g_3 and g_4 may differ considerably from g_1 and g_2 in respect to phase. It is customary to normalize these terms in respect to g_1 as unity. In quantities so normalized, denoted by capital letters, the observation G_s may be written

$$G_s = 1 + G_2 + G_3 e^{i\theta_3} + G_4 e^{i\theta_4} \tag{1}$$

where G_2 is independent of the station location. The phase angle θ_3, which is essentially the phase of the local ocean tide with respect to G_2, may be large or small.

The value of G_2 (close to 0.160) has been accurately determined on the basis of knowledge of the earth's interior obtained largely from modern seismology (Alsop and Kuo, 1964; Alterman et al., 1959; Longman, 1963; Molodensky, 1953; Takeuchi, 1950). The magnitude of G_3 varies greatly with distance from the large sea tides, but almost always is less than 0.15, and generally less than 0.05. The anomaly G_4 is small; it is generated in some places by large local ocean loads. The G_4 term can be omitted from consideration for the present.

Appropriate numerical values of G_s, G_1, G_2, G_3, can be introduced into Eq. (1), denoting their respective relative errors by ϵ_s, ϵ_1, ϵ_2, ϵ_3. The most unfavorable case can be assumed in which the phase angle, $\theta_3 = 0$. When $\theta_3 = 90°$, the error ϵ_3 is independent of ϵ_1 and ϵ_2. Equation (1) then becomes

$$(1.16 \pm 0.05)(1 \pm \epsilon_s) = 1 \pm \epsilon_1 + 0{\cdot}16(1 \pm \epsilon_2) \pm 0.05(1 \pm \epsilon_3) \tag{2}$$

From Eq. (2), the following error relation is obtained in which the ambiguous signs have been chosen to provide the largest error ϵ_3, the most unfavorable case:

$$\epsilon_3 = 24.2\epsilon_s + 20\epsilon_1 + 3.2\epsilon_2 \tag{3}$$

Equation (3) emphasizes the degree to which relative errors ϵ_3 in the geophysically meaningful part of a gravity tide measurement may be multiplied by the presence of small errors in the associated quantities. For example, let ϵ_s be an instrumental scale calibration error, $\epsilon_s = 0.005$; let $\epsilon_1 = 0.005$ be the fractional error in the calculated theoretical tide for a rigid earth; and let $\epsilon_2 = 0.01$ be the fractional error in G_2 associated with the adopted earth model and the calculation of G_2. From these error values, taken to be all of the same sign to obtain the most unfavorable case, the result is

$$\epsilon_3 = 0.121 + 0.10 + 0.032 = 0.253$$

Thus the geophysically pertinent residual value, G_3, may have a 25% error, resulting chiefly from 0.5% errors in the instrument's calibration constant and in the computation of the theoretical rigid earth tide. (These two major errors apply also to tiltmeters. The strainmeter measures tidal strain components directly and is free of errors of type ϵ_1.)

It is apparent that the optimum role of earth tide meters is not in measuring the global earth tide but rather local perturbations, those caused by local loads. As will be described, ocean tides of large amplitude furnish such loads. Large and rapid changes in the barometric pressure also furnish load variations of usable size (M. Caputo, private communication, 1971). At special locations that are only slightly disturbed by ocean tides, measurements of global G_2 can profitably be made, as will be shown for the South Pole. When the aim is the evaluation of the local effects, G_3 and G_4, the quality of the observations must be very high, and as described, modern instrumentation meets the need. Theoretically, the slowing of the earth's rotation rate should be demonstrable by earth tide observations, but because of the limited accuracy and incomplete geographic coverage of the observations, significant results on the slowing torque cannot be obtained.

Tidal Resonance by Coupling between Mantle and Core

There is one global tidal relationship which is being verified by earth tide observations. In theory (Jeffreys and Vicente, 1957a and b; Jobert, 1964; Melchior, 1966; Molodensky, 1961), a resonance effect in the coupling between the elastic mantle and the fluid core should occur at the sidereal period, 23.935 hours, of the earth's rotation, which is the period of the K_1 tide. The resonance effect should distinguish this tide from its major neighbor, O_1, of period 25.819 hours. Until the corrections for the K_1 and O_1 constituents of the ocean tides are better known, however, the significance of the results is not fully clear. Values deduced by Jeffreys-Vicente and Molodensky of tidal parameters for these two tides are listed in Table 1, for the several earth models used by these authors. The comparisons are presented in the form of differences and ratios of the O_1 and K_1 parameters. (In this table and elsewhere, we use the diminished numbers $G - 1 = h - 3/2k$ in place of G, and $1 - \gamma = h - k$ in place of γ because the diminished number characterizes the earth more sensitively, undiluted by the relative large constant, "1".)

Melchior (1966, p. 384, 385) lists differences, $\gamma(O_1) - \gamma(K_1)$, at 17 stations and $G(O_1) - G(K_1)$ at 18 stations which accord in sign and in approximate magnitude with the computed values listed in Table 1. He has also recently reported (personal communication, 1969) extensive tidal tilt measurements and gravity-tide observations from 7 stations which provide the respective mean values $R_\gamma = 1.28$, $R_G = 1.14$. For the 10 IGY stations at which both O_1 and K_1 were observed (Harrison et al., 1963), the mean R_G value was 1.38 (+ 0.11, − 0.54). No corrections for water tides were introduced. As better knowledge of ocean tide corrections is developed, the spread in the observed values of R_γ and R_G will hopefully decrease.

Table 1. Comparisons after Jeffreys-Vicente and Molodensky of parameters of the O_1 and K_1 tides

Author	Model no.	$\gamma(O_1)$	$\gamma(K_1)$	$\gamma(O_1)$-$\gamma(K_1)$	$R_\gamma = \frac{1-\gamma(O_1)}{1-\gamma(K_1)}$	$G(O_1)$	$G(K_1)$	$G(O_1)$-$G(K_1)$	$R_G = \frac{G(O_1)-1}{G(K_1)-1}$
Jeffreys-Vicente	1	0.654	0.714	- 0.060	1.21	1.224	1.183	0.041	1.22
	2	0.658	0.693	- 0.035	1.11	1.221	1.185	0.026	1.14
Molodensky	1	0.688	0.730	- 0.042	1.16	1.159	1.137	0.022	1.16
	2	0.686	0.727	- 0.041	1.15	1.164	1.143	0.021	1.15

OCEAN-EARTH TIDAL INTERACTIONS

The sea tides with their two major constituents of semidiurnal and diurnal period, and different geometric patterns, serve as unique sources for deforming the earth's crust and mantle. This of course has long been known, and an extensive literature records researches on this complex subject. Melchior (1966) provides a full and convenient documentation of these researches, and his bibliography lists selected studies of this subject. An additional general reference is Nowroozi et al. (1969). My purpose here is to illustrate different kinds of recent studies of ocean-earth tide interactions. Four examples will serve. The first study is chiefly concerned with very distant tides. In the second, the effects of ocean tides upon tidal gravity are traced systematically across a continent. In the third, the influence of extremely high tides of small areal extent and the effects of other neighboring tides surrounding a peninsula nearly as an island, are reported. In the fourth example, two types of studies are involved: (1) correlations are deduced between the observed tidal gravity constituents and the phases of the corresponding constituent of the ocean tide for a world-wide

Table 2. Gravity observation of wave M_2 before and after correction for ocean tides (after Pertsev, 1969)

	Observed $G_0 - 1$	Correction ΔG	Corrected $G_c - 1$
East European stations			
Pulkovo	.238	- .029	.209
Krasnaya Pakhra	.188	- .009	.179
Kiev	.195	- .018	.177
Poltava	.183	- .010	.173
Simferopol	.173	- .010	.163
Mean	.195 ± .011	- .015	.180 ± .008
Central Asian stations			
Talgar	.166	+ .023	.189
Tashkent	.143	+ .018	.161
Frunze	.145	+ .022	.167
Lanchjow	.148	+ .014	.162
Novosibirsk	.158	+ .034	.192
Mean	.152 ± .004	+ .022	.174 ± .007

Ratio of means	Observed	Corrected
East European ÷ Asiatic	1.28 ± .11	1.034 ± .09

Note.–Mean oceanic correction for Asian stations = 0.84 ± .14 μgals. Mean pertinent M_2 signal for Asian stations = 6.42 μgals. Correction = 0.84/6.42 = 13%.

distribution of observing stations and (2) ocean tidal loads known with exceptional confidence are used in interpreting tidal gravity observations with a unique gravimeter.

1. The first example comes from Pertsev (1969) who evaluated for 10 stations in eastern Europe and central Asia the effects of the world ocean tides situated at distances more than 1500 km from these stations. Pertsev's computed corrections indicate that "recordings of the gravimeter may be affected by sea tides in all parts of the earth" and that "the corrections are not negligible even for stations situated in the middle of the Asian continent." As to tiltmeters, the effects are much more local, but still "their recordings are affected by tides . . . no further than 80° from the station." This is a distance that includes almost half the earth.

His results for the gravimetric factor $G_2 - 1$, observed and corrected, are reproduced for the M_2 tide in Table 2. The upper half of the table refers to five east European stations; the lower half to five central Asian stations. The ratio of means at the bottom indicates that, before correction, the mean of $G_0 - 1$ for the five eastern European stations exceeded that for the five mid-Asiatic stations by 28%, which is several times the observational error. The ocean tide corrections reduce this excess to only 3%; that is, to zero within the observational uncertainty of 9%. The mean corrected value of $G_2 - 1$ for all 10 stations is 0.177, which is 10% greater than the global value 0.160 deduced from seismic results.

2. In continuation of a program concerning the spatial variation of tidal gravity (Kuo and Ewing, 1966), the first transcontinental tidal gravity profile across the United States has been completed (Kuo et al., 1970). The upper half of Fig. 1 shows the disposition of the nine semipermanent gravimetric stations along the 40th parallel and the co-tidal patterns of the Atlantic and the Pacific M_2 tides. The lower portion shows values of ΔG (in percent) and of the phase lag k (in degrees) compared with calculated values of ΔG and k (triangles). The gravimetric factors are high near the east coast and low at the west coast in comparison with those observed in the interior. "The observed values $\Delta\delta$ and k agree remarkably well with those calculated for the M_2 constituent. . . . Nevertheless, there is a considerable degree of uncertainty about the ocean tides on open oceans. Currently, the agreement between the observed deviations of the gravimetric factors and the phases and the calculated deviations due to the influence of ocean tides does substantiate the primary importance of the influence of ocean tides on tidal gravity. The small residual deviations of both the gravimetric factors and the phases, after subtracting the effects of ocean tides on tidal gravity, do not correlate with the major different geological provinces, such as the Interior Plains and the Rocky Mountains. These deviations may well result from the imperfect knowledge of the tidal characteristics on open oceans." (Kuo et al., 1970).

3. Nova Scotia, located on the continental margin but surrounded by seas almost as an island, with the famous high tides of the Bay of Fundy on its west coast, provides an unusual opportunity for studies of tidal tilts. This opportunity

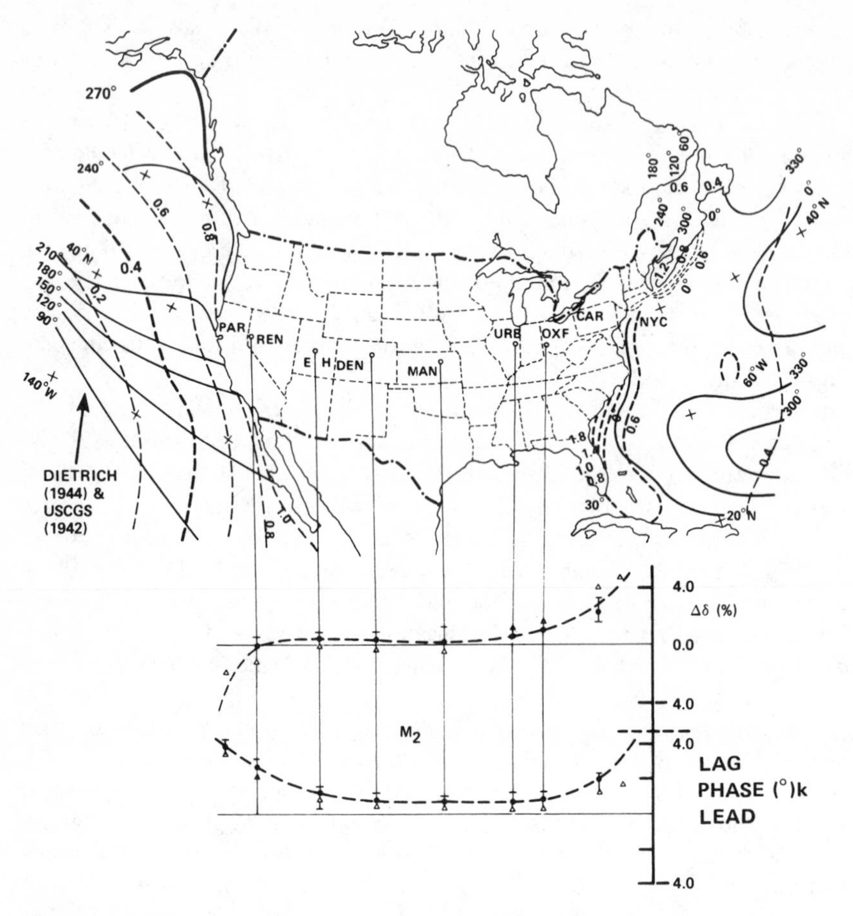

Fig. 1. Observed and calculated values of tidal gravity parameters along a transcontinental traverse. After Kuo et al. (1970).

has been taken by Lambert (1970). Figure 2 shows the major semidiurnal tide M_2 in the vicinity of Nova Scotia. In the Bay of Fundy, this tide attains amplitudes of 5 to 15 feet or more. In the Gulf of Saint Lawrence, there is an amphidrome and weak M_2 tides. In the Atlantic, moderate but of course extensive tides exist. The corresponding chart for the major diurnal tide K_1 (Fig. 3) is almost the inverse–weak tides in the Bay of Fundy, weak but more extensive tides in the Gulf of Saint Lawrence, and in the Atlantic, an amphidrome with nearly zero local tides. The land-locked tides, of course, are exceptionally well defined.

The solid circle marks the site (Rawdon) where tilt measurements were obtained during 8 months with two Verbaandert-Melchior horizontal pendulums. Because the station is at 45° latitude, it is theoretically free of the north-south diurnal tilt component in the solid earth tide, leaving purely

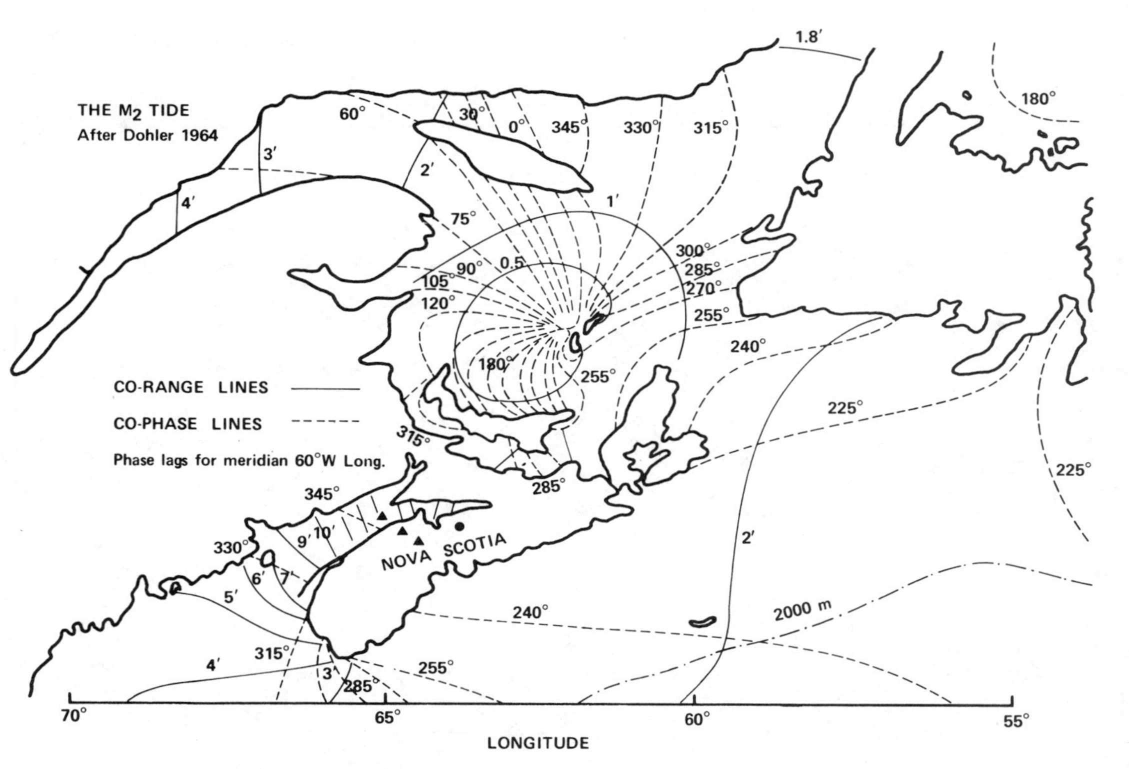

Fig. 2. Co-range and co-phase contours for M_2 tide near Nova Scotia.

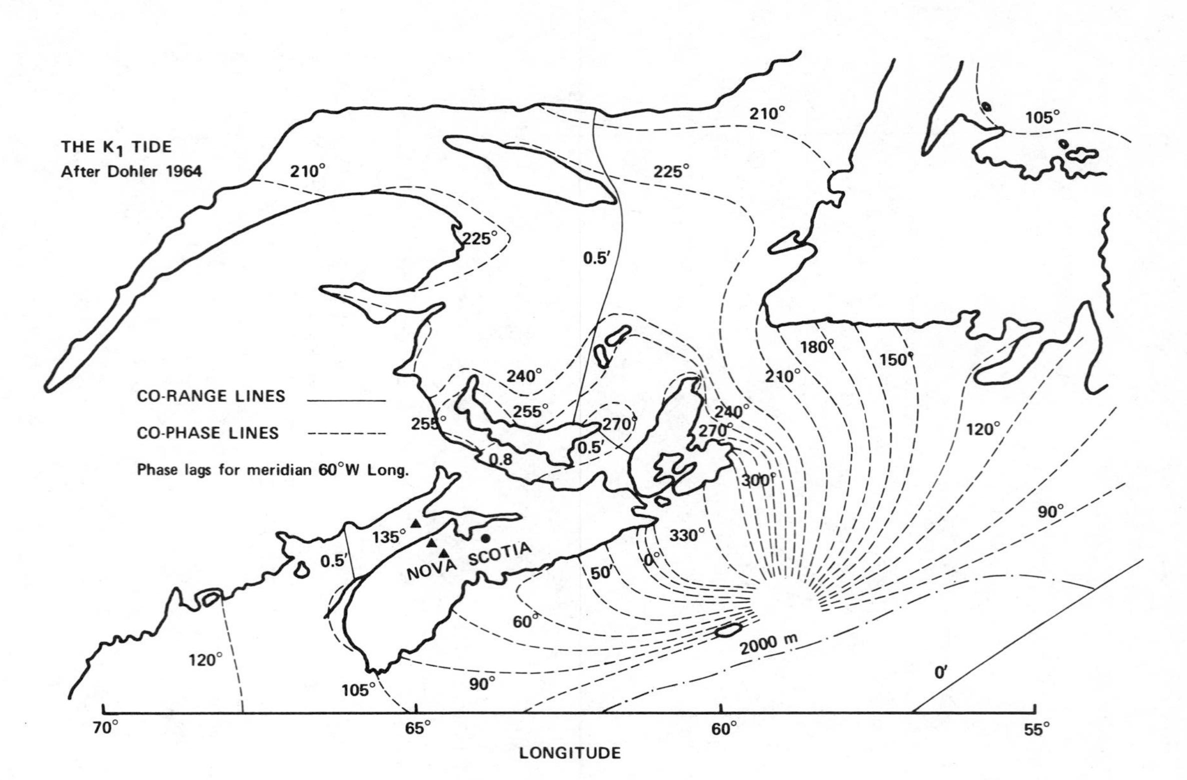

Fig. 3. Co-range and co-phase contours for K_1 tide near Nova Scotia.

secondary effects for observation in this component. The three triangles represent sites of observation with a LaCoste-Romberg tidal gravimeter, of which the two coastal stations were only exploratory. At the interior site (Berwick), 7 weeks of observations were obtained.

It was possible to separate the respective tidal loading effects caused by the three water areas. The local tilts produced by large loads in the Bay of Fundy can be explained by a two-layer local crustal model having the rigidities deduced for seismic models of the area. Lambert (1970) states that, "the tilting in phase with the more distant larger scale semidiurnal tides in the Atlantic Ocean is not compatible with laterally homogeneous models having rigidities suggested by seismic experiments. . . ." The continental (Appalachian) structure tilts independently in response to loading. "There is a preferred direction of tilt in response to loading in these areas. The gravity results on the whole confirm the tilt results." Apparently this area will continue to provide unusual opportunities for observing the regional yielding of the crust and upper mantle under known loads.

4. Excellent correlations between major local ocean tides and the corresponding gravity tides have been demonstrated recently by Farrell (1970), using gravity tide observations taken by E. A. Kraut and R. F. Forbes during the IGY with two LaCoste-Romberg meters (Harrison et al., 1963). Farrell's method of exhibiting these correlations is illustrated in Fig. 4, for the case of the M_2 tide at Bermuda. To normalize against the latitude effect, the observations are expressed by the usual gravimetric factor; i.e., as the ratio of the observed amplitude to that which would have been observed at the same location on a perfectly rigid oceanless earth. The observed gravimetric factor $\vec{G}_0$ is plotted in a polar vector diagram, with its observed temporal phase lag (here + 5.2°), behind the local equilibrium tide, written with positive sign. The common reference vector $\vec{G}_t$ for all tides has zero phase, and (except for K_1), amplitude 1.16. (For the K_1 tide, the value 1.14 for $|\vec{G}_t|$ was used.) The influence of loading (and of

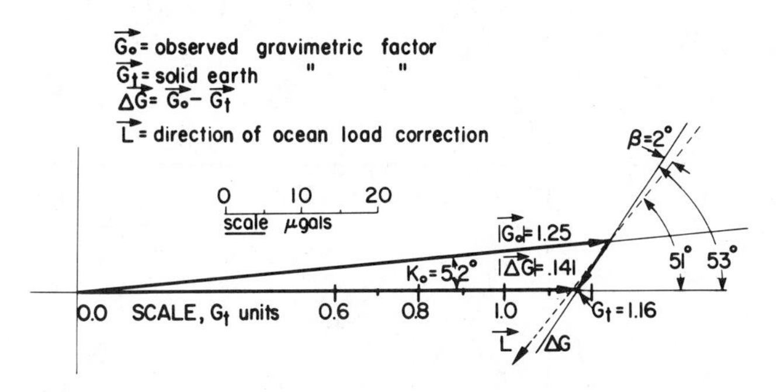

Fig. 4. Observed and corrected M_2 gravity tide at Bermuda. Polar vector plot showing difference in direction (β = 2°) between supposed correction vector ΔG and the local ocean load correction vector L. After Farrell (1970).

instrumental errors and unknown effects) appears in the vector difference $\vec{G}_0 - \vec{G}_t \equiv \Delta\vec{G}$, called the "anomaly." The phase angle of the anomaly is 53°, its amplitude = 0.14, which corresponds to 7.8 μgals on an absolute scale. Although the magnitude of the M_2 ocean load is not known, its temporal phase, 51°, with respect to the local equilibrium tide is available. In the present example, the difference β between the phase of the anomaly and the ocean load is $\beta \equiv 53 - 51 = 2°$. Thus the ocean tide accounts well for the phase of the observed M_2 anomaly.

Similar examination was made of the 16 tidal constituents at the 7 stations listed in Table 3. At the remaining five stations (Manila Observatory, New Delhi, Bunia, Lwiro [both on the equator in Central Africa], and Winsford, England), the discrepancies are not correlated in a simple way with ocean tides. For the seven stations listed in Table 3, the phase difference between the anomaly and the corresponding ocean tide has a nearly zero mean, with standard deviation ± 16°. The excellent correlations between the small observed anomalies and the respective phases of the local tides indicate that the portion of these anomalies attributable to differences in geologic structures among the sites must be small.

Knowledge of the ocean tides off the California coast has been much improved by the offshore measurements and the new theoretical analysis of Munk, Snodgrass, and Wimbush (MSW) (1970). One of their new charts (for the M_2

Table 3. Influence of ocean loads at seven UCLA gravity tide stations (after Farrell, 1970)

Station	Tide	Total load correction $\vec{\mid \Delta g \mid}$ μgals	Phase of observed gravity tide anomaly less phase of ocean load β
Bermuda	M_2	7.8	+ 2°
	S_2K_2	1.6	+ 3°
Glendora,	M_2	4.3	- 22°
Calif.	P_1K_1	4.8	+ 22°
Honolulu	M_2	3.4	+ 19°
	S_2K_2	3.3	+ 9°
	P_1K_1	2.2	+ 10°
Wake	O_1	2.6	+ 10°
	P_1K_1	2.5	+ 4°
Manila	S_2K_2	2.3	+ 21°
Observatory	O_1	2.1	- 20°
	P_1K_1	1.6	- 12°
Saigon	O_1	2.1	+ 24°
	P_1K_1	4.8	- 19°
Trieste,	M_2	3.6	- 16°
Italy	S_2K_2	1.3	+ 2°

Note.–Mean phase difference = + 2.3 ± 16°.

tide) appears in the lower right corner of Fig. 5, in comparison with three of the many older charts for this area. The differences among these representations clearly are large. Farrell has used the MSW charts to obtain gravity load corrections for the two tidal constituents K_1 and M_2, at La Jolla and UCLA. For La Jolla, O_1 and S_2 were estimated by the method of Lennon (1961). To obtain the effect at UCLA, the amplitude was assumed to be the same, and the phase was delayed as in MSW (Fig. 5). For La Jolla, two runs of 55 and 44 days duration with the superconducting gravimeter were used. For UCLA, the observations were of 485 days duration with two gravimeters (Slichter et al., 1964). The crustal yielding correction was computed for a flat multilayered gravitating earth. The earth model is that plotted by Backus and Gilbert (1970). The scale value of the La Jolla gravimeter was determined by matching the mean corrected M_2 amplitude of the two UCLA series, thus leaving all phase angles and the amplitudes of the other three constituents free.

Comparisons of observations of the constituents M_2, S_2, K_1, O_1 at La Jolla

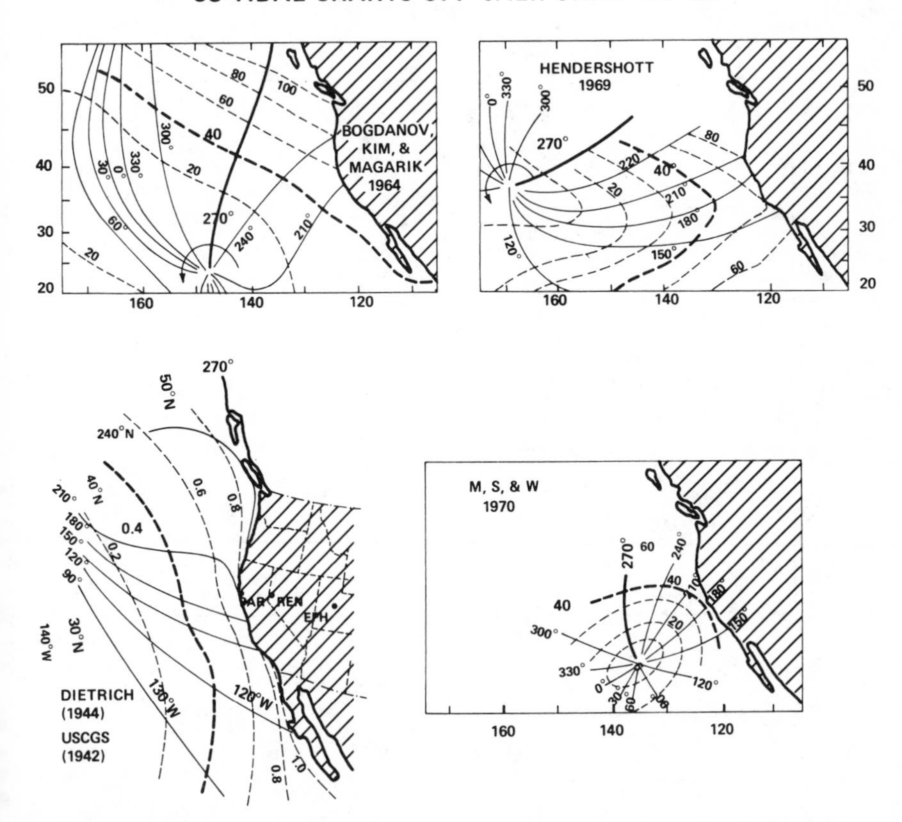

Fig. 5. Comparisons of co-tidal charts for M_2 tide off the California coast.

and UCLA, before and after correction for tidal loads, are provided by the phasor diagram, Fig. 6. The corrected G values are indicated by the points plotted at the tips of the correction vectors. The corrected amplitude of G_{M_2} at La Jolla has been normalized to that at Los Angeles. The large values for the S_2

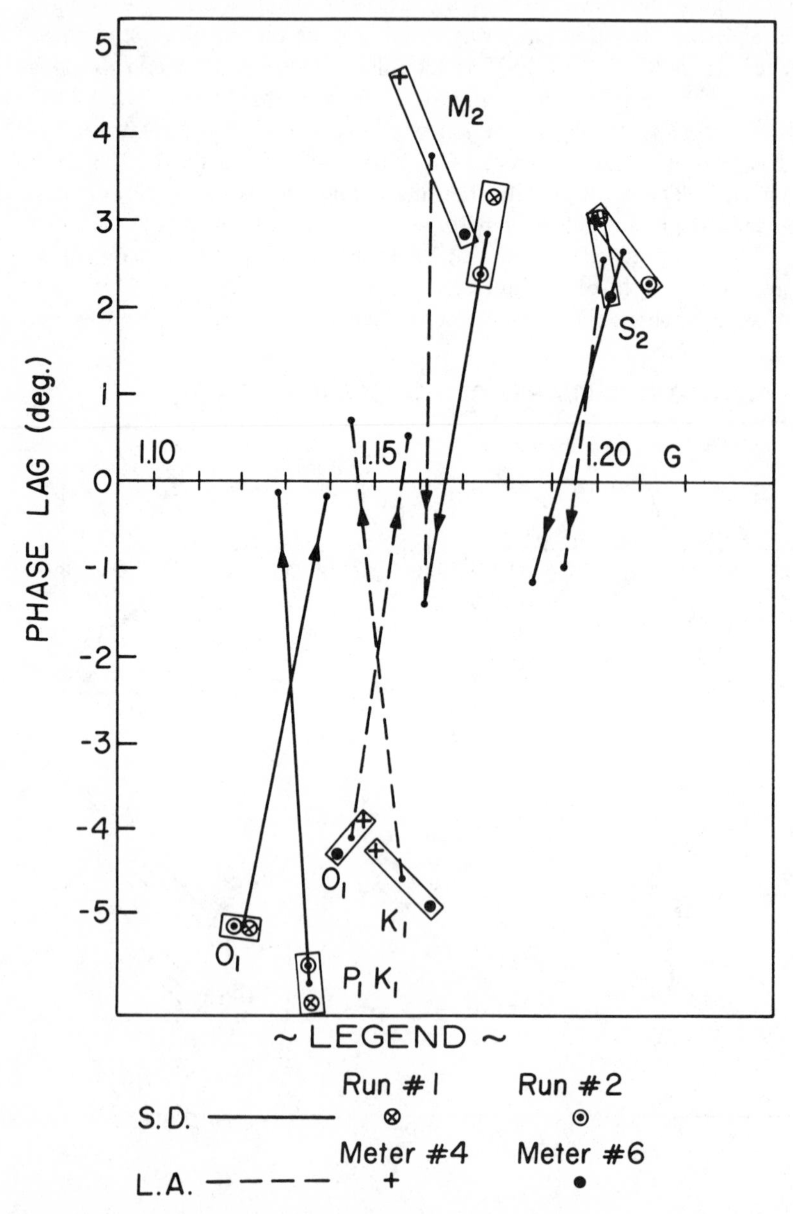

Fig. 6. Comparisons of observed and corrected gravity tides at UCLA, L.A., and La Jolla, S.D. (after Farrell and Prothero, personal communication, 1970).

tide are not understood. The corrected G factor for M_2 at Los Angeles is close to the theoretical 1.16, and that for K_1 is smaller than that for O_1 at both locations. The ratios R_G (see Table 1) for corrected values at Los Angeles and La Jolla, respectively, are 1.09 and 1.08; i.e., smaller than the theoretical values 1.15 to 1.22 listed in Table 1. The corrections introduced, however, are critical, for they essentially invert the values of these ratios. The success of the corrections in producing much closer clustering of the points is obvious and has encouraged more detailed examination of loading effects, such as those from Mission Bay and San Diego Bay.

INSTRUMENTATION

Earth tide instrumentation in essence provides the means of observing extremely small geophysical variations occurring at unusually small rates (changes of a few parts per billion at periods of a day or even weeks). Thus, these remarkable tiltmeters, gravimeters, and strainmeters have applications which will probably transcend their original purposes. An example of the extended field of application is the observation of the earth's free vibrations after great earthquakes. The extensive spectrum of these low-period vibrations was first observed with earth tide meters, and these meters continue to provide the prime observational information about the earth's free modes.

A developing application of geologic importance and instrumental challenge aims at the measurement of crustal movements in active tectonic regions. Such measurements at selected spots, made over a limited timespan of days or years, can reveal features of tectonic activity that escape the large-scale observations of geodesy or the long time scale of geology. This frontier is new and basic. In respect to instrumentation, success in observing these slow geologic deformations almost automatically implies corresponding excellence in earth tide recordings.

The chief new requirement is either increased stability of the instruments over times of a year or more or the ability meaningfully to calibrate and to correct a meter's slowly changing scale. A second desirable quality is portability to a degree that enables reasonable freedom of choice of sites within the selected geologic areas of interest, without demanding such rare circumstances as the presence of a long tunnel or a mine. In each type of meter (tiltmeter, gravimeter, and strainmeter), new designs are appearing that better meet these needs of tectonophysics. A few examples follow.

Tiltmeters

In detecting small tilts, the site and its environment is as important as the qualities of the instrument itself. A good tiltmeter site can hopefully be produced at tolerable cost wherever bedrock is near surface by boring a shallow hole to house the instrument. Several types of tiltmeters are now in service within drill holes. In Alaska, tiltmeters are in use at three locations, in bore holes 15 inches in diameter and 30 feet deep (E. Berg, personal communication, 1970). In Clausthal-Zellerfeld, West Germany, a two-component Askania pendulum has been under test in an 8-inch-diameter bore-hole about 50 meters deep (O. Rosenbach, personal communication, 1969). The small Hughes optical flat tiltmeter, about 4-inch-diameter by 8.5-inch height, developed by S. Hansen (1968) has received critical study in respect to its long-term stability by Harrison (1969) at the Poorman Mine site in Colorado. This small meter has promise of successful operation in small, relatively inexpensive drill holes for the study of slowly tilting geologic structures. Two component tilt meters obviously have the advantage of furnishing both the magnitude and the direction of tilt.

Gravimeters

Of the three main types of earth tide meters, the gravimeter is the most portable and makes least demands on the stability of its foundations. In the context of tectonic changes, the drift rates of conventional spring gravimeters are large–about 1 to 20 μgals per day in LaCoste-Romberg earth tide meters. Nevertheless, there are many tests that indicate that sensitive gravimeters are indeed competent to evaluate gravity differences among local sites with the precision required in monitoring the more rapid tectonic differential changes in elevation. In fact, a routine calibration of gravimeters consists in making repeated observations between two local test sites of accurately known gravity difference. In the case of the LaCoste-Romberg earth tide meters, such tests indicate that these comparisons, when extended to many dozens of repetitions, would establish statistically the differential gravity value within a fraction of a microgal; namely, a differential change in elevation of a fraction of a centimeter. Precise field observations of differential gravity among selected sites, repeated after a lapse of a few years, might reveal a significant pattern of gravity change, interpretable as a map of relative changes in elevation. I suggest that the conventional spring gravimeter, like the tiltmeter, has its special type of application in studies of secular change.

The superconducting gravimeter. A new type of gravimeter, in which the coventional spring for suspending the active mass is replaced by the magnetic flux associated with current flow in a superconductor, has been produced by Goodkind and Prothero (1971). This permanent magnetic field supports a 1-inch-diameter superconducting sphere at liquid helium temperatures, whose position is sensed and nulled by a capacitor-plate sensor and feedback electronic system of a type (Block and Moore, 1966; Weber and Larson, 1966) developed for conventional gravimeters. Observations of daily tides by this gravimeter are shown in Fig. 7. A power spectrum in the tidal frequency band, which reveals the large signal power to noise power multiples being obtained is seen in Fig. 8. Examples of these large factors over the major span of tidal periods are the following: fortnightly tide, M_f, 31; diurnal tide, K_1, 65,000; semidiurnal tide, M_2, 880,000; tridiurnal tide, M_3, 170. The noise performance is unusually good at the low frequencies. Whereas for a conventional gravimeter, the ambient

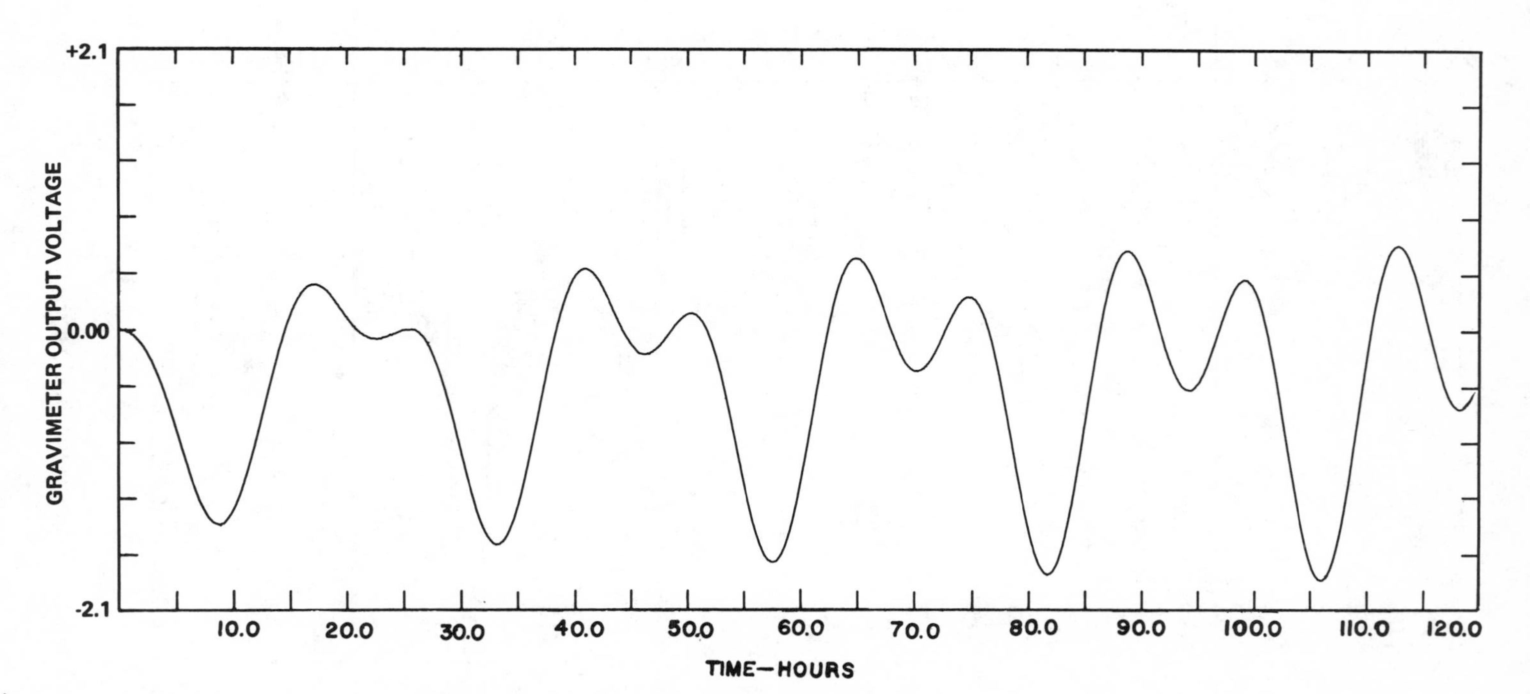

Fig. 7. Gravity tide observed with Goodkind-Prothero superconducting gravimeter.

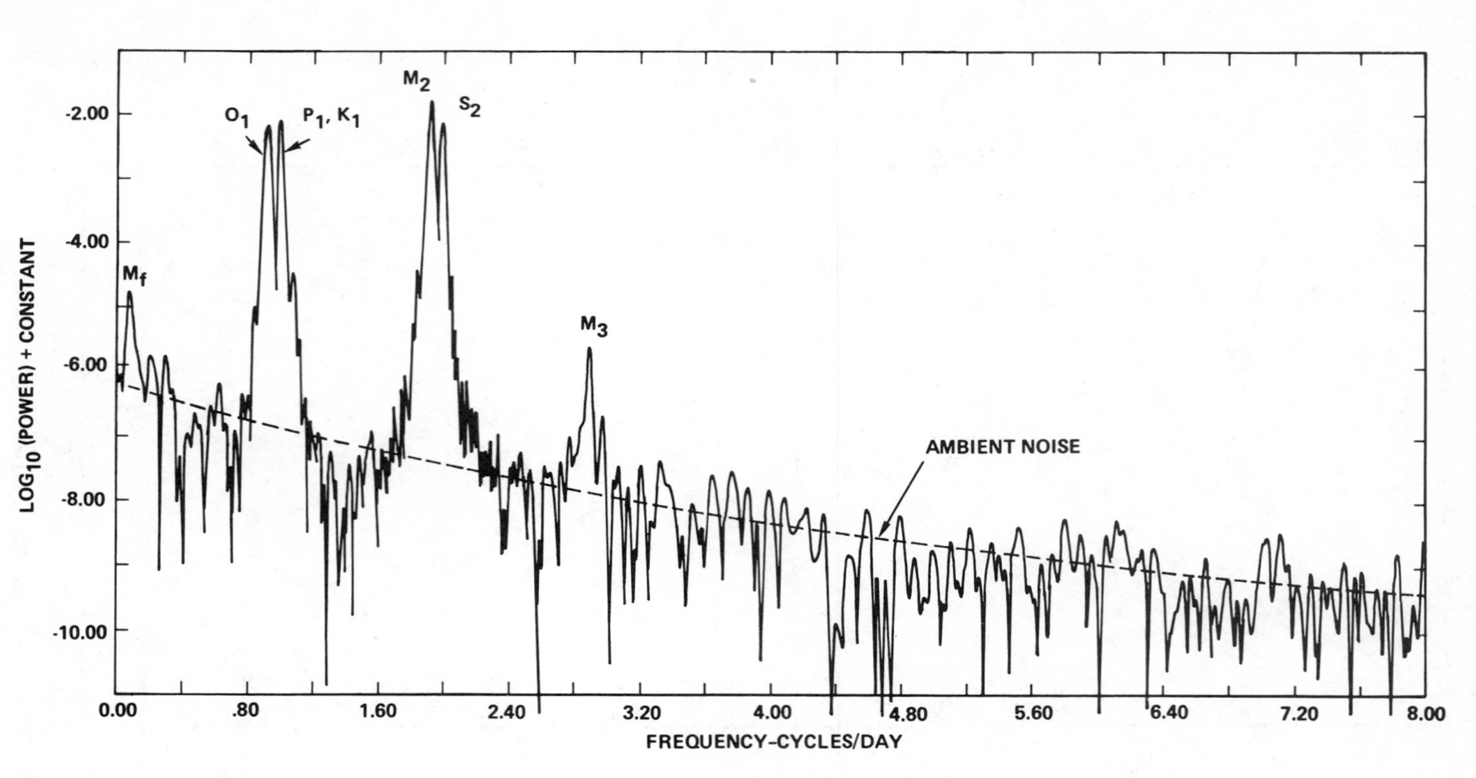

Fig. 8. Power spectrum of gravity tides observed with Goodkind-Prothero superconducting gravimeter.

noise power may increase about 4500-fold in the band 3.6 cpd toward 0.0 cpd, Fig. 8 indicates that this factor for the superconducting meter is about 95. The drift rate of the meter is also reported to be low, about 0.5 μgal per day.

Strainmeters

A bold concept in the design of strainmeters is that of Berger and Lovberg (1970), who have produced a laser beam strainmeter installed above ground at Camp Evans, near La Jolla, California, instead of underground in a tunnel as are other sensitive strainmeters. Figure 9 is an aerial view showing the 6-inch OD stainless-steel tube, 800 meters long, suspended on "A" frames. At either end is an instrument pier–a 2 by 2 by 10 foot column of black granite, grouted along its bottom 3 feet into holes 7 feet deep drilled through the top soil into the underlying conglomerate. A week's recordings of earth tides by this strainmeter is seen in Fig. 10. Solid curve represents the observed strain, the dotted curve theoretical strain computed with use of Longman's Love numbers (1963) up to P_7. The phase shift is produced largely by the M_2 ocean tide. Figure 11 is a recording of nuclear event "Benham," about 500 km away, which had a Richter magnitude 6.3.

The meter appears to be recording secular earth strains, in compression, at rates between 10^{-8} and 10^{-10} per day. It is planned to move the meter to a location on fresh granite, north of the Santa Rosa Mountains and about 15 miles east of Palm Desert, California, where conditions are expected to be favorable for observing tectonic strain.

In summary, all three main types of earth tide meters are being applied also to the problem of observing local tectonic deformations in progress, on the short time scale suitable for human beings.

OBSERVATIONS AT SOUTH POLE

The purposes of the current program of gravity tide observations at South Pole are (1) to observe long-period earth tides, which attain maximum amplitude at Pole and (2) to observe the earth's spheroidal free modes, which also have maximum amplitude there, and moreover are expected to be relatively free of the rotational fine structure observed in the lower latitudes. Three topics that have special relevance at this station will be briefly discussed: (1) special instrumental requirements; (2) an unusually small water tide correction; and (3) the analysis of records, which are short relative to the long tidal periods being recorded. In this analysis, the evaluation of instrumental variations and the identification of intervals of uniformity is critical. Results for a selected interval of observation in 1969 are reported.

Instrumental Requirements

Two general instrumental problems, not necessarily unique to the Pole station, were encountered there.

Fig. 9. Aerial view of Lovberg and Berger 800-meter laser strainmeter.

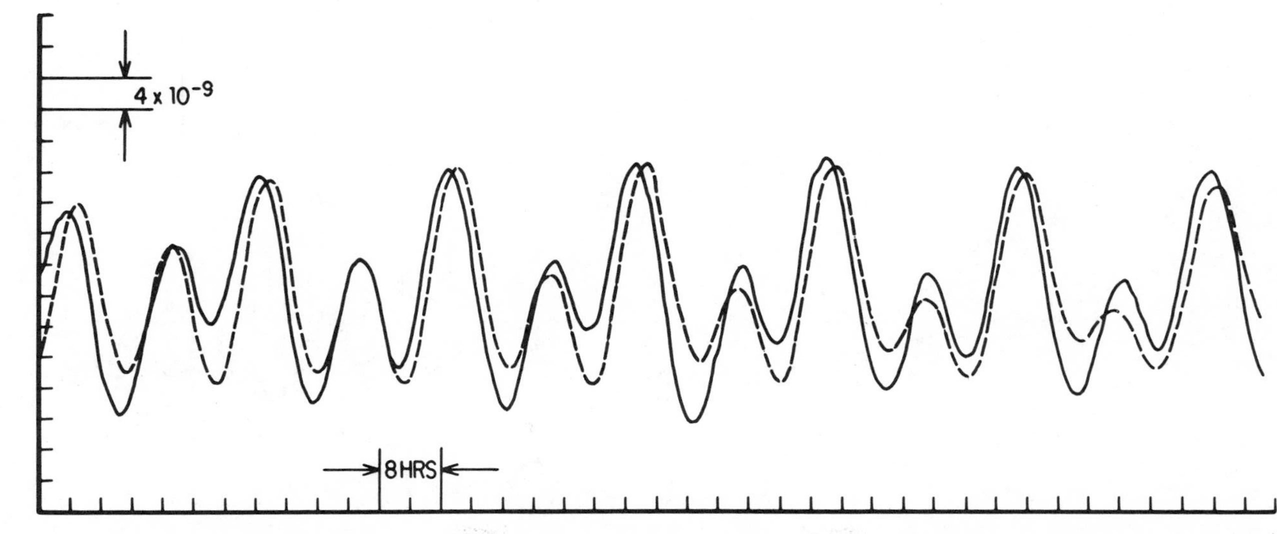

Fig. 10. Earth tides observed with Lovberg-Berger strainmeter. (Legend: Dotted curve = theoretical tide; solid curve = observed tide.)

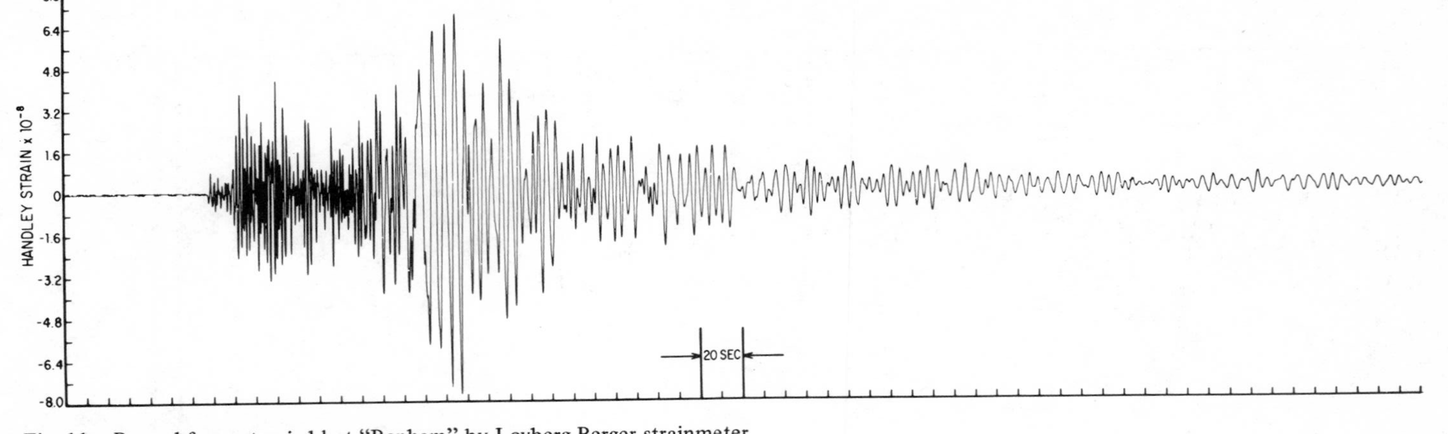

Fig. 11. Record from atomic blast "Benham" by Lovberg-Berger strainmeter.

1. The thick ice sheet, which moves about 100 feet a year, is subject to many small, irregular tilts, which produced excessive noise in the long-period tidal band. To neutralize tilting of the foundation, the meters have been suspended in the plumb line. Figure 12 is a schematic drawing illustrating the suspension system, and Fig. 13 shows the actual frame and suspended meter. A small chain is used as in a chemical chain balance to obtain fine level adjustments. The true level position of the beam is found by the use of test weights operated by solenoids, which can provide three known tilt-moments of either sign. Figure 14 shows the results of a test for true level. The theoretical parabolic variation of

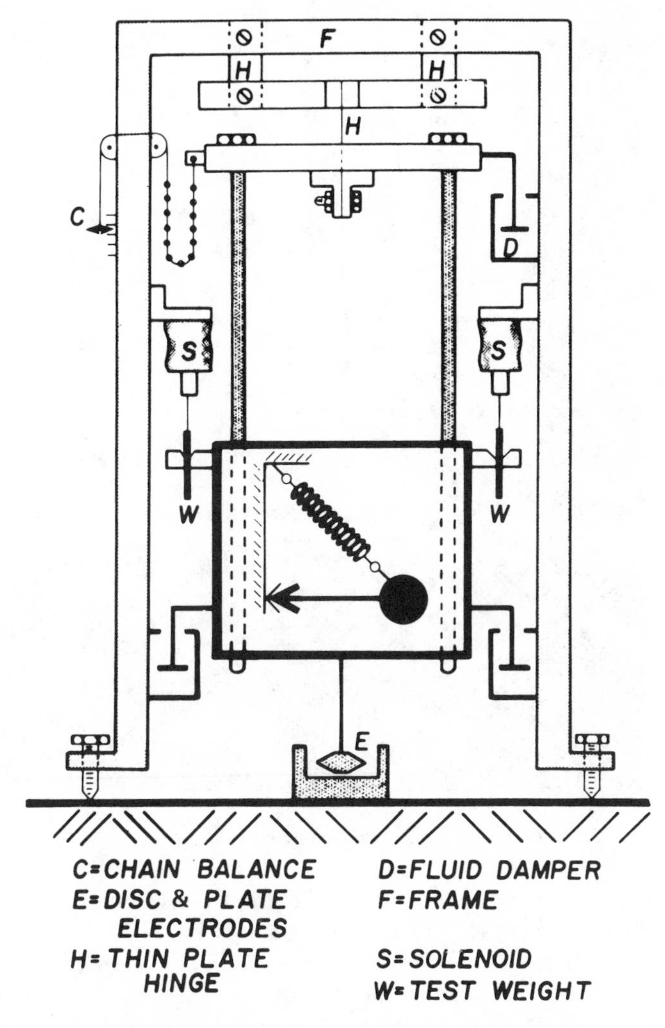

Fig. 12. Schematic drawing showing method of suspending and leveling tidal gravimeter.

Fig. 13. Photograph of LaCoste-Romberg earth-tide gravimeter suspended in anti-tilt frame.

apparent gravity with tilt angle θ from the vertical is obtained. This is more clearly evident in the rectified graph, $\sqrt{\Delta g} \sim \theta$. The intercept in this graph indicates that the meter beam is correctly leveled, within 5×10^{-6} rad. (From this position, an additional tilt of 10^{-5} rad. = 2 arc sec would introduce a false signal of only 0.10 μgal. If the initial error in beam level were 10^{-4} rad. = 20 arc sec, an increase of 10^{-5} rad. then would produce a significant error, 1 μgal.)

2. The local power supply was not of the desired reliability, nor of the required constancy in frequency and voltage. Stable 60-cycle, 110-volt power (about 500 watts for each gravimeter) was provided by transforming the 110 service voltage to 30 volts, rectifying with a zenode bridge, and trickle-charging a 24-volt automobile-type battery. This battery powered a Topaz inverter that furnished highly stable 110-volt, 60 cycle power. This system is constantly in circuit and will automatically supply the needed emergency power throughout several hours of main power outage.

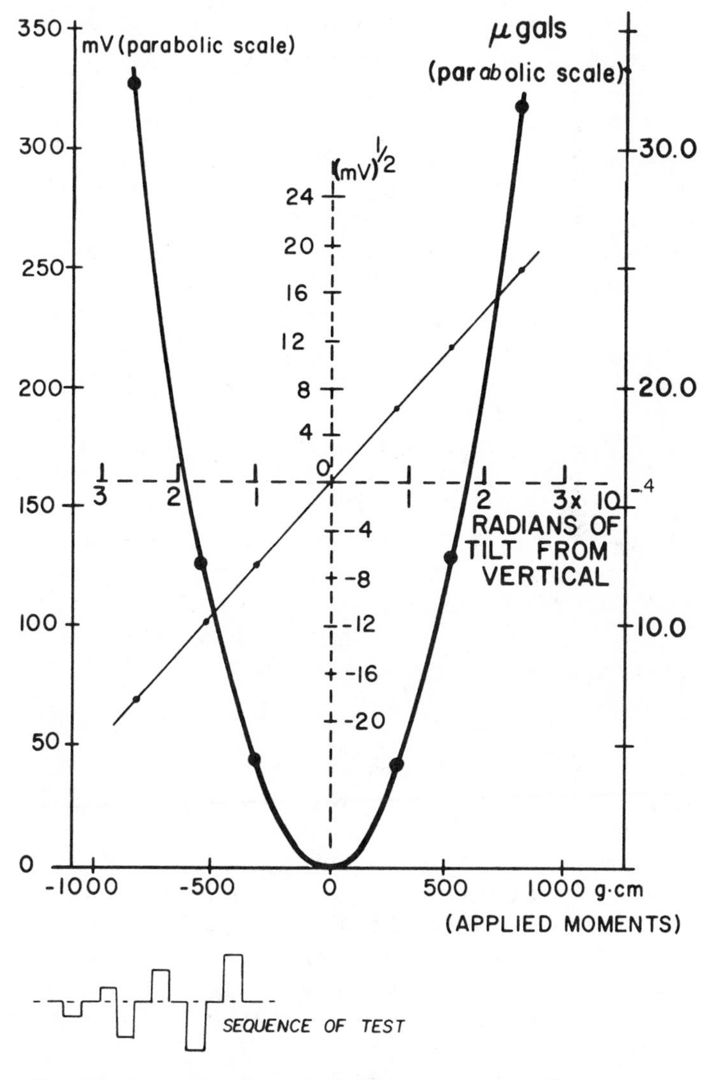

Fig. 14. Parabolic relations between changes in g due to tilt, and the tilt angle, θ (results of a test for true level).

Ocean Tide Corrections

The long-period ocean tides are very imperfectly known at present. The only analyses that succeed in distinguishing the fortnightly tide (M_f) above noise level appear to be those of Munk and Cartwright (1966) at Honolulu, those of Wunsch (1967) for a group of 16 stations in the Pacific and those of Cartwright (1968) for tides round north and east Britain. The amplitudes observed are not far from equilibrium, although the phase variations (Wunsch) indicate that the tide is substantially nonequilibrium. The estimate in Appendix A indicates that the gravitational influence at South Pole of equilibrium long-period ocean tides is small. In fact, the water correction at Pole is less than 1% of total tidal gravity there, and only 7% of the useful residual that corresponds to the reduced parameter $G - 1$. In contrast, at La Jolla, where the amplitude of the M_2 ocean tide is moderate (about 40 cm), the water correction is as large as 50% of the useful residual (see Fig. 6). Also significantly, the percentage correction to $G - 1$ for the M_2 ocean tide found by Pertsev (Table 2) even at the distant Central Asian stations, is *twice* that needed at South Pole for the long-period tides.

The percentage corrections for equilibrium water tides are the same for all the long-period tidal constituents. The observed amplitude of the fortnightly gravity tide, it should be noted, changes by the large factor 2.4 during a half cycle of the 18.61-year period of the precession of the lunar nodes. A maximum theoretical value, 21.1 μgals was attained in March 1969; the next minimum, 8.7 μgals, will occur in July, 1978. The percentage correction resulting from equilibrium tides remains constant. In absolute terms, the mean M_f amplitude at Pole is 14.9 μgals for $G = 1.16$, increased by 0.143 μgals by the computed ocean tide correction. The "useful residual" is $(G - 1)\, G^{-1}\, 14.9 = 2.06$ μgals.

Averaged in longitude over the full latitude circle, it seems unlikely that the true long-period tides will depart much from their equilibrium values. Their contribution to tidal gravity at the South Pole is then very small. Thus in theory, disregarding practical problems, the South Pole appears to be the ideal spot for determining global values of $G - 1$ for long-period tides.

Analysis of Short Records

During the long time intervals required for observing fortnightly and monthly tides, instrumental faults or failures may occur to interrupt the measurements. Sufficiently long records of high quality may be rare. The problem thus arises of extracting information from records that are too short for analysis by usual methods. Furthermore, during a fortnightly period, the drift rate of the gravimeter may change significantly. It is essential to determine this drift rate with reliability and to identify recording intervals during which the instrument's behavior appears to be unusually steady. Procedures for monitoring instrumental behavior in short records and for analyzing such records are described in Appendix B.

A record of a part of a selected 35-day interval of observations obtained by R.

Holbrook in 1969 at South Pole is seen in Fig. 15. The smooth curve represents theoretical tidal gravity on a rigid earth. The neighboring wavy curve is a programmed plot of 1-hour averages of the original data, from which drift has been removed. Riding on the long-period wave is a ripple of amplitude about 1 μgal and period 24 hours, which is better seen in the lower graph, from which the long-period tide has been removed. On a truly symmetrical earth, no daily tidal constituents should be present at South Pole. The ripple appears to represent asymmetry in the 24-hour ocean tide and will be the subject of a separate report.

The quality of the observations over the needed minimum time intervals of 3 or 4 weeks has been judged by the constancy of the linear and quadratic coefficients, b_j and c_j, respectively, in the interpolation formulae (described in Appendix B) for representing instrumental drift. A small value of the quadratic coefficient is especially desirable.

In Table 4, coefficients b_j and c_j and associated values of the gravimetric parameter $G - 1$ are shown, derived for observations for a 39-day interval, which includes that used in Fig. 15. This interval encompasses five half periods, $i = 1, 2, \ldots, 5$. For the three interior intervals, $i = 2, 3, 4$, two determinations of b_i and c_i may be obtained. These two, together with the corresponding values deduced for $G - 1$, are listed in rows A and B, columns 2, 3, 4. These two determinations of $G - 1$ differ only in respect to the estimated drift

Table 4. Instrumental drift characteristics and associated $G - 1$ values

Half wavelength no.		i	1	2	3	4	5
Time, t_i, of initial zero		Hours	4,352.6	4,513.8	4,725.1	4,888.7	5,008.1
$-b_j$	A	mv/hr	13.640	13.640	13.385	13.388	13.226
	B	mv/hr		13.385	13.388	13.226	
Ratio $\frac{\text{Values } A}{\text{Values } B}$				1.019	.9998	1.012	
$+c_j \times 10^3$	A	mv/hr^2	1.24	1.24	-.35	.14	1.35
	B	mv/hr^2		-.35	.14	1.35	
$(G-1)_i$	A		.107	.047	.177	.145	.116
	B			.109	.180	.093	
Mean $(G-1)_i$			.107	.078	.178	.119	.116
Means by pairs			.092	.128	.148	.117	
Mean of $G - 1$				$.121 \begin{smallmatrix} +.027 \\ -.029 \end{smallmatrix}$			

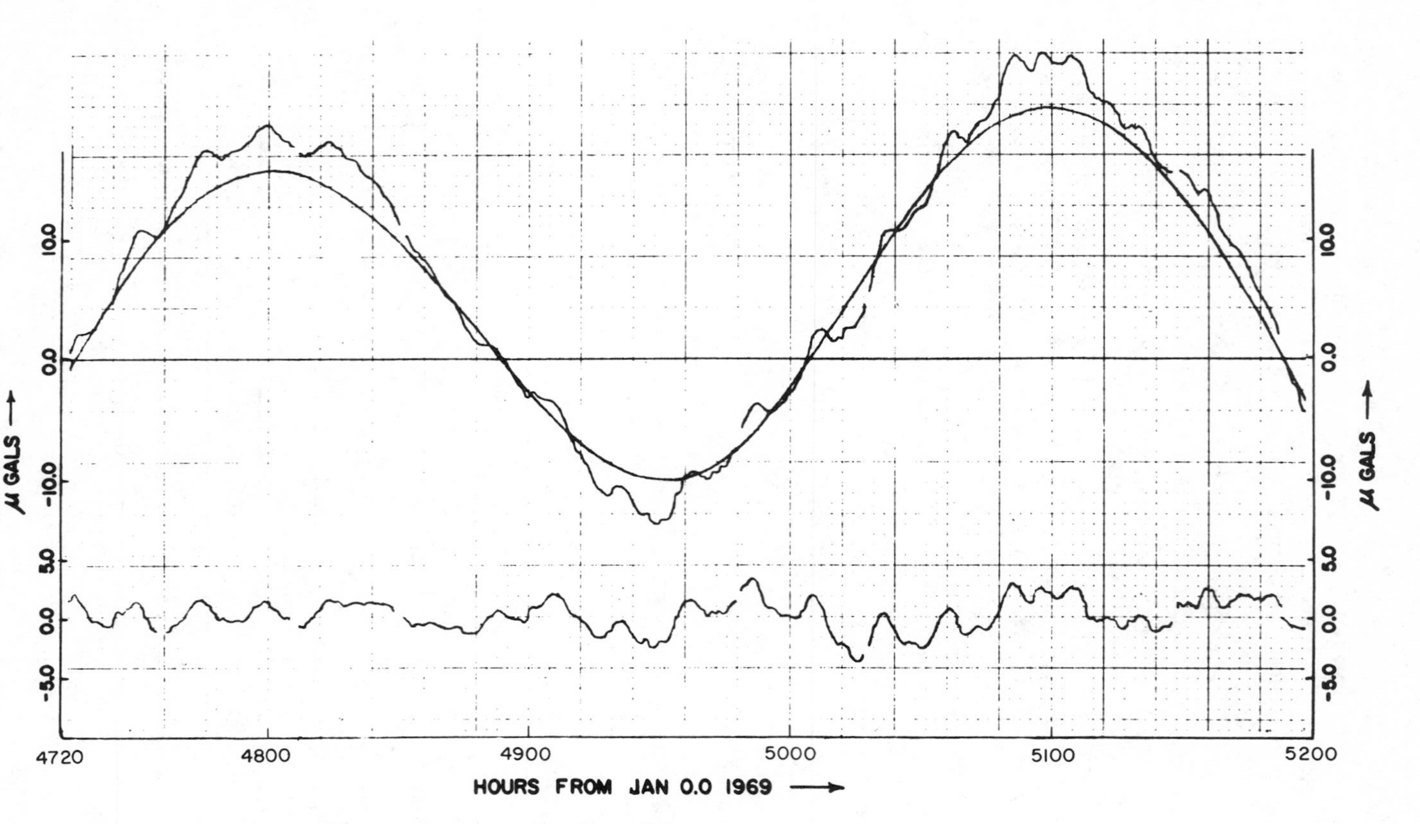

Fig. 15. Observed and theoretical long-period gravity tide at South Pole.

characteristic or reference zero of the gravimeter. The same observed readings were used in each case. In the columns $i = 2$, $i = 4$, the two values of the linear drift coefficient differ by 1 to 2%, and the differences in the curvature coefficients are large. The resulting $G - 1$ values differ by 50 to 100%. In column $i = 3$, the drift rates are the same within a few parts per 10,000, and the curvature values are very small. The two $G - 1$ values are also, necessarily, nearly the same. These examples clearly illustrate the large changes in $G - 1$ obtained from small percentage changes in the estimated drift rate of the meter. The value of $G - 1$ for the best interval, $i = 3$, reduced by 7% to correct for ocean tides, is 0.166. There is obvious need of much more extensive recordings of the higher quality data.

During the current year (1970), a second meter, #4, with low drift rate, ca. 1 μgal per day, has been put into service at Pole. Furthermore, the drift rate of the 1969 meter (#6) has become 1/3 less, and operational procedures have been improved. Cable reports indicate that better data are being obtained, which are expected to supersede those for 1969.

SUMMARY

Two of the major aspects of modern earth tide studies have been illustrated by means of examples: (1) the interaction of ocean and land tides and (2) new dual purpose instrumentation capable also of measuring secular tectonic changes. Modifications in the idealized theoretical earth tides produced by large known ocean tidal loads provide information about the mechanics of the region being deformed. Conversely, earth tide observations may offer controls upon the mapping of uncertain offshore ocean tides. Earth tide instrumentation is capable of providing information on a short time scale and at localized areas about secular changes in active tectonic regions.

At interior continental stations in the United States and in central Asia the correction for ocean tides is relatively small. This correction is remarkably insignificant at the South Pole station, which is therefore theoretically an ideal spot for measuring a global value of $G - 1$.

APPENDIX A

Correction to Tidal Gravity at South Pole Resulting from Equilibrium Ocean Tides

The gravity tide Z (positive, upward) and the height H of the equilibrium water tide are simply related, as follows, to the tide-producing gravitational potential U_2

$$Z = 2Ga^{-1}U_2 = 3.64 \times 10^{-9}U_2 \tag{A1}$$

$$H = \gamma g^{-1}U_2 = 6.98 \times 10^{-4}U_2 \tag{A2}$$

The numerical values are those for a spherical earth of radius $a = 6.371 \times 10^8$, surface gravity $g = 982.04$ (on a nonrotating earth), gravimetric factor $G = 1.16$, tiltometric factor $\gamma = 0.685$. For the long-period tides, U_{2L} is independent of the longitude of the observing point. It has the simple form (Bartels, 1957)

$$U_{2L} = -C\left(\frac{c}{R}\right)^3\left(\cos 2\delta - \frac{1}{3}\right)P_2(\cos\theta) \tag{A3}$$

where δ is the declination of the heavenly body, c the mean distance of the body from the earth, R its instantaneous distance, θ is co-latitude from South Pole, and the constant C (equivalent to G of Bartels) is

$$C = \frac{3}{4}fM\frac{\rho_1^2}{c^3} = 2.627 \times 10^4$$

where $f = 6.67 \times 10^{-8}$, M = mass of moon, and ρ_1 = earth's volumetric mean radius.

The distribution of the earth's ocean areas as a function of the co-latitude from South Pole is seen in Table 5. The number q_i in column 3 represents the fraction of the area in the co-latitude band between θ_i and θ_{i+1} (generally 10° wide) which is water covered.

The direct (downward) attraction Δg_i of a tide of height $H = -t_2P_2(\cos\theta)$ in the band θ_i, θ_{i+1} is

$$\begin{aligned}\Delta g_i &= -2\pi\rho f t_2 q_i \int_{\theta_i}^{\theta_{i+1}} \frac{1}{2}[2(1-\cos\theta)]^{-1/2}P_2(\cos\theta)\sin d\theta \\ &= -2\pi\rho f t_2 q_i \left\{u^{1/2}\left[P_2(\cos\theta) + 4u\cos\theta + \frac{16}{5}u^2\right]\right\}_{\theta_i}^{\theta_{i+1}}\end{aligned} \tag{A4}$$

Table 5. Water area and gravity at pole for indicated co-latitude bands

Index no., i	Co-latitude band, θ_i, θ_{i+1} (degrees)	Fractional area of sea, q_i	Δg_i (μgals)
0	< 15	0	0
1	15-20	.394	-1.75×10^{-2}
2	20-22.5	.715	- 1.47
3	22.5-30	1.000	- 5.40
4	30-40	.990	- 5.18
5	40-50	.962	- 2.14
6	50-60	.842	+ .02
7	60-70	.769	+ 1.54
8	70-80	.778	+ 2.60
9	80-90	.762	+ 2.82
10	90-100	.747	+ 2.67
11	100-110	.686	+ 1.76
12	110-120	.617	+ .85
13	120-130	.557	+ .01
14	130-140	.435	- .39
15	140-150	.414	- .70
16	150-160	.220	- .35
17	160-170	.500	- .61
18	170-180	.900	- .40
Total			-6.13×10^{-2}

where ρ = density of sea water = 1.025, $f = 6.67 \times 10^{-8}$, and $u \equiv 1/2\,(1 - \cos\theta)$. The values of the successive contributions Δg_i are listed in the last column of Table 5, for the case $t_2 = -2.86$ cm, which is the height of the M_f equilibrium tide [using Bartels' value $U_{M_f} = -(4{,}100)\, P_2(\cos\theta)$ in Eq. (A1)]. The total contribution is

$$\sum_{i+1}^{18} \Delta g_i = -0.061 \mu\text{gals}$$

The corresponding value at Pole of Z_{M_f} [Eq. (A1)] is

$$Z_{M_f} = (+3.64 \times 10^{-9})(-4{,}100) = -14.91 \mu\text{gals}$$

Of this, the part of geophysical significance is $(G-1)\,G^{-1}\,(-14.91) = -2.06$ μgals. Thus, the ratio of the direct water attraction to the useful signal is 0.061/2.06 = 0.03, so this correction is not of serious magnitude. Obviously, the same percentage correction applies to the major $P_2(\cos\theta)$ terms of all the long-period tides.

Minor additional corrections attributable to the water load are required.

Because the earth is not completely water covered, the tidal height t_2P_2 (cos θ) does not satisfy the requirement of conservation of water volume. To satisfy this condition, a small constant tide of amplitude t_1 independent of θ will be introduced, so $H = t_1 + t_2P_2$ (cos θ). The condition for conservation of water volume then determines t_1 through the relation

$$\Delta v = 0 = \sum_{i+1}^{18} q_i \int_{\theta_i}^{\theta_{i+1}} 2\pi a^2 [t_1 + t_2 P_2(\cos\theta)] \sin\theta \, d\theta \tag{A5}$$

which requires that $t_1 = 0.0442\, t_2 = -0.126$ cm.

The increase in downward gravity at Pole resulting directly from this t_1 tide is

$$\Delta g_1 = 2\pi\rho f t_1 \sum_{i=1}^{18} q_i \int_{\theta_i}^{\theta_{i+1}} \frac{1}{2} \sin\theta [2(1 - \cos\theta)]^{-1/2} d\theta \tag{A6}$$

$$= -0.035\mu\text{gals}$$

The computed direct tidal correction at Pole, $\Delta g_2 = -0.063$ μgals, is increased numerically by the elastic deformation of the earth and the change in station elevation produced by the tidal load. This increase for the present case is determined directly by the two known load deformation coefficients h'_2, k'_2 (Longman, 1963; Munk and MacDonald, 1960) provided that the load tide $H = -2.86P_2$ (cos θ), completely covers the globe. In that event, the direct correction should be multiplied by the factor $1 + h'_2 - 3/2k'_2 = 1.54$. To approximate this correction for the 70% water-covered globe, we apply this same factor, 1.54, to the reduced direct attraction of the actual ocean tide, -0.063 μgals, getting -0.097 μgals. This appears to overestimate the correction.

Similarly, correction for elastic deformation is needed in the case of the uniform tide t_1. For this purpose, this tide is regarded as uniform everywhere except over a circular disk 15° in radius that represents the Antarctic continent. Such a load (Caputo, 1962) produces at Pole an upward displacement of $+0.0036$ cm, thus increasing upward g by 0.011 μgals.

In summary, tidal gravity at Pole is changed by $-0.097 - 0.035 - 0.011 = -0.143$ μgals because of the equilibrium tide t_2P_2 (cos θ) and its associated uniform tide t_1 distributed over the ocean areas and because of the associated load deformations. In relation to the estimated geophysically significant tidal amplitude, -2.06 μgals, this ocean tide correction is only 7%.

APPENDIX B

Drift Correction and Analysis of Short Tidal Records

At the observing point, let tidal gravity on a rigid earth be

$$y(t) = y_s(t) + y_m(t) + y_0 \tag{B1}$$

Here $y_s(t)$ represents the contribution of the fortnightly and monthly tides only, which are henceforth designated by subscript s as the *signal* tides. The quantity $y_m(t) + y_0$ represents all the remaining (minor) tidal constituents of all periods–semiannual, annual, 18.6-year, trimonthly, and the constant tide y_0. The constant tides M_0 and S_0 and the eight principal minor tides (the 18.6-year tide, Sa, Ssa, Sta, MStm, Mtm, the tide of period 9.12068 days, and MSqm) were computed and removed from $y(t)$ to yield the signal portion of interest,

$$y_s(t) = y(t) - y_m(t) - y_0 \tag{B2}$$

which at known times, t_j, has zero value.

The decimated observed record is written in a form analogous to Eq. (B1)

$$r(t) = g_s(t) + g_m(t) + g_0 + n(t) \tag{B3}$$

where $g_s(t)$ and $g_m(t)$ are the signal and minor tides previously defined. The drift function $n(t)$ represents both instrumental drift and other types of noise, and g_0 is an arbitrary constant, which is especially so because a tide gravimeter is a variometer with no defined zero. The minor tides $g_m(t)$ may be satisfactorily removed by use of the relation

$$g_m(t) = 1.16y_m(t) \tag{B4}$$

The tides $g_s(t)$ may lag the theoretical tide $y_s(t)$ by a small time interval τ_0 resulting from the earth's imperfections in elasticity, but initially τ_0 will be omitted. Into Eq. (B3), let the substitution (B4) and $g_s(t) = Gy_s(t)$ be introduced to obtain

$$r(t) = Gy_s(t) + 1.16y_m(t) + g_0 + n(t) \tag{B5}$$

At the known zeros t_j of $y_s(t)$, Eq. (B5) provides known values of the drift function $n(t)$

$$n(t_j) = r(t_j) - g_0 - 1.16y_m(t) \tag{B6}$$

Improved precision for $r(t_j)$ in (B6) was obtained by averaging $r(t)$ over a 2-day interval centered at $t = t_j$.

The drift function $n(t)$ was approximated by successive quadratic interpolation formulas determined by the observations $\bar{r}(t_j)$ at three zones, t_{j-1}, t_j, t_{j+1}; i.e.,

$$n_j(x) = a_j + b_j x_j + c_j x_j^2 \tag{B7}$$

where x_j is the local time coordinate,

$$x_j \equiv t - t_j, \quad t_{j-1} - t_j < x_j \leq t_{j+1} - t_j \tag{B8}$$

In terms of the notations

$$\tau_j \equiv t_j - t_{j-1}, \quad D_j \equiv n(t_j) - n(t_{j-1}) \tag{B9}$$

the values of a_j, b_j, c_j are

$$a_j = \bar{r}(t_j) - g_0 \tag{B10}$$

$$b_j = (D_{j+1}\tau_j\tau_{j+1}^{-1} + D_j\tau_{j+1}\tau_j^{-1})(\tau_{j+1} + \tau_j)^{-1} \tag{B11}$$

$$c_j = (D_{j+1}\tau_{j+1}^{-1} - D_j\tau_j^{-1})(\tau_{j+1} + \tau_j)^{-1} \tag{B12}$$

In a record containing n zeros, there are $n - 2$ interior intervals for which quadratic interpolation (B8) applies. For the two end intervals, only a linear interpolation is determined. For each "half" of an interior interval, one obtains two solutions for the constants b and c. This provides both a check and insight about the quality of the record.

Using (B7), Eq. (B5) provides the following solution for the gravimetric factor, specific to each interval j (in which $t_{j-1} \leqslant t < t_{j+1}$, $x = t - t_j$)

$$\begin{aligned} G &= [r(t) - n_j(t) - 1.16y_m(t) - g_0][y_s(t)]^{-1} \\ &= [r(t) - r(t_j) - b_j x - c_j x^2 - 1.16y_m(t)][y_s(t)]^{-1} \\ &\equiv K_j(t)[y_s(t)]^{-1} \end{aligned} \tag{B13}$$

where $K_j(t)$ represents the known numerator.

These ratios should be weighted by factors proportional to the theoretical amplitude $y_s(t)$. The resultant weighted ratio evaluated for a half wavelength of the record, $t_j \leqslant t \leqslant t_{j+1}$, is, in fact, the ratio of the area under the observed half wavelength (corrected) tidal curve to the area under the corresponding rigid earth tidal curve; i.e.,

$$G_+ = \int_{t_j}^{t_{j+1}} K_j(t)\,dt \left[\int_{t_j}^{t_{j+1}} y_s(t)\,dt\right]^{-1} \tag{B14}$$

Similarly, the value G_- for the first half wavelength in the wave interval j, $t_{j-1} \leqslant t < t_{j+1}$, is obtained by substituting limits t_{j-1}, t_j for t_j, t_{j+1}, respectively.

To compensate for a potential slight error in the zero of reference, G values for adjacent positive and negative areas are averaged to obtain

$$G_j = \frac{1}{2}(G_- + G_+) \tag{B15}$$

REFERENCES

Alsop, L. E., and Kuo, J. T., Semidiurnal earth tidal components for various earth models, Ann. Geophys., **20**, 3, 286-300, 1964.

Alterman, Z., Jarosch, H., and Pekeris, C. L., Oscillations of the earth, Proc. Roy. Soc. (London), Ser. A., **252**, 80-95, 1959.

Backus, G., and Gilbert, F., Uniqueness in the inversion of inaccurate gross earth data, Phil. Trans. Roy. Soc. (London), Ser. A, **266**, 123-192, 1970.

Bartels, J., Geseitenkrafte (in German), Encyclopedia of Physics, edited by S. Flugge, pp. 734-774, Springer-Verlag, Berlin, 1957.

Berger, J., and Lovberg, R. H., A laser earth strain meter, Rev. Sci. Instr., **40**, 12, 1569-1575, 1969.

Block, B., and Moore, R. D., Measurements in the earth mode frequency range by an electrostatic sensing and feedback gravimeter, J. Geophys. Res., **71**, 18, 4361-4375, 1966.

Bogdanov, K. T., Kim, K. V., and Magarik, V. A., Numerical solution of tidal hydrodynamic equations for the Pacific area by means of BESM-2 electronic computer, Akademia Nauk SSSR, Trudy Instituta Okeanologii, **75**, 73-98, 1964.

Caputo, M., Tables for the deformation of an earth model by surface mass distributions, J. Geophys. Res., **67**, 1611-1616, 1962.

Cartwright, D. E., A unified analysis of tides and surges round north and east Britain, Phil. Trans. Roy. Soc. (London), Ser. A, **263**, 1134, 1-55, 1968.

Farrell, W. E., Gravity tides, Ph.D. thesis, University of California, San Diego, 1970.

Dietrich, G., Veroeffentl, Inst. Meersfrosch, Univ. Berlin, A41, 1944.

Dohler, G., Tides in Canadian waters, Marine Sciences Branch, Dept. Energy, Mines, and Resources, Publications, Ottawa, 1964.

Goodkind, J., and Prothero, W., Tidal measurements with a superconducting gravimeter, (in press), J. Geophys. Res., 1971.

Hansen, S., A highly stable geophysical tiltmeter, Trans. Am. Geophys. Union, **49**, 664, 1968.

Harrison, J. C., A preliminary report on tilt and gravity-tide measurements in the Poorman Mine near Boulder, Colorado, Technical Memorandum ERLTM-ESL 8, ESSA Research Laboratories, U.S. Department of Commerce, 1969.

Harrison, J. C., Ness, N. F., Longman, I. M., Forbes, R. F. S., Kraut, E. A., and Slichter, L. B., Earth-tide observations made during the International Geophysical Year, J. Geophys. Res., **68**, 1497-1516, 1963.

Hendershott, M. C., A numerical integration of Laplace's tidal equations in idealized ocean basin, Symp. on Math-Hydrodynamical Investigations of Physical Processes in the Sea Proc., Moscow, 1966.

Jeffreys, H., and Vicente, R. O., The theory of nutation and the variation of latitude, Monthly Notices, Roy. Astron. Soc., **117,** 142-161, 1957a.

______, The theory of nutation and the variation of latitude: The Roche model core, Monthly Notices, Roy. Astron. Soc., **117,** 162-173, 1957b.

Jobert, G., Théorie de la nutation et des marées diurnes d'après Molodensky (in French), Comm. 236, Ser. Geophys., **69,** de l'Observ. Roy. Belgique, 64-91, 1964.

Kuo, J. T., and Ewing, M., Spatial variation of tidal gravity, *in* The Earth Beneath the Continents, pp. 595-610, Geophys. Monograph 10, American Geophysical Union, Washington, D.C., 1966.

Kuo, J. T., Jachens, R. C., Ewing, M., and White, G., Transcontinental tidal gravity profile across the United States, Science, **168,** 3934, 968-971, 1970.

LaCoste, L. J. B., A simplification in the conditions for the zero-length-spring seismograph, Bull. Seism. Soc. Am., **25,** 176-179, 1935.

Lambert, A., The response of the earth to loading by the ocean tides around Nova Scotia, Geophys. J. Roy. Astron. Soc., **19,** 449-477, 1970.

Lennon, G. W., The deviation of the vertical at Bidston in response to the attraction of ocean tides, Geophys. J. Roy. Astron. Soc., **6,** 1, 64-84, 1961.

Longman, I. M., A Green's function for determining the deformation of the earth under surface mass loads, 2, Computations and numerical results, J. Geophys. Res., **68,** 485-496, 1963.

Melchior, P., Diurnal earth tides and the earth's liquid core, Geophys. J. Roy. Astron. Soc., **12,** 15-21, 1966.

______, The Earth Tides, Pergamon Press, Oxford, 1966.

Michelson, A. A., and Gale, H. G., The rigidity of the earth, Astrophys. J., **50,** 330-345, 1919.

Molodensky, M. S., Elastic tides, free nutation, and some problems of the earth's structure, Ann. Geophys. Inst., Acad. Sci. USSR, **N 19,** 146, 3-52, 1953.

______, The theory of nutation and diurnal earth tides, Comm. 188, Ser. Geophys., **58,** de l'Observ. Roy. Belgique, pp. 25-56, 1961.

Munk, W. H., and Cartwright, D. E., Tidal spectroscopy and prediction, Phil. Trans. Roy. Soc. (London), Ser. A, **259,** 533-581, 1966.

Munk, W. H., and MacDonald, G. J. F., The Rotation of the Earth, Cambridge University Press, Cambridge, England, 1960.

Munk, W. H., Snodgrass, F., and Wimbush, M., Tides off-shore: Transition from California coastal to deep-sea waters, Geophys. Fluid Dynamics, **1,** 161-235, 1970.

Nowroozi, A., Kuo, J. T., and Ewing, M., Solid earth and oceanic tides recorded on the ocean floor off the coast of northern California, J. Geophys. Res., **74,** 2, 605-614, 1969.

Pertsev, B. P., The effect of ocean tides upon earth-tide observations, paper presented at the Sixth International Symposium on Earth Tides, International Association of Geodesy, Strasbourg, 1969.

Slichter, L. B., MacDonald, G. J. F., Caputo, M., and Hager, C. L., Report of earth tide results and of other gravity observations at UCLA, Comm. 236, Ser. Geophys., **69,** de l'Observ. Roy. Belgique, pp. 124-130, 1964.

Takeuchi, H., On the earth tides of the compressible earth of variable density and elasticity, Trans. Am. Geophys. Union, **31,** 651, 1950.

U.S. Coast and Geodetic Survey, Tidal harmonic constants, Reports PH-1 and TH-2, 1942.

Weber, J., and Larson, J. V., Operation of LaCoste-Romberg gravimeter at sensitivity approaching the thermal fluctuation limits, J. Geophys. Res., **71,** 24, 6005-6009, 1966.

Wunsch, C., The long-period tides, Rev. Geophys. **5,** 4, 447-475, 1967.

IMPLICATIONS OF PLATE TECTONICS

13
PLATE TECTONICS

by **D. P. McKenzie**

The theory of plate tectonics now provides a remarkably successful kinematic theory of the present tectonic activity of the earth and is described in some detail. The origin of the forces that move the plates is by no means clear, though a surprising number of suggestions can be excluded using very simple arguments.

THE THEORY

Like most scientific theories, plate tectonics evolved gradually over a long period. Work on the gravity field in the 19th century had suggested that the outer 100 km of the earth behaved like an elastic plate overlying a more deformable layer below. Indeed the model for the long term mechanical properties of the earth described by Fisher (1889) is remarkably similar to the one outlined below.

The modern theory was perhaps first used explicitly by Hess and Maxwell (1953) when they discussed the bathymetry of the western Pacific. They believed that the trenches are sites of plate consumption and recognized that the plate boundary joining the northern end of the Tonga trench to the southern end of the Marianas trench must be an enormous left lateral strike slip fault, which would now be called a transform fault. Though Wilson (1965) recognized the essential postulates of the modern theory and Bullard, Everett, and Smith (1965) published the basic mathematical theory, the connection between the two was not recognized until 1967 when McKenzie and Parker (1967) and Morgan (1968) independently published the present theory. Le Pichon (1968) then showed that the new theory of plate tectonics applied on a global scale, and Isacks et al. (1968) reviewed an enormous mass of seismic information, which they showed was also in agreement with the new ideas. Later work has extended the original theory, but has not suggested that the basic assumptions are in error. Indeed such a demonstration now seems unlikely, and it therefore appears that plate tectonics provides a reliable kinematic description of global tectonics.

Though there is much in common between the concepts of the new theory and the earlier ideas of sea floor spreading (Hess, 1962) and continental drift (Wegener, 1929), there is a considerable difference in the emphasis given to certain observations. The essential postulate of plate tectonics is that the earth's surface may be divided into a number of rigid spherical caps whose boundaries are the seismic belts of the world (Fig. 1). It is the relative motion between these caps, or plates, which causes earthquakes. Thus the division of the world into plates depends on the absence of a seismic deformation, and the success of Le Pichon's calculations (1968) suggests that no such boundaries exist. The relative motion between any two rigid spherical caps on a sphere can be described by an angular velocity vector passing through the center of the sphere. This result is a consequence of Euler's theorem, which states that any finite or infinitesimal displacement of a rigid cap on the surface of a sphere can be described by a rotation about some axis through the center of the sphere (Wigner, 1959). The major difference between plate tectonics and the earlier theories is this emphasis on rigidity and of relative motion between plates. Also the plate motions are no longer believed to be closely related to the mantle motions below, so that there is no longer any difficulty in understanding how the African and Antarctic plates can be almost entirely surrounded by ridges. Plate boundaries are important only because they are the sites of plate creation and destruction, and are also the places where it is easiest to measure the direction and magnitude of the relative velocity vector between plates. Though the relative movements across plate margins have attracted an enormous amount of interest, they are unlikely to be related to large-scale mantle motions. The absence of major deformation, though not of large stresses, within the plates themselves is perhaps even more important than the deformation now taking place along their margins.

Certain important features of the new theory have caused confusion. There is no known difference between the motion and behavior of plates mostly made of continental lithosphere, such as the Eurasian plate (Fig. 1), and those that consist almost entirely of oceanic material, such as the Pacific plate. There is, however, a remarkable difference in the distribution of seismicity where plate boundaries cross continental regions (Fig. 1). In the oceans, the earthquakes are confined to very narrow strips, whose width in many places still has not been resolved. On the continents, however, the earthquakes are spread over a wide belt; such as the Alpide or Himalayan region. Though certain speculative ideas

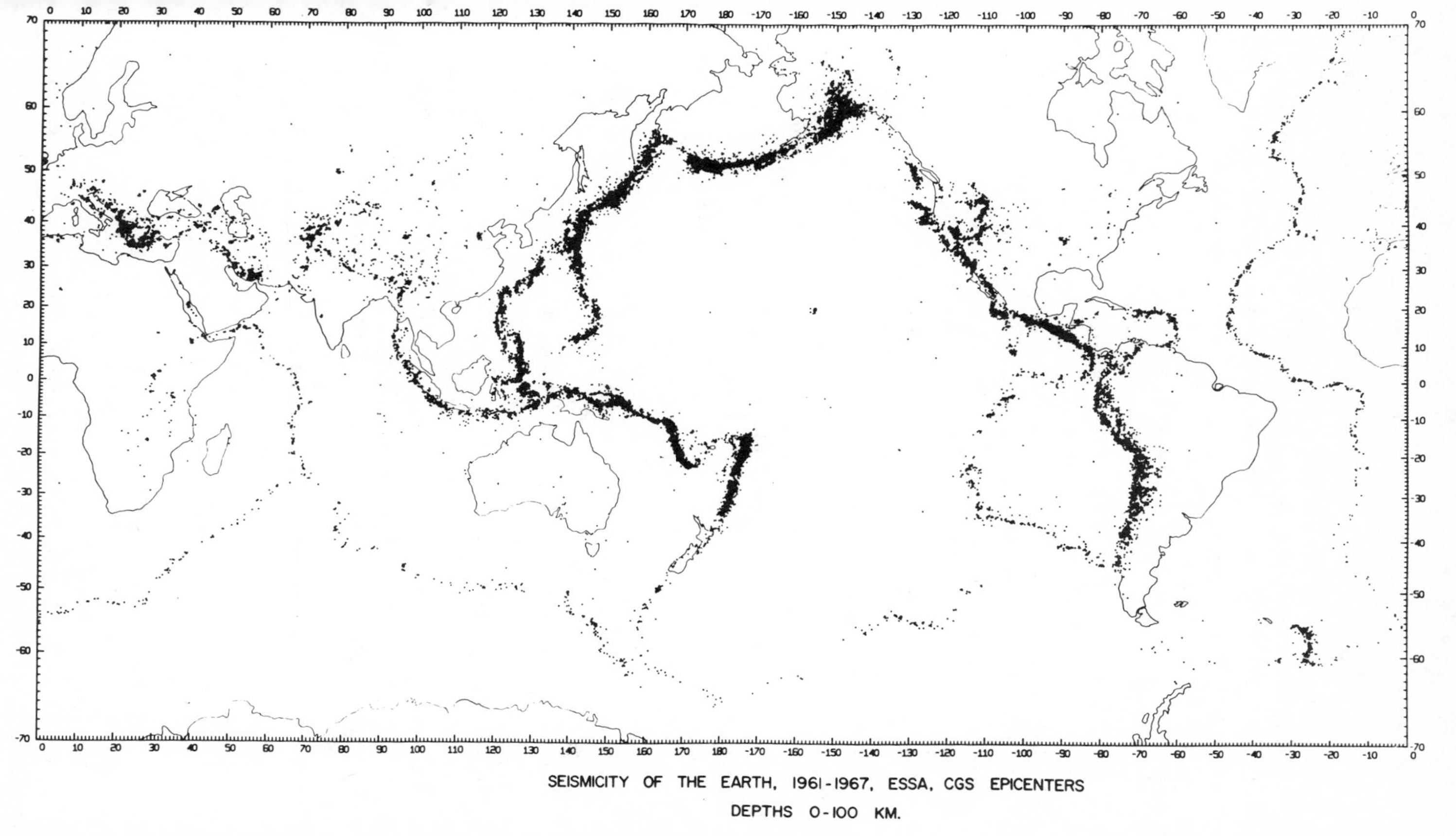

Fig. 1. Earthquake epicenters whose foci were shallower than 100 km and which occurred between 1961 and 1967 (Barazangi and Dorman, 1969). In oceanic areas the seismicity is confined to narrow strips, whereas in continental regions it is spread over a broad zone.

have been suggested to explain this difference (McKenzie, 1969a), its origin and importance is still not properly understood. In view of this difference, it is surprising that plate boundaries bear little relationship to continent-ocean boundaries and that most plates consist of parts of both oceans and continents. It is for this reason that the theory is better called plate tectonics, rather than continental drift or sea floor spreading.

The other confusion that still appears in the literature is concerned with the importance of relative motion. This difficulty is clearly brought out by Talwani (1969) and by Irving and Robertson (1969), all of whom accept the new theory, but still believe that even though the plates do not define an absolute frame of reference, their margins do. Francheteau and Sclater (1970) have used Le Pichon's poles (1968) of relative motion to demonstrate that neither ridges nor trenches form such a frame, since they are all in relative motion. It is therefore clear that any frame can be used as a reference frame and that no special frame is favored by the observations. For instance the rotational axis and the Greenwich Meridian form such a frame, and all plate motions can be referred to it. This choice shows that polar wandering is an unnecessary concept in plate tectonics, a point first made by Munk and MacDonald (1960). It may, however, still be a useful one in certain circumstances provided it is carefully defined.

Another important change of emphasis that still causes confusion is concerned with stresses and displacements on plate boundaries. It has been common geological practice to describe normal faults as tensional faults, and other thrusts as compressional ever since Anderson's (1951) work on the relationship between the orientation of the axes of principal stress and the directions of faulting. Though his arguments do apply to the formation of a fault in an initially homogeneous material, they are not valid in most geological situations. The very existence of plate margins and seismic belts shows that the relative motion between plates is taken up along weak zones in the lithosphere whose position changes slowly if at all with time. Under these conditions there is very little relationship between the type of faulting and the orientation of the principal stresses (McKenzie 1969b). The nature of the faulting is intimately related to the relative motion between the plates and the strike of the plate boundary, or to the displacement rather than the stress field. Normal faulting is therefore referred to as extensional, since it is the consequence of the separation of two plates, whereas overthrust faulting is the consequence of the consumption of lithosphere, which must take place whenever two plates move toward each other. Since the type of faulting is not closely related to the stress field, very little is known about the orientation of the principal stress axes within the plates.

Though the concepts of plate tectonics seem to describe the tectonics of the earth remarkably well, they are clearly an idealization of the real world. In practice, plates are elastic and can take up small variations of velocity between different parts of the same plate. Earthquakes are the result of the equalization of these variations in displacement. Small anaelastic displacements must also occur within all plates, and though all such events could be described as earthquakes and the slip planes as plate boundaries, the resulting description would be useless because of the number of plates that would then be required.

A more interesting effect that could possibly be important is the result of the earth's ellipticity. Plate motions on a spherical earth can occur by rotations about any axes without plate deformation, since the curvature of a sphere is everywhere the same. On an elliptical earth, however, all rotations about axes other than the axis of symmetry of the oblate spheroid must produce deformation of the plates. The present ellipticity must produce strains of the order of one part in a hundred in plates the size of the Pacific plate when they move through 90° of latitude. As a result of this deformation, a restoring force must act on all plates or they attempt to change their orientation or latitude with respect to the rotational pole. Observations of present plate motions and palaeomagnetic observations of rocks of Silurian age and younger do not suggest that this effect has been important, though it does have interesting consequences.

The original discussions of the concepts of plate tectonics were principally concerned with boundaries between two plates and with the description of the relative motion between them as an angular velocity about some axis through the earth's center. Though McKenzie and Parker (1967) made an attempt to discuss points where three plates meet, they were not especially successful. Both they and Morgan (1968) made use of the equation that relates the relative angular rotation vectors of three plates surrounding any such point:

$$ {}_A\boldsymbol{\omega}_B + {}_B\boldsymbol{\omega}_C + {}_C\boldsymbol{\omega}_A = 0 \tag{1} $$

where ${}_A\boldsymbol{\omega}_B$ is the angular velocity vector of plate B relative to plate A. The sign convention used throughout this paper takes ${}_A\boldsymbol{\omega}_B$ to be positive if it is in a right-handed screw sense when viewed looking outward from the center of the earth. Equation (1) simply described the obvious statement that the relative motion between plates A and B must be the same whether it is obtained directly or by comparing the motion of A and B with respect to C. If the vector from the center of the earth to the point where A, B, and C meet is $\mathbf{r}_0$, then:

$$ {}_A\boldsymbol{\omega}_B \times \mathbf{r}_0 + {}_B\boldsymbol{\omega}_C \times \mathbf{r}_0 + {}_C\boldsymbol{\omega}_A \times \mathbf{r}_0 = 0 \tag{2} $$

If ${}_A\mathbf{v}_B$ is the vector velocity between A and B at a point a distance ϵ from $\mathbf{r}_0$ on the boundary between A and B, then:

$$ {}_A\mathbf{v}_B = {}_A\boldsymbol{\omega}_B \times \mathbf{r}_0 + 0(\epsilon) $$

hence

$$ {}_A\mathbf{v}_B + {}_B\mathbf{v}_C + {}_C\mathbf{v}_A = 0 + 0(\epsilon) \tag{3} $$

Equation (3) formed the basis of the recent discussion by McKenzie and Morgan (1969) of the evolution and movement of points where three plates meet. Their results will not be discussed here in detail, since the problem is somewhat complicated. In general such points, called triple junctions, move with respect to all three plates and can produce apparent changes in the relative motion between plates which must be carefully distinguished from true changes in the relative motion. The same results have recently been obtained in a more general form by McKenzie and Parker (1971).

MEASUREMENT OF RELATIVE MOTION BETWEEN PLATES

The instantaneous relative motion between two plates can now be measured in a variety of ways. Most methods give either the velocity or its direction, and only geodetic methods or direct observation can give both. The most accurate and widely used method of obtaining the magnitude of the rate of relative motion depends on the Vine-Matthews hypothesis, which perhaps should now be called a theory (Vine and Matthews, 1963). They suggested that the linear magnetic anomalies observed over most ridges are formed by sea floor spreading, or plate creation, combined with reversals of the main field. Their suggestion is now generally accepted (Vine, 1966; Heirtzler et al., 1968) and permits the age of any part of the ocean floor to be determined from the characteristic magnetic anomalies observed above it. The reversal time scale has now been reliably established by magnetic measurements on well-dated volcanic rocks (Cox, 1969; Dalrymple et al., 1967). The spreading rate ν_a on most ridges can therefore be measured to an accuracy of about 0.1 cm/yr simply by comparison of calculated and observed magnetic profiles. Such a comparison can only give the true rate of separation between plates if the angle between the strike of the ridge and the direction of motion between the plates is known. The true separation rate is given by $\nu_a/\cos\phi$, where ϕ is the angle between the normal to the ridge and the direction of motion between plates. Some confusion occurs because all known ridges produce plates on either side at an equal rate, and therefore their axes move at the same velocity relative to either of the plates. This symmetry has resulted in the spreading rate of ridges usually being given as $\nu_a/2$ rather than ν_a. Also some authors assumed that $\phi = 0°$, a result that is approximately true for many but by no means all ridges. Prominent exceptions are the Reykyanes ridge, south of Iceland, with $\phi \simeq 30°$, the Juan da Fuca ridge off Oregon with $\phi \simeq 20°$, and the ridge in the center of the Gulf of Aden. Though various suggestions have been made to explain the symmetry of all ridges and to account for the fact that ϕ generally differs little from zero, none have yet been generally accepted.

Magnetic lineations can only be used to measure the rate of separation of two plates along an extensional boundary. The rate of approach of the two plates on either side of a trench must be determined by other methods. The most powerful of these is at present the measurement of the seismicity of such plate margins. This technique has been developed by Brune (1968) who used the surface wave magnitudes of earthquakes on transform faults and trenches to determine the rate of slip between plates. Unlike the short period body wave magnitude, the surface wave magnitude of an earthquake depends only on the product of the area over which slip takes place multiplied by the displacement that occurred on the fault. If the area of the plate boundary is known, the slip rate can be obtained from the surface wave magnitudes of earthquakes in a time interval that is sufficiently long to average out the statistical variations of earthquake energy release. The length of the interval depends on the slip rate; shorter periods can be used for boundaries between rapidly moving plates. Some of Brune's results (1968) agree with rates obtained by using other methods and suggest that his calculations are reliable for boundaries with high slip rates.

The most obvious but least general method of measuring both the direction and rate of motion between plates is by direct observation by triangulation. This method has been attempted in California (Hofmann, 1968; Whitten, 1956) across the San Andreas fault and in Japan. It is not accurate unless the observing stations are situated on solid rock at some considerable distance (for instance 100 km) from the fault on either side. If the stations are close to the plate boundary, displacements between them occur only after earthquakes. Displacements of 5 meters are not uncommon on faults after large earthquakes and can occur only once in a hundred years if the slip rate is 5 cm/yr. Careful measurements are not available over such long periods anywhere, and it is therefore essential to use stations at a considerable distance from the fault. The same objection does not apply to Brune's method (1968) if the plate boundary used is some thousands of kilometers long, since the displacements will occur at different places at different times, and hence the average activity will reflect the average slip rate.

The last and probably least reliable method of measuring the rate of motion between plates moving toward each other depends on an observation of Isacks et al. (1968). They noticed that the time taken for a piece of plate to move from the earth's surface down to the deepest point on the plane containing intermediate and deep focus earthquakes is about 10 m.y., if Le Pichon's slip rates are correct. Since the thermal time constant of the lithosphere is about 10 m.y., the most probable explanation of their observation is that earthquakes can only occur in cold material of the sinking slab and not within the more widespread and hotter material elsewhere in the upper mantle (McKenzie, 1969a). This suggestion is consistent with Isacks and Molnar's observation (1969) that the intermediate and deep focus earthquakes occur within the sinking slabs and not by slip between the slabs and the surrounding mantle. If, therefore, the relative motion between two plates has not changed during the last 10 m.y., the distance from the surface to the deepest earthquakes measured down the dip of the slab can be used to obtain the velocity of relative motion normal to the trench. In practice, this method generally involves too many assumptions to be reliable, since if the motion between two plates is known from other evidence to have remained constant for a long period, this evidence can be used to obtain their relative velocity. The method can, however, sometimes be used to determine the lower limit on the consumption velocity when the history of the plate movements is not known.

The direction of relative motion between the plates on either side of a plate boundary is generally easier to measure than the rate of motion. The most accurate method is to determine the strike of active transform faults where they occur on the plate boundaries. These faults must always be parallel to the motion direction between plates, since otherwise they would not be lines of pure slip but would also involve plate creation or destruction as well as slip. These structures are obvious and easy to map wherever they offset the axes of ridges, since they produce remarkable straight slots that may extend 2 km below the rest of the sea floor. How this bathymetry is produced is not clear, but it does permit the motion direction to be determined to an accuracy that is often better than 5°.

It is unfortunate that transform faults that offset trenches are much less easy to recognize. The principal difficulty is that trenches, unlike ridges often have a large component of strike slip across them. This difficulty is well illustrated by the Aleutian trench (McKenzie and Parker, 1967), which is pure overthrust with no strike slip at its eastern end and develops an increasing component of strike slip faulting going westward until the boundary becomes a transform fault at its western end where it joins Kamchatka. Such faults can only be recognized by studies of fault plane solutions of earthquakes that occur on them and, therefore, do not themselves determine the directions of motion.

The most powerful general method of studying the relative motion of two plates is by using fault plane solutions of earthquakes. The technique now used for obtaining fault plane solutions was developed by Stauder (Stauder, 1962; Stauder and Bollinger, 1966a, b) and by Sykes (1967). An excellent description of the technique has been given by Honda (1962), who also discusses the theory on which the method depends.

The method depends on the P and S wave radiation patterns from earthquakes and on the availability of a world-wide collection of long-period seismograms from any earthquake. Figure 2 shows the P wave radiation pattern from a strike slip earthquake on a vertical fault. In directions such as B and E, the initial motion of the ground is away from the focus of the earthquake and is called compressional motion. In other directions, such as A and F, the initial P wave motion is toward the source and is known as a dilatational motion. Figure 2 shows that there are two planes that divide the radiation field into dilatational and compressional quadrants. The general theory of dislocations requires these two planes to be at right angles to each other, and also requires one of them to coincide with the fault plane (Honda, 1962). The theory also shows that there is no method of deciding which plane is the fault plane and which plane is the other plane at right angles to it, called the auxiliary plane. A choice of fault plane can be made only if additional information is available. If, for instance, the earthquake produced slip on a fault which was observed on the surface, the

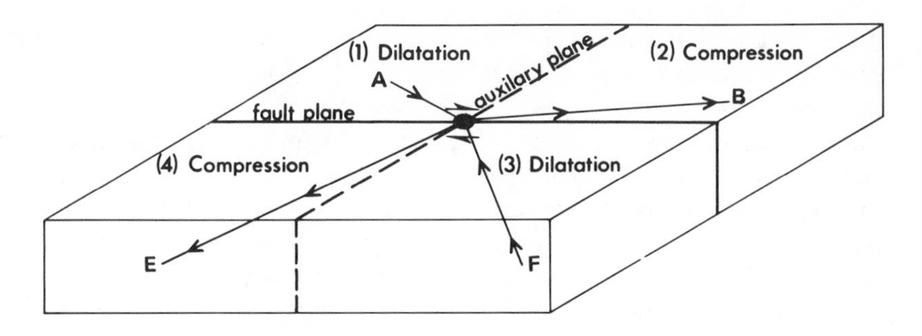

Fig. 2. The radiation field from a strike-slip earthquake. The arrows on the rays mark the initial direction of motion of the earth.

strike of the fault must agree with that of the fault plane and so must the sense of motion. Other methods of deciding which plane is the fault plane are the distribution of aftershocks and the direction of progation of the initial break. Though the dip and strike of the fault plane are of considerable interest, the direction of slip between the sides of the fault is of more general importance, since it can be used to obtain the direction of relation motion between the plates on each side of the fault. Figure 2 shows that the auxiliary plane is vertical and at right angles to the fault plane if the slip is horizontal. It is easy to see that the line of intersection between the fault plane and the auxiliary plane, called the null vector, is at right angles to the direction of slip in the fault plane. Since the two planes are at right angles, this result requires the slip vector in the fault plane to be the normal to the auxiliary plane. If this normal is taken to be the unit vector $\mathbf{u}_1$ and the normal to the fault plane to be $\mathbf{u}_2$, then the null vector is in the direction $\mathbf{u}_1 \times \mathbf{u}_2$. The horizontal projection of the slip vector in the fault plane then is $\mathbf{u}_H = \mathbf{u}_1 - (\mathbf{u}_1 \cdot \mathbf{a}_z)\mathbf{a}_z$, where $\mathbf{a}_z$ is a unit vector in the vertical direction. $\mathbf{u}_H$ is the horizontal projection of the relative motion vector between the two plates, and since $\mathbf{u}_1$ is the normal to the auxiliary plane, the strike of $\mathbf{u}_H$ may be obtained most easily by adding $90°$ to the strike of the auxiliary plane. In Fig. 2, the auxiliary plane is vertical, therefore its normal is horizontal and the slip direction between the two sides has the same strike as the fault plane. This result is true only for horizontal slip on vertical faults. In the past, great importance has been attached to the orientation of the so-called pressure and tension axes, generally labeled P and T of fault plane solutions. These axes lie in the plane containing the normals to the fault and auxiliary planes, and they bisect the angles between them. Thus, the two axes are $(\mathbf{u}_1 + \mathbf{u}_2)/\sqrt{2}$ and $(\mathbf{u}_1 - \mathbf{u}_2)/\sqrt{2}$ when written as unit vectors. The P axis is taken to be in the dilatational quadrant, the T in the compressional. Though under certain conditions (McKenzie, 1969b) the orientation of the P and the T axes is related to the orientation of the principal stress axes, these conditions are rarely satisfied by surface fractures, and therefore the orientation of these axes is not in general important for shallow shocks.

The radiation pattern in Fig. 2 is cumbersome to draw, and for this reason is generally represented in two dimensions by imagining a small sphere to be drawn centered on the focus of the earthquake, and then projecting the points where the rays cut the lower hemisphere of the sphere into a horizontal plane using either an equal area or stereographic projection. Figure 3 shows such a plot for the radiation pattern of the Turkish earthquake of July 22, 1967, with two mutually orthogonal planes drawn to separate the dilatations shown by open circles from the compressions shown by solid circles. This earthquake produced right-handed strike slip motion on an east-west fault, and therefore the east-west plane is the fault plane and the north-south the auxiliary plane. It is also possible

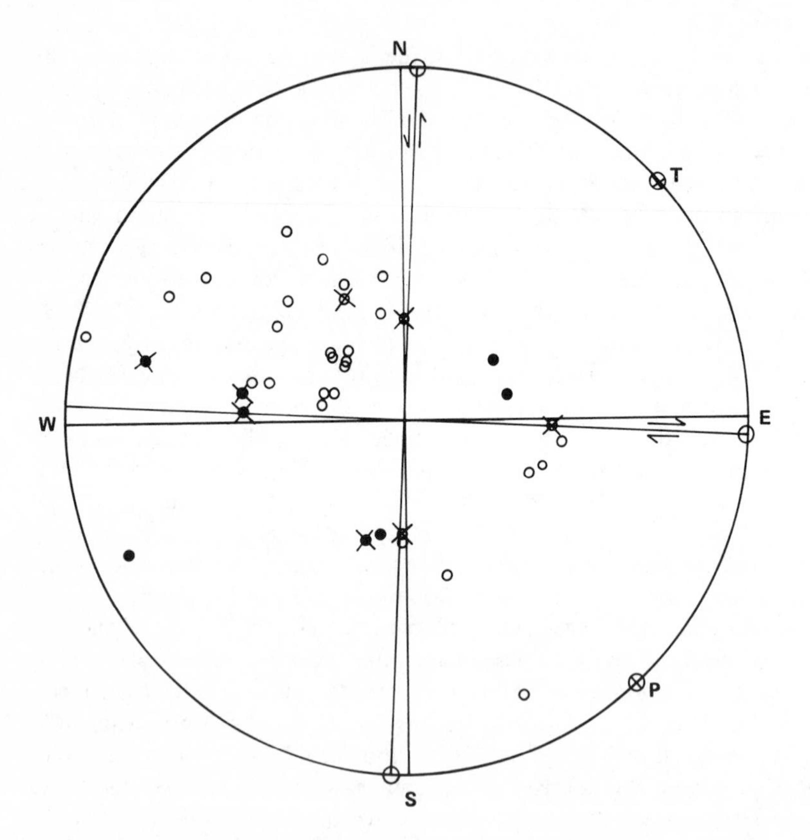

Fig. 3. Strike-slip fault plane solution for the earthquake of July 22, 1967, in Western Turkey. Dilatations are marked with open circles, compressions with solid circles. Points marked with a cross were believed to lie close to the fault plane or the auxiliary plane from the character of the record alone. The larger open circle on the fault plane marks the normal to the auxiliary plane, and similarly, the circle on the auxiliary plane is the normal to the fault plane. This earthquake produced slip on an east-west fault, and therefore the choice between the two planes is not in doubt.

to use the concepts of plate tectonics to choose between the two planes, since the slip vector between two plates cannot vary rapidly over short distances, whereas the fault plane or plate boundary can. Thus, if the normals to both planes from several earthquakes are plotted on a map, one direction will remain approximately constant whereas the other may vary widely and remain at right angles to the plate boundary. This method of using fault plane solutions to determine the relative motion between plates was first used by McKenzie and Parker (1967) on Stauder and Bollinger's (1964, 1966a, b) fault plane solutions in the North Pacific. They also resolved the ambiguity between the planes of the fault plane solutions in this region using the argument outlined above. Their technique has since been widely applied to all three types of plate boundary (Banghar and Sykes, 1969; Isacks et al., 1968; McKenzie, 1970a; Molnar and Sykes, 1969) and is perhaps the most general method for studying plate motions yet devised.

The direction of the motion between two plates which occurs in an earthquake can also sometimes be obtained by direct triangulation in a region that had previously been surveyed. This powerful method can reveal many of the details of the deformation of an area that cannot be obtained by other techniques, but is unfortunately restricted to well-surveyed continental areas. In practice, this at present means Japan and the western United States. The method is therefore more used to study the complicated way in which faults slip (Scholz et al., 1969) than for observing plate motions.

The cheapest and most direct method of measuring the direction of relative motion between two sides of a fault is to observe the relative displacements of the two sides after a major earthquake (Allen and Smith, 1966; Ambraseys, 1963; Ambraseys and Zatopek, 1969). This method has been widely used in poorly surveyed areas and is very valuable in such regions. The surface displacements on some faults, however, are strongly controlled by the mechanical behavior of less competent shallow layers, and it is important to compare the results of direct observations with fault plane solutions or geodetic measurements wherever possible.

Various methods have been used to combine all measurements of the direction and magnitude of the relative motion between two plates to obtain the angular velocity vector (Le Pichon, 1968; Morgan, 1968). The most obvious method is trial and error, using some function of the difference between the observed and calculated rates and directions as a measure of the goodness of fit. Most search techniques of this type suffer from the problem of local minima. The angular velocity obtained by them fits the observations better than other choices little different in direction or magnitude. There is generally no method of deciding whether the best fit is the absolute best fit or merely a local best fit. This problem has been discussed by McKenzie and Sclater (1971) who showed that the fitting function F', closely related to F defined by

$$F = \sum_i (\boldsymbol{\omega} \cdot \mathbf{u}_i)^2$$

where $\mathbf{u}_i$ are the vector directions of motion along the plate boundary, may be minimized with respect to $\boldsymbol{\omega}$ subject to the condition

$$\boldsymbol{\omega} \cdot \boldsymbol{\omega} = 1$$

Once the direction of $\boldsymbol{\omega}$ has been obtained in this way, its magnitude is easily obtained from measurements of the magnitude of the velocity vector.

All the methods described so far can measure the relative motion between two plates, but not the relative motion between one plate and the earth's North rotational pole. Since longitude can only be defined by reference to some arbitrarily chosen point on one plate, the component of the angular velocity vector describing the motion of one plate relative to the rotation pole that points along the rotational axis of the earth can be arbitrarily chosen and can be set equal to zero. Thus, the motion of one plate A relative to the rotational pole P taken to be fixed can always be described by an angular velocity vector ${}_P\Omega_A$ passing through the equator. If the angular velocity between A and another plate B is known, then the angular velocity of B relative to the pole ${}_P\Omega_B$ is easily obtained:

$${}_P\boldsymbol{\Omega}_B = {}_P\boldsymbol{\Omega}_A + {}_A\boldsymbol{\omega}_B - ({}_A\boldsymbol{\omega}_B \cdot \mathbf{a}_z)\,\mathbf{a}_z \tag{4}$$

where $\mathbf{a}_z$ is a unit vector along the rotational axis. The last term on the right is included to cause ${}_P\Omega_B$ to pass through the equator. Since the definition of longitude is arbitrary, any vector parallel to $\mathbf{a}_z$ can be added to ${}_P\Omega_B$ without affecting the rate or direction of motion between plate B and the pole.

No measurements of the instantaneous rate and direction of movement of the pole relative to any plate have yet been published. Such a measurement could in principle be obtained from the observations of the International Latitude Service. Unfortunately, at least three of the contributing observatories are sufficiently close to plate margins to cause difficulties in interpretation. Perhaps the observations of Bureau de l'Heure could be used for this purpose, or perhaps the orbit of a drag free satellite will be sufficiently stable for a long period of time to determine small motions of the tracking stations relative to the rotational axis. Since the relative motions between plates are already known with some accuracy (Le Pichon, 1968), the astronomical observations can be reduced using Le Pichon's poles to determine the relative motion between one plate and the pole. Thus, many of the problems that faced astronomical tests of continental drift (Munk and MacDonald, 1960) will not confuse this measurement.

It is clear from Eq. (4) that polar wandering is an unnecessary concept in plate tectonics. It may, however, still be a useful one for the same reason as the concept of the angular velocity of the earth is useful: if the motion of the pole relative to any plate is very much faster than the motion between plates, it is clearly convenient to discuss relative motions between plates in a different way from their relative motion with respect to the earth's rotational pole. If the angular velocity of the pole P relative to a plate A is ${}_A\Omega_P$, then the velocity and direction of polar motion ${}_A\mathbf{v}_P$ is:

$$ {}_A\mathbf{v}_P = a({}_A\Omega_P \times \mathbf{a}_z) \tag{5}$$

where $\mathbf{a}_z$ is a unit vector along the rotational axis and a is the radius of the earth. It is convenient to define a new vector ${}_A\mathbf{V}_P$ by:

$$ {}_A\mathbf{V}_P = A_A\, {}_A\mathbf{v}_P \tag{6}$$

where A_A is the total area of the plate A. This definition means that small but rapidly moving plates do not confuse the definition as they would without the factor A_A. Two situations are shown in Fig. 4, where the arrows are the vectors ${}_1\mathbf{V}_P \ldots {}_N\mathbf{V}_P$ drawn from the north pole to represent the polar wandering directions of all N plates that cover the earth. In Fig. 4*a*, polar wandering is not a useful concept, whereas in 4*b*, it is. In mathematical terms, polar wandering is useful if and only if:

$$ \sum_{n=1}^{N} |{}_n\mathbf{V}_P| \gg \sum_{n=1}^{N} |({}_n\mathbf{V}_P - \mathbf{V}_m)| \tag{7}$$

where

$$ \mathbf{V}_m = \frac{1}{N} \sum_{n=1}^{N} {}_n\mathbf{V}_P \tag{8}$$

Thus, the usefulness or otherwise of the concept of polar wandering can only be decided when the relative motion between all the major plates and the pole has been determined. At present, this cannot be done for the instantaneous poles, though it can probably now be decided for the finite displacements since the Cretaceous.

There are far fewer methods of measuring the finite displacements between plates. The magnetic lineations and their offsets on transform faults are very useful for this purpose, but they are only formed along extensional plate boundaries. The relative position of two plates on either side of such a boundary can be obtained at some period simply by fitting together the magnetic lineations on either side of the present ridge axis, in the same way that Bullard et al. (1965) or McKenzie et al. (1970) did using the continental edges. Few lineations, however, have yet been surveyed in sufficient detail, and therefore little use has yet been made of this technique.

The other method of reconstructing the relative positions of plates at earlier

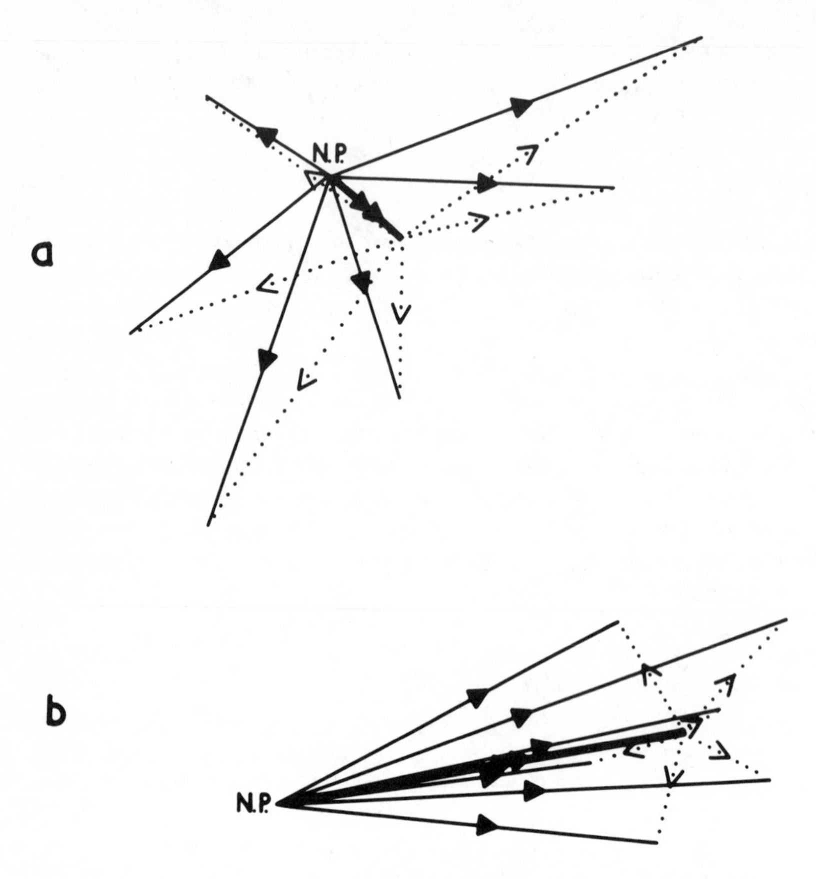

Fig. 4. The thin solid lines with arrows represent the vector velocities $_n\mathbf{v}_P$ of the North Pole, marked N.P., with respect to each of six plates, in each case multiplied by the area of the plate. These six plates are imagined to cover the whole earth's surface. The thicker line with two arrows shows $\mathbf{V}_m = 1/6 \sum_{n=1}^{6} {_n\mathbf{V}_P}$ and the dotted lines show $_n\mathbf{V}_P - \mathbf{V}_m$ for each of the six plates. If polar wandering is to be a useful concept then

$$\sum_n |{_n\mathbf{V}_P}| \gg \sum_n |({_n\mathbf{V}_P} - \mathbf{V}_m)|$$

or the rate of relative motion of all plates relative to the pole must be very much reduced by a particular choice of polar wandering direction and velocity. This condition is satisfied by (*b*) but not by (*a*).

times is by the use of palaeomagnetic observations. These give the latitude and orientation of a plate with respect to the earth's rotational pole, but cannot give the relative longitude between two points on different plates in relative motion. This ambiguity can often be removed by careful use of sea floor features

(McKenzie and Sclater, 1971), and then the combination of the two methods provides a powerful technique for obtaining the positions of plates at times since the Cretaceous.

Most palaeomagnetic measurements are made on red beds or igneous rocks and involve careful thermal and alternating current washing to remove any unstable component of the magnetization (Irving, 1964). Many measurements are generally made on the same formation and the results averaged to remove the effect of secular variations. This technique is accurate and reliable if sufficient care is taken, but it is restricted to rocks exposed on land. Since half the earth is covered by ocean, this restriction is very important. Two methods are available for determining the direction of magnetization of parts of oceanic plates, and hence their latitude and orientation at a certain time.

The first depends on the magnetic anomaly produced by seamounts (Vacquier and Uyeda, 1967; Vine and Matthews, 1963). A seamount is chosen on a piece of sea floor that does not possess magnetic lineations, and a careful magnetic and bathymetric survey made. The seamount is then assumed to be uniformly magnetized in a direction that is determined by least squares fitting of a calculated magnetic anomaly to that observed. The fit is often surprisingly good. The seamount then has to be dated. Though the age of the sea floor on which it sits is often known, this age puts only an upper bound on the age of the seamount. The seamount must therefore be dredged in the hope of obtaining datable fossils. Even if the dredging is successful, the age of the seamount is still not well determined unless the fossil ages are close to that of the sea floor. Thus, the assumptions and uncertainties of this method are somewhat large, though it has been used by Vacquier and Uyeda (1967) with considerable success in the north Pacific.

The other method that can be used in oceanic areas depends on the shape of the magnetic lineations. The magnetization of any piece of oceanic crust occurred on the ridge axis where it was formed and remains unchanged during later plate movements. The shape of the magnetic lineations depends on the direction of magnetization of the rock that produces them and, hence, on the latitude and orientation of the ridge that formed them, as well as on their present position and orientation. If the magnetic anomalies were originally formed by a rapidly spreading ridge, they are often entirely undisturbed by topographic effects. Such a profile is shown in Fig. 5, taken from McKenzie and Sclater (1971). The profile above the observed profile was calculated assuming that the anomalies were formed at their present latitude and orientation. The agreement between it and the observed profile is not good, but becomes so if the anomalies are produced by a ridge at 40°S by NE-SW spreading. This method does not give an accurate pole position, but it does give an accurate date.

These palaeomagnetic methods have been applied to most major plates, and it appears that all except Antarctica have moved a considerable distance northward since the Cretaceous. The least certain of these measurements are those for the Pacific (Vine, 1968), but if future work supports the preliminary studies that have already been carried out, polar wandering is unlikely to be a useful concept to describe these motions.

Finally, there are two methods that can be used to reconstruct the original arrangement of fragments of one plate which are now widely dispersed. Both methods depend on the absence of relative motion between the fragments in their original positions. The first method is well known and depends on fitting continental edges (Bullard et al., 1965; Smith and Hallam, 1970). A different and perhaps more physical criterion for the goodness of fit has been used by McKenzie et al. (1970), though it gives essentially the same results.

The other method depends on palaeomagnetic observations and was first pointed out by Graham et al. (1964). They noticed that the polar wander curves of two fragments of an old plate should be superimposable during the time when the fragments were joined. McElhinney (1967) and Creer (1968) have both used

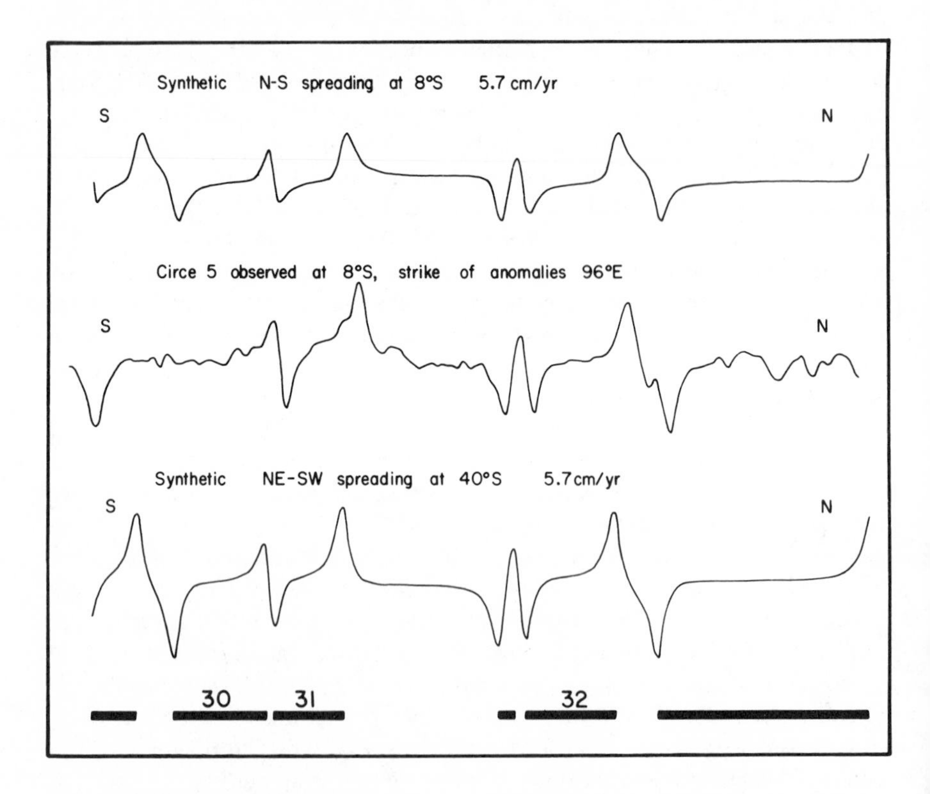

Fig. 5. Magnetic anomalies observed south of Ceylon by leg 5 of the Circe expedition of Scripps Institute of Oceanography. The uppermost profile shows the shape the anomalies would have if they had been generated at the latitude and orientation that they now possess, whereas the lowest profile shows the corresponding anomalies generated by a ridge at 40°S with NE-SW spreading and since drifted to their present positions. The lowest profile agrees much better with that observed then does the uppermost and shows that this piece of sea floor has moved northward with India since the Cretaceous (McKenzie and Sclater, 1971).

this method to reconstruct the original positions of South America and Africa and have obtained results that agree very well with Bullard and his coworkers' reconstruction. It is, however, extremely important not to attempt to superimpose the polar wander paths of two plates in relative motion, since the resulting reconstructions will then be nonsense (McElhinney et al., 1968).

The discussion above shows that instantaneous relative plate motions are well determined and the theory is well understood, since it involves vectors alone. The same is not true of finite rotations of plates, which appear to be rather complicated, and whose motion must be described by matrices. Considerable simplifications should be possible, perhaps by the use of group theory (Wigner, 1959). The discussion below is principally concerned with other geophysical phenomena that appear to be related to the motion of the plates.

THE THERMAL STRUCTURE OF PLATES

The various methods of studying plate motions already outlined show that oceanic plates are formed by the upwelling of hot mantle material along the axes of oceanic ridges and plunge back down into the mantle beneath island arcs. The processes by which the plates are created and destroyed cannot be studied by the methods described in the last section, nor can their thickness be estimated.

Since plates are formed from the mantle by upwelling and then cooling of the new material, a region of high heat flow through the ocean floor exists on either side of the axes of oceanic ridges. The width of this region is controlled by the time taken for a slab of hot material to cool. This time τ is governed by the thickness l of the plate that is formed:

$$\tau = \frac{\rho C_p l^2}{\pi^2 \kappa} \tag{9}$$

(Carslaw and Jaeger, 1959). The ρ is the density of the plate material, C_p its specific heat at constant pressure, and κ its thermal conductivity. Therefore, if the plate is being formed at velocity v on each side of the ridge axis, the high heat flow region will extend approximately $v\tau$ on either side of the ridge. The heat flow observations on various ridges spreading at different rates suggest that $l \sim 50$ km (McKenzie, 1967). A recent careful analysis of heat flow and topography in the Pacific by Sclater and Francheteau (1970) shows that a value of 80 km is in better agreement with the observations.

The equation that governs the temperature within a plate moving with constant velocity is:

$$\rho C_p v \frac{\partial T}{\partial x} = \kappa \left(\frac{\partial^2 T}{\partial x^2} + \frac{\partial^2 T}{\partial z^2} \right) + H \tag{10}$$

The x is measured from the ridge axis and z from the base of the plate. The rate of radioactive heat generation in a unit volume is H, and it can be neglected for oceanic plates. Analytic solutions of Eq. (10) are in general possible only if v does not depend on x or z. This is clearly the case within the ridge plate, though it is not true where the plate is being formed beneath the ridge. Fortunately, the temperature structure of most plates is governed by the convection of heat horizontally by the moving plate and not by the details of the processes beneath ridge axes. Thus, for most plates the horizontal conduction of heat can be neglected, and the temperature within them is governed by the balance between heat convected horizontally from the ridge axis and heat lost by vertical conduction to the sea floor. Thus Eq. (10) reduces to:

$$\rho C_p v \frac{\partial T}{\partial x} = \kappa \frac{\partial^2 T}{\partial z^2} \tag{11}$$

which is a good approximation for most plates. If the base of the plate is at a temperature T_0, and the plate has thickness l, Eq. (11) can be reduced to a dimensionless equation by substituting

$$\begin{aligned} T &= T_0 T' \\ z &= l z' \\ x &= l R x' \end{aligned} \tag{12}$$

where R is the Thermal Reynolds number:

$$R = \frac{\rho C_p v l}{\kappa} \tag{13}$$

to give:

$$\frac{\partial T'}{\partial x'} = \frac{\partial^2 T'}{\partial z'^2} \tag{14}$$

Equation (14) has the same form as the one-dimensional time dependent heat conduction equation, whose solutions have been discussed in some detail by Carslaw and Jaeger (1959). The solution that satisfies the boundary conditions

$$\begin{aligned} T &= 0 \quad \text{on} \quad z = l\,, \; x \neq 0 \\ T &= T_0 \quad \text{on} \quad z = 0 \quad \text{and on} \quad x = 0 \end{aligned}$$

is easily shown to be

$$T = T_0 \left[1 - \frac{z}{l} + 2 \sum_{n=1}^{\infty} \frac{(-1)^{n+1}}{n\pi} \exp\left(-\frac{n^2 \pi^2 x}{Rl}\right) \sin \frac{n\pi z}{l} \right] \tag{15}$$

The heat flow through $z = l$ is then obtained by differentiation:

$$-\kappa \left(\frac{\partial T}{\partial z}\right)_{z=l} = \frac{\kappa T_0}{l} \left[1 + 2 \sum_{n=1}^{\infty} \exp\left(-\frac{n^2 \pi^2 x}{Rl}\right) \right] \tag{16}$$

Both (15) and (16) have the interesting property that the temperature and heat flow depend only on x/R, or x/v, and not on x directly. Since x/v is the age of this part of the plate, these equations show that the temperature structure depends only on the age of that part of the plate.

It is also of interest to calculate the total heat lost by plate formation. In the absence of plate creation, only the first term on the right of (16) would remain. Thus, the extra heat E lost by the earth because of plate creation is:

$$E = 2 \int_0^{\infty} \left[-\kappa \left(\frac{\partial T}{\partial z}\right)_{z=l} - \frac{\kappa T_0}{l} \right] dx$$

$$= \frac{2 T_0 \rho C_p v l}{\pi^2} \sum_{n=1}^{\infty} \frac{1}{n^2}$$

per unit length of ridge axis per second. But

$$\sum_{n=1}^{\infty} \frac{1}{n^2} = \frac{\pi^2}{6}$$

and therefore:

$$E = \frac{\rho C_p v l T_0}{3} \tag{17}$$

This result does not agree with that obtained by a different method by McKenzie and Sclater (1969), who obtained $\rho C_p v l T_0$. The difference is caused by a slight change in the method of calculating the excess heat flow. If

$$T_0 = 1200°\text{C}, \quad l = 80 \text{ km}$$

$$\rho = 3 \text{ gm/cm}^3, \quad K = 5 \times 10^{-3} \text{ cal/cm deg C sec}$$

$$C_p = .25 \text{ cal/gm deg C}$$

then the total heat lost by the earth through plate creation corresponds to about 30% of that lost by steady conduction through the ocean floor outside the high heat flow region.

Though (11) is an adequate approximation for most ridges, it is of some interest to examine the conditions under which it forms a poor approximation. If H is neglected in (10) and the same substitutions (12) are made, (10) becomes:

$$\frac{\partial T'}{\partial x'} = \frac{\partial^2 T'}{\partial z'^2} + \frac{1}{R^2}\frac{\partial^2 T'}{\partial x'^2} \tag{18}$$

substitution of $e^{\alpha x'} \sin n\pi z'$ gives

$$\frac{\alpha^2}{R^2} - \alpha - n^2\pi^2 = 0$$

If

$$R >> n\pi$$

then

$$\alpha \simeq -n^2\pi^2\left(1 - \frac{n^2\pi^2}{R^2}\right) \tag{19}$$

The temperature structure is governed by the term with $n = 1$, thus the condition that must be satisfied before the difference between the solutions to (18) and (14) can be neglected is

$$1 >> \frac{\pi^2}{R^2} \tag{20}$$

or using (13)

$$v \gtrsim 0.1 \text{ cm/yr}$$

Thus if the separation rate between the plates is less than about 2 mm/yr the temperature structure is controlled by horizontal conduction of heat and no longer only depends on the age of the piece of the plate concerned. It is perhaps of interest to obtain a similar result in a different way, which shows the physical processes involved. Where the plate is formed by upwelling of mantle material, the temperature of the rising matter is controlled by the temperature of the sea floor only when it is close to it. Some estimate of the depth to which the temperature of the upwelling material is controlled by the cold sea floor may easily be obtained. The material moves upward everywhere at a constant velocity V and is cooled from T_0 to 0 by the time it reaches the sea floor at $z = 0$, where it is removed. The equation that governs the temperature within the rising material is

$$\frac{\rho C_p V}{\kappa} \frac{dT}{dz} = \frac{d^2 T}{dz^2} \tag{21}$$

whose solution is

$$T = T_0 (1 - e^{z/\lambda}) \tag{22}$$

where

$$\lambda = \frac{\kappa}{\rho C_p V}$$

Equation (22) shows that the temperature to a depth of $\sim \lambda$ is controlled by upper boundary condition. If $V \sim$ 4 cm/yr, $\lambda \sim$ 500 meters. This result shows that the cold crust on the axis of a rapidly spreading ridge like the East Pacific Rise is very thin, and explains why the earthquakes on the axes of rapidly spreading ridges are generally too small to be detected by the Standard World Wide Network. Equation (22) also shows that the temperature structure at a depth greater than λ will be the same beneath all ridges and will be governed by the temperature of the mantle beneath the plates. As this mantle material rises, it expands adiabatically and probably partially melts at a depth of perhaps 20 km. The extent of partial melting will be governed by the temperature and composition of the upper mantle and not by more complicated crustal effects. Therefore, provided $\lambda \lesssim 20$ km, the composition of the basalts erupted along the ridge axis should be constant everywhere. Gast (1968) has used a similar but slightly more complicated model to explain the constant composition of the tholeitic basalts erupted everywhere on the ridge axes (Engel and Engel, 1964) and also the composition of the alkali basalts of which most seamounts consist.

Equation (22) can also be used for another purpose. If the plates are produced by upwelling within a horizontal distance l of the ridge axis, equal to their thickness, then the initial temperature structure of the plates will be constant if $\lambda \ll l$. When $\lambda \sim l$ it will be partly controlled by heat loss to the cold sea floor even on the axis of the spreading ridge. This is the same condition as that given by (20) and shows that the temperature structure beneath ridges spreading at less than 1 mm/yr should be different from all ridges spreading faster than this, and therefore the composition of the magma erupted from them should not be the standard tholeite of most oceanic ridges. Only two ridges with such slow spreading rates have yet been discovered: the East African Rift Valley and the western end of the Azores-Gibraltar ridge. Since the rocks erupted along the Rift Valley may have assimilated some continental material, the Azores-Gibraltar Ridge seems a more useful place to test these ideas by dredging. It is perhaps significant that carbonatite lavas have been erupted near both these ridges.

The mean oceanic heat flow through the plates away from the ridges is 1.1 to 1.2×10^{-6} cal/cm^{2}sec and is remarkably constant everywhere. This heat is not generated by radioactive decay within the plate but is conducted to the surface from the bottom of the lithosphere. Since the heat flow is constant, the transfer of heat to the base of the plates must be much more rapid than the conduction of heat through them. Such transfer is most easily achieved by convection, but if this is to be effective, the plates cannot move with the mantle below them. The mean heat flow through the continental plates is less well determined but is probably between 1.5 and 1.6×10^{-6} cal/cm^{2}sec. The difference between this value and the deep oceanic mean agrees well with the calculated rate of radioactive heat generation of continental rocks. There is no longer any reason to believe that the mean continental and oceanic heat flows are the same. This result was originally obtained because the excess heat flow resulting from plate creation was included in the average heat flow. This should not be done because there is no corresponding process on the continents.

Calculations similar to those already described have been carried out by McKenzie (1969a) to obtain the temperature structure in the sinking slab beneath island arcs. When the plate is first thrust into the mantle beneath an island arc, the temperature structure is the mean oceanic temperature structure provided that the ridge that produced the plate is not close to the island arc. If this is not the case, the calculations of the temperature can still be carried out and show that it is similar to that of a plate whose thickness is $\sim \pi (L\kappa/\rho C_p v)^{1/2}$, where L is the separation between the ridge and the trench and v is the rate at which the plate is being created on one side of the ridge. Provided the ridge and the trench are separated by a distance greater than $\sim \rho C_p l^2 v/\pi^2 \kappa$, where l is the plate thickness, the temperature is easily calculated from (14) and the boundary conditions that the temperature on the upper and lower surfaces is T_0 everywhere, and that the temperature on $x = 0$ is given by:

$$T = T_0 \left(1 - \frac{z}{l}\right)$$

The temperature is then given by:

$$T = T_0 \left[1 + 2 \sum_{n=1}^{\infty} \frac{(-1)^n}{n\pi} \exp\left(- \frac{n^2 \pi^2 x}{Rl} \right) \sin \frac{n\pi z}{l} \right] \quad (23)$$

where

$$R = \frac{\rho C_p v l}{\kappa} \quad (24)$$

v is the velocity of relative motion between the plates and perpendicular to the island arc, and x is the distance from the surface measured in the plane of the earthquakes. A section of the isotherms within the sinking slab is shown in Fig. 6, taken from McKenzie (1969a). He used

$$v = 10 \text{ cm/yr}, \quad \rho = 3 \text{ gm/cm}^3$$

$$\kappa = 10^{-2} \text{ cal/cm deg C sec}, \quad l = 50 \text{ km} \quad (25)$$

$$C_p = 0.25 \text{ cal/gm deg C}$$

The values of κ and l are least well determined, and Sclater and Francheteau (1970) have suggested that values of:

$$\kappa = 5 \times 10^{-3} \text{ cal/cm deg C sec}, \quad l = 80 \text{ km} \quad (26)$$

fit the heat flow and topography of ridges better than those on (25). If these values are used to calculate the isotherms, the cold regions extend to even greater depths than those in Fig. 6. McKenzie (1969a) also showed that there is an intimate relationship between the temperature at the center of the sinking slab and the occurrence of intermediate and deep focus earthquakes. Figure 7 shows the projection of the earthquakes beneath the Tonga-Kermadec-New Zealand region onto a vertical plane, with the lowest tips of the isotherms calculated from plate tectonics (Fig. 6) for comparison. The relationship between the two is striking and does not depend on the correctness or otherwise of the parameters in (25).

The most important assumption in this calculation of the isotherms is that the mantle outside the slab can be taken to be isothermal. If the mantle is moving as fast as the sea floor, the variation of temperature with depth must be almost adiabatic, or about 1°C/km. Thus, the temperature outside the slab probably increases by about 600°C from 80 km to 680 km where earthquakes cease. This increase cannot be included in the calculations simply by applying a steadily

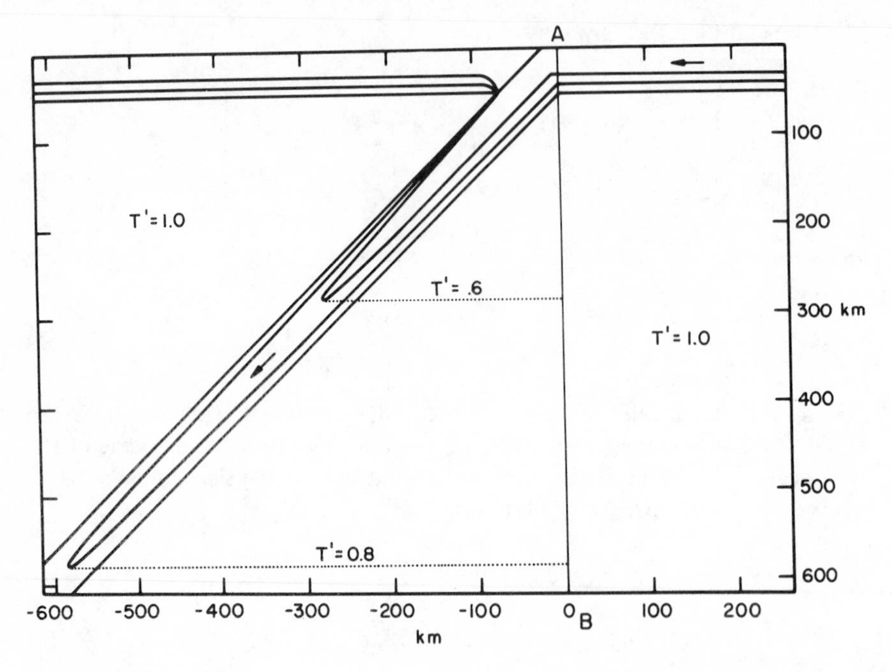

Fig. 6. Dimensionless temperature T' beneath island arcs when the consumption velocity is 10 cm/yr (McKenzie, 1969a).

increasing temperature as a boundary condition on $z = 0$ and $z = l$ because the material of which the slab is made is also adiabatically compressed and therefore heated as it sinks. The problem can, however, be solved by the use of potential temperature (McKenzie, 1970b), and the resulting temperature is given by:

$$T = \exp\left[\frac{(x \sin\phi - z\cos\phi)}{h}\right] \times \left[\theta_1 + 2(\theta_1 - 273) \sum_{n=1}^{\infty} \frac{(-1)^n}{n\pi} \exp\left(-\frac{n^2\pi^2 x}{Rl}\right) \sin\frac{n\pi z}{l}\right] \tag{27}$$

where h is the scale height of the adiabatic temperature:

$$h = \frac{C_p}{\alpha g}$$

where ϕ is the dip of the sinking slab, α the thermal expansion coefficient, and θ_1 the absolute temperature of the mantle below the plates. The resulting

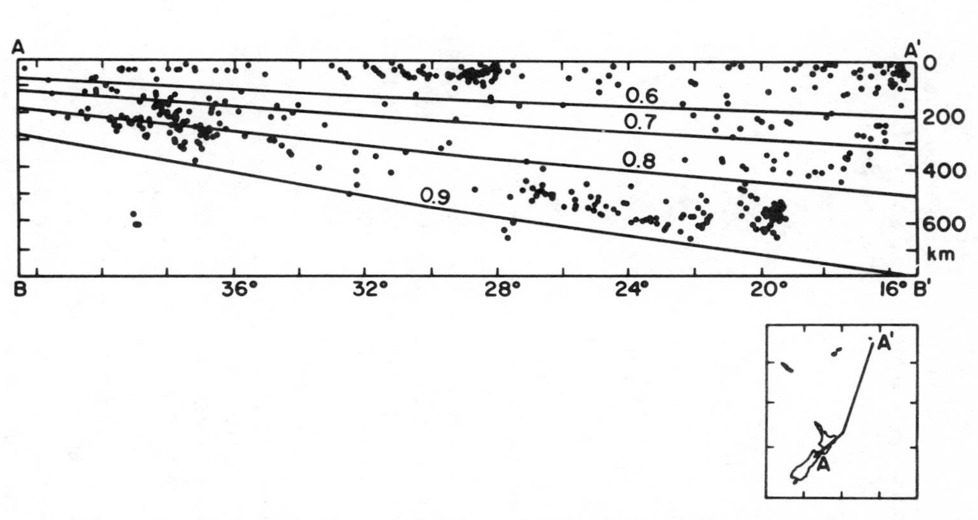

Fig. 7. Projections of foci of earthquakes and isotherms onto the vertical plane AB in Fig. 6 in the Tonga Fiji Kermadec New Zealand region (see inset). Except for six anomalous deep earthquakes, intermediate and deep focus activity ceases at about $T' = 0.85$ (McKenzie, 1969a).

isotherms are similar to those from (23) though they do not extend to such great depths. There is, however, one remarkable difference between the behavior of an incompressible and a compressible slab. Equation (27) shows that the temperature difference between the center and the outside of the slab can increase as it sinks if:

$$\frac{\sin\phi}{h} \gtrsim \frac{\pi^2}{Rl}$$

Using the value for κ and l given by (26) and $\alpha = 4 \times 10^{-5}\ {}^\circ\mathrm{C}^{-1}$ gives

$$v \gtrsim \frac{30}{\sin\phi}\ \mathrm{cm/yr}$$

Since phase changes may increase the local value of α by an order of magnitude, this effect could well be important in certain parts of the slab. Since the density contrast between the slab and the mantle outside is proportional to the temperature contrast, this effect will give rise to a convective instability. Though there appear to be no obvious physical reasons why such convection is not important in the mantle, it nevertheless does not appear to control the plate motions.

This discussion of the shape of the isotherms in terms of an extremely simple model shows that the temperature structure beneath ridges and trenches is controlled by plate motions. One important, though more enigmatic, heat flow anomaly also appears to be related to the plate motions, but in a considerably more complicated way. Vacquier et al. (1966) noticed that the mean heat flow behind the island arcs of the northwest Pacific was about twice that through the deep oceans. McKenzie and Sclater (1968) pointed out that no simple explanation of this anomaly existed, and indeed they could not devise a mechanism that was able to sustain the observed heat flow for a geologically long period. McKenzie (1969a) did finally suggest a mechanism, but there is at present no other evidence to support his ideas.

THE MECHANICAL BEHAVIOR OF PLATES AND SINKING SLABS

This subject differs from the others discussed in this study because most of the work involved was stimulated by gravity observations and the concept of isostasy in the 19th century. Classical work was done by Thomson and Tait (1883), Darwin (1908), Love (1926), and others in a notation that is now difficult to follow because it uses neither vectors nor matrices. Also, many functions that are now known by standard names and letters were at that time generally expressed as infinite series. An excellent description of some of this work is to be found in Jeffreys (1929). Recent work (McKenzie 1967, Walcott,

1970) has principally been concerned with applying certain parts of the 19th century theory to specific geophysical problems. One important obvious result that appears to be little known is that all gravity anomalies must be supported by shearing forces. Therefore the absence of a local or regional gravity anomaly over oceanic ridges (Talwani et al., 1965) implies that the material beneath them is not supporting shearing stresses. McKenzie and Sclater (1969) used this result to calculate the shape of ridges as a function of the spreading rate. A similar result may be obtained from (15) by balancing columns of lithosphere:

$$e(x) = \frac{\rho_0 \alpha T_0 l}{2(\rho_0 - \rho_\omega)} \sum_{k=0}^{\infty} \frac{8}{\pi^2 (2k+1)^2} \exp\left[-\frac{(2k+1)^2 \pi^2 x}{Rl}\right] \tag{28}$$

where $e(x)$ is the elevation of the ridge above the deep sea floor, ρ_0 is the density of the lithosphere at 0°C, and ρ_ω that of sea water. Sclater and Francheteau (1970) have used a similar expression to show that the calculated shape agrees excellently with the bathymetry if values for l and κ given by (26) are used, rather than those of McKenzie (1967). Equation (28) shows that:

$$e(0) = \frac{\rho_0 \alpha T_0 l}{2(\rho_0 - \rho_\omega)} \tag{29}$$

Therefore the height of the ridge axis above the deep sea floor does not depend on the spreading rate. This result has long been known from bathymetric surveys, and it now appears to have a simple explanation in terms of plate tectonics.

Most topographic features on the earth differ from ridges in having a topographic gravity anomaly, even if they are compensated on a regional scale. The magnitude and wavelength of such anomalies gives a direct estimate of the magnitude of the shearing stresses involved (McKenzie, 1967). Since it is now known that the shear stresses in plates can be at least as large as 1 kb (Wyss, 1970), even the large gravity anomalies over trenches can easily be supported by the strength of the lithosphere. This argument therefore suggests that an elastic sheet overlying a viscous fluid will provide a good model for the long-term mechanical behavior of plates. Walcott (1970) has used this method to determine the thickness and rigidity of the plates from the deformation produced by surface loads. He has obtained thicknesses of 75 and 110 km for oceanic and continental plates, respectively, using thin plate theory. Though it appears doubtful that the deformations involved are on a sufficient scale to justify the use of thin plate theory, Walcott's results would probably be little affected by the use of the full theory of thick plates. Thus, the results from gravimetric and topographic measurements obtained so far are in excellent agreement with the model developed to account for the heat flow observations.

The stress field within sinking plates is of considerably greater interest. Several arguments now suggest that most intermediate and deep focus earthquakes are produced by the formation of fractures with the sinking slab at stresses close to the yield stress. It is perhaps even more remarkable that the internal friction within the material must be rather small. Though various suggestions have been made to explain these observations (Griggs and Baker, 1969; Raleigh and Paterson, 1965; Savage, 1969), the mechanism that permits the release of elastic energy is still not properly understood.

Recent discussions of the fault plane solutions of intermediate and deep focus earthquakes (Isacks and Molnar, 1969; Wyss, 1970) show that they must occur within the sinking slab and not by friction between the slab and the mantle outside. Figure 8, taken from Isacks and Molnar (1969), shows the orientation of the *P*, *T*, and null vectors and also that of the normals to the fault and auxiliary planes in relation to the sinking slab. The solutions are taken from earthquakes in many island arcs and are plotted in equal area projection with the hemisphere of the projection orientated to make the vector pointing down the dip of the seismic zone appear in the center of the projection. Thus, vectors pointing down the dip will appear at the center of the circles. The projection is also orientated with respect to the slab in such a way as to make the strike of the slab vertical in Fig. 8. The advantage of this particular choice of projection is that it shows clearly any preferred orientation of any of the vectors of the fault plane solutions with respect to the sinking slab. If, for instance, these earthquakes were produced by slip between the slab and the mantle, one plane of the fault plane solution should be parallel to the slab. Its normal would then be normal to the slab and, hence, would always plot on the outside circle of the plot at a point 90° from the line showing the strike of the slab. Figure 8*d* shows the normals to the fault planes and auxiliary planes, and it is clear that the poles from both groups of earthquakes are not normal to the dipping slab. Unfortunately, it is not yet possible to determine which plane is the fault plane and which the auxiliary for any intermediate or deep focus shock, and therefore nothing is yet known about any preferred orientation of slip planes within the slab. It will, however, be surprising if large structural weaknesses such as oceanic transform faults do not control failure within the slab in their vicinity. It is, therefore, clear from Fig. 8*d* that the intermediate and deep focus earthquakes occur within the sinking slab. This result is not surprising, since Fig. 7 shows that their occurrence is controlled by the temperature at the center of the slab. A more surprising result is the striking relation between the orientation of the *P* axes of deep shocks and *T* axes of intermediate ones, and the dip of the slab. Since McKenzie (1969b) has shown that failure on pre-existing faults does not give *T* and *P* axes whose orientation is closely related to the stress field, this result of Isacks and Molnar's suggests that both groups of shocks are produced by failure of an originally homogeneous material. Even under these conditions, there will be a considerable scatter of *P* and *T* axes about the directions of greatest and least principal stress unless the coefficient of internal friction is

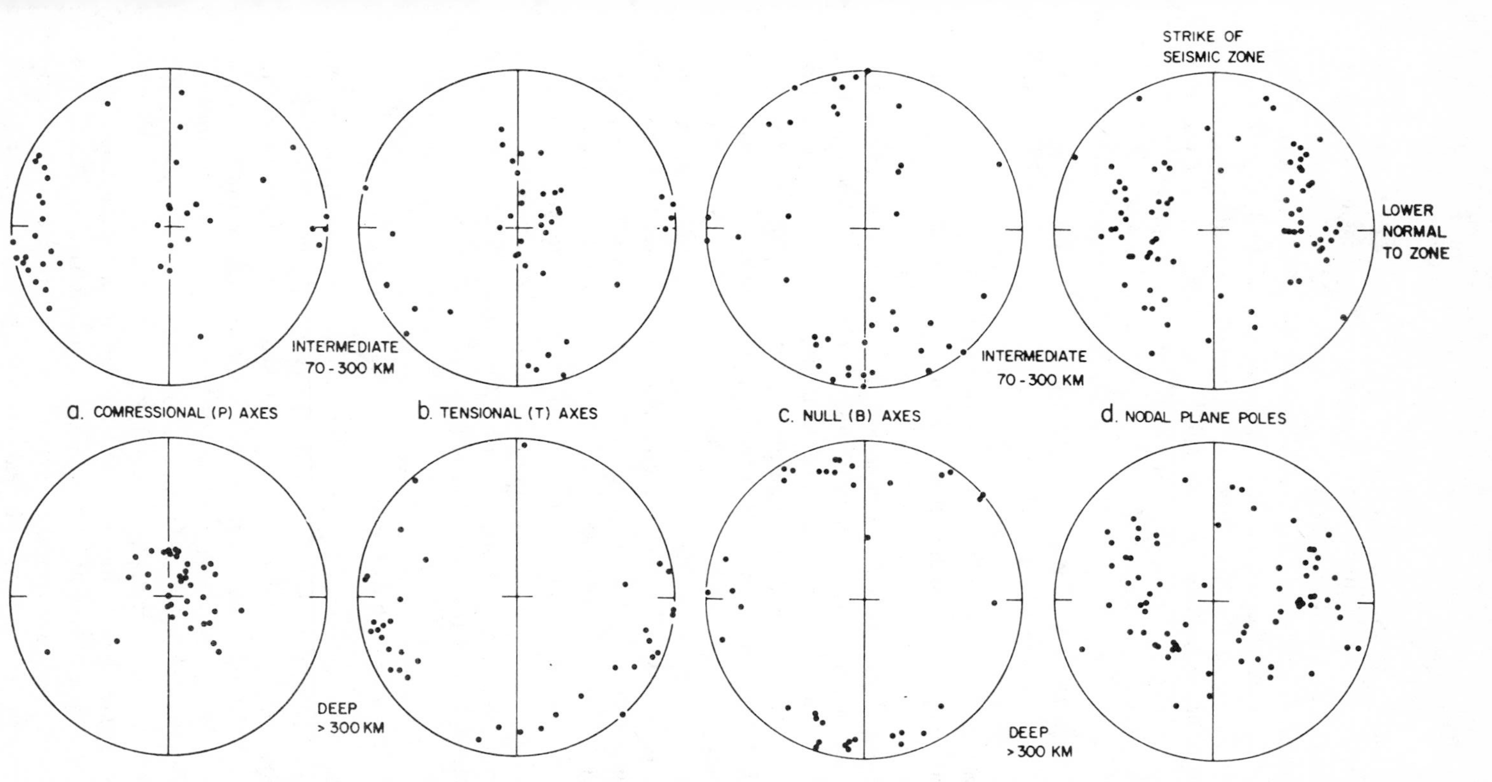

Fig. 8. The pressure P, tension T null axes and normals to fault and auxiliary planes of a number of intermediate and deep focus earthquakes chosen from uncomplicated planar parts of sinking slabs. The lower hemisphere of the sphere is shown in equal area projection. In all cases the equator of the projection sphere is taken in the plane of the normal and the strike of the dipping plane, and the position of the strike in this plane is marked by the straight line. All vectors pointing down the dip of the slab plot in the center of this projection (Isacks and Molnar, 1969).

small (McKenzie, 1969b). These two surprising results suggest that the faulting that occurs during intermediate and deep focus earthquakes is different and simpler than that during shallow shocks.

The stress conditions within the sinking slab are also important because they may provide the force required to move the plates (Elsasser, 1967; Isacks and Molnar, 1969). Since the distribution of temperature within the plates has already been obtained, it is simple to calculate the resultant stress f within the plates if their buoyancy is entirely supported by stresses transmitted through the plate and not be viscous traction on its surface (McKenzie, 1969a):

$$f = \frac{g\alpha\rho_0 T_0 lR}{24} \sin\phi \tag{30}$$

where ϕ is the dip of the slab. Substitution of the values for l, T_0 and κ given by Sclater and Francheteau (1970) gives $f \simeq 10$ kb if $\phi = 45°$. This value exceeds the yield stress of most rocks even when the experiments are carried out in the laboratory and last for a short time compared with that of geological processes. Though such order of magnitude arguments show that stresses sufficiently large to produce fractures could exist within sinking slabs, it is difficult to show that they are not balanced by viscous forces. An independent argument does, however, support these estimates. Wyss (1970) has used long- and short-period seismic waves to determine the apparent stresses involved in intermediate and deep focus earthquakes. He obtains the highest apparent stresses of about 400 bars at a depth of 100 km, and estimates that the true shear stress involved is about 3 kb. These new results on the stress conditions and temperature of the material where intermediate earthquakes occur suggest that the failure processes are strongly dependent on temperature and occur only if the shearing stress approaches the yield point. Suggested mechanisms of such earthquakes had not predicted either of these results and should perhaps be re-examined.

THE THICKNESS OF THE PLATES

When plate tectonics was first put forward there seemed no reason to believe that the base of the plates was the top of the low-velocity zone or, indeed, any other obvious boundary within the earth. McKenzie (1967) suggested that the plates became decoupled at a depth of 50 km and at a temperature of 550°C. His argument depended on heat flow observations and has been questioned by later authors (Sclater and Francheteau, 1970; Turcotte and Oxburgh, 1969). They argued that the topography of the ridges and temperature of the basalts erupted along their axes required the plates to be about 80 km thick with a temperature at their base of at least twice the value suggested by McKenzie. It now appears that these arguments are correct, and therefore value of the thermal conductivity must be nearer to 0.005 cal/°C cm sec, rather than 0.01 cal/°C cm sec used by McKenzie. This is perhaps the least well known of the parameters involved, especially at high temperatures (Fujisawa et al., 1968). A plate thickness of between 70 and 80 km now seems to agree best with the observations and agrees remarkably well with the depth to the low velocity in oceanic areas recently established by Kanamori and Abe (1968). Mechanical decoupling therefore appears to take place at the top of the low-velocity zone. This result emphasizes the importance of careful measurements of the thickness of continental plates and the depth to the low-velocity zone, such as Walcott (1970) and Archambeau et al. (1969) have carried out. Such measurements should establish whether the decoupling in these regions is also at the top of the low-velocity zone.

THE DRIVING MECHANISM

Various simple conditions must be satisfied by any mechanism that sustains plate motions. The mean annual seismic energy release, principally from shallow shocks, was 6×10^{24} ergs/yr between 1907 and 1955 (Gutenberg, 1956). If recent estimates of seismic efficiency (Wyss, 1970) are correct, then the annual elastic energy release during this period must be about 6×10^{25} ergs/yr. Since the mechanisms of shallow earthquakes shows that they are the result of plate motions, this energy is a lower bound to the energy required to keep the plates in motion. Most of the more speculative hypotheses that have been suggested are quite incapable of satisfying this requirement for geological periods of time. Thermal convection alone can do so without difficulty. Since the annual heat loss is $\sim 10^{28}$ ergs, the convection mechanism that converts heat into mechanical work need only be about 0.5% efficient for this purpose. For this reason, most of the work that has been so far carried out on the driving mechanisms (Elsasser, 1967; Lliboutry, 1969; McKenzie, 1968a; Turcotte and Oxburgh, 1967) has been concerned with some type of thermal convection, though generally of geometry very different from that simple cellular convection studied by Runcorn (1965).

Elsasser (1967) and Lliboutry (1969) have both suggested rather simple mechanisms by which the plate motions can be maintained. The driving force in Elsasser's model is the buoyancy force exerted by the sinking slab on the surface plate. McKenzie (1969a) has discussed the difficulties in accounting for observed plate movements by this mechanism, and his objections will not be restated. Lliboutry's mechanism is attractive for the same reason as that of Elsasser's: it avoids the complicated theory of thermal convection in a medium with unknown mechanical properties. Unfortunately, it suffers from many of the same defects as Elsasser's in an even more extreme form. Lliboutry argued that the hydrostatic pressure resulting from the elevation of the ridges above the surrounding sea floor must produce a force on the lithosphere acting away from them. It is easy to show that the mean pressure ΔP exerted on the lithosphere in this way is:

$$\Delta P = \frac{1}{2} g(\rho_0 - \rho_\omega) e \tag{31}$$

where e is the elevation of the ridge above the deep sea floor and ρ_0 and ρ_ω are the densities of the lithosphere and sea water, respectively. Thus, the total work done W by this force in a year is

$$W = 2\Delta P v l L \tag{32}$$

where $2v$ is the mean separation rate of all the world's extensional boundaries of total length L. Substitution of

$$e = 2 \text{ km} \qquad l = 80 \text{ km}$$

$$L = 50{,}000 \text{ km}$$

gives:

$$W = 2 \times 10^{25} \text{ ergs/yr} \tag{33}$$

This estimate of the energy release is unlikely to be in error by more than perhaps 50% since the least well-known parameter is l, the thickness of the lithosphere. Thus Lliboutry's mechanism cannot provide the energy released by shallow earthquakes. Even if the annual energy release is less than current estimates by a factor of 3, there is another important objection to Lliboutry's suggestion. The driving force provided by any ridge is the same, regardless of the spreading rate or size of plate being added to. Thus, small plates with an extensional boundary should move faster than large plates, since the driving force is constant and the frictional resistance between the plate and the mantle presumably increases with the size of the plate. The observations of global plate motions (Le Pichon, 1968) show the opposite effect, since the fastest relative motion between two plates occurs between the largest plates. These arguments do not favor Lliboutry's suggestion, nor even a combination of his and Elsasser's driving mechanisms, since the motion of small plates should still be faster than those of large ones.

These arguments show that any convective forces must act through traction on the base of the plates rather than only by pulling from trenches or pushing from ridges. At present, nothing is known about the circulation in the mantle which moves the plates, principally because of the success of the plate tectonics. Ridges are no longer believed to exist only on top of the rising limb of a convection cell, nor trenches only over the sinking limb. Apart from plate motions themselves, only the external gravity field and mean oceanic heat flow at present appear to be the result of circulation within the mantle. McKenzie (1968a) believed that deep earthquakes and the slope of the deep sea floor were also the result of mantle circulation, rather than plate creation and destruction. Intermediate and deep focus earthquakes, however, have now been demonstrated to be the consequence of plate consumption (Isacks et al., 1968; Oliver and Isacks, 1967) and the model for the thermal structure of the spreading sea floor (McKenzie and Sclater, 1969) appears to account for the slope of the sea floor (Sclater and Francheteau, 1970). These two phenomena should therefore be removed from the list of observations that can be used to test any theory of convection.

There is a considerable difference between the present status of the convection problem and that of the sea floor spreading 5 years ago. There has always been geological and seismological evidence for movements of the earth's crust on an enormous scale, and thus many relevant observations could be used to test the theory of plate tectonics. The other major difference is the complexity of the underlying physics. Plate tectonics is a kinematic theory that depends on simple ideas developed originally by Euler. In contrast, the equations that govern convection in a medium with variable viscosity have no known analytic solutions and have been studied very little. For these reasons, work on the convection problem will probably be beset by fierce controversies and slow progress.

Perhaps the most important of the existing arguments is concerned with the constitutive equation that governs flow within the mantle. Orowan (1965), Weertman (1970), and others believe that the creep of rocks within the mantle cannot be described by a viscosity, principally because they believe that the creep rate must be rate limited by the movement of dislocations. Gordon (1965) and McKenzie (1968b) disagree and believe that the creep rate depends linearly on the stress because the rate of deformation is governed by diffusion and not by dislocation movement. The reason for this disagreement is that no creep experiments have been carried out on silicates with a grain size of about 1 mm at high temperature and low stress. Orowan and Weertman have based their arguments on experiments with metals, which McKenzie believes are misleading. Gordon and McKenzie have used the results from high temperature creep experiments on ceramics with very small grain size, which Weertman believes must behave in a quite different way to specimens containing larger grains. This disagreement can only be resolved by careful experiments on silicates.

The other controversy concerns depth to which convection extends. Munk and MacDonald (1960), MacDonald (1963), and McKenzie (1966) have used the nonhydrostatic equatorial bulge of the earth to estimate the viscosity of the lower mantle. They believed that as the earth decelerates in response to tidal friction, its figure lags behind the equilibrium figure because of the high viscosity of the lower mantle. The basis for this argument has been questioned by Goldreich and Toomre (1969), who showed that the observations of the gravity field could be explained equally well by polar wandering. Polar wandering will take place unless the greatest principle axis of the moment of inertia is orientated along the rotational axis. Since the ellipticity is produced by the rotation itself and will therefore follow the pole, it must not be included in the calculation of the moment of inertia. The rate of polar wandering must be controlled by the viscosity and the angle between the rotational axis and the axis of greatest moment of inertia. Polar wandering must be defined with some care, since the variations in density that control the orientation of the principal axes of the moment of inertia tensor are probably related to convection in the mantle and, therefore, are not static features. If, however, the rate of polar wandering is rapid compared with the plate motions and convective velocities, the situation shown in Fig. 4*b* applies. Since the polar wandering velocities would have to be considerably greater than 20 cm/yr and such rates are not supported by palaeomagnetic evidence, it appears unlikely that polar wandering of the type discussed by Goldreich and Toomre (1969) has taken place. This difficulty does not mean that the greatest principal axis of the moment of inertia tensor will not move toward the rotation axis, but it does mean that this movement cannot usefully be described as polar wandering.

There is at present no obvious method of deciding in favor of either MacDonald's (1963) and McKenzie's (1966) model or of that of Goldreich and Toomre (Kaula, 1969), though there is now some slight evidence in favor of a highly viscous lower mantle. Isacks and Molnar (1969) found that sinking slabs that did not extend to a depth of 600 km or deeper generally possessed intermediate earthquake mechanisms that showed that the slab was in tension. If, however, the slab penetrated below this depth, the mechanisms at all depths showed that the slab was in compression along its entire length. This critical depth approximately corresponds to the expected position of a major phase change from tetrahedrally to octahedrally coordinated silicon. This is not the olivine-spinel phase change, but from spinel to perhaps a strontium plumbate structure (Anderson, 1967). For theoretical reasons (McKenzie, 1966), the denser phase is expected to have greater resistance to deformation, and Isacks and Molnar's (1969) results appear to support these arguments. It is therefore still unclear whether the lower mantle is involved in convection, though the evidence is at present more against than in favor of this suggestion. At present, the post glacial uplift of the Canadian Shield appears to be the only observation that may decide between these two models.

Therefore, the most useful model for discussing convection in the mantle at present appears to be a plane layer of fluid heated from below and also internally, with a viscosity that decreases rapidly with temperature. The surface boundary layer consists of the plates, and it may or may not be convenient to consider them independently from the rest of the convection problem. All calculations should be compared with the mean energy released each year by earthquakes, the plate velocities, the mean oceanic heat flow, and the satellite gravity field.

CONCLUSION

Plate tectonics started as an instantaneous theory of the tectonics of the ocean basins. It now appears that the theory provides a useful description of continental tectonics and a basis for the discussion of the evolution of the plates. Palaeomagnetic measurements are easily included within the framework of the present theory and provide one of the best methods of studying plate evolution over long periods of time.

This success is in marked contrast to progress in the convection problem in the last 5 years. Little progress has been made in understanding the mass motions in the mantle, which must move the plates. The problem does, however, now appear to be well posed and tractable using present large computers.

REFERENCES

Allen, C. R., and Smith, S. W., Parkfield earthquakes of June 27-29, Monterey and San Luis Obispo counties, California. Pre-earthquake and post-earthquake surficial displacements. Bull. Seism. Soc. Am., **56**, 966, 1966.

Ambraseys, N. N., The Buyin-Zara (Iran) earthquake of September 1962. A field report, Bull. Seism. Soc. Am., **53**, 705, 1963.

Ambraseys, N. N., and Zatopek, Z., The Mudurnu Valley, West Anatolia, Turkey, Earthquake of 22 July 1967, Bull. Seism. Soc., Am., **59**, 521, 1969.

Anderson, D. L., Phase changes in the Upper Mantle, Science, **157**, 1165, 1967.

Anderson, E. M., The Dynamics of Faulting, Oliver & Boyd, Edinburgh, 1951.

Archambeau, C. B., Flinn, E. A., and Lambert, D. G., Fine structure of the Upper Mantle, J. Geophys. Res., **74**, 5825, 1969.

Banghar, A. R., and Sykes, L. R., Focal mechanisms of earthquakes in the Indian Ocean, J. Geophys. Res., **74**, 632, 1969.

Barazangi, M., and Dorman, J., World seismicity maps compiled from ESSA Coast and Geodetic Survey epicenter data, 1961-1967, Bull. Seism. Soc. Am., **59**, 369, 1969.

Brune, J. N., Seismic moment, seismicity, and rate of slip along major fault zones, J. Geophys. Res., **73**, 777, 1968.

Bullard, E. C., Everett, J. E., and Smith, A. G., The fit of the continents around the Atlantic, Phil. Trans. Roy. Soc. (London), Ser. A, **258**, 41, 1965.

Carslaw, H. S., and Jaeger, J. C., Conduction of heat in solids, Oxford Univ. Press, London, 1959.

Cox, A., Geomagnetic reversals, Science, **163**, 237, 1969.

Creer, K. M., Arrangement of continents during the Palaeozoic era, Nature, **219**, 41, 1968.

Dalrymple, G. B., Cox, A., Doell, R. R., and Grommé, C. S., Pliocene geomagnetic polarity epochs, Earth Planet. Sci. Letters, **2**, 163, 1967.

Darwin, G. H., On the stresses caused in the interior of the Earth by the weight of continents and mountains, *in* Scientific Papers, vol. 2, p. 459, 1908.

Elsasser, W. M., Convection and stress propagation in the Upper Mantle, Princeton Univ. Tech. Rept. 5, 1967.

Engel, A. E. J., and Engel, C. G., Igneous rocks of the east Pacific rise, Science, **146**, 477, 1964.

Fisher, O., Physics of the Earth's Crust. MacMillan, New York, 1889.

Francheteau, J., and Sclater, J. G., Comments on paper by E. Irving and W. A. Robertson, "Test for Polar Wandering and Some possible Implications." J. Geophys. Res., **75**, 1023, 1970.

Fujisawa, H., Fujii, N., Mizutani, H., Kanamori, H., and Akimoto, S., Thermal diffusivity of Mg_2SiO_4, Fe_2SiO_4 and NaCl at high pressures and temperatures, J. Geophys. Res., **73**, 4727, 1968.

Gast, P. W., Trace element fractionation and the origin of tholeiitic and alkaline magma types, Geochim. Cosmochim. Acta., **32**, 1057, 1968.

Goldreich, P., and Toomre, A., Some remarks on polar wandering, J. Geophys. Res., **74**, 2555, 1969.
Gordon, R. B., Diffusion creep in the Earth's mantle, J. Geophys. Res., **70**, 2413-2417, 1965.
Graham, K. W. T., Helsley, C. E., and Hales, A. L., Determination of the relative positions of continents from palacomagnetic data, J. Geophys. Res., **69**, 3895, 1964.
Griggs, D. T., and Baker, D. W., The origin of deep-focus earthquakes, *in* Properties of Matter, p. 11, John Wiley & Sons, New York, 1969.
Gutenberg, B., Great earthquakes 1896-1903, Trans. Am. Geophys. Union, **37**, 608, 1956.
Heirtzler, J. R., Dickson, G. O., Herron, E. M., Pitman, W. C., and Le Pichon, X., Marine magnetic anomalies, geomagnetic field reversals, and motions of the ocean floor and continents, J. Geophys. Res., **73**, 2119, 1968.
Hess, H. H., History of the ocean basin, *in* Petrologic Studies, pp. 599-620, Buddington Memorial Volume, Geol. Soc. Am., 1962.
Hess, H. H., and Maxwell, J. C., Major structural features of the southwest Pacific: A preliminary interpretation of H. O. 5484, bathymetric chart, New Guinea to New Zealand. Proc. Seventh Pacific Sci. Congr., New Zealand, **2**, 14, 1953.
Hofmann, R. B., Geodimeter fault movement investigations in Calif., Calif. Dept. of Water Resources, Bull., **116**, 45, 1968.
Honda, H., Earthquake mechanism and seismic waves. J. Phys. Earth. (Tokyo), **10**, 1-97, 1962.
Irving, E., Palaeomagnetism, John Wiley & Sons, New York, 1964.
Irving, E., and Robertson, W. A., Test for polar wandering and some possible implications, J. Geophys. Res., **74**, 1026, 1969.
Isacks, B. L., and Molnar, P., Mantle earthquake mechanisms and the sinking of the lithosphere, Nature (London), **223**, 1121, 1969.
Isacks, B. L., Oliver, J., and Sykes, L. R., Seismology and the new global tectonics, J. Geophys. Res., **73**, 5855, 1968.
Jeffreys, H., The Earth. 2d edition, Cambridge Univ. Press, London, 1929.
Kanamori, H., and Abe, K., Deep structure of island arcs as revealed by surface waves, Bull. Earth. Res. Inst., **46**, 1001, 1968.
Kaula, W. M., Comment on Paper by P. Goldreich and A. Toomre, "Some Remarks on Polar Wandering," J. Geophys. Res., **74**, 1568, 1969.
Le Pichon, X., Sea-floor spreading and continental drift, J. Geophys. Res., **73**, 3661, 1968.
Lliboutry, L., Sea floor spreading, continental drift and lithosphere sinking with an Asthenosphere at melting point, J. Geophys. Res., **74**, 6525, 1969.
Love, A. E. H., Some Problems of Geodynamics, Cambridge Univ. Press, London, 1926.
MacDonald, G. J. F., The deep structure of oceans and continents, Rev. Geophys., **1**, 587, 1963.
McElhinney, M. W., The Palaeomagnetism of the Southern Continents–A Survey of Analysis UNESCO Symposium on Continental Drift, Montevideo, in press, 1971.
McElhinney, M. W., Briden, J. C., Jones, D. L., and Brock, A., Geological and geophysical implications of palaeomagnetic results from Africa. Rev. Geophys., **6**, 201, 1968.
McKenzie, D. P., The viscosity of the lower mantle, J. Geophys. Res., **71**, 3995, 1966.
______, Some remarks on heat flow and gravity anomalies, J. Geophys. Res., **72**, 6261, 1967.
______, The influence of the boundary conditions and rotation on convection in the Earth's mantle, Geophys. J., **15**, 457, 1968a.
______, The geophysical importance of high temperature creep, *in* The History of the Earth's Crust, Proceedings of NASA conference, Princeton Univ. Press, Princeton, N. J., 1968b.
______, Speculations on the causes and consequences of plate motions, Geophys. J., **18**, 1, 1969a.

______, The relation between fault plane solutions for earthquakes and the directions of the principal stresses, Bull. Seism. Soc. Am., **59**, 591, 1969b.

______, The plate tectonics in the Mediterranean region, Nature (London), **226**, 239, 1970a.

______, Temperature and potential temperature beneath island arcs, Tectonophys., **10**, 357, 1970b.

McKenzie, D. P., Molnar, P., and Davies, D., Plate tectonics of the Red Sea and East Africa, Nature (London), **226**, 243, 1970.

McKenzie, D. P., and Morgan, W. J., The evolution of triple junctions, Nature (London), **224**, 125, 1969.

McKenzie, D. P., and Parker, R. L., The North Pacific: An example of tectonics on a sphere, Nature (London), **216**, 1276, 1967.

______, Plate teçtonics in Ω space, in preparation, 1971.

McKenzie, D. P., and Sclater, J. G., Heat flow inside the island arcs of the northwestern Pacific, J. Geophys. Res., **73**, 3173, 1968.

______, Heat flow in the eastern Pacific and sea floor spreading, Bull. Volcanologique, **33**, 101-118, 1969.

______, The evolution of the Indian Ocean, Geophys. J., in press, 1971.

Molnar, P., and Sykes, L. R., Tectonics of the Caribbean and Middle America regions from focal mechanisms and seismicity, Bull. Geol. Soc. Am., **80**, 1639, 1969.

Morgan, W. J., Rises, trenches, great faults, and crustal blocks, J. Geophys. Res., **73**, 1959, 1968.

Munk, W. H., and MacDonald, G. J. F., The rotation of the Earth. Cambridge Univ. Press, Cambridge, 323 pp., 1960.

Oliver, J., and Isacks, B., Deep earthquake zones, anomalous structures in the upper mantle, and the lithosphere, J. Geophys. Res., **72**, 4259, 1967.

Orowan, E., Convection in a non-Newtonian mantle, continental drift, and mountain building, Phil. Trans. Roy. Soc. (London), Ser. A, **258**, 284, 1965.

Raleigh, C. B., and Paterson, M. S., Experimental deformation of serpentinite and its tectonic implications, J. Geophys. Res., **70**, 3965, 1965.

Runcorn, S. K., Changes in the convection pattern in the Earth's mantle and continental drift: Evidence for a cold origin of the Earth, Phil. Trans. Roy. Soc. (London), Ser. A, **258**, 228, 1965.

Savage, J. C., A possible explanation of the orientation of fault planes for deep focus earthquakes (abstract), annual meeting of the Seismological Society of America, 1969.

Scholz, C. H., Wyss, M., and Smith, S. W., Seismic and Aseismic Slip on the San Andreas Fault, J. Geophys. Res., **74**, 2049, 1969.

Sclater, J. G., and Francheteau, J., The implications of terrestrial heat flow observations on current tectonic and geochemical models of the crust and upper mantle of the Earth, Geophys. J., **20**, 509, 1970.

Smith, A. G., and Hallam, A., The fit of the southern continents, Nature (London), **225**, 139-144, 1970.

Stauder, W., The focal mechanism of earthquakes, Adv. Geophys., **9**, 1, 1962.

Stauder, W., and Bollinger, G. A., The *S*-wave project for focal mechanism studies, earthquakes of 1962, Bull. Seism. Soc. Am., **54**, 2199, 1964.

______, The focal mechanism of the Alaska earthquake of March 28, 1964, and its aftershock sequence, J. Geophys. Res., **71**, 5283, 1966a.

______, The *S*-wave project for focal mechanism studies, earthquakes of 1963, Bull. Seism. Soc. Am., **56**, 1363, 1966b.

Sykes, L. R., Mechanism of earthquakes and nature of faulting on the mid-oceanic ridges, J. Geophys. Res., **72**, 2131, 1967.

Talwani, M., Plate tectonics and deep sea trenches (abstract), Trans. Am. Geophys. Union, **50**, 180, 1969.

Talwani, M., Le Pichon, X., and Ewing, M., Crustal structure of the mid-ocean ridges 2. Computed model from gravity and seismic refraction data, J. Geophys. Res., **70**, 341, 1965.

Thomson, W., and Tait, P. G., Treatise on natural philosophy. Cambridge Univ. Press, London, 1883.

Turcotte, D. L., and Oxburgh, E. R., Finite amplitude convection cells and continental drift, J. Fluid Mech., **28**, 29, 1967.

______, Convection in a mantle with variable physical properties, J. Geophys. Res., **74**, 1458, 1969.

Vacquier, V., Uyeda, S., Yasui, M., Sclater, J., Corry, C., and Watanabe, T., Studies of the thermal state of the Earth. The 19th paper: Heat flow measurements in the northwestern Pacific, Bull. Earthquake Res. Inst., **44**, 1519, 1966.

Vacquier, V., and Uyeda, S., Palaeomagnetism of nine seamounts in the Western Pacific and of three volcanoes in Japan, Bull. Earthquake Res. Inst., **45**, 815, 1967.

Vine, F. J., Spreading of the ocean floor: New evidence, Science, **154**, 1405, 1966.

______, Palaeomagnetic evidence for the northward movement of the north Pacific Basin during the past 100 m.y. (abstract), Trans. Am. Geophys. Union, **49**, 156, 1968.

Vine, F. J., and Matthews, D. H., Magnetic anomalies over oceanic ridges, Nature (London), **199**, 947, 1963.

Walcott, R. I., Flexural rigidity, thickness and viscosity of the lithosphere, J. Geophys. Res., **75**, 3941, 1970.

Weertman, J., Creep strength in the earth's mantle, Rev. Geophys. and Space Physics, **8**, 145, 1970.

Wegener, A., The origin of continents and ocean basins, 1929, translated by J. Biram, 246 pp., Dover Publications, New York, 1966.

Whitten, C. A., Crustal movement in California and Nevada, Trans. Am. Geophys. Union, **37**, 393, 1956.

Wigner, E. P., Group theory, Academic Press, New York and London, 1969.

Wilson, J. T., A new class of faults and their bearing on continental drift, Nature (London), **207**, 343, 1965.

Wyss, M., Stress estimates for South American shallow and deep earthquakes, J. Geophys. Res., **75**, 1529, 1970.

14

THE SINKING LITHOSPHERE AND THE FOCAL MECHANISM OF DEEP EARTHQUAKES*

by **David T. Griggs**

A multitude of recent reliable earthquake mechanism solutions in island arc areas (largely by the Lamont group) yield a consistent set of three types: (1) shallow normal faults, (2) underthrust faults at depths of 10 to 100 km, and (3) intermediate and deep solutions indicating compression (or occasionally extension) parallel to the dip of the Benioff seismic zone. The first seems correctly explained by the Stauder hypothesis that bending of the foundering lithospheric plate causes graben-type normal faulting on its upper surface. The second is explained by the new Global Tectonics as resulting from the shearing interaction between the foundering plate and its passive neighbor. Conventional fracture and stick-slip behavior, facilitated by high water pressure from the sediments entrained by the downgoing plate, suffice to explain these two classes of earthquakes, though precise definition of their origin must await more detailed knowledge of their character and a more definitive model of the interplate reaction zone.

No satisfactory explanation of the intermediate and deep-focus mechanisms has yet been advanced, to my knowledge. The nature and the environment of these earthquakes is intimately bound up with the thermal and mechanical history of the sinking lithospheric slab. Because

*Paper No. 797 of the Institute of Geophysics and Planetary Physics, University of California, Los Angeles, California.

of its large thermal time constant and rapid rate of descent, the slab is expected to preserve its integrity to great depths consistent with the seismic zones. The wealth of seismic, gravity, and heat flow observations in island arcs tempts theoretical attack on the nature of the sinking slab. McKenzie, Minear and Toksöz, and the writer have explored this problem, with comparable yet different results. McKenzie and Griggs tend to get relatively cool slab centers persisting to great depths, while the model of Minear and Toksöz leads to a slab that becomes hotter than the ambient geotherm at depths comparable to those of deep earthquakes. The principal difference between these two sets of models is the great shear strain heating assumed in the Minear-Toksöz model.

If the olivine-spinel transition occurs in the mantle at approximately 365 km (Anderson, 1967) and has the properties given by Akimoto and Fujisawa (1968), then the depth at which this transition occurs in the sinking lithosphere is a sensitive function of its thermal behavior. This and other mantle transitions may play a role in intermediate and deep-focus earthquakes.

The strength of the sinking lithosphere is calculated, showing how the slab thins and weakens as it sinks. This calls in question the stress-guide hypothesis of the origin of down-dip compression in deep earthquakes.

PREFACE

The origin of deep earthquakes seems to be paradoxical. They occur only where there is reason to believe that the lithosphere is plunging into the mantle. When the rate of this motion was first realized, it became clear that such known processes as dehydration embrittlement and weakening could cause shear fractures at even the maximum earthquake depths. Such fractures were expected to occur roughly parallel to the surface of the sinking plate, as in the underthrust earthquakes occurring in the Aleutians. Recent focal plane solutions for intermediate and deep earthquakes seem to point to shear surfaces inclined $\sim 45^\circ$ to the sinking plate, with the stress axis of greatest compression (or occasionally greatest extension) parallel to the dip of the slab. If these quakes are of simple mechanical origin, they would seem to imply that the colder center of the slab serves as a stress guide (Elsasser, 1967), though the origin of the compressive stresses is uncertain.

The origin of shallow earthquakes will not be pursued in this paper for two reasons: (1) shallow earthquakes are being investigated in detail by many seismologists, and it would be beyond the scope of this paper to do justice to this extensive work, and (2) Crustal rock motions seem everywhere consistent with mechanisms of conventional fracture and stick-slip behavior facilitated at depth by high fluid pressure. There thus seems to be no problem of mechanisms, although the details of these processes will only become clear as more intensive studies are made.

Since deep earthquakes seem to be intimately related to processes within the sinking lithospheric slab, the first part of this paper examines in an exploratory way the character of the slab. The second part comments on the earthquake mechanics.

NATURE OF THE SINKING LITHOSPHERE

Introduction

It has been dimly perceived for decades that the crust and upper mantle are underthrust at the oceanic trenches, and the discovery of the deep earthquake zones extending downward from the trenches lent credence to this idea. Sykes' detailed study (1966) of the Tonga and Kermadec zones first showed that these earthquakes occur in a very narrow zone ($<$ 20 km thick), which is for the most part nearly planar and inclined at roughly 45° to the horizontal.

Oliver and Isacks (1967) showed that these earthquake zones have anomalously low attenuation of S waves and 1 to 2% higher P and S wave velocities. From these facts, they deduced that the upper part of the mantle under the ocean (the lithosphere) sinks or is thrust down at the trenches and extends in recognizable form at least as far down as the deepest earthquakes.

Isacks, Oliver, and Sykes (1968), hereafter called IOS, argued that this behavior is characteristic of trenches and showed that the down-dip length along the earthquake zone is roughly proportional to the sinking velocity derived for that trench from the rigid plate model of the earth's surface motion (LePichon, 1968).

This author has reasoned (Griggs, 1968, 1969) (1) that the lithosphere is similar to the underlying asthenosphere except that it is cooler, which accounts for its higher seismic velocity and lower attenuation, and (2) that the limiting depth of earthquakes is that point at which the lithosphere, which is heated as it sinks, becomes too hot for seismic instability (fracture?) to occur. Below this depth, all deformation occurs by flow.

Related Work

Two recent papers have dealt with the nature of the sinking lithospheric plates (McKenzie, 1969; Minear and Toksöz, 1970). McKenzie extended his analysis of the heat flow anomalies associated with the ridges (McKenzie, 1967) to the solution of the temperature distribution in the sinking lithosphere. He also derived the streamlines, the stresses, and the strain heating in the mantle above and below the sinking plate. He then considered the broad problem of mantle convection and the role played in this process by the sinking plates.

Minear and Toksöz did a numerical two-dimensional heat flow calculation of the sinking lithosphere and the surrounding mantle, taking into account adiabatic compression, radioactivity, and phase changes. Estimates of the effect of strain heating in the slab boundary, surface heat flux, gravity anomalies, and seismic travel time in the vicinity of the slab are presented.

Because of uncertainties in the values of the large number of parameters and the mathematical difficulties in handling the nonlinear real-earth processes, these analyses (and mine also) can only hope to throw additional light on the probable trend and the relative importance of key factors, not to solve the problem. They may, however, point the way toward future geophysical exploration of significance and toward refinements to be sought in the mathematical and calculational techniques.

McKenzie's analysis (1969) is made tractable by assuming that the temperature gradient in the mantle below the lithosphere is adiabatic everywhere except in the slab and that mantle flow is characterized by a constant Newtonian viscosity. The latter assumption is probably apt for the upper mantle below the low-velocity zone because it is very hot (near the melting point) and the stresses are low. I believe the former assumption is grossly in error and that the mantle more nearly follows the melting point gradient below the low-velocity zone. The thermal history of the slab is not sensitive to the assumed geotherm, however, and McKenzie's analysis gives a very good qualitative picture of the nature of the temperature distribution in the slab and most of the gross effects that he considers.

Minear and Toksöz (1970) used a 20- by 20-km grid and translated the boundary temperatures 60 km with respect to the slab in each step of the calculation. They adopted MacDonald's chondritic mantle radioactivity and temperature gradient (1963). Adiabatic heating and heating resulting from passing through phase transitions are introduced.

Strain heating in the mantle at the slab boundaries is treated parametrically. They apparently assume constant shear stress over the whole length of the slab rather than conform to McKenzie's model, which has a maximum near the top and diminishes several-fold downward. Typical shear stresses assumed (2500 bars in Fig. 11) are very much higher than McKenzie's average stress of about 100 bars. The strain heating in Minear and Toksöz' model is thus very much greater and even causes inversion in the slab. The slab at depth is hotter than the surrounding mantle.

I believe McKenzie's values of strain heating in the mantle are about right. Certainly, in the model of the earth which I use, neither the drag stresses nor the strain heating could approach Minear and Toksöz' values for reasons that will become clear later.

Calculational Model of the Sinking Lithosphere

Attention is restricted to the interior of the downgoing slab. For simplicity, I employ a one-dimensional heat flow calculation of the most rudimentary sort (Carslaw and Jaeger, 1959, p. 470). Because of the fact that the transverse temperature gradient is vastly greater than the longitudinal, one-dimensional heat flow is a good approximation except at very low sinking velocities. This may be qualitatively verified by comparing my isotherms with those of McKenzie (1969).

The basic difference equation is:

$$T_j = R_j + w(R_{j-1} - 2R_j + R_{j+1}) \tag{1}$$

where T is the new temperature, R the old, the subscript j denotes the grid point number across the slab, and $w = \kappa\Delta t/\Delta x^2$. κ is the diffusivity, Δt the time interval between each set of calculations and Δx the distance between grid points across the slab. For calculational stability, w must be less than 0.5, and is given the value 0.316. A grid of 27 points across the slab (25 active cells) suffices to reveal with the requisite fidelity the thermal structure and also the fine structure in the resultant gravity anomaly. Duplicate calculations with 50 active cells and hence with a time interval reduced fourfold show very small differences in the thermal structure. The κ is taken as 0.01 cm^2/sec. Radiative conductivity is of no consequence except very near the borders of the slab at depth and then only if the opacity is 10 cm^{-1} or less. It is neglected in these calculations, but it may not be negligible for calculations on the mantle.

Adiabatic heating is added at each calculational step as follows (Jeffreys, 1929, p. 139):

$$\frac{dT}{dZ} = g\,\frac{\alpha T}{c_p} \tag{2}$$

where Z = depth, α = coefficient of themal expansion, and c_p = specific heat. The value of α is of first-order importance for calculating the gravity anomalies and the stresses resulting from the excess density of the slab. Values of α and α/c_p versus depth are shown in Fig. 1. These are taken from Birch (1952) and seem as good as any in the later literature.

Only a single phase change was explored in this model–the olivine-spinel transition presumed by Anderson (1967) to occur at a depth of 365 km. As will be seen, the *P-T* slope of this transition has a pronounced effect on the density structure of the slab. Table 1 shows the transition parameters that were used in the calculations. Transition *B* is consistent with Akimoto and Fujisawa (1968). Its slope is so nearly parallel to the geotherm that the transition occurs at shallow depth in the center of the cold slab. For this reason, other slopes were considered. Transitions *D, E,* and *F* correspond to a mantle composition of 50% olivine.

The melting temperature within the mantle is approximated by:

$$T_{mp} = 1273\left(1 + \frac{P}{15}\right)^{0.25} \tag{3}$$

where P is the pressure in kilobars. This is shown in Fig. 2 faired to 1100°C at

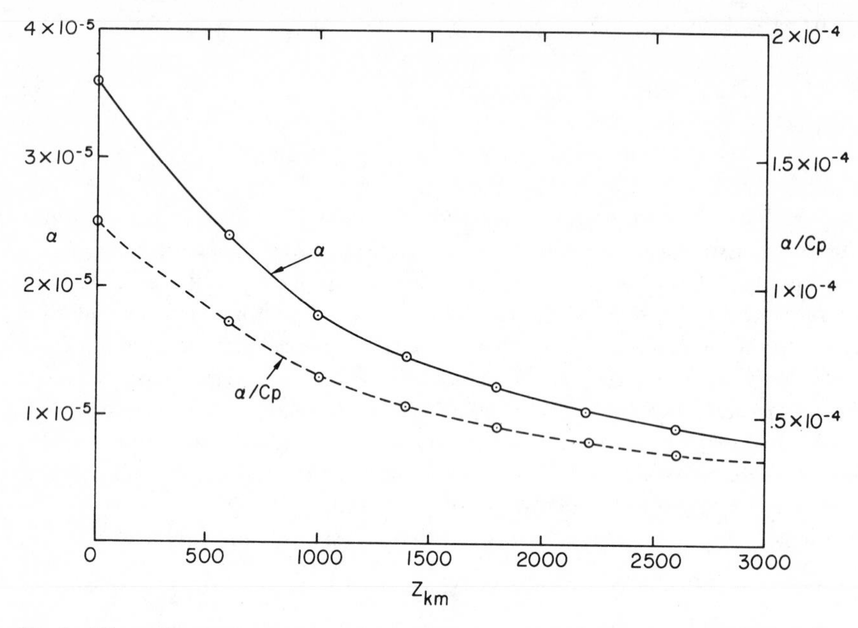

Fig. 1. Thermal expansion coefficient α/c_p versus depth in the earth. From Birch (1952).

$P = 0$. At low pressure, this corresponds to the solidus of peridotite (lherzolite) as determined by Ito and Kennedy (1967) and Kushiro, Syono, and Akimoto (1968). At higher pressures, it is adjusted by a Simon-type fit to a melting temperature of 4000°K at the core boundary. While this is functionally wrong, the result corresponds to the best guesses of my colleagues (principally that of G. C. Kennedy, 1970, oral communication) as to the temperature of first melting in the mantle. The importance of melting temperature in these calculations is twofold: (1) it is used to find β (the ratio of temperature in the slab to the melting temperature at that pressure) and (2) it sets an upper limit for the rise of temperature resulting from strain heating, etc.

The principal factor governing the choice of the undisturbed mantle geotherm

Table 1. Six sets of parameters used for the olivine-spinel transition in the calculations

Transition	ρ_{sp}/ρ_{ol}	dT/dP (°C/kb)	T_0 ($P = 0$)	ΔT
A	1.10	30.0	-1747°C	35°C
B	1.10	15.7	-2	70
C	1.10	20.0	-527	50
D	1.05	15.7	-2	35
E	1.05	20.0	-527	25
F	1.05	30.0	-1747	17.5

is my belief that the mantle temperature must closely approach or equal the temperature of first melting in the low-velocity zone. A wide variety of geotherms were explored, and it was found that the results of the calculations were not very sensitive to the shape of the geotherm, provided it approached the melting point at depths comparable to that of the low-velocity zone. The geotherm used in the calculations reported here was:

$$\left.\begin{array}{ll} T = 1340\left[1 - \exp\left(-\dfrac{Z}{45}\right)\right] + 1.7Z & 0 < Z < 215 \text{ km} \\ T = T_{mp} - \dfrac{(Z - 215)}{4} & Z \geq 215 \text{ km} \end{array}\right\} \quad (4)$$

where Z is the depth in kilometers and T_{mp} from Eq. 3 is here used in °C. This is shown in Fig. 2, together with the adiabat from 215-km depth. This geotherm would give a surface heat flow of 1.6 μcal/cm^{2}sec if the surface conductivity were 0.005 cal/cm deg C sec. It will be noticed that this geotherm corresponds to a Lachenbruch (1968) steady-state model with radioactivity decreasing exponentially from a surface value of 3×10^{-13} cal/cm^{3}sec and an e-folding length of 45 km. Actual upper mantle temperatures, however, must be largely determined by their cooling history as they spread away from the ridges. The

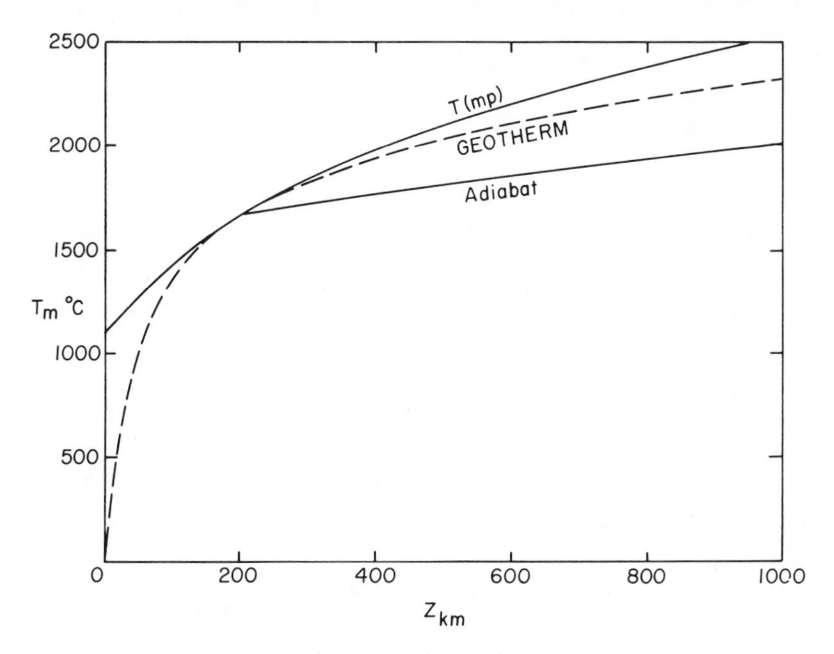

Fig. 2. Solidus melting temperature, *T*(*mp*), geotherm, and adiabat used in most calculations, versus depth.

geotherm chosen is intended to approximate such cooling after transport from ridge to trench.

Pressure and density in the mantle were calculated as follows:

$$\left.\begin{aligned} P &= g\int \rho dZ \\ \rho &= \rho_0\left[1 + \int (a - 2bP)\, dP - \int a dT\right], \end{aligned}\right\} \quad (5)$$

where $a = 8 \times 10^{-4}$ and $b = 1 \times 10^{-6}$, from the two-constant fit for the compressibility of olivine (Clark, 1966, p. 131). Density versus depth is shown in Fig. 3 for the two types of olivine-spinel transitions (Table 1), each of which is presumed to be gradational over a depth range of 35 km. At depths greater than 200 km, these densities are lower than those commonly estimated for the outer mantle, suggesting that the compressibility used in these calculations is too low. Calculations with other equations of state have shown, however, that the effects to be discussed below are not sensitive to the magnitude of the compressibility, but only to the general shape of the ρ versus Z curve.

We are now ready to describe the calculational procedure. It is assumed that the slab slips down through undisturbed mantle. This is clearly not correct, but

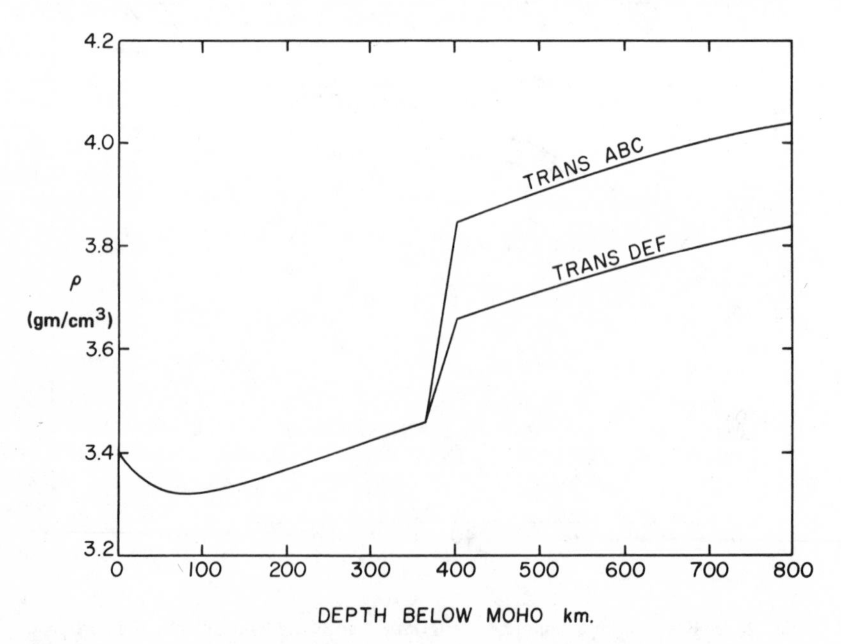

Fig. 3. Density versus depth for the six sets of olivine-spinel transitions (Table 1).

is akin to McKenzie's assumption (1969) of an adiabatic mantle. As he points out, there is no obvious alternative to numerical solution of the general nonlinear convection equations, which is beyond my capability.

Values are chosen for sinking velocity (parallel to the slab), V, angle of descent (assumed constant throughout the sinking), ψ, slab thickness, D, and type of transition. The details of the bending of the lithosphere are ignored, since it was found that these have little effect on the thermal structure in the slab except at very low velocity. The transverse grid points are initially assigned the temperature of the mantle at their depths. The grid is then translated parallel to the slab a distance ΔY–the distance the slab would have traveled in time Δt ($\Delta Y = Vw\Delta x^2/\kappa = V\Delta x^2/10$ km, where V is in cm/yr). The upper and lower boundary temperatures are set to the corresponding mantle temperatures. The heat flow calculation is done by Eq. (1), and adiabatic heating is added for each grid point. Radioactive heating is negligible except at very low velocities, and so it is neglected. Pressure and density are calculated for each grid point and each equivalent point in the undisturbed mantle. The contribution of the density difference in each cell to the gravity anomaly is calculated. The contribution of these density differences to the down-slab extensile stress is also calculated. The minimum temperature in the slab is found, and $\beta = T/T_{mp}$ is calculated for that point. The calculation is then iterated and the behavior of the slab is observed to some chosen depth. Temperature and density at each grid point (if 25 cells, or every other grid point, if 50 cells are used) are printed out, so that the details of the process may be observed. Cumulative gravity anomalies, extensile stress, and diagnostic data are printed out every five time steps.

At each time step, a test is made to see if any point in the slab has passed the threshold for the transition and the density, pressure, and temperature are appropriately incremented if this condition is met, until the transition is complete. This is done separately, of course, for the slab grid points and the corresponding mantle points, since the transition occurs earlier in the cooler parts of the slab.

Isotherms and Transition Elevation in the Slab

The thermal structure in the slab for a typical calculation is shown in Fig. 4. Similarity between this and McKenzie's calculation (1969, Fig. 3) is readily apparent. The upward extension of the transition within the slab is also shown for the two extreme transition slopes considered. The early history of the slab is dominated by the slowness of thermal diffusion. The time required for changes in the boundary temperatures to be felt appreciably at the center of a slab is (Carslaw and Jaeger, 1959, p. 98, Fig. 10a):

$$\tau \simeq \frac{D^2}{30\kappa} \simeq \frac{D^2}{1000} \tag{6}$$

where D is the total slab thickness in km and τ is in millions of years. IOS show

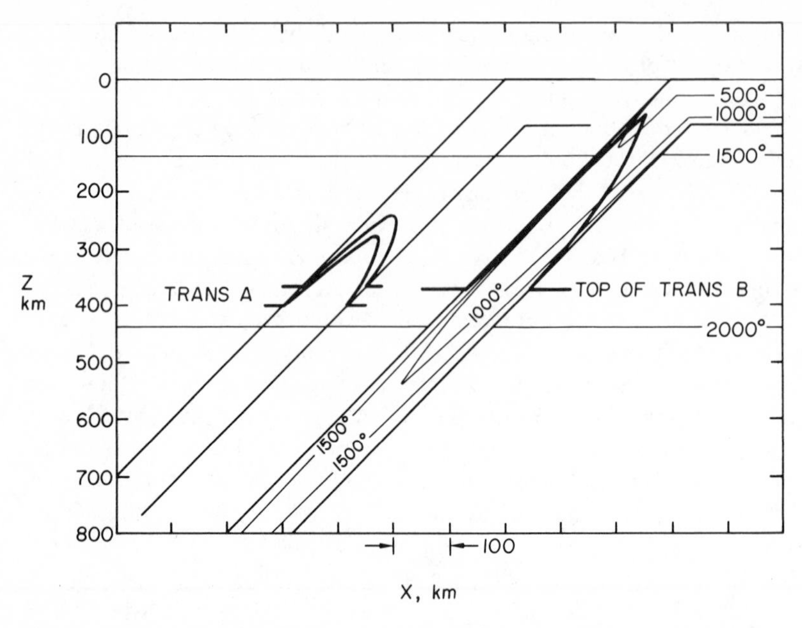

Fig. 4. Isotherms and transition boundaries for typical calculations. Lithosphere thickness (D) = 80 km, ψ = 45°, ν = 8 cm/yr, ρc_p = 1.0, adiabatic heating and transition heating included.

that the sinking time to the deepest earthquake is about 10 m.y., so if the thickness is greater than 100 km, the center of the slab will be nearly as cold as it was initially, except for adiabatic heating and transition heating. In a 50-km-thick slab, however, the central temperature will have increased to about half the elevation of the boundary temperature.

It follows that the maximum elevation of an olivine-spinel transition such as B or D within the slab is insensitive to slab thickness because the sinking time to the transition boundary is small compared to the thermal time constant. Values for transitions A and B (Table 1) are given in Table 2. Figure 5 shows temperature versus pressure profiles through a slab and the intersection with the various transitions for various lengths, L, down the slab.

I have dwelt on this elevation of the olivine-spinel transition in more detail

Table 2. Depth to uppermost point of olivine-spinel transition in slab at ν = 8 cm/yr

Transition	Slab thickness, kilometers			
	50	75	100	150
B	47	47	47	47
A	225	220	218	216

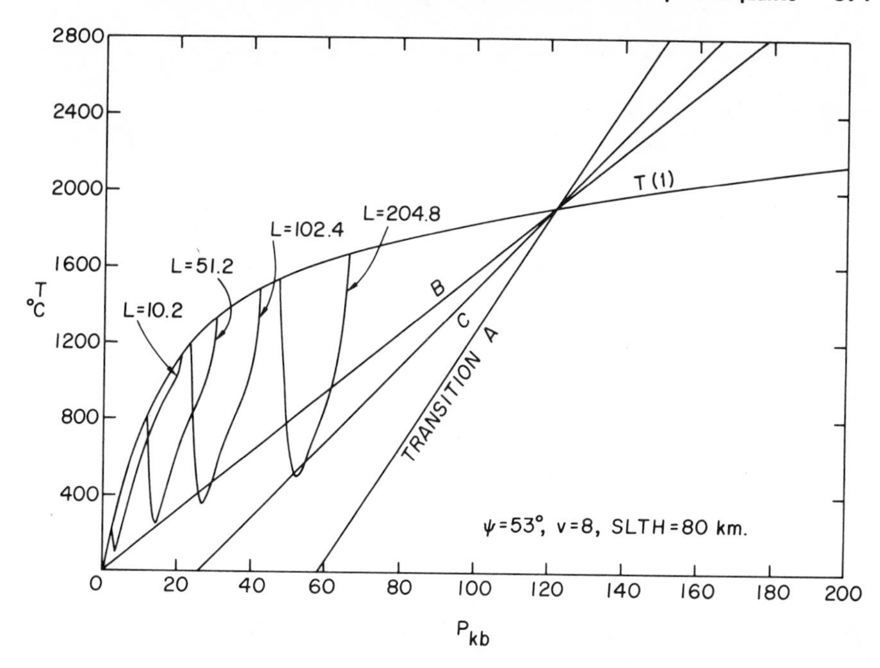

Fig. 5. Thermal sections through lithosphere at different values of length down the slab (L) in km, plotted as a function of ambient pressure. Intersection with transition boundaries illustrates degree of elevation of transition within the sinking lithosphere. $T(1)$ is temperature at slab boundary, ψ is angle of descent, v is velocity of sinking in cm/yr, and SLTH is slab thickness.

than Minear and Toksöz (1970) because this seems to be an effect capable of exploration by seismic methods. Since Anderson (1967) and others now have strong evidence that such a transition exists, it seems appropriate to mount special studies to determine the shape of the high-density, high-velocity region within the sinking slab.

Maximum Depth of Earthquakes

We turn next to the limiting depth of earthquakes. McKenzie (1969) assumes that this limit occurs when the minimum temperature in the slab exceeds some critical value. Testing this against the variations of sinking velocity along the Tonga-Kermadec trench zone, he shows that the limit corresponds to heating of the center of the 50-km-thick slab in his model to 85% of his adiabatic mantle temperature (McKenzie, 1969, Fig. 4). I believe it is physically more reasonable to assume that the limiting depth is that point at which the minimum slab temperature exceeds a critical fraction β_c of the melting temperature.

IOS (their Fig. 16) give the length down the slab to the deepest earthquake versus sinking velocity for trenches all over the world. This seems to be the best data to use for a test of the limiting β hypothesis. Figure 6 shows the results from 30 calculations at three different angles of sinking and 10 velocities from 1

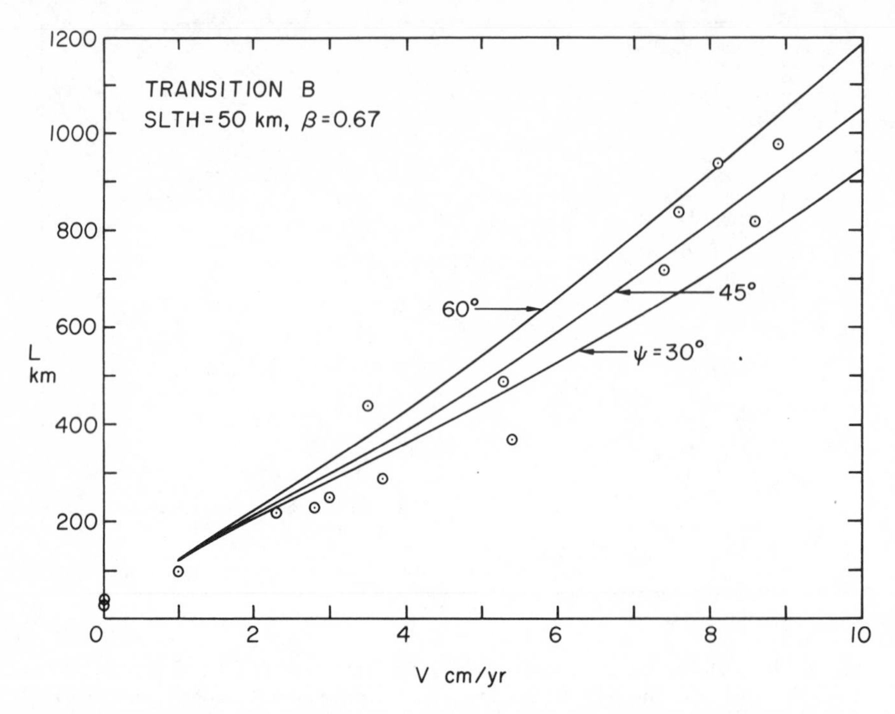

Fig. 6. Length down-dip in slab to the point where the minimum temperature in the slab reaches β = 0.67 versus velocity of sinking, for three different angles of dip. Points are from IOS (Fig. 16) excluding points of widest scatter. This shows that the critical β hypothesis of the limiting depth of earthquakes fits the observations within the scatter of the data, and that it fits better than the 10-m.y. straight line of IOS, which extends from L = 0, ν = 0, to L = 1000, ν = 10.

to 10 cm/yr, for $\beta_c = 2/3$, and the IOS data with points of widest scatter omitted. It is seen that the 45° curve fits the data somewhat better than the 10-m.y. straight line of IOS. I do not have sufficient data on the dip of the zones to check whether the quality of the fit would be improved or not by correcting each point for dip. Unfortunately, a wide variety of model parameters will give about equally good fit to the IOS data, so this can not serve as a discriminant of model validity.

The variation of β with slab thickness is shown in Fig. 7, together with the amplitude of the gravity anomaly produced by the slab. The nature of the transition affects β very little, but drastically affects the magnitude of the gravity anomaly, because of the elevation of the transition in the slab. Density contours for a typical calculation illustrate this effect (Fig. 8).

Gravity Anomalies

The characteristic shape of the gravity anomaly due to the slab is shown in Fig. 9 (see also Minear and Toksöz, 1970, Fig. 14). This anomaly is partly compensated by the characteristic negative anomaly associated with the trenches caused by the sinking lithosphere. Taking Tonga as the model, a symmetric negative anomaly is created by mirroring the oceanic-side bathymetry about the trench axis and assuming a density contrast of 1.8 (Talwani, Worzel, and Ewing, 1961). Sykes (1966, Fig. 9) shows that the trench is offset about 50 km from the top of the sinking slab as revealed by earthquakes. This symmetric anomaly is shown in Fig. 9 with the 50-km offset. The sum of this and the slab anomaly is a broad positive anomaly of large magnitude, which must be regionally compensated.

The problem of regional compensation is very complicated, since it involves time-dependent strength and non-Newtonian plastic flow of the crust and

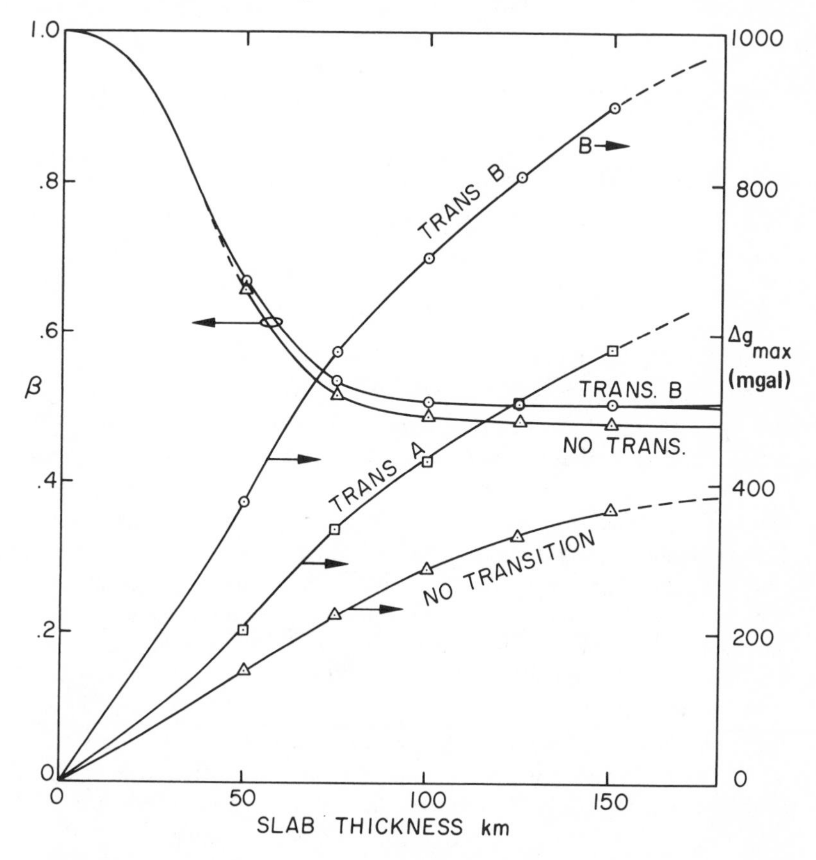

Fig. 7. Ratio of minimum temperature in slab to the melting temperature (β) and the maximum value of the gravity anomaly component attributable to the slab alone (Δg_{max}) versus slab thickness. Arrows indicate the appropriate ordinate scale. L = 820 km, ν = 8 cm/yr, ψ = 45°. Illustrates increase in gravity anomaly attributable to elevation of transition in slab. As slab thickness increases, thermal conduction asymptotically approaches zero, in which case β is determined solely by the initial temperature plus adiabatic and transition heating.

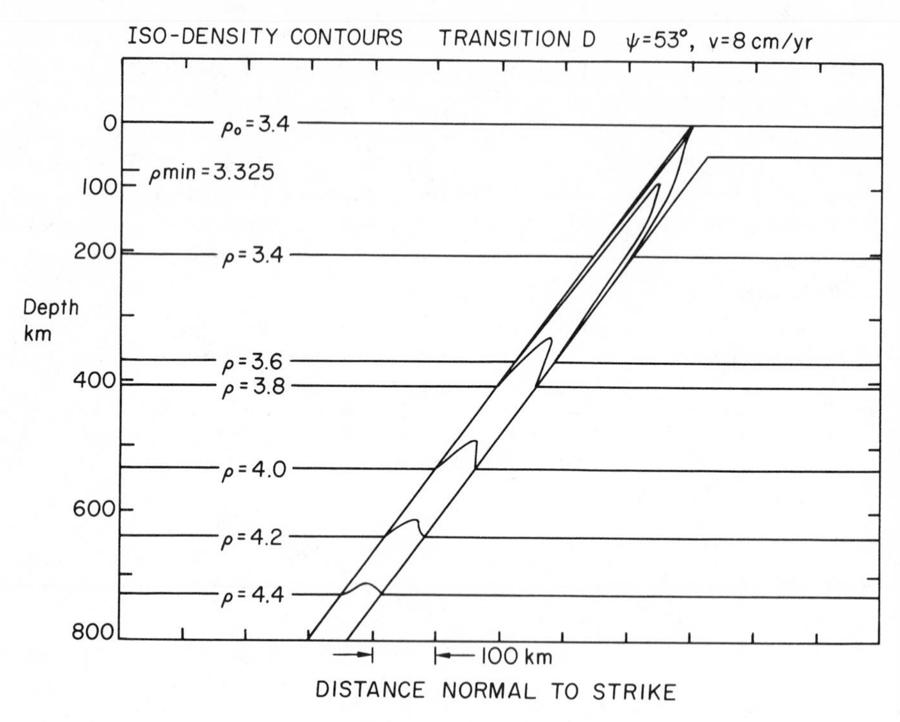

Fig. 8. Distribution of density, ρ, in gm/cm^3, in a typical calculation.

lithosphere plus flow of the asthenosphere (McConnell, 1968). Based on McConnell's and other comparable analyses of isostatic adjustment, it seems that a fairly good approximation of regional compensation in this case would be achieved by a $\cos^2(\pi X/\lambda)$ deviation from the geoid, with $\lambda \cong 2000$ km, where X is now the horizontal distance from the center of effective mass of the sum of slab and trench anomalies. Such a regional compensation anomaly is shown in Fig. 9, adjusted so that the net regional anomaly is zero. The sum of these three anomaly components yields my best estimate of the observable free-air anomaly attributable to the sinking lithosphere and the trench alone.

Net anomalies derived in this way are shown in Fig. 10 for three different types of transition, several lithosphere thicknesses, and three angles of dip of the sinking slab. If there were no transition, or a transition with large dT/dP, a lithospheric thickness of 100 km would fit the observed anomaly well. If there were a transition with the slope of Akimoto and Fujisawa (1968), then the slab thickness must be less, but also the effect of dip becomes more important (Fig. 10*d*). The dip of the Benioff zone at 20 to 25°S, where the anomalies were measured, is about 53° (Sykes, 1966, Fig. 9). The net anomalies with this dip are shown in Fig. 11 for transitions *B* and *D*. Of these, transition *D* with a lithosphere thickness of 50 km best fits the observed data. Thus, it seems possible to explain the island arc anomalies without resort to the assumed tapered transition zone adduced by Talwani et al. (1961).

Kanamori and Press (1970) conclude from seismic evidence that the thickness of the lithosphere is close to 70 km, and that the bottom of the lithosphere is probably at the solidus, because of the sharp decrease in rigidity observed. If the lithosphere differs from the underlying upper mantle material only in being cooler from surface conduction, as presumed here, then the thickness of the lithosphere should increase as it moves from the ridges to the trench. The magnitude of this effect will bear investigation, both from seismic observations and from theory of mantle currents. As it stands, the finding of Kanamori and Press is consonant with the model presented here.

Strain Heating in the Slab

Strain heating can at most raise the boundary temperature to the melting point. Because the geotherm is so near the melting point at depth in my model, this has little effect on the temperature in the interior of the slab. A sample calculation showed that the central temperature in the slab is only raised from 1410 to 1433°C (0.668 to 0.677 β) when the boundary temperature was assumed to be at melting throughout (D = 50 km, ν = 8 cm/yr; transition B, L = 800 km). This is consistent with the strain heating calculated by McKenzie, (1969), but far less than that assumed by Minear and Toksöz (1970).

Since heating by boundary drag is unimportant in our model, let us estimate

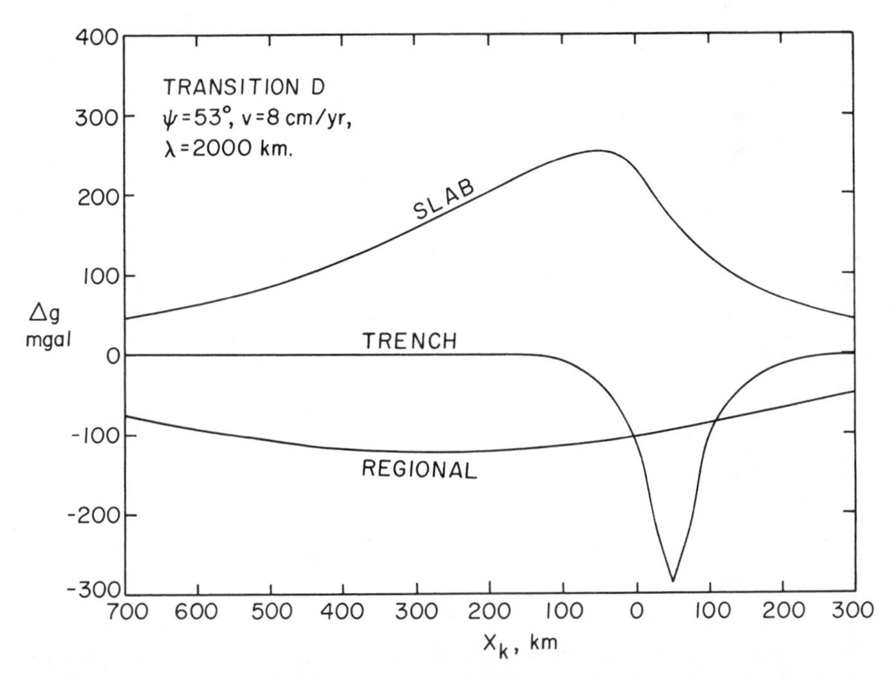

Fig. 9. Gravity anomaly components attributable to slab, symmetric trench, and regional compensation. Surface distance X_K measured normal to trench axis from the point at which the projection of the upper surface of the slab intersects the surface (X_K = 0).

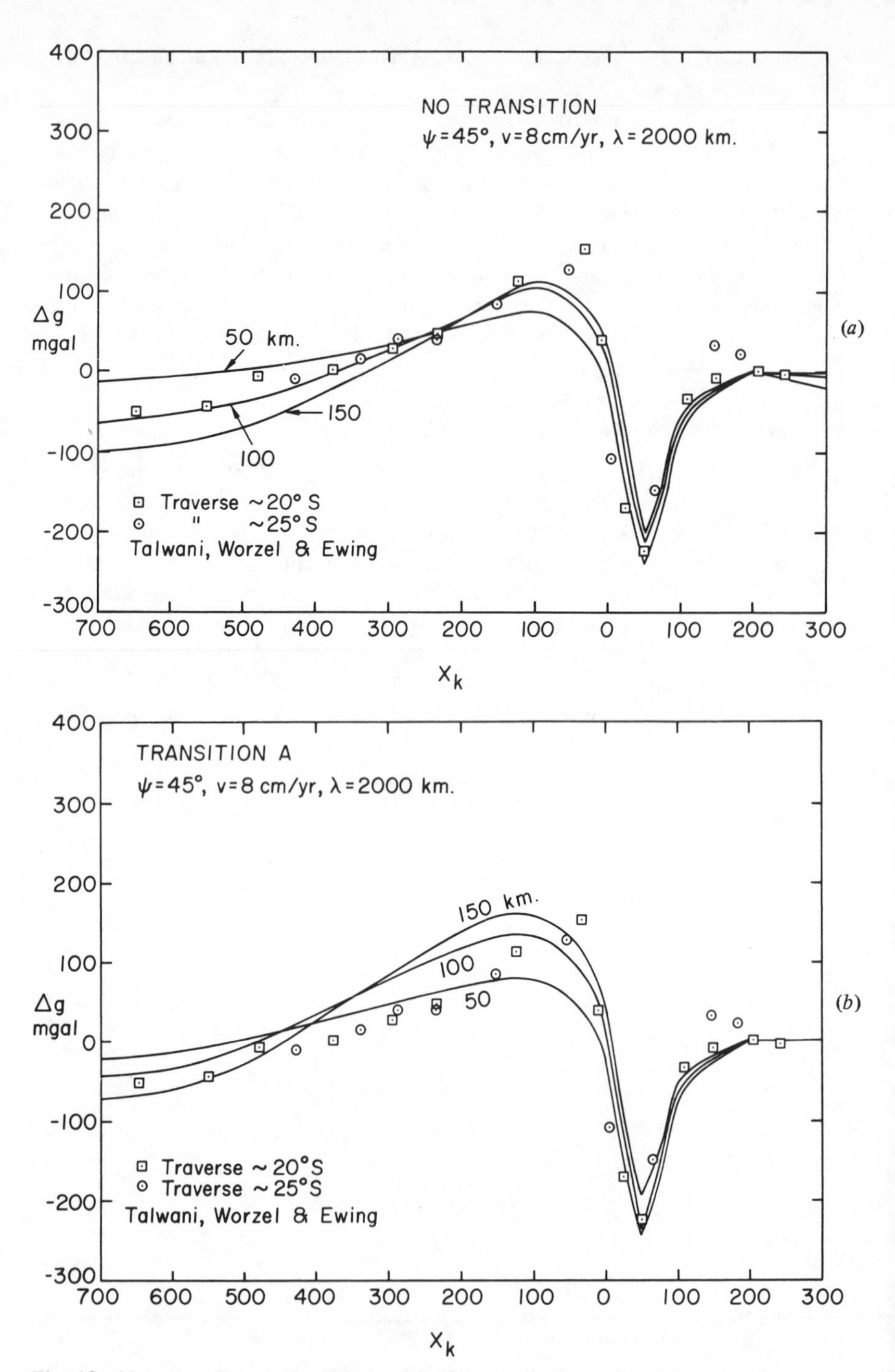

Fig. 10. Net anomalies attributable to slab plus trench plus regional compensation compared with observed gravity anomalies (Talwani et al, 1961). Effects of transitions, lithosphere thickness, and dip angle are illustrated.

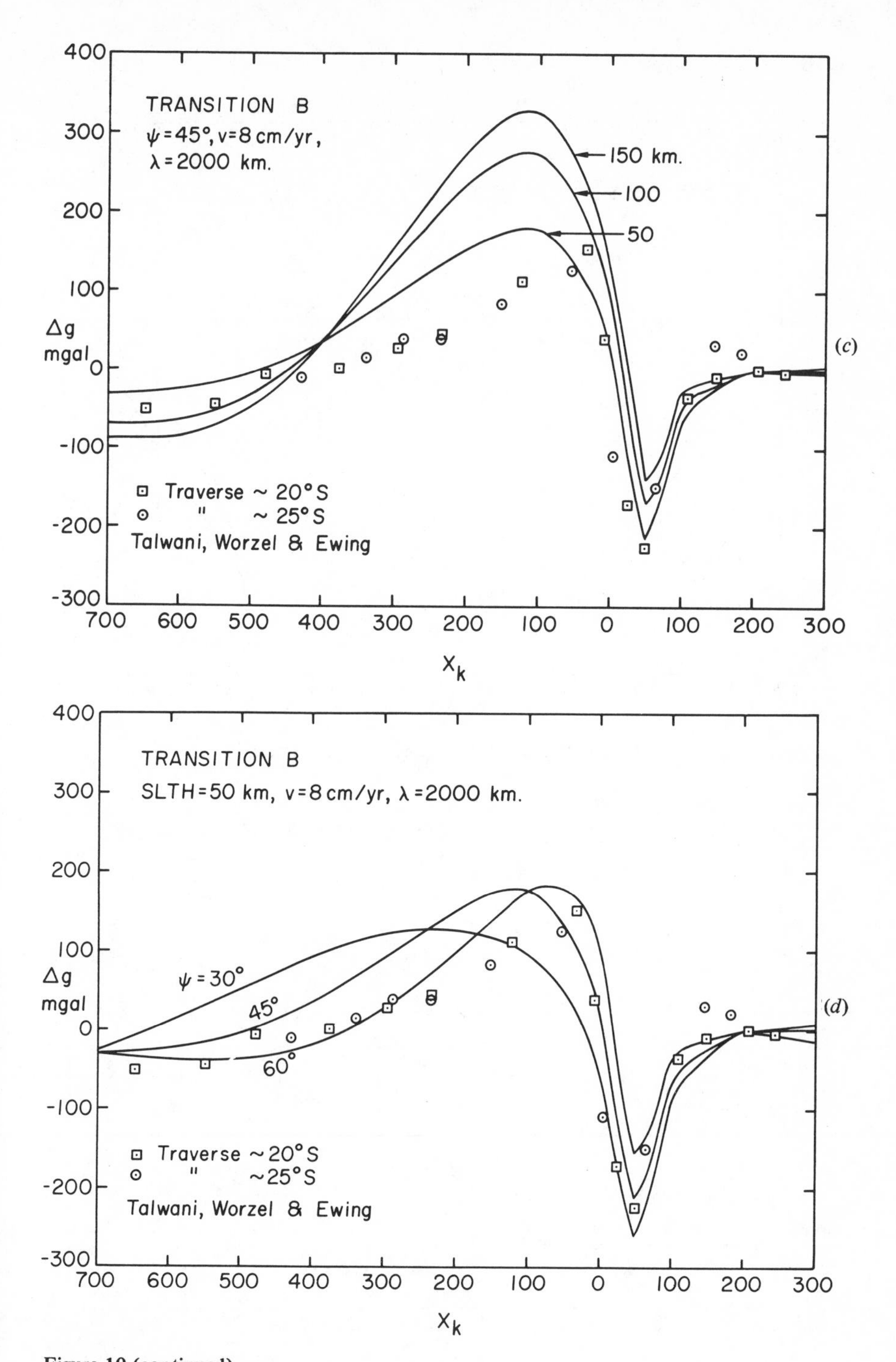

Figure 10 (continued).

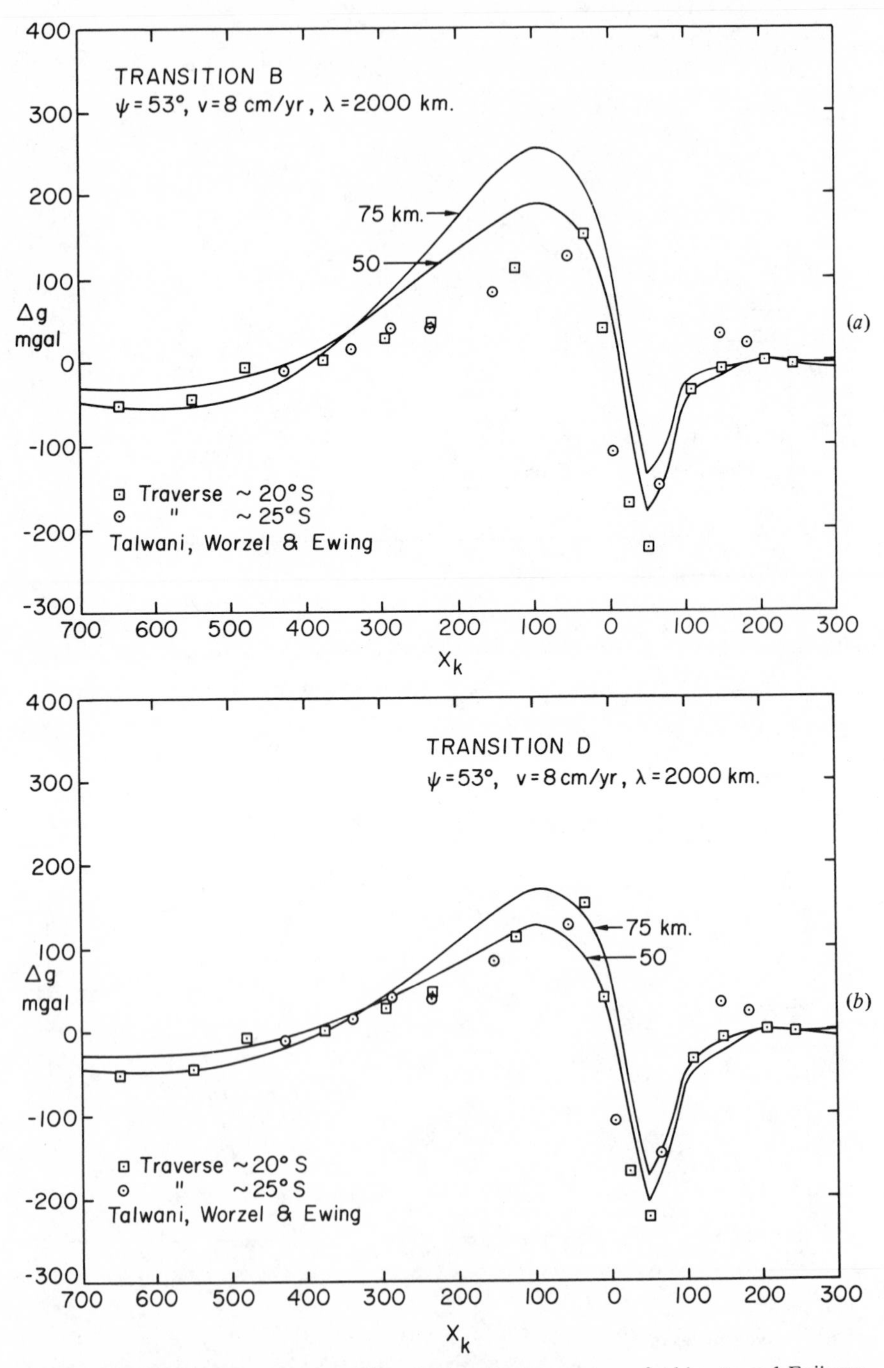

Fig. 11. Net anomalies with transitions having the parameters of Akimoto and Fujisawa (1968) and dip angle of Sykes (1966) compared with observed anomalies (Talwani et al, 1961). (*a*) Mantle 100% olivine composition. (*b*) Mantle 50% olivine.

the maximum strain heating that could be produced *within* the slab. The flow law applicable to the interior of the slab must be of the Weertman form:

$$\dot{\gamma} = a\tau^n \exp\left(-\frac{E}{RT}\right) = a'\tau^n \exp\left(-\frac{G}{\beta}\right) \tag{7}$$

where $\dot{\gamma}$ is strain rate, τ shear stress, T temperature, β is T/T_{mp}, and α, n, E, R, G are constants.

The integration of this over the length of the slab proved very difficult, and an upper limit estimate is made by assuming that all the work done by the sinking lithosphere is converted to internal heat in the slab. The work done is σDL where σ is the extensile stress described above and L is the length of the slab. The maximum slab heating is:

$$\Delta T = \frac{\sigma DL}{DL\rho c_p} = 23.9\sigma\,(^\circ C)$$

For a typical case (D = 50 km, ν = 8 cm/yr $\psi = 53^\circ$, Transition D), σ = 9.14 kb, $\Delta T = 218^\circ$C. This is to be compared with the total heating of the slab by conduction: $\Delta T' = 1466^\circ$C. Hence the maximum strain heating of the slab is only 15% of the conduction heating.

This is a substantial overestimate of the actual strain heating, since much of the work must go to pulling the lithosphere horizontally ($\sigma_l \sim$ 1-3 kb) some goes into the flow field (σ_f from McKenzie $\sim$ 2 kb) and much of the remainder will be lost by conduction. It is concluded that strain heating is not of great importance in this model of lithosphere sinking.

It may be noticed that the boundary shear stress of 2.5 kb assumed by Minear and Toksöz (1970, Fig. 11, ν = 8 cm/yr, D = 160 km) is 25 times McKenzie's calculated average shear stress and 5 times the maximum value available from my calculation above if all the sinking energy went into heating the slab.

EARTHQUAKE MECHANISMS

If deep earthquakes were of the shear fracture type, as seems indicated by all studies of the focal mechanisms, I expected that they would reflect the shearing flow associated with the mantle currents (the shear surfaces would be predominantly coincident with the upper surfaces of the sinking slab). Discovery by IOS, Isacks, Sykes, and Oliver (1969), Isacks and Molnar (1969), and Fitch and Molnar (1970) that focal mechanisms involve predominantly an axis of maximum compressive stress or occasionally maximum extensile stress down-dip in the sinking slabs requires some other mechanism. Elsasser's (1967) conception of the lithosphere as a stress guide has been appealed to by most of these authors. It is suggested by them that the slab hits a more resistant zone at depth, causing down-dip compression at the bottom, which may grade upward into extension if the resistance is sufficiently small. The extension is sometimes believed sufficient to pull apart the lithosphere, accounting for seismic gaps in some Benioff zones.

Raleigh and Kirby (1970) advance persuasive arguments that Weertman power law flow characterizes behavior of the lithosphere. Hot creep experiments on dunite by Post and Blacic (unpublished data) in our laboratory yield a preliminary flow law as follows:

$$\dot{\epsilon} = 13.2\sigma^{5.9} \exp\left(-\frac{66700}{RT}\right)$$

where $\dot{\epsilon}$ and σ are compressive strain rate in $\sec^{-1}$ and stress in kb, respectively. Converting to shear strain and shear stress and noting the Weertman observation (1970) that the effects of pressure may be included by expressing temperature in units of β (Raleigh and Kirby, 1970); we obtain:

$$\dot{\gamma} = 788\tau^{5.9} \exp\left(-\frac{66700}{RT}\right)$$

$$\dot{\gamma} = 788\tau^{5.9} \exp\left(-\frac{24.5}{\beta}\right) \tag{8}$$

where $\dot{\gamma}$ is shear strain rate in $\sec^{-1}$ and τ is shear stress in kb.

This flow law allows us to calculate the shear stress at any point in the cooler interior of the slab corresponding to a chosen shear strain rate. It does not apply to the mantle below about 100 km nor to the hot outer regions of the slab where Nabarro-Herring flow is to be expected. A strain rate of 10^{-14} $\sec^{-1}$ seems appropriate for an illustrative calculation. At this strain rate, the slab could, for instance, be pulled apart in about 3 m.y. Figure 12 shows stress contours in a typical slab calculation with this strain rate and the flow law of Eq. (8). These represent the effective strength distribution within the slab for this strain rate.

It is seen that, viewed as a stress guide, the sinking lithosphere resembles a tongue getting thinner and weaker as it dips into the mantle. We now face the question: how can such a thin tongue if subject to resistance in the nether regions yield so as to exhibit shear fractures of predominantly horizontal strike and dipping at angles of about 45° to the slab trend? Further, how can such fractures occur over most of its length from near surface to the limiting depth of earthquakes?

One might expect buckling in such a thin tongue subject to resistance near its tip. This could cause short wavelength folding, with maximum shear on the limbs of the folds. This is one farfetched solution to the enigma of the observed focal mechanisms. If this were the case, one would expect to see evidence for these folds in earthquake distributions and in the distribution of focal mechanisms. There is no trace of it in Sykes (1966) earthquake distributions in

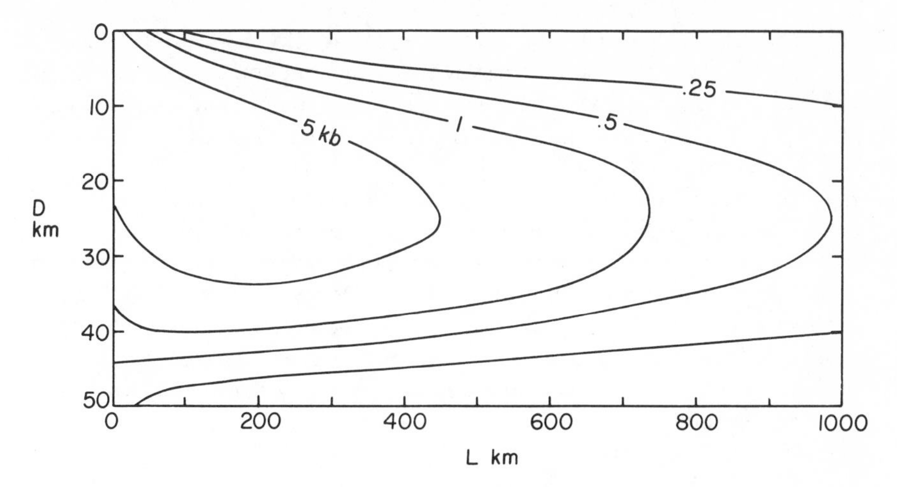

Fig. 12. "Strength" distribution within sinking lithosphere. Note exaggerated transverse scale (D). Values plotted are stress (calculated from preliminary laboratory flow law from hot creep of dunite at a strain rate of 10^{-14}/sec and appropriate values of β) calculated throughout the sinking slab. Transition D, 50-km slab, ψ = 53°, ν = 8 cm/yr. Note progressive weakening and/or mechanical thinning of slab as it sinks from surface (left) to depth of deepest earthquakes ($L \simeq 800$ km).

the Tonga-Kermadec region, though the apparent thickening toward the bottom might suggest some such mechanism. As yet, good focal mechanism solutions are too sparse to check this hypothesis.

Another problem is that the effective length of the strong "tongue" is roughly proportional to the velocity of sinking. "Resistance" in the mantle must occur at roughly equal depth everywhere, so that such resistance should be felt much more by long, high-velocity tongues than by short slow tongues. There seems to be no evidence of such an effect.

Other related problems are associated with the finding of extension axes down-dip in the slabs at odd places and the apparent separation in zones of no earthquakes. Clearly, the slab can be subject to extension whenever the resistance to downward motion is sufficiently small. The slab can readily be pulled apart by gradual necking with increased heating.

With an assumed flow law, the total mechanical behavior of the slab can, in principle, be numerically calculated for any set of assumed boundary conditions.

I believe that shear instabilities already known in the laboratory may suffice to explain deep earthquake phenomena once the puzzling geometry of the sources is understood. The slab center is sufficiently cool at the depths of the deepest earthquakes so that dehydration embrittlement and fracture (Raleigh and Paterson, 1965) may occur, so this is one possible mechanism.

Another type of mechanism is suggested by our hot creep experiments. In the extensive work on dunite deformation done in our laboratory first by Blacic and now by Post (unpublished data), shear instabilities akin to hot creep fracture are common. Characteristically, these seem to develop from zones of distributed shear that progressively become narrower. At the same time, the creep rate (at constant stress) rises, so that the local strain rate is increasing roughly exponentially with time, until there is a sudden stress drop. Our understanding of the phenomena involved is at this stage inadequate to know whether these hot-creep instabilities have the characteristics requisite for deep focus earthquakes, but this phenomenon appears to be a promising candidate.

One wonders what the role of transitions is in deep earthquakes. We have done hundreds of deformation experiments in which we transgressed the stability fields of two reconstructive transitions (quartz-coesite and calcite-aragonite) and have noted no mechanical effects equivalent to shear fracture or any other indication of sudden stress drop resulting from the transitions. We cannot, however, exclude the possibility that some such phenomenon exists and may play a role in deep earthquakes.

CONCLUSIONS

The thermal structure in the sinking lithosphere according to my calculations is similar to that of McKenzie and unlike Minear and Toksöz' calculations (1970) with strain energy added. I believe Minear and Toksöz have greatly overestimated the magnitude of strain heating in the boundary layer of the sinking slab. The effects of adiabatic heating, radioactive heating, latent heat of transitions, and radiative conductivity are second order compared to the effects resulting from variations in the assumed thickness of the lithosphere and the rate of descent. (This is true in both Minear and Toksöz' calculations (1970) and in mine.)

The hypothesis is advanced that the limiting depth of earthquakes is that depth at which the minimum temperature in the slab exceeds a critical value of $\beta(T/T_{mp})$. This agrees with the data on maximum depth of earthquakes within the rather wide scatter of the observations. This hypothesis seems physically more reasonable than McKenzie's assumption of a critical temperature.

The loci of mantle transitions within the sinking lithosphere is sensitive to the dT/dP of the transition. An olivine-spinel transition with the Akimoto and Fujisawa slope (1968) would rise within a rapidly descending lithosphere to a level above the ambient low-velocity zone. Since the density change is large (10%), the actual location of the transition within the sinking slab should be determinable by appropriate studies of body wave transmission from deep earthquakes if an adequate network of local seismometers can be established. This would provide important additional information as to the nature of mantle transitions above 600 to 700 km depth.

The gravity anomaly comprised only of that calculated for the slab, the contribution of a *symmetric* trench and plausible regional compensation, is remarkably close to the observed anomaly across the Tonga trench. If one could solve the problem of determining the nature of regional compensation, then using the known bathymetry and crustal structure, it should be possible to reduce greatly the uncertainty as to lithosphere thickness and the loci of transitions within the sinking slab. It might even be possible to derive outer mantle temperatures with some precision. I favor McKenzie's idea of counter-flow on the landward side of the trench, which would aid in explaining the high heat flow in this region and would also explain the net east-west gravity assymmetry at Tonga, when the known crust and bathymetry are included with the compensated slab anomaly.

The thickness of the lithosphere cannot be uniquely determined from present calculations. It can vary from perhaps 50 to 100 km depending on the nature of the actual transitions in the mantle.

Deep focus earthquakes can be explained by either or both of two mechanisms now known in the laboratory: (a) the dehydration-embrittlement and weakening of Raleigh and Paterson (1965) and (b) hot creep instability as found by Blacic and Post.

The geometry required by recent focal mechanism solutions for deep earthquakes (down-dip compression) presents problems. It is shown that the stress-guide concept has difficulties because strengthwise the sinking slab is a long thin tongue, thinning and weakening from the surface down. Further, if the compression were a result of resistance at some depth, there should be a highly nonlinear relationship between maxim depth of earthquakes and sinking velocity rather than the linear relationship that is observed.

The role of transitions, if any, in deep earthquakes is unknown.

The energy released by the sinking lithosphere is of the order of 10% of the energy given off by the total surface heat flow of the earth, according to my calculations. Hence, this must be an important part of the driving potential for worldwide mantle currents and plate tectonics. It seems paradoxical, however, that focal mechanism solutions of earthquakes in these sinking slabs indicate predominantly down-dip compression. If the sinking slabs were driving the mantle and plate motions, I would expect down-dip extension to predominate.

ACKNOWLEDGMENTS

R. W. Post, T. Tullis, I. C. Getting, L. Knopoff, W. M. Kaula, and L. B. Slichter provided valuable discussions. D. W. Baker and E. Nyland helped with program and computer bugs. This work was supported in part by NSF Grant GA-1394.

REFERENCES

Akimoto, S., and Fujisawa, H., Olivine-spinel solid solution equilibria in the system Mg_2SiO_4-Fe_2SiO_4, J. Geophys. Res., **73**, 1467-1479, 1968.

Anderson, D. L., Phase changes in the upper mantle, Science, **157,** 1165-1173, 1967.

Birch, F., Elasticity and constitution of the earth's interior, J. Geophys. Res., **57**, 227-286, 1952.

Carslaw, H. S., and Jaeger, J. C., Conduction of Heat in Solids, Oxford, 1959.

Clark, S. P. (Ed.), Handbook of Physical Constants, Memoir 97, Geol. Soc. Am., pp. 587, 1966.

Elsasser, W. M., Convection and stress propagation in the upper mantle, Tech. Rept. 5 for NSG-556, Princeton, 1967.

Fitch, T. J., and Molnar, P., Focal mechanisms along inclined earthquake zones in the Indonesia-Philippines region, J. Geophys. Res., **75**, 1431-1444, 1970.

Griggs, D. T., Invited lectures (no abstracts), NAS fall meeting 1968, and Am. Geophys. Union, 1969.

Isacks, B., and Monar, P., Mantle earthquakes mechanisms and the sinking of the lithosphere, Nature, **223**, 1121-1124, 1969.

Isacks, B., Oliver, J., and Sykes, L. R., Seismology and the new global tectonics, J. Geophys. Res., **73**, 5855-5899, 1968.

Isacks, B., Sykes, L. R., and Oliver, J., Focal mechanisms of deep and shallow earthquakes in the Tonga-Kermadec region and the tectonics of island arcs, Bull. Geol. Soc. Am., **80**, 1443-1470, 1969.

Ito, K., and Kennedy, G. C., Melting and phase relations in a natural peridotite to 40 kilobars, Am. J. Sci., **265**, 519-538, 1967.

Jeffreys, H., The Earth, 2d. ed., Cambridge University Press, London, 1929.

Kanamori, H., and Press, F., How Thick is the Lithosphere? Nature, **226**, 330-331, 1970.

Kushiro, I., Syono, Y., and Akimoto, S., Melting of a peridotite nodule at high pressures, and high water pressures, J. Geophys. Res., **73**, 6023-6030, 1968.

Lachenbruch, A. H., Preliminary geothermal model of the Sierra Nevada, J. Geophys. Res., **73**, 6977-6990, 1968.

LePichon, X., Sea-floor spreading and continental drift, J. Geophys. Res., **73**, 3661-3697, 1968.

MacDonald, G. J. F., The deep structure of continents: Reviews of Geophysics, **1**, 587-665, 1963.

McConnell, R. K., Jr., Viscosity of the mantle from relaxation time spectra of isostatic adjustment, J. Geophys. Res., **73**, 7089-7105, 1968.

McKenzie, D. P., Some remarks on heat flow and gravity anomalies, J. Geophys. Res., **72**, 6261-6274, 1967.

______, Speculations on the consequences and causes of plate motions, Geophys. J., **18**, 1-32, 1969.

Minear, J. W., and Toksöz, M. N., Thermal regime of a downgoing slab and new global tectonics, J. Geophys. Res., **75**, 1397-1420, 1970.

Oliver, J., and Isacks, B., Deep earthquake zones, anaomalous structure in the upper mantle, and the lithosphere, J. Geophys. Res., **72**, 4259-4275, 1967.

Raleigh, C. B., Kirby, S. H., Strength of the upper mantle, *in* The minerology and petrology of the upper mantle, edited by B. A. Morgan, Mineral. Soc. Am. Special Paper 3, pp. 113-121, 1970.

Raleigh, C. B., and Paterson, M. S., Experimental deformation of serpentinite and its tectonic implications. J. Geophys. Res., **70**, 3965-3985, 1965.

Sykes, L. R., The seismicity and deep structure of island arcs, J. Geophys. Res., **71**, 2981-3006, 1966.

Talwani, M., Worzel, J. L., and Ewing, M., Gravity anomalies and crustal section across the Tonga Trench, J. Geophys. Res., **66**, 1265-1278, 1961.

Weertman, J., The creep strength of the earth's mantle, Rev. Geophys. and Space Physics, **8**, 145-168, 1970.

15

GLOBAL GRAVITY AND TECTONICS*

by **William M. Kaula**

The improved resolution of the new solution for the earth's gravity field by Gaposhkin and Lambeck (1971) results in a significant change in our knowledge of the relationship of the gravity to tectonics in the southern oceans. Large positive anomaly areas are now located along the ocean rises, the sole exception in the oceans being Hawaii. This correlation of positive anomalies with rises is enhanced for representations that are isostatic and residual to a fifth-degree figure.

The trench and island arc belts are also predominantly positive. The paradox of both tension and compression features being areas of mass excess near the surface may be the consequence of the lithosphere acting primarily as a free boundary to vertical stresses in the former case and a fixed boundary in the latter, as well as both being stagnation points in mantle flow.

The commonest negative anomaly features are the ocean basins, always located to the flanks of the rises. The basins may be regions of maximum horizontal acceleration (Lagrangian) in the asthenospheric flow, or there may be a downflow of a denser component from the flow between the rise to the basin.

Correlation of negative anomalies on land with glaciated areas is marked, particularly isostatic anomalies residual to the fifth degree. The

*Publication No. 835, Institute of Geophysics and Planetary Physics, University of California, Los Angeles, California.

Antarctic negative is much too large to account for by glacial melting; it appears to require a deficiency of the asthenospheric flow associated with the outward migration of ocean rises. The Himalayas-Turkestan negative seems well explained by great thicknesses of low-density crust carried by stiff lithosphere, which to some extent displaces asthenospheric material.

INTRODUCTION

This paper is a continuation of Kaula (1969), which attempted a tectonic classification of the main features of the earth's gravitational field. On the basis of magnitude and extent of mean gravity anomalies for 5° squares, 19 areas on the earth were selected as markedly positive, 14 as markedly negative, 10 as exceptionally mild. Other geological and geophysical data for each of these 43 areas were examined. On the basis of certain patterns of correlation, 11 types of areas were defined. It was found that, where characteristics of different types appeared, certain characteristics were dominant over others. In general, characteristics associated with positive anomalies were dominant over those associated with mild or negative anomalies, and characteristics associated with recent tectonics dominant over those associated with ancient. The 11 types in order of dominance, with sign and a leading example of each given in parentheses, were: trench and island arc (+, Indonesia-Philippines), Cenozoic oceanic flood basalts (+, Iceland-North Atlantic), Cenozoic orogeny with Quaternary extrusives (+, Caucasus), Quaternary glaciation (–, Canadian Shield), vigorous ocean rise (0, southeast Pacific), current orogeny without extrusives (–, Himalayas), ocean basin (–, Somali-Arabian), continental basin (–, Parnaiba Basin), pre-Cenozoic orogeny (0, eastern United States), continental shield (0, Brazilian Shield), pre-Cenozoic ocean flood basalts (0, Darwin Rise). The strongest correlation found was between positive gravity anomalies and Quaternary volcanism. Positive correlation of gravity anomalies with topography residual to a fifth-degree figure was almost universal. The extent to which the different area types relate to the global tectonics inferred from paleomagnetic and seismic data varied from strong (trench and island arc, vigorous ocean rise) to negligible (Cenozoic oceanic flood basalts, Quaternary glaciation). The lack of systematic correlation between temperature indicators (heat flow, P_n velocities, seismic station delays) and gravity anomalies indicated that horizontal variations in petrology are significant.

Since Kaula (1969), there has been a major improvement in the determination of the global gravity field by Gaposhkin and Lambeck (1971). In this paper, we first examine this improved determination, and attempt to transform it so as to be most useful for geophysical interpretation. Then we hypothesize mechanisms by which the main features of the gravitational field are maintained.

THE DATA

Figure 1 of this paper differs from Fig. 1 of Kaula (1969) in four significant respects:

1. The new determination of the gravity field by Gaposhkin and Lambeck (1971) is used. This analysis is based primarily on the orbits of 21 artificial satellites and secondarily on mean gravity anomalies for 5° by 5° squares covering 56% of the earth.
2. The gravimetry is the same as that used by Kaula (1966), but the manner of combination of data in effect gives higher weight to the satellites than in the 1966 analysis.
3. The results are given in the form of spherical harmonic coefficients of the potential, complete through the 16th degree, (plus 33 coefficients of higher degree to which satellite orbits are sensitive), rather than area means. Hence, the resolution, or shortest half wavelength represented, is about 11° or 1200 km.
4. The free air anomalies in Fig. 1 are referred to the figure of hydrostatic equilibrium, an ellipsoid of flattening 1/299.8, in accord with the explanation of Goldreich and Toomre (1969) for the excess oblateness.

There are two major effects of these changes:

1. The improved resolution results in the breakup of the two largest features in the southern oceans. The large area of mild anomaly in the South Pacific is now resolved into two negative areas with a positive area between, the former over basins and the latter along the East Pacific Rise. In the area between Africa and Antarctica, a single large positive feature centered in the "vee" between the two rises is now divided into two positive features over the rises and an area of mild anomaly between. In general, most of the ocean rises are now positives, rather than "mild" features.
2. The use of the hydrostatic flattening results in the intensification of the negative anomalies in the glaciated areas near the poles: at the South Pole to an extent that is much greater than can be imputed to glacial loading.

Lesser effects are the appearance of the highest Himalayas as a small positive belt; the reduction of an overlap of the southeast Indian Ocean rise by the south Australian basin negative anomaly; the emphasis of the positive belt from the Carpathians to Iran; the reduction of the East Mediterranean negative; and the reduction or removal of positive features in areas of slight recent tectonic activity in northeast USSR, the Central Pacific, and Australia.

Figure 2 is the corresponding isostatic anomaly map, using the spherical harmonic expansion of the Airy-Heiskanen 30-km crust isostatic correction calculated by Uotila (1962). As usual, oceanic maxima and continental minima are enhanced in the isostatic map, but significant change in the pattern would occur only if the compensation were placed at, or below, asthenospheric depths.

As previously pointed out (Kaula, 1967), the correlation of gravity with topography is poor for the fifth and lower degrees. On the other hand, Hide and Malin (1970) have recently shown that the low degree harmonics of the gravity

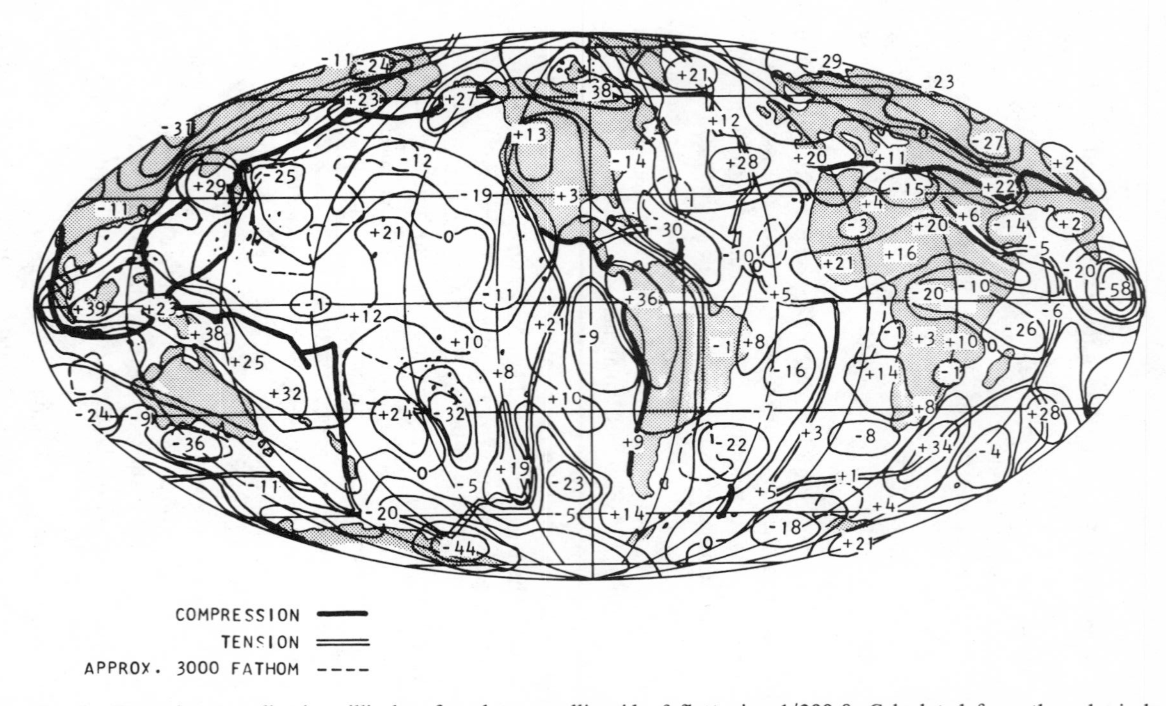

Fig. 1. Free air anomalies in milligals referred to an ellipsoid of flattening 1/299.8. Calculated from the spherical harmonic coefficients of the gravitational field of degrees 2 through 16 of Gaposhkin and Lambeck (1971). (Non-zero contours enclosing only one value have been omitted on all figures.) Global tectonic lines of compression and tension from Isacks et al. (1968), and major basins indicated by approximate 3000-fathom line on all figures.

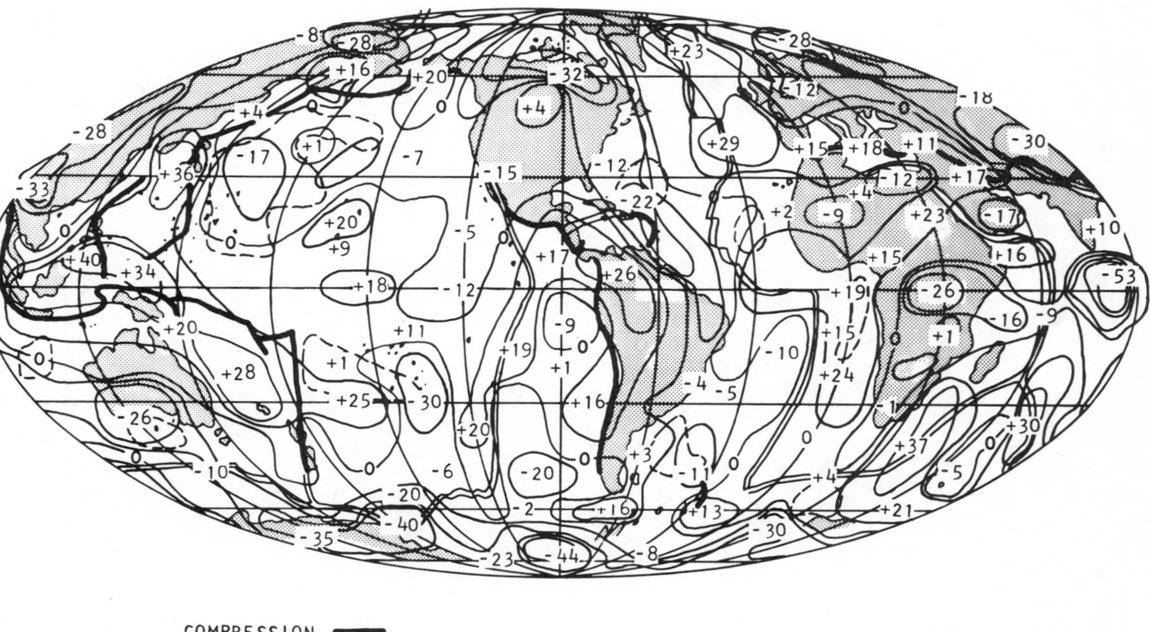

Fig. 2. Isostatic anomalies in milligals referred to an ellipsoid of flattening 1/299.8. Airy-Heiskanen compensation with nominal crustal thickness of 30 km. Calculated from Fig. 1, less the spherical harmonic coefficients for the isostatic correction of degrees 2 through 16 of Uotila (1962).

field have a high correlation with the corresponding harmonics of the magnetic field, provided that the latter is rotated 160° eastward. The obvious application of these facts for the purpose of interpreting upper mantle and crustal phenomena is to use a residual field. Figure 3 is the free air anomaly field calculated from spherical harmonic coefficients of degrees 6 through 16, and Fig. 4 is the corresponding isostatic anomaly field. In the four successive representations of Figs. 1 through 4, the correlation of ocean rises with positive anomalies appears more and more emphasized.

As discussed in Kaula (1969), it seems appropriate to analyze the gravity field in terms of reasonably contiguous blocks of anomaly x area, since, by the half-space application of Gauss's theorem, this quantity is directly proportional to excess mass, which in turn is a primary measure of the stresses required. Table 1 gives the 30 largest blocks in terms of free air anomalies referred to the hydrostatic figure, while Table 2 gives the 25 largest blocks in terms of isostatic anomalies referred to the fifth-degree figure. Of the 12 question marks in Table 4 of Kaula (1969), about 10 seem to be resolved. The greatest question remaining is the great negative over Antarctica; it is too large by more than a factor of 3 to be attributable to the loss of ice in recent geologic time (O'Connell, 1971).

The types given in Tables 1 and 2 are those used in Kaula (1969), with some obvious modifications.

INTERPRETATION

The principal inference from the large areas of postglacial uplift is, of course, the existence of an asthenosphere: a relatively plastic layer in the upper mantle, 80 to 400 km or more deep. Of the seven or eight major feature types, the glaciated areas are alone in being transient, with a decay time on the order of a few thousand years (O'Connell, 1971).

Since the lithosphere is not capable of supporting elastically the necessary stresses for features thousands of kilometers in extent (McKenzie, 1967), the other broad departures in the earth from equilibrium must entail flow in the asthenosphere. The asthenosphere is stiff enough, however, and the thermal conductivity of the earth is poor enough (the Prandtl number is large), that it is generally agreed that the flow is essentially *steady state* (Turcotte and Oxburgh, 1967, 1969). As indicated by magnetic reversal patterns, the present pattern of tectonic motion has persisted for about 10 m.y. (Heirtzler et al., 1968).

In a steady-state flow system, to maintain a mass excess in a particular region, there must be effectively a Lagrangian deceleration of matter entering the region and an acceleration of matter leaving it; the converse must apply to a region of mass deficiency. Mathematically, this condition requires that for the volume containing the mass excess, the surface integral

$$-\int_{\text{surface}} \rho \frac{\partial \mathbf{v}}{\partial t} \cdot d\mathbf{n} = \int_{\text{surface}} \rho \mathbf{v} \cdot \nabla \mathbf{v} \cdot d\mathbf{n} > 0 \tag{1}$$

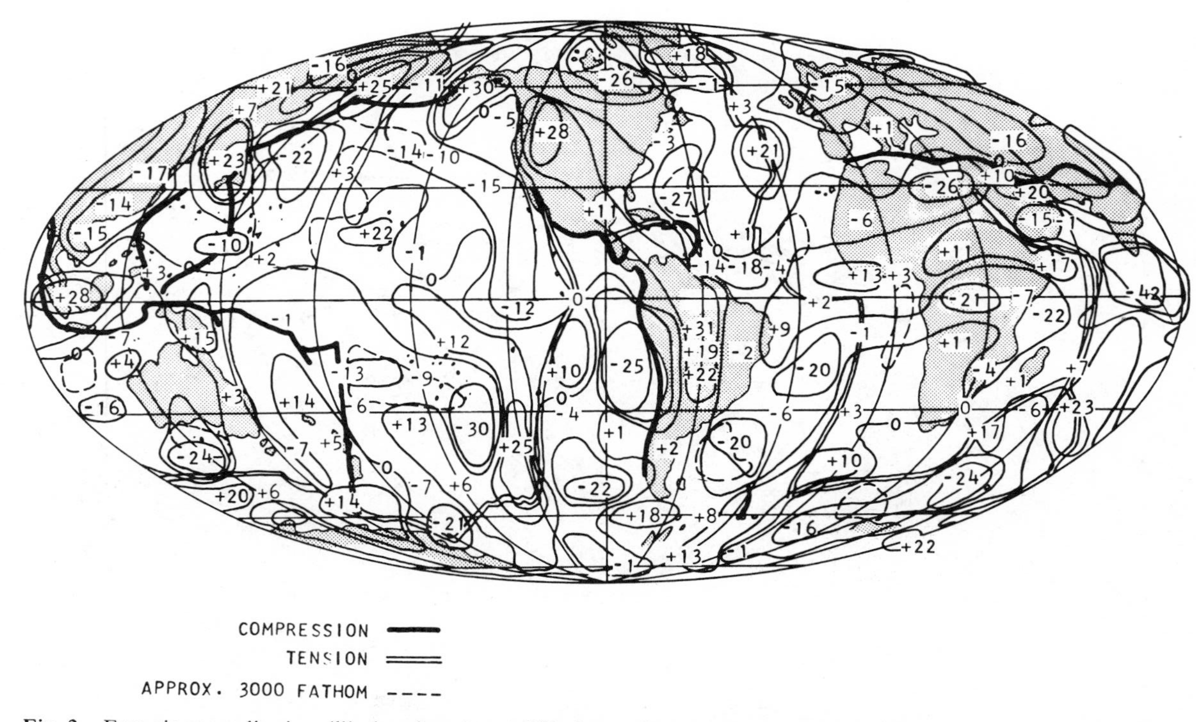

Fig. 3. Free air anomalies in milligals referred to a fifth-degree figure. Calculated from the spherical harmonic coefficients of the gravitational field of degrees 6 through 16 of Gaposhkin and Lambeck (1971).

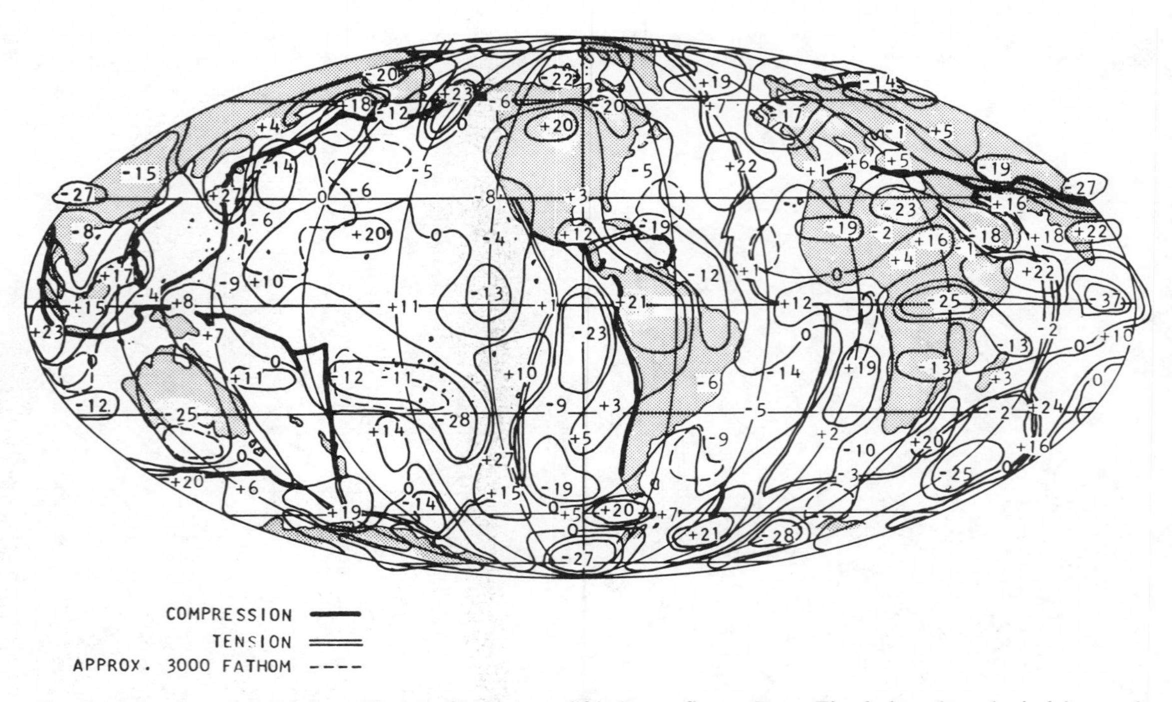

Fig. 4. Isostatic anomalies in milligals referred to a fifth-degree figure. From Fig. 3, less the spherical harmonic coefficients for the isostatic correction of degrees 6 through 16 of Uotila (1962).

where ρ is density, $\mathbf{v}$ is velocity, and $\mathbf{n}$ is the outward drawn unit vector normal to the surface. The uppermost boundary of this volume for purposes of gravity anomaly interpretation should be a segment of a bounding equipotential, such as the geoid. Since the compressibility of upper mantle material is slight, these "decelerations" must be accomplished by: (1) the piling up of material at the surface; (2) the replacement of less dense by more dense material at an interior interface; (3) thermal contraction; (4) transition to a denser phase; or (5) petrological fractionation in which a less dense component is left behind. The reverse of one or more of these processes is needed to accomplish an "acceleration." It is to be emphasized that mechanisms such as (3), (4), and (5) do not affect the gravity field directly by increasing the density, but rather by inducing mass transfers in accord with Eq. (1).

If the asthenosphere is a relatively thin layer, then the obvious direction to transfer matter so as to affect the external gravitational field is lateral. Vertical transfers, however, are not to be ruled out: an upward displacement of material making the density higher than the average at a shallow level, balanced by a mass deficiency at considerable depth (below about 200 km) could account for a gravity excess. From the formula for the potential arising from a spherical harmonic surface distribution of mass (Kaula, 1968, p. 67), we have for a mass excess of $\Delta\rho h$ of width L compensated at depth D:

$$\Delta g \approx 2\pi^2 G \frac{D}{L} \Delta\rho h \tag{2}$$

Then to say the data are satisfied by isostatic compensation at great depth, however, is to beg the question as to the response of the asthenosphere to the stresses that must necessarily exist at the intervening levels.

The greater effectiveness of lateral transfer also suggests that stagnation points (regions where the flow changes direction, such as ocean rises or trench and island arcs) will tend to be regions of gravity excess, while regions dominated by horizontal flow will tend to be regions of gravity deficiency.

The relationship of gravity anomalies to the flow system depends considerably on the boundary conditions. In a system of thermal convection, if the boundary is rigid, then an upcurrent is associated with a negative anomaly, because of its lower density (Runcorn, 1965). If the upper boundary is free, however, while the lower boundary remains fixed, then an upcurrent is associated with a positive anomaly because the effect of the mass pushed up at the surface outweighs the density effect (McKenzie, 1968, Pekeris, 1935).

In the case of the real earth, the question becomes to what extent the lithosphere (the layer of relative strength) and the crust (the lower density, uppermost layer of the lithosphere) act as a part of the convective flow, and to what extent they act as a restraining boundary to the flow. Manifestly, they act both roles to differing degrees in different parts of the earth. The lithosphere can even be simultaneously a rigid boundary for horizontal forces in being able to

Table 1. **Areas of exceptional gravity anomaly, defined as having an area x *free air* anomaly referred to the *hydrostatic* figure more than 50 mgal x 10^6 km^2 (1.17 x 10^{21} gm) in absolute magnitude, and absolute anomaly of more than 10 mgal throughout the area**

General location	Area (10^6 km^2)	Free air anomaly x area (mgal x 10^6 km^2)	Mean free air anomaly (mgal)	Type
Positive features				
1. Sumatra-Philippines-Solomons	18.9	+ 461.	+ 24.	Arc
2. Andes-W. Amazon Basin	9.4	206.	22.	Arc-Orogenic
3. Solomons-Tonga-Kermandec	10.4	178.	17.	Arc
4. Mid-Indian Rise-Indian Antarctic Rise	9.4	175.	19.	Rise
5. Crozet Plateau-S. Madagascar Rise	6.3	117.	19.	Rise
6. Mexico-N.W. Colombia	7.3	107.	15.	Arc
7. Carpathians-Turkey-Iran	7.2	103.	14.	Orogenic
8. Hawaii	6.1	96.	16.	Shield
9. Azores Plateau	5.8	95.	16.	Rise
10. Japan-Bonins	4.6	92.	20.	Arc
11. Atlas-Iberia-W. Mediterranean	5.5	86.	16.	Arc-Orogenic
12. West Africa-Guinea Basin	6.2	85.	14.	Orogenic
13. Ahaggar-Tibesti-Nigeria	4.7	77.	16.	Orogenic

14. Greenland-Iceland-Norwegian Sea	4.6	76.	16.	Rise
15. East Pacific Rise, N. of Easter	4.2	64.	15.	Rise
16. Walvis Rise-S.W. Africa	4.5	+ 58.	+ 13.	Rise
Negative features				
1. Antarctica	22.4	- 511.	- 23.	Glac.-Basin?
2. Siberian Platform-Turkestan-Himalayas	15.1	289.	19.	Glac.-Orogenic
3. North Canada	10.1	218.	22.	Glaciated
4. N. American-Guiana Basins	11.7	212.	18.	Basin
5. Somali Basin-Central Indian Ocean	6.2	193.	31.	Basin
6. N. Pacific Basin-N.E. Pacific Slope	12.3	175.	14.	Basin
7. W. Australian Shield-S. Australian Basin	4.6	121.	26.	Basin
8. N.W. Pacific Basin	5.3	116.	22.	Basin
9. Wharton Basin	5.6	98.	17.	Basin
10. N.W. Siberia-Aleutian Basin	5.6	87.	16.	Glac.-Basin
11. Society Islands-S.W. Pacific Basin	3.2	69.	22.	Basin
12. Congo-Kenya	5.0	60.	12.	Basin-Rift
13. Argentine Basin	3.6	54.	15.	Basin
14. Chile Rise-Pacific Antarctic Basin	2.8	- 50.	- 18.	Basin

Table 2. Areas of exceptional gravity anomaly, defined as having an area × *isostatic* anomaly referred to a *fifth-degree* figure more than 50 mgal × 10^6 km^2 (1.17 × 10^{21} gm) in absolute magnitude, and absolute anomaly of more than 10 mgal throughout the area

General location	Area (10^6 km^2)	Isostatic anomaly × area (mgal × 10^6 km^2)	Mean isostatic anomaly (mgal)	Type
Positive features				
1. Southeast Pacific Rise	6.6	+ 135.	+ 20.	Rise
2. Mid-Indian-Amsterdam-Naturaliste Ridge	6.1	100.	16.	Rise
3. Borneo-Sumatra-Cocos	5.3	98.	18.	Arc
4. Indian Peninsula-Bay of Bengal	4.7	82.	17.	Sediment?
5. S.E. Indian and MacQuarrie Rises	5.1	76.	15.	Rise
6. North Atlantic-Arctic Ocean	7.1	73.	10.	Glaciated(?) Rise
7. Azores Plateau	4.7	73.	16.	Rise
8. Indian.-Antarctic, Gaussberg Ridges	4.5	72.	16.	Rise
9. N. Andes-W. Amazon Basin	5.0	72.	14.	Arc-Orogenic
10. N.E. Georgia-S. Sandwiches-Mid-Atlantic	4.3	68.	16.	Rise-Arc

11. Japan-Bonins	4.0	66.	16.	Arc
12. Carlsberg Ridge-Gulf of Aden	3.7	65.	18.	Rise
13. South Alaska	3.6	65.	18.	Arc-Orogenic
14. Walvis Rise	4.5	+ 61.	+ 14.	Rise
Negative features				
1. Himalayas-China	8.0	- 141.	- 18.	Orogenic
2. Antarctica	8.1	140.	17.	Glaciated Basin?
3. Laccadives-Ceylon-Mid-Indian Ocean	5.0	130.	26.	Basin?
4. N. Canada-Greenland	7.5	123.	16.	Glaciated
5. Australian Shield-S. Australian Basin	7.7	113.	15.	Basin
6. N. Amer.-Guiana-Parnaiba Basins	7.3	96.	13.	Basin
7. Galapagos-Peru Basin	6.1	96.	16.	Basin
8. Society-Tuamotu-Austral Seamount	6.8	96.	14.	Basin?
9. Congo-Kenya	3.7	74.	20.	Basin-Rift
10. E. Crozet Basin-Kerguelen	2.6	53.	20.	Basin
11. N.W. Europe	3.5	- 51.	- 15.	Glaciated

act as a rigid plate in tectonic motions and a free boundary for vertical forces in not resisting convective upthrusts. The extent to which a particular portion of the lithosphere acts as a free or rigid boundary depends on its temperature, size of feature, rate of motion of material into and out of a feature, and composition, particularly its water content. The situation may be further complicated by steady surface transfers of matter: erosion and sedimentation.

How a boundary acts in the range between perfectly free and perfectly rigid depends on both (1) its elastic properties (its rigidity and thickness) and (2) its plastic properties (most simply expressed as a decay time in response to a transient loading, dependent on dimensions of the loading and stress as well as creep properties of the material). Under small stresses, the decay time of the lithosphere is very long: it is effectively acting as an elastic layer in areas of post glacial uplift. Under greater stress, however, such as in the major areas of mass excess, the effective decay time may be much shorter because of the nonlinear dependence of strain rate on stress (Weertman, 1970), as evidenced by the seismicity of these regions. Qualitatively, for both elastic and plastic behavior, we should expect that the thicker, the colder, the less hydrous the lithosphere is in a particular region, the more it will behave like a rigid boundary. Quantitatively, however, we should expect that in some cases it may be difficult even to infer the correct sign of the gravity anomaly.

The flow system for a body that has boundaries that are partly rigid, partly free, would be a difficult problem to treat rigorously. However, we might expect that usually the nature of the local boundary conditions would predominate in determining the characteristics of a particular region. We shall apply this assumption in the analysis of feature types.

Of the 11 gravity anomaly area types proposed in Kaula (1969), 6 appear to be associated with current internal activity in the earth. We shall discuss these 6 (somewhat modified) in an order suggested by their apparent relationship to the global tectonic pattern: (1) active ocean rises, (2) oceanic shield basalts still active in Quaternary, (3) basins, (4) trench and island arcs currently active, (5) current orogeny without extrusives, and (6) Cenozoic orogeny with extrusives in Quaternary.

Active Ocean Rises

The indication from the new data that these areas are generally of positive gravity anomaly is consistent with their being free boundaries over upcurrents in a convective system. Their well-known characteristics of high heat flow, shallow depth in the ocean, thin sediments, large scale volcanism, frequent moderate earthquakes, lack of a distinct Moho, and prevalence of intermediate seismic primary velocities in the range 7.2 to 7.7 km/sec are generally taken to indicate that the rises are the sites of upwelling and spreading out in a convective cycle. The intensity and uniformity of heating is apparently sufficient to prevent this mass imbalance from being large. The small temperature gradients entailed are the expected consequence of a strong temperature dependence of viscosity (Tozer, 1967; Turcotte and Oxburgh, 1969).

Oceanic Shield Basalts

With the improved data, all major oceanic positive areas appear to be associated with spreading centers except one: Hawaii. Hawaii appears to be the buildup of an appreciable mass excess by extrusive activity off the rise. This buildup is in spite of a sinking of the crust, as pointed out by Menard (1969). An approach to isostatic adjustment is also suggested by depths to the Moho somewhat greater than the oceanic average (Drake and Nafe, 1968). Apparently the lithosphere has cooled sufficiently to cause a lag in the attainment of equilibrium. This notion is corroborated by the relatively low heat flow. The existence of such a feature indicates that the asthenospheric flows that generate the required pressure do not necessarily have a simple and direct relationship to the lithospheric plate motions (McKenzie, 1969).

Hawaii is unique in that it falls midway in the 10,000-km stretch from the East Pacific Rise to the trench and island arc along the west Pacific margin: the only region in the world where the rise-continent distance exceeds 5000 km. It is also the only region in the world where the rise-basin distance exceeds 3500 km.

Basins

This commonest of the major features always occurs somewhere to the flanks of ocean rises. Landward of the basin, however, may be a trench and island arc, an orogenic belt, or a relatively quiescent continent on the same tectonic plate. This suggests that the nature of the flows associated with basins depends more on where the material came from than where it is going.

The direct source of the negative isostatic anomaly is most likely that the crust carried along in the sea floor spreading is thicker than compatible with the depth of the basin; a Moho deeper by less than a kilometer is adequate to account for the average isostatic anomaly of – 14 mgal.

The underlying cause, of course, is asthenospheric withdrawal, which results in the 3-km topographic drop from the rise to the basin. Such a drop could be caused by either (1) a horizontal acceleration in the asthenosphere or (2) a downflow of a denser component of the asthenospheric material.

A horizontal acceleration in the asthenosphere at a distance from the rise equal to the typical rise-basin distance (1500 to 3500 km) seems implausible, given that the velocity at the upper boundary is maintained constant by the lithosphere. If we assume a two-dimensional flow, neglect the effect of temperature gradients on the flow, then adopting the customary stream function form (Batchelor, 1967, p. 76)

$$u = \frac{\partial \psi}{\partial y}, \qquad v = -\frac{\partial \psi}{\partial x} \tag{3}$$

where (x, y) and (u, v) are, respectively, the horizontal and vertical position and velocity (positive downward); we obtain the biharmonic equation

$$\nabla^4 \psi = 0 \tag{4}$$

for which a solution is (Batchelor, 1967, p. 225)

$$\begin{aligned} \psi(r, \theta) &= r f(\theta) \\ f(\theta) &= A \sin\theta + B \sin\theta + C\theta \sin\theta + D\theta \cos\theta \end{aligned} \tag{5}$$

where r is distance from the origin and θ is the angle from the x axis. Make the boundary conditions

$$\begin{aligned} u &= 0, \quad v = -v_0, \quad x = 0 \\ u &= u_0, \quad v = 0, \quad y = 0 \end{aligned} \tag{6}$$

i.e., the x axis is the asthenospheric-lithospheric boundary, u_0 is the spreading velocity, and the y axis is the vertical center of the plume of constant velocity v_0. The solution then is

$$\begin{aligned} u &= u_0[1 - f(\xi)] + v_0 g(\xi) \\ v &= v_0[c(\xi) - 1] - u_0 d(\xi) \end{aligned} \tag{7}$$

where

$$\xi = \frac{y}{x} = \tan\theta \tag{8}$$

and

$$f(\xi) = \frac{\pi}{2} q \left(\frac{\xi}{1 + \xi^2} + \tan^{-1}\xi \right) - q \frac{\xi^2}{1 + \xi^2}$$

$$g(\xi) = q \left(\frac{\xi}{1 + \xi^2} + \tan^{-1}\xi \right) - \frac{\pi}{2} q \frac{\xi^2}{1 + \xi^2}$$

$$c(\xi) = \frac{\pi}{2} q \left(\frac{\xi}{1 + \xi^2} - \tan^{-1}\xi \right) + q \frac{\xi^2}{1 + \xi^2} + 1$$

$$d(\xi) = q\left(\frac{\xi}{1+\xi^2} - \tan^{-1}\xi\right) + \frac{\pi}{2}q\frac{\xi^2}{1+\xi^2} \tag{9}$$

$$q = \frac{4}{\pi^2 - 4} = 0.678\ldots \tag{10}$$

The condition imposed by Eq. (1) necessary to make the rise a mass excess and the basin a mass deficiency is that

$$\frac{\partial u}{\partial x} > 0 \tag{11}$$

or

$$u_0 \frac{\partial f}{\partial \xi} - v_0 \frac{\partial q}{\partial \xi} > 0 \tag{12}$$

Developing f and g in series of ξ, for small ξ:

$$v_0 < \frac{\pi}{2}\left(1 + \frac{4-\pi^2}{\pi}\xi + \cdots\right)u_0 \tag{13}$$

In words, the process of lithospheric formation has to consume more than a certain fraction of the material brought up by the vertical flow v_0 if the lithosphere of velocity u_0 is to cause an acceleration in the asthenosphere by dragging the asthenosphere along with it.

The settling out of a denser component would be expected in a multicomponent laterally moving flow that was cooling. A 3-km drop requires much more than thermal contraction, however. If there is an appreciable negative gravity anomaly as well, then Eq. (2) indicates that the settling out cannot just be immediately below the basin lithosphere, but must be either (1) at several 100 km depth below the basin or (2) between the ocean rise and the basin. The process (1) could be induced by the phase transitions of olivine and pyroxene, which occur at depths of 300 to 600 km, while the process (2) might be facilitated by gabbro-to-eclogite transitions at shallower depths.

The sharpness of the crust-mantle boundary, with the 7.2 to 7.7 km/sec gap in velocities (Drake and Nafe, 1968), makes it impossible for the crust to be directly involved in causing the negative anomaly. If the crust is being carried passively along (as suggested by the lack of seismicity, volcanism, or disturbance of the sea floor), then it is hard to understand why it is thicker under the basins than on the flanks of the rises, as emphasized by Le Pichon (1969). Could it be that consolidated sediments are mistaken for basement rock? Sedimentation itself is a secondary process in explaining the gravity anomaly pattern, more a result than a cause: if the thick sediments were the driving force, then the isostatic anomalies in ocean basins would be positive, rather than negative.

Possibly includable in the category of basins caused by behavior of the lithosphere as a free boundary over flows with horizontal accelerations or settling out of denser components are two land features, Antarctica and the Congo Basin. Antarctica is an extremely large feature–large enough to require an unique explanation.

Trench and Island Arcs

The now generally accepted model of McKenzie (1969) and others of a colder, denser oceanic lithospheric slab being thrust down under a less dense but stiffer continental margin fits a simple notion of the gravity pattern: the dominant feature is the broad positive anomaly associated with the denser downthrust slab, while the secondary feature is the narrow negative belt associated with the trench caused by tensile cracking along the downward breaking line. This simple picture is based on the assumption that the applicable boundary condition of the convective flow is more "rigid" than "free": in other words, the time scale of the process is short enough that the strength of the continental lithosphere (and perhaps the oceanic lithosphere as well) significantly resists being pulled down by the downcurrent. This general idea that resistance to flow combined with densification creates positive gravity anomalies applies not only to the boundary layer, but also to deeper strata: the downthrust slab could in part be supported by stiffer matter below the asthenosphere, as suggested by Isacks and Molnar (1969) and others from seismic data.

The association of the downthrust slab with positive anomalies also suggests that the driving cause is a push from above rather than withdrawal from below. Whether this "push" is the gravitationally caused sinking of the denser oceanic lithosphere, the pressure of the spreading sea floor behind it, or the viscous drag by the sublithospheric flow, does not seem resolvable from the gravity data.

Current Orogeny Without Extrusives

The hypothesis of McKenzie (1969) that purely continent versus continent compression results in folding rather than downthrust because of the excessive bouyancy of the thicker crust is appealing as an explanation for the strongly negative gravity anomalies associated with the Asian part of the Alpide belt. The resulting pileup of lower density material results in a mass deficiency in the short run because the stiffness of the lithosphere containing low density crust enables it to push out of the way higher density asthenospheric material. In the longer run, however, the trend from "rigid" to "free" boundaries is expressed by the forcing upward of the lithosphere; geologic and geodetic indications are that the Himalayas-Turkestan complex is currently rising (Artyushkov and Mescherikov, 1969; Gansser, 1964).

The thick layers of sedimentary and metamorphic rocks constituting the upper part of the Himalayas have existed since Precambrian times. The resulting excess of radioactive material combined with low thermal conductivity will lower crustal densities. Furthermore, there may be a contribution to the negative anomaly by erosion, as corroborated by the positive features over the corresponding sedimentation basins, the Bay of Bengal and the Arabian Sea, which appear in Fig. 4.

It is possible that the foregoing suggested mechanisms are all quantitatively insufficient and that an asthenospheric withdrawal is necessary.

Cenozoic Orogeny with Extrusives

These mountain-building areas, listed as orogenic among the positive features in Table 1, are of more limited extent and closer to isostatic equilibrium. Most are associated with compressive belts of the global tectonic system, but this is not entirely so. The reason why they differ from the Himalayas-Turkestan complex in being positive may be (1) the lack of the pre-existing great thicknesses of sedimentary and metamorphic rocks or (2) the presence in the eastern Mediterranean of oceanic crust that can be "consumed" or "subducted" (McKenzie, 1970). They may also be the continental equivalents of Hawaii to some extent: the coincidence of weak features in the lithosphere with regions of excess pressure and heat in a convective system that is not directly related to surface features. Most of these areas have positive seismic delay residuals, suggesting high temperatures to considerable depth.

DISCUSSION AND CONCLUSIONS

The gravity data now appear to be quite reconcilable with the dependence of plate tectonics on mantle convection inferred from other phenomena associated with the mid-ocean rises and the compressive belts (Isacks et al., 1968). Gravity still suffers, however, from its traditional ambiguity in being insufficient to infer the exact mechanism, such as whether the ocean basin negatives are caused by horizontal acceleration or downflow of a denser component.

The greatest feature not readily related to the global tectonic system is the Antarctic negative, much too large to be explained by glacial melting. Antarctica is five-sixths surrounded by ocean rises. Hence, either the spreading rises are migrating away from Antarctica, or Antarctica is a sink for lithospheric material. The latter seems ruled out by the complete absence of the seismicity expected with the destruction or folding of lithosphere. Given that the rises are migrating away, Antarctica must be a mass deficiency because there is not an asthenospheric flow to match the lithospheric spread: i.e., the condition of Eq. (13) applies over most of the rises around Antarctica.

Other features not well explained by the global tectonic pattern are the gravity excesses associated with extrusive flows that occur away from the ocean rises and trench and island arcs, in both oceanic and continental areas. These features appear to require higher temperatures in the asthenosphere generating excess pressures, together with weaknesses in the lithosphere allowing the extrusions. It is, however, difficult to choose whether the resulting net mass excess is a consequence of sufficient overall strength in the lithosphere to support the extruded load or of behavior as a free boundary over a horizontal deceleration or an upcurrent, the reverse of the processes that appear necessary to account for the ocean basins.

Anticipated properties of the mantle convective system that need to be better related to the gravity field are the stress- and temperature-dependence of the effective viscosity, the horizontal temperature gradients arising from variations in radiogenic heating and the contributions to driving the system by fractionations and phase transitions. All these properties are important, of course, to solution of the entire global tectonic problem.

ACKNOWLEDGMENT

I am grateful to E. M. Gaposhkin and Kurt Lambeck for providing their results in advance of publication. This work has been supported by NSF Grant GA-10963.

REFERENCES

Artyushkov, E. V., and Mescherikov, Y. A., Recent movements of the earth's crust and isostatic compensation, *in* The Earth's Crust and Upper Mantle, Am. Geophys. Union Monograph **13**, edited by P. J. Hart, pp. 379-390, 1969.

Batchelor, G. K., An Introduction to Fluid Dynamics, Cambridge Univ. Press, London, 615 pp., 1967.

Drake, C. L., and Nafe, J. E., The transition from ocean to continent from seismic refraction data, *in* The Crust and Upper Mantle of the Pacific Area, Am. Geophys. Union Monograph **15**, edited by L. Knopoff, C. L. Drake, and P. J. Hart, pp. 174-186, 1968.

Gansser, A., Geology of the Himalayas, John Wiley (Interscience), London, 289 pp., 1964.

Gaposhkin, E. M., and Lambeck, K., Earth's gravity field to the sixteenth degree and station coordinates from satellite and terrestrial data, J. Geophys. Res., **76**, 4855-4883, 1971.

Goldreich, P., and Toomre, A., Some remarks on polar wandering, J. Geophys. Res., **74**, 2555-2567, 1969.

Heirtzler, J. R., Dickson, G. O., Herron, E. M., Pitman, W. C., III, and LePichon, X., Marine magnetic anomalies, geomagnetic field reversals, and motions of the ocean floor and continents, J. Geophys. Res., **73**, 2119-2136, 1968.

Hide, R., and Malin, S. R. C., Novel correlations between global features of the earth's gravitational and magnetic fields, Nature, **225**, 605-609, 1970.

Isacks, B., and Molnar, P., Mantle earthquake mechanisms and the sinking of the lithosphere, Nature, **223**, 1121-1124, 1969.

Isacks, B., Oliver, J., and Sykes, L. R., Seismology and the new global tectonics, J. Geophys. Res., **73**, 5855-5899, 1968.

Kaula, W. M., Test and combination of satellite determinations of the gravity field with gravimetry, J. Geophys. Res., **71**, 5303-5314, 1966.

______, Geophysical implications of satellite determinations of the earth's gravitational field, Space Sci. Rev., **7**, 769-794, 1967.

______, An Introduction to Planetary Physics: the Terrestrial Planets, John Wiley & Sons, New York, 490 pp., 1968.

______, A tectonic classification of the main features of the earth's gravitational field, J. Geophys. Res., **74**, 4807-4826, 1969.

Le Pichon, X., Models and structure of the oceanic crust, Tectonophysics, 7, 385-401, 1969.

McKenzie, D. P., Some remarks on heat flow and gravity anomalies, J. Geophys. Res., **72**, 6261-6273, 1967.

______, The influence of the boundary conditions and rotation on convection in the earth's mantle, Geophys. J. Roy. Astron. Soc., **15**, 457-500, 1968.

______, Speculations on the consequences and causes of plate motions, Geophys. J. Roy. Astron. Soc., **18**, 1-32, 1969.

______, Plate tectonics of the Mediterranean region, Nature, **226**, 239-243, 1970.

Menard, H. W., Growth of drifting volcanoes, J. Geophys. Res., **74**, 4827-4837, 1969.

O'Connell, R. J., Pleistocene glaciation and the viscosity of the lower mantle, Geophys. J. Roy. Astron. Soc., in press, 1971.

Pekeris, C. L., Thermal convection in the interior of the earth, Monthly Notices Roy. Astron. Soc., Geophys. Supp., **3**, 343-367, 1935.

Runcorn, S. K., Changes in the convective pattern in the earth's mantle and continental drift: evidence for a cold origin of the earth, Phil. Trans. Roy. Soc. London, Ser. A, **258**, 228-251, 1965.

Tozer, D. C., Towards a theory of thermal convection in the earth's mantle, *in* The Earth's Mantle, edited by T. F. Gaskell, pp. 325-353, Academic Press, New York, 1967.

Turcotte, D. L., and Oxburgh, E. R., Finite amplitude convection cells and continental drift, J. Fluid Mech., **28**, 29-42, 1967.

______, Convection in a mantle with variable physical properties, J. Geophys. Res., **74**, 1458-1474, 1969.

Uotila, U. A., Gravity anomalies for a model earth, Ohio State Univ. Dept. Geodetic Sci. Tech. Rept., **37**, 15 pp., 1962.

Weertman, J., The creep strength of the earth's mantle, Rev. Geophys. Space Phys., **8**, 145-168, 1970.

16

TECTONICS INVOLVED IN THE EVOLUTION OF MOUNTAIN RANGES

by **James Gilluly**

Although no adequate driving mechanism has thus far been presented, the "new global tectonics" is consistent with many features of the ocean floors and of the circum-Pacific oceanic-continental interfaces. It seems appropriate to examine some of the salient records of earlier continental evolution in the light of similar plate tectonics mechanisms. Because of the persistent loss of geologic record with time, this examination mainly considers Cenozoic and Mesozoic continental history.

Much, but by no means all, of the tectonic activity of western North America seems readily to be accommodated by a scheme of plate tectonics. Sedimentary volumes, both on and offshore of the Pacific Coast, and metamorphic and plutonic histories contrasting with those of the eastern seaboard, together with striking differences in structure and in metamorphic facies, all agree in suggesting a formerly active Benioff zone off the Pacific coast. The hypothesis of a former extension of the East Pacific Rise beneath the continent is examined, with contradictory but largely opposed evidence. In any event the spreading of the Basin and Range province has been far less than is indicated for the ocean floor by the magnetic patterns associated with the Rise.

Although many features of the Mesozoic tectonics of California seem to fit a scheme of subduction zones, it is certain that the presently active tectonism there cannot be so related. Yet, it is fully as active today as it was during the Mesozoic. Many other features of Cenozoic tectonism elsewhere in the country are also impossible to relate to plate tectonics and must be attributed to movements in the immediately underlying mantle.

The history of the Alps and Himalayas proper can be fitted into a scheme of plate tectonics, but the demonstrable oceanization of the Mediterranean in Miocene and Pliocene time seems to demand localized mantle movement, many oroclinal folds, difficult to reconcile with rigid plates, and either subcrustal erosion or the subduction of vast masses of sial, formerly present, into the depths of the mantle. The history of the Alboran Sea, just east of Gibraltar, and of the Betic and Riff Ranges, shows that the reconstruction of pre-drift geography by Bullard and his colleagues cannot be reconciled with the tectonic history either before or after the onset of drift. Spain lay at least 400 km relatively north of the position they indicate. Similarly, since the beginning of the Jurassic the Atlantic shelf off the coast of North America has prograded at least 200 km, so that the present 500-fathom contour used in the Bullard reconstruction is far to the east of the pre-drift contour. Persuasive as their fit appears, it is in part purely fortuitous.

Neither do plate tectonics adequately account for many other Cenozoic tectonic features of southern Europe, such as the uplift of the Spanish Meseta, and Plateau Central, the Vosges and Schwarzwald, the uplift of the Carpathians and Caucasus and many other features. Likewise plate tectonics do not account for the great ranges and depressions of the Tien Shan, Kirgiz, and northern Afghanistan. The deep focus earthquakes of Tibet can be referred to the Indus-Tsangpo subduction zone but hardly to any farther north.

In summary, the theory of plate tectonics seems applicable to many of the earth's greatest ranges—especially those that expose considerable volumes of ophiolites probably derived from pieces of oceanic crust—but other great ranges seem difficult or impossible to derive through such a mechanism. Intraplate deformation is at least as active in the continental segments as Menard has shown it to be in the oceanic.

INTRODUCTION

The many dramatic advances in oceanography and geophysics during the past decade have produced compelling evidence (Dickson et al., 1968; Heirtzler et al., 1968; Isacks et al., 1968; LePichon and Heirtzler, 1968; Morgan, 1968; Orowan, 1964; Pitman and Heirtzler, 1966; Pitman et al., 1968; and Vine, 1966) that much of the current tectonic activity of the earth is related to the convergence of a few large crustal plates that are moving away from the oceanic ridges.

Although no generally accepted mechanism for driving these movements has yet been suggested, their current reality seems undeniable.

Most of the evidence for these movements has been derived from the magnetic patterns of the sea floor and from seismology, supported in some degree by deep sea drilling and dredging. A highly persuasive correlation of seismic activity with the current tectonics of the circum-Pacific island arcs and volcanic chains provides strong support for the model (Fig. 1). It seems appropriate, therefore, to examine the tectonic evolution of other orogenic regions and earlier geologic times to test the model's general applicability. Because of persistent loss of geologic record with time (Gilluly, 1969b), the patterns of Mesozoic and Cenozoic orogenies are especially considered, with only minor reference to the earlier, more fragmentary record.

WESTERN NORTH AMERICA

Much of the tectonic activity of western North America during the Mesozoic and early Cenozoic eras seems to fit neatly into a pattern of plate tectonics somewhat like that of the present.

The vast discrepancy between the volumes of Mesozoic and younger sediment off the Atlantic and Pacific shores has been noted (Gilluly, 1969a; Gilluly et al., 1970). Off the Atlantic, at least 10×10^6 km^3 of sediment are reasonably referred to the weathering and erosion of the continental United States; off the Pacific, so much of the 2×10^6km^3 of offshore sediment is pyroclastic that the total land-derived material probably does not exceed half, about 1×10^6 km^3. The terrigenous material in the Atlantic is thus probably 10 and certainly 6 times as voluminous as that in the Pacific, though the drainage areas tributary to the two oceans have been roughly equal since Triassic time. Off the eastern coast most of the sediment is of Jurassic and Cretaceous age, for the pre-Jurassic erosion surface of the Piedmont has been traced seismically beneath coastal plain, continental shelf, and continental rise to a junction with the oceanic crust (Drake et al., 1959; Ewing et al., 1939), and the Tertiary coastal plain sediments are relatively thin. Off the Pacific Coast, however, the magnetic anomalies suggest that the oceanic crust is of mid-Tertiary or younger age (Heirtzler et al., 1968); the overlying sediment cannot, of course, be older. Many other features support this interpretation.

The mild epeirogenic uplift of the Atlantic seaboard has returned relatively little post-Triassic sediment to the continent, according to a careful estimate, less than 2×10^5 km^3 (Fig. 2). Huge volumes of Mesozoic and Cenozoic marine sedimentary rocks have been incorporated into the continent from the Pacific during the orogenies of Mesozoic to Holocene time. The Franciscan Group of California, of Jurassic and Cretaceous ages, has been estimated to comprise a volume of nearly 1.5×10^6 km^3 (Bailey et al., 1964), an amount fully equal to that of all the sediment still remaining off the coast. The coeval Great Valley sequence is about as voluminous, and the Tertiary and Quaternary rocks of the

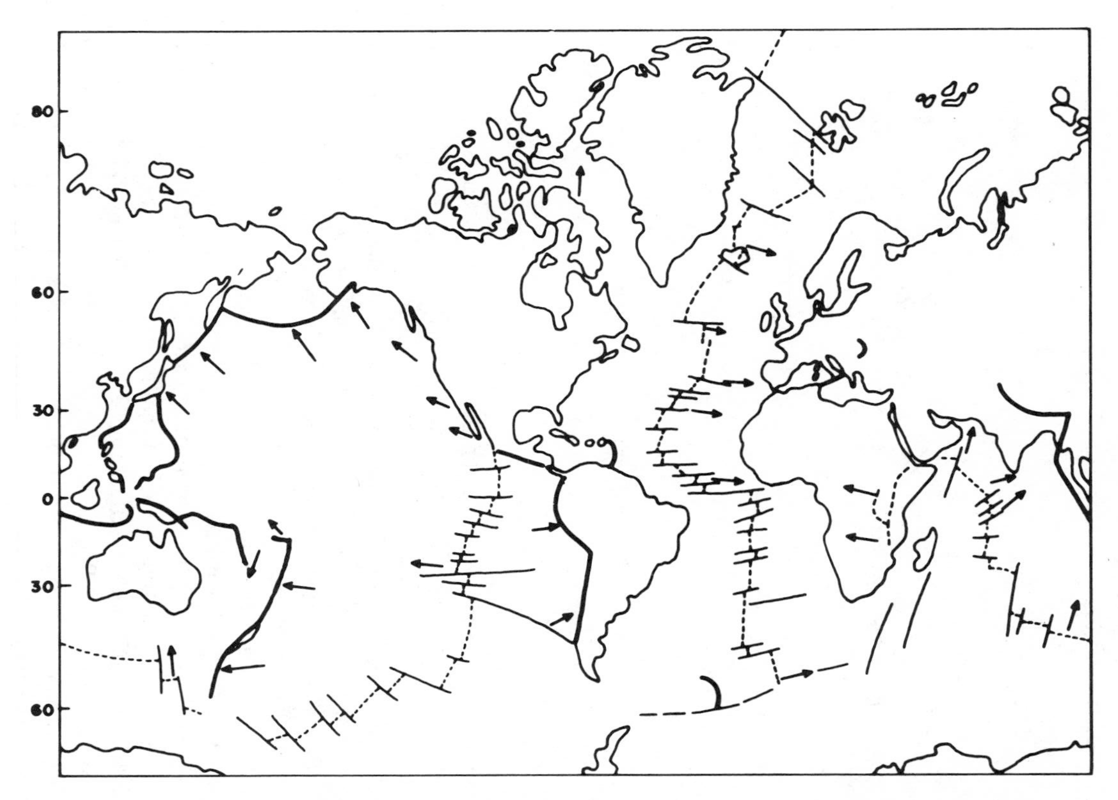

Fig. 1. Summary of slip vectors (arrows) derived from earthquake mechanics studies. Dotted lines, oceanic ridges; heavy lines, inferred zones of subduction. After Isacks et al. (1968).

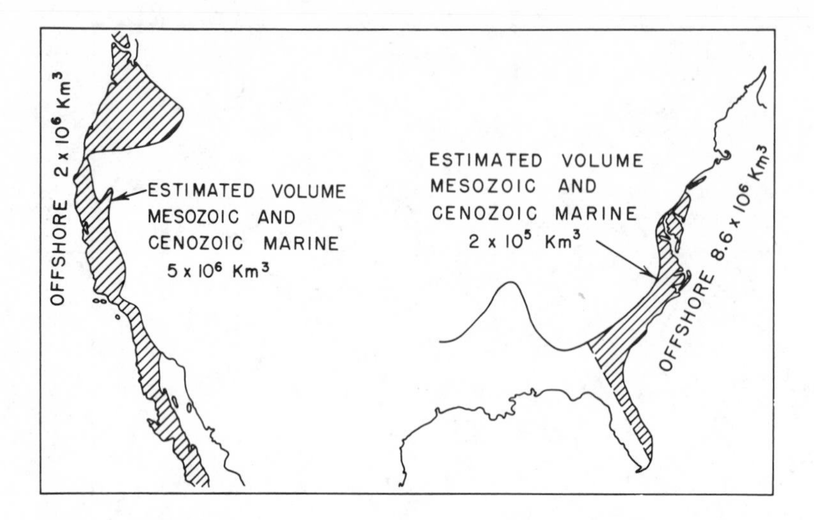

Fig. 2. Estimated volumes of Mesozoic and Cenozoic rocks on the Atlantic and Pacific seaboards.

Los Angeles and Ventura Basins, the Great Valley, Coast Ranges, and Willamette and Puget Sound troughs are comparable or greater masses (Gilluly et al., 1970). All these deposits, incorporated into the Pacific margin, together would perhaps account for half the difference in volume of the offshore sediments of the two coasts. Yet a very large volume of Mesozoic and younger sediments, together with Paleozoic deposits that must once have been delivered to the Pacific, are still to be accounted for.

If the Atlantic began to open in the Triassic (Wilson, 1963), and North America then began to drift westward, overriding the oceanic crust that was then moving eastward from an offshore East Pacific Rise, the overriding may have been on an inclined zone such as Benioff (1954, 1955, 1959) suggested now underlies western South America. Such a mechanism seems to be operating now beneath Japan (Hess, 1948; Kuno, 1966). Several noteworthy features of the geology of western North America may be explicable by such a process.

For example, the plutonic activity of the Mesozoic was on a scale far greater than that of any comparable span of Phanerozoic time in the western Americas (Gilluly, 1963, 1969a; Hamilton, 1969; Knopf, 1955). Taking North and South America together, the Mesozoic plutons cover more than 3 times the area of the Paleozoic plutons and more than 12 times the area of the Cenozoic plutons (Gilluly, 1969a). Even though part of the disparity in area of outcrop might be explained by the burial of some Paleozoic plutons beneath younger rocks and by the failure of erosion to unroof the Cenozoic plutons proportionately, the difference is so very large as strongly to suggest a significant discontinuity in igneous history. It is probably not merely fortuitous that Mesozoic plutonism was widespread throughout the circum-Pacific ring.

If a Benioff zone had been activated beneath western America during the early Mesozoic, some of the water-saturated Paleozoic strata that must then have lain offshore might have been dragged, along with their underlying or adjoining oceanic crust, far down the zone into the mantle. We must infer the existence of such pre-Mesozoic rocks from the present exposures of Paleozoic rocks in the Santa Lucia (Trask, 1926) and Klamath regions (Davis, 1968; Irwin, 1960) and of Precambrian rocks in the San Gabriel Mountains of southern California (Silver et al., 1960) virtually at the present seaboard. With a drift rate of 2 or 3 cm/yr and a 45° dip of the Benioff zone, some deposits might have been carried inland about 200 km and to a comparable depth in about 15 m.y. During this time, they could have been heated, modestly by conduction from the mantle, but very notably by friction in the zone (Oxburgh and Turcotte, 1968)–a factor that, according to Shaw's analysis (1969), would have been highly significant in magma generation.

The deposits would thus be expected to reach anatectic temperatures rather promptly, as has been independently suggested by Hamilton (1969). The onset of the tremendous Mesozoic plutonism of western America, from the Peninsular Ranges of Baja California through the Sierra, the Idaho batholith, and on northward through British Columbia, Yukon, and Alaska, might thus have been related to the onset of the current pattern of continental drift. Other plutons of the circum-Pacific mountains may have originated in the same way. If so, drift began virtually simultaneously throughout the Pacific Basin.

If such a Benioff zone continued intermittently active until mid-Tertiary time, it might interrelate several features of California geology; The Jurassic-Cretaceous Franciscan Group of California lies to the west of the zone of major plutons and thus would overlie a shallower segment of the postulated Benioff zone than the plutons do. The Franciscan eclogites are of the sort derived from metamorphism of basalt, not from the mantle (Coleman et al., 1965). Associated with the eclogites are widespread, though small, masses of glaucophane schist, aragonite marble, and serpentine, an association indicating unusually high pressures and relatively low temperatures for its formation (Coleman and Lee, 1962, 1963; Coleman et al. 1965; Ernst, 1960, 1963). It is suggested that this association may have been formed from deep sea sediments and associated lavas that were dragged down a Benioff zone to depths of 30 or 40 km so rapidly as not to have had time to heat to high temperatures before being squeezed toward the surface and quenched as we see them today. Tectonic overpressure could not have sufficed for the blueschist facies; deep burial was required.

The serpentine associated with these rocks may represent fragments of the mantle (peridotite hydrated by water from the wet sediments) torn from the upper mantle as the rocks were squeezed upward. An objection to this suggestion is, of course, the high density of blueschist, eclogite and peridotite, which would surely tend to oppose their upward motion. Perhaps the serpentine helped to float the denser rocks associated with it. In any event, this same objection confronts any scheme of emplacing peridotite, either solid or molten, into the sialic crust, yet we see it there and locally in abundance.

Squeezing upward of the whole mass of variably metamorphosed sedimentary rocks, pillow lavas, and mantle fragments would be expected to produce the chaotic arrangement and abrupt changes in metamorphic grade that characterize the "mélange structure" of Hsu (1968), so widespread in the Franciscan terranes. A similar interpretation has also been made by Hamilton (1969). It seems significant that many other blueschist terranes of the Pacific rim (in Kamchatka, New Caledonia, New Zealand, the Philippines, and Central America) are found where seismic data suggest the existence of Benioff zones.

The sedimentary and even the volcanic rocks of the ocean floor are of course far less dense than the rocks of the mantle and should tend to rise in diapirs through the mantle rather than being dragged to great depths. This tendency to rise would be opposed, however, by the traction of the descending plate, so that while much of the lighter crustal rocks would perhaps be scraped off the descending plate to become incorporated in the lower crust, some parts of the mass might nevertheless be carried to great depths. The volume relations in the Alps, later discussed, suggest that this is so; the volume of the Alpine roots is far less than the volume of sial carried down on subduction zones there.

If the plutonic and tectonic features mentioned are indeed a result of plate tectonics governed by crustal spreading from the Mid-Atlantic Ridge and East Pacific Rise, two aspects demand discussion: (1) the East Pacific Rise is no longer offshore of the continent between Cape Mendocino and the tip of Baja California and (2) a Benioff zone is not now active anywhere off the North American coast, yet orogeny is still as active there as ever it was.

Both the San Cayetano and Oak Ridge Faults in the Santa Clara Valley of California are actively offsetting terrace gravels, as is the Aliso Canyon Fault farther east. The San Gabriel Mountains are rising at the rate of 50 cm per century (Gilluly, 1949). It is very doubtful whether the tectonic activity was greater at any time during Mesozoic or Cenozoic time than it now is.

Is there a subcontinental extension of (or proxy for) the East Pacific Rise now affecting inland tectonics?

Menard (1964) traced the East Pacific Rise toward the Gulf of California, before it disappeared. He suggested that it may continue beneath the Basin and Range province, known to be an area of crustal extension (Gilluly, 1963, 1970) and of higher than average heat flow (Simmons and Roy, 1969), as would be expected if it overlay a belt of rising mantle. This suggestion has been endorsed by many (Cook, 1966; Hamilton, 1969; Vine, 1966; Warren et al., 1969). Bostrom (1967) suggested a different subcontinental extension (through the Los Angeles Basin, the Great Valley of California, the Willamette and Puget Sound Valleys, and western Columbia) but without evidence of either high heat flow or crustal extension along this trend.

It is true that large parts of the Basin and Range province resemble structurally many sections of the East African Rift system in Ethiopia, Kenya, Uganda, and

Tanzania, which is thought by many to be a continental extension of the Carlsberg Ridge in the Indian Ocean (Fig. 3). Neither in the African Rift region nor the Basin and Range province do the faults trend in rigid parallelism as they do along the Mid-Atlantic Ridge, nor are they distributed along a narrow band. The Basin and Range province is by no means linear, as is the East Pacific Rise, where it is well developed. Along the Mexican border the province is nearly 1000 km broad and it extends far to the south through Sonora and Chihuahua to a point well below the latitude of Cape San Lucas. To the north it splits, sending a long prong northward into Colorado, while to the west, beyond the Colorado Plateau, a segment about 300 km broad connects the southern area with a 700-km-wide region in Utah and Nevada. This splitting is comparable to that between the eastern and western rift systems in Africa but is quite different from the recognized patterns of the oceanic ridge systems.

These contrasts between suboceanic and continental regions of tension may be a result of the greater inhomogeneities of continental as compared with oceanic crust. Both the Basin and Range province and the East Pacific Rise are areas of high heat flow (Menard, 1964; Simmons and Roy, 1969; Warren et al., 1969), though again, the continental pattern is not linear, as is the oceanic.

Another argument favoring the extension of the East Pacific Rise beneath the continent is that the magnetic pattern offshore can be interpreted as recording a subcontinental source of crustal spreading (Heirtzler et al., 1968).

Opposed to the idea of an extension of the East Pacific Rise beneath the continent is the great contrast in structures of the Basin and Range province with those of the East Pacific Rise. The Rise, unlike the Mid-Atlantic Ridge, is not characterized by extensive normal faults, whereas the characterizing feature of the Basin and Range province is the abundance of normal faults. These intracontinental faults are by no means closely parallel, as are those of the Mid-Atlantic Ridge, but splay out in irregular patterns, at least to the south of the Canadian border. It is questionable whether the Rocky Mountain trench belongs to the Basin and Range system, for it seems to have been formed in latest Cretaceous and early Tertiary time (Leech, 1966), whereas Basin and Range faulting is much younger. (Incidentally the southern rift valleys of Africa are far older (Permian) than those of Kenya and Uganda (Miocene and younger) according to De Swardt (1965).) Furthermore, the Rocky Mountain Trench is straight for a distance far greater than any fault of the Basin and Range province proper.

The map pattern of the Basin and Range faults is also unlike that of the suboceanic rise or ridges. The average trend of the Basin and Range faults swings from north-south in southern Colorado and eastern New Mexico through west-northwest in southern Arizona, nearly north-south again in Nevada and Utah, to northwest in Idaho and Montana and nearly due west in Oregon (Fig. 4), a pattern that finds no parallel in the suboceanic fault patterns. In fact, Fuller and Waters (1929) describe an elliptical graben in southern Oregon completely surrounded by normal faults, demanding crustal extension in all

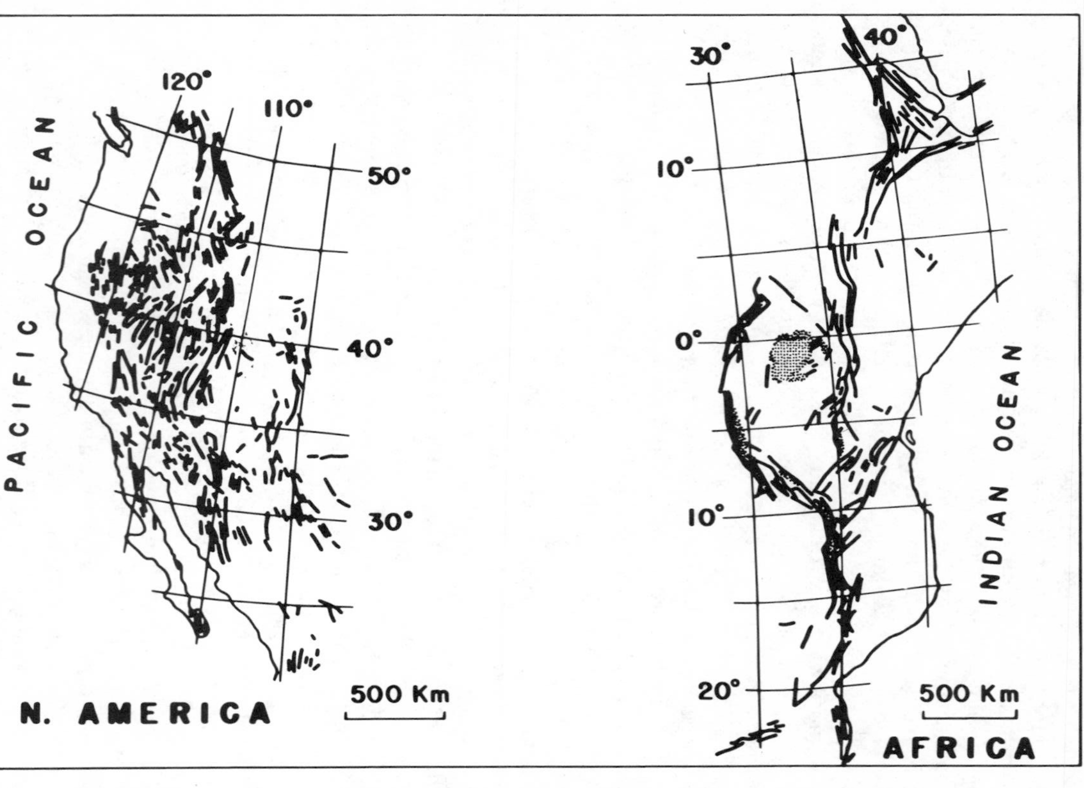

Fig. 3. The fault pattern of the Great Basin compared with that of the African Rift region. Note the similarity between the split pattern around the Colorado Plateau and that around the Lake Victoria block. The closer spacing of the American faults is probably accounted for by the more detailed mapping.

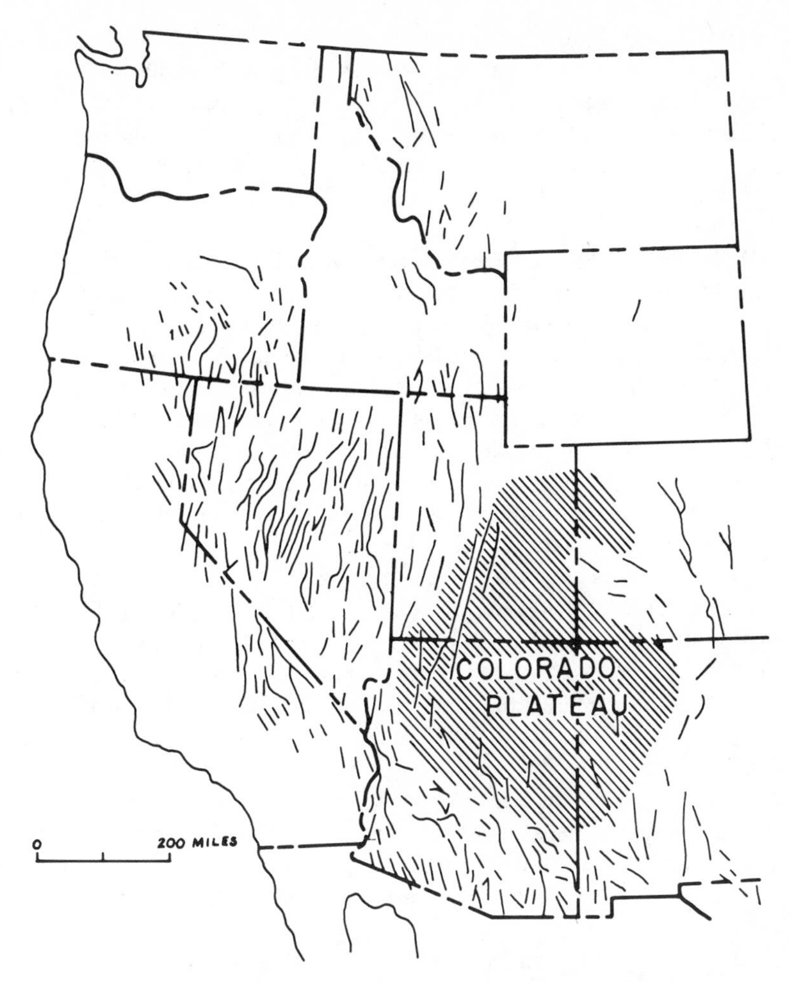

Fig. 4. The Basin Range type faults of the western United States. From the tectonic map of the United States, 1961, published by the U.S. Geological Survey and the American Association of Petroleum Geologists.

azimuths. Thus, if the Rise does lie beneath the continent, the reaction of the overlying continental crust has been drastically different from that of the oceanic crust.

It may be significant that, after the late Mesozoic plutonism, the next most active time of igneous activity in western America was in the Oligocene and Miocene, when Basin Range faulting also began (Nolan, 1943). Perhaps this was the time when the continent overran the East Pacific Rise, or when the Rise was moved beneath the continent on a transform fault, where it could both supply heat for the igneous activity and spread the crust apart, as is recorded in the normal faulting. Warren and others (1969) have suggested that the Rise was overrun during the Cretaceous, but it seems clear that the metamorphism of the Franciscan was going on at that time, so the postulated Benioff zone was still active at the continental border. The volcanic and plutonic activity of Oligocene and Miocene time, however, was by no means restricted to the Basin and Range province, as is demonstrated by the plutonism of the northern Cascades (Smith and Mendenhall, 1900) and the extrusion of the vast plateau lavas of the northwest.

McKenzie and Parker (1967) present a strong argument that the East Pacific Rise does not extend north of the Central American trench, being terminated by transform fault movement along the San Andreas fault (assumed to extend down the Gulf of California). A possible difficulty with their persuasive argument is that the San Andreas may lie some distance inland from the coast of Sonora, rather than following down the Gulf, so that it would not be in contact with the Rise to form a triple junction (L. T. Silver, oral communication, 1970). In any event, the magnetic patterns in the Gulf of California do not support a simple extension of the San Andreas fault down the Gulf; several offsets are required (Chase et al., 1970; Larson et al., 1968).

The suggestions that the San Andreas fault is a tranform fault has become widely accepted. There are, however, reasons for skepticism. It seems to me significant that no other transform faults extend into the continental crust from the ocean basins except in India. For example, the Mendocino fracture zone fails to offset continental structures even as old as Late Devonian (the Antler orogenic belt) on its strike to the east, and the Basin Range faults lying on its trend extend indifferently across its projection (Gilluly, 1963, 1970). It is true that the Murray fracture zone projects along strike into the Transverse Ranges the only east-trending structure for many miles either north or south, and henc a relation that I once thought possibly significant (Gilluly, 1970), but it has been impossible to trace the fracture zone across the continental shelf (Von Huene, 1968).

John and Maurice Ewing (1967) have suggested that, after a time of relatively slow movement, sea floor spreading began to speed up in the Miocene. Perhaps this speed-up came about after the Rise was overrun by the drifting continent, an event postulated by Orowan (1964), because the westward movement of the continent away from the Mid-Atlantic Ridge was no longer opposed by the oceanic crust moving eastward from an offshore East Pacific Rise. Either such an overrun or an offset of the Rise on a transform fault, as postulated by McKenzie and Parker (1967) and others would eliminate the postulated Benioff zone inasmuch as part of the new crust generated at the Rise would be moving with the continent, not against it as before.

A very different interpretation is that of Palmer (1968), who thinks that the magnetic pattern of the oceans indicates a halt in the drifting between 70 and 10 m.y. before the present and that the continent overran the East Pacific Rise in mid-Jurassic time. This suggestion seems most improbable, as it leaves both the blue schist metamorphism of the Franciscan and the earlier plutonism of the Mesozoic unrelated to the ocean-continent interface.

It seems highly improbable that the Rise does lie beneath the Basin and Range province; the tectonics and volcanism of that region more likely result from activity of the mantle underlying the main continental plate and not from interaction of two plates.

If, despite the drastic differences in pattern and alignment of the faults and in the patterns of heat flow between continent and ocean rise, the East Pacific Rise nevertheless exists beneath the Basin and Range province, this may account for the high heat flow, the burst of volcanic activity (the great ash flows of the southern and eastern parts of the province are chiefly of Oligocene and younger age), and the tearing apart of the crust to give rise to the innumerable normal faults that dissect the area. It might also account for the abnormally thin crust of the province (Fig. 5) reported by Pakiser and Steinhart (1964). The crust may have been thinned by subcrustal erosion (Gidon, 1963; Gilluly, 1955, 1963), or by metamorphic flowage resulting from the high temperatures, or both. The crustal material dragged from the Basin and Range province may have gone east to thicken the crust beneath the Colorado Plateau and west to form the root of the Sierra. Both the Sierra root and the Plateau were developed in late Tertiary time, as both areas stood very much lower until the mid-Tertiary and both are in virtual isostatic balance at present.

Whatever the driving mechanism for the crustal spreading that is responsible for the still active normal faulting in the province, the spreading of the continent is proceeding much more slowly than the estimated spreading of the East Pacific Rise. In the roughly 35 m.y. since the Basin Range faulting began, the province has spread only about 50 km on the average and only about 90 km as a maximum (Gilluly, 1970), whereas, if the spreading had been equal to that of the Rise, it should have been about 1300 km at the rates suggested by Heirtzler et al. (1968). (See Fig. 6) Perhaps this is an argument that the continental spreading is by metamorphic flowage, brought about principally by high temperatures rather than by traction exerted by movements of the underlying mantle.

On the other hand, the difference in spreading rate may be accounted for by a very weak coupling between the new "oceanic" crust forming at the Rise and the older continental crust beneath which it is growing. Ample evidence shows that the continental and oceanic crustal segments are at present virtually decoupled at the coast, and that no Benioff zone is now active there (Gilluly, 1963, 1970). If the drifting is still going on, as seems very likely, the continental plate is moving over the oceanic on a surface at a depth measured in scores, rather than hundreds of kilometers.

Yet it seems most unlikely that the depth of the sliding surface is extremely shallow; it must lie within the mantle, rather than at or above the Moho and at sufficient depth within the mantle to allow most of the petrogenesis of rocks that reach the surface to go on above it. The metallogenic province characterized by copper and molybdenum ore deposits has been furnishing these metals in virtually the same longitudes of the Cordillera from Precambrian (Jerome, Arizona) through Jurassic (Bisbee, Arizona), Eocene (Ajo, Arizona; Santa Rita,

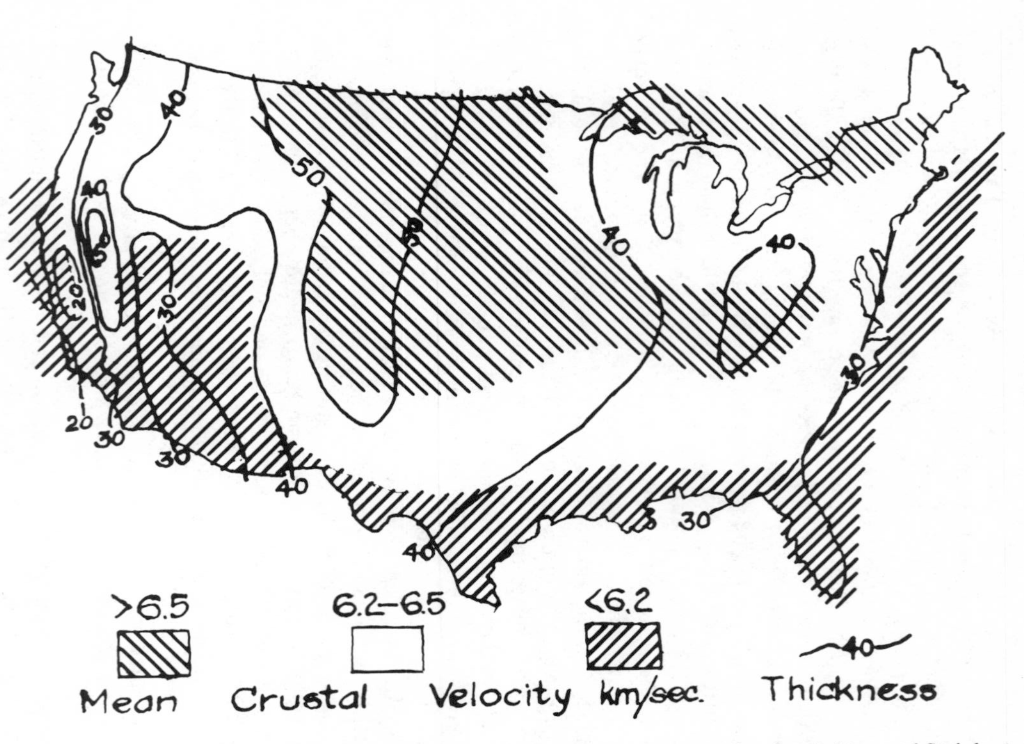

Fig. 5. Crustal thickness and elastic properties of the crust and upper mantle, after Pakiser and Steinhart (1964). Note low crustal velocities and thickness in the Basin and Range area.

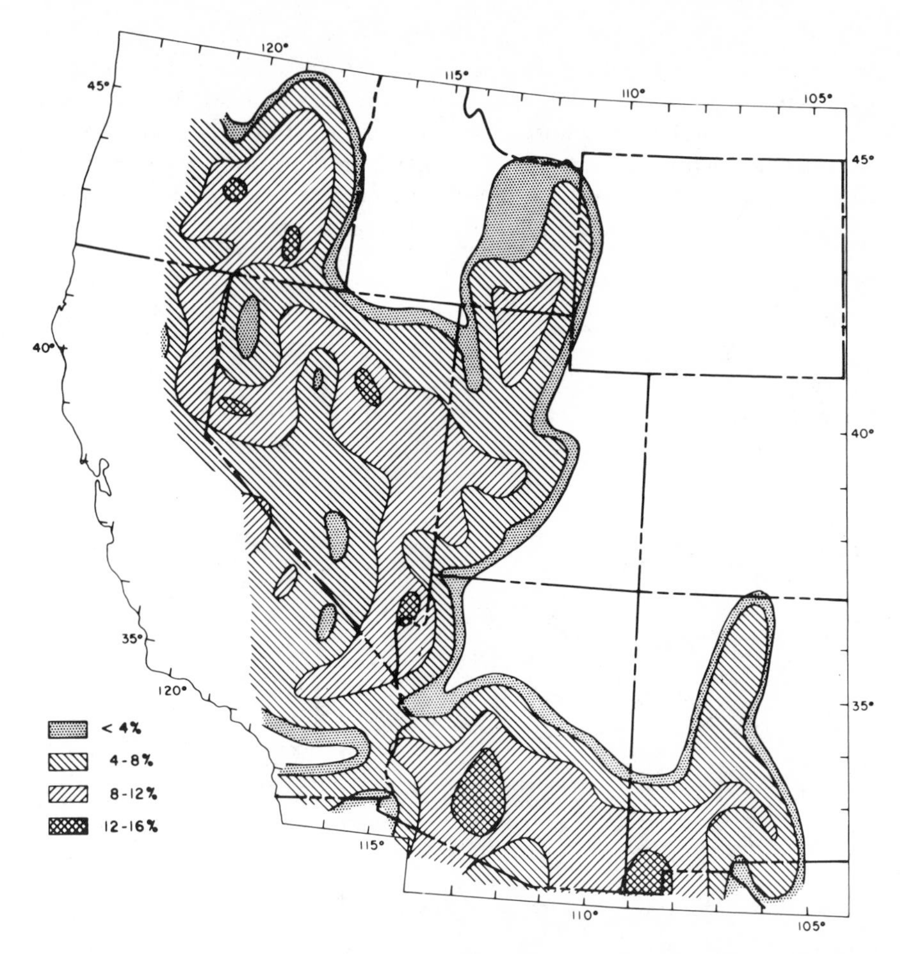

Fig. 6. Percentage of areal expansion in each square degree of the Basin and Range province resulting from normal faulting, on the assumptions that the faults dip at 60° and have an average throw of 5 km.

New Mexico), and Oligocene (Bingham, Utah) time. Spurr's "Great Silver Channel" (1923), though not demonstrably active for so long, also suggests that the upper mantle is coupled to the continental crust. One need not accept a strict linearity of these metallogenic features, but it seems clear that the mantle underlying them is distinctly different than, say, the mantle beneath the Great Plains, and it has been so for hundreds of millions of years (Fig. 7).

For this and many other reasons, it seems most unlikely that any activity along a Mesozoic to mid-Tertiary Benioff zone can have had any significant influence on either the tectonic or igneous processes of the central and eastern Cordillera. The early Jurassic thrusting in the Hawthorne region of western Nevada was directed from north to south (Ferguson and Muller, 1949), so that any direct influence of east-west compression was lacking (Fig. 8).

To judge of the possible relation between the Jurassic and younger thrusts farther east and the postulated Benioff zone along which the great plutons to the west were being developed, we may turn to the geology of South America where the "type" Benioff zone is now active. Here the most easterly of the Andean structures are active at a distance of 500 km from the coast in Peru (Ham and Herrera, 1963), at 750 km in Bolivia (Sonnenberg, 1963), and Argentina (Herrero-Ducloux, 1963). It is thus possible that the post-Permian pre-Lower

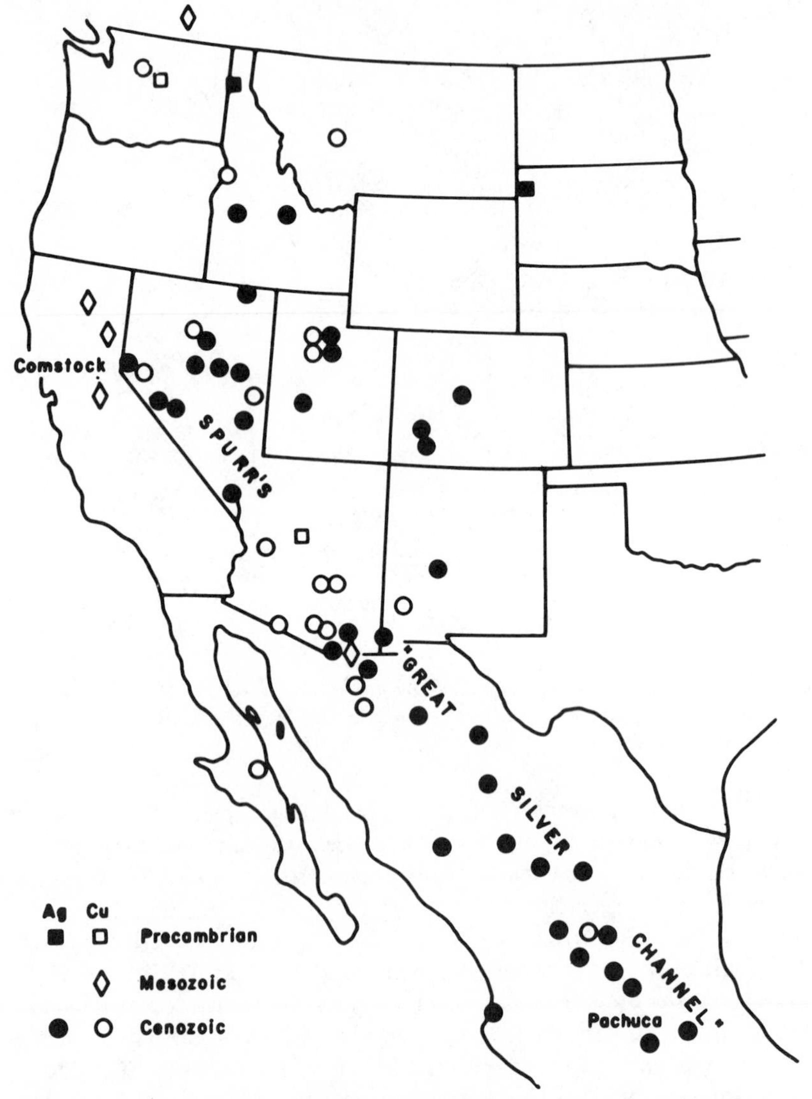

Fig. 7. Silver and copper deposits of the western United States and northern Mexico, showing Spurr's (1923) "Great Silver Channel."

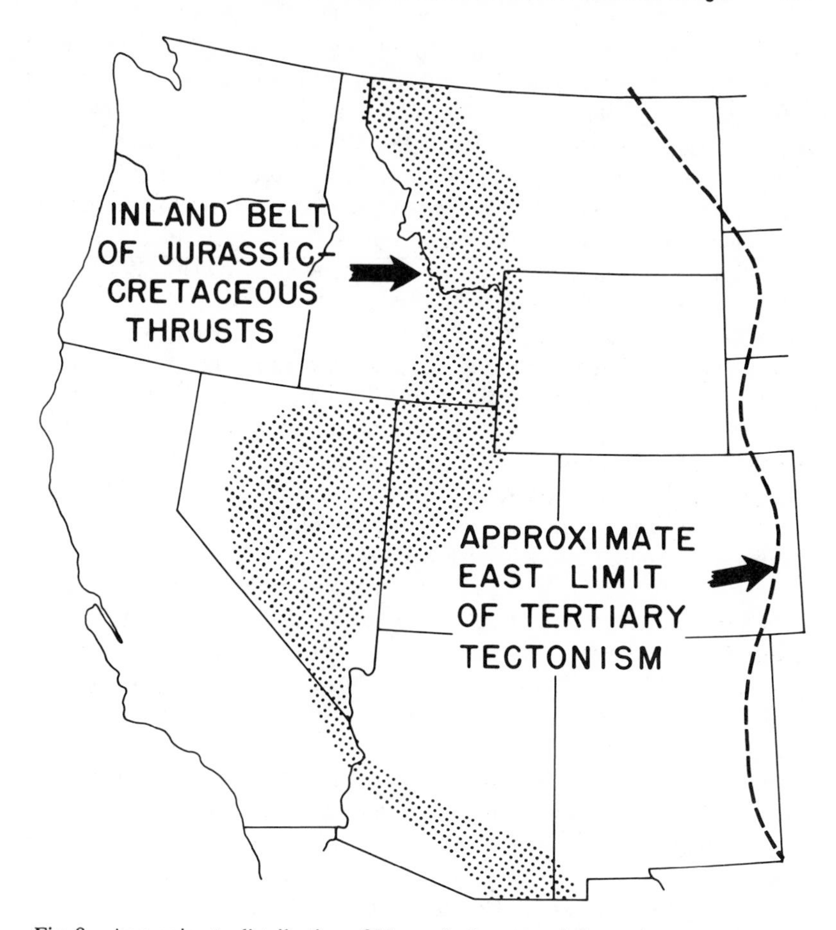

Fig. 8. Approximate distribution of Mesozoic thrusts and Cenozoic tectonism.

Cretaceous deformation at Eureka (Nolan et al., 1956) and the Jurassic plutons of central Nevada lying at about equal distances from the postulated Benioff zone to the west may have been similarly related. The deformation in easternmost Nevada (Lee et al., 1970) and western Utah (Nolan, 1935) and still more so the Jurassic thrusting in Idaho (Armstrong and Oriel, 1965), however, seem considerably too far east to have been related to any Benioff zone cropping out to the west of the Sierra Nevada and the Idaho batholith; such a zone would surely have lain at a depth beneath eastern Idaho far too great to have influenced the thin-skinned décollements there, which nowhere involved basement rocks.

If, during the Jurassic, the probable Benioff zone cropping out at the continental border lay too deep to have influenced the tectonics of eastern Nevada and Idaho, it must have lain still deeper beneath the more easterly areas of active Cretaceous tectonism. By Cretaceous time, the Benioff zone was relatively still farther west than it had been during the Jurassic, while the wave of Cordilleran tectonism was moving progressively eastward. The eastward march of tectonic activity can be followed through the Cretaceous of eastern Nevada (Drewes, 1958, 1967, 1970; Misch and Hazzard, 1962; Woodward, 1964); through western Utah (Armstrong, 1968; Christiansen, 1952; Crittenden, 1961; Eardley, 1944; Nolan, 1935; Spieker, 1946); eastern Idaho and Wyoming (Armstrong and Oriel, 1965; Rubey, 1955; Veatch, 1907) to the Eocene of the eastern Cordillera (Darton, 1906a, 1906b; Hewett, 1920; Pierce, 1941) and to the Oligocene of the southern Rockies (Burbank and Goddard, 1937).

This eastward march of tectonism thus extended to distances of more than 1500 km from the coast; if the Benioff zone dipped comparably to that beneath South America, it would have lain at a depth of more than 1000 km at the time of the Tertiary deformation in Colorado and Montana. It must be recalled that most trench-associated fault zones dip at angles of 45 to 60° (Benioff, 1949; Sykes, 1966), exceptionally at 70° to vertical (Denham, 1969). If a Benioff zone did govern the tectonic activity of the central and eastern Cordillera, it would surely have had a dip far lower than any of those now active. Such an assumption is purely special pleading.

The Benioff zone now active beneath South America does not seem to be causing any notable tectonic activity in central Brazil, Uruguay, and central Argentina. Yet, these areas are no farther from the outcrop of the currently active Benioff zone of South America than Utah was from the Cretaceous continental margin. A Cretaceous North American Benioff zone like the modern South American one would lie at the mantle depths far greater than those of any known deep focus earthquakes if it were projected far enough east to underlie the eastern Cordillera. To attribute the Cretaceous and Tertiary deformations and plutonism of the eastern Cordillera to activity on such a zone seems entirely gratuitous. It must not be forgotten that most of the uplift (as distinct from the tectogenesis) of the eastern Cordillera was post-middle Miocene, by which time the East Pacific Rise had been presumably overrun or had died out as an active source of eastward crustal motion.

Whatever the mechanism of the virtually mid-continental tectonics of much of the North American Cordillera, it seems impossible to attribute it to active plate tectonics and Benioff zones. Many other tectonic features seem equally independent. For example, the currently active uplifts and deformations of the Palos Verdes Hills (Woodring et al., 1946), Santa Clara Valley (Bailey and Jahns, 1954), and other parts of the Transverse Ranges (Gilluly, 1949) of California are proceeding along east-west axes, not at all consistently with the magnetic patterns offshore. Similarly, the northwest-striking uplifts of the Cascade anticlines in central Washington are rising along axes highly divergent from the trends of the magnetic patterns in the sea to the west (Waters, 1955). Farther back in time, the east-west-trending Uinta Range was being elevated simultaneously with the active eastward thrusts taking place immediately to the north in southern Wyoming (Osmond, 1965). At the same time, the strongly arcuate Axial Basin-Grand Hogback uplift in Colorado was forming as a continuation of the Uinta uplift to the East. In Wyoming, the east-trending Owl Creek Mountains (Darton, 1906a) and Sweetwater uplift were forming at the same time as the northwest-trending Big Horn Mountains (Darton, 1906a). It seems difficult to account for these relatively small-scale variations in surficial tectonic patterns by appeal to plate movements at a depth of many hundred kilometers that a contemporaneously active Benioff zone must have had, this far from the continent-ocean interface.

Still another group of tectonic features seem to have formed without any direct influence of plate tectonics. Among these are the great epeirogenic uplifts of the Colorado Plateau, the Great Plains, and the high plateaus of the Rocky Mountain states and provinces. Along the Appalachian-Ouachita trends (Fig. 9), subsidence of thousands of meters took place beneath the Mississippi embayment and the Texas coastal plain while the Solitario and Marathon regions and the Ouachita Mountains were being elevated. The uplift of the old Appalachian Mountains in post-Triassic time seems impossible to relate to any plate tectonics active at that time; so too, are the uplifts of the Adirondacks and Labrador. We are prone to forget that the intracontinental New Madrid and Charleston earthquakes were almost surely of greater magnitude than any of the California quakes, even though we have no seismographic records of them.

In short, plate tectonics and Benioff zones active at different times through

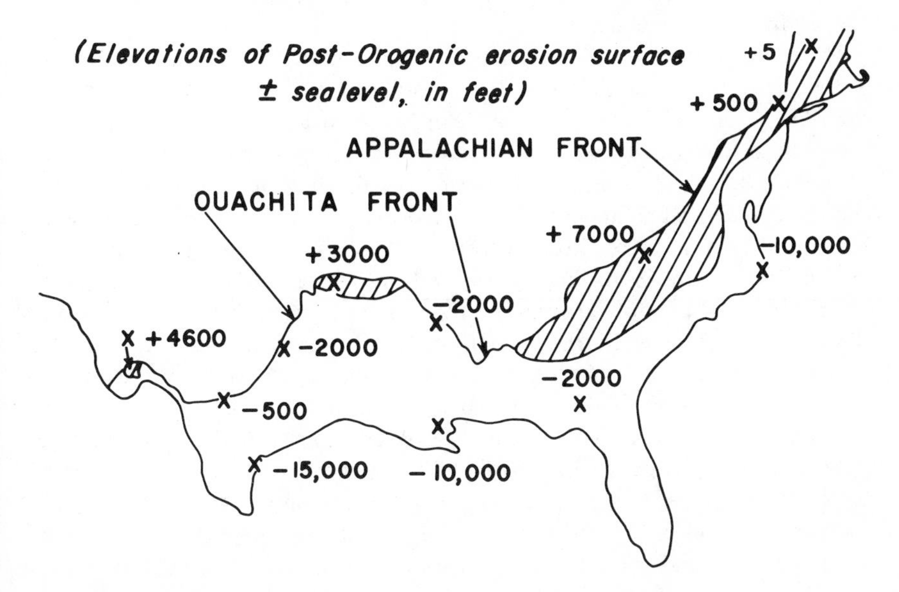

Fig. 9. Cenozoic deformation of the Appalachian-Ouachita orogenic belt.

the Mesozoic and Cenozoic may account for many tectonic features of western North America. They do not by any means explain them all; nor do they account for the widespread crustal deformation farther east. Many structures developed within the continental plate, not at or near any recognizable plate boundary.

Menard (1969) has pointed out that great vertical changes have also gone on in the midst of oceanic plates at the Shatsky, Manihiki, Solomons, Eauropik, and New Guinea Rises in the Pacific and at the Bermuda, Corner, Rio Grande, and Argentine Rises in the Atlantic. These midplate rises, must, as Menard noted, be attributed to local mantle motion not connected directly with the larger scale plate tectonics. The same conclusion must be drawn from the deformations of the interiors of the continental plate of North America; they are completely independent of the tectonics at the plate borders and must be attributed to movements in the underlying mantle.

THE ALPS

The Alps are the best known mountain range on earth; their history should provide a significant test of the plate tectonic model (Fig. 10).

Aside from the autochthonous and parautochthonous massifs, composed of Permian and older rocks folded during the late Paleozoic, the Alps are made up of Mesozoic and early Tertiary strata forming three great groups of decken: the Helvetides, the Pennides, and the Austrides. Each of these groups represents a distinct paleogeographic province. These provinces originally covered broad areas; their strata have now been compressed into a region only a fraction the size of the depositional domain.

The strata of the Helvetide decken were laid down on a continental shelf in the Tethyan sea on the south coast of Hercynian Europe. Unlike such a shelf as that off the present Atlantic coast of North America (but like that of the proto-Appalachians), it was separated from the emergent part of the continent by a wide epeiric sea. Most of the Helvetide sediments are carbonates, but the subordinate detrital components were derived, not from the northern continent, but from ridges in the sea floor or from an island arc to the south (Heim, 1922; Kraus, 1951; Staub, 1924; Trümpy, 1960). None of the Helvetide strata are deep sea deposits, so it is not surprising that the basement upon which they were laid down is sialic, as is seen in the Mont Blanc, Aar, Gotthard, and other massifs.

The Pennide decken came from a region farther south. Much of their strata were derived from islands from which sediment was spilled out in every direction—some onto shelves (Trümpy, 1960) and some into water of probable oceanic depths (Cadisch, 1953; Staub, 1924; Steinmann, 1926), as is suggested by the association of radiolarian cherts, aptycus limestones, condensed "starved" sections, and lack of benthonic fossils. Depth changes were many, abrupt, and along varying axes. The area from which the Pennide decken came was several hundred kilometers across but despite this width and the great depth of the sea

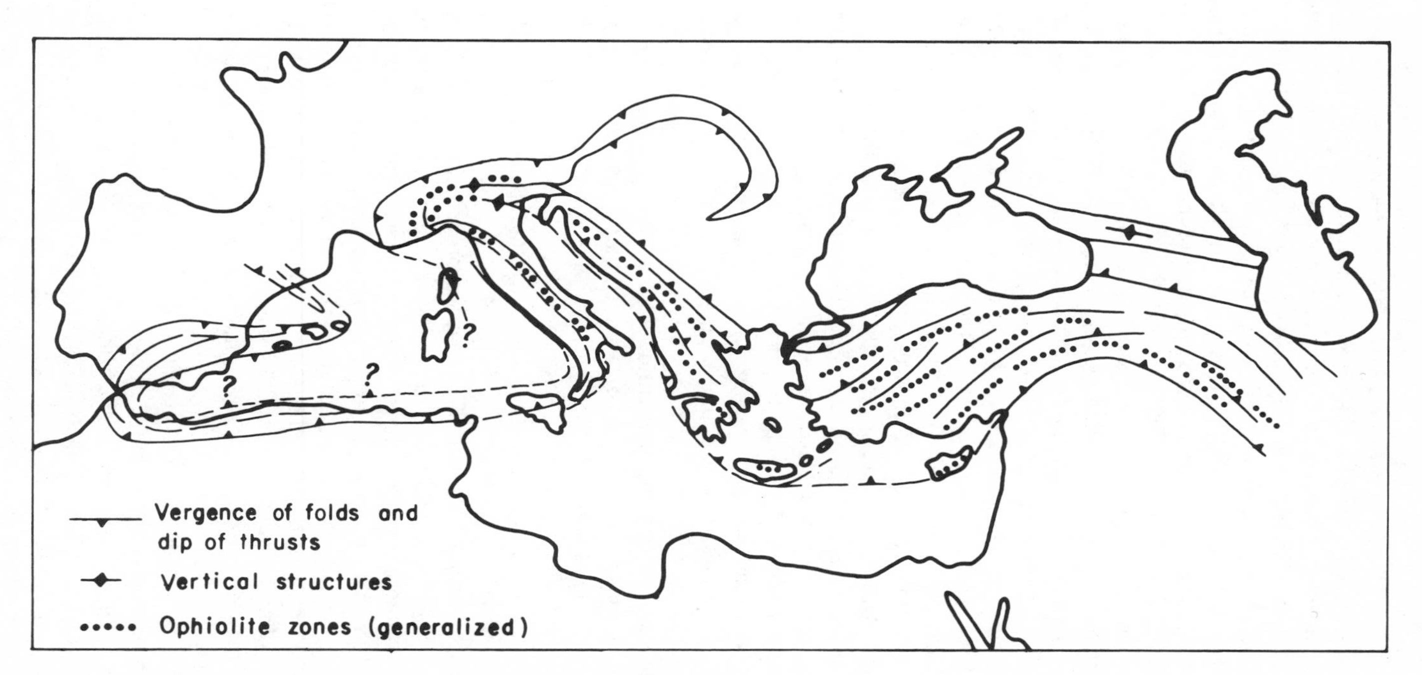

Fig. 10. Structural trends of the Mediterranean and its associated ranges.

in places, the floor on which the strata were laid down was only in small part oceanic crust; most of it, like the basement of the Helvetide decken, was sialic, as is shown by the granitic gneisses that comprise most of the Pennide root zone.

Submarine lavas, commonly associated with chert and locally accompanied by peridotite, serpentine and gabbro (the ophiolite suite) abound in the Pennide decken (Steinmann, 1926; Trümpy, 1960). Contrary to earlier interpretations, Trümpy, does not believe that the ophiolites are localized along thrust surfaces. Nevertheless, the ophiolites of the Ivrea zone are clearly associated with the Insubric shear zone. If others are not so localized, it is improbable that they represent oceanic crust dragged off along a fossil Benioff zone. The eclogite locally associated with the ophiolites was probably metamorphosed during the Alpine tectogenesis and is not derived directly from the mantle. In this respect, the Alps seem to differ greatly from the other Mediterranean ranges farther east, in Greece (Aubouin et al., 1963; Moores, 1969), Crete (Bear, 1966), and Turkey (Blumenthal, 1963), where the ophiolitic rocks seem largely to be pieces of the oceanic floor and are uniformly dragged out on shear zones.

The Austride decken, structurally the highest in the Alps, consist not merely of Mesozoic and Cenozoic strata but also of Paleozoic strata and great volumes of crystalline pre-Mesozoic rocks, in extremely complex patterns. The sediments were considered by Staub (1924) to be shelf deposits on the south shore of the Tethyan sea; they were clearly derived from the south. Many statigraphic gaps testify to local uplifts and erosion. The earliest orogenic disturbances of the Alpine cycle (the pre-Gosau orogeny of the Austrian Alps, of probable Turonian age) was confined to the Austride decken. The thrusts formed at this time were buried beneath later Cretaceous and Tertiary strata before the main episode of Alpine tectogenesis in the Oligocene. As with the lower decken groups, the basement of the Austride decken was sialic orthogneiss and related rocks, now exposed in the extremely thick Silvretta and Oetztal decken in eastern Switzerland and Austria.

The closing of the wide Tethyan sea proceeded throughout the Eocene and Oligocene, with subduction of huge volumes of material along the root zone in the southern Alps. That great volumes of crust had to be disposed of at depth was pointed out long ago by Ampferer (1906). Though the sea was in part of truly oceanic depth, we know from the exposed massifs and root zones associated with the Helvetide and Pennide decken and from the great volumes of crystalline rocks in the Austrides that the crust involved was sialic through most of the width of the Tethyan basin. We may therefore assume a crustal thickness averaging 30 km.

The Alps now average perhaps 150 km in breadth. The crustal shortening involved has been estimated by Trümpy as not less than 50% (1960), thus 150 km. Goguel (1963) puts the shortening at 200 km, which he considers "a moderate estimate." Heim (1922) estimated 65% shortening (280 km); Cadisch (1953) 76% (460 km), and Staub (1924) thought that since the Permian Africa has approached the Schwarzwald by 1500 km. Not all this distance (which was based on the inference that the Gondwana glaciation implies a polar position for central Africa before drift began) was thought to be represented by local shortening in the Alps themselves. It seems clear, however, that at least 150 km and perhaps more likely 350 and 400 km of shortening has taken place. The Tethyan Sea was as wide as the present Black or Caspian Seas; it was not a narrow trough.

The present sialic root of the Alps, part of it perhaps 70 km deep (Knopoff et al., 1966), can represent at most about 300 km^3 in volume per km of strike length. The mountain mass standing above the sea involves at most 250 km^3. An average of fully 2000 km^3, but surely not more than 3000 km^3, has been eroded, but much of this remains in the Molasse deposits on the north flank and in the deep fill of the Po Basin on the south. We can thus account for about 3500 km^3 of crust per km of strike length of the range. Assuming the original sialic crust of the Tethyan region was 30 km thick, the shortening implies that a minimum of 4500 km^3 must be accounted for, and a more probable figure is perhaps 10,000 km^3. Thus at least 1000 km^3 and more probably 6000 km^3 of crustal material has vanished into the depths, down a subduction zone along the Alpine root zone. Kraus (1954) and Gansser (1968) have suggested the Insubric line marks the scar of the subduction zone.

The Insubric line is represented by a mylonite zone that now dips vertically. It may once, however, have dipped at more moderate angles as do most presently active Benioff zones. Unlike the currently active subduction zones of the island arcs, South America and Tonga, which involve oceanic crust and overlying sediments, the material missing along the Insubric line was almost wholly sial. The zone of Ivrea, with its peridotites and other ultrabasic rocks probably represents a slab of the crust and upper mantle of a basin between islands of the Pennine Sea, which was dragged off along the Insubric line (Giese, 1968; Kaminski and Menzel, 1968) as the crust descended the subduction zone.

The last tectonic movements in the Alps, except, of course, for the still continuing uplift in response to erosional unloading, took place in Miocene and Pliocene time, with the backfolding of the southern Alps and the folding of the Jura Mountains. Neither of these events seems to have involved any notable subduction, and it is especially difficult to attribute the Jura décollement to the operation of any Benioff zone.

THE OCEANIZATION OF THE MEDITERRANEAN

Much of the sedimentary material of the Alpine decken was derived from the south (Bourcart, 1962; Heim, 1922; Staub, 1924; Trümpy, 1960), from the area of the present Ligurian Sea, which was also the source of most of the strata of the northern Apennines. During the Miocene and Pliocene, the relief of the region was reversed, with the beginning of the formation of the Mediterranean. The sialic crust that formerly underlay the area of the Ligurian Sea was swept northward and eastward to form the Alps and northern Apennines, thus baring the mantle to accumulate an oceanic crust (Bourcart, 1962; Glangeaud et al., 1966; Kraus, 1960, 1962).

Farther west, although most of the Mesozoic sediments now exposed on Majorca and Minorca were derived from an uplift in Catalonia, much of the Tertiary strata are continental and derived from a land immediately to the northeast, an area that began to break up during the Miocene and is now sunk to depths of more than 2000 meters beneath the sea (de Booy, 1969; Colom and Escandell, 1962; Glangeaud, 1966; Pannekoek, 1969; Ritsema, 1969).

Still farther west, the site of the present Alboran Sea, just east of Gibraltar, was occupied by a substantial land mass that was the source area of deposits laid down both to the north, in the Betic geosyncline of Spain, and to the south, in the Riff geosyncline of Algeria and Morocco (Durand-Delga et al., 1962; Durr et al., 1962; Egeler and de Booy, 1962; Kieken, 1962). These geosynclines, originally connected as they are today, were squeezed during Tertiary orogeny and bent into the orocline of Gibraltar. Obviously, an orocline demands a sialic crust flexible about steep axes and accordingly not as usually conceived of, a result of plate tectonics.

Incidentally, the history of the Alboran Sea not only illustrates once more the oceanization of former land areas, but also it shows clearly that the very persuasive pre-drift reconstruction of the conjunction of Africa, Europe, and North America by Bullard, Everett, and Smith (1965) is in part fortuitous. The fit of South America and Africa suggested by this reconstruction has been all but proven by Hurley et al. (1966), but the shortening of at least 2° and perhaps 4° of latitude involved in the deformation of the Riff and Betic cordilleras shows that, before drift, the gap between Europe and Africa at Gibraltar was much wider than it is now or than it is shown on the Bullard et al. compilation. Prior to drift and until Miocene time, the Alboran land mass was shedding sediment both north and south; it did not become oceanized until Pliocene time. Thus both pre- and post-drift history are inconsistent with the suggested reconstruction (Fig. 11).

A considerable pre-drift gap between Spain and Africa is also supported by the floral contrasts between Europe and the southern continents prior to drift, contrasts from which the idea of Gondwanaland originally gained its strongest support (Wegener, 1929). It will be noted that the Bullard, Everett, and Smith reconstruction allowed no room for Central America; when this former wider gap is taken into account there is ample room. It should in passing be noted that the present 500 fathom isobath (on which the fit was made) lies well to the east of the pre-Jurassic isobath along the east coast of North America from Cape Hatteras northeastward (data from Drake, Ewing, and Sutton, 1959). The Mesozoic and Cenozoic prograding of the continental shelf has been at least 200 km, so that here again an impressive fit is deceptive.

The Riff Mountains extend on strike to the east through the coastal Atlas of Algeria, and though submerged for a stretch beneath the Sicilian Channel, continue on strike across northern Sicily, through Calabria, to join the

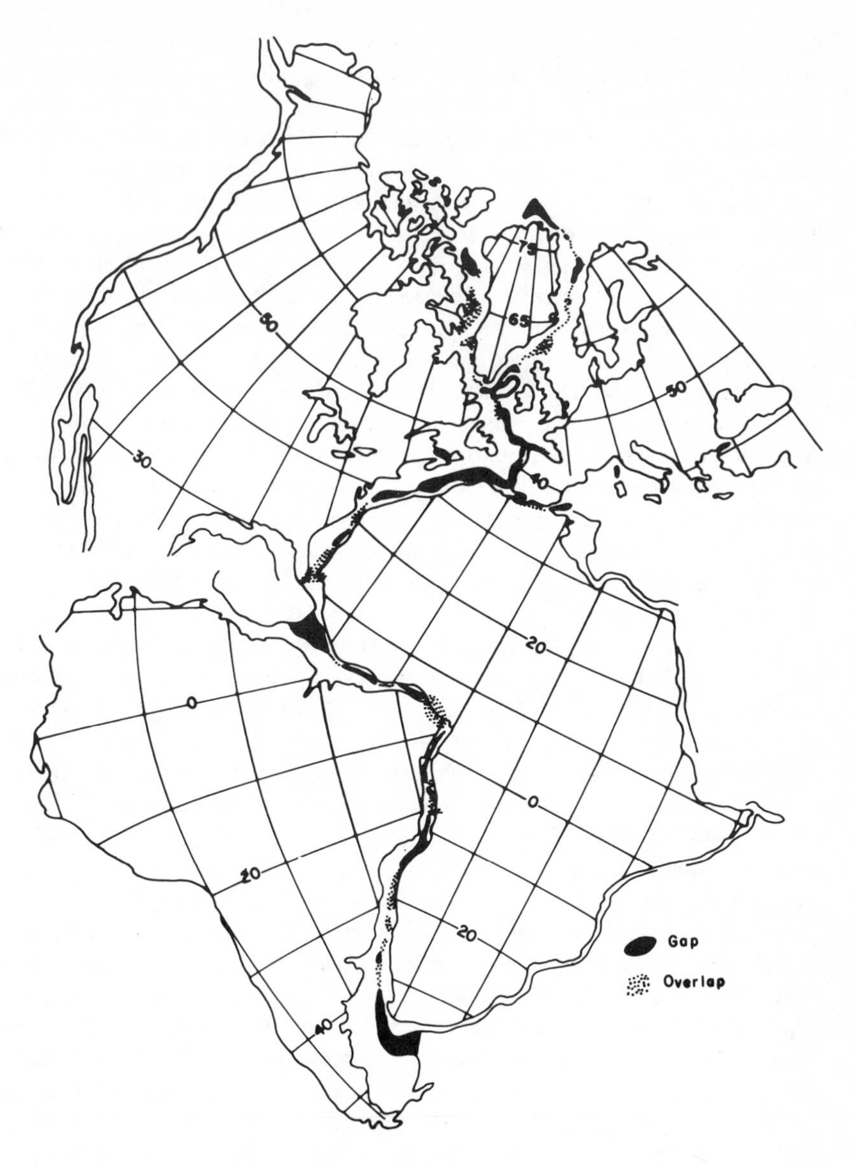

Fig. 11. Fit of all the continents around the Atlantic at the 500 fathom contour. After Bullard, Everett, and Smith (1965).

Apennines in an arc comparable to that of the western Alps. All the detrital rocks of this range were derived from the Tyrrhenian Sea and more westerly parts of the present Mediterranean (Glangeaud, 1966; Kieken, 1962; Kraus, 1960). A more decisive example of oceanization of former sialic land to present great depths and its conversion to an oceanic crust (Payo, 1967) could hardly be found, though any subduction zones along which crust was absorbed are mostly submarine. Perhaps there were none and the oceanization was caused wholly by subcrustal erosion.

It seems highly probable that the Betic Range extended similarly through the Balearic Islands and swung in a comparable arc through eastern Sardinia to join the western Alps. Its former continuations have disappeared beneath the Ligurian and western Mediterranean Seas by oceanization. Menard (1967) and Berry and Knopoff (1967) have shown that the crust beneath the western Mediterranean is intermediate between a normal continental and a normal oceanic one, whereas that of the central Mediterranean greatly resembles that of a continent.

Farther east in the Mediterranean, the folded chains of Yugoslavia and Greece swing southeastward to connect with the chains of Crete and Anatolia, with huge parts of their formerly continuous extent drowned beneath the Aegean and eastern Mediterranean Seas. Here, more than anywhere else in the Mediterranean, are huge masses of ultramafic and accompanying volcanic bodies that seem clearly to be remnants of a former sea floor, exposed because their cover has disappeared down subduction zones (Aubouin et al., 1963; Bear, 1966; Blumenthal, 1963).

The geology of the Mediterranean strongly supports the hypothesis of a northerly movement of Africa with respect to Europe, narrowing and finally destroying the old Tethys Sea. The present Mediterranean is only indirectly a remnant of Tethys; in large part, it is a new sea, formed by the oceanization of large areas formerly occupied by sialic continental rocks. The Alpine root zone preserves a record, in the Insubric line, of a former subduction zone along which vast quantities of sial have been dragged downward into the mantle. This structure seems to fit well into a scheme of plate tectonics. The younger oceanization farther south, southeast, and southwest, however, must have been generated by many much smaller convection cells in the mantle underlying the Mediterranean Sea. This oceanization may have been in large part caused, not by subduction but by subcrustal erosion, leaving a crust of intermediate properties. The volume of sial that has disappeared into the mantle by both processes is huge, indeed. The hypothesis advanced by McKenzie (1970), that the Mediterranean tectonics are to be explained by the interactions of many small rigid plates seems quite inconsistent with the many young oroclinal folds and also with the geologic history. The present Mediterranean is only in small part a remnant of Tethys, most of it is new sea formed by sweeping off formerly overlying sial.

As in North America, by no means all the tectonic features of southern Europe are to be explained by plate tectonics. The Tertiary uplift of the Spanish Meseta, of the Plateau Central, the Vosges, and Schwarzwald, and the Oligocene faulting of the Rhine graben seem difficult to associate with plate tectonics. The Carpathians, an obviously continuous feature with the Alps, have nothing comparable in their degree of shortening and lie wholly within the continent, with no evidence of oceanic influence.

SOME ASIATIC RANGES

The great Caucasus Range, though of approximately the same trend as the Alps, shows no signs of being associated with any subduction zone; on the contrary, it seems to have come about by uplift alone, with little or no crustal shortening being involved. There are no ophiolites (Khain and Milanovsky, 1963). If plate tectonics were involved in the formation, it must have been very indirectly; wholly subcrustal mantle convergence seems more probable in view of the trivial shortening of the crust.

The great Himalayan Range is not a geosynclinal range (Gansser, 1964). Though it is in small part composed of Tethyan geosynclinal rocks, by far the largest part is made of gneiss and other crystalline rocks identical with those of the Indian Precambrian Shield to the south, most of which had lain with little disturbance from Precambrian to Miocene time. (In some respects the relation resembles that in the northern Cordillera of the United States, where the Belt Supergroup, of geosynclinal thickness, lay virtually undisturbed from late Precambrian through most of Cretaceous time. These examples show that no close correlation is necessary between thick accumulation of sediment and orogeny.)

The Indian shield Precambrian rocks, with the overlying Talchir tillite of Permian age, extends well north of the Himalayas proper; most of the Tethyan rocks lie north of the main range (Gansser, 1964; 1966). Like the Alps, the Himalyas were formed by the collision of Gondwanaland and Eurasia, but the sea floor that must once have lain between the continents is but poorly represented in the thrust sheets. Contrary to the relations in the Alps, the thrust sheets are rooted at the north; Asia has overthrust India or India has underthrust Eurasia, rather than the converse. They are separated from the shield to the south by the alluvium of the Gangetic plain and by a thick series of Late Cenozoic continental sandstone and shale comprising the Siwaliks. The Siwaliks are entirely composed of detritus derived from the shield, not from the Tethys strata far to the north, and not, like the molasse of the Alpine foreland, from the immediately adjacent thrust sheets, but from areas to the east (and southeast) (Gansser, 1964).

Sheets of crystalline rocks many kilometers thick were thrust south for many miles over the Siwaliks and their crystalline basement. The crustal shortening has been estimated by Gansser as perhaps 500 km. Only the most northerly thrust sheets carry any Tethyan strata. The highest sheets root vertically along the line of the Indus and Tsang-po, on the north side of the Himalayan divide (a line along which vast masses of ophiolite are smeared out and along which also huge masses of rock blocks from a wide range of strata are embodied in other strata both as a kind of wildflysch and as olistostromes). This is clearly the principal subduction zone along which the crustal material involved in the shortening was eliminated (Fig. 12).

In sharp contrast with western America, where the Pacific fracture zones (with the possible exception of the Murray) definitely do not extend into the continent, the two great syntaxes of the Himalayas lie on the Quetta and Arakan Yoma lines, which are direct continuations of the Owen Fracture zone (Murray Ridge and Ninety East Ridge, respectively) on the Indian Ocean floor (Gansser, 1966). Between these two great structures (the Ninety East Ridge is the longest straight lineament on earth), the Indian Ocean floor has moved northward at least 500 km, perhaps considerably more, with the subcontinent of India ahead of it, to collide with and underthrust the Asian continent. The Quetta line shows large left-lateral displacement; in contrast, right-lateral displacement seems very

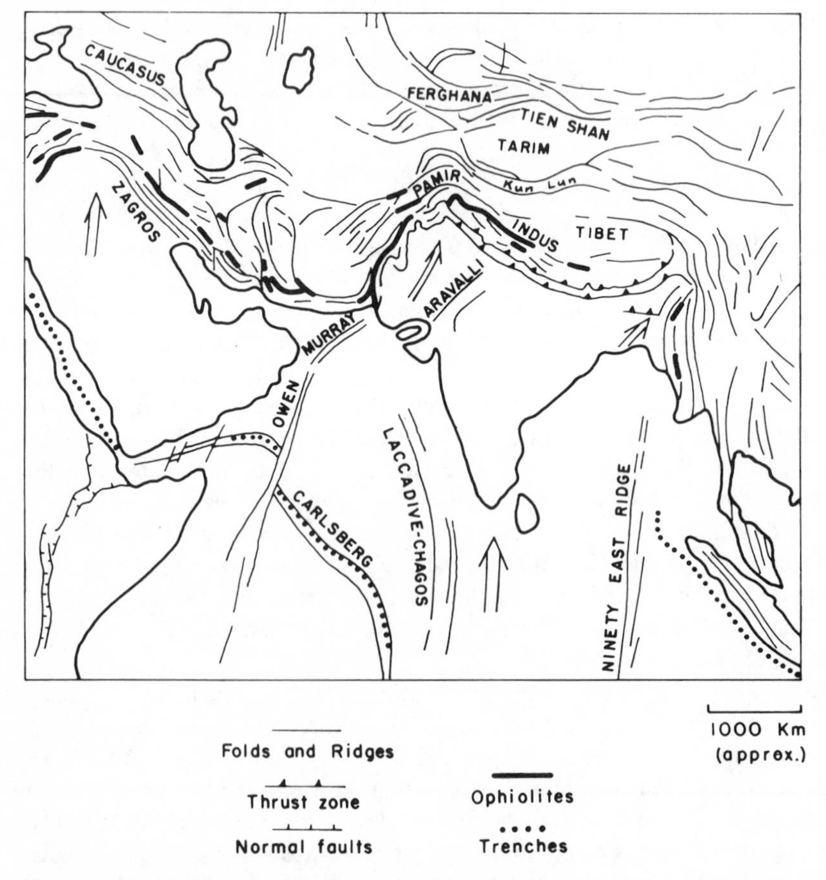

Fig. 12. The relations of the Indian peninsula to the Himalaya Range and the Indian Ocean ridges. After Gannser (1966).

probable along the Arakan Yoma trend, but the geology is so poorly known that this cannot be demonstrated.

Gansser (1966) has also pointed out other interesting features of south Asian geology. The tremendous thrusts of the Himalyas do not continue on anything like the same scale either to the west of the Quetta line in Baluchistan nor to the east of the Arakan Yoma line in Burma. It is as though the Indian block had been pushed under its northern extension to form the main range and had then carried the Himalayan block and Tibetan plateau northward for more than 1000 km beyond the Baluchistan ranges. This is also suggested by the fact that to the north of the Tibetan block the Kun Lun Range contains an ophiolite belt suggesting another subduction zone.

Although many active volcanic fields lie both east and west of the Himalaya block, none are present in the area of the great thrusts; evidently the north-south compression was too great to allow magmas to reach the surface. Although crustal shortening began in both Alps and Himalyas during the Upper Cretaceous, the Alpine tectogenesis was completed along most of the range during the Miocene, whereas the major thrusting of the Himalayas was Pliocene and Pleistocene.

In summary, it seems clear that the Himalayan tectogenesis is readily accommodated into a pattern of plate tectonics similar to that now active. Perhaps the rise of the Tibetan plateau (the greatest on earth) came about because the crustal material that disappeared down the Indus-Tsangpo line was dragged down beneath it, floating it up isostatically. The chiefly Mesozoic sediments of the plateau extend north to the Kun Lun Range, which is a facies boundary between Tethys and Angara sediments and whose ophiolites suggest deep-seated deformation.

Again, as in North America and Europe, many features of Asian tectonics are difficult to fit into a single tectogenetic pattern. The Tien Shan, Kirgiz, and northern Afghan ranges lie too far inland to be directly influenced by sea floor spreading or by subduction zones such as are apparently now active.

SUMMARY

Space does not permit a discussion of many other areas affected by Mesozoic and Cenozoic tectonics. It seems that most, if not all, the many ranges with associated ophiolites may reasonably be attributed to movements resembling the currently active plate motions, though the sharp recurvature of the western Alps and northern Apennines seems to require a flexibility of the sial that makes the term "plate tectonics" signify something other than was originally intended. It also seems, however, that many ranges have formed far from the sea and within a plate rather than at a plate boundary. The plate tectonics idea, satisfactory as it seems to be for many ranges, seems quite inapplicable to the eastern Cordillera of North America, to the Caucasus, and to most of the ranges of interior Asia. It goes without saying that the uplift of Ruwenzori, in the midst of the western

Rift of East Africa (an area of tension) demands a very different mechanism. As with most generalizations in geology, there are enough deviations from the concept of plate tectonics to assure plenty of work for future generations of geologists and geophysicists.

ACKNOWLEDGMENTS

My thanks are due to John C. Maxwell and to William R. Dickinson for their very helpful criticisms of the original manuscript. They are of course not responsible for any remaining errors of fact or of interpretation.

REFERENCES

Ampferer, O., Uber das Bewegungsbild von Faltengebirge: Jahrbuch der k. k. Reichsanstalt, H. 3 and 4, 539, 1906.

Armstrong, R. L., Sevier orogenic belt in Nevada and Utah, Geol. Soc. Am. Bull., **79,** 429, 1968.

Armstrong, F. C., and Oriel, S. S., Tectonic development of Idaho-Wyoming thrust belt, Bull. Am. Assoc. Petrol. Geologists, **49,** 1847, 1965.

Aubouin, J., Brunn, J. H., Celet, P., Dercourt, J., Godfiaux, L., and Mercier, J., Esquisse de la géologie de la Grèce: Livre à la mémoire du Prof. P. Fallot, Soc. Géol. France, **2,** 583, 1963.

Bailey, T. L., and Jahns, R. H., Geology of the Transverse Range province, Southern California, California Div. Mines Geol. Bull., **170,** 83, 1954.

Bailey, E. H., Irwin, W. P., and Jones, D. L., Franciscan and related rocks and their significance in the geology of Western California, California Dept. Nat. Resources, Div. Mines and Geology Bull., **183,** 177 pp., 1964.

Bear, L. M., The evolution and petrogenesis of the Troodos complex, Ann. Rept. Geol. Survey Dept., Cyprus, Nicosia, Cosmos Press, **26,** 1966.

Benioff, H., Seismic evidence for the fault origin of oceanic deeps, Geol. Soc. Am. Bull., **60,** 1837, 1949.

______, Orogenesis and deep crustal structure: Additional evidence from seismology, Geol. Soc. Am. Bull., **65,** 385, 1954.

______, Seismic evidence for crustal structure and tectonic activity, *in* Crust of the Earth, A Symposium, edited by A. Poldervaart, Geol. J. Soc. Am. Spec. Paper 62, 61, 1955.

______, Circum-Pacific tectonics, *in* The Mechanics of Faulting, with Special Reference to the Fault Plane Work—A Symposium, edited by J. H. Hodgson, Canada Dominion Observatory Pub. 20, 395, 1959.

Berry, M. J., and Knopoff, L., Structure of the upper mantle under the western Mediterranean Basin, J. Geophys. Res., **72,** 3613, 1967.

Blumenthal, M. M., Le systeme structural du Taurus sud-Anatolien: Livre à la mémoire du Prof. Paul Fallot, Soc. Géol. France, **2,** 611, 1963.

de Booy, T., Repeated disappearance of continental crust during the geological development of the western Mediterranean area, Verhand. Kon. Nederland. Geol. Mijnbouwk, Gen., **26,** 79, 1969.

Bostrom, R. C., Ocean ridge system in northwest America, Bull. Am. Assoc. Petrol. Geologists, **51,** 1816, 1967.

______, The ocean ridge system in the northeast Pacific Ocean, Pacific Geology, **1,** 1, 1968.

Bourcart, J., La Mediterranée et la Révolution du Pliocene: Livre à la mémoire du Prof. P. Fallot, Soc. Géol. France, **1,** 103, 1962.

Bullard, E. C., Everett, J. E., and Smith, A. G., The fit of the continents around the Atlantic: Phil. Trans. Roy. Soc. London, **1088**, 41, 1965.

Burbank, W. S., and Goddard, E. N., Thrusting in Huerfano Park, Colorado, and related problems of orogeny in the Sangre de Cristo Mountains, Geol. Soc. Am. Bull., **48**, 931, 1937.

Cadisch, J., Geologie der Schweizer Alpen, Basel, Wepf, 2d ed., 480 pp., 1953.

Chase, C. G., Menard, H. W., Larson, R. L., Sharman, G. F., III, and Smith, S. M., History of sea-floor spreading off Baja California: Geol. Soc. Am. Bull., **81**, 491, 1970.

Christiansen, F. W., Structure and statigraphy of the Canyon Range, central Utah, Geol. Soc. Am. Bull., **63**, 717, 1952.

Coleman, R. G., and Lee, D. E., Metamorphic aragonite in the glaucophane schists of Cazadero, California, Am. J. Sci., **260**, 577, 1962.

______, Glaucophane-bearing metamorphic rock types of the Cazadero area, California, J. Petrology, **4**, 260, 1963.

Coleman, R. G., Lee, D. E., Beatty, L. B., and Brannock, W. W., Eclogites and eclogites, their differences and similarities, Geol. Soc. Am. Bull., **76**, 483, 1965.

Colom, G., and Escandell, B., L'évolution du géosynclinal Baléare: Livre à la mémoire du Prof. P. Fallot, Soc. Géol. France, **1**, 125, 1962.

Cook, K. L., Rift system in the Basin and Range Province, *in* The World Rift System, Geol. Survey of Canada paper 66-14, 246, 1966.

Crittenden, M. D., Magnitude of thrust faulting limits in northern Utah, U.S. Geol. Survey Prof. Paper 424D, D 128, 1961.

Darton, N. H., Geology of the Owl Creek Mountains: U.S. 59th Cong., 1st session, Senate Doc. 219, 48 pp., 1906a.

______, Geology of the Big Horn Mountains: U.S. Geol. Survey Prof. Paper 51, 129 pp., 1906b.

Davis, G. A., Westward thrust faulting in the south-central Klamath Mountains, Geol. Soc. Am. Bull., **79**, 911, 1968.

Denham, D., Distribution of earthquakes in the New Guinea-Solomon Islands region, J. Geophys. Res., **74**, 4290, 1969.

Dickson, G. O., Pitman, W. C., III, and Heirtzler, J. R., Magnetic anomalies in the South Atlantic and ocean-floor spreading, J. Geophys. Res., **73**, 2087, 1968.

Drake, C. L., Ewing, M., and Sutton, G. H., Continental margins and geosynclines–The east coast of North America north of Cape Hatteras, *in* Physics and Chemistry of the Earth, edited by L. H. Ahrens, F. Press, K. Rankama, and S. K. Runcorn, **3**, 110, 1959.

Drewes, H. D., Structural geology of the southern Snake Range, Nevada, Geol. Soc. Am. Bull., **69**, 221, 1958.

______, Geology of the Connors Pass Quadrangle, east-central Nevada, U.S. Geol. Survey Prof. Paper 557, 93 pp., 1967.

______, Tertiary tectonics of the White Pine-Grant Range region, east-central Nevada, and some regional implications–A discussion, Geol. Soc. Am. Bull., **81**, 319, 1970.

Durand-Delga, M., Hottinge, L., Marcais, J., Mattauer, M., Milliard, Y., and Suter, G., Données actuelles sur la structure du Rif: Livre à la mémoire du Prof. P. Fallot, Soc. Géol. France, **1**, 399, 1962.

Dürr, S., Hoeppener, R., Hoppe, P., and Kockel, F., Géologie des montagnes entre le Rio Guadalhorce et le Campo de Gibraltar (Espagne meridional), Livre à la mémoire du Prof. P. Fallot, Soc. Géol. France, **1**, 209, 1962.

Eardley, A. J., Geology of the north-central Wasatch Mountains, Utah, Geol. Soc. Am. Bull., **55**, 819, 1944.

Egeler, C. G., and de Booy, T., Signification tectonique de la presence d'eléménts du Bétique de Malaga dans la Sud-Est des Cordillères bétiques avec quelques remarques sur les rapports entre Bétique de Malaga et Subbétique: Livre à la mémoire du Prof. P. Fallot, Soc. Géol. France, **1**, 155, 1962.

Ernst, W. G., Glaucophane stability and the glaucophane schist problem (abstract), Washington Acad. Sci. J., **50**, 2, 1960.
______, Petrogenesis of glaucophane schists, J. Petrology, **4**, 1, 1963.
Ewing, M., Woollard, G. P., and Vine, A. C., Geophysical investigations in the emerged and submerged Atlantic Coastal Plain, Pt. 3., Barnegat Bay, N.J. section, Geol. Soc. Am. Bull., **50**, 257, 1939.
Ewing, J. L., and Ewing, M., Sediment distribution on the mid-ocean ridges with respect to spreading of the sea floor, Science, **156**, 1590, 1967.
Ferguson, H. G., and Muller, S. W., Structural geology of the Hawthorne and Tonopah quadrangles, Nevada, U.S. Geol. Survey Prof. Paper 216, 55 pp., 1949.
Fuller, R. E., and Waters, A. C., The nature and origin of the horst and graben structure of southern Oregon, J. Geology, **37**, 204, 1929.
Gansser, A., Geology of the Himalayas, Interscience, New York, 289 pp., 1964.
______, The Indian Ocean and the Himalayas: A geological interpretation, Eclogae Geol. Hevetiae, **59**, 831, 1966.
______, The Insubric line: A major geotectonic problem, Schweiz. Min. Petr. Mitt., **48**, 123, 1968.
Gidon, P., Courants magmatiques et évolution des continents, Masson et Cie, Paris, 154 pp., 1963.
Giese, P., Die Struktur der Erdkruste im Bereich der Ivrea-zone. Ein Vergleich verschiedener seismischer Interpretationen und der Versuch einer petrographisch-geologischen Deutung, Schweiz. Min. Petr. Mitt., **48**, 261, 1968.
Gilluly, J., The distribution of mountain-making in geologic time, Geol. Soc. Am. Bull., **60**, 561, 1949.
______, Geologic contrasts between continents and oceans, *in* Crust of the Earth, edited by A. Poldervaart, Geol. Soc. Am. Special Paper 62, 7, 1955.
______, The tectonic evolution of the western United States, Geol. Soc. London Quart. J., **119**, 133, 1963.
______, Oceanic sediment volumes and continental drift, Science, **166**, 992, 1969a.
______, Geologic perspective and the completeness of the geologic record, Geol. Soc. Am. Bull., **80**, 2303, 1969b.
______, Crustal deformation in the western United States, *in* The Megatectonics of Continents and Oceans, p. 47, edited by H. Johnson and B. L. Smith, Rutgers University Press, New Brunswick, N.J., 47, 1970.
Gilluly, J., Reed, J. C., Jr., and Cady, W. M., Sedimentary volumes and their significance, Geol. Soc. Am. Bull., **81**, 353, 1970.
Glangeaud, L., Les grandes ensembles structureaux de la Méditerranée occidentale d'après les données de Géomede 1, Compt. Rend. Acad. Sci. Paris, 262, Ser. D., 2405, 1966.
Glangeaud, L., Alinat, J., Polveche, J., Guillaume, A., and Leenhardt, O., Grandes structures de la mer Ligure; Leurs évolutions et leurs relations avec les chaines continentales: Bull. Soc. Géol., France, Ser. 7, 8, 921, 1966.
Goguel, J., L'interpretation de l'arc des Alpes Occidentales, Bull. Soc. Géol. France, Ser. 7, 5, 20, 1963.
Ham, C. K., and Herrera, L. J., Jr., Role of subandean fault system in tectonics of eastern Peru and Ecuador, *in* Backbone of the Americas, Amer. Assoc. Petrol. Geologists Mem., **2**, 47, 1963.
Hamilton, W., Mesozoic California and the underflow of the Pacific mantle, Geol. Soc. Am. Bull., **80**, 2409, 1969.
Heim, A., Geologie der Schweiz, Leipzig, C. Tauchnitz, 3 vols. 1919-1922.
Heirtzler, J. R., Dickson, G. O., Herron, E. M., Pitman, W. C., III, and LePichon, X., Marine magnetic anomalies, geomagnetic field reversals, and motions of the ocean floor and continents: J. Geophys. Res., **73**, 2119, 1968.

Herrero-Ducloux, A., The Andes of western Argentina, *in* Backbone of the Americas, Am. Assoc. Petrol. Geologists Mem., **2**, 16, 1963.
Hess, H. H., Major structural features of the western North Pacific: An interpretation of H. O. 5485 Bathymetric Chart, Korea to New Guinea, Geol. Soc. Am. Bull., **59**, 417, 1948.
Hewett, D. F., The Heart Mountain overthrust, Wyoming, J. Geology, **28**, 536, 1920.
Hsu, J. K., Principles of mélanges and their bearing on the Franciscan-Knoxville paradox: Geol. Soc. Am. Bull., **79**, 1063, 1968.
Hurley, P. M., Melcher, G. C., Rand, J. R., Fairbairn, H. W., and Pinson, W. H., Rb-Sr whole-rock analyses in northern Brazil correlated with ages in west Africa, Geol. Soc. Am. Program 1966 Annual Meeting, 100, 1966.
Irwin, W. P., Geologic reconnaissance of the northern Coast Ranges and Klamath Mountains, northern California, with a summary of the mineral resources, California Div. Mines Geol. Bull., **179**, 80, 1960.
Isaaks, B., Oliver, J., and Sykes, L. R., Seismology and the new global tectonics, J. Geophys. Res., **73**, 5855, 1968.
Kaminski, W., and Menzel, H., Zur Deutung der Schwereanomalie des Ivrea-Körpers, Schweiz. Min. Petr. Mitt., **48**, 255, 1968.
Khain, V. E., and Milanovsky, E. E., Structure tectonique du Caucase d'après les données modernes: Livre à la mémoire du Prof. P. Fallot, Soc. Géol., France, **2**, 663, 1963.
Kieken, M., Les traites essentiels de la géologie Algérienne: Livre à la mémoire du Prof. P. Fallot, Soc. Géol., France, **1**, 545, 1962.
Knopf, A., Bathyliths in time, *in* The Crust of the Earth, edited by A. Poldervaart, Geol. Soc. Am. Special Paper 62, 685, 1955.
Knopoff, L., Mueller, S., and Pilant, W. L., Structure of the crust and upper mantle in the Alps from the phase velocity of Rayleigh waves, Seism. Soc. Am. Bull., **56**, 1009, 1966.
Kraus, E., Die Baugeschichte der Alpen, Berlin, Akademie Verlag, **2**, 470 pp., 1951.
______, Neue Gedanken zur Entstehung der Alpen, Eclogae Geol. Helvetiae, **47**, 61, 1954.
______, Beobachtungen and Gedanken zur Geologie von Sardinien und Korsika, Zeit. Deutsche Geol. Gesell., **112**, 75, 1960.
______, Le probléme de l'espace en tectonique dans la région méditerranéen: Livre à la mémoire du Prof. P. Fallot, Soc. Géol., France, **1**, 117, 1962.
Kuno, H., Lateral variation of basalt magma across continental margins and island arcs, Bull. Volcanol., **29**, 195, 1966.
Larson, R. L., Menard, H. W., and Smith, S. M., The Gulf of California: A result of ocean-floor spreading and transform faulting, Science, **161**, 781, 1968.
Lee, D. E., Marvin, R. F., Stern, T. W., and Peterman, Z. E., Modification of potassium-argon ages by Tertiary thrusting in the Snake Range, White Pine County, Nevada, U.S. Geol. Survey Prof. Paper 700D, D92, 1970.
Leech, G. B., The Rocky Mountain Trench, *in* The World Rift System, Geol. Survey Canada Paper 66-14, 307, 1966.
Le Pichon, X., and Heirtzler, J. R., Magnetic anomalies in the Indian Ocean and sea-floor, J. Geophys. Res., **73**, 2101, 1968.
McKenzie, D. P., Plate tectonics of the Mediterranean Region, Nature, **226**, 239, 1970.
McKenzie, D. P., and Parker, R. L., The North Pacific: An example of tectonics on a sphere, Nature, **216**, 1276, 1967.
Menard, H. W., Jr., Marine Geology of the Pacific, McGraw-Hill, New York, 271 pp., 1964.
______, Transitional types of crust under small ocean basins, J. Geophys. Res., **72**, 3061, 1967.
______, Elevation and subsidence of oceanic crust, Earth Planet. Sci. Letters, **6**, 275, 1969.
Misch, P., and Hazzard, J. C., Stratigraphy and metamorphism of late Precambrian rocks in central northeastern Nevada and adjacent Utah, Am. Assoc. Petrol. Geologists Bull., **46**, 289, 1962.

Moores, E. D., Petrology and structure of the Vourinous ophiolitic complex of northern Greece: Geol. Soc. Am. Spec. Paper 118, 74 p., 1969.
Morgan, W. J., Rises, trenches, great faults and crustal blocks: J. Geophys. Res., **73,** 1959, 1968.
Nolan, T. B., The Gold Hill mining district, Utah, U.S. Geol. Sur. Prof. Paper 177, 172 pp., 1935.
______, The Basin and Range Province in Utah, Nevada and California, U.S. Geol. Survey Prof. Paper 197-D, 141, 1943.
Nolan, T. B., Merriam, C. W., and Williams, J. S., The stratigraphic section in the vicinity of Eureka, Nevada, U.S. Geol. Survey Prof. Paper 276, 77 pp., 1956.
Orowan, E., Continental drift and the origin of mountains, Science, **146,** 1003, 1964.
Osmond, J. C., Geologic history of the site of Uinta Basin, Utah, Am. Assoc. Petrol. Geologists Bull., **49,** 1957, 1965.
Oxburgh, E. R., and Turcotte, D. L., Problem of high heat flow and volcanism associated with zones of descending mantle convective flow, Nature, **218,** 1041, 1968.
Pakiser, L. C., and Steinhart, J. S., Explosion seismology in the western hemisphere, Research in Geophysics, **2,** 123, MIT Press, Cambridge, Mass., 1964.
Palmer, H., East Pacific Rise and westward drift of North America, Nature, **220**, 341, 1968.
Pannekoek, A. J., Uplift and subsidence in and around the western Mediterranean since the Oligocene: A review, Verhandl. Kon. Ned. Geol. Mijnbouwk, **26,** 53, 1969.
Payo, G., Crustal structure of the Mediterranean Sea by surface waves: Pt. A., Group velocity, Seism. Soc. Am. Bull., **57,** 151, 1967.
Pierce, W. G., Heart Mountain and South Fork thrust, Park County, Wyoming, Am. Assoc. Petrol. Geologists Bull., **25,** 2021, 1941.
Pitman, W. C. III, and Heirtzler, J. R., Magnetic anomalies over the Pacific-Antarctic Ridge, Science, **154,** 1164, 1966.
Pitman W. C. III, Herron, E. M., and Heirtzler, J. R., Magnetic anomalies in the Pacific and sea-floor spreading: J. Geophys. Res., **73,** 2069, 1968.
Ritsema, A. R., Seismic data of the west Mediterranean and the problem of oceanization, Verhandl. Kon. Ned. Geol. Mijnbouwk, **26,** 105, 1969.
Rubey, W. W., Early structural history of the overthrust belt of western Wyoming and adjacent States, *in* Wyoming Geological Association Guidebook, 10th Ann. Field Conf., Green River Basin, 125, 1955.
Shaw, H. R., Rheology of basalt in the melting range, J. Petrology, **10**, 510, 1969.
Silver, L. T., McKinney, C. R., Deutsch, S., and Bolinger, J., Precambrian age determinations of some crystalline rocks of the San Gabriel Mountains of southern California (ab.): J. Geophys. Res., **65,** 2522, 1960.
Simmons, G., and Roy, R. F., Heat flow in North America, *in* The Earth's Crust and Upper Mantle, edited by P. J. Hart, p. 78, Am. Geophys. Union Geophys. Monograph, 13, 1969.
Smith, G. O., and Mendenhall, W. C., Tertiary granite in the northern Cascades, Geol. Soc. Am. Bull., **11,** 223, 1900.
Sonnenberg, F. P., Bolivia and the Andes, *in* Backbone of the Americas, Am. Assoc. Petrol. Geologists Mem., **2,** 36, 1963.
Spieker E. M., Late Mesozoic and early Cenozoic history of central Utah, U.S. Geol. Survey Prof. Paper 205D, 117, 1946.
Spurr, J. E., The Ore Magmas, McGraw-Hill, New York, vol. 2, 458, 1923.
Staub, R., Der Bau der Alpen, Beitr. zur Geol. Karte der Schweiz, 272 pp., 1924.
Steinmann, G., Die ophiolithischen Zonen in den Mediterranean Kettengebirgen: 14th Internat. Geol. Congress, Compte Rendu, pt. 2, 637, 1926.
Sykes, L. R., The seismicity and deep structure of island arcs, J. Geophys. Research, **71,** 2981, 1966.

Trask, P. D., Geology of the Pt. Sur quadrangle, California, California Univ. Dept. Geol. Sci. Bull., **16,** 119, 1926.

Trümpy, R., Paleotectonic evolution of the central and western Alps, Geol. Soc. Am. Bull., **71,** 843, 1960.

Veatch, A. C., Geography and geology of a portion of southwestern Wyoming, U.S. Geol. Survey Prof. Paper 56, 178 p., 1907.

Vine, F. J., Spreading of the ocean floor: New evidence, Science, **154,** 1405, 1966.

Von Huene, R., Geologic structure between the Murray fracture zone and the Transverse Ranges, Geol. Soc. Am. Abstracts for 1968, 308, 1968.

Warren, R. E., Sclater, J. G., Vacquier, V., and Roy, R. F., A comparison of terrestrial heat flow and transient geomagnetic fluctuations in the southwestern United States, Geophysics, **34,** 463, 1969.

Waters, A. C., Geomorphology of south-central Washington, illustrated by the Yakima East quadrangle, Geol. Soc. Am. Bull., **66,** 663, 1955.

Wegener, A., Die Entstehung der Kontinente und Ozeane, Fourth ed., translated by John Biram, Dover Press, New York, 246 pp., 1929.

Wilson, J. T., Continental drift, Sci. Am., **208,** 86, 1963.

Woodring, W. P., Bramlette, M. N., and Kew, W. S. W., Geology and paleontology of the Palos Verdes Hills, California, U.S. Geol. Survey Prof. Paper 207, 145 p., 1946.

Woodward, L. W., Structural geology of central-northern Egan Range, Nevada, Am. Assoc. Petrol. Geologists Bull., **48,** 22, 1964.

17
HISTORY OF THE OCEAN BASINS

by **H. W. Menard**

The ocean basins as we see them are largely the result of the horizontal and vertical motion of large rigid crustal plates during the last 200 m.y. The principal modifiers have been vulcanism within the plates and deposition of continental denudation products. Much of the history of the relationship between adjacent plates can be reconstructed from the topography, magnetic anomalies, crustal structure, and vulcanism associated with plate edges.

None of these edge phenomena indicate the motion of the plates relative to a fixed frame of reference such as the pole of rotation. Such information can be derived with only limited assumptions from the paleomagnetic poles of marine volcanoes, and the trend of the equatorial belt of pelagic sediment. The former suggest that the Pacific plate has moved some tens of degrees of latitude north since Cretaceous time, and the latter seem to suggest that it has remained relatively fixed in latitude but has moved a little north. These differences can be resolved if most of the motion occurred before the deposition of the pelagic sediment and if the remainder was largely latitudinal in the equatorial region. Such motion appears compatible with both paleomagnetism of continents and several aspects of marine geology.

INTRODUCTION

Geologists have been mapping the land for almost 150 years by actually climbing mountains, swimming rivers, breaking trail through forest, and doing whatever was necessary to look at outcrops and collect rock samples. Aided in later years by their colleagues in geochemistry and geophysics, they found a very complicated world with a history that adds another dimension of complexity. The continents are at least 3.5 b.y. old, and they have repeatedly gone through episodes of subsidence below sea level, sedimentation, deformation and mountain building, and erosion to sea level. Sedimentary rocks are as much as 10 km thick and their history is obscured by lack of fossils, facies changes, faulting, overturned folding, metamorphism, and intrusion and engulfment in igneous rocks. Observations beyond counting are available from continental geology, and many are unified by broad hypotheses but usually by more than one hypothesis. This indicates that the wealth of painstakingly acquired observations simply has not been adequate to solve many of the major problems of the history of the earth. Solving millions of small problems does not necessarily solve big ones at the same time.

Mapping of the sea floor has also been underway for 150 years but, largely through the circumstance that the ocean is deep, almost everything else about marine geology has been different from the work on land. It has cost much more to work at sea; on a logarithmic scale, oceanography is about half way between mapping the land and landing on the moon. Consequently, the number of scientists involved has been small and there has been some premium on multiple, integrated, and efficient observations. Perhaps most important, marine geologists could easily work at a distance from the sea floor, but only recently and with great difficulty could they look at outcrops and collect rock samples. Thus, they have been unable to make detailed observations and could not be confused by them. It is not surprising, therefore, that the history of the ocean basins appears remarkably simple, coherent, and easy to integrate compared to the land. The extant ocean basins appear no more than 0.2 b.y. old, although the water is older and therefore other basins existed before to hold it. In most places, basins have a thin, uniform crust covered with only a few hundred meters of undeformed sediment. Except for island-building vulcanism and gradual subsidence, hardly anything seems to affect the vast floor of the deep sea except at the edges. We know from plate tectonics that parts of the sea floor are in relative motion at rates of centimeters per year but this is not easy to detect within the oceanic crustal plates. How startingly different are those parts of the sea floor where something is happening, namely, those parts that are now recognized as plate edges. There the sea floor is obviously deformed.

Along the trailing edges of plates, there are earthquakes, high heat flow, a thin oceanic crustal layer or none at all, no sediment, fresh volcanic rock, and commonly an exceptionally large magnetic anomaly and a deep median valley. To make it even more obvious that the trailing edge is different from the placid interior of a plate, it is elevated several kilometers to form one-half of a midocean ridge. In sea floor spreading and plate tectonics, we have a paradigm that successfully explains the major features that occur and are formed at the trailing edges of plates. One aspect of the paradigm is that many of the features cannot form anywhere except at these trailing edges. Thus, when we find them in the middle of plates, or even near the leading edges, we have reason to believe that they formed in a regular sequence at a trailing edge and migrated within the plate as it moved along. Much of the elucidation of the history of ocean basins, therefore, is reduced to mapping and dating these undeformed trailing edge phenomena such as linear hills and mountains and parallel magnetic anomalies, and the very distinctive and generally perpendicular ridges, troughs, and scarps of fracture zones.

The leading edges of oceanic plates are crumbled into mountain ranges or plunge down into the mantle to form island arcs and a Benioff zone. Either way they are lost to the marine geologist and become the concern of land geologists and seismologists, respectively. The geological history of the island arcs and folded mountains can be reconstructed and compared with the history of the adjacent oceanic plate. This is being done in many places, and it appears that the simple history measured from ships provides a broad synthesis into which the detailed observations of the land can be integrated. Moreover, the modern island arcs and fold mountains have ancient equivalents that can still be identified in the geological record. Thus, the simple history of the youthful ocean basins may help unravel much of the history of the ancient, complex continents.

Plate tectonics and many other aspects of geophysics are dealt with elsewhere in this volume. Therefore, I shall address myself to new aspects of marine geology interpreted in the light of plate tectonics in order to avoid duplication.

DRIFT OF THE PACIFIC PLATE RELATIVE TO THE EQUATOR

The data of sea floor spreading and plate tectonics describe relative plate motion but hardly ever indicate anything about "absolute" motion relative to a frame of reference such as the equator or the poles of rotation. We can, however, derive some information about latitudinal motion from other sources, namely, the stratigraphy of the equatorial Pacific and the paleomagnetism of oceanic volcanoes. Given such latitudinal motion, it becomes possible to use plate tectonics to deconvolute continental drift and obtain the position of the major tectonic plates in the past. In this section, we analyze the evidence for latitudinal motion and test the results by deconvolution.

We begin with marine stratigraphy. The accumulation of calcareous ooze on the deep sea floor depends on a balance between productivity in the surface water and solution in the bottom water. Empirically, it has been established that the equatorial waters are highly productive and calcareous ooze lies in a band along the equator in the eastern Pacific (Arrhenius, 1952). For this reason, Arrhenius about 15 years ago suggested that the equatorial band could be used to identify the ancient equator when technology made drilling possible. This has now occurred and his prediction is confirmed. Meanwhile, much has been learned about bottom solution and marine stratigraphy. Calcium carbonate spheres dissolve at any depth in the ocean but much more rapidly below 3700 meters in the Pacific (Peterson, 1966). The only extensive areas above this depth are on the crests of midocean ridges. Below this depth, little accumulates but red clay and it at a rate of only a few percent of that for calcareous ooze (Goldberg and Koide, 1961). Therefore, almost all pelagic sediment is deposited on the crests of midocean ridges and is preserved as the ridges spread and gradually sink at rates of 2 to 9 cm/10^3 years (Menard, 1969b).

Consequently, the equatorial productive zone is matched by thick accumulation of sediment only where it intersects the crest of the East Pacific Rise. It is well to remember that any part of the sea floor with linear magnetic anomalies was at one time at the crest of a midocean ridge. Thus calcareous oozes, drilled on the flank of a midocean ridge may have been deposited when they were on the shallow crest.

The JOIDES holes 69 to 75 give a north-south profile of the stratigraphy of the Pacific approximately along longitude 140°W. Holes #74 and #75 bottomed in basalt, and the sediment directly above it at #74 is Middle Eocene (C. Von der Borch, personal communication, 1969). From the topography and the orientation of magnetic anomalies in the region, it appears that the crust at this locality is at about magnetic anomaly 17 to 18 and thus is about 42 to 46 m.y. old (Heirtzler et al., 1968). The belt of equatorial sediment is readily apparent in the preliminary results of the drilling (Figs. 1 and 2).

The belt can be identified from Eocene time to the present. During Upper Miocene and later time, rapid sedimentation has been centered on the equator and has built an actual sea floor ridge along it (Fig. 1). The maximum thickness of sediment, however, lies progressively north of the equator with greater age in Oligocene, and Lower and Middle Miocene time. With regard to plate motion, this suggests no latitudinal motion at this locality for the last 10 m.y., or in other terms, since magnetic anomaly #5. From 10 to 37 m.y. ago, corresponding to anomalies #5 to #13, the latitudinal movement of the plate was to the north. The change in plate motion 10 m.y. ago suggested by the equatorial sediment is simultaneous with the change in direction of relative motion observed in the northeastern Pacific (Menard and Atwater, 1968), which suggests that the motion of the Pacific plate can be determined even for relatively short periods.

With only preliminary and incomplete results from JOIDES, however, this seems premature. A constant motion is therefore assumed for the last 37 m.y., and this gives an average rate of 1 cm/yr to the north at this longitude. Little more can be determined from the JOIDES data, and we must turn to the thickness of the equatorial belt indicated by subbottom profiling (Ewing et al, 1968). The sediment near the equator consists of several components. The very thick accumulation northeast of New Guinea is associated with shallow water and mid-plate rises (Menard, 1969b). So is the area off Central America. The remaining thick accumulations form a long thin curving band extending from 80°W to 105°E with interruptions in the deep central Pacific and at the very shallow but youthful crest of the East Pacific Rise (Fig. 3). Presumably this is the sediment deposited at the equator. Although it is a weaker assumption, it

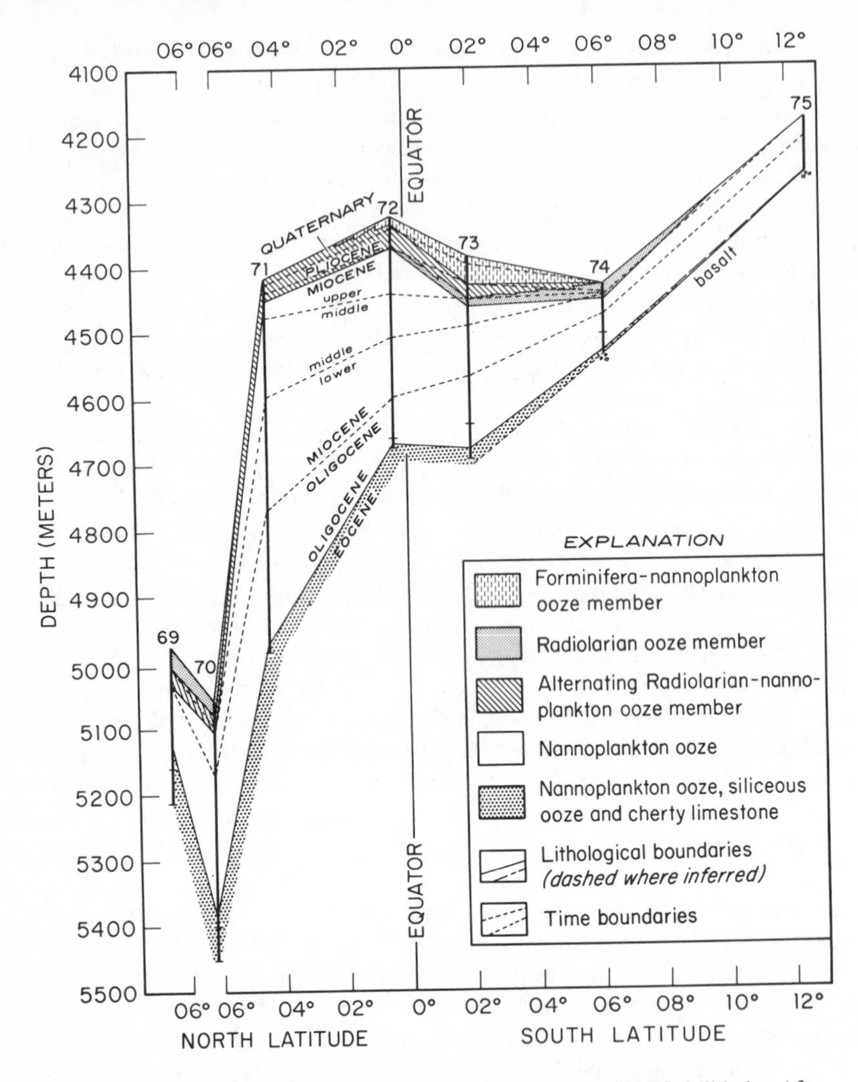

Fig. 1. Stratigraphy of the eastern equatorial Pacific from JOIDES drill holes. After C. Von der Borch (personal communication, 1969).

seems reasonable that the thickest part corresponds to the locus of the equator when that part of the crust was relatively young and drifting under the productive waters. If so, the thickest sediment is not the same age everywhere, but it does mark the former position of the equator.

At present, there is no reason to believe that the Pacific plate has been fragmented in the equatorial region in about the last 150 m.y. Consequently, the plate has moved rigidly relative to the equator and its average motion can be described in terms of a pole of relative rotation, which I shall call an "Euler pole" after the mathematician. The rotation relative to the equator has been such as to produce a curved line trending generally south of east on a mercator projection. The plate motion that would produce this configuration is a northerly (clockwise) rotation of the Pacific plate at 1.9×10^{-7} deg/yr around an Euler pole at the equator and 95°W longitude. This Euler pole at present is at the intersection of the East Pacific Rise and the equator and hence the thick equatorial sediment farther east is on a different plate. Nevertheless, this sedimentary belt has rotated around the same Euler pole at 95°W relative to the equator, but the motion has been southerly (counterclockwise).

These records of latitudinal motion put no restrictions on longitudinal motion except that it is either due east or west. That is, both the belt of sediment and the Euler pole can drift in longitude, but the pole cannot drift in latitude. These data, if correctly interpreted, eliminate the possibility, for example, that the Pacific plate is rotating relative to the equator around the Euler pole for the Pacific and America plates. Instead, this latter Euler pole is itself moving relative

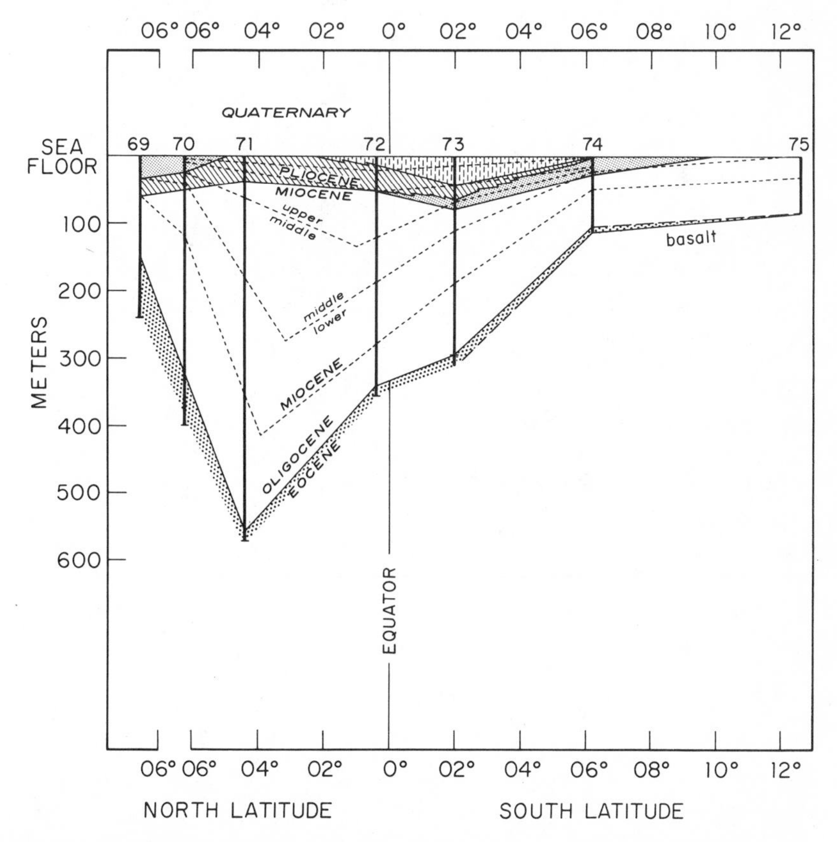

Fig. 2. Diagrammatic stratigraphy of the eastern equatorial Pacific normalized to a sea floor of constant depth to show shifting locus of thickest sediment at different times.

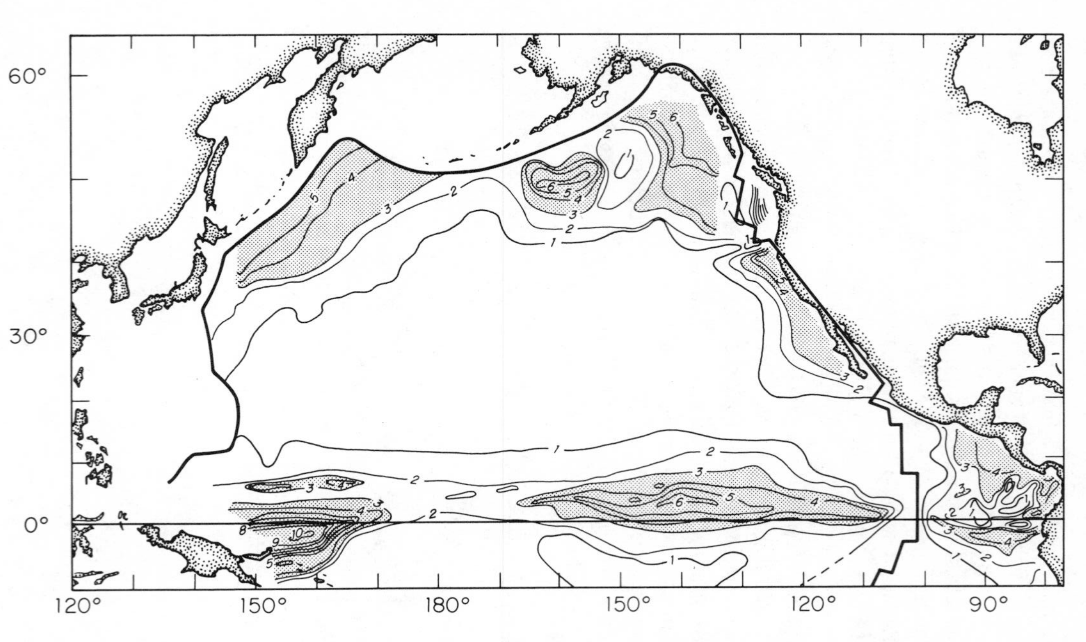

Fig. 3. Thickness of sediment in 100's of meters assuming sound velocity in sediment equivalent to water. After Ewing et al. (1968).

to the equator. Armed with this fact, we can construct a paleogeographic map for 37 m.y. ago or in early Oligocene time (Fig. 4). The construction is as follows:

1. Accept the Euler poles and rotation rates computed by Le Pichon (1968) for modern plates.
2. Rotate the Euler poles of Le Pichon in accordance with the rate and direction of motion of the Pacific plate relative to the equator.
3. Rotate the various plates around their new poles to their positions 37 m.y. ago.

No independent tests have been devised for this deconvolution. The marine stratigraphic record is not yet known in sufficient detail, and the paleomagnetic record for almost all continents suggests little change in latitude. This in itself gives a negative check in that the deconvolution also indicates little variation in latitude for any continent except Australia. Tests will emerge, however, as the deconvolution is extended backward. A test of the assumptions is given by the degree of convergence of magnetic anomalies which are 37 m.y. old and thus were at spreading centers at that time. The now split anomalies should fit together exactly if the assumption of constant rotation is correct. The fit in the Indian Ocean for anomaly #13 could hardly be improved upon, suggesting the assumptions are reasonable for the Africa and India plates for this period. The fit between the Africa and America plates in the South Atlantic is not as satisfactory but the plate edges as indicated by anomaly #13 overlap by only a few degrees of longitude. For the Antarctic, the fit is hopeless for Le Pichon's rotation of 10.8×10^{-7} deg/yr. This is hardly surprising because the Euler pole is quite near to the magnetic anomalies and therefore the spreading rate is prone to errors. Accordingly, a new rate of rotation has been calculated to give a fit of the Pacific and Antarctic plate edges 64 m.y. ago at the beginning of Cenozoic time. The rate of 6.1×10^{-7} deg/yr gives an excellent fit for 27 m.y. ago as well (Fig. 4).

PALEOMAGNETISM OF SEAMOUNTS

Before discussing the history of ocean basins as indicated by the movement of the equatorial sediment belt for the last 37 m.y., it seems worthwhile to see if other criteria indicate earlier motion relative to the equator. The paleomagnetism of small volcanic seamounts in the Pacific implies just such a motion (Uyeda and Richards, 1966; Vacquier, 1962; Vacquier and Uyeda, 1967). Similar evidence of drift relative to the magnetic pole and, thus, by the usual assumptions, relative to the geographical equator comes from the shape of linear magnetic anomalies in the Great Magnetic Bight of the northeastern Pacific (Vine, 1968). The observations consistently show a large northerly motion of the Pacific plate since the volcanoes and the magnetic bight were formed (Fig. 5). The age and thus the history of the motion is less known. Groups of seamounts in the western Pacific and near Hawaii have been dated by dredging

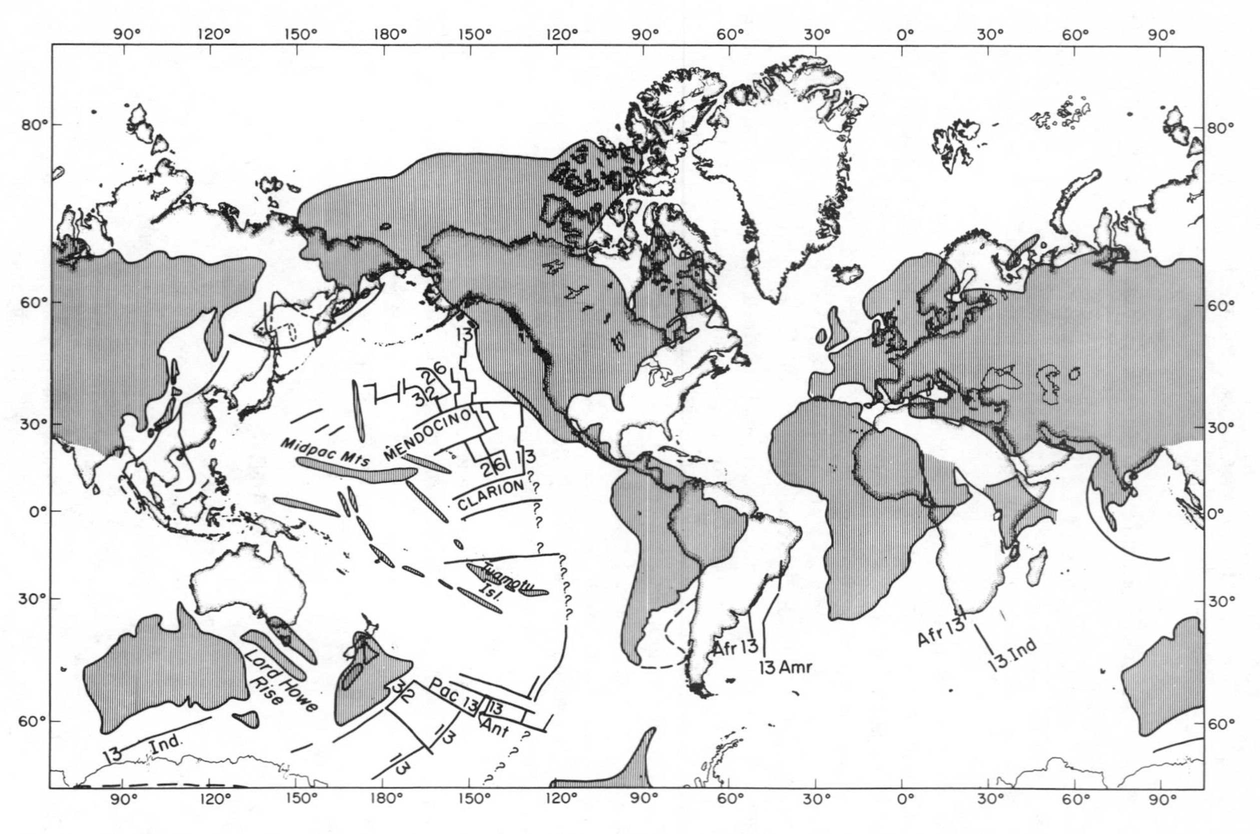

Fig. 4. The distribution of continents and some oceanic features 37 m.y. ago in very early Oligocene time.

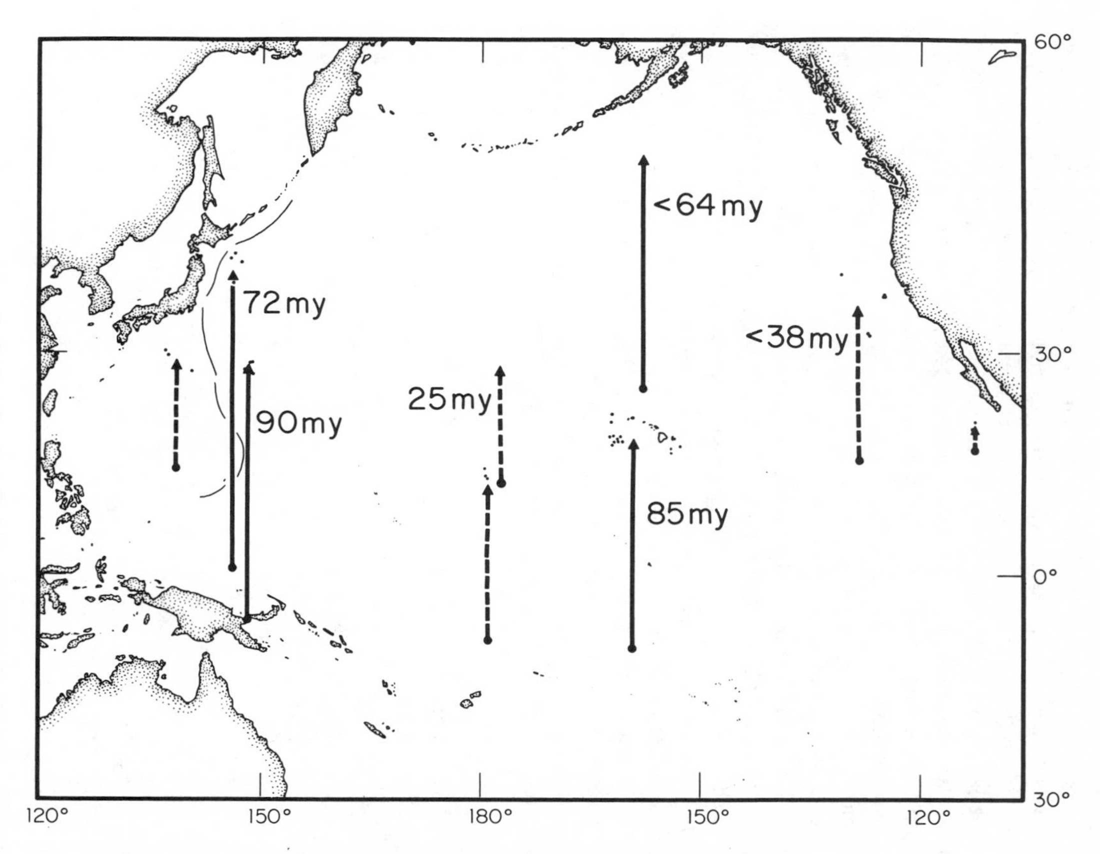

Fig. 5. Indicated latitudinal movement of sea floor features from various sources.

the tops and the age of the magnetic bight is known by comparison with the dated anomalies of the South Atlantic (Heirtzler et al, 1968). Some of the indicated motions are puzzling. Midway Atoll appears to have moved 15° north in the last 25 m.y., and the group of seamounts have an indicated motion even greater in less than 38 m.y. The indicated Midway motion (Vine, 1968) is completely incompatible with the indicated motion of the equatorial sediment belt and is dismissed here as an unresolved puzzle. The seamounts off California lie on a crust with linear magnetic anomalies, and thus the maximum age is known. On the other hand, the linearity of the surrounding magnetic anomalies in itself poses difficulties in determining the paleomagnetic poles.

The various sea floor features are attached to the Pacific plate, and once they were all formed, they moved rigidly around some Euler pole. Figure 6 shows the motion of the various older features compared to the motion that would result from a rotation of 29° to the north around an Euler pole at the equator and 95°W longitude. The correspondence of the observed and calculated motions encourages the assumptions that the Pacific plate has so rotated since the formation of the visible Great Magnetic Bight was completed 64 m.y. ago. The indicated Euler pole is the same as for the plate motion deduced from equatorial

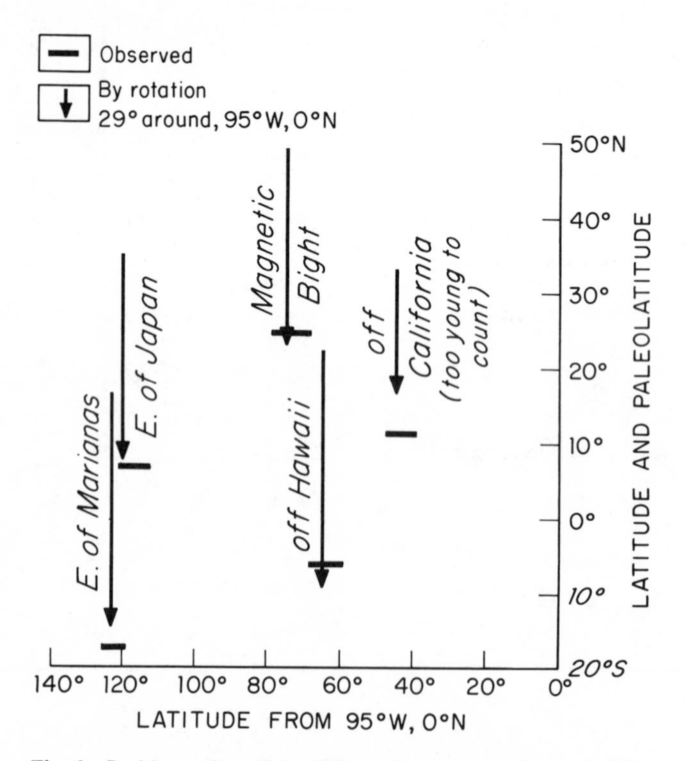

Fig. 6. Positions of sea floor features 64 m.y. ago compared with calculated positions assuming a plate rotation.

sediment. Thus, the rotation of 7° in 37 m.y. can be readily subtracted from that of 29° in 64 m.y. to give an average rate of rotation between 37 and 64 m.y. ago of 8.2×10^{-7} deg/yr. The motion may not have been uniform during the period, and in the reasonable extreme, it may all have occurred in the million years just before 37 m.y. ago. If so, it was more than 20 times faster than any relative motion known at present (Le Pichon, 1968). The only reason for suggesting it is that it accounts reasonably well for the motion of the seamounts of California as well as the older ones. This would be an extraordinary event in plate tectonics just at the end of Eocene time. It is also possible to account exactly for the latitudinal motion of the California seamounts and the others by a model with slow rotation from the formation of the magnetic bight to the formation of the California seamounts, then very rapid rotation for a million years, then slow rotation through mid-Tertiary time and no rotation for the last 10 m.y. Such an elaborate model is hardly warranted by the small number of observations, and an average motion will be assumed between 64 and 37 m.y. ago. Even so, three rates of rotation are identified and a gradual slowing of the northerly motion.

We can now construct a paleogeographic map of the world at the beginning of Eocene time, 64 m.y. ago, by another deconvolution of relative plate motions (Fig. 7). The only apparent check of this deconvolution by marine geophysics is that in the south Pacific anomaly #26 for the Pacific and Antarctic plates fits together in the map. This fit, however, was only accomplished by a rotation around the Euler pole of the Pacific-Antarctic at 6.1×10^{-7} deg/yr rather than at the faster rate estimated by Le Pichon. The only check on this rate, therefore, is that it also produces a fit of anomaly #13 as previously discussed.

The deconvolution can most readily be tested by comparing the latitudes of the continents with those determined by paleomagnetism (Fig. 8). During the last 100 m.y., Canberra and Bombay have had large latitudinal motions, the motion of the present south pole has been moderate, and Salisbury and Brasilia have moved less than 10° (Creer, 1965). The positions of the cities by our deconvolution agree within the scatter of the points with all three of the larger motions. The paleolatitudes for the continents with little latitudinal motion are, however, not in such good agreement with the curves based on continental data. Even so, the maximum indicated error is 15° and the continental observations are much too few to define the curves precisely. In sum, it appears that the motion of the Pacific plate relative to the equator is compatible with the latitudinal motions of the continents.

Inasmuch as the deconvolution is based on the rotations of rigid plates, we can also reconstruct relative motions in longitude. Absolute longitude is meaningless, so for convenience we assume that the Euler pole for the Pacific plate relative to the equator is fixed in longitude. This means that the points with an Euler latitude of zero are also fixed in longitude, and it also determines the changes in longitude of points at other Euler latitudes. It does not specify what happens to the edges of the plate because they can build outward at spreading centers or be

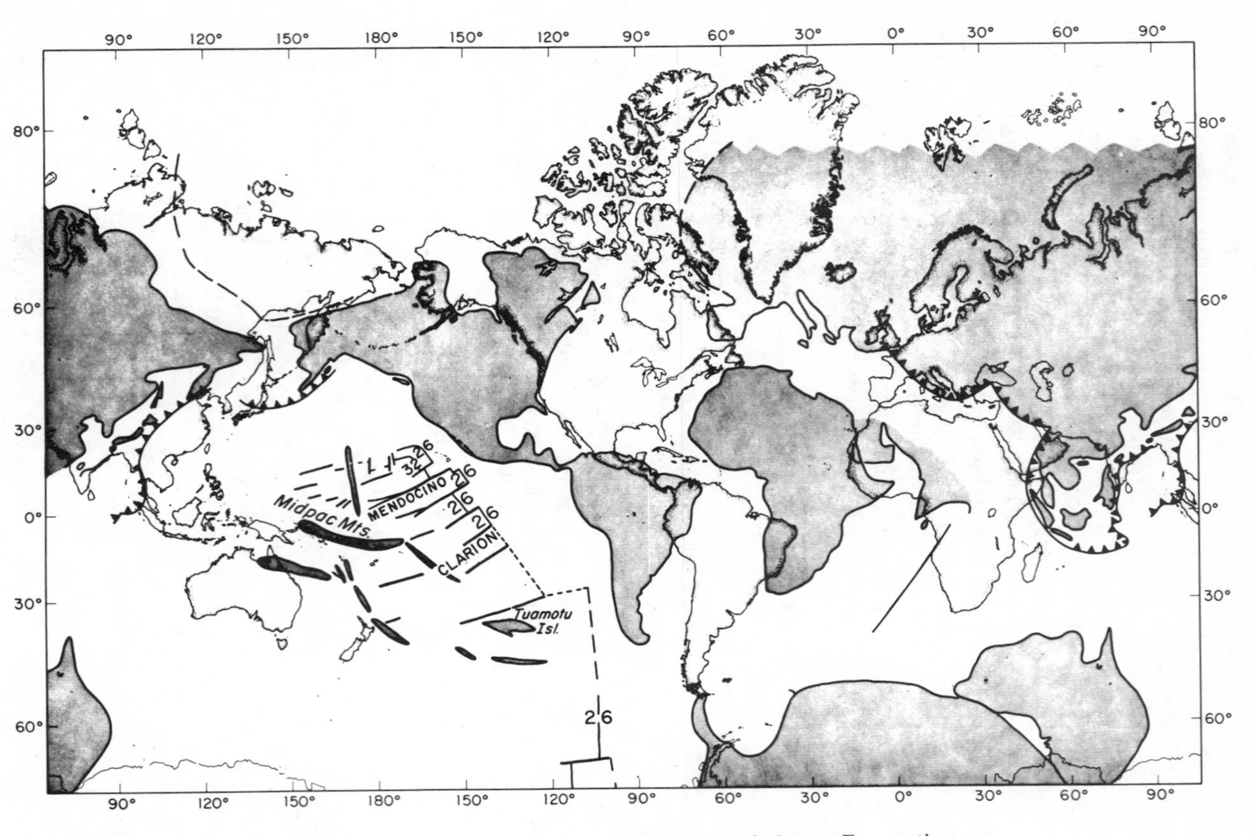

Fig. 7. The distribution of continents and some oceanic features 64 m.y. ago in lowest Eocene time.

eaten inward at trenches. In fact, the only boundary that appears even roughly fixed in longitude is the crest of the East Pacific Rise near 20°S. Consider the motion of Pacific plate anomaly #26, which marks the position of the trailing edge of the plate at the time it was formed. The part of the anomaly that was south of the paleoequator has been rotating northwest, and the part north of the paleoequator has rotated northeast. At the same time, the plate edge in the South Pacific has built toward the east and so has the plate edge in the North Pacific. Indeed, it is the universal characteristic of active plates that they grow larger on their trailing edges. The only noteworthy feature here is that the orientation and motion of the edge are known. The other tectonic plates have motions relative to the Pacific, which determine the actual boundary phenomena. For the Pacific plate alone, however, the resultant of rotation and outbuilding has been a conspicuous change in the orientation of the trailing edge.

In the plate tectonics paradigm, continents are displaced and occasionally rifted or joined together but since the breakup of Gondwanaland, changes in the ocean basins have been far more spectacular. The total area of the basins is constant if the area of the earth and of continents are constant, it is only the shape and area of individual basins that change. During the last 64 m.y., the Atlantic, particularly the South Atlantic, has widened by a major fraction. The distance between Africa and India has also increased with the opening of the Carlsberg Ridge, and the distance from India to Antarctica has increased enormously. In contrast, the distance from Africa to Australia has hardly changed at all in the last 37 m.y. and the change was small between 64 and 37 m.y.

In the Pacific, the changes have been highly varied depending on the part of the perimeter involved. For the period of interest, the basin originally consisted of three major plates, the Pacific plate, *P*, to the southwest, the Farallon plate, *F*, to the southeast, and an "Aleutian" plate, *A*, to the north, as shown in Fig. 8*a*. Some 64 m.y. ago, the Aleutian plate perhaps extended from the Mariana Trench or the Philippine Islands to not far off North America. The Farallon plate formed the Pacific boundary of the remainder of North and South America. All the rest of the Pacific probably was the Pacific plate although possibly some minor plates existed in Melanesia.

Since that time, the Aleutian plate has been entirely consumed in the Aleutian and probably the Kurile trenches. The Farallon plate moved northwest relative to the America plate until it extended from Alaska to Patagonia (Fig. 8*b*). During the entire time it was very narrow or, in other terms, the crest of the East Pacific Rise was near the continental margin. This means that the geography was very similar to that off Mexico or Oregon at present. The crest of the ridge was shallow and collected carbonate oozes. Much of the boundary between the America and Farallon plates was a trench or young mountain range that formed a trap and little sediment reached the Farallon plate. Even if it did, it was largely dammed by the elevated ridge crest as occurs off Washington at present. Some

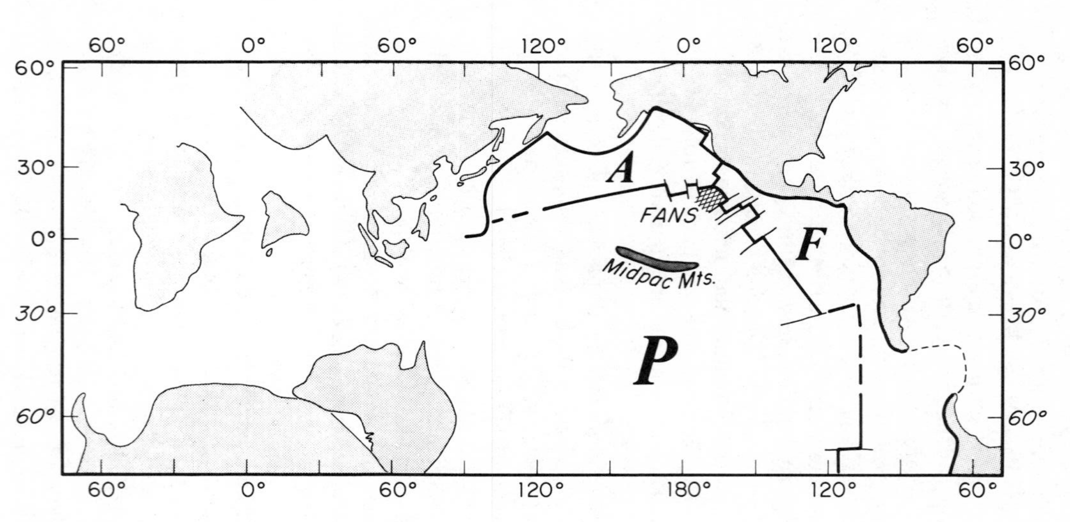

Fig. 8*a*. Schematic map of Pacific plates 64 m.y. ago.

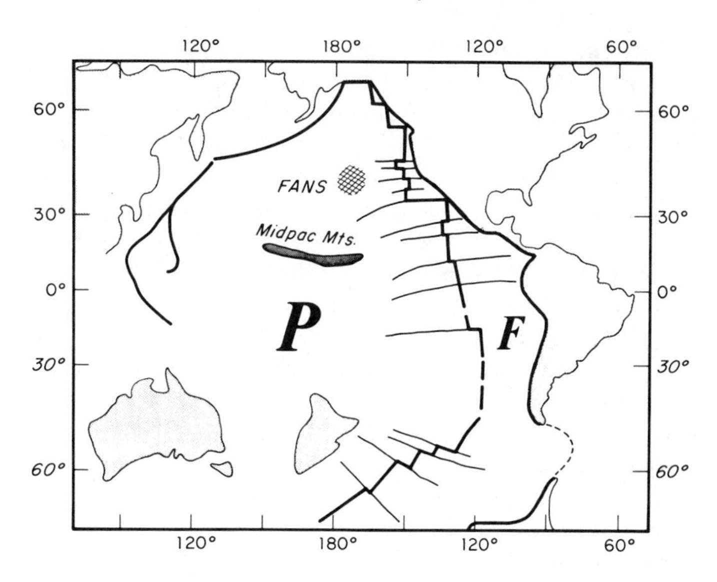

Fig. 8*b*. Schematic map of Pacific plates 37 m.y. ago.

sediment under ideal circumstances could bypass both plate margins just as it does now off Oregon and Washington. Thus, a combination of carbonate oozes and turbidites on the Farallon plate moved into the American trenches. The structure too probably was like that off Mexico. The lithosphere was still thin when it plunged on the Benioff zone, and consequently it warmed quickly and did not plunge to great depths (Atwater, personal communication, 1969; McKenzie, 1967).

The crest of the East Pacific Rise is offset by many long ridge-ridge transforms. This was true in the past as well although the length of the transforms varied and new ones appeared and others were destroyed because of changes in plate motion (Atwater and Menard, 1970; Menard and Atwater, 1968, 1969). On the other hand, the west coast of North America is relatively straight. Thus the Farallon plate, narrow everywhere, was extremely so in a few places such as between the Pioneer and Murray fracture zones when they were active transform faults. The relative motion between the Farallon and America plate was converging rather than entirely parallel to their mutual boundary. Consequently, the easternmost projections of the rise one by one intersected the America plate, and the Farallon plate was broken into sections in a complex but orderly manner (Chase et al, 1970; Larson et al, 1968; McKenzie and Morgan, 1969). In this way, the Farallon plate largely disappeared in the North Pacific just like the Aleutian plate.

In the South Pacific it also broke up because several plates now form the east flank of the East Pacific Rise. It expanded, however, rather than disappearing into a trench. Between South America and the East Pacific Rise, the Chile Rise appeared in another complex series of events (Herron and Hayes, 1969). It either split the Farallon plate lengthwise or originated at the boundary between the America and Farallon plates. In any event, the Farallon plate in the area is now preserved between the crests of the East Pacific and Chile rises and is growing on both sides. This will continue until either the Chile or East Pacific rise intersects a trench when the Farallon plate will begin to experience marginal effects in accordance with its relative motion.

The Pacific plate has been growing at the expense of the Aleutian and Farallon plates but not as rapidly as it is being consumed by the Asia and India plates. Across the North Pacific little change has occurred. The eastern edge of the plate has built outward about as much as the western edge is consumed. A southeasterly line from the Mariana Trench to the crest of the East Pacific Rise, however, is shorter by about 3000 km than it was 64 m.y. ago. The whole basin is about the same width along this line, but the Farallon and other plates now take up more space than in the past. The greatest change in the Pacific plate is along a line from the Aleutian Trench to New Guinea. Despite the disappearance of the Aleutian plate, this distance has shortened by 4000 km as the northeasterly movement of Australia and New Guinea has encroached upon the Pacific plate. An enormous amount of the southwestern part of the Pacific plate apparently disappeared into the complex island arcs of Melanesia.

We can now consider marine geological tests of the motion of the Pacific plate relative to the equator and of some tectonic effects that have occurred at its edges. The effects to be evaluated are the absence of a pre-Oligocene sediment belt, the absence of thick coral reefs on the Midpac Mountains, and the origin of the ancient abyssal plain south of the Aleutian Trench. The formation of a thick belt of sediment at the equator requires rapid calcareous sedimentation, shallow water to preserve the sediment, and slow drift of the crust across the equator. All these conditions have existed for the last 37 m.y., and as much as 600 meters of sediment have accumulated within a few degrees of the equator. Why is the sediment not similarly thick to the north? Calcareous pelagic foraminifera and coccoliths were abundant in earlier times and oceanic circulation, reasonably similar to the present, presumably gave high productivity at the equator. Likewise, the ancient East Pacific Rise crest was elevated by the same process as at present, and presumably to about the same depth, or perhaps a little less because of growth of the ocean (Menard, 1969b; Rubey, 1951). We have found, however, that the angular velocity at which the crust rotated past the equator was about four times as great before Oligocene time as after. Thus, the expected maximum would be about 150 meters and in most places it would be less. This agrees very well with the observed thickness in the North Pacific (Fig. 3). It will be apparent that the Pacific plate moves only very slowly relative to the equator near the Euler pole for the motion. Thus, an enormous thickness of sediment at this pole might be expected. It has not accumulated for the simple reason that the trailing edge of the Pacific plate has been far to the west of the Euler pole until quite recently. During most of Tertiary time the Euler pole has been over the Farallon plate or one of its southern fragments. Consequently, the expected thick sediment has disappeared into the American trenches.

The Midpac Mountains are a long chain of enormous guyots with flat summits that were once islands and then shallow banks. On them a few patches of Upper Cretaceous reef corals accumulated (Hamilton, 1956) but only Wake Island became an atoll. Why are they not all atolls? A possibility is suggested by Fig. 8. When the patches of coral reef formed, the Midpac Mountains were mostly a little south of the equator, which is a favorable location for atoll development. At present, atolls would form if the mountains were shallow banks. In the eastern Pacific atolls do not exist even in equatorial waters because the crust is too young for them to form (Menard, 1969a). Even if old crust existed, however, there might not be any atolls because coral reefs of any kind are virtually absent. Indeed the entire biota is highly impoverished (Durham, 1963), and this is generally attributed to the enormous distance separating the eastern islands from the up-drift biota of the central Pacific. The pelagic larval forms normally float far enough to populate any suitable island, but they can only very rarely live the time necessary to span the eastern Pacific. If they do make a lodgement, disease or a predator may eliminate the species and repopulation may never occur. The possibility of difficulties of populating and repopulating the Midpac Mountains has been suggested (Menard and Ladd, 1963), but the idea seemed to conflict with what is now a favorable location.

Figure 8*a* suggests that the circumstances may have actually been highly unfavorable because, when they were shallow banks, the Midpac Mountains were in the eastern rather than the central Pacific. The Farallon plate was narrow, and the mountains were not far west of the eastern edge of the Pacific plate. To the west of the mountains was the combined Pacific-Indian ocean, which apparently extended to Africa except for a few islands along the equator (Fig. 7). The islands of Indonesia were far west of their present position and those of Melanesia were far to the south. Thus, the Midpac Mountains were in an unfavorable environment rather like an extreme example of what now obtains in the eastern Pacific.

As an aside, we may consider the bearing of this paleogeography on the existence of an ancient, subsided midocean ridge, the Darwin Rise, in the western Pacific (Hess, 1965; Menard, 1964). The Darwin Rise, if correctly proposed, has as yet unmapped bilaterally symmetrical magnetic anomalies with the youngest crust in the middle like any active ridge. To the south and also the east of it, the East Pacific Rise has developed. The process whereby one ridge turned into another may have been the same as that which caused the Chile Rise to form within the Farallon plate or the crest of the East Pacific Rise to jump east and leave a dead ridge between the Murray and Molokai fracture zones (Menard and Atwater, 1969). Thus, although the existence of the Darwin Rise remains to be tested, a mechanism by which it could have died has been shown to exist elsewhere.

The final aspect of plate motion and tectonic edge effects to be examined here is related to the source area of the ancient turbidites of the Aleutian "abyssal plain" south of the Aleutian Trench (Hurley, 1959). Subbottom profiles indicate a thick lens of sediment (Ewing et al, 1968; Fig. 3) in this region. Very careful analysis of high resolution subbottom profiles reveals a terrane of channels, levees, and coalescing fans overlain by a relatively uniform blanket of pelagic ooze (Hamilton, 1967). The age of the blanket can only be estimated from typical sedimentation rates, but it probably began to accumulate no later than middle Tertiary time and perhaps much earlier. The subbottom profiling data are not adequate to do more than suggest an isopach map of the deposit. It thins to the west, south, and southeast but the northern edge is unknown; consequently, the source area is not identifiable by this technique. The obvious source is Alaska, which is separated from these fans only by the narrow Aleutian Trench. The trench, however, has been an impenetrable sediment barrier since at least late Tertiary time during which an enormous amount of the Aleutian and Pacific tectonic plates has disappeared into the trench. The position of the continents and the fans 37 m.y. ago illustrates the importance of the crust that has been lost (Figs. 4 and 8). At that time, the fans were more than 1000 km from Alaska and almost as distant from any other continental source.

The fans lie on magnetic anomaly #26, and therefore are no more than 64 m.y. old, at which time the paleogeography was hardly favorable for their accumulation (Fig. 8*a*). A trench and spreading center separated the continental source from the side of deposition. The total distance was small, however. The ridge crest was very near the continent, perhaps somewhat nearer than suggested by Figure 8*a*. Thus, the paleogeography was very similar to that now existing off Mexico or Oregon and Washington. In the latter environment, the vast flux of sediment from the Columbia River overwhelms the tectonic relief and Cascadia Channel is funneling sediment onto fans west of the active Juan de Fuca and Gorda ridges. Perhaps the Aleutian deep sea fans formed in the same way from deposits of a large river draining what is now Mexico.

CONCLUSIONS

From the stratigraphy of equatorial pelagic sediment and the paleomagnetism of seamounts, it appears that the Pacific tectonic plate has been rotating northward with respect to the equator around an Euler pole at 95°W and on the equator. The average rotation was relatively fast from 64 to 37 m.y. ago. It was slower for 27 m.y., and apparently stopped 10 m.y. ago.

With regard to the Pacific basin, the calculated motion offers an explanation for the distribution of both thick and thin sediment, paleomagnetism of most but not all seamounts, and the absence of atolls rising from the guyots of the Midpac Mountains. On the speculative assumption that relative plate motions calculated by Le Pichon (1968) have been constant for all of Teritary time, the positions of plates relative to the equator can be deconvoluted. The deconvolutions are compatible with paleomagnetism on continents with rapid latitudinal motion, but may be less so with others that were less mobile.

REFERENCES

Arrhenius, G., Sediment cores from the East Pacific, Rept. Swed. Deep Sea Exp., **5**, Fasc. 1, 1952.

Atwater, T. M., and Menard, H. W., Magnetic lineations in the Northeast Pacific, Earth Planet. Sci. Letters, **7**, 445-450, 1970.

Chase, C. G., Menard, H. W., Larson, R. L., Sharman, G. F., III, and Smith, S. M., History of sea-floor spreading west of Baja California, Bull. Geol. Soc. Am., **81**, 491, 1970.

Creer, K. M., Paleomagnetic data from the Gondwanic continents, *in* A Symposium on Continental Drift, edited by P. M. S. Blackett, Sir Edward Bullard, and S. K. Runcorn, p. 27, Philos. Trans. Roy. Soc. London, Ser. A., 258, 1965.

Durham, J. W., Paleogeographic conclusions in light of biological data, *in* A Symposium on Pacific Basin Biogeography, p. 355, Bishop Museum Press, Honolulu, 1963.

Ewing, J., Ewing, M., Aitken, T., and Ludwig, W. J., North Pacific sediment layers measured by seismic profiling, *in* The Crust and Upper Mantle of the Pacific Area, edited by L. Knopoff, C. L. Drake, and P. J. Hart, p. 147, Geophysical Monograph No. 12, Am. Geophys. Union, Washington, D.C., 1968.

Goldberg, E. D., and Koide, M., Geochronological studies of deep sea sediments by the ionium/thorium method, Geochim. Cosmochim. Acta, **26**, 417, 1961.

Hamilton, E. L., Sunken Islands of the Mid-Pacific Mountains, 97 pp., Geol. Soc. Am. Mem., 64, Baltimore, 1956.

______, Marine geology of abyssal plains in the Gulf of Alaska, **72**, 4189, 1967.

Heirtzler, J. R., Dickson, G. O., Herron, E. M., Pitman, W. C., III, and LePichon, X., Marine magnetic anomalies, geomagnetic field reversals, and motions of the ocean floor and continents, J. Geophys. Res., **73**, 2119, 1968.

Herron, E. M., and Hayes, D. E., A geophysical study of the Chile Ridge, Earth Planet. Sci. Letters, **6**, 77, 1969.

Hess, H. H., Mid-oceanic ridges and tectonics of the sea floor, *in* Submarine Geology and Geophysics, edited by W. F. Whittard and R. Bradshaw, p. 317, Proc. Seventeenth Symp. Colston Soc., Butterworths, London, 1965.

Hurley, R. J., The geomorphology of abyssal plains in the northeast Pacific Ocean, Ph.D. thesis, University of California, Los Angeles, 1959.

Larson, R. L., Menard, H. W., and Smith, S. M., The Gulf of California: A result of ocean-floor spreading and transform faulting, Science, **161**, 781, 1968.

Le Pichon, X., Sea-floor spreading and continental drift, J. Geophys. Res., **73**, 3661, 1968.

McKenzie, D. P., Some remarks on heat flow and gravity anomalies, J. Geophys. Res., **72**, 6261, 1967.

McKenzie, D. P., and Morgan, W. J., The evolution of triple junctions, Nature, **244**, 125, 1969.

Menard, H. W., Marine Geology of the Pacific, 271 pp. McGraw-Hill, New York, 1964.

______, Growth of drifting volcanoes, J. Geophys. Res., **74**, 4827, 1969a.

______, Elevation and subsidence of oceanic crust, Earth Planet. Sci. Letters, **6**, 275, 1969b.

Menard, H. W., and Atwater, T. M., Changes in direction of sea-floor spreading, Nature, **219**, 463, 1968.

______, Origin of fracture zone topography, Nature, **222**, 1037, 1969.

Menard, H. W., and Ladd, H. S., Oceanic islands, seamounts, guyots and atolls, *in* The Sea, **3**, edited by M. N. Hill, p. 365, Interscience, New York, 1963.

Peterson, M. N. A., Calcite: Rates of dissolution in a vertical profile in the Central Pacific, Science, **154**, 1542, 1966.
Rubey, W. W., The geological history of sea water, Bull. Geol. Soc. Am., **62**, 1111, 1951.
Uyeda, S., and Richards, M., Magnetization of four Pacific seamounts near the Japanese islands, Bull. Earthquake Res. Inst. Tokyo Univ., **44**, 179, 1966.
Vacquier, V., A machine method for computing the magnetization of a uniformly magnetized body from its shape and a magnetic survey, Proceedings Benedum Earth Magnetism Symposium, p. 123, Pittsburgh, 1962.
Vacquier, V., and Uyeda, S., Paleomagnetism of nine seamounts in the Western Pacific and of three volcanoes in Japan, Bull. Earthquake Res., Inst. Tokyo Univ., **45**, 815, 1967.
Vine, F. J., Paleomagnetic evidence for the northward movement of the North Pacific Basin during the past 100 m.y. (abstract), Trans. Am. Geophys. Union, **49**, 156, 1968.

REGIONAL GEOPHYSICS

18

REGIONAL VARIATIONS IN GRAVITY*

by **G. P. Woollard**

1 by 1° Bouguer gravity anomaly maps for each of the continental areas are examined in terms of apparent anomalous relations to elevation and correlated with isostatic anomaly and seismic crustal data where available. It is found that in most areas anomalous gravity is related to the mean velocity of the crust and upper mantle. There is a corollary correlation with crustal thickness such that where the crust has a high mean velocity, it is of greater than normal thickness for the surface elevation and the gravity values have a positive bias. Conversely, where the crust has a subnormal mean velocity, it has a subnormal thickness for the surface elevation and the gravity values have a negative bias. These results suggest that there are few actual departures from isostatic equilibrium and that in most areas abnormalities in gravity are related in large measure to subsurface changes in crustal density. There are areas such as eastern Canada where the gravity values have a negative bias because of an apparent time lag in crustal rebound following deglaciation. There may be major mass inequalities in the upper mantle other than those normally recognized. These are suggested by the subnormal gravity field associated with southern India and the abnormal gravity field associated with much of western Europe and Australia, but as yet, there is no seismic confirmation of such disturbing masses at depth.

*Hawaii Institute of Geophysics Contribution No. 331.

INTRODUCTION

The recognition of isostasy as a natural phenomenon whereby variations in surface mass distribution as defined by regional changes in surface elevation are compensated by changes in mass distribution of opposite sign at depth represented one of the significant discoveries in earth science. Despite the accumulation of an extensive body of knowledge concerning the subsurface mass distribution in recent years, it cannot be said that there is as yet an understanding of the isostatic mechanism. Seismic refraction measurements have shown that major changes in regional surface elevation are in general accompanied by changes in crustal thickness as defined by the depth of the Mohorovicic seismic velocity discontinuity.

This relation, when first established, suggested that the concept of a floating crust supported by a denser mantle substratum, which is embodied in the Airy theory of isostasy, was correct. As the number of seismic refraction crustal measurements has increased, however, and data have been obtained from many different areas, it has become apparent that the original concept of a simple relation between surface elevation and crustal thickness based on the assumption of a uniform density differential between the crust and mantle does not exist; that the crust and upper mantle vary laterally in composition and density; that there is no uniform pattern of crustal layering; and that in certain areas there appear to be deeper mass distributions in the upper mantle that also play a role in the establishment of isostasy. These observations have given credence to the concept embodied in the Pratt model of isostasy; namely, that surface elevation and gravitational compensation are dependent on variations in mean density above some level at depth where all crustal columns exert equal pressure.

It is to be noted, however, that where the compensating mass distribution is made up of a number of point sources with varying density differences at varying depths, their integrated gravity effect at the surface of a spherical earth will not correspond to that of the Pratt isostatic model even though the depth of isostatic compensation is the same. This is because the Pratt isostatic model assumes a single point source with the mean density difference for the entire column concentrated at the center of mass of the compensating column. Because gravitational attraction varies as the square of the distance to a point source, it would be fortuitous for the Pratt model to give the same value as the integrated value from several point sources at different depths.

This dependence of the compensation value on a changing pattern of mass distribution at depth is probably the reason there is no apparent universal depth of compensation under the Pratt concept of isostasy. This was early recognized by Bowie (1917), who on the assumption that the depth of compensation assumed which would yield a minimum isostatic gravity anomaly was the most nearly correct one, found that the most probable depth of compensation varied on a regional basis in the United States. In some regions, the value was about 56 km and in others it was around 130 km.

The most recent quantitative study of the isostatic mass distribution (Dorman and Lewis, 1970), and its application to the United States (Lewis and Dorman, 1970), defines a four point mass distribution for the compensation extending down to a depth of 400 km. This isostatic model is of interest in that it brings out clearly the possible contributions of deep seated density changes in explaining the relationship between regional variations in surface elevation and observed regional changes in Bouguer gravity anomalies.

The analysis of the data for the United States by Lewis and Dorman (1970) is germane to the present study in that the results define what appears to be a more realistic "norm" than current isostatic models for evaluating abnormalities in the regional Bouguer anomaly pattern as expressed by 1 by 1° average Bouguer anomaly values. The only questions concerning this isostatic model are:

1. The reality of the compensation depth of 400 km, since a three point source extending to 250 km gives results that are much the same as those obtained using a 400 km depth and a four point source.
2. The high degree of smoothing incorporated in the regional elevation and Bouguer anomaly values.

The relation obtained by Lewis and Dorman for their preferred model (400-km depth with a four point source) for the gravitational compensation associated with regional changes in elevation (wavelengths greater than 410 km) can be written as $BA' = -0.12\ \Delta h' + 0$ where $\Delta h'$ is the regional elevation in meters. This expression is very similar to that derived by Woollard (1968) for evaluating abnormalities in Bouguer anomaly values on a regional basis in the United States. The writer's expression is based on the assumption that areas of 3 by 3° size are compensated and there is a linear relation between the 1 by 1° Bouguer anomaly at the center of each 3 by 3° square and the average elevation for the 3 by 3° square. This relation originally proposed by Wilcox (1967) varies for different areas, but in the United States can be written as $BA' = -0.12\ \Delta h' + 4$ where $\Delta h'$ is the average elevation for the 3 by 3° square and BA' pertains to the center 1 by 1° square. Where the elevation of the center 1 by 1° square differs significantly from that of the 3 by 3° area, a terrane correction is applied. This terrane correction term is $C = -0.013\ (h_1 - h_3)$ where h_1 and h_3 are the respective elevations in meters for the center 1 by 1° square and the 3 by 3° square. The principal difference between the above expression, other than the 4-mgal zero elevation intercept value, and that based on Lewis and Dorman's analysis is in the regional elevation value. In the highest portion of the Colorado Rocky Mountains centered at about 38°30′N latitude, 106°30′W longitude, the 3 by 3° average elevation value is ≈ 2520 meters and that determined by Lewis and Dorman ≈ 2370 meters. This difference of 150 meters in regional elevation results in a 13-mgal difference in the predicted Bouguer anomaly values. The Lewis and Dorman relation gives $BA' = -285$ mgal and that based on the average 3 by 3° elevation − 298 mgal. Although this example represents a "worst" case, it points up the significance of the size of area that is believed to be completely compensated. In this connection, the space domain plots of Lewis and Dorman (1970), which show the dependence of the vertical Bouguer anomaly gradient on distance, indicate zero dependence is obtained at distances beyond about 150 km. This suggests that the critical size area is in fact about 3 by 3° in size. The fact that the Lewis and Dorman elevation values are less than the 3 by 3° average values in the Rocky Mountains and greater than the 3 by 3° average values in low lying areas, such as the Mississippi embayment, suggests that the filter used by Lewis and Dorman in determining the regional Bouguer anomaly and elevation values gives results that in effect are for an area that is considerably larger than 3 by 3° in size.

That the Lewis and Dorman isostatic model does give somewhat different compensation values from those based on the Airy-Heiskanen or Pratt-Hayford concepts of isostasy can be shown for the same sector of the Rocky Mountains already considered. The Lewis and Dorman value as indicated is + 285 mgal. The Airy-Heiskanen value is ≈ + 278 mgal for H = 30 km, and the Pratt-Hayford value for H = 113.7 km is ≈ + 273 mgal. A Bouguer anomaly value of about + 298 mgal is predicted using the empirical relation of the 1 by 1° versus 3 by 3° anomaly values; this relation, although not based on an isostatic model, has proved to be reliable to better than 10 mgal over 65% of the time even without considering the terrane effect, changes in geology, or changes in crustal and upper mantle parameter values.

These uncertainties, plus the marked difference in the relation of average Bouguer anomaly values to elevation values for 3 by 3° size areas on the different continents, with consequent differences in Bouguer anomaly values that amount to over 80 mgal for the same elevation (Woollard, 1969b), point up the fact that the isostatic compensation mass distribution below the crust mantle interface is probably as heterogeneous as that now defined by seismic measurements for the crust. This conclusion appears to be verified by the regional differences in upper mantle structure above the 400-km seismic discontinuity defined by travel time studies of seismic waves from large explosions as well as earthquakes. See, for example, Anderson (1967), Archambeau et al. (1969), Green and Hales (1968), and Lewis and Meyer (1968) for regional differences in upper mantle structure in the United States.

Woollard (1959, 1962, 1966, 1968, 1969b, 1970) in studying the interrelationship between surface elevation, geology, gravity, and seismic crustal data has come to the conclusion that except for areas where there is tectonic or former glacial displacement of the crust, there is isostatic equilibrium and, in general, an interdependence of surface elevation, crustal thickness, crustal and upper mantle composition and density values as reflected in the seismic velocity values. Changes in the values of the latter parameters result in anomalous changes in gravity because of: (1) relatively short wavelengths for changes in the composition of the lithosphere, (2) earth curvature, and (3) the dominant gravitational effect of the near surface mass distribution reflecting the density of the crust over that of deeper mass distributions, especially that associated with the crustal root, contributing to isostatic compensation. It is because the gravitational effect of deep seated changes in mass are only partially realized at the surface within the confines of a given lithosphere unit (Woollard, 1968) that the near surface effect is dominant. This effect, although analogous to the surface topographic effect that results in positive free air anomalies where there are hills and negative free air anomalies where there are valleys, is not removed by the Bouguer anomaly reduction as is the topographic effect because it is of subsurface origin.

The effect of lateral changes in subsurface density and mass distribution is considerable, and in areas of short wavelength change in values of crustal seismic parameters, the effect can give rise to local gravity anomalies exceeding 50 mgal.

With very short wavelength changes, such as are obtained with an intrusive mass, the gravity effect can exceed 100 mgal. The Koolau caldera on the island of Oahu, for example, is characterized by a local Bouguer anomaly "high" of + 115 mgal (Woollard, 1951). Seismic reflection and refraction studies of this area (Furumoto et al., 1965) show the top of the disturbing mass is about 2 km below sea level and that the associated seismic velocity is 7.6 km/sec. This represents an extreme case, but it illustrates the importance of the crustal contribution to anomalous gravity where there are lateral inhomogeneities in crustal composition.

The effect is not known of mass inhomogeneities below the Moho other than those associated with upper mantle structure above the 400 km seismic discontinuity, and analysis of them by Lewis and Dorman (1970) suggests they are largely compensated by changes in density differential, at depths of 60, 120, 250, and 400 km. In particular, there is no understanding as yet of the significance of the large regional changes in gravity derived from the orbital perturbations of earth-orbiting artificial satellites. It is clear, however, that the external gravity field sensed at satellite height is an integration of the surface gravity field as has been shown by Strange (1966), Talwani and LePichon (1969), Woollard and Khan (1970), as well as others. Whether large-scale regional changes in gravity reflect regional compositional changes in the crust and upper mantle, or variations in mass below the lithosphere in the asthenosphere and induced by sea floor spreading, as postulated by Moberly and Khan (1969), or variations in mass induced by lateral changes in temperature associated with convection below the lithosphere, as suggested by Lee and Uyeda (1965), or stress induced by thermal variations at depth as suggested by McKenzie (1967), or phase transformations at the base of the asthenosphere as postulated by Magnitsky and Kalashnikova (1970), or a combination of these factors is as yet not established. The only factors contributing to anomalous gravity that do appear to be well established are those of relatively near surface origin, those above about a 400-km depth.

There appear to be other correlations of anomalous gravity in some areas with tectonic setting, geologic age, tectonic history, and vulcanism, as suggested by various investigators (Drake and Nafe, 1968; Strange, 1966; Kaula, 1969, and this volume), and if these factors as well as changes in the seismic parameters of the crust and upper mantle could be related to heat flow, a strong case could be made for the thermal regime of the upper mantle being the primary cause of anomalous gravity. There is, however, no simple one to one relation between heat flow and anomalous gravity on either a local basis or a regional basis. Lee and MacDonald (1963) as well as Lee and Uyeda (1965), for example, examined heat flow and satellite-defined gravity data for a possible correlation on a global basis using spherical harmonic representations of the data, and while these investigators found there is a suggestion that in places heat flow is high where gravity is low, there is no general confirmation of this relation. On a more local basis it is known that the oceanic rises representing centers of crustal spreading and high heat flow are areas of excess gravitational attraction. Similarly, it is found that the western province of the United States that is characterized by high heat flow and anomalously low mantle velocities has an essentially normal gravity field, and if anything excess gravity, if the data are incorporated in a global spherical harmonic analysis as was done by Strange (1966).

Kaula (1969 and this volume) in his global analysis of the satellite-defined gravity field, as did Strange (1966), found that there are no unique geological associations with regional anomalous variations in gravity as defined by satellite data, although it does appear possible to make some generalizations as to types of geologic areas that usually have no anomalous gravity field, positive values or negative values. The writer, looking at the problem from a more restricted point of view, asked what is the gravity field where we have seismic measurements of crustal thickness, and concluded (Woollard, 1968) that where the crust is thick and has a higher than normal mean velocity, gravity is positive and that where the crust is thin and has a low mean velocity, gravity is negative. As these two types of areas also appear to correspond respectively to areas of crustal subsidence (basins) and uplifts (domes), it appears that there is some process that is operative at shallow depths that results in significiant changes in crustal mass distribution above and below the Moho. As changes in mantle velocity are also noted to correspond to changes in mean crustal velocity, the observations suggest the phenomenon involves both the upper mantle and the crust. Serpentinization as proposed by Hess (1955) could offer an explanation since this results in a change in density and seismic velocity in the same sign and is a reversible process. Another possibility is a phase transformation analogous to that of basalt and eclogite as proposed by Kennedy (1959). As will be discussed later, however, there are certain areas where other mechanisms appear to be needed to explain the relations noted.

The problem of the significance of anomalous variations in gravity is, therefore, a complex one that has still to be resolved, and at this stage, it appears best to try and better define the actual complexity of the problem by establishing what correlations appear to be of a universal nature and areas where it appears other factors are involved. As one step toward this objective, Woollard (1969b) examined how free air and Bouguer anomaly values vary as a function of surface elevation for 1 by 1° and 3 by 3° size areas in different parts of the world. Previous studies of the relation of gravity anomalies to either surface elevation or crustal thickness had been based on point values and on global compilations of data (Demenitskaya, 1959, 1967; Woollard, 1959) in an attempt to establish average relations.

The large scatter in values obtained using this approach not only made it difficult to formulate any meaningful relation between Bouguer anomaly values and elevation, but also implied the existence of such large geological and tectonic effects that there was reason to think there were systematic differences in gravity field associated with the different continents. To a certain extent, this was indicated by satellite-derived gravity data, but for the most part the satellite data indicated it was the ocean basins that were characterized by marked regional differences in gravity field rather than the continents. The most recent free air anomaly map based on satellite data in combination with surface data (see Kaula, this volume) does indicate, however, there are significant regional differences in gravity on the continents, and these correlate well with those defined by 3 by 3° averages of surface data. That there should be differences between Kaula's values and the 3 by 3° average values is to be expected as a satellite scans an effective area of about 10 by 10° rather than 3 by 3°. Kaula also uses a reference ellipsoid with a polar flattening of 1/299.8 rather than the international ellipsoid with a polar flattening of 1/297. With these limitations in mind, it is of interest that Kaula's highest continental free air anomaly value (+ 36 mgal) occurs in the Andes Mountains of South America, which is the area where Woollard (1969b) finds the 3 by 3° free air anomaly values are highest for any elevation, and that one of Kaula's lower free air anomaly areas (– 20 mgal) is in equatorial central Africa, where the 3 by 3° free air anomaly values indicate subnormal gravity for all elevations. The respective 3 by 3° free air anomaly values for an elevation of 1000 meters are + 41 mgal and – 6 mgal. Any allowance for the difference in mean elevation of the two areas would significantly increase the difference between the two values since the Andes have a higher elevation and a vertical free air anomaly gradient that is 16% higher than that in equatorial Africa.

The choice of 1 by 1° average values to express the regional Bouguer anomaly field in the present paper was made in order to suppress the "noise" from very short wavelength anomaly changes caused by local changes in geology, and yet retain the basic Bouguer anomaly pattern associated with regional changes in elevation, geology, orogenic belts, major tectonic features, and changes in crustal and upper mantle parameter values. The 3 by 3° average values bring out regional differences in values. These, however, are not to be confused with compensation values although it appears that areas of 3 by 3° size represent the minimum size area that is compensated. Averaging on a 3 by 3° areal basis, however, tends to over-suppress the gravitational effect of even major tectonic features such as the Sierra Nevada Mountains whose associated Bouguer anomaly pattern has a half wavelength of ≈ 2°. On an anomaly of this size, the reduction in total amplitude using 1 by 1° averages is about 10% whereas using 3 by 3° averages gives a reduction in total amplitude of about 70%. A 4 by 4° average would eliminate all evidence of the feature.

GENERAL COMMENTS ON FACTORS OTHER THAN ELEVATION AFFECTING BOUGUER GRAVITY ANOMALY VALUES

An inspection of any well-controlled gravity anomaly map covering an area of 10^5 km^2 (4 x 10^4 mi^2) or more will show that anomaly closures exceeding 10 mgal have half wavelength (peak to trough width) values ranging from ≈ 20 km to more than 200 km. In shield areas of exposed basement crystalline rocks where anomalies of many wavelengths are found, the short wavelength anomalies can be identified with changes in basement lithology, particularly granite and ultrabasic intrusives and small horsts and grabens. The long wavelength anomalies in such areas appear to be related more to metamorphic rock provinces and areas of former orogenic tectonic disturbance or else major horsts or grabens or batholithic intrusions. There are special cases such as Fenno-Scandinavia and eastern Canada where there is also an apparent inherited effect from Pleistocene ice loading, which gives very long wavelength negative anomalies.

Other causes of long wavelength anomaly changes can be related to changes in crustal and upper mantle parameter values, which may or may not be associated with mountain ranges and changes in regional elevation. In the United States, for example, Pakiser and Steinhart (1964) have shown that there are major differences in crustal and upper mantle parameter values associated with the high plains area east of the Rocky Mountains and the basin and range area west of the Rocky Mountains. The Rocky Mountains themselves have no distinctive "crustal root," but lie in a transition zone between the two flanking areas of anomalous crust.

To illustrate this point for two other areas, consider the two crustal profiles shown in Figs. 1 and 2. Figure 1 is based on the data of Ewing et al. (1966) in the Maritime Provinces of Canada across the northern Appalachian tectonic province. All stations have essentially the same surface elevations (near sea level), and the thickness of the crust varies from 31 to 46 km with a half wavelength value of about 500 km (300 miles). It is to be noted that there is a marked correlation between crustal thickness, mantle velocity, mean crustal velocity, and the composition and structure of the crust. These variations in implied mean crustal density are also reflected in the free air gravity anomaly values, which average about +10 mgal where the crust is thick and −15 mgal where the crust is thin. As the area is well inside the continental shelf, there is no bias from the continental edge effect.

The crustal section shown in Fig. 2 is across the island of Oahu in the Hawaiian Islands and is based on the work of Shor and Pollard (1964) and Furumoto et al. (1968). This section shows that south of the Hawaiian Ridge the structure and thickness of the crust as well as the mantle velocity is distinctly different from

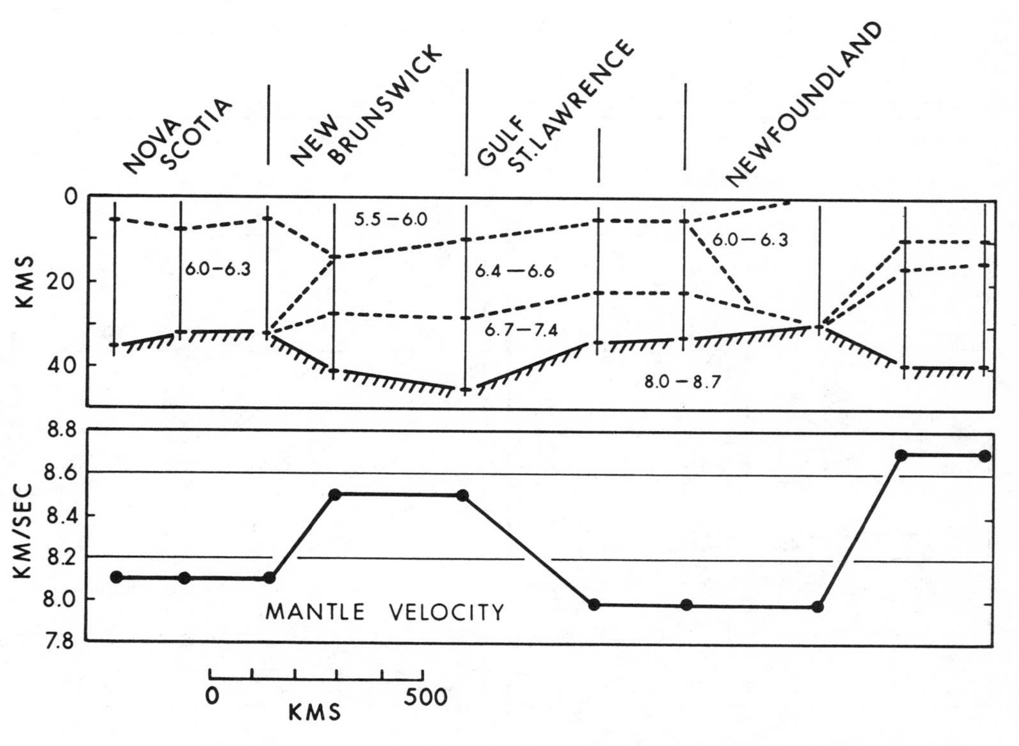

Fig. 1. Crustal section across the Maritime Provinces of Canada. After Ewing et al. (1966).

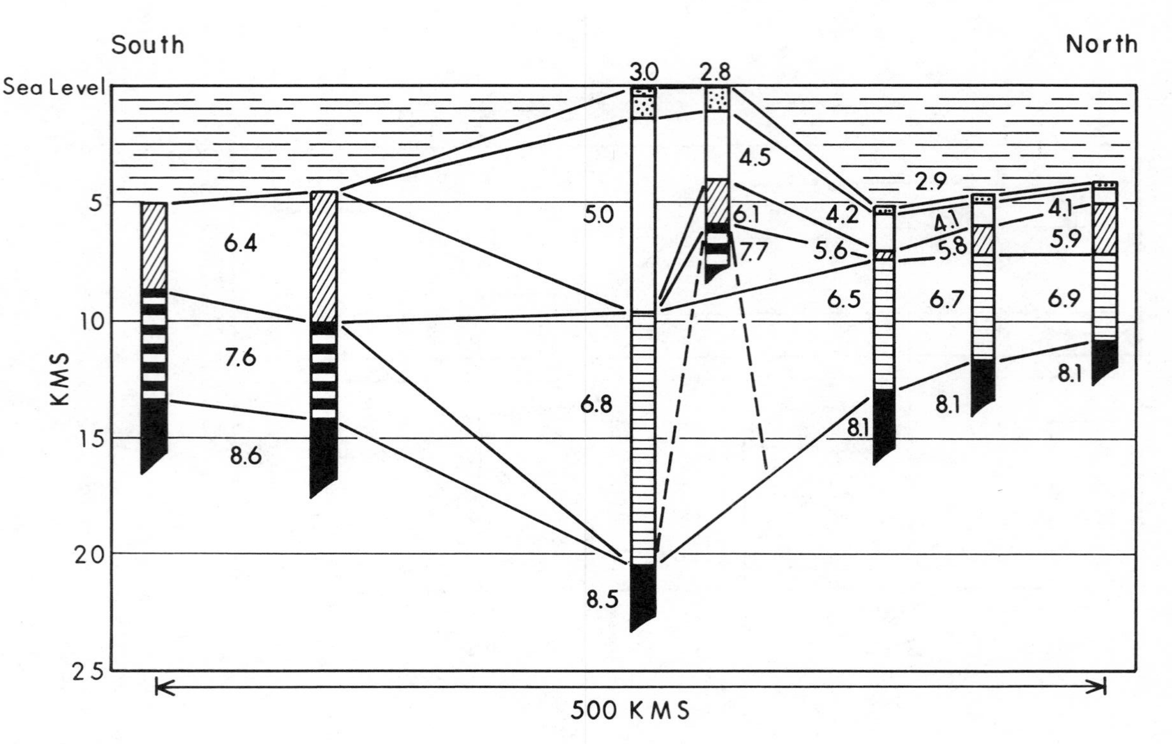

Fig. 2. Crustal section across the Hawaiian Ridge at island of Oahu.

that north of the ridge. It also shows that there is a systematic transition in crustal layering on the north side of the ridge and that the basal (6.8 km/sec) crustal layer thickens by approximately 5 km beneath the ridge. The Pratt-Hayford isostatic anomaly values with H = 113.7 km progressing from north to south are: 0; – 46 mgal in the trench; + 42 mgal on the ridge; + 14 mgal immediately south of the ridge; and + 10 mgal at the southernmost crustal section. The total wavelength of the structure in this case is about 500 km (300 miles). The crustal column shown bringing 7.7 km/sec material to within 6 km of sea level is over an intrusive rift. As indicated earlier, similar high velocity values were obtained at a depth of only 2 km below sea level in the Koolau caldera on the windward coast of Oahu (Furumoto et al., 1965) with local changes in gravity in excess of + 110 mgal.

These two sections not only illustrate the heterogeneous nature of the crust and upper mantle and the scale of change, but also suggest that in general (the Hawaiian Ridge is an exception) there is an intimate relation between the composition of the crust and that of the upper mantle such that where the upper mantle velocity is high the crust is thick and has a high mean velocity. The gravity effect, as described, varies significantly with the width of the disturbing mass as well as with density as implied by the seismic velocity values. The Hawaiian Ridge appears to represent a special geochemical problem related to volcanism, for it appears that at least 5 km of former mantle material was converted to crustal material and that the crustal structure that must have existed above the 6.8 km/sec layer prior to volcanism has been destroyed. The only other alternative is that the data taken were not adequate to define the structure of the upper crust. Since similar results were obtained by Shor and Pollard (1964) off the island of Maui, this does not appear to be the explanation.

THE 1 BY 1° ANOMALY PATTERN IN THE UNITED STATES

Figure 3 shows the 1 by 1° Bouguer anomaly pattern in the United States. It corresponds in gross pattern to that portrayed in the actual Bouguer Anomaly Map (Woollard and Joesting, 1964), but there are many features of narrow width, such as the Mid-Continent Gravity High, which are removed by averaging on this scale. Because the Bouguer anomalies incorporate the effect of all changes in subsurface mass contributing to the isostatic compensation of the surface topographic mass, anomalous changes in crustal and upper mantle mass are difficult to recognize except where there is not obvious correlation with surface elevation. It is to be noted that East of the Rocky Mountain Front, which coincides roughly with the – 100 mgal contour extending as a broad arc northward from the Big Bend border of Texas with Mexico, the pattern is mostly anomalous. Even in the Appalachian Mountains, it is only in the southern Appalachian region defined by a – 60 mgal closure that this orogenic belt is reflected in the 1 by 1° anomaly values. The fact that the apparent Rocky Mountain effect extends eastward beyond the physical boundary of these

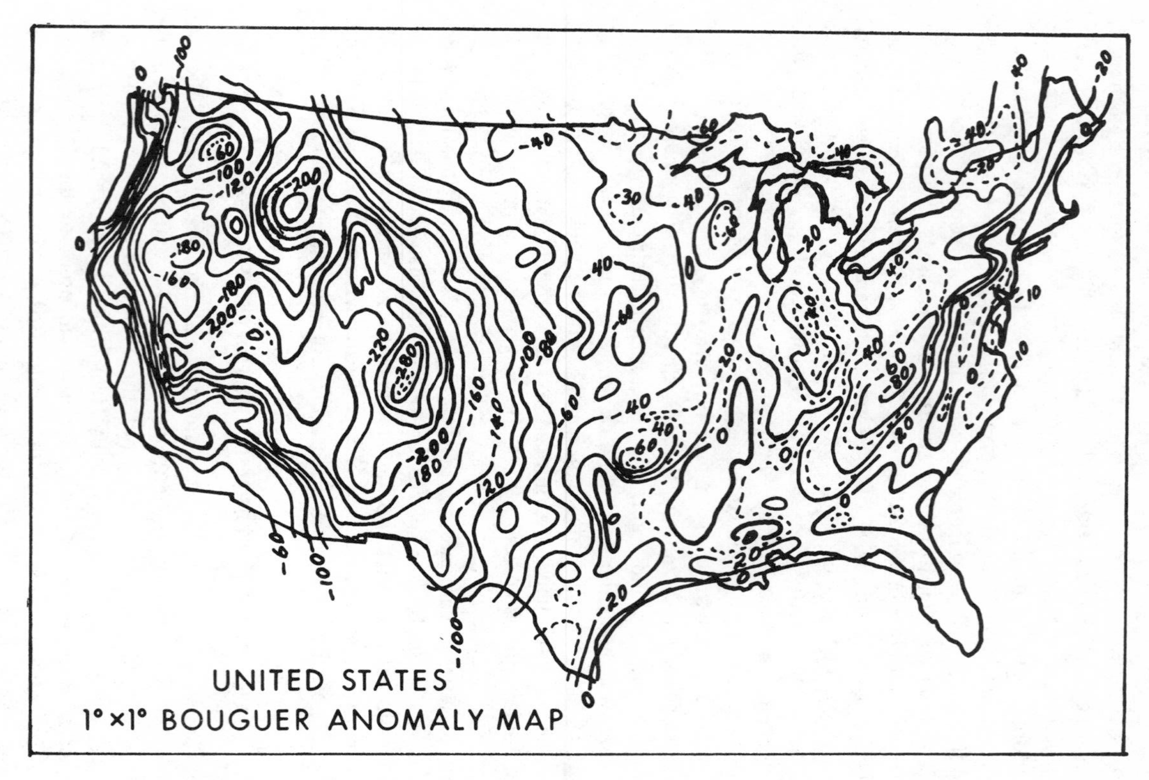

Fig. 3. 1 by 1° Bouguer anomaly map of the United States.

mountains, as reflected in the conformable configuration of the – 60 to the – 100 mgal contour, is in part related to the thick section of sediments in the basins along the Rocky Mountain front, in part to the fact that compensation of the topographic mass is probably on the basis of the average elevation of 3 by 3° size areas rather than 1 by 1° size areas, and in part by the fact that about 25% of the compensation is realized from distant sources.

That the pattern is anomalous west of the Central Rocky Mountains–Colorado Plateau block also appears to be clearly indicated by the configuration of the – 200 mgal contour. The pattern of anomalous gravity, however, is best portrayed in the isostatic anomaly map shown in Fig. 4. This map, based on actual values using the Airy-Heiskanen concept of isostasy, with a standard sea level column with H = 30 km, explains why there is such a wide scatter in 1 by 1° Bouguer anomalies for the same elevation. Most of the pattern of positive and negative isostatic anomalies in the eastern half of the country appears to be related to former orogenic belts and presumably represents changes in crustal and upper mantle parameter values similar to those shown in Fig. 1 for the Maritime Provinces.

Certainly, the broad positive anomaly area shown adjoining the Canadian border at about 105°W longitude is an area of extra thick, high-velocity crust with a high-velocity upper mantle. It is also an area where there is seismic evidence for layering below the M-discontinuity. In contrast to the Appalachian orogenic province, the Rocky Mountain region shows a highly fractured anomaly pattern in which individual basins, uplifts, horsts, grabens, and batholiths are conspicuous as well as features having no surface geologic counterpart. It is only along the west coast that the gravity anomaly pattern shows an overall agreement with orogenic trends similar to that noted in the Appalachian region.

In Fig. 5, a different type of isostatic representation is shown. This is a plot of depths of compensation under the Pratt-Hayford concept of isostasy that yield minimum isostatic anomalies. It is based on a regional distribution of 1100 gravity stations as established by the U.S. Coast and Geodetic Survey (Duerksen, 1949). In gross form, it is seen that the pattern east of the Rocky Mountains conforms to that portrayed in the Airy-Heiskanen isostatic anomaly map (Fig. 4). In general, where a deep depth of compensation (113.7 km) is indicated, the Pratt-Hayford isostatic anomaly values are positive, and where a shallow depth of compensation (56 km) is indicated, the isostatic anomaly values are negative. The surprising thing is that so few areas appear to give minimum anomalies for an intermediate (96-km) depth of compensation and that few areas appear to be insensitive to the depth of compensation used.

The relation between the sign of the isostatic anomaly and depth of compensation can be explained by the fact that the observed gravity values and free air anomalies are affected by changes in crustal density and that a fixed crustal density is used in the Bouguer and isostatic reductions. As the derived compensating density is dependent on the surface elevation and the length of the compensating column, cancelation of the Bouguer anomaly, which incorporates

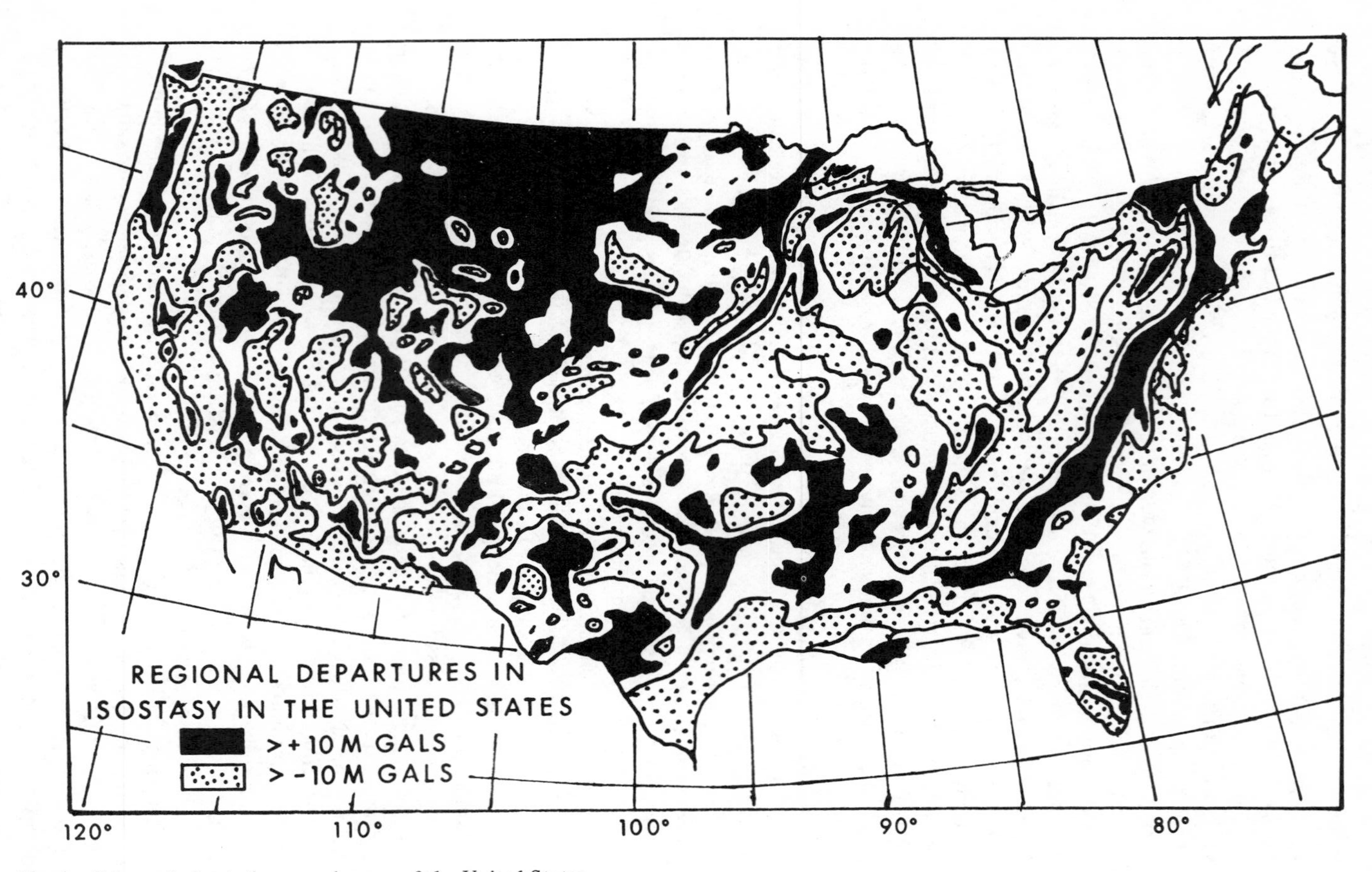

Fig. 4. Schematic isostatic anomaly map of the United States.

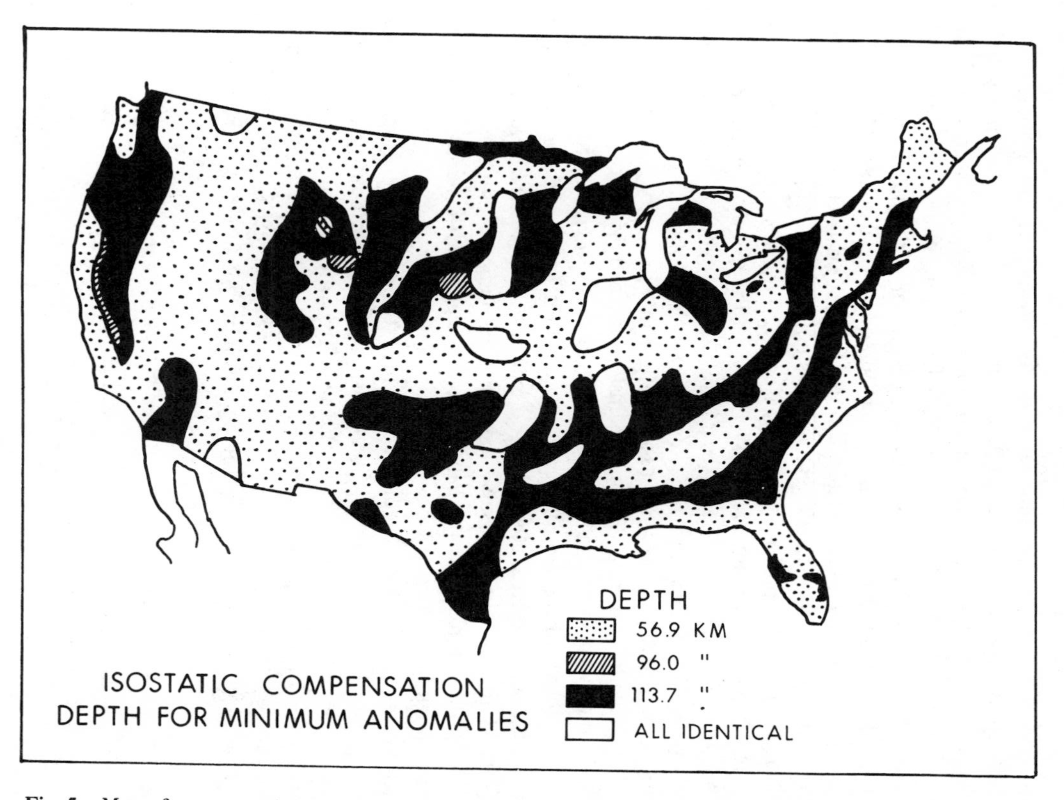

Fig. 5. Map of compensation depth yielding minimum isostatic anomalies in the United States.

any bias in the free air anomaly, will be dependent on the length of the compensating column. The fact that over most of the country a shallow depth of compensation yields minimum isostatic anomalies suggests most of the compensating mass is associated with the configuration of the *M*-discontinuity.

That there are significant differences in crustal and upper mantle parameter values in the United States has been pointed out by Woollard (1962), Pakiser and Steinhart (1964), Pakiser and Robinson (1966) as well as others. Woollard (1968) has also shown that there is a systematic relationship between mantle velocity, crustal parameter values, and isostatic anomaly values. The magnitude of the gravity effect is illustrated by Fig. 6, which is a plot of average isostatic anomalies as a function of reported Moho velocity values. Corresponding mean crustal velocity values and the elevation of the Moho for changes in mantle velocity in the United States are as shown in Table 1.

These results corroborate the conclusion reached by Pakiser and Steinhart (1964) that where there is an anomalous increase in either crustal velocity or mantle velocity, there is an increase in crustal thickness, and that where there is an increase in both crustal velocity and mantle velocity, there is marked increase in crustal thickness. Conversely, these authors indicate that subnormal crustal and mantle seismic velocity values are associated with subnormal values of crustal thickness.

Although Pakiser and Steinhart conclude that if the observed changes in seismic velocity values have a direct correspondence in density values there is isostatic equilibrium, it is evident that there are isostatic anomaly values associated with these areas of anomalous crustal and upper mantle velocity and that the sign of the anomaly is related to the anomalous density. This does not necessarily imply a lack of isostatic equilibrium, but rather is a consequence of

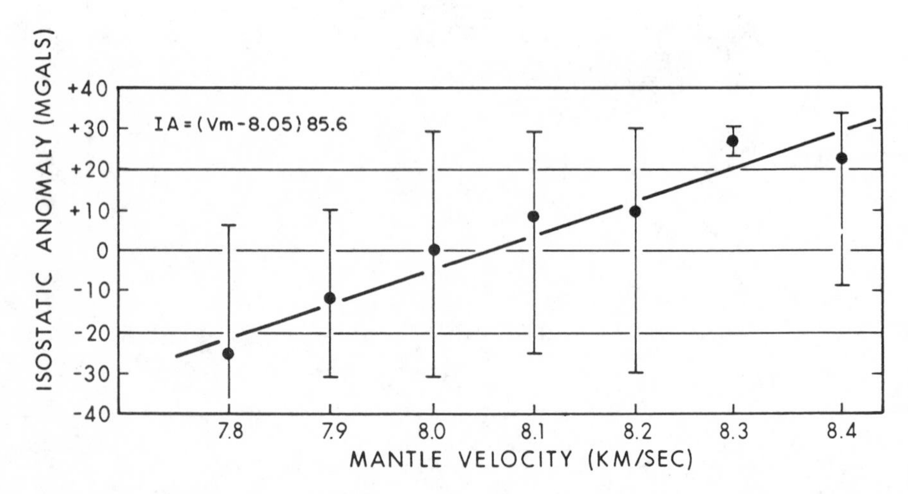

Fig. 6. Relation of isostatic anomaly values to variations in mantle velocity in North America.

Table 1. Crustal and Moho velocities according to elevation of Moho

Moho velocity (km/sec)	Mean crustal velocity (km/sec)	Elevation of Moho (km)
7.7	6.1	- 25
8.0	6.4	- 36
8.3	6.6	- 46

the dominant gravitational effect of crustal density over that of the crustal root where the lithospheric blocks are of finite width. If this were not the case, then all areas with a thicker than normal crust for the surface elevation would be characterized by negative isostatic anomaly values because of the excess crustal root. As it is, only graben areas and recently glaciated areas exhibit this characteristic. All other areas of excess crustal thickness have positive anomaly values.

Figure 7 shows that there are definite indications of a correlation between the elevation of the Moho with geologic age of the basement complex. As shown by Pakiser and Robinson (1966), there is also a correlation between the depth of the Moho and the relative percentages of granitic and mafic materials comprising the crust as reflected in the seismic velocity structure of the crust. Beliayevsky (1969) has also shown that variations in crustal thickness and structure in eastern Europe and Asia can be classified on the basis of age and tectonic

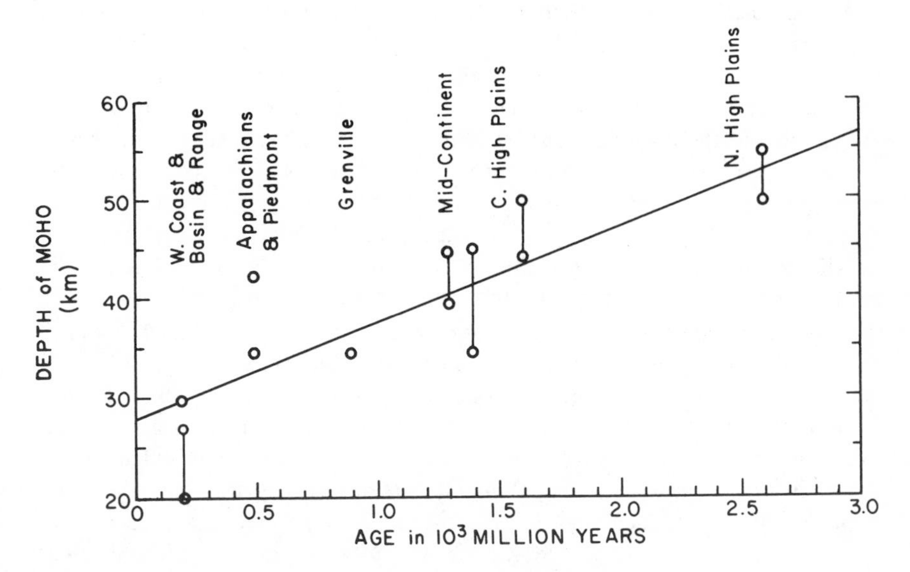

Fig. 7. Relation of depth of mantle to geologic age in North America.

characteristics. Woollard (1969a) in studying these relationships in North America finds that whereas the thickness of the crust is dependent on crustal structure and composition, and there is, on the average, a relation to geologic age, there is no distinctive crustal model associated with the different geological provinces, but rather regional and local changes whose cause is not always obvious.

Where there are indications of sub-Moho layering, the depth of such layering varies significantly as shown in Fig. 8. As with crustal data, there is at first glance no pattern of consistent relationships. By isolating individual parameters as a function of the sub-*M*-velocity, however, it is found that the bulk of the data do suggest that there are certain systematic relationships. These as shown by Woollard (1970) are as follows:

1. There is a dependence of the *M*-velocity value on that of sub-*M*-velocity.
2. The depth of the sub-*M*-discontinuity is related to the sub-*M*-velocity value.
3. The depth of the *M*-discontinuity is related to the sub-*M*-velocity value.

These relationships are similar to those noted between the velocity of the *M*-discontinuity and the velocity of the crust and the depth of the mantle, and therefore represent a downward continuation of the near surface pattern. It is to be noted, however, that where there is sub-*M*-layering, there appears to be no pronounced dependence of the depth of the *M*-horizon on the *M*-velocity value such as is found in general. This suggests that in areas where the sub-*M*-horizon occurs, it is more important than the *M*-horizon in defining the compensation associated with changes in surface elevation. To test this hypothesis, a plot was made of the relation between surface elevation and the depth of the sub-*M*-horizon. This plot (Fig. 9) shows that except for two points, which are erratic, the data do suggest there is a relation between surface elevation and the depth of the sub-*M*-horizon. The large scatter in values is not surprising since there is no more reason to suppose that there is a constant density contrast between the sub-*M*-layer and the overlying lithosphere than between the *M*-horizon and the overlying crust. The fact that the data do suggest there is in general a systematic relationship between surface elevation and the depth of the sub-*M*-horizon (admittedly poorly defined) is significant in that it brings out the probable importance of deeper mass distributions in the isostatic mechanism, which for the most part have not received too much attention because of uncertainties in interpreting seismic wave arrivals from distant sources. Even these studies, however, appear to support the above results based on local seismic refraction studies. For example, Lewis and Meyer (1968) define a seismic column for the mid-continent region based on nuclear explosions that shows a discontinuity at 70 km with a velocity of 8.2 km/sec. Green and Hales (1968), using similar data for the same region, show a discontinuity at approximately 90 km with a velocity 8.4 km/sec. Both interpretations may well be real for the respective areas examined, and the fact that the 70-km discontinuity has a lower velocity (8.2 km/sec) than the 90-km discontinuity (8.4 km/sec) is in agreement with the general conclusion that the depth of a seismic discontinuity, other than those in the crust, is related to its velocity.

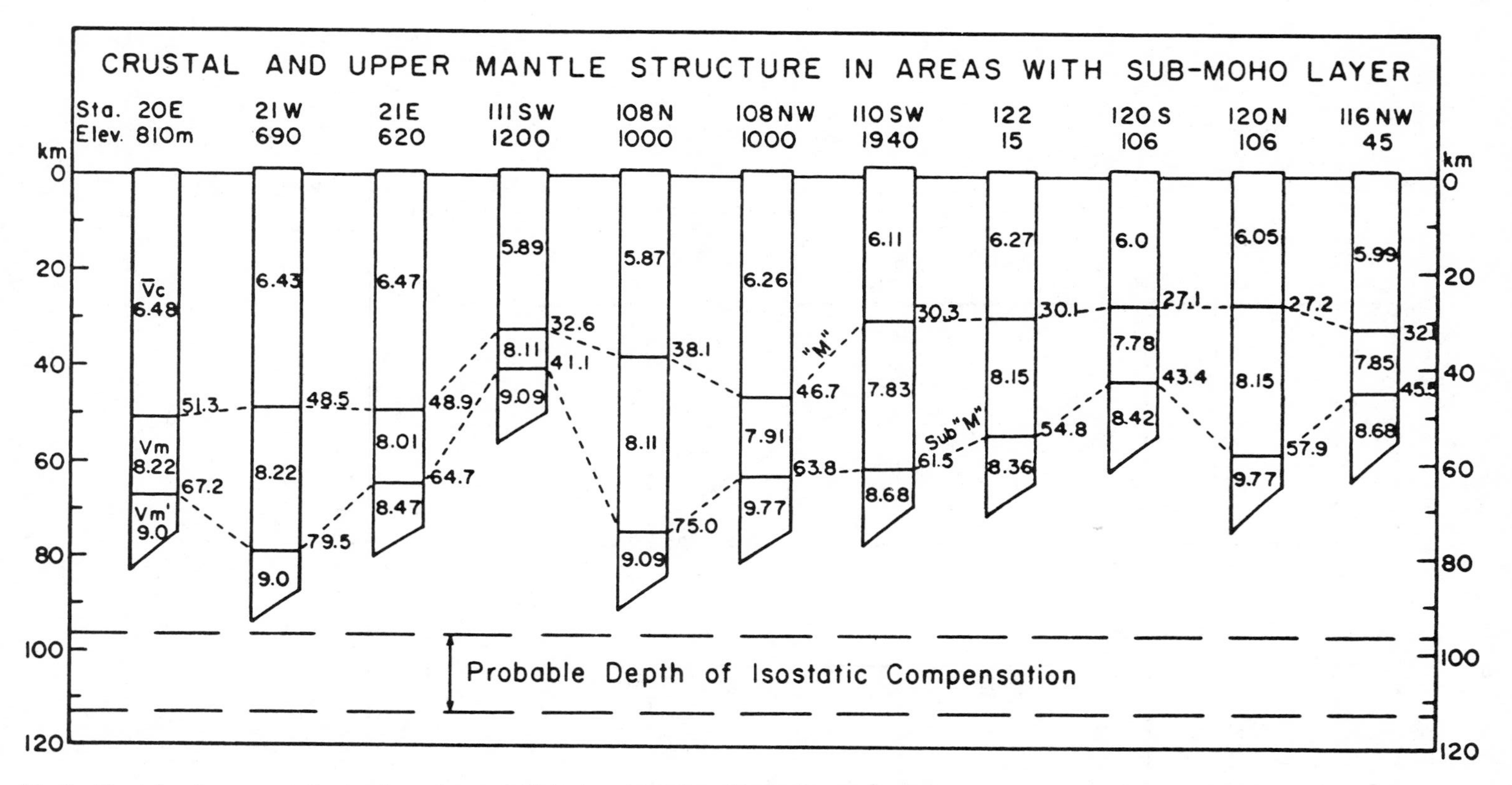

Fig. 8. Crustal and upper mantle structure where sub-Moho layering is found in the United States.

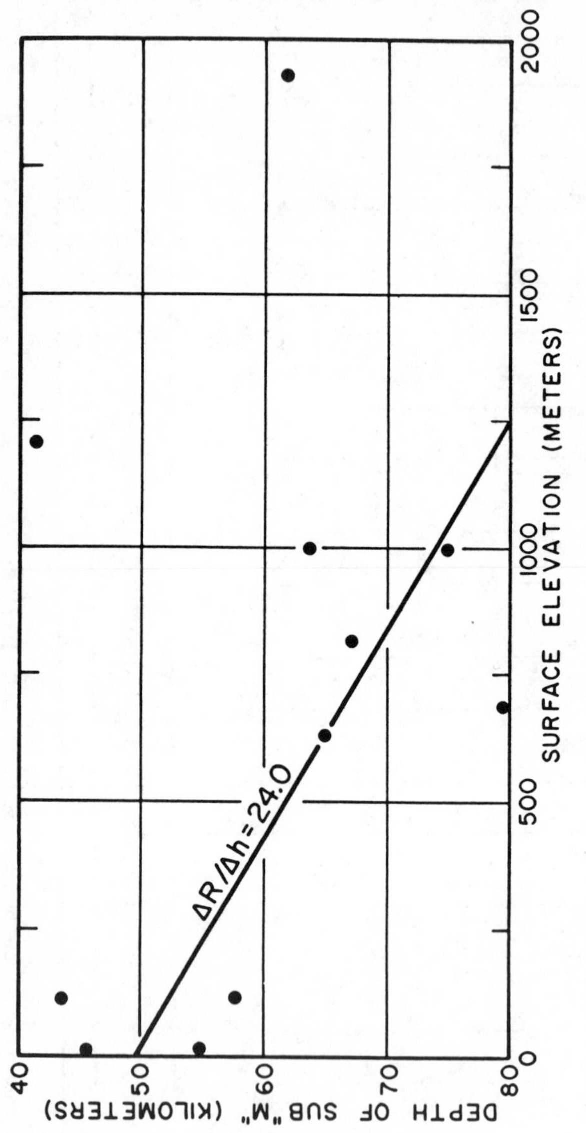

Fig. 9. Relation of sub-Moho layer depths to surface elevation in the United States.

The implications of the above and the fact that mean crustal velocity values and crustal structure are in general dependent on the M-velocity value are that we are dealing with a continuous vertical system of inter-related changes in the physical properties of the lithosphere at least down to the low-velocity zone at a depth of about 100 km. This of course assumes that there is a direct relationship between seismic velocity and density and that the suites of rocks represented have essentially the same mean atomic weight. If this is the case, however, there is a cumulative mean density bias with depth in areas of anomalous crustal velocity. Changes in the geometry of the crustal mass distribution are therefore only part of the isostatic mechanism, and in order to have isostatic equilibrium in areas of anomalous crustal and upper mantle velocity, there have to be either compensating differences in mass distribution of opposite sign at greater depths or significant differences in surface elevation.

In this connection, it appears significant that neither Lewis and Meyer (1968) nor Green and Hales (1968) find a low-velocity zone in the crust in the mid-continent region, whereas both Green and Hales (1968) and Archambeau et al. (1969) as well as Johnson (1967) and Anderson (1967) find a well-defined low-velocity zone at a depth ranging from 70 to 100 km in the western part of the country which is characterized by a thin crust and a low-velocity mantle. Except for the low-velocity zone, the deep-velocity structure defined by these various investigators is not markedly different down to the postulated olivine-spinel transformation at a depth of about 400 km. It should be noted, however, that Green and Hales (1968) indicate this change is actually at about 450 km rather than 400 km in the western states.

With normal relations between velocity and density, it would thus appear that the western states have a markedly subnormal density compared to the mid-continent region down to at least 400 km. It can be argued that this conforms to the basic Pratt concept of isostasy that the surface elevation is related to the mean density above the compensation depth since the elevation of the mid-continent area is significantly lower than that in the western states. If the mid-continent seismic model also applies to the high plains area east of the Rocky Mountains where the elevation is not significantly different from that in the Basin and Ridge area, there is a problem. In the high plains, the crust and upper mantle are both characterized by high velocity values, implying a high density, and it is difficult to see how the two areas can be in isostatic equilibrium without postulating that the low velocity values in the Basin and Range area at some point really represent a high density rather than a low density. This would imply that there is a change comparable to that for a change in Mg-Fe ratio that would decrease the modulus of rigidity and velocity and raise density. The existence of compensation is indicated by the low amplitude of the isostatic anomalies. That the compensation is not deep seated (below 100 km) is suggested by the fact that the minimum isostatic anomalies under the Pratt-Hayford concept of isostasy in the Basin and Range area indicate a shallow depth of compensation as shown in Fig. 5.

The only direct evidence for a high density mass at shallow depth in the Basin and Range area is the high-velocity "lid" defined by Green and Hales (1968) at a depth of about 90 km, immediately overlying the low-velocity zone. This layer, however, is only 10 km thick, and its velocity (8.3 km/sec) appears to be too low to give material with a density high enough to provide the required compensation. Although other investigators of this area do not find this high-velocity layer, its presence or absence does not alter the basic fact that the mean velocity of the entire lithosphere down to at least 400 km is subnormal, and that if there is a normal relation of velocity to density there should be a significant deficiency in gravity field and indications of a deep depth of compensation in the Pratt-Hayford isostatic anomaly values. To argue that the areas are not in isostatic equilibrium is not reasonable in view of their large size, which is in excess of 5 by 5°.

A separate, but possibly related, problem is the dependence of crustal thickness and mean crustal velocity on the velocity of the *M*-discontinuity, and of the *M*-velocity value on that of the sub-*M*-layer where it is found. As indicated earlier, the relations are such that where the crust is thickened through the addition of a high-velocity basal layer, the mantle velocity is higher and there is an apparent decrease in density contrast between the crust and mantle as evidenced by surface subsidence. It is difficult to see how either serpentinization or a phase transformation would affect the crust and mantle in the same sign unless the process is one that has its origin at greater depth and travels as a wave through both the upper mantle and crust with presumably pressure and temperature controlled differential effects in the same sign where there are compositional changes.

Magnitsky and Kalashnikova (1970) have proposed a somewhat analogous mechanism to explain crustal uplift and subsidence, and they attribute the changes in surface elevation to changes in the thickness of the asthenosphere, which they identify with the low-velocity zone. The causative mechanism is related to heat flow and its effect on the olivine-spinel transformation zone at around 400-km depth.

Whatever the process, it appears to be reversible, affects the crust and mantle in the same sign, results in changes in surface elevation and gravity, and also is able to operate in opposite sign in adjacent areas over the same time span. This is suggested by basins and adjacent uplifts such as the Michigan Basin and the Wisconsin uplift, which developed over the same period in geologic time. Although crustal seismic data are lacking for Michigan, the positive isostatic gravity anomaly values over Central Michigan (Fig. 4) suggest the area is characterized by a high-density and high-velocity crust. This appears to be substantiated by the map of the depth of the Moho prepared by Woollard (1968), which shows that the – 40-km contours spread in this area and suggests there is a closure centered over the Michigan Basin with a value in excess of – 40-km. The adjacent Wisconsin uplift is characterized by negative isostatic anomaly values as shown in Fig. 4; the mantle has a lower velocity than adjacent areas, and the depth of the Moho in the central area is less than − 35 km. As geologic evidence from well data indicates, the center of the Michigan Basin migrated eastward with time to its present position; presumably the causative mechanism also migrated eastward with time.

The writer has no pat explanation for these observations other than that it appears they are related to thermal convection.

The principal value in calling attention to these relations at this time is to define a problem, not provide the solution, which obviously is of a geochemical nature.

With the above relations in mind as determined in the United States, we can proceed to examine more limited data in other areas.

1 BY 1° RELATIONS IN CANADA

Figure 10 shows the 1 by 1° Bouguer anomaly map for Canada. The marked negative anomaly area in eastern Canada is not related to elevation so much as to a probable time lag in the isostatic rebound of the crust following removal of the Pleistocene ice cap. Although this area corresponds also to the location of the Laurentian highlands, these terminate at about latitude 55°N, whereas the negative anomaly pattern as defined by the − 40 mgal contour extends up to about latitude 60°N. Innes et al. (1968) in discussing the gravity relations and seismic crustal results obtained in the Hudson Bay region note that there is no apparent correlation between the Bouguer anomaly pattern and crustal thickness, although it is stated that the anomaly pattern correlates with structures within the crust as observed in the exposed geology that are related to Precambrian orogeny. The observed variations in crustal thickness based on preliminary results for a section from Churchill to the Ottawa Islands (94°W to 80°W longitude between 58 and 59°N latitude) show the M-discontinuity rising eastward as an undulating surface from about − 39 km at Churchill to near − 27 km at about 83°W longitude and then sinking to − 39 km beneath the Ottawa Islands. The preliminary results show no evidence of layering within the crust, and the mean crustal velocity is given as 6.3 km/sec with a mantle velocity of 8.3 km/sec. It is difficult to see why the long wavelength gravity anomaly pattern with half wavelength values of about 200 km and amplitude values of about 40 mgal should have no expression in the seismic velocity values for the basement surface and upper crust or why there is no correlation between the anomaly pattern and the change of approximately 12 km in crustal thickness with uniform crustal and mantle velocity values. As deduced by the writer (Woollard, 1969b) from studies of the change in Bouguer anomaly values with elevation in Canada, Innes et al. (1968) attribute about − 25 mgal to the effect of former ice loading in this area.

The marked excess in gravity defined by the − 20 mgal contour in the high plains area at about 100°W longitude represents an extension of the positive isostatic anomaly block in the United States. Here, as in the United States, it is

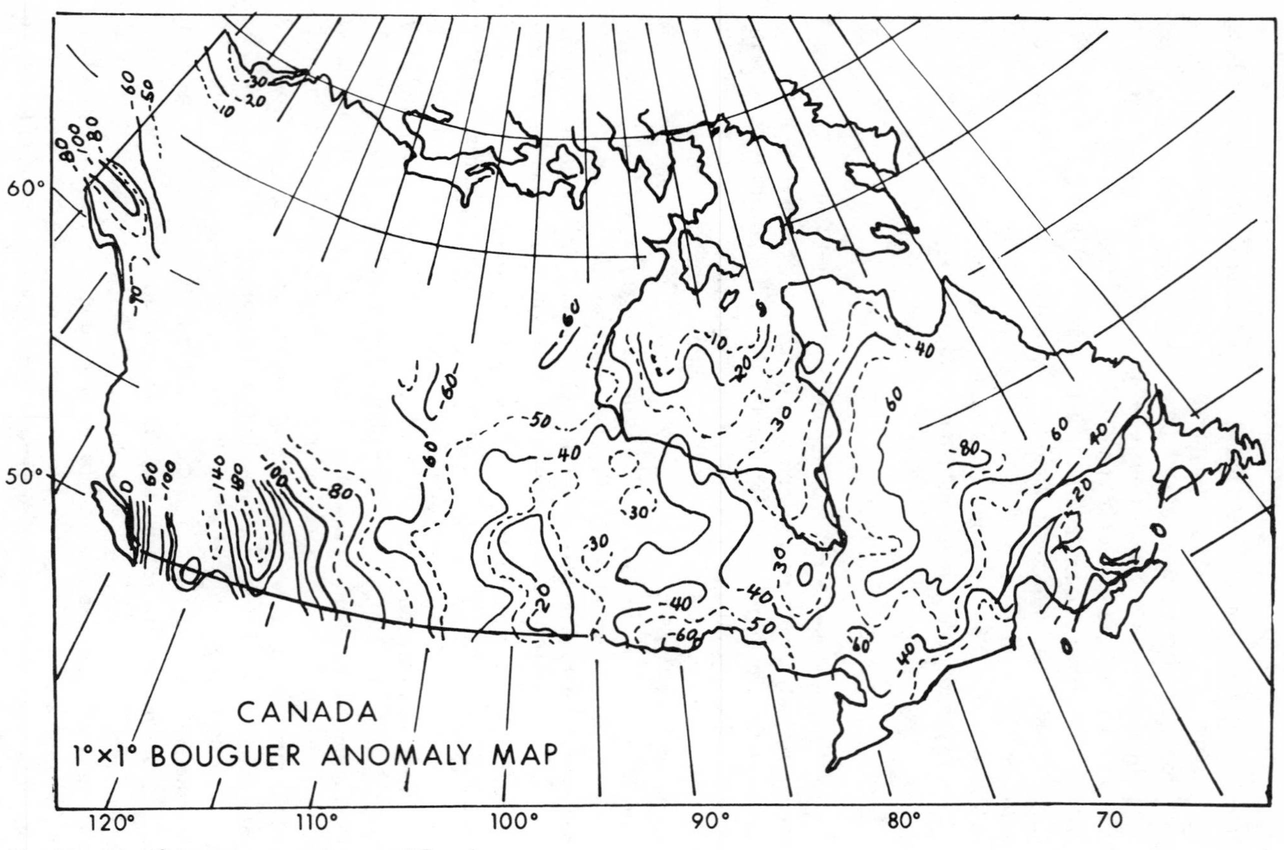

Fig. 10. 1 by 1° Bouguer anomaly map of Canada.

probable that there is a thick, high-velocity crust. Farther west at about 110°W longitude, which is part of the same positive anomaly area, Cummings et al. (1962) found the crust to have a thickness of about 47 km, which is abnormal for the surface elevation of around 900 meters. The crust here is characterized by a high mean velocity (6.8 km/sec) and a high mantle velocity (8.3 km/sec). Immediately north of the 20 mgal closure at 100°W longitude, where the gravity contours are – 40 to – 50 mgal, the crust appears to be of subnormal thickness, 31 to 34 km for the surface elevation of around 270 meters. The crustal velocity is very close to normal (6.4 to 6.5 km/sec), but the mantle velocity is subnormal (7.9 km/sec).

The Rocky Mountains region of southern Canada poses a problem analogous to the Basin and Range area in the United States in that the gravity field appears to be near normal for the surface elevation, but there is no corresponding thickening of the crust but rather a thinning of the crust. The *M*-horizon is at a depth of only about – 30 km according to White and Savage (1965) with a mantle velocity of 7.8 km/sec. White et al. (1968) indicate these conditions characterize all of the area westward from Banff, Alberta, to the Coast Range.

On the other hand, on Vancouver Island where the anomalies are positive, the crust appears to be abnormally thick (51 km) and is composed predominantly of material with a velocity of 6.7 km/sec.

The relations in Canada therefore appear to substantiate those found in the United States in terms of crustal and upper mantle provinces and the relation of gravity to variations in crustal and upper mantle parameters except for the Hudson Bay region, which represents a set of conditions not found in the United States.

1 BY 1° BOUGUER ANOMALY VALUES IN ALASKA

The 1 by 1° Bouguer anomaly map of Alaska is shown in Fig. 11. The anomaly pattern in southern Alaska in the Coast range area has, if anything, a negative bias as indicated by a low vertical Bouguer anomaly gradient of – 59 mgal per 1000 meters (in the United States the value is – 109 mgal per 1000 meters) and spreading between the + 40 and – 20 mgal contours. This appears to be related to a batholithic intrusion. Both Woollard et al. (1960) and Hales and Asada (1966) found the crustal velocity to be subnormal (6.1 to 6.2 km/sec) with a depth to the *M*-horizon of 35 to 36 km and a near normal mantle velocity of 8.0 km/sec. A profile based on the same shot point (Skagway) cutting across the southern end of the Wrangel Mountains, where there is a steep anomaly gradient, showed a significantly higher velocity for the crust (6.45 km/sec) and a depth to the *M*-discontinuity of about 41 km. In the zone of the – 100-mgal closure shown in Fig. 11, the crustal velocity was again low (6.1 km/sec) and the depth of the Moho, 36 km.

In the Prince William Sound area where the – 60 mgal contour shows a large re-entrant at about 148°W longitude, the same authors have also reported seismic

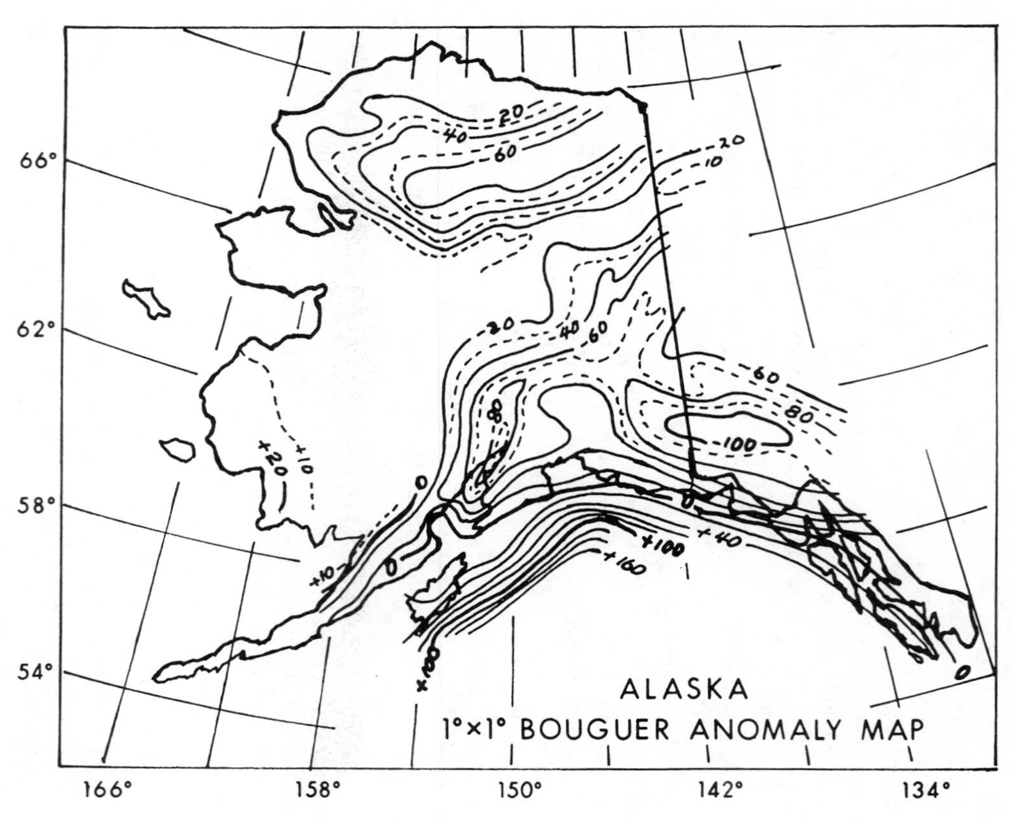

Fig. 11. 1 by 1° Bouguer anomaly map of Alaska.

results, and although there are some differences in detailed interpretations, the overall results obtained are again in general agreement. At the coast in the Prince William Sound area, the mean crustal velocity exceeds 6.8 km/sec and the depth of the Moho is about 53 km with a mantle velocity of about 8.3 km/sec. Inland from the same shot point across the St. Elias range, the crust is somewhat thinner (48 km) and again abnormally thick for the surface elevation. Along the coast towards the Kenai Peninsula, the seismic interpretations differ. Woollard et al. (1960) found a depth to the *M*-discontinuity of about 46 km with a three layer high-velocity crust, and Hales and Asada (1966) indicate a crustal thickness of only 35 km with a single layer 6.0 km/sec crust. This difference in interpretation of the same data results from the weight given second arrivals in defining crustal layering and the fact that Woollard et al. recognized the presence of a major crustal fault with a significant time offset of the data beyond 152 km. They also allowed for the effect of the Cook Inlet graben on arrivals at the outer end of the profile using magnetic depth estimates for the thickness of the sediments present.

There are no seismic data for the Brooks Range area in northern Alaska where a – 60-mgal closure is indicated, but on the basis of the positive bias shown in the plot of anomaly values as a function of elevation both here and in adjacent Canada, it is probable that the crust has a high density.

1 BY 1° BOUGUER ANOMALY PATTERN IN MEXICO

The 1 by 1° Bouguer anomaly pattern for Mexico is shown in Fig. 12, and the pattern is distinctive both because it is only partially related to the elevation pattern and because of obvious distortions in pattern. The fact that the Sierra Madre Oriental forming the eastern boundary of the Plateau of Mexico does not stand out in the gravity closures as does the Sierra Madre Occidental, which is marked by a – 200-mgal closure, suggests there is a real difference in crustal structure and composition beneath these two mountain ranges. The only crustal seismic data are those of Meyer et al. (1961) on the east side of the Sierra Madre Occidental south from Durango, which indicate a 43-km four layer crust with a mean velocity of 6.5 km/sec overlying mantle material with a velocity of 8.4 km/sec. All velocity values, however, are apparent rather than actual values as the profile was not reversed. The fact that the area is characterized by negative Pratt-Hayford isostatic anomalies of about – 20 mgal for a compensation depth of 113.7 km and the depth of the *M*-horizon is subnormal for the elevation and Bouguer anomaly suggests the actual velocity values may be somewhat lower than those reported.

Although it is not possible to attach seismic crustal significance to the sharp flexures in the contours between 100 and 104°W longitude, these do correspond to changes in topographic and geological provinces. The broad flexure in contours in northeast Mexico appears to be related to the Torreon-Monterrey fault, which displaces surface geologic features some 250 km and carries out into the Gulf of Mexico where it is evident in sparker measurements.

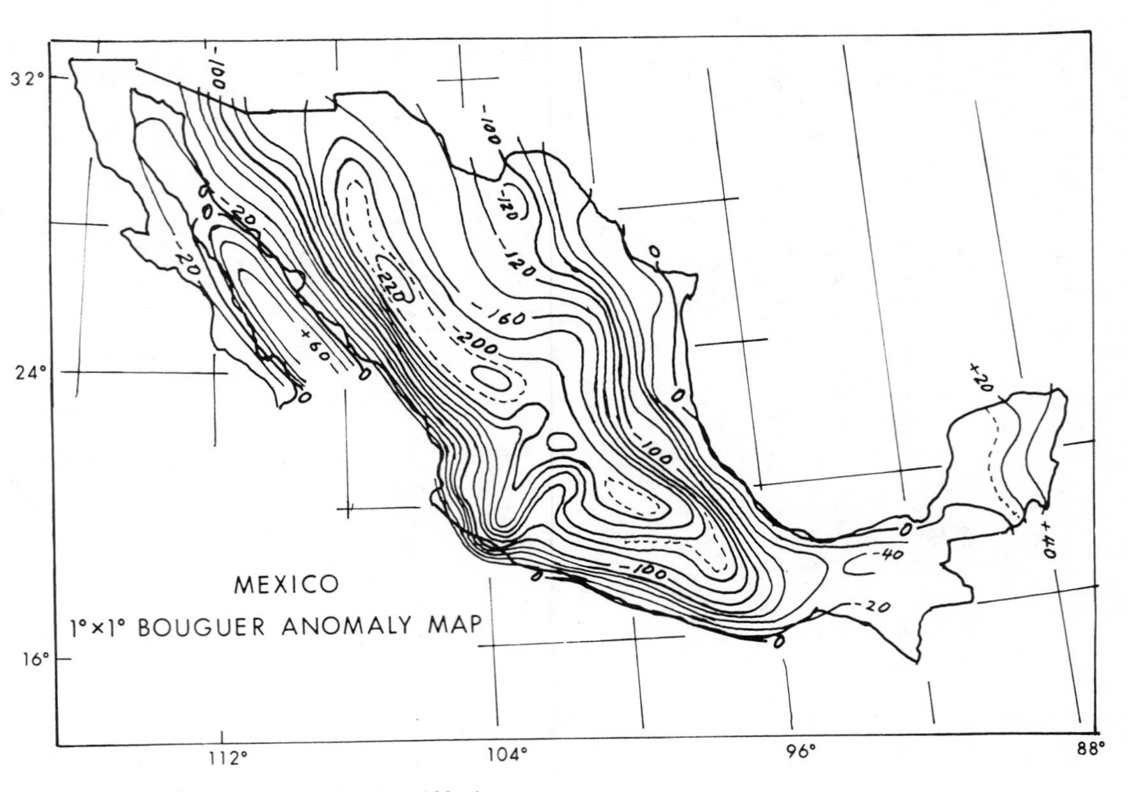

Fig. 12. 1 by 1° Bouguer anomaly map of Mexico.

The pattern of positive Bouguer anomaly values associated with the Gulf of California, which is believed to represent a rift and an extension of the East Pacific Rise loci of crustal spreading, is a definite area of crustal thinning and change in crustal composition. Phillips (1964) reports depths to the Moho as shallow as 7 km and Moho velocity values as low as 7.5 km/sec in this area.

1 BY 1° BOUGUER ANOMALY PATTERN IN SOUTH AMERICA

The gravity data available for South America are relatively sparse, and the seismic crustal data are even more limited. Figure 13 shows the 1 by 1° Bouguer anomaly map. The anomaly pattern in Colombia is related to apparent major horsts and grabens which divide the northern Andes Mountains into three distinct tectonic elements.

The – 300 mgal contour indicating a closure of this magnitude in northern Chile and Bolivia is associated with the Altiplano. This is the only area for which there are seismic data and they are of very low quality. A preliminary interpretation of these data which were taken by the Carnegie Institution of Washington (Woollard, 1960), suggests that in Peru the crust has a mean velocity of 6.4 km/sec and a thickness of 64.9 km beneath the Altiplano at an elevation of 4300 meters with a poorly determined mantle velocity of 8.0 km/sec. On the west flank of the Altiplano at an elevation of 1520 meters, near Arequipa, Peru, the crustal thickness indicated is 51.7 km. Both areas are characterized by positive Pratt-Hayford isostatic anomalies of + 10 and + 30 mgal, respectively.

Farther south in Chile at an elevation of 4500 meters the crust has a mean velocity of 6.6 km/sec and a thickness of 70.3 km. On the west flank of the mountains at an elevation of about 1520 meters the mean velocity is 6.5 km/sec and the crustal thickness is 56.5 km. As in Peru, the mantle velocity is poorly defined as about 8.0 km/sec. A subsequent interpretation of the data for the measurement near Arequipa, Peru (Carnegie Institution of Washington, 1969) indicates a crustal thickness of 39.5 km. This same group also reports new results taken in the Altiplano region of Bolivia in 1968. These indicate the crust has the following section:

	Thickness (km)	Velocity (km/sec)
	10	5.0
	15	6.0
	35	6.6
Total	60	8.0

These results are similar in both velocity values and total thickness to those obtained by Woollard (1960) for the adjoining Altiplano region of Peru. Other than the difference of about 5 km in total thickness, the most significant differences are in the thickness of the surface layer, which is believed to be associated with sedimentary rocks, and that of the upper crustal layer. The section obtained by Woollard based on the 1957 measurements is as follows:

	Thickness (km)	Velocity (km/sec)
	4.1	5.3
	21.2	6.2
	39.6	6.7
Total	64.9	8.0

The significance of the obvious positive bias in anomaly values in northern Argentina adjoining the Chaco region of Paraguay is not known, but presumably it is related to a high-density crust.

1 BY 1° BOUGUER ANOMALY PATTERN IN EUROPE

The 1 by 1° Bouguer anomaly pattern in western Europe is shown in Fig. 14. There is a marked positive bias suggested over most of the area, which is brought out more clearly in the schematic isostatic anomaly map based on Airy-Heiskanen anomalies shown in Fig. 15. This map shows that positive values predominate with negative values occurring in association with the Carpathian Mountains, the High Alps, the Baltic lowlands and Scandinavia, the Iberian Plateau, the eastern Pyrenees Mountains, and the northern Apennine Mountains paralleling the Po Valley.

A summary portrayal of the depth of the mantle, mean crustal velocity and age of the basement complex as prepared by Meissner (1970) is shown in Fig. 16. An inspection of the relations shown in Fig. 16 shows that as in the United States, crustal thickness appears to be definitely related to mean crustal velocity and within limits also geologic age.

Closs (1969) has also given a summary presentation of crustal profiles obtained in Europe. This summary includes results obtained in the Alps that are not covered in Meissner's summary. On the section extending from Regensberg, West Germany, to Vicenza, Italy, the crust thickens from about 30 to about 35 km beneath the Alps. Earlier results (Closs and Labrouste, 1963) for a profile between Grenoble, France, and Turin, Italy, indicate a marked change in crustal composition and thickness in crossing from the negative isostatic anomaly area of the French Alps to the positive isostatic anomaly area of the Italian (Graian) Alps. A more recent study by Berckheimer (1968) shows that this change in crustal composition is characteristic of the area as suggested by the gravity anomaly pattern.

Similarly, Sollogub (1969) has summarized refraction results in southeastern Europe that are only partially covered by Meissner's summary. He reports a

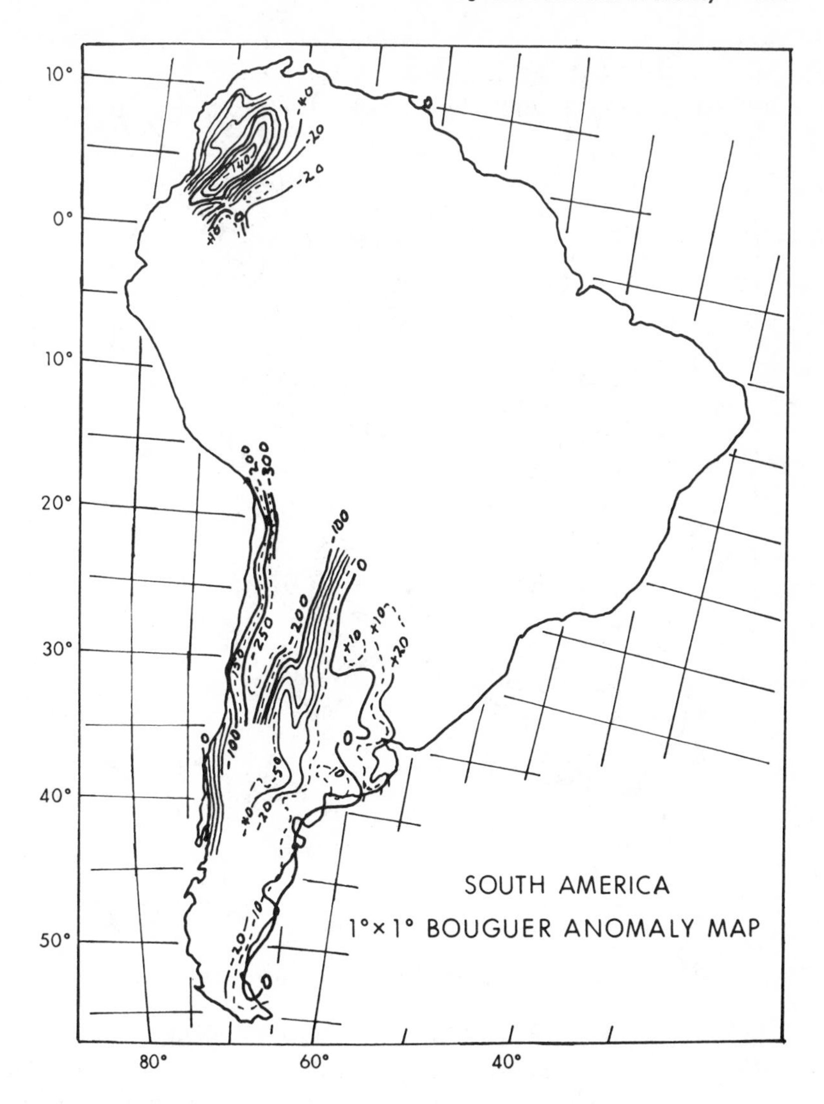

Fig. 13. 1 by 1° Bouguer anomaly map of South America.

30-km, high-velocity crust at Bari on the Italian Adriatic coast, which thickens to about 44 km beneath the Dinaric Alps of Yugoslavia. In the vicinity of Szeged, the crust has a low mean velocity and the thickness is only about 28 km, but beneath the Transylvanian Alps it thickens again to about 53 km with a marked increase in mean crustal velocity. Although gravity data are not available for most of this profile, it is significant that the available isostatic anomaly data indicate positive values at both Bari and in the Transylvanian Alps where the crust has a high mean velocity.

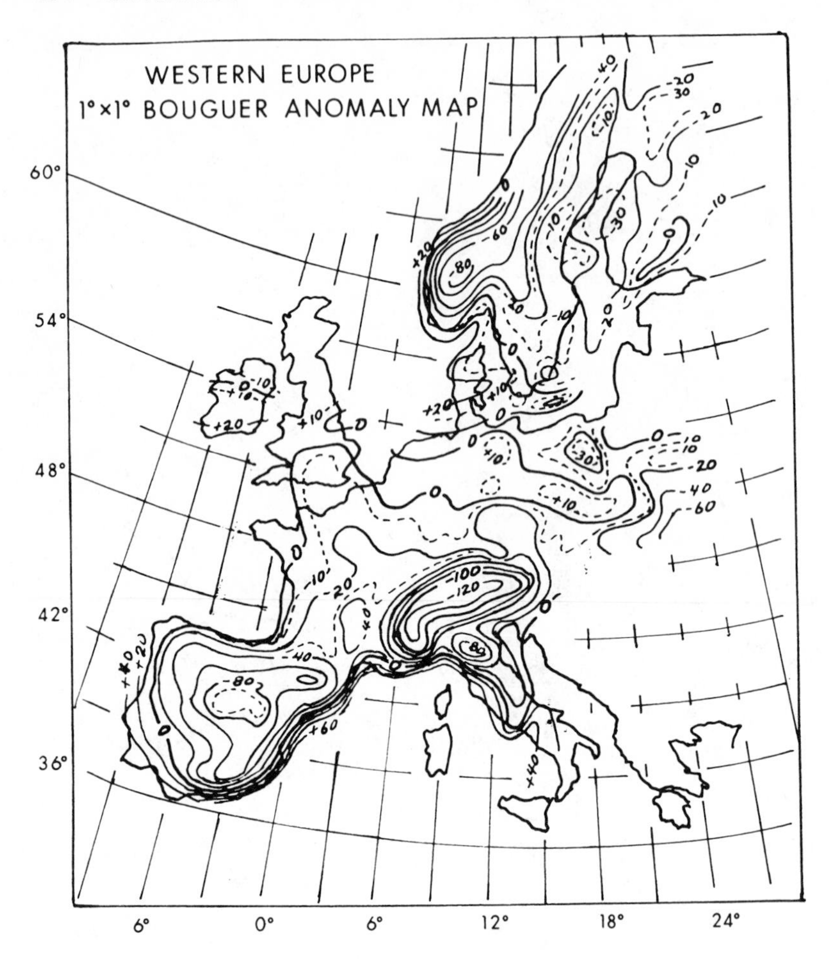

Fig. 14. 1 by 1° Bouguer anomaly map of Western Europe.

In summary, the pattern of abnormal gravity in Europe appears to bear the same relation to the seismic parameters of the crust as noted in North America. The correlation with mantle velocity variations is not as systematic, but whether this represents actual conditions or experimental error cannot be readily ascertained.

1 BY 1° BOUGUER ANOMALY PATTERN IN INDIA

The 1 by 1° Bouguer anomaly pattern in India is shown in Fig. 17, and it is seen that central India north of 20°N latitude appears to have an overall positive bias, whereas the area south of 20°N latitude has a negative bias. The schematic isostatic anomaly map based on Pratt-Hayford anomalies with a compensation depth of 113.7 km shown in Fig. 18 indicates that there are several significant

Fig. 15. Schematic isostatic anomaly map of Europe and North Africa.

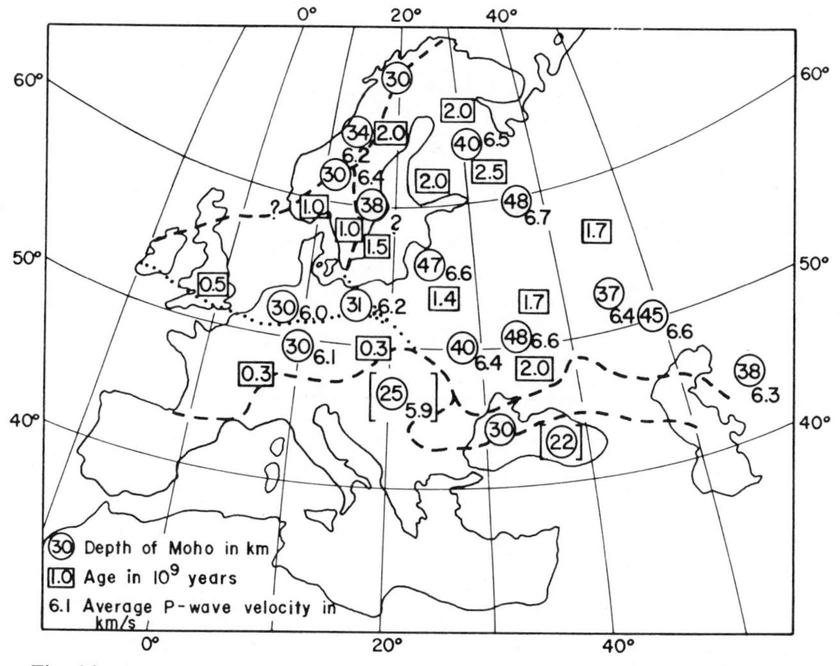

Fig. 16. Summary on crustal parameters and age variations in Europe. Prepared by Meissner (1970).

geological associations: positive isostatic anomalies characterize the central India mountain ranges and also the Thar desert depression extending northward from the coast at about 73°E longitude. Positive anomalies are also associated with the Himalaya Mountains, and there are marked negative anomalies in association with the Ganges valley. As the Ganges valley is known to be a deep trough filled with clastic material, there may be no crustal correlation other than a local downwarping of the crust or a crustal graben. The other anomalous areas appear to be related to changes in crustal and possibly upper mantle parameters. As there are no seismic measurements, these postulates cannot be verified.

Although the Deccan plateau with an extensive cover of flood basalts embraces much of southern India, there is no obvious expression of this geologic feature in the gravity field, which is uniformly subnormal over southern India. The cause of this negative bias in gravity in southern India is not known, but since it

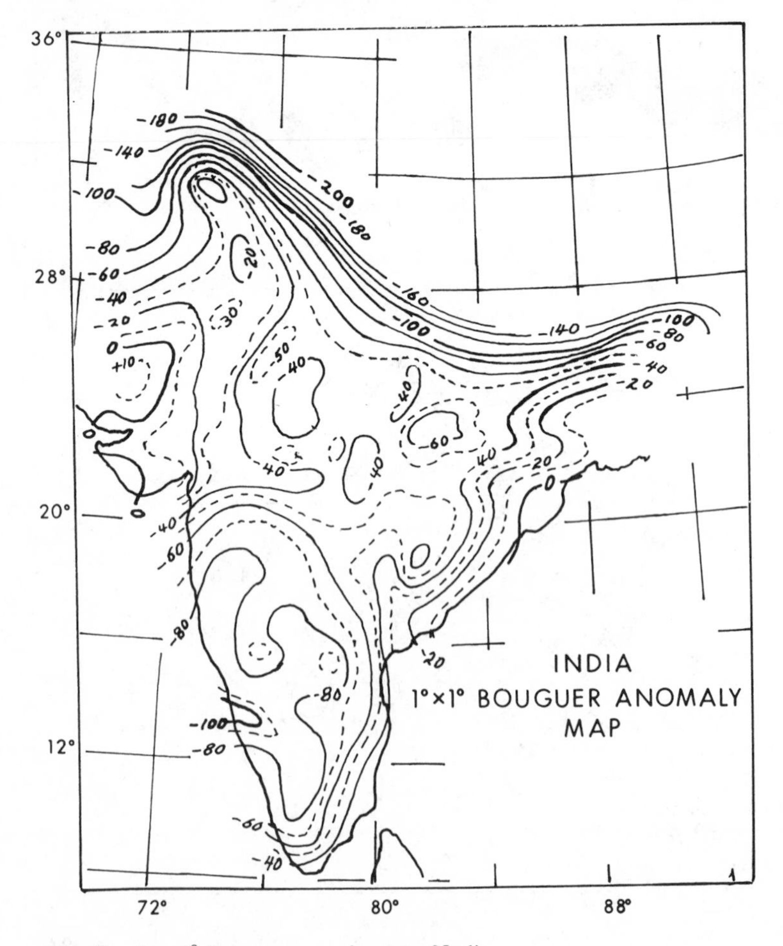

Fig. 17. 1 by 1° Bouguer anomaly map of India.

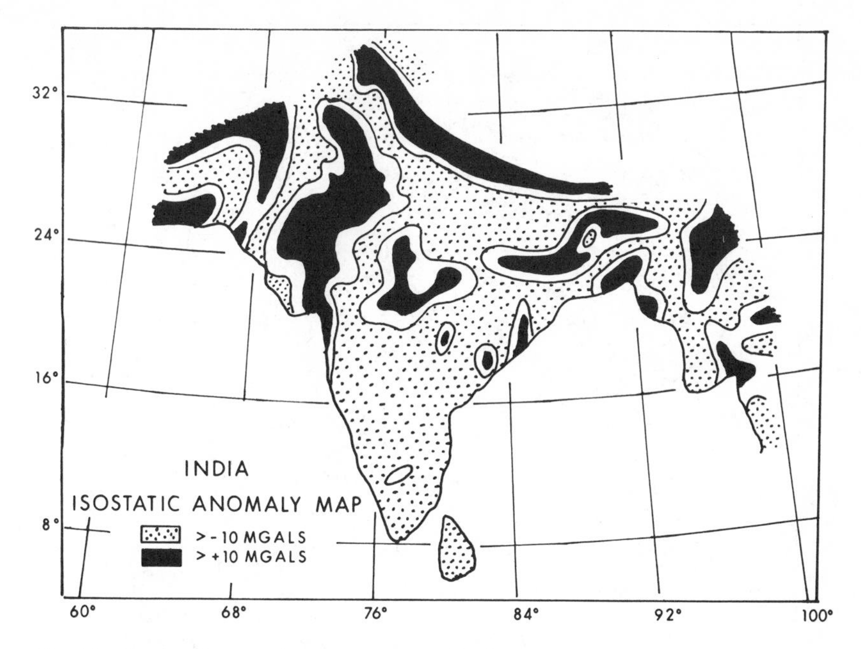

Fig. 18. Schematic isostatic anomaly map of India.

extends southward into the Indian Ocean as shown by Le Pichon and Talwani (1969) and is also evident in the satellite-derived gravity field, it appears that it probably has its origin in the mantle.

1 BY 1° BOUGUER ANOMALY PATTERN IN AFRICA

The 1 by 1° Bouguer anomaly map for Africa is shown in Fig. 19, and as is shown in the isostatic anomaly map of Europe (Fig. 15), much of north Africa appears to have a positive bias. As has been pointed out by Woollard (1969b), however, plots of average 1 by 1° Bouguer anomalies versus corresponding elevation values indicate that there is no overall pattern of positive bias as in Europe, but rather marked regional variations, which are also evident in the 3 by 3° average anomaly values as shown in Fig. 20.

The only seismic crustal measurements are those reported by Hales and Sacks (1958) for the area around Johannesburg, which lies near the center of the – 140-mgal closure defined in Fig. 19. Although these measurements show a variation in mean crustal velocity from about 6.1 to 6.3 km/sec and there is a corresponding change in crustal thickness from 36 to 42 km, there is no absolute correlation with Pratt-Hayford isostatic anomaly values, which range from – 10 to + 20 mgal as found in the United States. This can be attributed to the + 20-mgal bias in the values for South Africa. In general, the crustal thickness is subnormal for the surface elevation, which varies between 1300 and 1600

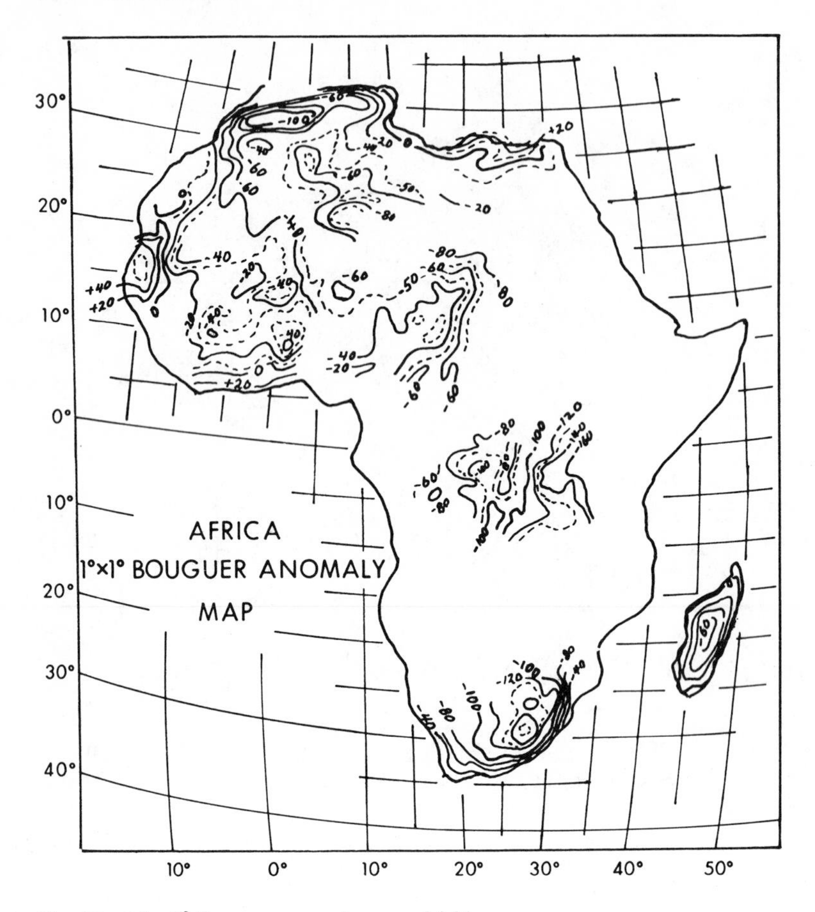

Fig. 19. 1 by 1° Bouguer anomaly map of Africa.

meters. In view of the low crustal velocity values, this would appear to be a normal relation.

The cause of the marked difference of about 60 mgal between northern Africa and central equatorial Africa with south Africa lying about half way between the two as shown in Fig. 20 is not known.

1 BY 1° BOUGUER ANOMALY PATTERN IN AUSTRALIA

The 1 by 1° Bouguer anomaly map of Australia is shown in Fig. 21. Except for the Macdonald Mountains in central Australia marked by a – 100-mgal closure, much of the continent appears to have a positive bias. This bias, however, is not a constant one, but appears to vary significantly from area to area. As a result, there is about 100-mgal variation in 1 by 1° Bouguer anomaly values for an elevation of 500 meters. Although much of this large spread in 1 by 1° anomaly

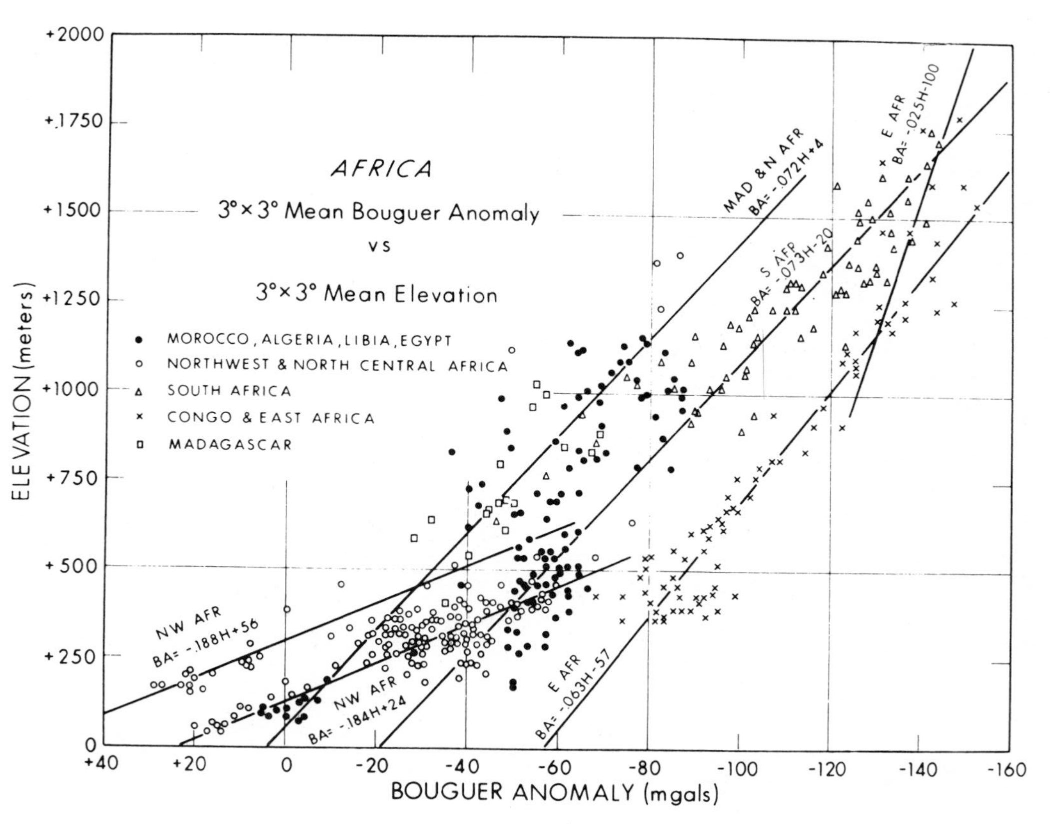

Fig. 20. Relation of 3 by 3° Bouguer anomalies to elevation in Africa.

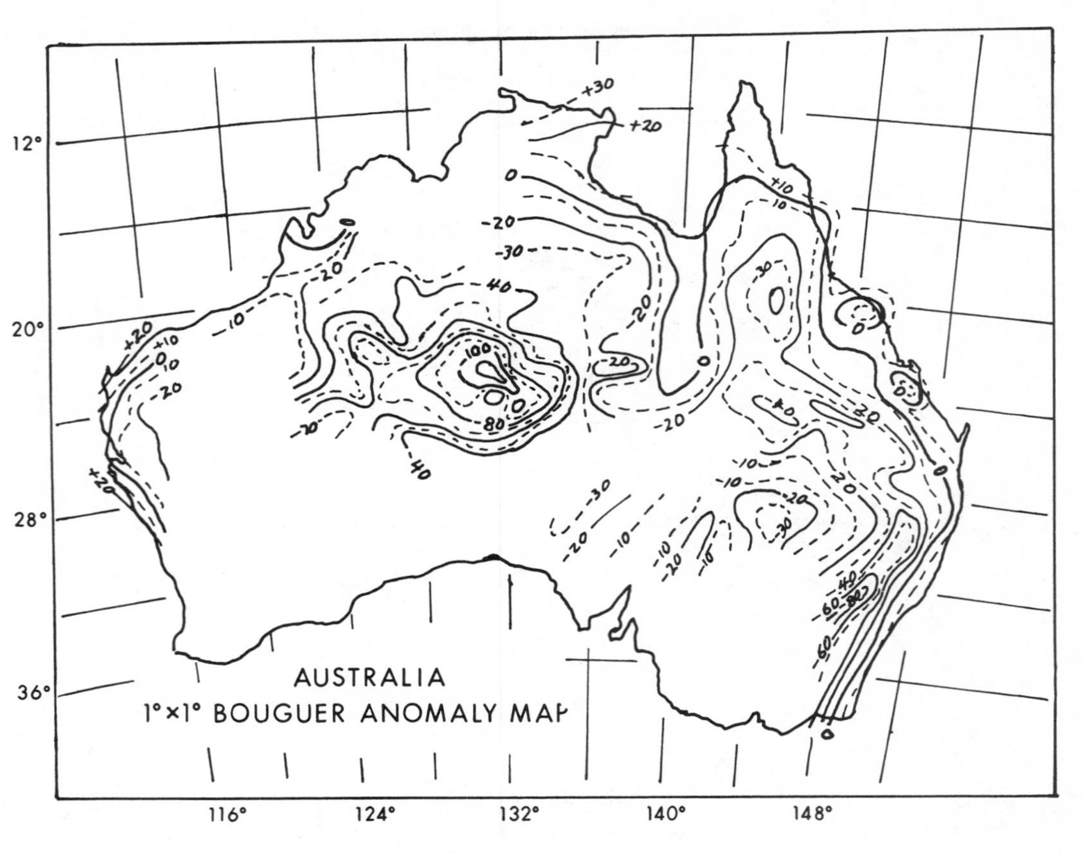

Fig. 21. 1 by 1° Bouguer anomaly map of Australia.

values is related to local changes in geological structure rather than regional changes in crustal and upper mantle parameters, there are apparent regional changes in the mantle and crust as indicated by the fact that the 3 by 3° Bouguer values show two groupings of data separated by about 40 mgal at 500-meters elevation. This difference is emphasized in the 3 by 3° free air anomaly values, which show vertical free air anomaly gradients of −64 mgal per 1000 meters in the central area and + 91 mgal per 1000 meters in the eastern area.

Although there have been seismic crustal measurements reported (Bolt et al., 1958; Doyle et al., 1959) for both eastern and western Australia, these, while in different geographic areas, define similar crustal conditions. The elevations are similar (625 and 300 meters) and there are similar Pratt-Hayford negative isostatic anomalies of − 15 to − 30 mgal. The values of crustal thickness are much the same (37.0 and 37.4 km), and in both areas a single layer low velocity (5.9 to 6.0 km/sec) crust was found. More recent unpublished reflection studies in southern Australia by the Bureau of Mineral Resources show a similar thin crust of about 34-km thickness in the Ballarat region of southeastern Australia where the marked positive vertical free air anomaly gradient suggests there is a lack of crustal compensation.

SUMMARY

Where there are seismic crustal measurements, it appears that the pattern of change in gravity 1 by 1° anomaly values on the continents can be explained in most areas by changes in crustal and upper mantle parameter values. Few areas appear to be out of actual isostatic equilibrium, and in most, any bias in the gravity anomaly values appears to be related to changes in mean crustal density as expressed in mean velocity values. There is a suggestion that crustal density, composition, and thickness can be correlated with geologic age and because of this, there is in general a correlation between isostatic gravity anomaly values and geologic age. The age pattern within the basement crystalline rock complex is not simple, however, and other factors as orogeny, crustal uplift, and subsidence as well as major intrusive bodies and volcanism also influence crustal composition, structure and thickness, and consequently, the associated gravity field.

There are areas of major changes in crustal and upper mantle parameter values where the underlying velocity pattern suggest there is a density bias over a considerable depth range, and the overall isostatic compensation appears to conform to that of the Pratt concept of isostasy. In most areas though where there is seismic control, the compensation appears to be achieved primarily at the depth of the *M*-discontinuity. Contributions to the compensation from deeper mass discontinuities, while undoubtedly present, appear to be of smaller magnitude in terms of their gravitational effect. The interrelation of crustal parameter values to the velocity of the mantle and the general dependence of the depth of upper mantle seismic discontinuities on their velocity, at least above the velocity reversal at about 100 km, suggest a process that affects velocity and density from within the upper mantle up into the crust. The process appears to be reversible leading to a thinning of the crust, a decrease in seismic velocity values, surface uplift, and negative isostatic anomalies in one phase. In opposite phase, the process leads to a thickening of the crust, an increase in seismic velocity values, surface subsidence, and positive isostatic anomalies. As both phases appear to have been operative at the same time in the geologic past in adjacent areas of crustal uplift and subsidence, it would appear to be thermally controlled and a consequence of thermal convection.

Some orogenic belts, such as the Appalachian tectonic province, exhibit a pattern of negative isostatic anomaly values flanked by positive anomaly values. In the Alps where this anomaly pattern is quite pronounced, seismic data indicate it is associated with a marked change in mean crustal velocity values and crustal composition. Other orogenic belts as the Rocky Mountains, show no distinctive gravity pattern such as characterizes the Appalachian Mountains, and seismic data suggest there is no local root beneath the Rocky Mountains, such as is found beneath the Sierra Nevada Mountains. Although most orogenic mountain ranges appear to be isostatically compensated regardless of whether there is or is not seismic evidence of a crustal "root," some, as the Great Dividing Range in eastern Australia, appear to be only partially compensated as evidenced by a marked positive vertical free air anomaly gradient. In general, shield areas are characterized by a number of lenticular anomaly patterns that apparently are related to former Precambrian orogenic belts. In these areas, the gravity control appears to lie in the upper portion of the crust, but the seismic evidence of associated variations in crustal thickness is not always consistent.

Regional long wavelength changes in gravity field can be explained in areas of former ice caps by a time lag in the isostatic rebound of the crust following deglaciation. Areas such as southern India, however, appear to be regionally anomalous because of as yet unknown mass distributions that are probably located in the upper mantle. A similar explanation is required to explain why each continent appears to represent a special gravity domain with resulting bias in gravity anomaly values and the vertical gravity anomaly gradient. These regional gravity domains, however, are not always of continental size, and in North America and Africa there appear to be three regional areas of anomalous gravity in which the relation of free air and Bouguer gravity anomalies to elevation varies significantly.

These differences are significant, as is indicated by the fact that they affect both absolute gravity values and the vertical gravity gradient. Because of these differences, 3 by 3° average Bouguer gravity anomaly values on the different continents show a spread of about 80 mgal for a median continental elevation of 1000 meters. Since it is on this pattern of varying base relationships that geologic and crustal effects are superimposed, the same disturbing mass distribution will be characterized by different anomaly values on the different continents, and as indicated, in some cases different portions of the same continent.

ACKNOWLEDGMENTS

The writer would like to acknowledge the help extended by many oil companies and other groups in making data available for this study, and in particular the cooperation of the Australian Bureau of Mineral Resources who have systematically released new data each year as it was taken. He is also indebted to the U.S. Air Force Chart and Information Center for data and support to evaluate and adjust the data collected to the international gravity standard. Particular thanks are also due to Dr. Rolf Meissner for permission to use his summary diagram on crustal and age relationships in western Europe prior to publication.

REFERENCES

Anderson, D. L., Phase changes in the Upper Mantle, Science, **157,** 3793, 1165-1173, 1967.

Archambeau, C. B., Flinn, E. A., and Lambert, D. G., Fine structure of the Upper Mantle, J. Geophys. Res., **74,** 5825-5866, 1969.

Bolt, B. A., Doyle, H. A., and Sutton, D. J., Seismic observations from the 1956 atomic explosions in Australia, Geophys. J., **1,** 135-145, 1958.

Beliayevsky, N. A., The problem of relationship between geological structures and deep-seated crustal structures, Tectonophysics, **7,** 403-406, 1969.

Berckheimer, H., Topographic des Ivrea-Korpers abgeleit aus seismischen und gravimetrischen Daten, Sonderdruch Schweis Mineral Petrog. Mittel, **48,** 235-246, 1968.

Bowie, W., Investigation of Gravity and Isostasy, U.S. Coast Geod. Surv. Spec. Pub., 40, 195 pp., 1917.

Closs, H., Explosion seismic studies in western Europe, *in* the Earth's Crust and Upper Mantle, Am. Geophys. Union Geophys. Monograph 13, edited by P. J. Hart, pp. 178-188, 1969.

Closs, H., and Labrouste, Y., Recherches seismologiques dans les Alpes occidentales du moyen de grandes explosions en 1956, 1958 et 1960, Seismologie, **12,** 241-250, 1963.

Cummings, G. L., Garland, G. D., and Vozoff, K., Seismological Measurements in Southern Alberta, Tech. Report submitted to U.S. Air Force, Cambridge Research Laboratory, Office of Aerospace Research, Bedford, Mass., 1962.

Carnegie Institution of Washington, Year Book 67, Annual Report of Director, Department of Terrestial Magnetism, 1969.

Demenitskaya, R. M., The method of research of the geological structure of the crystalline mantle of the Earth, Soviet Geol., **1,** 92-112, 1959.

______, Crust and Mantle of the Earth, Nedra, 280 pp., 1967.

Dorman, L. M., and Lewis, B. T. R., Experimental isostasy, 1, Theory of the determination of the Earth's isostatic response to a concentrated load, J. Geophys. Res., **75,** 3357-3365, 1970.

Doyle, H. A., Everingham, I. B., and Hogan, T. K., Recording of large explosives in southeastern Australia, Austral. J. Phys., **12,** 222-230, 1959.

Drake, C. L., and Nafe, J. E., The transition from ocean to continent from seismic refraction data, *in* the Crust and Upper Mantle of the Pacific Area, Am. Geophys. Union Geophys. Monograph 12, edited by L. Knopoff, C. L. Drake, and P. J. Hart, pp. 174-186, 1968.

Duerksen, J. A., Pendulum Gravity Data in the United States, U.S. Coast Geod. Surv. Spec. Pub., 244, 217 pp., 1949.
Ewing, G. N., Dainty, A. M., Blanchard, J. E., and Keen, M. J., Seismic studies on the eastern seaboard of Canada, 1, The Appalachian System, Can. J. Earth Sci., **8,** 89-110, 1966.
Furumoto, A. S., Thompson, N. J., and Woollard, G. P., The structure of the Koolau Volcano from seismic refraction studies, Pacific Sci., **19,** 306-314, 1965.
Furumoto, A. S., Woollard, G. P., Campbell, J. F., and Hussong, D. M., Variation in the thickness of the crust in the Hawaiian Archipelago, *in* the Crust and Upper Mantle of the Pacific Area, Am. Geophys. Union Geophys. Monograph 12, edited by L. Knopoff, C. L. Drake, and P. J. Hart, pp. 94-111, 1968.
Green, R. W. E., and Hales, A. L., The travel times of *P* waves to 30° in the central United States and Upper Mantle structures, Bull. Seism. Soc. Am., **58,** 267-290, 1968.
Hales, A. L., and Asada, T., Crustal structure in coastal Alaska, *in* the Earth Beneath the Continents, Am. Geophys. Union Geophys. Monograph 10, edited by J. S. Steinhart and T. J. Smith, pp. 420-432, 1966.
Hales, A. L., and Sacks, I. S., Evidence for an intermediate layer from crustal structure studies in the eastern Transvaal, Geophys. J., **2,** 15-25, 1958.
Hess, H. H., Serpentines, Orogeny and Epiorogeny in Crust of the Earth, Geol. Soc. Am. Sp. Paper 62, pp. 391-408, 1955.
Innes, J. J. S., Goodacre, A. K., Weston, A. A., and Weber, J. R., Gravity and isostasy in the Hudson Bay region, *in* Science, History and Hudson Bay, vol. 2, Can. Dept. Energy, Mines and Resources, pp. 703-727, 1968.
Johnson, L. R., Array measurements of *P* velocities in the Upper Mantle, J. Geophys. Res., **72,** 6309-6325, 1967.
Kaula, W. M., A tectonic classification of the main features of the Earth's gravitational field, J. Geophys. Res., **74,** 4807-4826, 1969.
______, Global gravity and tectonics, (in this volume), 1971.
Kennedy, G., The origin of continents, mountain ranges and ocean basins, Am. Sci., **47,** 494-504, 1959.
Lee, W. H. K., and MacDonald, G. J. F., The global variation of terrestrial heat flow, J. Geophys. Res., **68,** 6481-6492, 1963.
Lee, W. H. K., and Uyeda, S., Review of heat flow data, *in* Terrestrial Heat Flow, Am. Geophys. Union Geophys. Monograph 8, pp. 87-190, edited by W. H. K. Lee, 1965.
Le Pichon, X., and Talwani, M., Regional gravity anomalies in the Indian Ocean, Deep Sea Res., **16,** 263-274, 1969.
Lewis, B. T. R., and Dorman, L. M., Experimental isostasy, 2, An isostatic model for the USA derived from gravity and topographic data, J. Geophys. Res., **75,** 3367-3386, 1970.
Lewis, B. T. R., and Meyer, R. P., A seismic investigation of Upper Mantle structure to the west of Lake Superior, Bull. Seism. Soc. Am., **58,** 565-596, 1968.
Magnitsky, V. A., and Kalashnikova, I. V., Problem of phase transitions in the Upper Mantle and its connection with the Earth's crustal structure, J. Geophys. Res., **75,** 877-885, 1970.
McKenzie, D. P., Some remarks on heat flow and gravity anomalies, J. Geophys. Res., **72,** 24, 6261-6273, 1967.
Meissner, R., Some results from deep seismic soundings in Europe (abstract), EOS Am. Geophys. Union Trans., **51,** 356, 1970.
Meyer, R. P., Steinhart, J. S., and Woollard, G. P., Central plateau of Mexico, *in* Explosion Studies of Continental Structure, Carnegie Inst. Wash. Pub. 622, pp. 199-226, 1961.
Moberly, R., and Khan, M. A., Interpretation of the sources of the satellite-determined gravity field, Nature, **223,** 263-267, 1969.
Pakiser, L. C., and Robinson, R., Composition of the continental crust as estimated from seismic observations, *in* the Earth Beneath the Continents, Am. Geophys. Union Geophys. Monograph 10, edited by J. S. Steinhart and T. S. Smith, pp. 620-626, 1966.

Pakiser, L. C., and Steinhart, J. S., Explosion seismology in the Western Hemisphere, *in* Research in Geophysics, vol. 2, Solid Earth and Interface Phenomena, MIT Press, edited by H. Odishaw, pp. 123-147, 1964.

Phillips, R. P., Seismic refraction studies in Gulf of California, *in* Marine Geology of Gulf of California, Mem. 3, Am. Assoc. Petrol. Geol., edited by T. H. van Andd and G. G. Shor, Jr., pp. 90-121, 1964.

Shor, G. G., Jr., and Pollard, D. D., Mohole site selection studies north of Maui, J. Geophys. Res., **69**, 1627-1637, 1964.

Sollogub, V. B., Seismic crustal studies in southeastern Europe, *in* the Earth's Crust and Upper Mantle, Am. Geophys. Union Geophys. Monograph 13, edited by P. J. Hart, pp. 189-194, 1969.

Strange, W. E., Comparisons with surface gravity data, *in* Geodetic Parameters for a 1966 Smithsonian Institute Standard Earth, vol. 3, Smithsonian Astrophys. Obser. Sp. Pub. 200, pp. 15-20, 1966.

Talwani, M., and LePichon, X., Gravity field over the Atlantic Ocean, *in* the Earth's Crust and Upper Mantle, Am. Geophys. Union Geophys. Monograph 13, edited by P. J. Hart, pp. 341-351, 1969.

White, W. R. H., and Savage, J. C., A seismic refraction and gravity study of the Earth's crust in British Columbia, Bull. Seism. Soc. Am., **55**, pp. 463-486, 1965.

White, W. R. H., Bone, M. N., and Milne, W. G., Seismic refraction surveys in British Columbia 1941-1966. A preliminary interpretation, *in* the Crust and Upper Mantle in the Pacific Area, Am. Geophys. Union Geophys. Monograph 12, edited by L. Knopoff, C. L. Drake, and P. J. Hart, pp. 85-93, 1968.

Wilcox, L., A method for predicting Bouguer anomaly values (unpublished), paper presented at the Int. Assn. Geod. and Geophys. General Reunion, Lucerne, Switzerland, 1967.

Woollard, G. P., A gravity reconnaissance of the island of Oahu, Trans. Am. Geophys. Union, **32**, 358-368, 1951.

______, Crustal structure from gravity and seismic measurements, J. Geophys. Res., **64**, 1521-1544, 1959.

______, Seismic crustal studies during the IGY, Part 2, Continental program, Trans. Am. Geophys. Union, **41**, 351-355, 1960.

______, The Relation of Gravity Anomalies to Surface Elevation, Crustal Structure and Geology, Univ. Wis. Geophys. and Polar Res. Center Rept. 62-9, 350 pp., 1962.

______, Regional isostatic relations in the United States, *in* the Earth Beneath the Continents, Am. Geophys. Union Geophys. Monograph 10, edited by J. S. Steinhart and T. J. Smith, pp. 557-594, 1966.

______, The interrelationship of the crust, the Upper Mantle and isostatic gravity anomalies in the United States, *in* the Crust and Upper Mantle of the Pacific Area, Am. Geophys. Union Geophys. Monograph 12, edited by L. Knopoff, C. L. Drake, and P. J. Hart, pp. 312-341, 1968.

______, A Regional Analysis of Crustal Structure in North America, Hawaii Inst. Geophys. Tech. Rept. 69-12, 55 pp., 1969a.

______, Regional variation in gravity, *in* The Earth's Crust and Upper Mantle, Am. Geophys. Union, Geophys. Monograph 13, edited by P. J. Hart, pp. 320-340, 1969b.

______, Evaluation of the isostatic mechanism and role of mineralogic transformations from seismic and gravity data, Phys. Earth Planet. Interiors, **3**, 484-498, 1970.

Woollard, G. P., Joesting, H. R., A Bouguer Gravity Anomaly Map of the United States. U.S. Geol. Surv. Spec. Map, 1964.

Woollard, G. P., and Khan, M. A., A review of satellite derived figures of the geoid and their geological significance, Pacific Sci., **24**, 1, 1-28, 1970.

Woollard, G. P., Ostenso, N. A., Thiel, E., and Bonini, W. E., Gravity anomalies, crustal structure and geology in Alaska, J. Geophys. Res., **65**, 1021-1037, 1960.

19
CONTINENTAL HEAT FLOW

by **Robert F. Roy, David D. Blackwell, and Edward R. Decker**

*To my mind, no single outstanding problem is more important for an understanding of geological processes than the problem of the distribution of temperature throughout the upper 100 km or so, as dependent upon such factors as time, sedimentary cover, and deformation; and probably no geological or geophysical problem has resisted solution more stubbornly. The principal difficulty is our ignorance of the distribution of the radioactive heat producing elements. We have no difficulty in finding possible distributions, consistent with our meagre observational data. But none of the possible distributions really imposes itself as conclusive. The differences between the possible distributions with regard to temperatures are very great, and probably entirely different geological processes would have to be invoked for some of the more extreme types. Thus, every conceivable method of introducing further restrictions on the possible distributions requires careful study.**

The correlation of heat flow with the radioactivity of surface rocks, as first proposed by Francis Birch, has done much to clarify the patterns of continental heat flow and, as an additional dividend, has raised interesting questions about the vertical distribution of radioactive elements. It now seems possible to establish the regional heat flow pattern for a whole continent with a few dozen measurements properly located with respect to knowledge of basement radioactivity.

*Birch (1947a), p. 793.

Results of measurements in the United States are presented in the form of a map of reduced heat flow. In most of North America east of the Rocky Mountains and north of the Gulf Coast geosyncline, large variations in heat flow at the surface are attributed entirely to variations in the radioactivity of the upper crust with a uniform flux from the lower crust and upper mantle. This portion of North America has been stable since the Triassic or longer. It is suggested that the same relationship between heat flow and radioactivity will be found on stable portions of other continents. The most prominent feature on the map is the large region of high flux generally following the North American Cordillera. Smaller regions of abnormally low heat flow are found in the Sierra Nevada and Peninsular Ranges. High values are found in the Franciscan block east of the San Andreas fault and normal values in the Salinian block west of the San Andreas fault. Transition zones between provinces have been studied in six places. All are narrow (less than 100 km wide), implying a shallow depth to partially molten upper mantle in the high heat flow provinces, with cold roots under the regions of abnormally low heat flow.

Considering these facts in terms of the concepts of plate tectonics, we feel it is possible to explain zones of abnormally low heat flow on the ocean side of regions of high heat flow, if the low heat flow is attributed to transient cooling by recently overridden portions of a cold oceanic plate. Subsequent warming of the subsiding block to mantle temperatures combined with heat sources in the oceanic plate would lead to partial melting in a few tens of millions of years. Upward convection of partially molten material would soon result in the high temperatures near the base of the crust that are necessary to explain the heat flow distribution observed at the surface.

INTRODUCTION

Birch's discussion in 1947 properly emphasized the importance of the distribution of radioactive heat-producing elements to understanding the temperature field in the outer part of the earth and the major tectonic processes that are driven by thermal energy. Heat flow at the surface is the observational boundary condition that all acceptable thermal models must satisfy. Throughout his career, Birch has placed a high priority on obtaining additional measurements of heat flow and investigating the various factors that must be considered in reducing raw heat flow data to a common base suitable for regional comparisons. These investigations have included analyses of the corrections required for subsurface temperature effects of topography (Birch, 1950), flowing water in drill holes (Birch, 1947b), climatic variations (Birch, 1948, 1954a), geologic evolution (Birch, 1950; Birch, Roy, and Decker, 1968), the thermal properties of major rock types (Birch and Clark, 1940), and the significance of combined studies of heat flow and radioactivity (Birch, 1947a, 1950, 1954b; Birch, Roy, and Decker, 1968; Roy, Blackwell, and Birch, 1968).

As a result of the work of the last decade, we now believe that there is sufficient information to provide preliminary answers to some of the important questions raised by Birch. The effects of climate, culture, topography, and hydrology can usually be regarded as secondary corrections to individual measurements of heat flow, whereas variations of geologic history and basement radioactivity are more fundamental. Examples of a few secondary corrections are given in the following section, but our aim in the bulk of the study is to summarize the broad scale implications of results obtained from recent studies of heat flow and radioactivity on the continents.

MEASUREMENTS IN SHALLOW HOLES

Prior to the early 1960's, most heat flow measurements were made in tunnels, oil wells, and holes drilled for mineral exploration. These measurements were few in number; each was treated with considerable care; and out of these studies came the standard corrections for topographic irregularities, uplift, erosion, climatic changes, drilling disturbances, and the like. Experience in working with data from holes drilled for mineral and oil exploration suggested that the majority were badly disturbed in the upper few hundred meters with irregular, even negative, gradients, which rendered that portion of the hole unusable for heat flow measurements. Thus attention focused on cored exploration holes deeper than 200 meters, where long segments of "undisturbed" (at least consistent) temperatures could be obtained and the upper portions could be ignored. With the advent of drilling for heat flow studies this disturbed zone received closer scrutiny, and it can now be demonstrated that most of the problems arise from three basic causes: (1) movement of ground water up or down a hole between previously unconnected fracture systems or aquifers; (2) changes in the mean annual surface temperature resulting from activities of man that cause transient disturbances to underground temperatures; (3) temperature anomalies at the surface resulting from contrasts in vegetation that lead to steady-state disturbances to underground temperatures.

An example of disturbed temperatures resulting from water movement in a hole is shown in Fig. 1. In this case, there was artesian flow at a rate of approximately 30 gal/min. This problem can usually be overcome by installing casing and filling the annulus between pipe and hole with a chemical grout or cement.

Figure 2 shows the subsurface temperatures in a part of Cambridge, Massachusetts, where most of the buildings were constructed between 1910 and 1920. If we assume that the heated basements led to an increase in the mean annual surface temperature of 5°C, the undisturbed temperatures can be estimated using a solution from Carslaw and Jaeger (1959, p. 321, 322). The average of the mean annual air temperatures for six weather stations surrounding

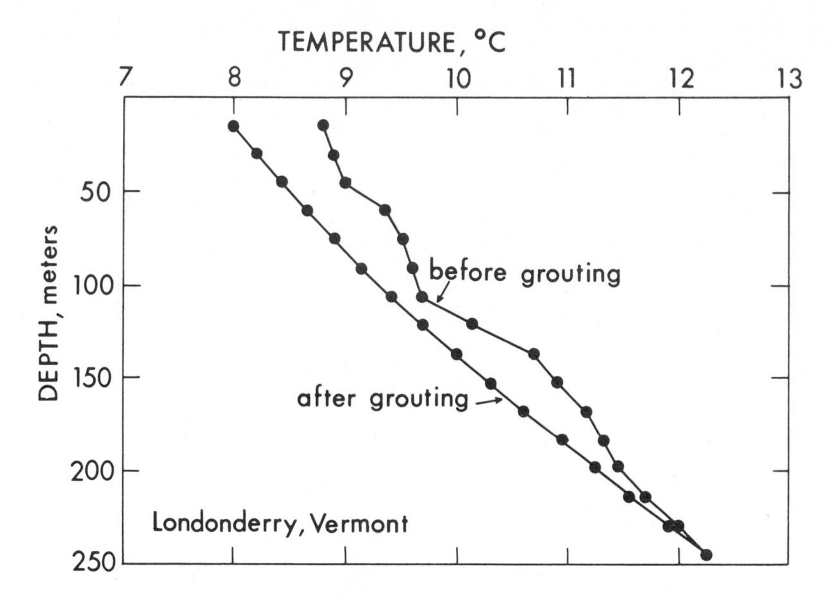

Fig. 1. Temperatures measured in a drill hole at Londonderry, Vermont. Before grouting, the hole had an artesian flow of approximately 30 gal/min.

Cambridge (Bedford, Blue Hill, Boston Airport, Chestnut Hill, Reading, and Weston) corrected to the elevation of the Cambridge site assuming an adiabatic lapse rate of 4.5°C/km (Birch, 1950) is 9.7°C. The mean annual soil temperature, estimated by substituting 0°C for the months with snow cover, is about 10.3°C, which agrees fairly well with the corrected temperatures extrapolated to the surface. We have observed similar temperature depth curves in the cities of Concord, New Hampshire, and Glens Falls, New York. Other human activities such as clearing forests and planting crops (or natural phenomena such as forest fires) cause similar disturbances, which may be significant to depths of 100 meters or more depending on time and the diffusivity of the rock. In most cases it is not possible to make accurate corrections for these transient effects; the only solutions are to avoid such areas or drill deeper holes.

Figures 3 and 4 show the effects of steady-state contrasts in surface temperatures resulting from variations in ground cover. The hole at Blodgett Forest was drilled at the edge of a small clearing in an extensive forest of virgin pine. The vegetation in the clearing consists mainly of small broadleafs reflecting the different soil that has devloped over an erosional remnant of an andesite flow. If the mean annual surface temperature of the clearing is approximately 0.5°C higher than in the surrounding forest, the temperature in the upper part of the temperature-depth curve is easily explained using methods described by Lachenbruch (1957). A similar disturbance, although opposite in sign, is found at Loomis, California, (Figs. 5 and 6) where the hole was drilled near a small

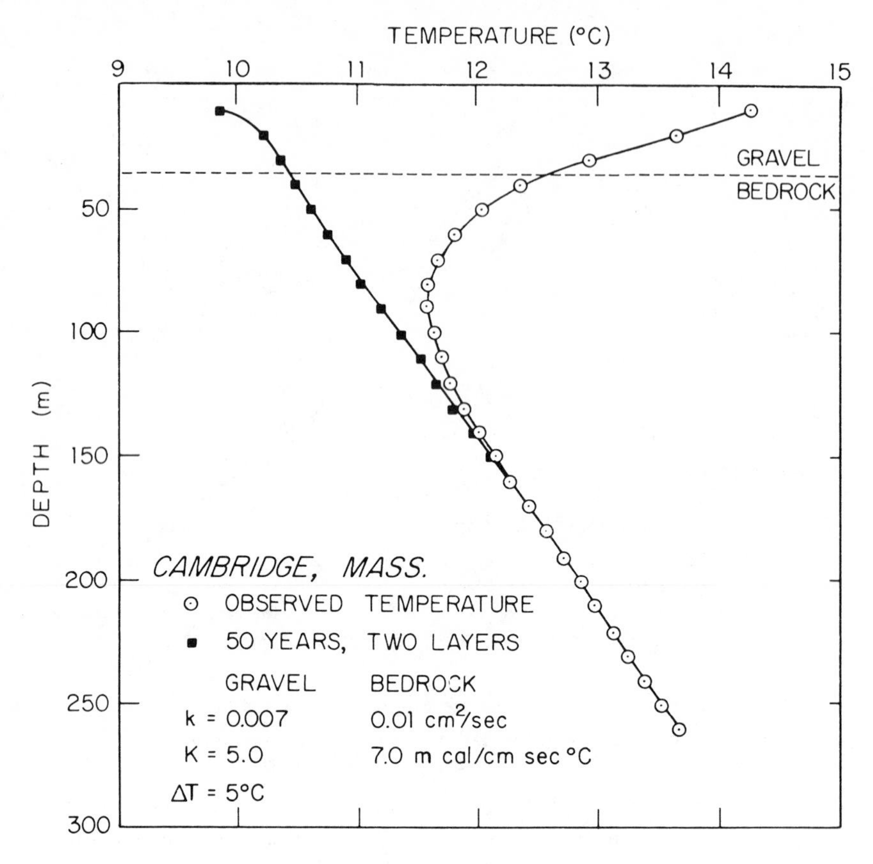

Fig. 2. Observed temperatures and estimated undisturbed temperatures in a drill hole at Cambridge, Massachusetts. Correction assumes that a sudden increase in mean annual surface temperature of 5°C took place 50 years ago as a result of building construction.

stream with a fringe of trees traversing open grasslands. As with the transient case, the uncertainties in corrections are uncomfortably large, and it is best to avoid such problems when locating drill sites.

Hualapai Mountains, Arizona (Fig. 7), is one of many localities where there is no evidence of human activity or contrasts in vegetation, and topographic relief is small. There is no significant change in gradient between 50 and 250 meters.

Details of these and other effects of climate, culture, hydrology, and topography will be presented elsewhere; the point we wish to emphasize here is our conclusion that with careful site selection, reliable heat flow values can be measured in holes 100 to 150 meters deep by conventional methods.

Oceanographic Techniques on Land

Sass, Munroe, and Lachenbruch (1968) have demonstrated that heat flow can be measured at shallow depths (under 200 meters) in unconsolidated sediments

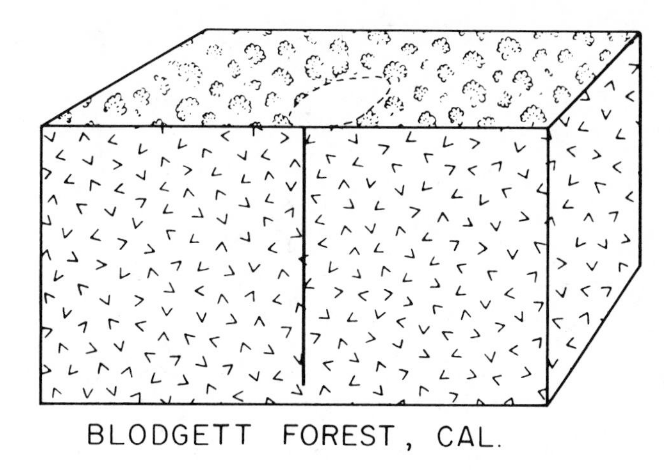

Fig. 3. Block diagram showing contrast in natural vegetation near the drill site at Blodgett Forest, California.

by utilizing closely spaced temperature and conductivity measurements within short cored intervals, similar to methods used by oceanographers. Strictly oceanographic techniques have been applied by Hart and Steinhart (1965), Von Herzen and Vacquier (1967), Steinhart, Hart, and Smith (1969), and Sclater,

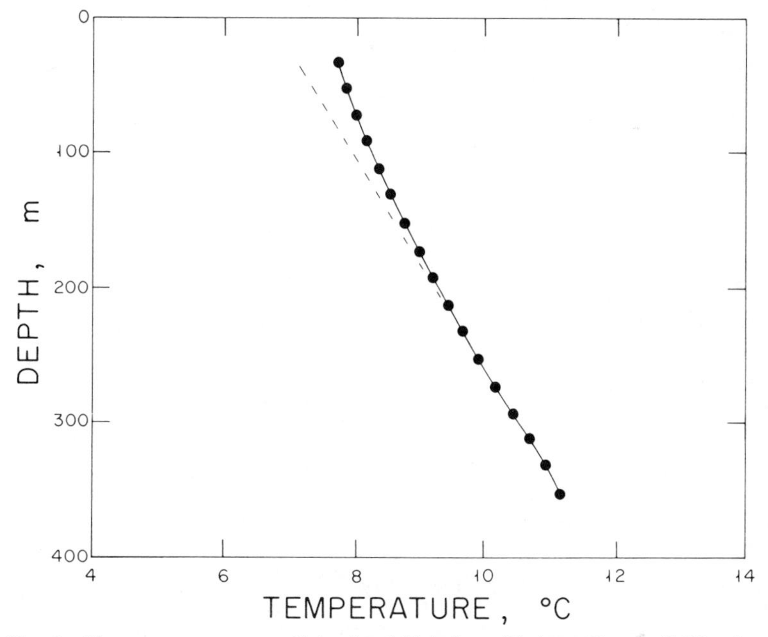

Fig. 4. Temperatures measured in the drill hole at Blodgett Forest, California. Dashed line represents temperatures corrected for a steady-state temperature anomaly of +0.5°C in the clearing.

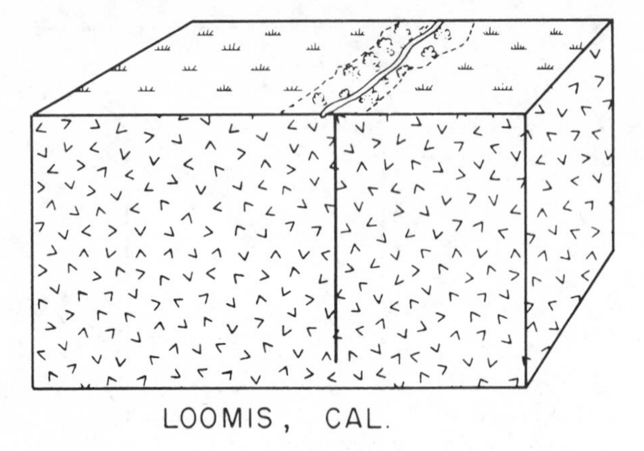

Fig. 5. Block diagram showing contrast in natural vegetation near the drill site at Loomis, California.

Vacquier, and Rohrhirsch (1970) to large deep lakes with isothermal bottom water. The advantages of working in large, deep lakes with oceanic techniques are limited by the small number of such lakes, their locations, and the need for large vessels with heavy hoisting equipment.

Measurements in small, meromictic lakes (Diment, 1967; Johnson and Likens,

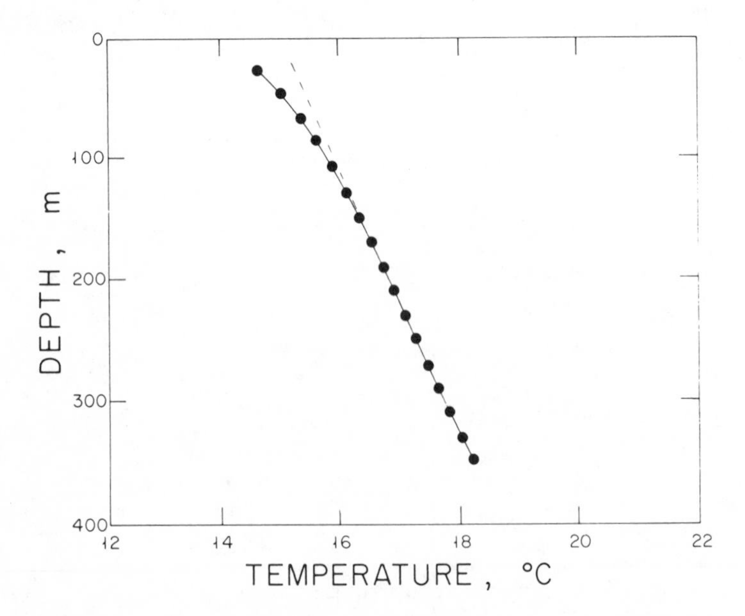

Fig. 6. Temperatures measured in the drill hole at Loomis, California. Dashed line represents temperatures corrected for a steady-state temperature anomaly of $-1°C$ near stream and trees.

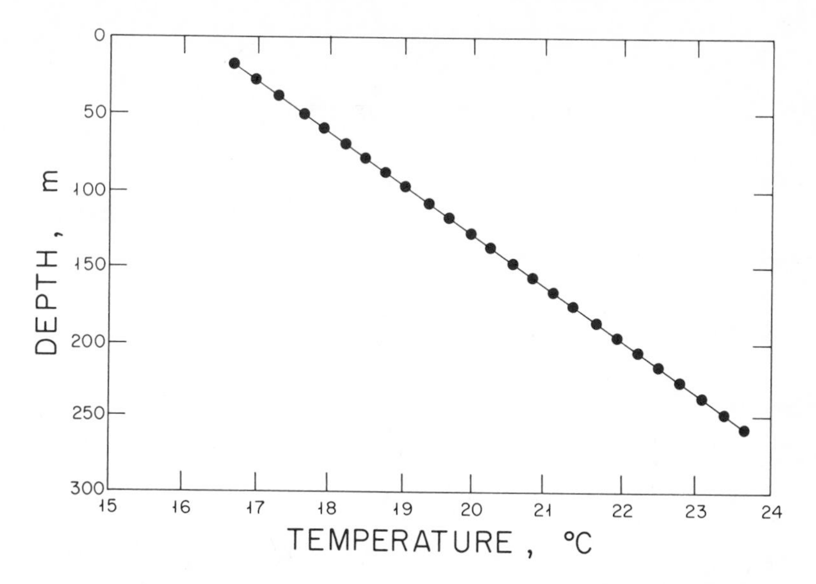

Fig. 7. Temperatures measured in a drill hole in the Hualapai Mountains, Arizona. There are no contrasts in natural cover or evidences of human activity (except for drilling) at this site. We have observed many similar temperature depth curves in the western United States where the natural vegetation appears to be in equilibrium with the environment.

1967; Reitzel, 1966) have large, uncertain corrections for temperature differences at the edges that render them unreliable for studies of regional heat flow, useful as they are for limnological purposes (Likens and Johnson, 1969).

Lakes of moderate size (2 km or more in diameter) are abundant if not ubiquitous, have much smaller corrections for the "warm rim effect," and show promise for quick inexpensive measurements, particularly in northern latitudes where it is possible to work from a stable ice surface during the winter (Williams and Roy, 1970). Although these lakes do not enjoy the advantage of constant temperatures in the bottom water, the annual temperature cycle is rapidly attenuated by the low diffusivity of the mud (about 0.002 cm^2/sec) and becomes negligible below about 15 meters for a temperature range of 10°C at the mud-water interface (Ingersoll, Zobel, and Ingersoll, 1954, p. 47). Utilizing piston coring techniques developed by limnologists (Wright, 1967; Wright, Livingstone, and Cushing, 1965), we have been able to penetrate more than 20 meters of lake mud with nearly complete core recovery. The holes are cased with plastic pipe and temperatures measured after transient disturbances have subsided. Conductivities are measured on the core with needle probes (Von Herzen and Maxwell, 1959). Details of this technique will be presented elsewhere (Williams et al., 1971).

Figures 8 and 9 show the data from measurements at Elk Lake, Minnesota, and Figure 10 the results of measurements in lakes and drill holes near the

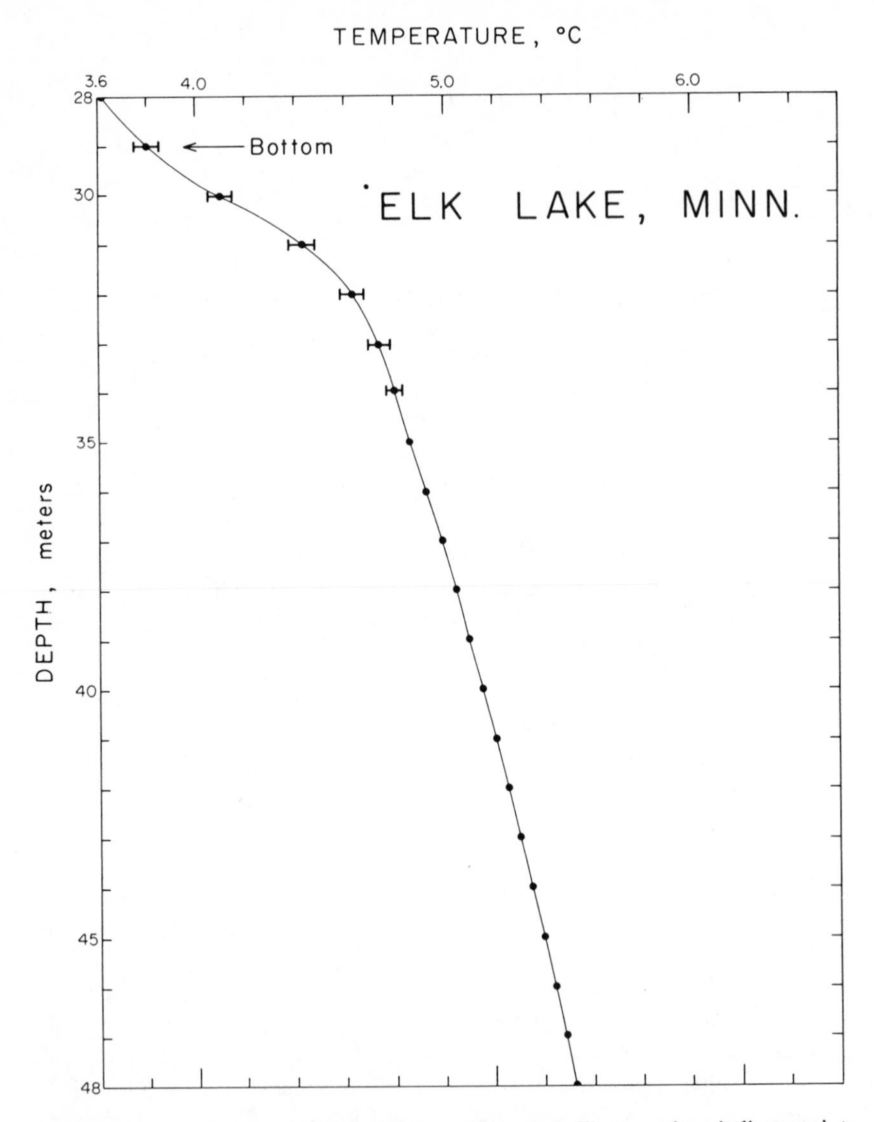

Fig. 8. Temperatures measured at Elk Lake, Minnesota. The error bars indicate points where temperatures were unstable (±0.1°C); at the points without error bars temperatures did not change more than 0.001°C in 5 min. (transducer time constant about 3 sec.).

Mid-Continent Gravity High. With the lake technique still unproved by comparison with nearby conventional measurements, it is premature to speculate about the heat flow on the Mid-Continent Gravity High beyond noting that it does not appear to be extraordinarily low as in the Sierra Nevada.

VERTICAL DISTRIBUTION OF RADIOACTIVITY

The first observation of the linear relation between heat flow and radioactive heat generation of plutonic rocks was made by Francis Birch and reported by

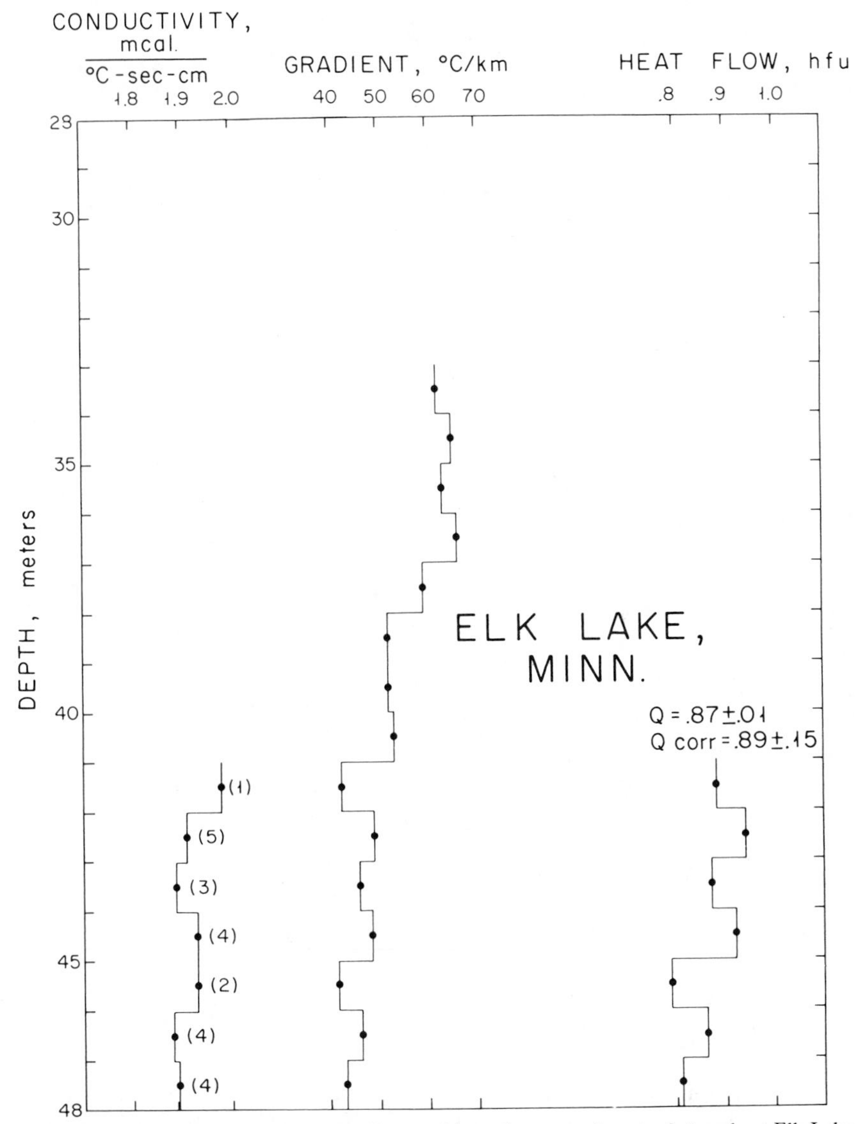

Fig. 9. Thermal conductivity, gradient and heat flow over 1-meter intervals at Elk Lake, Minnesota. The numbers in parentheses near the conductivity curve are the number of needle-probe measurements made in that interval. Q is the uncorrected heat flow computed by the resistance integral method (Bullard, 1939). Q corr includes corrections for warm rim effect, recent sedimentation, and refraction.

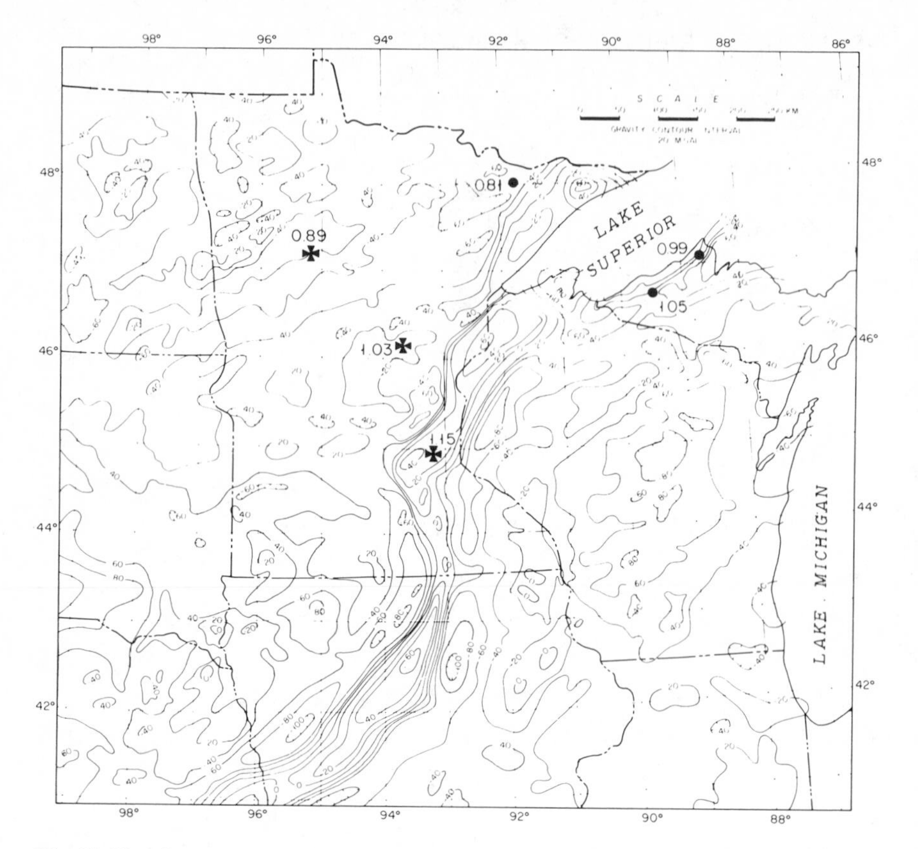

Fig. 10. Heat flow measurements near the Mid-Continent Gravity High. The gravity contours are after Woollard and Joesting (1964). The Maltese cross indicates measurements in lakes by Williams et al. (1971). The solid dots represent the locations of conventional measurements in drill holes reported by Roy, Decker, Blackwell, and Birch (1968). A large number of measurements in Lake Superior (nearly 180) by Steinhart, Hart, and Smith (1969) could not be plotted on this scale. They show a few measurements as low as 0.6 in western Lake Superior with the remainder of the lake between 0.8 and 1.3.

Birch, Roy, and Decker (1968). This relation is expressed by

$$Q = \mathbf{a} + \mathbf{b}A \tag{1}$$

where Q is the heat flow at the surface in units of 10^{-6} cal/cm^2 sec; A is the heat production of the surface rocks in units of 10^{-13} cal/cm^3 sec; **a** is the intercept value of heat flow measured in rocks of zero heat production; and **b** is the slope of the line relating Q and A and has the dimension of depth. Throughout the remainder of the paper the units of Q and A will be those cited above and will be omitted from the text.

The relation expressed by Eq. (1) has led to major advances in our understanding of the vertical distribution of heat sources, our ability to calculate more reliable temperature depth profiles, and our ability to accurately map lateral variations of heat sources in the upper mantle. The following sections will discuss these aspects of modern geothermal studies and their implications as to the distributions of subsurface temperatures, heat sources, and the other geological and geophysical parameters of the crust and upper mantle beneath continents.

Two vertical distributions of radioactivity have been proposed that are consistent with the observed linear relation between heat flow and heat generation (Birch, Roy, and Decker, 1968; Lachenbruch, 1968; Roy, Blackwell, and Birch, 1968): Model (1), A constant to the depth given by the slope of the straight line; Model (2), A exponentially decreasing with depth according to the relation $A = A_0 e^{-x/\mathbf{b}}$ where A_0 is a constant, **b** is the slope of the straight line, and x is depth. The intercept **a** in the model with constant heat generation is the heat flow from below the radioactive layer (approximately 7 to 10 km thick). For the model with exponentially varying heat generation, the constant **a** is the heat flow from the mantle. The relationship between heat production and heat flow, whatever the model used to explain it, implies that beneath large intrusive bodies the whole crust has been affected and the original distribution of U and Th modified in some regular way.

The conclusion that the vertical distribution of uranium and thorium can be inferred from a study of the surface heat flow and heat production is an extra bonus from recent heat flow studies and has important geophysical and geochemical ramifications. In geophysics a definite model for the vertical distribution of heat sources allows a much more accurate calculation of subsurface temperatures with resulting implications to the physical properties of rocks in the crust and upper mantle and possible phases involved. An interesting geochemical implication is that similar models might apply to other trace, or even major, elements. For example, Zartman and Wasserburg (1969) extended the regular variation of heat producing elements, inferred from the heat flow data, to the trace elements Sr, Rb, and Pb and were able to explain some of the apparent inconsistencies in the distribution of isotopes in the Rb-Sr, U-Pb, and Th-Pb systems studied in surface rocks. Thus, it is clear that the implications of a heat production versus depth model are so broad that more detailed studies of vertical U and Th distribution at suitable locations are of great importance.

Roy, Blackwell, and Birch (1968) have cited the evidence for uniform radioactive heat production over vertical ranges of about 1 or 2 km. Lachenbruch (1968) and Roy, Blackwell, and Birch (1968) both note, however, that the linear relationship between heat flow and heat production also are consistent with an exponential decrease of radioactivity with depth. Vertically decreasing radioactivity has been observed in a few deep (3 km) drill holes (Lachenbruch and Bunker, 1969), one survey of surface samples (Dolgushin and Amshinsky, 1966), and there is some evidence (Decker, unpublished) that an exponential relationship may exist between observed surface radioactivity and estimated mean erosion over four batholiths of Silver Plume granite in the Colorado Front Range (Fig. 11). The decreases of radioactivity in the drill holes are not clearly exponential (Lachenbruch and Bunker, 1969) and the **b** values (1 to 2 km) calculated from the Front Range and other surface data are all significantly lower than those (7 to 10 km) determined for the observed heat flow-heat production lines. Moreover, all other observational data are ambiguous because the accuracy (± 10%) of present heat generation measurements is roughly equivalent to the magnitude of the analytically predicted decreases with depth (10%/km; Lachenbruch and Bunker, 1969).

The recent combined studies of radioactivity and heat flow have provided us with the first reliable means for accurately mapping variations of temperature and heat flux in the upper mantle, but we believe that present radioactivity information is inconclusive and that the actual vertical distribution of heat production implied by Eq. (1) remains a very significant geochemical problem.

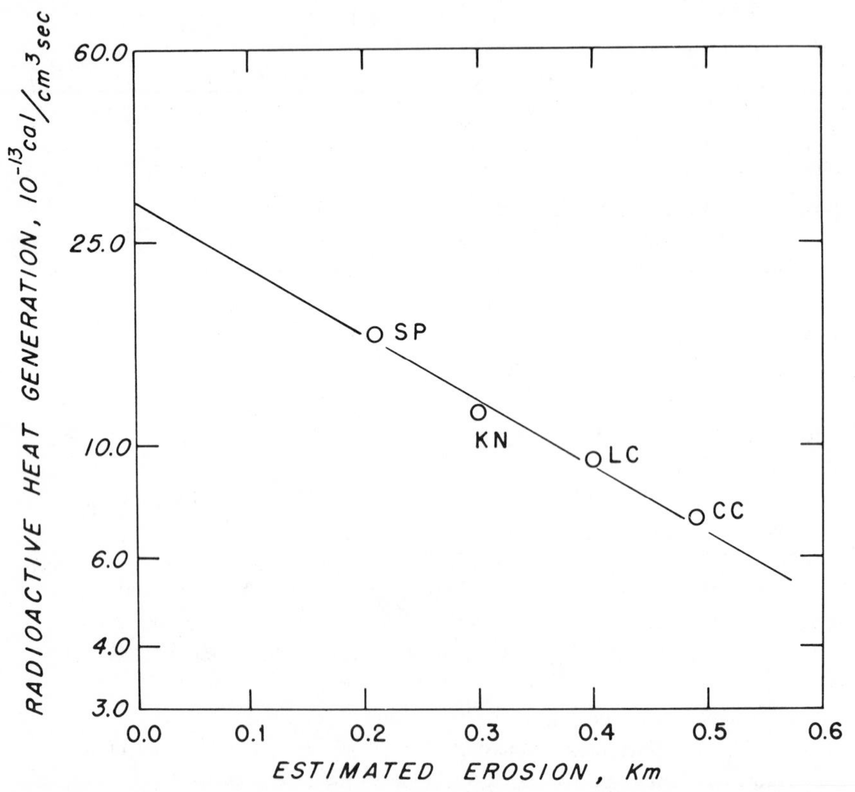

Fig. 11. Radioactive heat generation as a function of mean erosion over four batholiths of Silver Plume type granite in the Colorado Front Range. SP, KN, LC, and CC represent the Silver Plume, Kenosha, Log Cabin, and Cripple Creek batholiths, respectively. Radioactive heat generations are after Phair and Gottfried (1964). Estimated mean erosions by Decker (unpublished).

HEAT FLOW DISTRIBUTION

Regardless of the distribution of radioactivity preferred to explain Eq. (1), the intercept value **a** is obviously the parameter of importance in an investigation of regional variations in heat flow from the mantle. From a study of the heat flow and heat production in plutonic rocks in the United States, Roy, Blackwell, and Birch (1968) defined a heat flow province on the basis of its characteristic relationship between heat flow and heat production and identified three provinces: the eastern United States, where **a** = 0.8 and **b** = 7.5 km; the Basin and Range province, where **a** = 1.4 and **b** = 9.4 km; the Sierra Nevada, where **a** = 0.4 and **b** = 10 km.

From the broad scale geophysical point of view, the large variations of the intercept value **a**, which reflect variations in heat flow from the mantle, are of more interest than the slope **b**, which varies much less from province to province and probably reflects variations in the geochemistry of the upper crust. With the relative importance of **a** and **b** in mind, we have prepared a "reduced" heat flow map of the United States (Fig. 12). A reduced heat flow value is **a** calculated from Eq. (1) transposed to $\mathbf{a} = Q - \mathbf{b}A$. Points in the eastern United States, Basin and Range province, and Sierra Nevada reduce, of course, to 0.8, 1.4, and 0.4, because they were used to determine **a** and **b** for these regions initially. New data points in these and other regions have to be examined with care because there may be subprovinces with different slopes (**b**) but the same intercept (**a**) within the major provinces. In new regions with only a few data points such as the Peninsular Ranges of southern California, the Salinian block, the northern Cascades, etc., we have taken **b** = 10 km in calculating values of reduced heat flow.

Figure 12 is a map of the major physiographic and heat flow provinces in the United States. The locations for which we have reduced values (Roy, Blackwell, and Birch, 1968; Roy, Blackwell, and Decker, unpublished) are indicated except in the Sierra Nevada and Peninsular Ranges where the data are too closely spaced to be shown. The physiographic provinces make convenient units for discussion as heat flow and physiographic boundaries often seem to be close to one another.

All available measurements of terrestrial heat flow in the western United States and adjacent portions of the Pacific Ocean are plotted in Fig. 13. A similar map of the present distribution of heat flow in the eastern United States has been presented by Diment (this volume). Although the heat flow contours drawn in Fig. 13 show that the western United States is characterized by large areas of high and low regional heat flow, the map of reduced flux (Fig. 12) is more useful and clearly indicates that the regional heat flow patterns are related to significant variations of flux from the upper mantle.

Eastern United States

The reduced heat flow values in the United States east of the Rocky Mountains are all 0.8 ± 0.1. New data points in this region since the summary by

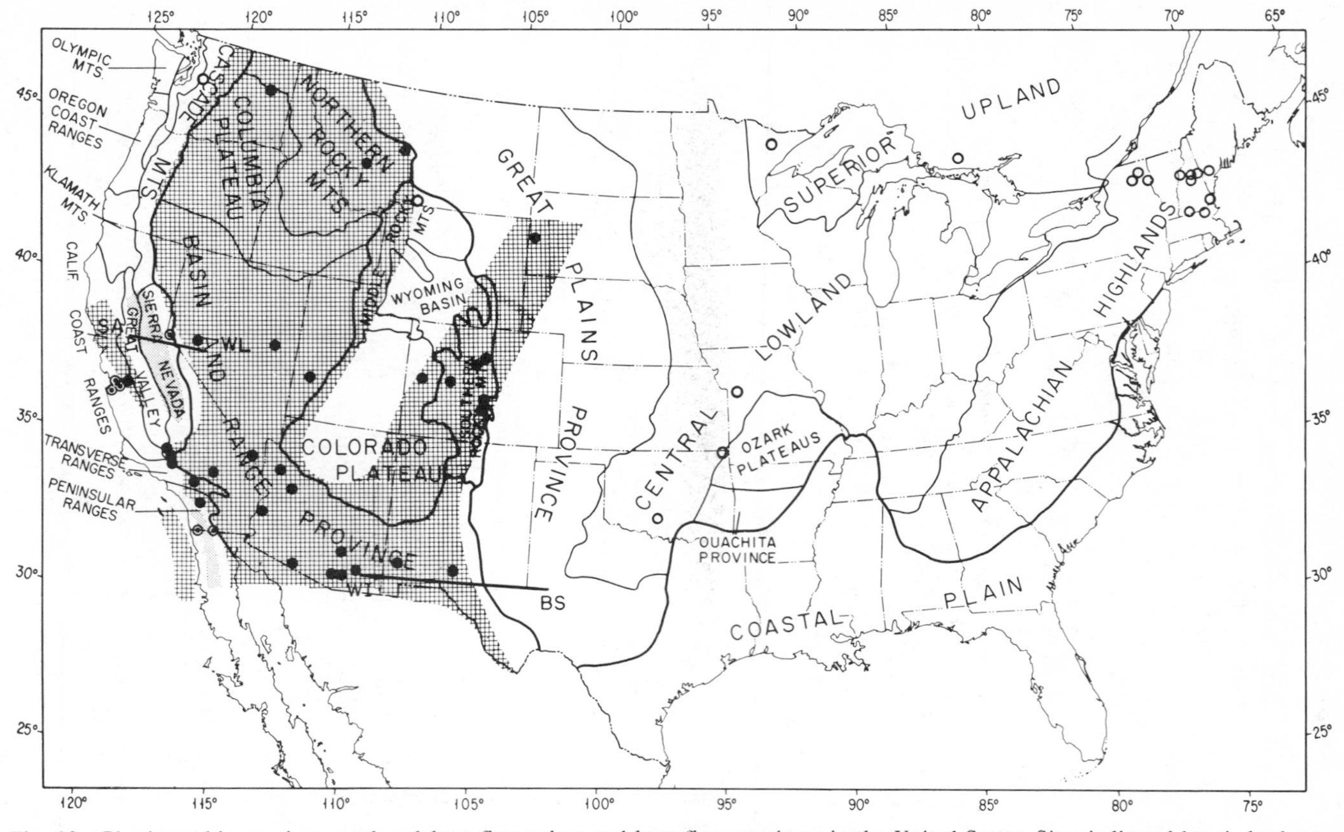

Fig. 12. Physiographic provinces, reduced heat flow values and heat flow provinces in the United States. Sites indicated by circles have reduced heat flow values of 0.8 ± 0.1; dotted circles, values between 0.9 and 1.3; solid dots values greater than 1.3. The extent of the regions of high reduced heat flow are designated with a square pattern, those with low reduced heat flow are designated by a dot pattern. The low values in the Sierra Nevada and many determinations in the Pacific Coast provinces could not be plotted because of the small scale of the map. The lines SA-WL and WI-BS are the locations of Figs. 16 and 17.

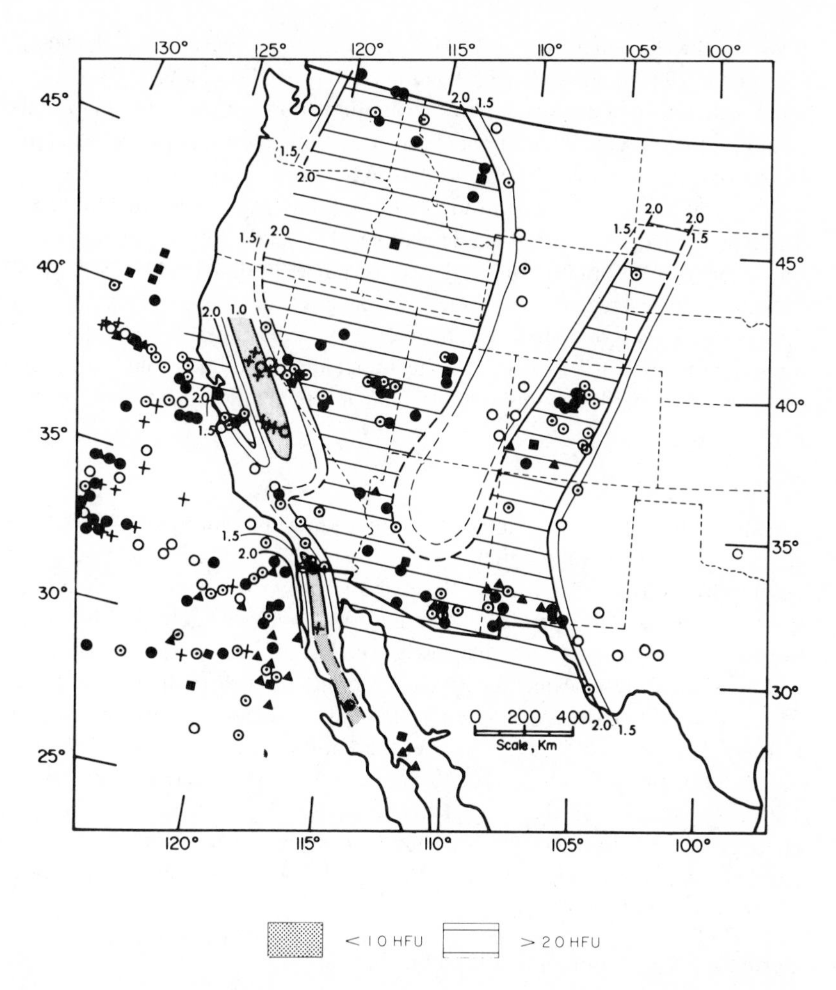

Fig. 13. Measurements of heat flow in the western United States and adjacent portions of the eastern Pacific Ocean. The contours delineate regions of high and low heat flow with average values of flux that would be measured in rocks with surface radioactivity within the range of granodiorite. Published heat flow data are from Benfield (1947), Herrin and Clark (1956), Clark, (1957), Foster (1962), Von Herzen (1964), Spicer (1964), Lee and Uyeda (1965), Lachenbruch, Wollenberg, Greene, and Smith (1966), Burns and Grim (1967), Vacquier, Sclater, and Correy (1967), Costain and Wright (1968), Roy, Decker, Blackwell and Birch (1968), Sass, Lachenbruch, Greene, Moses, and Monroe (1968), Henyey (1968), Warren, Sclater, Vacquier, and Roy (1969), Blackwell (1969), Decker (1969). Unpublished data by Roy, Blackwell, and Decker are also included. Pluses represent heat flow values in the range 0 to 0.99; open circles, 1.0 to 1.49; dotted circles, 1.5 to 1.99; solid circles, 2.0 to 2.49; solid triangles, 2.5 to 2.99; solid rectangles, > 3.0. A few measurements in the Pacific Coast Provinces have not been plotted because of lack of space on this scale.

Roy, Blackwell, and Birch (1968) are in southern Ontario (Sass, Killeen, and Mustonen, 1968) and southern Oklahoma (Blackwell, unpublished). Outside of the Northeast, data are sparse; however, other provinces may still be found; for example, there are no measurements of heat flow, reduced or unreduced, in the Gulf Coast geosyncline. There are no indications, however, of differences in mantle heat flow of the magnitude of those found in the western United States. The Appalachians have been stable geologically since early in the Mesozoic and the other regions for even longer. The crust ranges in thickness from about 30 to 50 km and the average *P* wave velocity in the lower crust is generally high (7.0 to 7.5 km/sec). *Pn* velocities are high (8.1 – 8.3 km/sec), and there is no low-velocity zone for *P* waves in the upper mantle (Green and Hales, 1968). There is only a moderate low-velocity zone in the mantle for *S* waves (Brune and Dorman, 1963).

Basin and Range

Reduced heat flow values in the Basin and Range province are 1.4 ± 0.2 except for some very high values that we attribute to local thermal anomalies (Roy, Blackwell, and Birch, 1968). The Basin and Range province has been tectonically active throughout the Cenozoic. The deformation is characterized by generally northtrending normal faults, many with vertical displacements of 3 km or more, and province-wide volcanic activity. Hot springs abound in the province and the east and west borders are presently the sites of bands of earthquake activity (Woollard, 1958). The crust is thinner than normal for the average elevation and the *Pn* velocities are only 7.8 to 7.9 km/sec (Diment, Stewart, and Roller, 1961; Eaton, 1963, 1966; Healy, 1963; Hill and Pakiser, 1966; Pakiser, 1963). There is an upper mantle low-velocity zone for *P* waves between about 60 and 170 km (Archambeau, Flinn, and Lambert, 1969), and *S* waves are hardly transmitted at all (Wickens and Pec, 1968).

Northern Rocky Mountains–Columbia Plateau

The Northern Rocky Mountains are composed of Precambrian, Paleozoic, and Mesozoic miogeosynclinal sediments in the east and Paleozoic and Mesozoic eugeosynclinal sediments in the west. The sedimentary rocks are intruded by many large bodies of Mesozoic plutonic rocks such as the Colville, Loon Lake, Kaniksu, Idaho and Boulder batholiths. The Cenozoic history is marked by several episodes of normal faulting and graben formation and by moderate amounts of volcanism. Hot springs are abundant in the southern part of the province and the eastern boundary is a zone of active earthquakes. Data obtained by Blackwell (1969, unpublished) show the same average surface and reduced heat flow as in the Basin and Range. The crust in the Northern Rocky Mountains is thin and the *Pn* velocities are 7.8 to 7.9 km/sec (Asada and Aldrich, 1966; Steinhart and Meyer, 1961; White and Savage, 1965; White, Bone, and Milne, 1968). Thus the Northern Rocky Mountains province is an extension of the Basin and Range heat flow province with a similar crustal structure and Cenozoic geologic history.

The Columbia Plateau physiographic province has a diverse geology. In much of the province, Cenozoic volcanic rocks cover in varying thicknesses a Mesozoic bedrock similar to that in the Basin and Range and Northern Rocky Mountains. In at least part of the Columbia Plateau proper and in part of the Snake River Plains, Miocene, and Pliocene and Pleistocene basalts, respectively, make up much if not all of the crust (Hill, 1963; Hill and Pakiser, 1966). Sparse data suggest a high heat flow similar to that in the Northern Rockies and Basin and Range province. Thus Blackwell (1969) has combined the Basin and Range, Northern Rockies, and Columbia Plateau into a single geophysical province called the Cordilleran Thermal Anomaly Zone.

Southern Rocky Mountains

Previously unpublished heat generation and heat flow data for seven sites in the Southern Rocky Mountains are plotted in Fig. 14. Neither the slope (10 km) nor the intercept (1.3) of the line calculated from these data is statistically different from the line calculated for the Basin and Range province. So at the present time, until more data in both areas either confirm or deny this equivalence, we consider the Southern Rocky Mountains to be a prong of the Basin and Range heat flow province. The Southern Rocky Mountains are composed of Precambrian granite and metamorphic rocks extensively faulted and uplifted in the Cenozoic and intruded and covered in places by late Cretaceous to Recent igneous rocks (Curtis, 1960). The long-period magnetic variation studies in the area by Porath, Oldenberg, and Gough (1970) indicate an upwarping of the low electrical resistivity layer similar to that beneath the Basin and Range province, but much smaller in lateral extent.

Although the geophysics and the Cenozoic geology of the Southern Rocky Mountains resemble those of the Basin and Range in many ways, there is a major difference in the crustal thickness. The crust is about 50 to 55 km thick and is thus much thicker than in the Basin and Range (Jackson and Pakiser, 1965; Ryall and Stuart, 1963), but the crust does appear to be 5 to 10 km thicker in the adjacent Great Plains even though elevations are lower (Jackson, Stewart, and Pakiser, 1963). The difference in crustal thickness between the two high heat flow areas demonstrates that high heat flow is not confined to regions with a thin crust.

The boundaries of the zone of high heat flow extending through the Southern Rockies are shown in Fig. 12 and 13 as are the data on which the boundaries are based. Blackwell (1969) inferred an extension of the zone of high heat flow at least to the Black Hills in South Dakota.

Sierra Nevada

All reduced heat flow values in the Sierra Nevada Mountains are within 10% of 0.4, only one-half that found in the eastern United States. As presently known, the province is only about 150 km wide by 600 km long, but is still of crucial importance as it is, at the present stage of heat flow investigations, unique on the earth. The bedrock of the range is composed of Paleozoic and Mesozoic

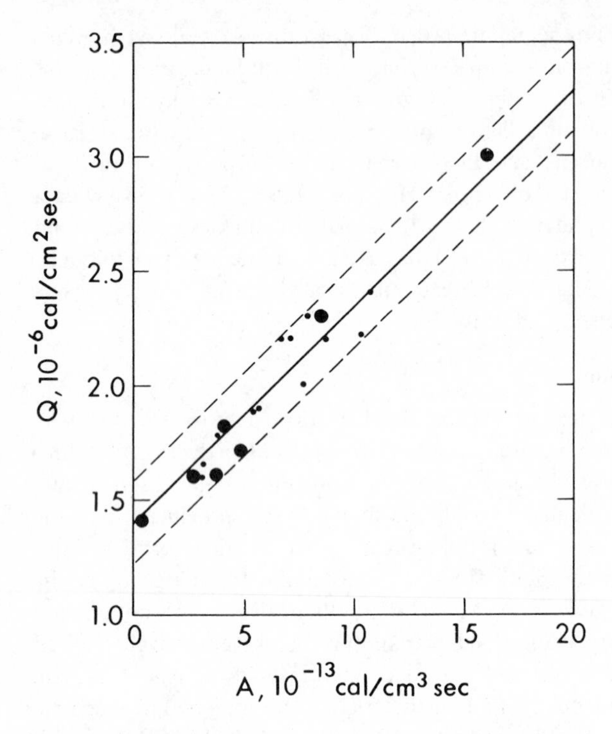

Fig. 14. Heat flow and radioactive heat production data for the Southern Rocky Mountain region. Large dots are Southern Rocky Mountain points; small dots are points in the Basin and Range province (Roy, Blackwell, and Birch, 1968). The solid line is the relationship between heat flow and heat production determined by Roy, Blackwell, and Birch (1968) for the Basin and Range province. Heat flows in the Southern Rocky Mountains are from Birch (1950), Roy, Decker, Blackwell and Birch (1968), Decker (1969), and Decker, unpublished. Heat generations after Phair and Gottfried (1964), Phair (personal communication, 1966), and Rogers (personal communication, 1969).

geosynclinal sediments and volcanics that were introduced by middle and late Mesozoic granitic rock on a grand scale. The range has been relatively quiescent during the Cenozoic although the east flank was uplifted in the late Cenozoic. The crust is 50 km thick under the crest and thins uniformly westward to about 20 km beneath the foothills (Eaton, 1963, 1966). Independent evidence for low temperatures in the upper mantle is not available, but investigations such as determinations of the electrical resistivity structure in the mantle and seismic wave velocities are obviously of crucial importance to our understanding of the heat flow anomaly.

Colorado Plateau—Wyoming Basin—Middle Rocky Mountains

This group of physiographic provinces has been investigated by a total of only seven heat flow determinations of which only two have reduced values. A reduced value in the Beartooth Plateau is about 0.9, while one at the eastern edge of the Colorado Plateau is about 1.3. Unreduced values of 1.3 (Sass, Lachenbruch, Greene, Moses, and Monroe, 1968) in the Wind River Range and 1.2 in the central Colorado Plateau may indicate a region of normal intercept values. A high unreduced value was measured in the Absaroka volcanic region. Our inferences on the heat flow are shown in Figs. 12 and 13.

Most of the area in these provinces is inferred to be normal based on: (1) the few heat flow data; (2) the conclusions of Porath, Oldenberg, and Gough (1970) that the low-resistivity layer is found at about the same depth beneath the Great Plains and the Colorado Plateau; (3) suggestions that the crustal structure changes abruptly crossing from the southern Colorado Plateau into the Basin and Range (Warren, 1969) and from the Basin and Range into the Middle Rocky Mountains (Willden, 1965); (4) evidence presented by Julian (1970) that the Colorado Plateau is structurally similar to eastern North America. The crust is about 40 km thick in the Colorado Plateau, and although the *Pn* velocity is only 7.8 km/sec (Roller, 1965), travel times from the Gasbuggy shot indicate higher average upper mantle velocity beneath the Colorado Plateau than beneath the provinces to the west.

The heat flow pattern is particularly hypothetical in northern Wyoming and indeed the two zones of high heat flow might connect instead of remaining separate as we have indicated on Figs. 12 and 13.

Pacific Coast Provinces

The heat flow along the Pacific coast is variable on a much smaller scale than in the provinces to the east of the Sierra Nevada. Determinations in granitic rocks in the Peninsular Ranges of southern California and Baja California (Roy and Brune, unpublished data) have reduced values of 0.6 to 0.7 near the center of the range rising sharply to 1.2 near the edge of the Imperial Valley. Although the reduced values in the center of the Peninsular Ranges are not as low as in the Sierra Nevada, we believe they are extremely important to the tectonic interpretation and have indicated a narrow zone of abnormally low heat flow for this region on Fig. 12 and 13.

In the granitic rocks of the Salinian block west of the San Andreas fault, reduced heat flow values are slightly above normal, 1.0 to 1.1 (Roy, Brune, and Henyey, in preparation). In contrast, the Franciscan block east of the San Andreas fault has high heat flow at the surface, 2.2 to 2.4 (Sass, Monroe, and Lachenbruch, 1968; Roy, Brune, and Henyey, unpublished data). If we take A = 2.8 for the Franciscan (Wollenberg, Smith, and Bailey, 1967) and $\mathbf{b}$ = 10 km, the reduced heat flow would be 1.9 to 2.1, approximately 0.6 higher than the Basin and Range. The extent of this region of extraordinarily high heat flow is not well known; on Figs. 12 and 13 we have extended it to the north to include the region near Clear Lake where the geothermal steam fields are an obvious indication of high heat flow.

There are no measurements of heat flow, reduced or unreduced, in the Pacific Coast provinces between Menlo Park, California, where Sass, Monroe, and Lachenbruch (1968) have reported a value of 2.2 and the northern Cascades where Blackwell (unpublished) has a reduced value of 1.0. More measurements are urgently needed in this region of complex interaction between continental and oceanic plates (Atwater and Menard, 1970; McKenzie and Morgan, 1969; Silver, 1969).

Heat Flow Provinces Outside the United States

Roy, Blackwell, and Birch (1968) suggested that the linear relationship between heat flow and heat production found in the eastern United States is the characteristic curve for *normal* continental heat flow in regions that have been tectonically stable for at least 100 to 200 m.y. The data they cited for Australia and the recently published data from three sites in Paleozoic plutons in New South Wales, Australia (Hyndman, Jaeger, and Sass, 1969), and two sites in sediments above Precambrian granitic rocks in the Canadian Shield (Lewis, 1969; Sass, Killeen, and Mustonen, 1968) lend strong support to that conclusion (see Fig. 15). The extension of this hypothesis to other continents gains support from the low values of flux measured in diorite and serpentinite (which almost certainly have low values of heat production) at Virtasalmi and Nivala, Finland (Puranen, Järvimäki, Hämäläinen, and Lehtinen, 1968), which are in the range (0.65 to 0.81) of the reduced heat flows in the eastern United States.

In his classic paper on heat flow in South Africa, Bullard (1939) measured a heat flow of 1.52 in the Dubbeldevlei borehole, which penetrated "Old Granite" of the Transvaal. Based on radium determinations on granites from the Cape Province by Immelman (1934), he estimated a heat production of 7.4×10^{-13} cal/cm^3 sec. This point plots slightly above the line for stable continental regions. It is perhaps more appropriate to combine the heat productivity of the Cape Granites with the heat flow determinations (1.21 to 1.45) by Gough (1963) in the southern Karoo. These points straddle the line characteristic of stable regions.

Thus far the only locality outside the United States that plots close to the Basin and Range line is the Snowy Mountains in Australia (Sass, Clark, and Jaeger, 1967). This region is tectonically similar to the Basin and Range province with normal faulting, Tertiary vulcanism and late P wave arrivals (Cleary, 1967; Hyndman, Jaeger, and Sass, 1969).

One of the important objectives of future studies of heat flow is to test the hypothesis that the same relationship exists between heat flow and heat production on other continents. If this relationship can be well established, it will strengthen the conclusion that all continental crust has developed by upward differentiation of sialic material from the underlying mantle.

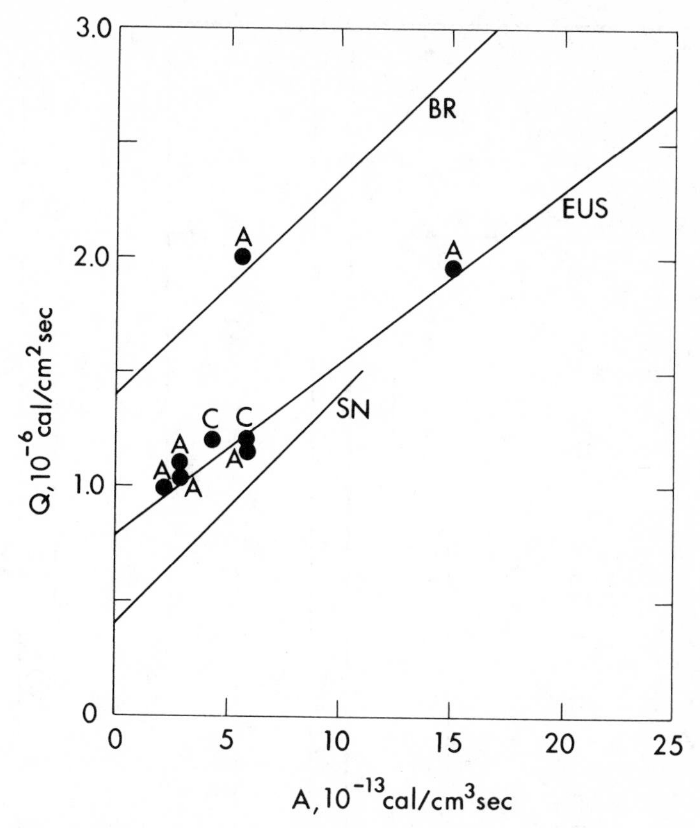

Fig. 15. Heat flow and heat production relations for the Basin and Range (BR), eastern United States (EUS), and Sierra Nevada (SN). Data points labeled A are from Australia, C from Canada.

HEAT FLOW TRANSITIONS

Zones of transition between the heat flow provinces previously discussed have been investigated in a few places. In all cases, the transition regions are narrow compared to the widths of the provinces. As examples, two profiles will be discussed: one across the Basin and Range-eastern United States transition (Fig. 16) and one across the Sierra Nevada-Basin and Range transition (Fig. 17).

Basin and Range–Great Plains Transition

Heat flow data along a profile from Wilcox, Arizona, to Big Spring, Texas, are shown in Fig. 16; the location of the profile (WI-BS) is indicated on Fig. 12. Data are from Warren, Sclater, Vacquier, and Roy (1969), Roy, Decker, Blackwell, and Birch (1968), Decker and Birch (in preparation). Although reduced heat flow values are available at only two sites, it appears that the high heat flux in the Basin and Range province extends at least as far as Cornudas,

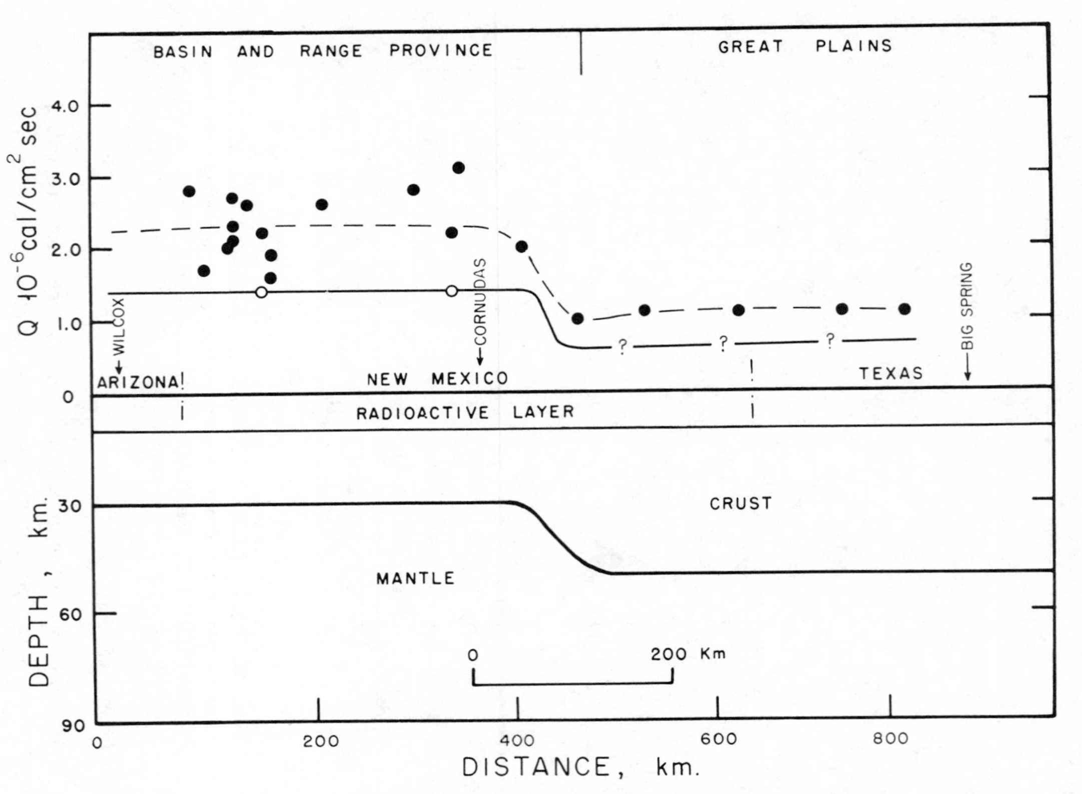

Fig. 16. Summary of the heat flow and other geophysical data for a profile from Wilcox, Arizona, to Big Spring, Texas (see line WI-BS, Fig. 12). The solid circles are heat flow values measured at the surface and the open circles are "reduced" heat flow values. The solid line represents the reduced heat flow from the lower crust and upper mantle.

New Mexico, where Schmucker (1964) put the eastern edge of the "Texas Anomaly" in his study of the long-period magnetic variations. The heat flux is 0.9 about 100 km to the south and 50 km to the east near Van Horn, Texas. A heat flow of 1.1 was found by Herrin and Clark (1956) at Carlsbad, New Mexico, about 120 km due east of Cornudas. That value also was found at several other sites in the southern Great Plains (Herrin and Clark, 1956). Although no reduced values are available in the Great Plains, it is clear that the transition occurs within a horizontal distance of 50 to 100 km and slightly within the Basin and Range physiographic province. The amplitude of the anomaly is 0.6, the difference between the intercept values for the two heat flow provinces.

Although only one estimate for the crustal thickness (30 to 35 km) has been published for the Basin and Range province in southwestern New Mexico (Stewart, Stuart, Roller, Jackson, and Mangan, 1964), the crust is inferred to be of similar thickness as far east as Cornudas (Fig. 16). The crustal thickness beneath the western Great Plains (50 km) is from a profile published by Stewart and Pakiser (1962). The transition in crustal structure is assumed to coincide with the heat flow transition.

The new data presented here demonstrate that the transition is more abrupt than assumed by Warren, Sclater, Vacquier, and Roy (1969), hence their model no longer fits the data and a shallower source or sharper rise in the isotherms is necessary.

Sierra Nevada--Basin and Range Transition

Heat flow data along a profile (line SA-WL, Fig. 12) at 39°N from Sacramento, California, to Fallon, Nevada, are plotted in Fig. 17 (Roy, Decker, Blackwell, and Birch, 1968; Blackwell and Roy, in preparation). Both surface and reduced values indicate a simple transition; however, we believe that, unlike the previous profile, there are two anomalies and that the transition must be compound. Relative to the eastern United States as the standard for *normal* heat flow, there is a negative anomaly of 0.4 in the Sierra Nevada and a positive anomaly of 0.6 in the Basin and Range so that the total anomaly is about 1.0.

The center of the transition in heat flow occurs about 50 km inside the physiographic boundary of the Basin and Range province and coincides with the rapid thinning of the crust from 50 km under the Sierra Nevada to 25 km beneath the Basin and Range (Eaton, 1963; 1966). A north trending zone of earthquake activity extends about 100 km into the Basin and Range province from the physiographic boundary (Ryall, Slemmons, and Gedney, 1966).

The maximum distance for the transition in heat flow of about 1.0 is 100 km. The observed anomaly curve shown in Fig. 17 can be fitted only to transient models because the shape of the curve is not similar to that produced by a steady source buried 20 to 50 km. Our preferred explanation is that there is an instantaneous sink in the crust and upper mantle beneath the Sierra Nevada and an instantaneous or continuous source in the upper mantle (less than 50 km

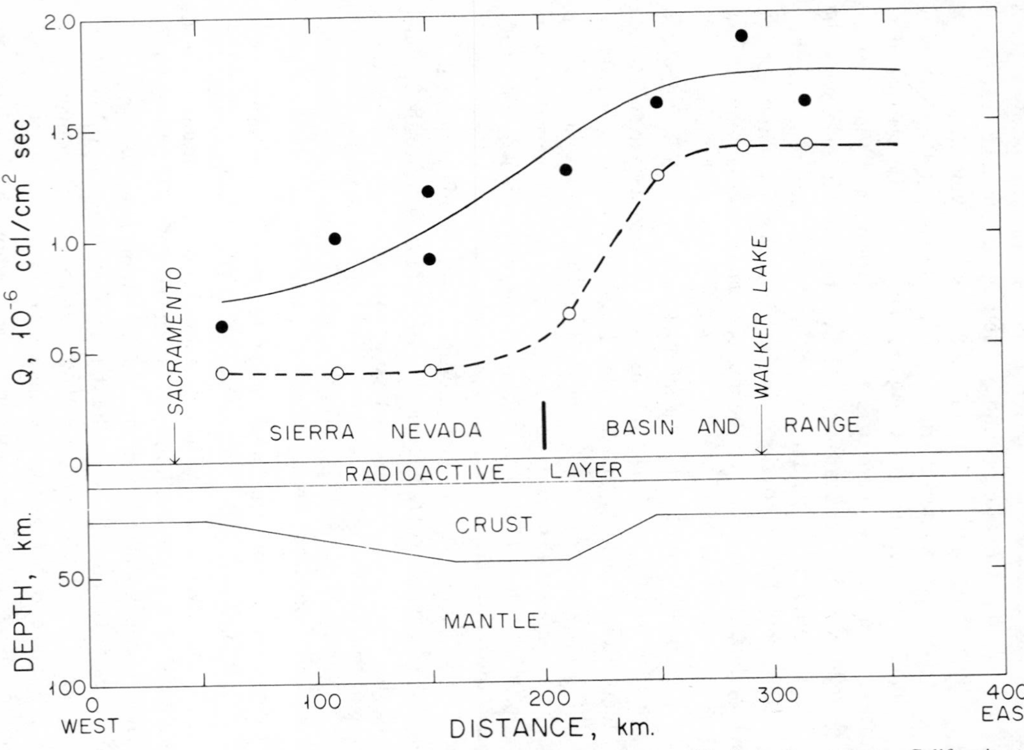

Fig. 17. Summary of heat flow and other geophysical data for a profile from Sacramento, California, to Walker Lake, Nevada (see line SA-WL, Fig. 12). Solid circles are heat flow values measured at the surface and open circles are "reduced" heat flow values. The dashed line represents the reduced heat flow from the lower crust and upper mantle.

deep) beneath the Basin and Range. The ages necessary to explain the present day data are 5 to 20 m.y.

Other Transitions

Data on transitions in regional heat flow between the Peninsular Ranges and the Basin and Range, the Salinian and Franciscan blocks, and the Southern Rockies and Colorado Plateau (Roy and Brune, unpublished; Decker, unpublished) also suggest characteristic widths of as little as 20 km to a maximum of 100 km. Hence, it is clear that the sources and/or sinks affecting the regional heat flow values are at a depth of less than 50 to 100 km beneath the surface (in the uppermost mantle or in some cases within the crust).

Variations in electrical conductivity structure of the upper mantle are in excellent general agreement with the results of surface heat flow measurements (Porath, Oldenberg, and Gough, 1970; Schmucker, 1964). If the boundary they map is an isotherm, however, it must be well below the source of the anomalous heat flux, which must be within a few tens of kilometers of the surface.

UPPER MANTLE SECTIONS

Our estimates on the distribution of temperatures and physical characteristics of the heat flow provinces in the western United States are summarized in Figs. 18, 19, and 20. These figures depict three east-west cross sections from the Pacific Ocean to the Great Plains: one typical of an east-west band of western North America from about 30 to 34°N latitude; one typical of a band from 36 to 40°N latitude; and a third typical of a band from 44 to 50°N latitude. The sections were drawn approximately along the 32, 38, and 47° parallels, and heat flow measurements within about 100 km of each side of the parallels were projected to the sections. The sections were chosen to be representative and to go through areas with the most data. The sections end at the Great Plains because published data suggest little variation in heat flow from the upper mantle in the eastern United States (Roy, Blackwell, and Birch, 1968). The pertinent assumptions made in preparing the cross sections will be summarized in three parts: heat flow, crustal and upper mantle seismic structure, and temperature calculations.

Heat Flow

The observed heat flow values have been plotted as solid dots and the trends indicated by a dotted line. This curve is the sort one might get from treating the observed values only, with no input of geological or other geophysical information. Reduced heat flow data were plotted as open circles and connected by a solid line. Aside from transition zones and the complex provinces along the Pacific Coast, all the points in the western United States are within $\pm 0.2 \times 10^{-6}$ cal/cm^2 sec of the intercept of values of either the Sierra Nevada or Basin and Range heat flow-heat production lines. The solid and dotted curves merge in

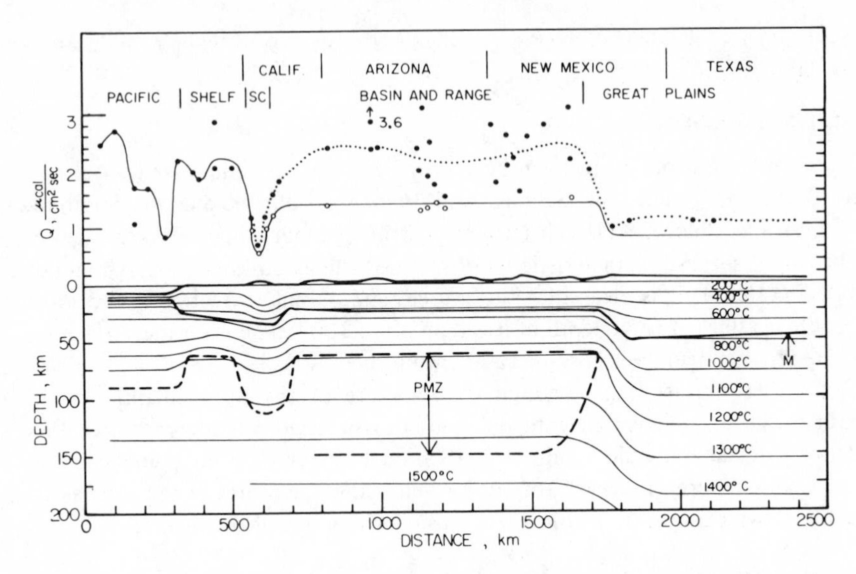

Fig. 18. Cross section across the southwestern United States and part of northern Mexico at approximately 32°N latitude. This cross section shows heat flow, crustal structure, and calculated temperatures. Solid dots represent observed heat flow at the surface and open circles, reduced heat flow. Data points have been projected to the section as much as 100 km. PMZ is the inferred Partial Melt Zone. *M* represents the Mohorovicic discontinuity. The region denoted by SC is the remarkable narrow zone of low heat flow in the Peninsular Ranges of Southern California, USA, and northern Baja California, Mexico. The temperature contours and PMZ boundaries west of the Basin and Range region are smoothed and on this scale do not attempt to fit the complex pattern of heat flow observed at the surface.

the oceans where as yet an analogous reduction for near surface heat sources has not been devised.

In contrast to the dotted line connecting the raw data, the solid line connecting the reduced heat flow values was drawn to emphasize a point previously made. At every heat flow transition where we have data, the transition is sharp and the total effect (within the limits of our precision) occurs within 100 km (the half widths are 50 km or less). Therefore, at each place on the cross sections where we infer a transition in heat flow from the mantle, the solid curve has been drawn to indicate a sharp transition, whether or not we have data for that specific transition. Thus the solid line consists of sections of almost constant mantle heat flow separated by narrow regions of rapid variations of heat flow. We repeat that for many of the transitions this is hypothesis rather than observation.

Crustal and Upper Mantle Structure

The crustal structure determined by seismic refraction surveys for the various heat flow provinces has already been mentioned. In a few places, it was necessary to extrapolate known crustal structure into unkown areas. When this extension was necessary the extrapolation was guided in large part by the mantle heat flow province involved. An example of this is the inference that a thin crust is characteristic of southwestern New Mexico; the crustal structure for the rest of the southern profile is based on published data from Warren (1969) and the references cited above. Crustal structure along the central profile has been extensively investigated and, in addition to the references already cited, is based on data presented by Hamilton, Ryall and Berg (1964).

Temperature Calculations

Temperatures were calculated for one- and two-dimensional models assuming steady-state conditions. The temperatures beneath the central portions of all heat flow provinces were calculated using one-dimensional models. Two-dimensional models were calculated across the transitions we have discussed in detail and the unstudied transitions that appear to be similar. The isotherms beneath the other unstudied transitions were calculated by interpolation between isotherms beneath the heat flow provinces on each side of the transition. Because of the uncertainities of temperatures calculated in this manner and because of the large variations of upper mantle heat flow in the Pacific Coast provinces, distributions of isotherms west of the Sierra Nevada are largely diagramatic.

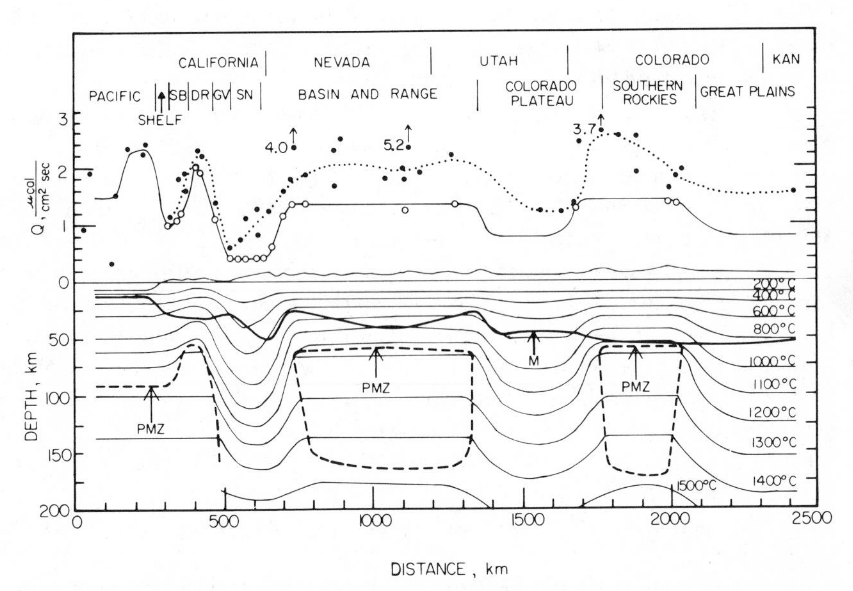

Fig. 19. Cross section across the western United States at approximately 38°N latitude. General notation and comments are the same as for Fig. 18. The abbreviations for the Pacific Coast provinces are: SB–Salinian Block, DR–Diablo Range, GV–Great Valley, SN–Sierra Nevada.

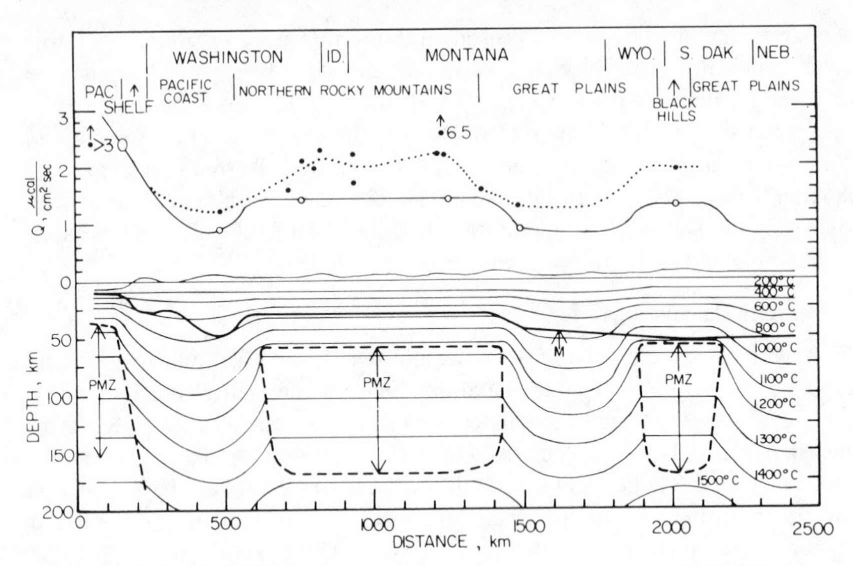

Fig. 20. Cross section across the western United States at approximately 47°N latitude. General notation and comments are the same as for Fig. 18.

Temperature depth curves for models with an upper layer of constant radioactivity as shown in Figs. 16 and 17 are given in Roy, Blackwell, and Birch (1968). The models shown in Figs. 18, 19, and 20 were calculated assuming four layers of different conductivity and two layers with exponentially decreasing heat sources with depth.

Layer 1 (the "granitic" layer) was assumed to have a constant conductivity of 6.5 mcal/cm sec °C and to extend to a depth equal to twice the slope of the heat flow, heat production curve for that province. Thus the layer varied in thickness from 20 km in the Sierra Nevada to 15 km in the Mid-Continent. At the bottom of layer 1 the heat production is assumed to be 0.68×10^{-13} cal/cm^3 sec. Layers 2, 3, and 4 were assumed to all be part of the same heat production layer with an A_0 of 0.68×10^{-13} cal/cm^3 sec and an exponential decrement of 80 km, thus 0.4 μcal/cm^2 sec was generated in this layer. In regions of high heat flow, an additional layer with high radioactivity was placed near the base of the crust to simulate a shallow heat source. In all models approximately 0.4 μcal/cm^{2}sec was assumed to come from below 400 km.

Layer 2 (the "gabbroic" layer) was assumed to extend from the base of layer 1 to the *M*-discontinuity, whose depth for a given province was taken from seismic refraction data. A constant conductivity of 5.0 mcal/cm sec°C was assumed for layer 2.

Layer 3 extends from the base of the crust to the 1000°C isotherm with a constant conductivity of 10 mcal/cm sec°C. Below the 1000°C isotherm the conductivity was assumed to increase 1.6 mcal/cm sec°C for each temperature increment of 100°C.

The model for temperatures beneath the Basin and Range province was also constrained such that the zone of partial melting was consistent with limits suggested by the upper mantle *P* wave profiles of Archambeau, Flinn, and Lambert (1969). Similar models were assumed for zones of partial melting beneath the Northern and Southern Rocky Mountains. The melting point curve assumed for determinations of temperature at the partial melt zone boundaries increases at a rate of 3°C/km from 1050°C at 1 bar and corresponds to the melting curve of a slightly wet peridotite (Ito and Kennedy, 1967; Kushiro, Syono, and Akimoto, 1968).

The two biggest gaps in the data are the Colorado Plateau and the Pacific Coast in the northwestern United States. The Colorado Plateau was assumed to have the same heat flow-heat production curve as the eastern United States, based on inferences from long period magnetic variations which suggest the same electrical conductivity structure as in the Great Plains (Porath, Oldenberg, and Gough, 1970). The heat flow and temperatures along the Pacific coast in the northwestern United States are assumed to be about the same as in the eastern United States, but could be somewhat lower if a slab of cold oceanic lithosphere is sinking beneath the coastline (Silver, 1969). The heat flow in this area and in the Colorado Plateau are the two most important areas for new data in the western United States.

DISCUSSION

So far our treatment of heat flow in the United States has been largely descriptive. In this section, we will consider briefly some thoughts on the origin of that heat flow pattern.

The most impressive feature on the map of reduced heat flow (Fig. 12) is the large region of high heat flow corresponding to the Basin and Range province (in a broad sense, i.e., Blackwell's Cordilleran Thermal Anomaly Zone). This region is similar in dimensions to the high heat flow in the Sea of Japan described by Vacquier et al. (1966). The superficial explanation of the high heat flow in the Basin and Range is that the upper mantle is partially molten near the base of the crust. The immediate cause of these high temperatures we attribute to local convection, i.e., diapiric intrusions of hot, solid and liquid material from deeper in the mantle to a position near the base of the crust as indicated in Figs. 18, 19, and 20.

Less impressive in areal extent on Fig. 12 but at least as important to the interpretation of recent tectonic history of the western United States are the low values of heat flow in the Sierra Nevada and Peninsular Ranges. There are many different types and origins of heat sources that could be invoked to explain high values of heat flow. The low values in the Sierra Nevada present a more difficult interpretational problem as heat sinks are less common than sources. Because there are fewer possible explanations, we believe the Sierra Nevada anomaly may be the key to explaining the origin of the mantle heat flow pattern in the western United States. So, what is the origin of the anomaly?

Several possibilities have been considered (Roy, Blackwell, and Birch, 1968). The model that envisioned complete depletion of the lower crust and upper mantle of radioactive heat sources seems unlikely at the present time. Recent heat flow determinations in the Mid-Continent suggest that heat flow values over basaltic basement such as the bedrock beneath the Mid-Continent Gravity High (Fig. 10) have values of 0.8 to 1.1 (Combs and Simmons, 1970; Williams and Roy, 1970; Blackwell, in preparation), which are the values expected if the parameters of Eq. (1) for the eastern United States apply. We have concluded that a transient heat sink is necessary and that the sink is low temperatures in the Sierra Nevada crust and uppermost mantle, which are a byproduct of recent sea floor spreading.

Current dogma on continental drift has the North American continent moving away from the Mid-Atlantic ridge toward the East Pacific Rise during much of the Cenozoic (Bullard, Everett, and Smith, 1965). Within the past few million years the continent and the East Pacific rise have closely approached, and the interaction of plates has become quite complicated. Papers too numerous to cite have discussed the nature of the interaction of the continental and oceanic plates and the resultant effect on the Cenozoic geologic history of the western United States (see, for example, Atwater and Menard, 1970; McKenzie and Morgan, 1969). We believe that the heat flow pattern described above is a definitive test of most of these hypotheses. Our general conclusions are that corollaries of the basic sea floor spreading hypothesis and plate tectonics explain the pattern we observe in the coastal areas almost uniquely and further inland as only one of several possibilities.

We summarize here briefly our hypothesis for the late Mesozoic and Cenozoic tectonic history of the western United States. During the late Mesozoic, a sinking slab of oceanic lithosphere existed at the site of the Franciscan terrain in western California (Ernst, 1970). Inland from the trench, about 100 to 200 km, the batholiths of the Peninsular Ranges, Salinian Block, Sierra Nevada, Klamath Mountains, Coast Ranges in Canada, etc., were being generated and emplaced in the crust. Here, incidentally, is geologic evidence from the observed low temperature assemblages in the Franciscan terrain and the high temperatures in the Sierra Nevada of a sharp transition in regional heat flow during the late Mesozoic. Near the beginning of the Cenozoic the direction, dip or rate of underthrusting changed so that the region of high heat flow shifted inland and the crust under the Sierra Nevada began to be cooled by the "cold conveyor belt" passing underneath, and the heat flow decreased to less than 0.4 $\mu cal/cm^2 sec$. The high heat flow in the Basin and Range was probably established with something near its present boundaries by early Oligocene.

High temperatures near the top surface of the underthrust plate have been attributed to friction, stress heating, shear strain heating, (McKenzie, 1969; McKenzie and Sclater, 1968; Minear and Toksöz, 1970; Oxburgh and Turcotte, 1968, 1970; Turcotte and Oxburgh, 1968). In addition to these mechanisms of mechanical-thermal energy conversion, we would like to suggest that radioactive heat sources in a slab of *undifferentiated* oceanic lithosphere may be important. If we take, for example, an oceanic plate with $A_0 = 0.7 \times 10^{-13}$ cal/cm^3 sec and **b** = 100 km (a popular dimension for slabs of oceanic lithosphere), the steady-state heat flow at the surface would be 0.7. Add to this a heat flow of approximately 0.4 from deep in the mantle, and we have the average heat flow observed in deep ocean basins (far from ridges) and the average heat flow on continents (excluding thermal anomaly regions such as the Basin and Range).

A 100-km-thick plate with the middle at a depth of 100 km and an internal heat generation of 0.7×10^{-13} cal/cm^3 sec would raise the surface heat flow by only 0.1 μcal/cm^2 sec in 10^8 years (Carslaw and Jaeger, 1959, p. 80). In 30 m.y., however, the temperature in the slab would rise about 200°C above the temperatures produced by heat conduction into a cold plate without radioactive heat sources. Partial melting of large portions of the descending slab thus becomes quite plausible. We find this suggestion attractive for three reasons: (1) it becomes possible to derive the intrusives and extrusives of quartz diorite to granodiorite composition now found at the surface directly from the mantle; (2) it is consistent with the low initial Sr ratios that are found in these rocks (Hedge and Peterman, 1969; Menzer and Jones, 1969); (3) it is the simplest explanation of the equality of oceanic and continental heat flow. The abrupt transitions between the high and low heat flow provinces are secondary effects because of rapid upward convection of the partially molten mantle material and do not reflect the distribution of sources actually responsible for the anomaly. Such effects would not be apparent in the first-order theories of mechanical-thermal energy conversion referenced above.

At some time in the middle Cenozoic, portions of the North American plate completely overrode the Cocos plate and abutted the Pacific plate (Atwater and Menard, 1970). The first such point was probably between the Mendocino and Murray fracture zones. At this time, the portion of the North American plate west of the San Andreas fault was welded to the Pacific plate and the San Andreas fault originated as a transform fault connecting the two remaining ends of the rise (Atwater and Menard, 1970). When the last vestige of the Cocos plate north of the southern tip of Baja California was trampled beneath the continent, the heat flow from the mantle beneath the Sierra Nevada and Peninsular Ranges ceased to be held down by the "cold conveyor belt" and the blocks began to warm up. The higher values in the Peninsular Ranges could be a result either of the fact that the block is narrower than the Sierra Nevada block (75 versus 150 km) and thus heated up more quickly or that the Peninsular Ranges are on a different time portion of the transient recovery.

We emphasize that whatever the explanation for the Sierra Nevada and Peninsular Range anomalies, the presence of these well-determined regions of low heat flow between the Basin and Range province and the coast rejects the hypothesis that the continent has simply drifted over the East Pacific Rise. There is no reasonable way we can devise to drop the heat flow to the present low values in the Penninsular Ranges and the Sierra Nevada subsequent to passing over an extensive rise system.

Attempts to extend the subduction model to explain the zone of high heat flow in the Southern Rocky Mountains and Black Hills encounter difficulties with the inferred zone of normal heat flow in the Colorado Plateau-Wyoming Basin-Middle Rocky Mountain region. There are only two values of reduced heat flow in this region (one of which is typically Basin and Range), 0.9 in the Beartooth Plateau and 1.3 near the eastern edge of the Colorado Plateau. If the Colorado Plateau is indeed a continuous zone of normal heat flow as inferred from the seismic and electrical resistivity data, then other mechanisms must be sought to explain the high mantle heat flow in the Southern Rocky Mountains. Clearly, the Colorado Plateau is one of the most important areas for further studies of heat flow and basement radioactivity in the United States.

Our general model implies that on the landward side of a subduction zone there will be a zone of low heat flow similar to the Sierra Nevada with a region of high heat flow still further inland. Some obvious areas to test this hypothesis are in the granitic plutons of the Klamath and Northern Cascade Mountains in the United States, the Sierra Madre del Sur in Mexico, and the coastal batholiths of Peru and Chile.

Note added in proof: Since the submission of this manuscript, a number of important papers on plate margin tectonics, subduction, and thermal structure of the western United States have appeared. Among these are: Atwater, 1970; Dickinson, 1970; Lipman et al., 1971; Porath et al., 1970; and Reitzel et al., 1970. The conclusions drawn in these and other papers, although based largely on petrology, plate tectonics, and electrical resistivity arguments, are generally in excellent agreement with the views expressed in this paper.

ACKNOWLEDGMENTS

We thank W. H. Diment, A. H. Lachenbruch, V. Rama Murthy, E. C. Robertson, and J. H. Sass for reading the manuscript and suggesting improvements. Various phases of the work were supported by the National Science Foundation (GP-701, GA-416, GA-715, GA-11351, GA-13553, GA-18450).

REFERENCES

Archambeau, C. B., Flinn, E. A., and Lambert, D. G., Fine structure of the upper mantle, J. Geophys. Res., **74**, 5825-5865, 1969.

Asada, T., and Aldrich, L. T., Seismic observations of explosions in Montana, *in* The Earth Beneath the Continents, Geophysical Monograph 10, edited by J. S. Steinhart and T. J. Smith, pp. 382-390, American Geophysical Union, Washington, D. C., 663 pp., 1966.

Atwater, T., Implications of plate tectonics for the Cenozoic tectonic evolution of western North America, Geol. Soc. Am. Bull., **81**, 3513-3536, 1970.

Atwater, T., and Menard, H. W., Magnetic lineations in the northeast Pacific, Earth Planet. Sci. Letters, **7**, 445-450, 1970.

Benfield, A. E., A heat flow value for a well in California, Am. J. Sci., **245**, 1-18, 1947.
Birch, F., and Clark, H., The thermal conductivity of rocks and its dependence upon temperature and composition, Parts I and II, Am. J. Sci., **238**, 529-558, 613-635, 1940.
Birch, F., Crustal structure and surface heat flow near the Colorado Front Range, Trans. Am. Geophys. Union, **28**(5), 792-797, 1947a.
______, Temperature and heat flow in a well near Colorado Springs, Am. J. Sci., **245**, 733-753, 1947b.
______, The effects of Pleistocene climatic variations upon geothermal gradients, Am. J. Sci., **246**, 729-760, 1948.
______, Flow of heat in the Front Range, Colorado, Bull. Geol. Soc. Am., **61**, 567-630, 1950.
______, Thermal conductivity, climatic variation, and heat flow near Calumet, Michigan, Am. J. Sci., **252**, 1-25, 1954a.
______, Heat from radioactivity, *in* Nuclear Geology, edited by H. Faul, pp. 148-174, John Wiley & Sons, New York, 414 pp., 1954b.
Birch, F., Roy, R. F., and Decker, E. R., Heat flow and thermal history in New York and New England, *in* Studies of Appalachian Geology: Northern and Maritime, edited by E. Zen, W. S. White, J. B. Hadley, and J. B. Thompson, Jr., pp. 437-451, Interscience, New York, 475 pp., 1968.
Blackwell, D. D., Heat flow in the northwestern United States, J. Geophys. Res., **74**, 992-1007, 1969.
Brune, J. N., and Dorman, J., Seismic waves and earth structure in the Canadian shield, Bull. Seism. Soc. Am., **53**, 167-210, 1963.
Bullard, E. C., Heat flow in South Africa, Proc. Roy. Soc. London, Ser. A, **173**, 474-502, 1939.
Bullard, E. C., Everett, J. E., and Smith, A. G., The fit of the continents around the Atlantic, Phil. Trans. Roy. Soc., **258**, 41-51, 1965.
Burns, R. E., and Grim, P. J., Heat flow in the Pacific Ocean off central California, J. Geophys. Res., **72**(24), 6239-6247, 1967.
Carslaw, H. S., and Jaeger, J. C., Conduction of Heat in Solids, 2d ed., Clarendon Press, Oxford, 1959.
Clark, S. P., Jr., Heat flow at Grass Valley, California, Trans. Am. Geophys. Union, **38**, 239-244, 1957.
Cleary, J., *P* times to Australian stations from nuclear explosions, Bull. Seism. Soc. Am., **57**, 773-781, 1967.
Combs, J., and Simmons, G., Heat flow in the north central United States (abstract), Trans. Am. Geophys. Union, **51**, 426, 1970.
Costain, J. K., and Wright, P. M., Heat flow and geothermal gradient measurements in Utah (abstract), Trans. Am. Geophys. Union, **49**, 325, 1968.
Curtis, B. F., Major geologic features of Colorado, *in* Guide to the Geology of Colorado, edited by R. J. Weimer and J. D. Haun, pp. 1-8, Geological Society of America, Rocky Mountain Association of Geologists, and Colorado Scientific Society, 1960.
Decker, E. R., Heat flow in Colorado and New Mexico, J. Geophys. Res., **74**, 550-559, 1969.
Dickinson, W. R., Relations of andesites, granites, and derivative sandstones to arc-trench tectonics, Rev. Geophys. Space Phys., **8**, 813-860, 1970.
Diment, W. H., The thermal regime of meromictic Green and Round Lakes, Fayetteville, New York (abstract), Trans. Am. Geophys. Union, **48**, 240, 1967.
Diment, W. H., Stewart, S. W., and Roller, J. C., Crustal structure from the Nevada Test Site to Kingman, Arizona, from seismic and gravity observations, J. Geophys. Res., **66**, 201-214, 1961.
Dolgushin, S. S., and Amshinsky, N. N., Uranium distribution in certain Altay granitoid intrusives, Geokhimiya, **9**, 1081-1086, 1966.

Eaton, J. P., Crustal structure from San Francisco, California, to Eureka, Nevada, from seismic-refraction measurements, J. Geophys. Res., **68**(20), 5789-5806, 1963.

______, Crustal structure in northern and central California from seismic evidence, Geol. Northern Calif., Calif. Div. Mines Geol. Bull., **190**, 419-426, 1966.

Ernst, W. G., Tectonic contact between the Franciscan Melange and the Great Valley Sequence–crustal expression of a Late Mesozoic Benioff zone, J. Geophys. Res., **75**, 886-902, 1970.

Foster, T. D., Heat flow measurements in the northwest Pacific and in the Bering Sea, J. Geophys. Res., **67**, 2991-2993, 1962.

Gough, D. I., Heat flow in the Southern Karoo, Proc. Roy. Soc. London, Ser. A, **272**, 207-230, 1963.

Green, R. W. E., and Hales, A. L., The travel times of *P* waves to 30° in the central United States and upper mantle structure, Bull. Seism. Soc. Am., **58**, 267-289, 1968.

Hamilton, R. M., Ryall, A., and Berg, E., Crustal structure southwest of the San Andreas fault from quarry blasts, Bull. Seism. Soc. Am., **54**, 67-77, 1964.

Hart, S. R., and Steinhart, J. S., Terrestrial heat flow: Measurement in lake bottoms, Science, **149**, 1499-1501, 1965.

Healy, J. H., Crustal structure along the coast of California from seismic-refraction measurements, J. Geophys. Res., **68**, 5777-5787, 1963.

Hedge, C. E., and Peterman, Z. E., Sr^{87}/Sr^{86} of circum-Pacific andesites (abstract), Geol. Soc. Am., Program Annual Meetings, Atlantic City, 96, 1969.

Henyey, T. L., Heat flow near major strike-slip faults in California, Ph.D. thesis, California Institute of Technology, Pasadena, 1968.

Herrin, E., and Clark, S. P., Heat flow in west Texas and eastern New Mexico, Geophysics, **21**, 1087-1098, 1956.

Hill, D. P., Gravity and crustal structure in the western Snake River Plain, Idaho, J. Geophys. Res., **68**, 5807-5819, 1963.

Hill, D. P., and Pakiser, L. C., Crustal structure between the Nevada test site and Boise, Idaho, from seismic-refraction measurements, *in* The Earth Beneath the Continents, Geophysical Monograph 10, edited by J. S. Steinhart and T. J. Smith, pp. 391-419, American Geophysical Union, Washington, D.C., 663 pp., 1966.

Hyndman, R. D., Jaeger, J. C., and Sass, J. H., Heat flow measurements on the southwest coast of Australia, Earth Planet. Sci. Letters, **7**, 12-16, 1969.

Immelman, M. N. S., A determination of the radium content of some South African granites, Phil. Mag., **17**, 1038-1047, 1934.

Ingersoll, L. R., Zobel, O. J., and Ingersoll, A. C., Heat Conduction with Engineering, Geological, and Other Applications, University of Wisconsin Press, 325 pp., 1954.

Ito, K., and Kennedy, G. C., Melting and phase relations in a natural peridotite to 40 kilobars, Am. J. Sci., **265**, 519-538, 1967.

Jackson, W. H., and Pakiser, L. C., Seismic Study of Crustal Structure in the Southern Rocky Mountains, U.S. Geol. Surv. Prof. Paper 525-D, pp. 85-92, 1965.

Jackson, W. H., Stewart, S. W., and Pakiser, L. C., Crustal structure in eastern Colorado from seismic-refraction measurements, J. Geophys. Res., **68**, 5767-5776, 1963.

Johnson, N. M., and Likens, G. E., Steady-state thermal gradient in the sediments of a meromictic lake, J. Geophys. Res., **72**, 3049-3052, 1967.

Julian, B. R., Regional variations in upper mantle structure in North America (abstract), Trans. Am. Geophys. Union, **51**, 359, 1970.

Kushiro, I., Syono, Y., and Akimoto, S., Melting of a peridotite nodule at high pressures and high water pressures. J. Geophys. Res., **73**, 6023-6029, 1968.

Lachenbruch, A. H., Three-Dimensional Heat Conduction in Permafrost Beneath Heated Buildings, U.S. Geol. Surv. Bull., **1052-B**, 51-69, 1957.

______, Preliminary geothermal model for the Sierra Nevada, J. Geophys. Res., **73**, 6977-6989, 1968.

Lachenbruch, A. H., and Bunker, C. M., Vertical distribution of crustal heat sources (abstract), Trans. Am. Geophys. Union, **50**, 316, 1969.

Lachenbruch, A. H., Wollenberg, H. A., Greene, G. W., and Smith, A. R., Heat flow and heat production in the central Sierra Nevada, preliminary results (abstract), Trans. Am. Geophys. Union, **47**, 179, 1966.

Lee, W. H. K., and Uyeda, S., Review of heat flow data, *in* Terrestrial Heat Flow, Geophys. Monograph 8, edited by W. H. K. Lee, pp. 87-190, American Geophysical Union, Washington, D.C., 1965.

Lewis, T., Terrestrial heat flow at Eldorado, Saskatchewan, Can. J. Earth Sci., **6**(5), 1191-1198, 1969.

Likens, G. E., and Johnson, N. M., Measurement and analysis of the annual heat budget for the sediments in two Wisconsin lakes, Lim. Oceanography, **14**, 115-135, 1969.

Lipman, P. W., Prostka, H. J., and Christiansen, R. L., Evolving subduction zones in the western United States, as interpreted from igneous rocks, Geol. Soc. Am. Abstracts with Programs, **3**, 148, 1971.

McKenzie, D. P., and Sclater, J. G., Heat flow inside the island arcs of the northwestern Pacific, J. Geophys. Res., **73**, 3173-3179, 1968.

McKenzie, D. P., Speculations on the consequences and cause of plate motions, Geophys. J. Roy. Astron. Soc., **18**, 1-32, 1969.

McKenzie, D. P., and Morgan, W. J., Evolution of triple junctions, Nature, **224**, 125-133, 1969.

Menzer, F. J., Jr., and Jones, J. A., Crustal evolution of the central Okanogan range, Washington (abstract), Geol. Soc. Am., Program Annual Meetings, Atlantic City, 148-149, 1969.

Minear, J. W., and Toksöz, M. N., Thermal regime of a downgoing slab and new global tectonics, J. Geophys. Res., **75**, 1397-1419, 1970.

Oxburgh, E. R., and Turcotte, D. L., Problem of high heat flow and vulcanism associated with zones of descending mantle convective flow, Nature, **218**, 1041-1043, 1968.

_____, Thermal structure of island arcs, Geol. Soc. Am. Bull., **81**, 1665-1688, 1970.

Pakiser, L. C., Structure of the crust and upper mantle in the United States, J. Geophys. Res., **68**, 5745-5756, 1963.

Phair, G., and Gottfried, D., The Colorado Front Range, Colorado, U.S.A., as a uranium and thorium province, *in* The Natural Radiation Environment, edited by J. A. S. Adams, and W. M. Lowder, pp. 7-38, Rice University Semicentennial Publications, 1964.

Porath, H., Oldenburg, D. W., and Gough, D. I., Separation of magnetic variation fields and conductive structures in the western United States, Geophys. J., **19**, 237-260, 1970.

Puranen, M., Järvimäki, P., Hämäläinen, U., and Lehtinen, S., Terrestrial heat flow in Finland, Geoexploration, **6**, 151-162, 1968.

Reitzel, J., Bottom temperature gradients in lakes with trapped sea water (abstract), Trans. Am. Geophys. Union, **47**, 181, 1966.

Reitzel, J. S., Gough, D. I., Porath, H., and Anderson, C. W., Geomagnetic deep sounding and upper mantle structure in the western United States, Roy. Astron. Soc. Geophys. Jour., **19**, 213-235, 1970.

Roller, J. C., Crustal structure in the eastern Colorado Plateau province from seismic-refraction measurements, Bull. Seism. Soc. Amer., **55**, 107-119, 1965.

Roy, R. F., Decker, E. R., Blackwell, D. D., and Birch, F., Heat flow in the United States, J. Geophys. Res., **73**, 5207-5221, 1968.

Roy, R. F., Blackwell, D. D., and Birch, F., Heat generation of plutonic rocks and continental heat-flow provinces, Earth Planet. Sci. Letters, **5**, 1-12, 1968.

Ryall, A., and Stuart, D. J., Travel times and amplitudes from nuclear explosions, Nevada test site to Ordway Colorado, J. Geophys. Res., **68**, 5821-5835, 1963.

Ryall, A., Slemmons, D. B., and Gedney, L. D., Seismicity, tectonism, and surface faulting in the western United States during historic time, Bull. Seism. Soc. Am., **56,** 1105-1135, 1966.

Sass, J. H., Clark, S. P., Jr., and Jaeger, J. C., Heat flow in the Snowy Mountains of Australia, J. Geophys. Res., **72,** 2635-2648, 1967.

Sass, J. H., Killeen, P. G., and Mustonen, E. D., Heat flow and surface radioactivity in the Quirke Lake syncline near Elliot Lake, Ontario, Canada, Can. J. Earth Sci., **5,** 1417-1428, 1968.

Sass, J. H., Munroe, R. J., and Lachenbruch, A. H., Measurement of geothermal flux through poorly consolidated sediments, Earth Planet. Sci. Letters, **4,** 293-298, 1968.

Sass, J. H., Lachenbruch, A. H., Green, G. W., Moses, T. H., Jr., and Monroe, R. F., Progress report on heat flow measurements in the western United States (abstract), Trans. Am. Geophys. Union, **49,** 325-326, 1968.

Schmucker, U., Anomalies of geomagnetic variations in the southwestern United States, J. Geomagnetism Geoelectricity, **15,** 193-221, 1964.

Sclater, J. G., Vacquier, V., and Rohrhirsch, J. H., Terrestrial heat flow measurements on Lake Titicaca, Peru, Earth Planet Sci. Letters, **8,** 45-54, 1970.

Silver, E. A., Late Cenozoic underthrusting of the continental margin off northernmost California, Science, **166,** 1265-1266, 1969.

Spicer, H. C., Geothermal gradients and heat flow in the Salt Valley anticline, Utah, Boll. Geofis. Teorica App., **6,** 263-282, 1964.

Steinhart, J. S., and Meyer, R. P., Explosion studies of continental structure, Carnegie Inst. Wash. Publ., **622,** 409, 1961.

Steinhart, J. S., Hart, S. R., and Smith, T. J., Heat flow, Carnegie Inst. Wash. Year Book 67, 360-367, 1969.

Stewart, S. W., and Pakiser, L. C., Crustal structure in eastern New Mexico interpreted from the GNOME explosion, Bull. Seism. Soc. Am., **52,** 1017-1030, 1962.

Stewart, S. W., Stuart, D. J., Roller, J. G., Jackson, W. H., and Mangan, G. B., Seismic propagation paths, regional travel times and crustal structure in the western United States, Geophysics, **29,** 178-187, 1964.

Turcotte, D. L., and Oxburgh, E. R., A fluid theory for the deep structure of dip slip fault zones, Phys. Earth Planet. Interiors, **1,** 381-386, 1968.

Vacquier, V., Sclater, J. G., and Corry, C. E., Studies of the thermal state of the earth. The 21st Paper: Heat-Flow, Eastern Pacific, Bull. Earthquake Res. Inst., **45,** 375-393, 1967.

Vacquier, V., Uyeda, S., Yasui, M., Sclater, J. G., Corry, C. E., and Watanabe, T., Studies of the thermal state of the earth, 19th paper; heat flow measurements in the northern Pacific, Bull. Earthquake Res. Inst., **44,** 1526-1535, 1966.

Von Herzen, R. P., Ocean-floor heat-flow measurements west of the United States and Baja California, Marine Geol., **1,** 225-239, 1964.

Von Herzen, R. P., and Maxwell, A. E., The measurement of thermal conductivity of deep-sea sediments by a needle-probe method, J. Geophys. Res., **64,** 1557-1563, 1959.

Von Herzen, R. P., and Vacquier, V., Terrestrial heat flow in Lake Malawi, Africa, J. Geophys. Res., **72,** 4221-4226, 1967.

Warren, D. H., Seismic refraction survey of crustal structure in central Arizona, Geo. Soc. Amer. Bull., **80,** 257-282, 1969.

Warren, R. E., Sclater, J. G., Vacquier, V., and Roy, R. F., A comparison of terrestrial heat flow and transient geomagnetic fluctuations in the southwestern United States, Geophysics, **34,** 463-478, 1969.

White, W. R. H., and Savage, J. C., A seismic refraction and gravity study of the earth's crust in British Columbia, Bull. Seism. Soc. Am., **55**(2), 463-468, 1965.

White, W. R. H., Bone, M. N., and Milne, W. G., Seismic refraction surveys in British Columbia, 1941-1966, a preliminary interpretation, *in* The Crust and Upper Mantle of the Pacific Area, Geophys. Monograph 12, edited by L. Knopoff, C. L. Drake, and P. J. Hart, pp. 81-93, American Geophysical Union, Washington, D.C., 1968.

Wickens, A. J., Pec, B., A crust-mantle profile from Mould Bay, Canada, to Tucson, Arizona, Bull. Seism. Soc. Am., **56**(6), 1821-1831, 1968.

Willden, R., Seismic-refraction measurements of crustal structure between American Falls Reservoir, Idaho, and Flaming Gorge Reservoir, Utah, U.S. Geol. Surv. Prof. Paper 525-C, 44-50, 1965.

Williams, D. L., and Roy, R. F., Measurement of heat flow in lakes by a combination of oceanic and continental techniques (abstract), Trans. Am. Geophys. Union, **51**, 425, 1970.

Williams, D. L., Roy, R. F., and Wright, H. E., Jr., Measurement of heat flow in lakes near the Mid-Continent Gravity High, Earth Planet. Sci. Letters, in press, 1971.

Wollenberg, H. A., Smith, A. R., and Bailey, E. H., Radioactivity of upper Mesozoic graywackes in the northern Coast Ranges, California, J. Geophys. Res., **72**, 4139-4150, 1967.

Woollard, G. P., Areas of tectonic activity in the United States as indicated by earthquake epicenters, Trans. Am. Geophys. Union, **39**(6), 1135-1150, 1958.

Woollard, G. P., and Joesting, H. R., Bouguer Gravity Anomaly Map of the United States, American Geophysical Union and U.S. Geological Survey, Arlington, Virginia, 1964.

Wright, H. E., Jr., A square-rod piston sampler for lake sediments, J. Sed. Petrology, **37**, 975-976, 1967.

Wright, H. E., Jr., Livingstone, D. A., and Cushing, E. J., Coring devices for lake sediments, pp. 494-520, *in* Handbook of Paleontological Techniques, edited by B. Kummel and D. M. Raup, W H. Freeman Co., San Francisco, California, 852 pp., 1965.

Zartman, R. E., and Wasserburg, G. J., The isotopic composition of lead in potassium feldspars from some 1.0 b.y. old North American rocks, Geochim. Cosmochim. Acta., **33**, 901-942, 1969.

20

SOME GEOPHYSICAL ANOMALIES IN THE EASTERN UNITED STATES

by **W. H. Diment, T. C. Urban, and F. A. Revetta**

The United States east of the 90th meridian is divided into three provinces in order to establish a framework for the analysis of some more local anomalies: a Western province west of the Cincinnati arch, a Central province including the Appalachian foreland and basin, and an Eastern province comprising the metamorphosed core of the Appalachians. The Western province is characterized by pronounced linear gravity features of high relief, a thick crust of high mean velocity, mildly negative *P* and *S* wave station anomalies, low attenuation of seismic waves, and average heat flow. In the Central province, gravity is generally low, becoming lowest toward the east edge of the province where the crust is thickest and heat flow lowest. *P* and *S* station anomalies are mildly to strongly negative and seismic wave attenuation is low. In the Eastern province, the crust is thinner and gravity, heat flow, attenuation, and station residuals are higher. Several problems of a more local character are considered: (1) Fourteen new heat flow determinations in New York and Pennsylvania suggest that heat flow is low in the deep part of the Appalachian geosyncline. This may be the result of widespread occurrence of radioactivity-impoverished anorthosites in the Precambrian basement or of greater erosion of the Precambrian basement in the vicinity of the heat flow low. (2) Analysis of additional gravity data over the Scranton gravity high, which extends 400 km from Albany, N.Y., to Harrisburg, Pa., and lies near the deepest part of Appalachian basin, suggests that this remarkably linear feature may be the result of mafic volcanics of late Precambrian age and that it represents a buried extension of the Blue Ridge of Maryland and Virginia. It is tentatively proposed that the northern end of this feature is terminated by a northwesterly trending right-lateral displacement of late Precambrian age.

INTRODUCTION

The present geological activity of the eastern United States does not offer the excitement of a lithospheric plate plunging under another along a Benioff plane or the rapid grinding of one plate against another along a system of transcurrent faults; nor is there active volcanism or continental rifting. No doubt, most or all of these processes were vigorously operative in the past. Today, we are left with a battered, contorted, and creaking remnant of the continental lithosphere that evolved over a long period of time and that somehow reflects these past processes. The challenge is to unravel the events and the influence of older fabric upon younger structure.

Nevertheless the eastern United States is not tectonically dead. Great earthquakes in the St. Lawrence River valley, Missouri, and South Carolina indicate release of considerable tectonic strain. The cause of these earthquakes, the extent of their fault systems, and their relation to older structure presents a challenge, particularly at a time when reactor siting requires a better system of seismic zoning.

We do not intend to make a comprehensive review or synthesis of the geophysical information but rather to present a progress report on our regional geophysical investigations in New York and Pennsylvania, with a few remarks concerning geophysical anomalies in similar geological environments of eastern North America.

It is appropriate to begin by dividing the eastern United States into provinces so as to establish a framework to which the more local anomalies can be related.

TECTONIC FRAMEWORK

In the exposed areas of the Canadian Shield and in the metamorphosed core of the Appalachian Mountain System, gravity and magnetic anomalies closely parallel many major geologic features. This can be clearly seen by comparing the Bouguer anomaly maps of the United States (Woollard and Joesting, 1964) and Canada (Dominion Observatory, 1957) with their respective geologic and tectonic maps. The correlations are even more striking when detailed gravity and magnetic maps of various regions are examined as, for example, within the bounds (35° to 39°N) of the Transcontinental Geophysical Survey (U.S. Air Force, 1968; Zietz et al., 1968), Maine (Kane and Bromery, 1968), Vermont and New Hampshire (Bean, 1953; Diment, 1968; Griscom and Bromery, 1968;

Joyner, 1963), New York (Simmons, 1964), and Massachusetts (Bromery, 1967). A generalized version of the Bouguer Gravity Anomaly Map of the United States (Woollard and Joesting, 1964) (Fig. 1) suggests that the United States east of the 90th meridian can be divided into at least three provinces on the basis of gravity anomalies alone.

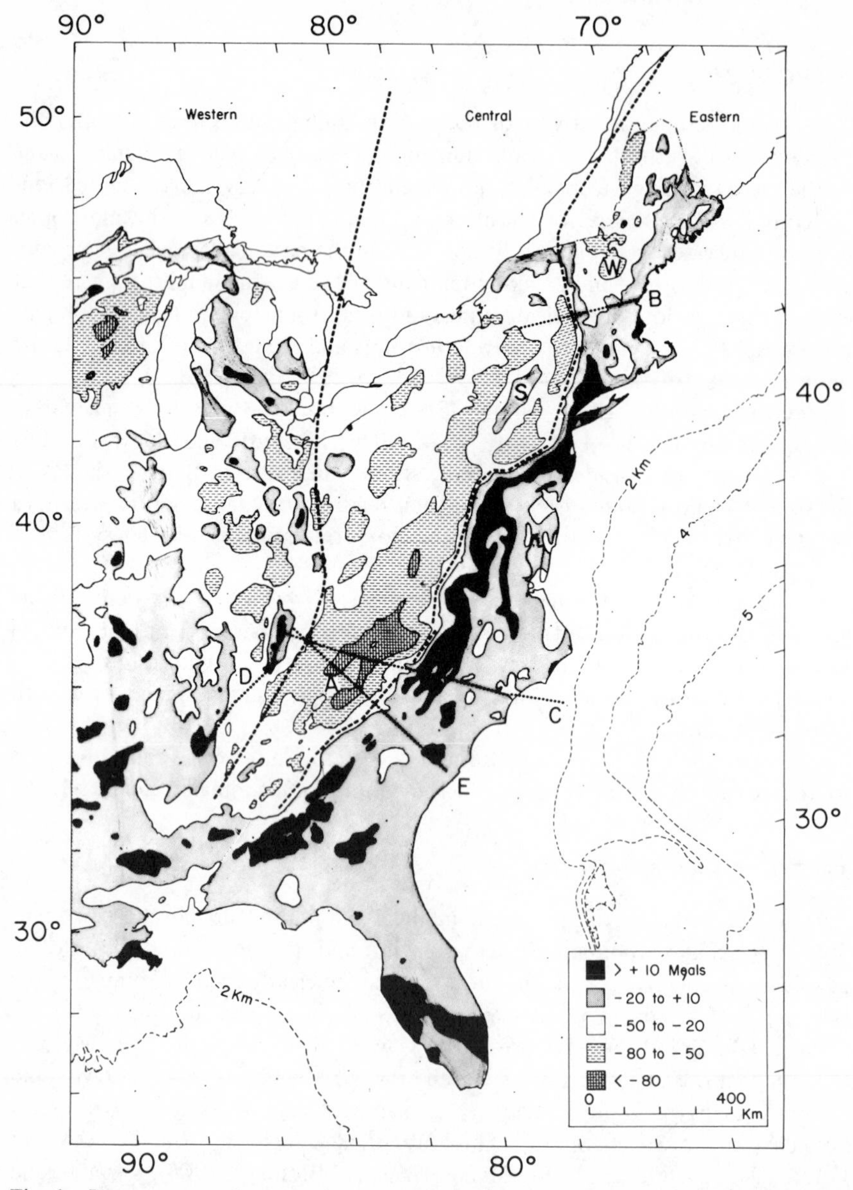

Fig. 1. Bouguer anomaly map of eastern United States. Generalized from Woollard and Joesting (1964). Dashed lines are province boundaries. Lines with letters indicate positions of gravity profiles or refraction profiles. Other letters refer to features discussed in text.

1. The Western province is characterized by large, linear gravity highs each with flanking gravity lows. This pattern is terminated by a rather poorly defined line (Fig. 1) which very roughly correlates with the Cincinnati and Finley arches (Cohee, 1962) and the southward extension of the Grenville front (Bayley and Muehlberger, 1968). McGinnis (1970) suggests that the gravity highs are largely the result of mafic intrusives emplaced in late Precambrian as a consequence of rifting and suggests that these sites were upwarped during late Precambrian and earliest Paleozoic and subsequently sank throughout the Paleozoic in response to isostatic forces. The eastern boundary of the province could be taken as the limit of the late Precambrian activity, although clearly such activity was present much farther east, as evidenced by the Catoctin greenstones of Maryland and Virginia and the Avalonian sequence of the Maritime Provinces (Rodgers, 1967).

2. The Central province, the next province to the east, includes the Appalachian Foreland, Valley and Ridge, and the southern part of the Blue Ridge provinces. The gravity relief is generally smaller in the Central province and Bouguer anomalies are more negative particularly near its eastern edge. Farther to the east an abrupt increase in gravity defines the eastern edge of the Central province. This boundary falls close to the western edge of the uplifted and metamorphosed core of the Appalachians.

Two distinct gravity highs within the Central province merit special consideration. The Scranton, Pennsylvania, gravity high (labeled *S* in Fig. 1) and the Ashville, N.C., gravity high (*A*) indicate anomalous structures within the province and will be considered in more detail in later sections. The northern and southern limits of the province also pose special problems (see, for example, King, 1959, 1964, 1969).

3. The Eastern province is characterized by more positive Bouguer anomalies. The highest values occur near its western edge and some 100 km to the east near or off the Atlantic coast. The intervening region generally exhibits lower anomaly values and, at least in places, this is a consequence of abundant felsic intrusives (Joyner, 1963) and perhaps unusually thick accumulations of eugeosynclinal sediments. The lower values (– 70 mgal) in the White Mountains of New Hampshire (labeled *W* in Fig. 1) are particularly notable in that they are by far the lowest in the eastern province and they occur in the region of the White Mountain magma series which are the youngest intrusives (Triassic-Cretaceous) of large extent in the Appalachians. The low values (– 50 mgals) are not restricted to the vicinity of these intrusives but do seem confined to areas of more abundant felsic intrusives, most of which are Paleozoic in age.

SEISMIC REFRACTION

The seismic refraction profiles within the bounds of the Transcontinental Geophysical Survey (Healy and Warren, 1969; Warren, 1968a, 1968b) and the results obtained by the time-term technique in the Mid-Atlantic states (James et al. 1968) suggest the following generalizations:

1. The crust in the Western province is generally thick and of high mean velocity, although considerable irregularity is evident among the various refraction profiles.

2. In the Central province the crust is thicker. The region of greatest thickness as determined by the time-term technique (James et al., 1968) is near the eastern edge of the province where the Precambrian basement is deepest and where gravity and heat flow are lowest.

3. In the eastern province the crust is thinner and the mean crustal velocities appear intermediate. Preliminary interpretations of two refraction profiles obtained by J. C. Roller and given in Warren's compilation (1968a) are shown in Fig. 2. Both are important and rather unusual. The top profile (*C*, Figs. 1, 2) extends from the edge of the Western province southeasterly across both the Central and Eastern provinces. It suggests that gravity change from the Central to the Eastern province can be accounted for largely by variations within the

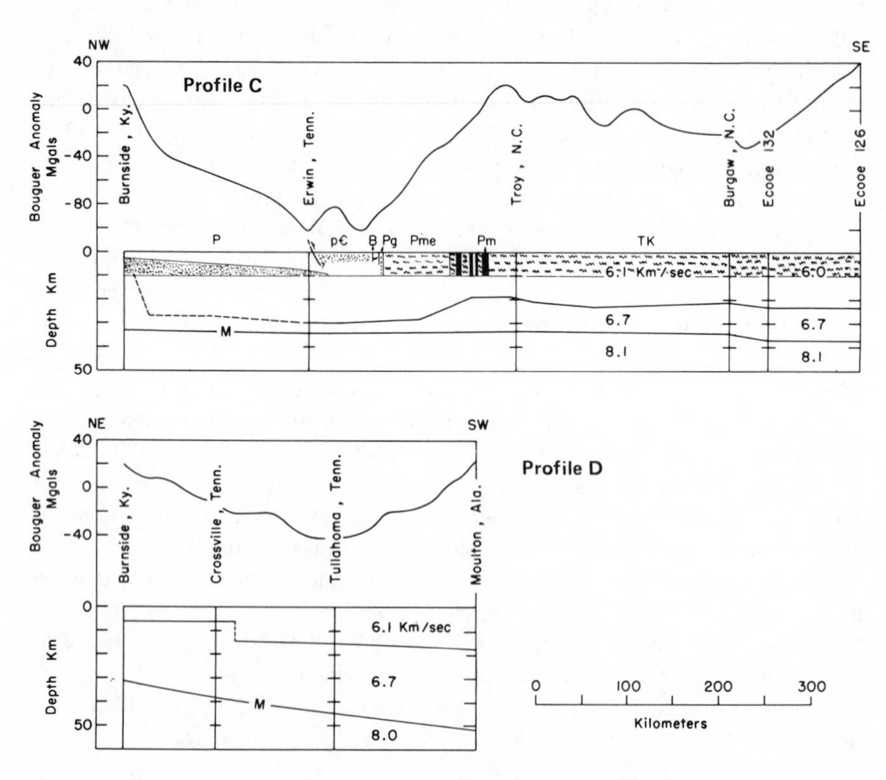

Fig. 2. Seismic refraction profiles. Redrawn from Warren (1968). Bouguer anomalies from Woollard and Joesting (1964). Geologic section generalized from a map of Wilden et al. (1968). Location of profiles is shown in Fig. 1. Geological symbols as follows: Pc, Precambrian; P, Paleozoic sedimentary rocks; Pe, metasedimentary rocks of Paleozoic and unknown age; Pg, Paleozoic granitic rocks; Pm, Paleozoic mafic rocks both intrusive and metavolcanics; TK, Tertiary and Cretaceous rock of the coastal plain; B, Brevard fault zone.

crust. This is in marked contrast to the results of James et al. (1968) in the region just to the north where relief on the *M*-discontinuity as determined from time-term analysis rather nicely explains the gravity anomalies. Whether the details of Roller's interpretation proves tenable or not, it does stress the fact that we should consider density contrasts within the crust for explanation of some major gravity anomalies. Moreover, the interpretations of James et al. (1968) involves data obtained in the central Appalachians, wheras Roller's profile is in the southern Appalachians. The two regions differ in several regards, and it is conceivable that real differences in gross crustal structure are present.

The second unusual profile (*D*, Figs. 1, 2) extends southwesterly along the axis of one of the linear gravity highs typical of the Western province. It indicates dense rocks of considerable thickness close to the top of the Precambrian surface. Late Precambrian mafic volcanics have been penetrated by the drill close to similar highs (Bayley and Muehlberger, 1968), but this is the only feature for which refraction information is available aside from the vast amount of information for the Lake Superior region (Steinhart and Smith, 1966). The interpretation that the *M*-discontinuity rises in the vicinity of the gravity high is interesting, but may not be a fact. Insertion of a "masked" layer of rather ordinary crustal velocity below the high-velocity material (mafic volcanics) and above the *M*-discontinuity would result in a flatter *M* or one that was depressed in the vicinity of the gravity high. This would make it easier to account for the lows that flank the highs in this province.

UPPER MANTLE VELOCITY AND ATTENUATION

During the past decade, it has become increasingly evident that body wave velocities vary laterally in the upper mantle. The information is of two kinds: (1) That obtained from *Pn* arrivals on short refraction profiles (several hundred kilometers) and (2) that obtained by examining *P* and *S* wave station residuals (station anomalies) at teleseismic distances appropriately corrected for azimuthal effects. It is supposed that station residuals are largely the result of velocity variations in the upper mantle. The two types of information yield similar results.

Herrin and Taggart (1962) first presented a map of *Pn* variations based on refraction measurements for the whole of the conterminous United States and latter modifications differ only in detail (Hales and Herrin, this volume). Several maps of *P* and *S* station anomalies have been presented by Cleary and Hales (1966) and Herrin (1969) of which those given by Hales and Herrin in this volume are typical. Solomon and Toksoz (1970) present maps of the lateral variation of the attenuation of *P* and *S* waves for the United States.

A principal result of these studies is that upper mantle velocities are low in a large part of the western United States where heat flow and seismic wave attenuation are high, and high in the east where heat flow and attenuation are low. The question remains as to whether such data are of use in contrasting lesser provinces such as those under consideration here (Pakiser and Steinhart, 1964).

Although the data are sparse, we note the following.

1. Upper mantle velocities seem highest in our Western and Central provinces with some indication that the highest velocities may be present in the Central province at midlatitudes. This is where the crust is thick and heat flow is low.

2. The mantle velocities seem a bit lower in the Eastern province, which comprises the metamorphosed and intruded core of the Appalachians. Here the mean heat flow and wave attenuation seem higher, at least in places, than in the Central province.

3. Although the contours of station anomalies are generally parallel to the Appalachian trends, there is a large departure in Pennsylvania. At first one might suppose that this results from the way the widely spaced data were contoured. Certainly, the data can be contoured so that the trends are parallel to the general structural trends, but then one is left with some large isolated closures. These may be the result of rather local structures such as that represented by the Scranton gravity high. (The Delhi station is on this gravity feature.) It does not seem entirely realistic, however, to explain the station anomalies on the basis of local features; the travel paths through rocks of anomalous velocity are rather short to give rise to anomalies of this size.

In summary, the mantle velocity and attenuation data seem to conform to the gravity provinces and there also appears to be a correlation with heat flow. The more local variations in velocity are intriguing and suggest that more dense control would prove rewarding.

HEAT FLOW

Birch et al. (1968), Roy et al. (1968a), and Lachenbruch (1968) have shown that heat flow (q) measured in igneous plutons bears a close relation to the measured radiogenic heat production (A) of the rocks in the plutons. A relation of the form $q = q^* + DA$ fits the data for a heat flow province surprisingly well. The parameter D varies little (7.5 to 10 km). The intercept q^* is interpreted as the heat flow coming from the lower part of the crust and upper mantle and is supposed to vary little within provinces of such dimensions as the eastern United States (~ 0.8 μcal/cm^2 sec), the Basin and Range (~ 1.4), and the Sierra Nevada (~ 0.4) (Roy et al., 1968a).

The preceding suggests that a regional heat flow map might be dominated by local contrasts in radiogenic heat production and that regional variations resulting from other causes might be completely masked. Nonetheless, we have contoured the heat flow data for the eastern United States (Fig. 3). Before examining the significance of the anomalies, a few comments on the accuracy of the data may be in order.

Each determination is subject to the usual uncertainties that have been treated in varying degrees by the individual investigators and summarized in part by Lee

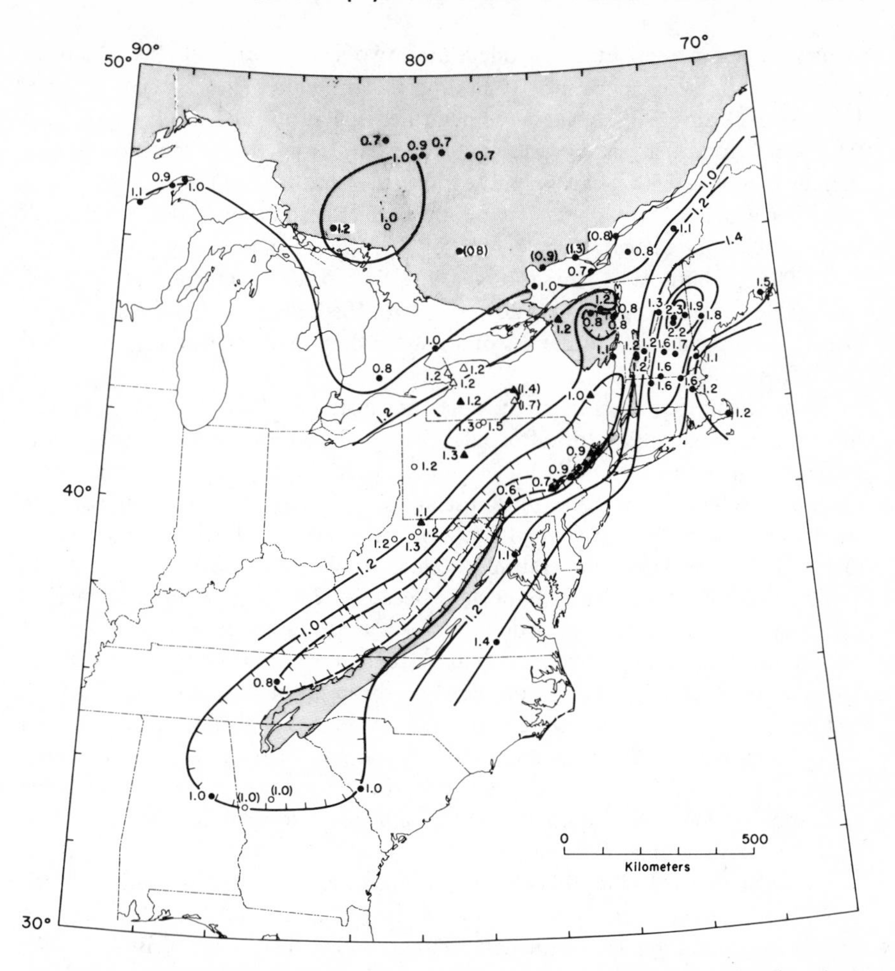

Fig. 3. Heat flow in eastern United States and Canada. Topographic corrections have been applied or are negligible. Questionable values are bracketed. Dots and triangles represent previously published values and this investigation, respectively. Open dot or triangle indicates thermal conductivity was estimated rather than measured (closed dot or triangle). Data from the following sources: Canada (Beck and Judge, 1969; Beck and Logis, 1964; Jessop, 1968; Judge and Beck, 1970; Leith, 1952; Misener et al., 1951; Sass et al., 1968; Saull et al., 1962); New England, New York, Alabama (Roy et al., 1968b); Pennsylvania and West Virginia (Joyner, 1960); Virginia (Diment and Werre, 1964; Diment et al., 1965b); South Carolina (Diment et al., 1965a); Tennessee (Diment and Robertson, 1963); Georgia (from Birch and Spicer, see Diment and Robertson, 1963); Michigan (Birch, 1954; Roy et al., 1968b).

and Uyeda (1965). The range in quality is large but difficult to quantify. Particularly suspect values are enclosed in brackets. It also appears that corrections for climatic change are significant at some localities where the holes are shallow or where the normal gradients are low because of highly conductive rocks.

Climatic effects might be divided into two categories: (1) the effects of Pleistocene climatic changes (> 10,000 years ago) and (2) more recent changes. Dansgaard et al. (1969) suggest temperature oscillations of periods of 120 and 940 years from analysis of oxygen isotope data from the Greenland ice cap. Periods as long as 2500 years might be inferred from the advances and retreats of mountain glaciers during the past 6000 years (Denton and Porter, 1970). All these oscillations appear to be so phased as to result in warming in the recent past. The amplitude of these oscillations is not well known, but it seems reasonable that they might be a degree or two. If so, they would have significant effects on the gradients to depths of a few hundred meters depending on period and diffusivity.

The gradients observed in many holes are distinctly low at a depth of 100 meters and apparently little affected below 200 meters. (See for an example Diment, 1965; Jessop, 1968; Judge and Beck, 1967; Sass et al., 1968.) Such disturbances could be caused by oscillations of periods less than about a thousand years (see also Cermak, 1970). The analyses of Lachenbruch et al. (1959, 1962, 1969) of temperature obtained from holes drilled in permafrost in arctic Alaska are particularly important because here temperature disturbances resulting from movement of ground water can safely be ignored. The form of Lachenbruch's temperature-depth curves and the time rate of change of temperature over an 8-year period clearly indicate a climatic warming of several degrees over the past century or more. It is evident that changes of such period do not significantly affect gradients to a depth greater than about 150 meters. Some heat flows, however, have been reported for holes in this depth range (Lubimova, 1964, Lee and Uyeda, 1965) and the results are most probably low by a significant amount (Diment, 1965).

The most plausible Pleistocene climatic histories examined by Birch (1948) produce maximum corrections of about + 3 deg C/km (Fig. 4, Table 1) for the depth ranges used in most of the determinations. The history (1) of Birch (1948) results in the lowest corrections and is shown in Table 1. These were applied to the values in Fig. 3 to obtain those in Fig. 5. History (2) contains no interglacial periods and results in an unrealistically high correction. To a first approximation, the correction to the heat flow is the product of the gradient disturbance and the conductivity of the rocks in which the heat flow was measured (see Lachenbruch, 1959, for the effect of stratified medium). Inasmuch as conductivities range from about 5 to 15 mcal/cm sec deg, heat flow corrections of 0.15 to 0.45 μcal/cm^2 sec would seem possible; however, most conductivities lie in the range 6.5 ± 1.5. Thus, the correction is only important in a relative sense when the conductivities are unusually high, as in quartzite, dolomite, salt, etc.

In an absolute sense, the correction may be significant because it raises the q^* values of Birch et al. (1968), Roy et al. (1968a), Lachenbruch (1968) by about 0.2 μcal/cm^2 sec. Moreover, the value of D increases about 5 percent for the eastern United States province of Roy et al. (1968a). This insignificant increase

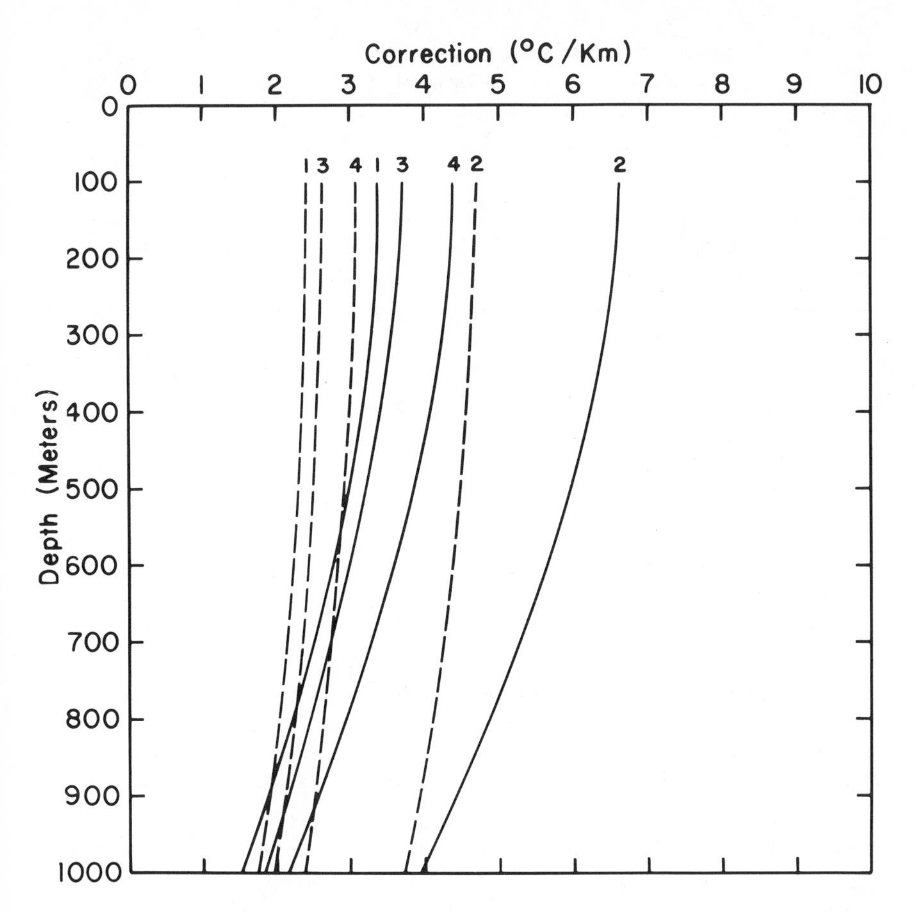

Fig. 4. Gradient corrections for the effect of Pleistocene climatic variations according to models examined by Birch (1948). Solid and dashed curves are for diffusivities of 0.01 and 0.02 cm^2/sec. Temperature histories are given in Table 1. Corrections applied to values in Fig. 3 to get those in Fig. 5 were obtained by using temperature history 1 and the diffusivity of the rocks for the depth interval in which the heat flow was determined.

is simply the result of the fact that higher radioactivity rocks are generally more quartz-rich and thus of higher thermal conductivity.

Whether a correction of the type constructed by Birch (1948) can be uniformly applied over the continents is doubtful. The chronology of Pleistocene events, however, seems world-wide, and the amplitudes of temperature oscillations deduced from oxygen isotope studies of deep sea cores (Emiliani, 1961) are nearly as large as those assumed by Birch (1948).

A map of the heat flows corrected for climatic change (Fig. 5) reveals patterns similar to the map of the uncorrected values, but there are some significant differences. The range of values is less, particularly within the Western province and between the Western province and the Canadian Shield. Indeed, additional

Table 1. Pleistocene climatic histories (after Birch, 1948) used in the calculation of corrections in Fig. 4

Time (thousands of years before present)	Temperature of surface (degrees Celsius relative to zero at present)			
	(1)	(2)	(3)	(4)
0-5	0	0	0	0
5-10	2	2	2	2
10-20	- 5	- 5	- 5	- 5
20-30	- 10	- 10	- 10	- 10
30-40	0	- 10	0	- 5
40-60	- 10	- 10	- 10	- 10
60-80	0	- 10	0	- 5
80-100	- 10	- 10	- 10	- 10
100-200	2	- 10	0	0
200-300	- 10	- 10	- 10	- 10
300-600	2	- 10	0	0
600-700	- 10	- 10	- 10	- 10
700-900	2	- 10	0	0
900-∞	0	0	0	0

measurements may show that there is no significant differences in the mean values between the Shield (or at least the Grenville province of the Shield) and our Central province.

Heat flow is low near the eastern edge of the Central province where gravity is lowest, the crust is thickest and *P* and *S* stations are mildly to strongly negative. We have no favored explanation for this anomaly but note that Birch et al. (1968) found anomalously low heat flow at three localities in the radioactivity impoverished Precambrian anorthosites of the Adirondack massif. This is shown as a closure in Fig. 3. A refraction correction would increase the anorthosite values by about 0.1 μcal/cm^2 sec, but this would not alter the general pattern. Large anorthosite bodies are rather common in the Grenville province of the Canadian Shield, and we might suppose that their frequency of occurrence is similar in the buried Precambrian of the Central province. If the low values are the result of anorthosite, we should not expect low heat flow to be continuous over such a large region, but would rather expect a series of isolated lows of limited extent.

Another steady-state model that could explain the heat flow low involves the removal of radioactivity by greater erosion of the Precambrian crust in the region of the heat flow low than to either side of it. This model requires that the Precambrian crust be thin in the vicinity of the heat flow low, yet this is where gravity and seismic information indicate that the total thickness of the crust is greatest. There are some 10 km of Paleozoic sediments where the crust is thickest, however. A significant fraction of this sediment is carbonate (Rodgers, 1968), which has low radioactivity (Clark, 1966, Wollenberg and Smith, 1970). Thus, we envision a situation where a radioactive part of the crust (the eroded Precambrian) has been replaced by less radioactive sediments (the carbonates).

Values exceeding 1.5 μcal/cm^2 sec (uncorrected for climatic change) are confined to New England and those exceeding 1.7 are restricted to plutons of the White Mountain magma series which are regarded as early Jurassic (Lyons and Faul, 1968) to Cretaceous (Foland et al., 1970). Most of the New England values were obtained in felsic plutons of radioactivity equal to or higher than their metasedimentary hosts; thus, their mean is considerably higher than that for the region. Values in the Precambrian gneisses of the Green Mountain anticlinorium and Chester dome and in the metasediments and the less

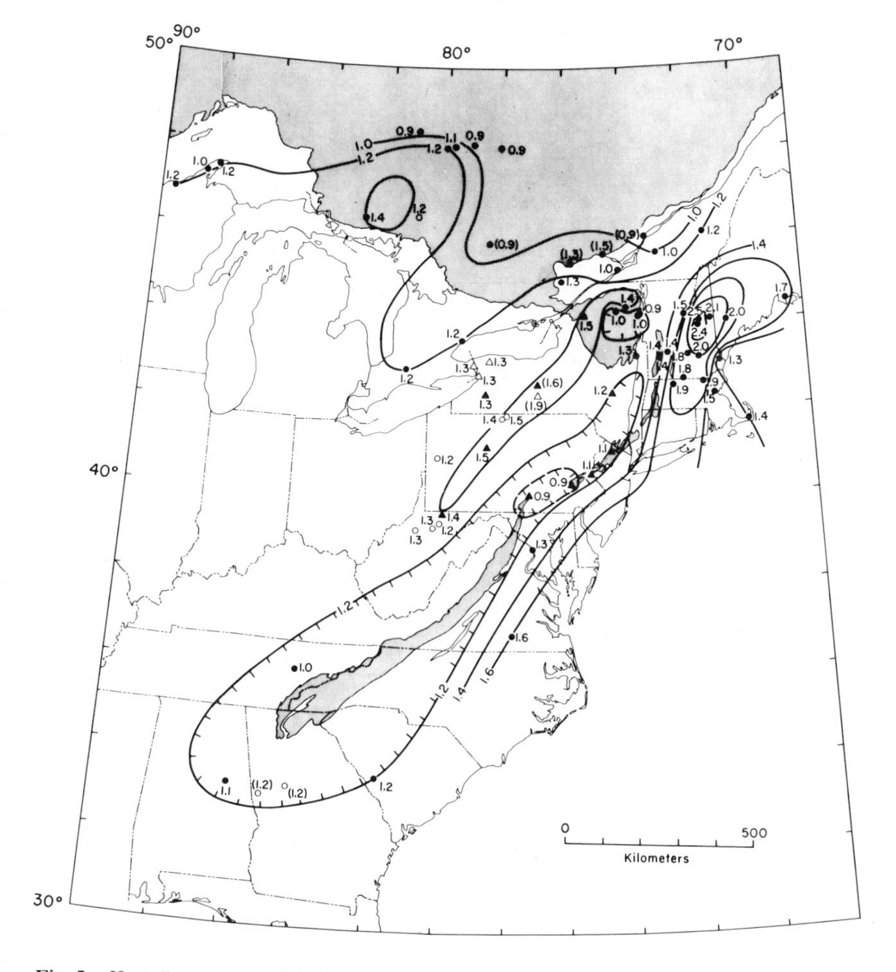

Fig. 5. Heat flow corrected for Pleistocene climatic variations.

radioactive plutons are all close to 1.2 (Birch et al., 1968, Fig. 1). We could suppose that the regional mean might be about 1.3.

The rocks of the White Mountain magma series are unusual. Although Birch et al. (1968) have shown that White Mountain magma series could be regarded as a late phase in the magmatic sequence that also involved the genesis of the mid-Paleozoic plutons, the White Mountain series appears to have no equivalents in type or time in the central and southern Appalachians in regions of extensive Paleozoic plutonism (Hadley, 1964); nor do there seem to be equivalents in the Maritime provinces. The rocks of the White Mountain magma series are, however, similar to the alkalic rocks of the Monteregian Hills, which also find no other expression in the Appalachians with the exception of a few alkaline dikes in Virginia and West Virginia (Dennison, 1970) and in the Mississippi embayment. It might also be noted that Bouguer anomalies are more negative (– 70 mgals) in the White Mountains than elsewhere in the Eastern province. All this suggests that the White Mountain series, and thus the high heat flows, are distinctly anomaluous within the Appalachian system and that other anomalous conditions might be expected in this region.

In summary, variations in heat flow in eastern North America are rather small except for the high heat flows obtained in plutonic rocks of high radiogenic heat production. Differences among the means for our three provinces and the Canadian Shield may not exceed 0.3 μcal/cm^2 sec. The Shield values may be the lowest, the Central and Western provinces about the same (see also unpublished data of Combs, 1970; Combs and Simmons, 1970), and the Eastern province the highest. The data are still extremely sparse and some of questionable reliability. No information has been published for such interesting regions as the Mississippi embayment, which has experienced more Tertiary activity than the rest of the region.

SCRANTON GRAVITY HIGH

Woollard and Joesting's Bouguer Gravity Anomaly Map (1964) of the United States shows a large elliptically shaped gravity high (labeled *S* in Fig. 1) extending from near Harrisburg, Pennsylvania, northeasterly to Albany, New York. This anomaly is especially interesting for it nicely fits in the area where the trend of the Appalachians changes from northerly in New England to near easterly in southeastern Pennsylvania. The anomalous mass responsible for this gravity high is somehow related to this marked recess in the Appalachian trends, and it was either formed as a consequence of the forces that produced the recess or its pre-existance controlled the formation of the recess.

In view of the importance of the Scranton gravity high as a possible indicator of the evolution of this part of the Appalachians and in view of the paucity of data upon which it was based, a much more detailed gravity survey was conducted (Revetta, 1970, Revetta and Diment, 1969) within the area with stations spaced at 5 km, roads permitting. Contrary to our expectations, the detailed survey revealed a much more regular feature (Fig. 6) than indicated by Woollard and Joesting's map. Indeed, the contours are remarkably linear for several hundred kilometers, and there is little indication of discontinuity except near the southern end.

Although we cannot explain the anomaly completely, we note the following.

1. The gravity high appears to be unique in the Central province, unless, perhaps, the more subdued and shorter Ashville gravity high (labeled *A* in Fig. 1) has a similar significance, which seems doubtful.

2. The anomaly is in, or just to the west of, the region where the miogeosynclinal deposits are thickest (Broughton et al., 1966, Fig. 3; King, 1968). There is, however, considerable uncertainity as to the exact configuration of the basement in this region for it is extremely deep and the stratigraphy of the lower Paleozoic is conjectural, and that of the late Precambrian supracrustal rocks (if any) is unknown.

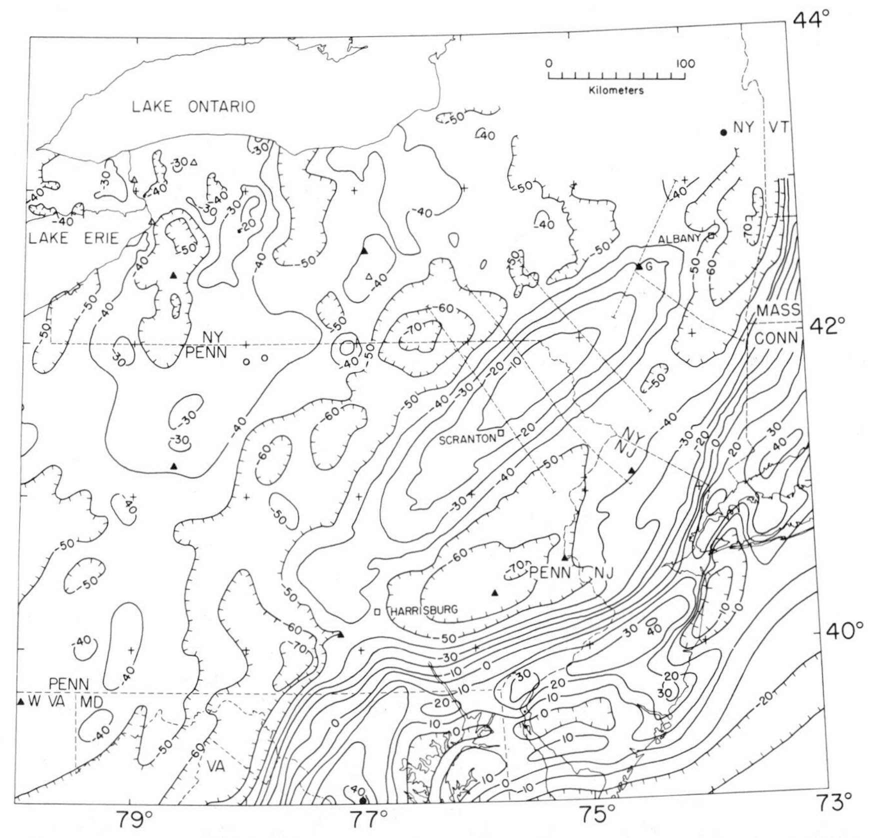

Fig. 6. Simple Bouguer anomaly map of Scranton gravity high. Dashed lines indicate locations of ground magnetometer traverses. Circles and triangles indicate positions of heat flow measurements (see Fig. 3). Gravity data in New York and eastern Pennsylvania are the result of this investigation; the remaining data from Woollard and Joesting (1964).

3. Aeromagnetic maps (Bromery and Griscom, 1967; U.S. Geological Survey, 1969) for the south central part of the anomaly exhibit low relief and small gradients. Several ground magnetometer traverses (Fig. 7), although of poor quality because of surface disturbances, suggest that the anomalous material is deep and that the Precambrian basement is depressed near the gravity high.

4. Katz's unreversed refraction profile (1955) from Milroy, Pa., to Palisades, N.Y. (a bit south of 41°N), which crosses the southern part of gravity high, yields *Pg* and *Pn* velocities of 6.04 and 8.21 km/sec and a crustal thickness of 33 ± 3 km assuming a one layer crust. Although the number of data points is not sufficient to reveal details of crustal structure, the crust could be regarded as a bit thin, or of high mean velocity, near the gravity high.

The results of James et al. (1968), obtained by the time-term technique cover the extreme southern end of the gravity high. The crust seems thickest just west of the gravity high and thins rapidly across it (James et al., 1968, Figs. 6 and 7). The latter could also be interpreted as increase in mean crustal velocity. Unfortunately, their work did not extend farther north, for it appears from the fragmentary data near the fringe of their survey that the technique would be valuable in elucidating the anomalous structure.

5. The gravity high is flanked by gravity lows that are, in part, clearly related to the high, and must be considered in any model used to explain the anomaly. The gravity lows could partly be the result of thick Paleozoic sediments, but from what is known of the configuration of the basement, it is unlikely that the sediments are the sole cause. Moreover, the sediments are well indurated. For example, the mean density of a 300-meter sand shale sequence at Gilboa (*G* in Fig. 6) is 2.7. The carbonates that dominate the Cambro-Ordivician sequence are more dense where much dolomite is present. A density contrast of more than 0.15 gm/cm^3 between the sediments and the Precambrian basement or the eugeosynclinal deposits to the east is unlikely.

If we admitted such a density contrast, however, and applied corrections to the gravity map proportional to the thickness of the sediments, the form of the gravity high would be significantly altered. First, the gradients on the southeast side of the high would be considerably less than those on the northwest side. Second, the Scranton gravity high would appear more nearly to be a continuation of the main gravity high that marks the western edge of the Eastern province of the central Appalachians of Virginia and Maryland.

6. The southeastern end of the gravity high is in the region where the strongly deformed lower Paleozoic rocks plunge eastward under the Carboniferous rocks of the anthracite region. The broad belt of Appalachian folds so prominent in the anthracite region fade out eastward into the Pocono Plateau. The Lackawanna syncline, which lies entirely within the Pennsylvania part of the gravity high, is deflected to the north near its eastern end. Rodgers (1964) has examined these structural anomalies and concludes: "Perhaps basement irregularities, already formed in the early Paleozoic, helped to control the distribution of these features as well as the culminations and depressions in the Valley and Ridge province [pp. 75-76]."

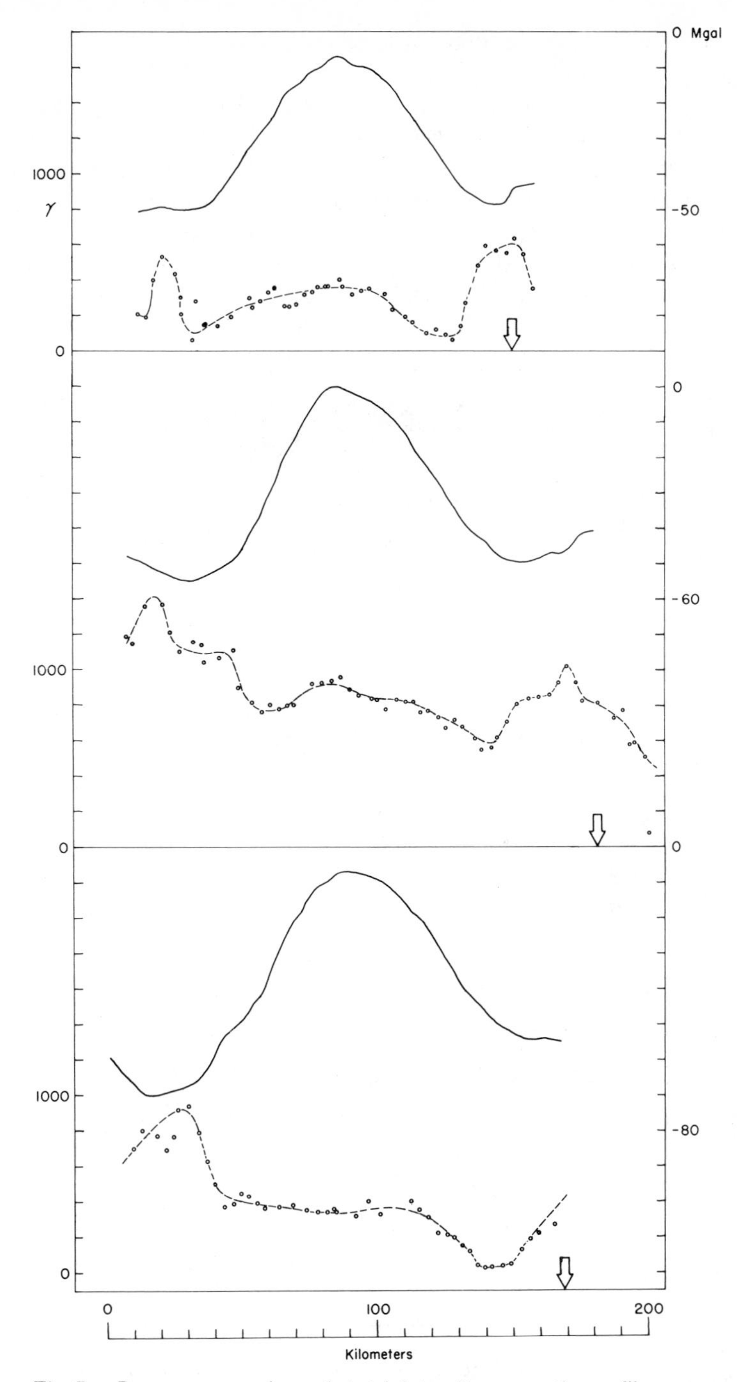

Fig. 7. Bouguer anomaly and total intensity magnetic profiles across Scranton gravity high. Ground magnetometer traverses (dashed lines with individual observations indicated) are of poor quality because of surface disturbances. A regional gradient of 4 gammas/km has been subtracted. Base is arbitrary, the arrows indicate the outcrop of the Silurian Shawangunk formation.

The preceding suggests (see also Wise and Werner, 1969) that the anomalous mass responsible for the Scranton gravity high may have been emplaced in late Precambrian time. Various classes of models are illustrated in Fig. 8. Indeed, perhaps this gravity feature is a continuation of the main gravity high of Maryland and Virginia. It is the result of a thick accumulation of mafic volcanics, such as the Catoctin greenstones, and if this is so, the problem remains as to why the Paleozoic structures cross over the earlier trend.

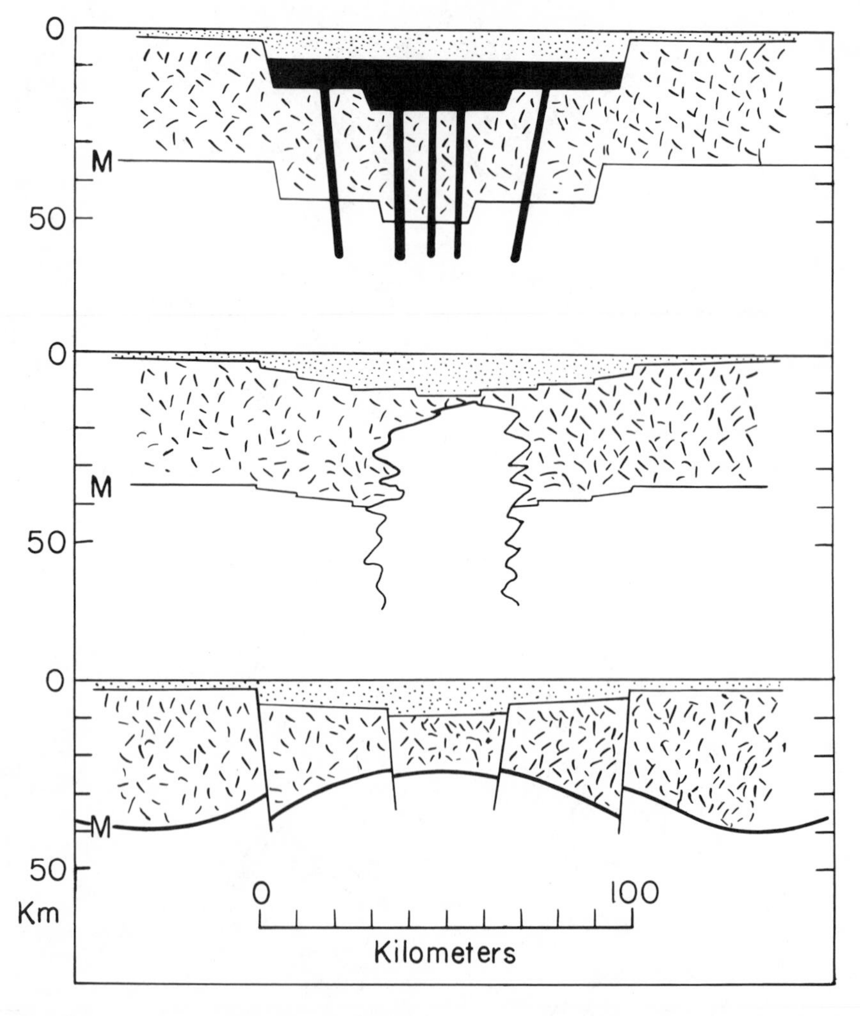

Fig. 8. Types of models that might explain the Scranton gravity high and its flanking gravity lows: (*a*) a basalt filled rift; (*b*) an intrusion from the asthenosphere consequent to a separation of the lithosphere; (*c*) a collapsed arch. The models are symmetrical about the axis of the gravity high but some asymmetry is required especially if the gravitational effect of the overlying sediments is removed.

Another matter that should be considered in connection with the origin of the gravity high is the right lateral fault proposed by Drake and Woodward (1963), Drake (1969), and Drake et al. (1968), which is supposed to extend along the Kelvin seamount chain, across the Appalachians near 40°N at the southern end of the Scranton gravity high, and thence southwest (line *D* of Fig. 12). If the gravity high existed before the faulting, another part of the Scranton gravity high might be expected somewhere to the southwest, but it cannot be identified at present. Perhaps the anomalous mass responsible for the gravity high was emplaced as a consequence of the right-lateral faulting. If so, the mechanics are not obvious. This and other information (Root, 1970) argues against the existence of this fault at least in Pennsylvania, but the case is not conclusive.

BOUNDARY BETWEEN CENTRAL AND EASTERN PROVINCES

The gravity boundary between these two provinces is sharply defined by an abrupt increase from less than – 50 mgals to more than – 20 mgals (Fig. 1). Although the path is sinuous, it appears continuous and rather closely parallels major structural trends from Alabama to the Gaspé Peninsula and perhaps across the Gulf of St. Lawrence into the Long Range of Newfoundland. In places there is a clear relation to geologic structure at the surface, but in others the relation is obscure or complex. In all localities, only eugeosynclinal deposits are found to the east of the boundary. These rocks were regionally metamorphosed and extensively plutonized and have been greatly uplifted relative to the miogeosynclinal rocks to the west. Although miogeosynclinal deposits predominate to the west, some eugeosynclinal rocks lie west of the boundary but were mainly thrust into this position from the east.

In central New England the boundary falls on the west flank of the Green Mountain anticlinorium (Fig. 9), but a little to the south the boundary lies over the anticlinorium (Diment, 1968). This lack of direct correspondence between gravity and geologic features along with the lack of suitable density contrasts at the surface suggests that the gravity anomalies owe their origin to deeper anomalous masses. We stress the fact that there is a rough correspondence between structure and gravity anomalies at the edge of the province and in particular that the gravity relief tends to increase with structural relief. Large overthrust masses such as the Taconic klippen (Rodgers, 1968, Zen, 1968) are in the gravity lows just to the west of the boundary.

A schematic cross section (Fig. 10) suggests one interpretation of the variation of gravity across the boundary. In this model, the abrupt increase in gravity at the boundary is regarded as the result of the displacement of the *M*-discontinuity and/or the dense lower part of crust. The decrease in gravity to the east is correlated with the increasing abundance of felsic intrusive rocks which are less dense than their metasedimentary hosts.

This model may apply to other parts of the Appalachians but only with considerable modification. In the central Appalachians of Maryland and northern Virginia, where the Precambrian rocks of the Blue Ridge are largely

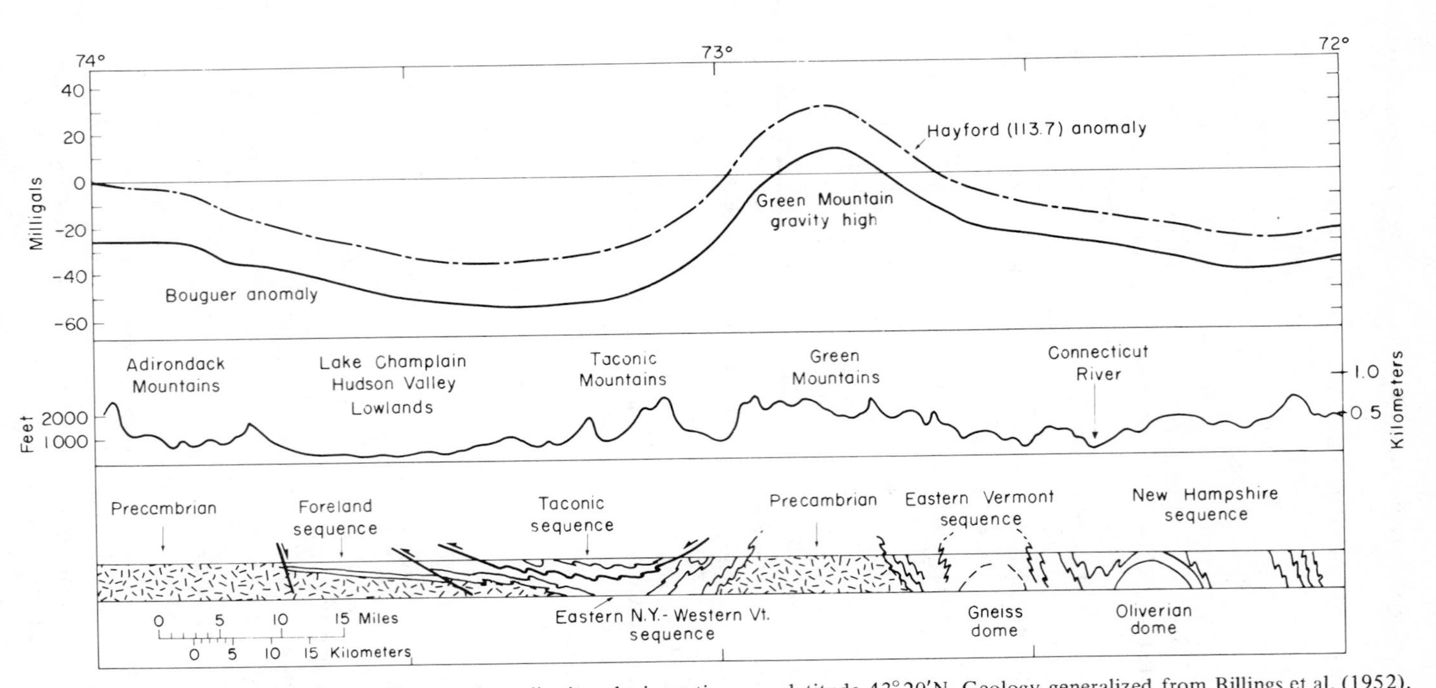

Fig. 9. Gravity and topography profiles and generalized geologic section near latitude 43° 20′N. Geology generalized from Billings et al. (1952). There is no vertical exaggeration in the geologic section. Figure modified from Diment (1968, p. 401).

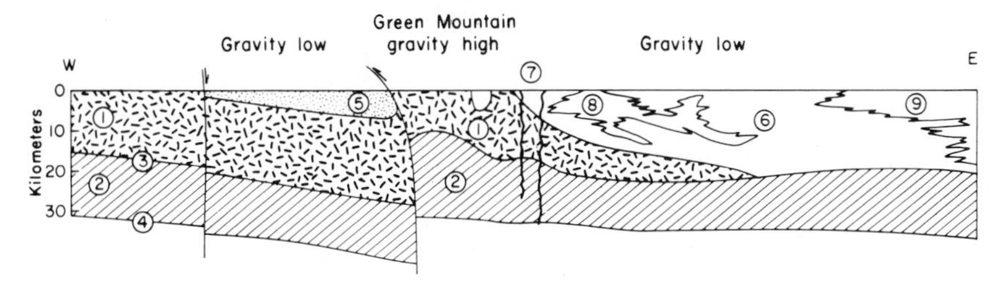

Fig. 10. Hypothetical schematic model to explain gravity anomalies across central New England. (1) Precambrian granitic "layer" of the continental crust. (2) Intermediate "layers" of continental crust or oceanic crust and part of oceanic mantle transformed to gabbro and or serpentinite. (3) Conrad "discontinuity." (4) Mohorovicic discontinuity. (5) Sedimentary rocks of the miogeosyncline. (6) Metasedimentary and metavolcanic rocks of the eugeosyncline. (7) Ultramafic rock (largely serpentinite). (8) Calc-alkalic syntectonic intrusive rocks. (9) Alkaline postorogenic intrusive rocks. Figure modified from Diment (1968).

autochthonous, the relation between gravity and structure is similar to that in central New England (Fig. 11). The gravity relief, however, is larger across the province boundary; more particularly gravity is higher in the Eastern province. A part of this is a consequence of thick sequences of late Precambrian volcanics such as the Catoctin greenstones (King, 1951, p. 128), and may also reflect the presence of dense intrusive rocks (Griscom, 1963). In the Central Appalachians, the seismic time-term analysis of James et al. (1968) indicates that most of the gravity relief in the vicinity of the province boundary can be explained by relief on the *M*-discontinuity (extremes 30 to 60 km). Their data are not sufficient, however, to examine the role of dense high-velocity rocks within the crust, nor is the control sufficient to suggest the type of structural boundary between the provinces.

South of central Virginia, the tectonic style changes as does the apparent relation between geology and the gravity field (King, 1964, 1969; Rodgers, 1964). The tectonic trends are more easterly, thrusts are more common, and a major part of the Precambrian of the Blue Ridge is allochthonous. The gravity province boundary falls to the east of the Precambrian mass of the Blue Ridge. Indeed, these Precambrian rocks are close to the axis of the great gravity low (– 100 mgals) to the west of the province boundary (Fig. 2). A major part of these Precambrian rocks are known to be allochthonous (King, 1964, 1968), and it might be tempting to suppose that the entire mass was thrust from the east into its present position and that the Precambrian rocks of the southern part of the Blue Ridge were emplaced in the same manner as the allochthonous masses of the northern Appalachians.

Note, however, that a small gravity high is located over the Precambrian outcrop (Fig. 2). This is the Ashville gravity high of Fig. 1. It increases in amplitude to the southwest of the line of Fig. 2 to attain a maximum amplitude of about 50 mgals. Does this anomaly have an analogue elsewhere in the

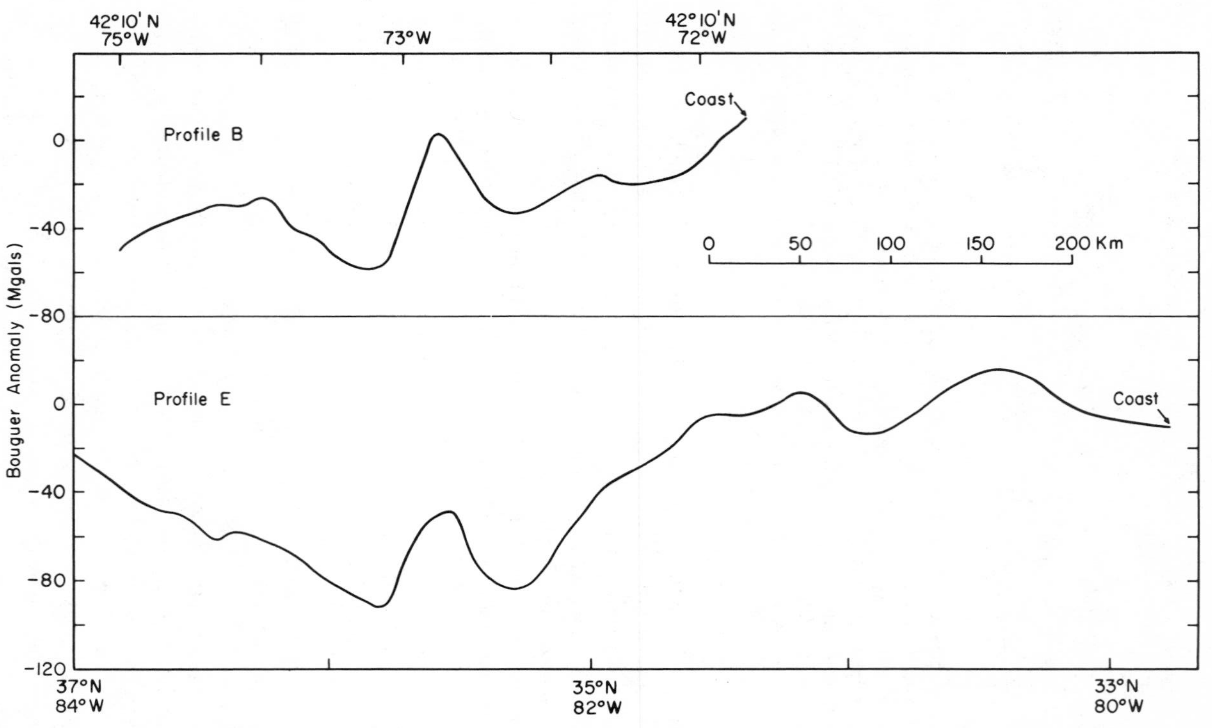

Fig. 11. Gravity profiles across central New England (B) and the Blue Ridge of the southern Appalachians (E) illustrating similarities in patterns. See Fig. 1 for location of profiles.

Appalachians? It most closely resembles the gravity high over the Precambrian rocks of the Green Mountains anticlinorium of Vermont (Fig. 9). Although the average Bouguer anomaly is about 40 mgals higher in Vermont (Fig. 11), this could be explained by a difference in average crustal thickness of some 2 km.

If the analogy between the Blue Ridge of the southern Appalachians and the Green Mountains is accepted, the boundary between the Central and Eastern Provinces should be reexamined. Indeed, perhaps two boundaries are involved rather than one (Fig. 12). Let the first boundary be the same as shown in Fig. 1 except that it is displaced to the west side of the Ashville gravity high in the southern Appalachians. This boundary is in the zone of intense Paleozoic thrusting (Logan's zone of Bird and Dewey, 1970) and is clearly a Paleozoic feature. Let the second boundary (*EF* of Fig. 12) be roughly at the western edge of the region where the Bouguer anomalies are more positive than 0 to + 10 mgals. This line strikes southwesterly along the coast of Maine into Massachusetts, then can be imagined to swing sharply west to the western edge of the Scranton gravity high. (Bouguer anomalies over much of this feature would be positive if allowance were made for the slightly less dense, but thick, sediments that overlie the basement in this region.) Then, the line continues southward roughly along the 0-mgal contour. This boundary, although highly conjectural, is interesting from several points of view.

1. In coastal Maine, it falls near the eastern limit of late Precambrian deformation as shown by Bird and Dewey (1970, Fig. 1). Farther to the south, it is along the western edge of the Scranton gravity high and the Blue Ridge of the central Appalachians. We have previously suggested that these two features are more or less continuous and that the high gravity values are in part the result of mafic volcanics of late Precambrian age. We suggest, therefore, that the boundary owes its origin to processes that occurred during late Precambrian or earliest Cambrian time. Perhaps the properties of the Precambrian crusts on either side of the boundary (*EF*) were sufficiently different so that their responses to Paleozoic deformations differed.

2. The shape of the boundary is quite similar to that of the continental shelf as defined by the 2-km isobath (Fig. 12), at least north of 34°N. This may be entirely fortuitous. The parallelism is rather good, however, and if a Precambrian age is accepted for the province boundary, we might suppose the continental margin is of equal antiquity. Drake et al. (1968) have proposed that certain structures of the margin date back at least to mid-Paleozoic time.

3. The sharp changes in trend of the boundary at the north end of Scranton gravity high could be taken to indicate a zone of right-lateral offset in Precambrian time (line *X* of Fig. 12).

Although most Paleozoic structures appear continuous across the cross-trending line, certain conditions north of this line are rather different from those to the south. In particular: (1) The Adirondack massif is just north of the line, whereas Paleozoic sediments thicken rapidly south of it. (2) A further extension of the line to the northwest falls close to the southern boundary of the Canadian

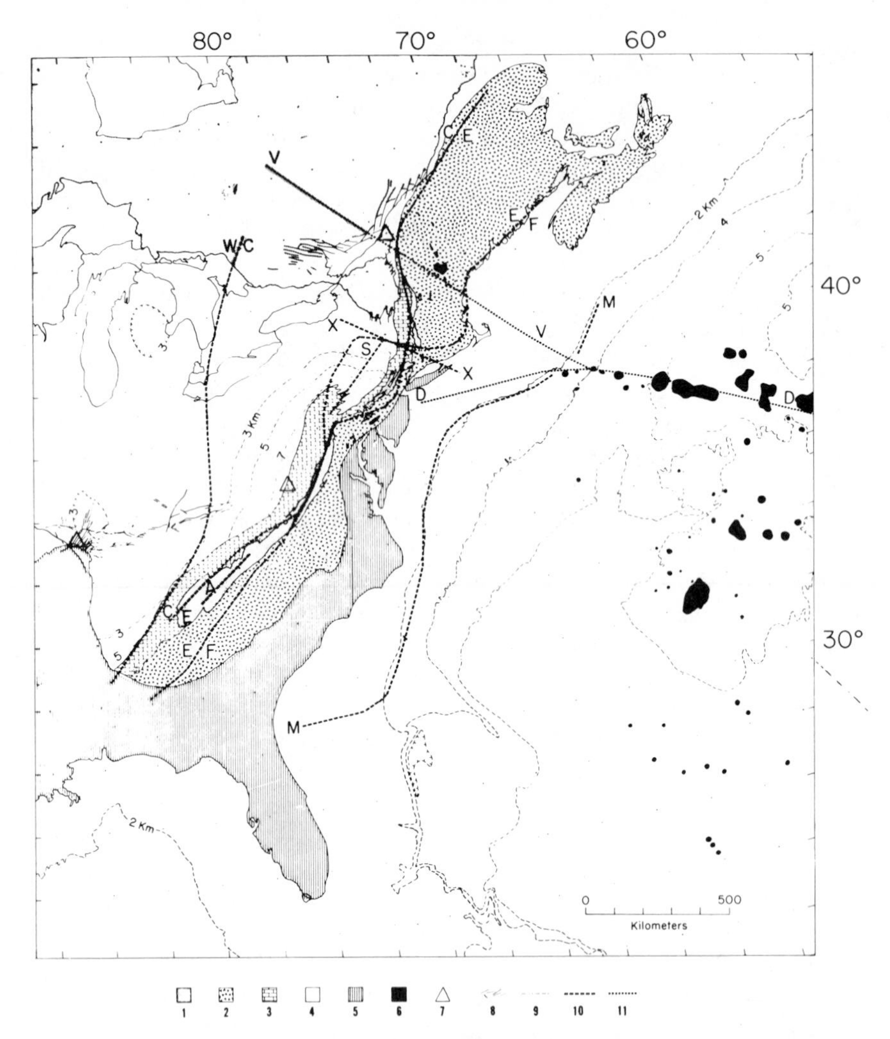

Fig. 12. Some tectonic features of eastern North America. Geologic features generalized from King (1969). (1) Precambrian rocks. (2) Paleozoic metasediments, metavolcanics, and plutonic rocks of core of the Appalachians. (3) Miogeosynclinal rocks. (4) Paleozoic platform deposits. (5) Cretaceous and later deposits of the coastal plane and Mississippi embayment. (6) On the land, post Paleozoic alkaline intrusive rocks; in the oceans, seamounts. (7) Areas of unusual concentrations of small alkaline Cretaceous and early Tertiary intrusions. (8) Fault systems in the platform deposits. (9) Depth contours: in ocean areas depth of water, in continental areas depth to Precambrian basement. (10) Province boundaries as discussed in text and line *M* which is the trace of the magnetic high that closely follows the continental margin except in the south where it changes form and turns inland (from Taylor et al., 1968). This magnetic high is continuous except at the intersection with the Kelvin seamount chain. (11) Cross-trending structural trends discussed in text.

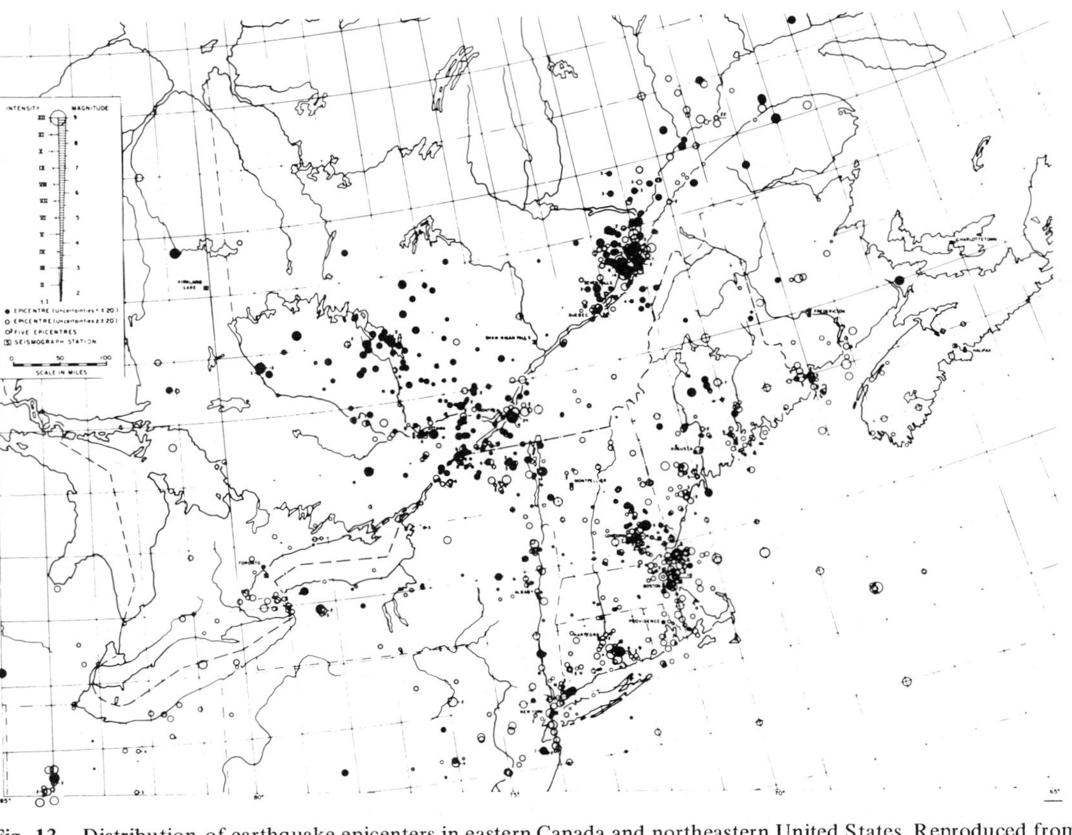

Fig. 13. Distribution of earthquake epicenters in eastern Canada and northeastern United States. Reproduced from Smith (1966).

Shield. (3) The gravity and structural trends of Vermont and Massachusetts are northerly whereas to the south the trends are northeasterly. (4) Moreover, the gravity relief is greater in southwestern New England than in northwestern New England. (5) The post-Paleozoic alkalic rocks of New England are all north of the cross-trending line.

Once one starts to draw cross-trending lines, it is difficult to stop, for there are signs of dislocations at many places in the Appalachians. The Kelvin seamount alignment (Cretaceous or older (Drake and Woodward, 1963) is particularly impressive and it is tempting to extend this line (*V* of Fig. 12) into the continent in the vicinity of the plutons of the White Mountain Magma series and the Monterigian Hills and to note the subparallelism of the faults of the Ottawa graben and possible southeasternly alignment of earthquake epicenters (Fig. 13) extending from the Canadian Shield through Montreal and Boston and out to sea. This line appears equally probable to that proposed by Drake and Woodward (1963) (line *D* of Fig. 12) or to the one that would be obtained by connecting our line *X* with the seamount alignment.

The southeasterly trending belt of epicenters is not well established. It is not difficult to imagine both a northwesterly and a northeasterly trend in the epicenters (Fig. 13). Isoseismal maps show both trends but favor the former (compare Fig. 3 and 6 of Smith, 1966). The northeasterly trend along the St. Lawrence River Valley has attracted particular attention (Hodgson, 1965; Woollard, 1969) because an alignment of small earthquakes through New York, Ohio, and Indiana suggests that the zone may extend into the New Madrid region of Missouri and adjacent states where great earthquakes have occurred. Perhaps this alignment of epicenters does indicate a tectonic zone of this extent. A northwesterly trend in the vicinity of line *V* is, however, nearly as impressive and suggests that we look more carefully for other alignments and entertain the existence of two sets of trends.

ACKNOWLEDGMENTS

We thank A. L. Baldwin, E. D. Saunders, Allan Cooper, K. A. and M. J. Goettel, and Marion Weaver for their assistance in the field and laboratory. Various aspects of the work were supported by the National Science Foundation (GA-1539, GA-947, GP-5225 and GJ-828) and by the New York State Museum and Science Service. We are also indebted to the U.S. Geological Survey for a generous loan of equipment. P. B. King, E. C. Robertson, R. F. Roy, and J. H. Sass reviewed the manuscript and suggested many improvements.

REFERENCES

Bayley, R. W., and Muehlberger, W. R., compilers, Basement Map of the United States, scale 1:2,500,000, U.S. Geological Survey, 1968.

Bean, R. J., Relation of gravity anomalies to the geology of central Vermont and New Hampshire, Bull. Geol. Soc. Am., **64,** 509-538, 1953.

Beck, A. E., and Judge, A. S., Analysis of heat flow data–1: Detailed observations in a single borehole, Geophys. J., **18,** 145-158, 1969.

Beck, A. E., and Logis, Z., Terrestrial heat flow in the Brent Crater, Nature, **201,** 383, 1964.

Billings, M. P., Thompson, J. B., Jr., and Rodgers, J., Geology of the Appalachian Highlands of east-central New York, southern Vermont, and southern New Hampshire, Geol. Soc. Am. Guidebook for field trips in New England, 65th Annual Meeting, 142 p., 1952.

Birch, F., The effects of Pleistocene climatic variations upon geothermal gradients, Am. J. Sci., **246,** 729-760, 1948.

______, Thermal conductivity, climatic variation, and heat flow near Calumet, Michigan, Am. J. Sci., **252,** 1-25, 1954.

Birch, F., Roy, R. F., and Decker, E. R., Heat flow and thermal history in New England and New York, *in* Studies of Appalachian geology: Northern and Maritime, edited by E-an Zen, W. S. White, J. B. Hadley, and J. B. Thompson, pp. 437-451, John Wiley & Sons, New York, 1968.

Bird, J. M., and Dewey, J. F., Lithosphere plate-continental margin tectonics and the evolution of the Appalachian orogen, Bull. Geol. Soc. Am., **81,** 1031-1060, 1970.

Bromery, R. W., Simple Bouguer Anomaly Map of Massachusetts, U.S. Geological Survey, Geophys. Inv. GP-612, 1967.

Bromery, R. W., and Griscom, A., Aeromagnetic and generalized geologic map of southeastern Pennsylvania, U.S. Geological Survey Geophys, Inv., GP-577, 1967.

Broughton, J. G., Fisher, D. W., Isachsen, Y. W., and Rickard, L. V., Geology of New York: A short account, Educational Leaflet 20, New York State Museum and Science Service, Albany, N.Y., 1966.

Cermak, V., Underground temperature and inferred climatic temperature of the past millennium (abstract), Trans. Am. Geophys. Union, **51**(4), 296, 1970.

Clark, S. P., Jr., Abundances of uranium, thorium and potassium, *in* Handbook of physical constants, Geol. Soc. Am., Mem. 97, pp. 521-541, 1966.

Cleary, J., and Hales, A. L., An analysis of the travel times of *P* waves to North American stations, in the distance range 28° to 82°, Bull. Seismol. Soc. Am., **56,** 467-489, 1966.

Cohee, G. V., Tectonic map of the United States, U.S. Geological Survey, Washington, D.C., 1962.

Combs, J., Terrestrial heat flow in northcentral United States, Ph.D. thesis, Massachusetts Institute of Technology, Cambridge, Massachusetts, 1970.

Combs, J., and Simmons, G., Heat flow in north central United States (abstract), Trans. Am. Geophys. Union, **51**(4), 426, 1970.

Dansgaard, W., Johnsen, S. J., Moller, J., and Langway, C. C., Jr., One thousand centuries of climatic record from Camp Century on the Greenland ice sheet, Science, **166,** 377-381, 1969.

Dennison, J. M., Effects of Tertiary instusions and the thirty-eighth parallel fracture zone in Virginia and West Virginia, Geol. Soc. Am. Abstracts with Programs, **2**(3), 205, 1970.

Denton, G. H., Porter, S. C., Neoglaciation, Sci. Am., **222,** 100-110, 1970.

Diment, W. H., Comments on a paper by E. A. Lubimova. Heat flow in the Ukrainian shield in relation to recent tectonic movements, J. Geophys. Res., **70,** 2466-2467, 1965.

______, Gravity anomalies in northwestern New England, *in* Studies of Appalachian Geology: Northern and Maritime, edited by E-an Zen, W. S. White, J. B. Hadley, and J. B. Thompson, Jr., pp. 399-413, John Wiley & Sons, New York, 1968.

Diment, W. H., and Robertson, E. C., Temperature, thermal conductivity and heat flow in a drilled hole near Oak Ridge, Tennessee, J. Geophys. Res., **68,** 5035-5048, 1963.

Diment, W. H., and Werre, R. W., Terrestrial heat flow near Washington, D.C., J. Geophys. Res., **69,** 2143-2149, 1964.

Diment, W. H., Marine, I. W., Neiheisal, J., and Siple, G. E., Subsurface temperature, thermal conductivity and heat flow near Aiken, South Carolina, J. Geophys. Res., **70**, 5635-3644, 1965a.

Diment, W. H., Raspet, R., Mayhew, M. A., and Werre, R. W., Terrestrial heat flow near Alberta, Virginia, J. Geophys. Res., **70**, 923-929, 1965b.

Dominion Observatory, Gravity Map of Canada, Gravity Division, Dominion Observatory, Ottawa, 1957.

Drake, C. L., Continental margins, *in* The Earth's Crust and Upper Mantle, Geophysical Monograph 13, edited by P. J. Hart, pp. 549-556, Amer. Geophys. Union, Washington, D.C., 1969.

Drake, C. L., and Woodward, H. P., Appalachian curvature, wrench faulting, and offshore structures, trans. N.Y. Acad. Sci., **26**, 48-63, 1963.

Drake, C. L., Ewing, J. I., and Stockard, H., The continental margin of the eastern United States, Can. J. Earth Sci., **5**(4, part 2), 993-1010, 1968.

Emiliani, C., Cenozoic climatic changes as indicated by the Stratigraphy and chronology of deep-sea cores of globeregina ooze facies, Ann. N.Y. Acad. Sci., **95**, 521-536, 1961.

Foland, K. A., Quinn, A. W., and Giletti, B. J., Jurassic and Cretaceous isotopic ages of the White Mountain magma series, Geol. Soc. Am., Abstracts with Programs, **2**(1), 19, 1970.

Griscom, A., Tectonic significance of the Bouguer gravity field of the Appalachians (abstract), Geol. Soc. Am. Special Paper 73, 163-164, 1963.

Griscom, A., and Bromery, R. W., Geologic interpretation of aeromagnetic data for New England, *in* Studies of Appalachian geology: Northern and Maritime, edited by E-an Zen, W. S. White, J. B. Hadley, and J. B. Thompson, pp. 425-436, 1968.

Hadley, J. B., Correlation of isotopic ages, crustal heating, and sedimentation in the Appalachian region, *in* Tectonics of the Southern Appalachians, edited by W. D. Lowery, pp. 33-46, V.P.I. Department of Geological Sciences, Mem. 1, Blackburg, Va., 1964.

Healy, J. H., and Warren, D. H., Explosion seismic studies in North America, *in* The Earth's Crust and Upper Mantle, Geophys. Monograph 13, edited by P. J. Hart, pp. 208-220, Am. Geophys. Union, Washington, D.C., 1969.

Herrin, E., Regional variations of *P*-wave velocity in the upper mantle beneath North America, *in* The Earth's Crust and Upper Mantle, Geophys. Monograph 13, edited by P. J. Hart, pp. 242-246, Am. Geophys. Union, Washington, D.C., 1969.

Herrin, E., and Taggert, J., Regional variations in *Pn* velocity and their effect on the location of epicenters, Bull. Sei. Soc. Am., **52**, 1037-1046, 1962.

Hodgson, J. H., There are earthquake risks in Canada, Can. Consult. Engineer, **7**, 42-51, 1965.

James, D. E., Smith, T. J., and Steinhart, J. S., Crustal structure of the Middle Atlantic States, J. Geophys. Res., **73**, 1983-2007, 1968.

Jessop, A. M., Three measurements of heat flow in eastern Canada, Can. J. Earth Sci., **5**, 61-68, 1968.

Joyner, W. B., Heat flow in Pennsylvania and West Virginia, Geophysics, **25**, 1229-1241, 1960.

______, Gravity in north central New England, Bull. Geol. Soc. Am., **74**, 831-858, 1963.

Judge, A. S., and Beck, A. E., An anomalous heat flow layer at London, Ontario, Earth Planet. Sci. Letters, **3**, 167-170, 1967.

Kane, M. F., Bromery, R. W., Gravity anomalies in Maine, *in* Studies of Appalachian geology: Northern and Maritime, edited by E-an Zen, W. S. White, J. B. Hadley, and J. B. Thompson, pp. 415-423, John Wiley & Sons, New York, 1968.

Katz, S., Seismic study of crustal structure in Pennsylvania and New York, Bull. Seis. Soc. Am., **75**(2), 303-325, 1955.

King, P. B., The Tectonics of Middle North America, Princeton Univ. Press, Princeton, N.J., 1951.

______, The Evolution of North America, Princeton Univ. Press, Princeton, N.J., 1959.

______, Further thoughts on the tectonic framework of the southeastern United States, *in* Tectonics of the Southern Appalachians, edited by W. D. Lowery, pp. 5-32, V.P.I. Department of Geological Sciences, Mem. 1, Blacksburg, Va., 1964.

______, Tectonic map of North America, U.S. Geological Survey, Washington, D.C., 1968.

______, The tectonics of North America–A discussion to accompany the Tectonic Map of North America, scale 1:5,000,000, U.S. Geol. Survey Prof. Paper 628, 1-94, 1969.

Lachenbruch, A. H., Periodic heat flow in a stratified medium with application to permafrost problems, U.S. Geol. Survey Bull. 1083-A, 1-36, 1959.

______, Preliminary geothermal model of the Sierra Nevada, J. Geophys. Res., **73,** 6977-6990, 1968.

Lachenbruch, A. H., and Brewer, M. C., Dissipation of the temperature effect of drilling a well in arctic Alaska, U.S. Geol. Surv. Bull. 1083-C, 73-109, 1959.

Lachenbruch, A. H., and Marshall, B. V., Heat flow in the Arctic, Arctic, **22**, 300-311, 1969.

Lachenbruch, A. H., Brewer, M. C., Greene, G. W., and Marshall, B. V., Temperatures in permafrost, *in* Temperature–Its measurement and control in science and industry, vol. 3, part I, pp. 791-803, Reinhold Publishing Corporation, New York, 1962.

Lee, W. H. K., and Uyeda, S., Review of heat flow data, *in* Terrestrial heat flow, Geophysical Monograph 8, edited by W. H. K. Lee, pp. 87-190, Amer. Geophys. Union, Washington, D.C., 1965.

Leith, T. H., Heat flow at Kirland Lake, Trans. Am. Geophys. Union, **33,** 435-443, 1952.

Lubimova, E. A., Heat flow in the Ukrainian shield in relation to recent tectonic movements, J. Geophys. Res., **69,** 5277-5284, 1964.

Lyons, J. B., and Faul, H., Isotope geochonology of the northern Appalachians, *in* Studies of Appalachian geology: Northern and Maritime, edited by E-an Zen, W. S. White, J. B. Hadley, and J. B. Thompson, Jr., pp. 305-318, John Wiley & Sons, New York, 1968.

McGinnis, L. D., Tectonics and gravity field in the continental interior, J. Geophys. Res., **70**(2), 317-332, 1970.

Misener, A. D., Thompson, L. G. D., and Uffen, R. J., Terrestrial heat flow in Ontario and Quebec, Trans. Am. Geophys. Union, **32,** 729-738, 1951.

Pakiser, L. C., and Steinhart, J. S., Explosion seismology in the western hemisphere, *in* Research in geophysics, Vol. 2, Solid Earth and interface phenomena, edited by H. Odishaw, pp. 123-145, M.I.T. Press, Cambridge, Mass., 1964.

Revetta, F. A., A regional gravity survey of New York and eastern Pennsylvania, Ph.D. thesis, University of Rochester, Rochester, N.Y., 1970.

Revetta, F. A., and Diment, W. H., A regional gravity survey of New York and eastern Pennsylvania, Geol. Soc. Am., Abstracts with Programs for 1969, pt. 7, 187, 1969.

Rodgers, J., Basement and no-basement hypotheses in the Jura and the Appalachian Valley and Ridge, *in* Tectonics of the Southern Appalachians, edited by W. D. Lowery, pp. 71-80, V.P.I. Department of Geological Sciences, Mem. 1., Blacksburg, Va., 1964.

______, Chronology of tectonic movements in the Appalachian Region of eastern North America, Am. J. Sci., **265,** 408-427, 1967.

______, The eastern edge of the North American continent during the Cambrian, *in* Studies of Appalachian geology: Northern and Maritime, edited by E-an Zen, W. S. White, J. B. Hadley, and J. B. Thompson, Jr., pp. 141-149, John Wiley & Sons, New York, 1968.

Root, S. I., Structure of the northern terminus of the Blue Ridge in Pennsylvania, Bull. Geol. Soc. Am., **81,** 815-830, 1970.

Roy, R. F., Blackwell, D. D., and Birch, F., Heat generation of plutonic rocks and continental heat flow provinces, Earth Planet. Sci. Letters, **5,** 1-12, 1968a.

Roy, R. F., Decker, E. R., Blackwell, D. D., and Birch, F., Heat flow in the United States, J. Geophys. Res., **73,** 5207-5221, 1968b.

Sass, J. H., Killeen, P. G., Mustonen, E. D., Heat flow and surface radioactivity in the Quirke Lake syncline near Elliot Lake, Ontario, Canada, Can. J. Earth Sci., **5,** 1417-1428, 1968.

Saull, V. A., Clark, T. H., Doig, R. P., and Butler, R. B., Terrestrial heat flow in the St. Lawrence lowland of Quebec, Can. Mining Met. Bull., **55**, 598, 92-95, 1962.

Simmons, G., Gravity survey and geologic interpretation, northern New York, Geol. Soc. Am. Bull., **75**, 81-98, 1964.

Smith, W. E. T., Earthquakes of eastern Canada and adjacent areas, Dominion Observatory Publ., **32**, 2, 87-121, 1966.

Solomon, S. C., and Toksöz, M. F., Lateral variation of attenuation of *P* and *S* Waves beneath the United States, Bull. Seis. Soc. Am., **60**, 819-838, 1970.

Steinhart, J. S., and Smith, T. J., editors, *in* The Earth beneath the continents: A volume of geophysical studies in honor of Merle A. Tuve, Geophys. Monograph 10, Amer. Geophys. Union, Washington, D.C., 1966.

Taylor, P. T., Zietz, I., and Dennis, L. S., Geologic implications of aeromagnetic data for the eastern continenal margin of the United States, Geophysics, **33**, 755-780, 1968.

U.S. Air Force, Transcontinental Geophysical Survey (35°-39°N) Bouguer Gravity Map from 74° to 87°W Longitude, compiled by U.S. Air Force Aeronautical Chart and Information Center, Misc. Geol. Inv. Map I-535-B, U.S. Geological Survey, Washington, D.C., 1968.

U.S. Geological Survey, Aeromagnetic map of the Harrisburg-Scranton area, northeastern Pennsylvania, U.S. Geol. Survey Map GP-669, 1969.

Warren, D. H., Transcontinental Geophysical Survey (35°–39°N) Seismic refraction profiles of the crust and upper mantle from 74° to 87°W longitude, Misc. Geol. Inv. Map. I-535-D, U.S. Geol. Survey, Survey, Washington, D.C., 1968a.

______, Transcontinental Geophysical Survey (35°–39°N), Seismic refraction profiles of the crust and upper mantle from 87° to 100°W longitude, Misc. Geol. Inv. Map I-535-D, U.S. Geol. Survey, Washington, D.C., 1968b.

Wise, D. U., and Werner, M. L., Tectonic transport domains and the Pennsylvania elbo of the Appalachians (abstract), Trans. Am. Geophys. Union, **50**(4), 313, 1969.

Wilden, R., Reed, J. C., and Carlson, J. E., Transcontinental Geophysical Survey (35°–39°N), Geological Map from the east coast of the United States to 87°W Longitude, Misc. Geol. Inv. Map I-535–C, U.S. Geol. Survey, Washington, D.C., 1968.

Woollard, G. P., Tectonic activity in North America as indicated by earthquakes, *in* The Earth's Crust and Upper Mantle, Geophys. Monograph 13, edited by P. J. Hart, pp. 125-133, Amer. Geophys. Union, Washington, D.C., 1969.

Woollard, G. P., and Joesting, H. R., Bouguer gravity anomaly map of the United States, U.S. Geol. Survey, 1964.

Wollenberg, H. A., and Smith, A. R., Radiogenic heat production in prebatholithic rocks of the central Sierra Nevada, J. Geophys. Res., **75**, 431-438, 1970.

Zen, E-an, Nature of the Ordovician orogeny in the Taconic area, *in* Studies of Appalachian Geology: Northern and Maritime, edited by E-an Zen, W. S. White, J. B. Hadley, and J. B. Thompson, Jr., pp. 129-140, John Wiley & Sons, New York, 1968.

Zietz, I., Stockard, H. P., and Kirby, J. R., Transcontinental Geophysical Survey (35°-39°N) magnetic and bathymetric map from 74° to 87°W Longitude, *in* Misc. Geol. Inv. Map I-535-A, U.S. Geol. Survey, Washington, D.C., 1968.

PHYSICAL PROPERTIES

21

PATTERNS IN ELASTIC CONSTANTS OF MINERALS IMPORTANT TO GEOPHYSICS

by **Orson L. Anderson**

The number and kind of crystals in mineralogy of potential interest to geophysical and geochemical theories of the earth is so large that it will be many years before a complete catalog of physical properties is available. While geophysicists await the completed catalog, empirical patterns of the physical constants have been proposed so that estimates can be made on the basis of a minimum amount of information (such as ambient density, chemical composition, and crystallographic structure). These empirical patterns have at least three applications: (1) they encourage geophysical speculation in the absence of data, (2) they help in the design of expensive experiments, and (3) they focus on the proper minerals to acquire for the impending experiments.

This paper gives: (1) a historical review of the concepts in patterns of elastic constant data starting with Birch's law (the linear relationship between ambient V_p and ambient density); (2) the nature of optimum experiments; (3) the impact of the present experimental situation upon theories of deformation at very high pressure.

BIRCH'S LAW

The connection between the seismic properties of the earth's interior and geochemical theories of the earth is made by use of laboratory data on elastic constants of pertinent minerals and rocks. It becomes important, therefore, to find how the elastic constants depend upon structural and atomic parameters. Conceivably many variables could affect the values of the elastic constants of a mineral.

Today Francis Birch is usually credited for demonstrating that only two variables are important for describing the compressional sound velocity of rocks and minerals. These variables are the density and the mean atomic weight (Birch, 1961). He suggested that the compressional sound velocity of silicates and oxides (at ambient pressure) is given by the empirical formula

$$V_p = \mathbf{a} + \mathbf{b}\rho_0 \tag{1}$$

and this relationship has come to be known as *Birch's law*. Least square solutions for the fitting of Eq. (1) to different classes of rocks indicated that **a** decreases as the content of iron increases. Birch's parameters for 15 granites are $V_p = -0.30 + 2.15\rho_0$ (km/sec), and for iron-rich rocks (ecolgite, diabase, and gabbros), $V_p = 1.00 + 2.67\rho_0$ (km/sec).

Equation (1) or its equivalent in other dimensions has been verified by many authors; and the utility of *Birch's law* as a tool in geophysical theories is widely recognized.

A linear relationship between V_p and ρ_0 was proposed by Nafe and Drake (1957) for ocean sediments, and their relation is still used in submarine geology. Equations like (1) are valid for highly porous-silicate materials.

The linear relationship between the sound velocity and density also holds for the earth's interior. (In this case silicates are under high pressure and temperature.) Figure 1 shows how the data of density and sound velocity (from one solution of the earth's interior) tend to cluster around a straight line such as given by (1). The value of (1) to theories of the earth's interior is obvious.

While (1) is well known in geophysical literature, it is not typical of empirical formulas found in the solid-state literature. For a sequence of semiconductor compounds having the same crystal structure, a well-known relationship exists between the bulk modulus and the inverse fourth power of the lattice constant. This relationship was apparently found by Bridgman (1923) who showed that to a good approximation the bulk modulus varies inversely as the product of the specific volume and interatomic distance

$$K_0 \approx \frac{1}{V_0 r_0} = \frac{1}{(r_0)^4} \tag{2}$$

This rule depends upon the assumption of an energy inverse with distance between the nearest neighbors. An example of how this rule is applied is given by Fig. 6 of the article by Cline et al. (1967).

Equation (2) is valid only for a particular function of the repulsive potential. If $\nu(r)$ is the repulsive potential, the actual relationship between K_0 and r_0 is given

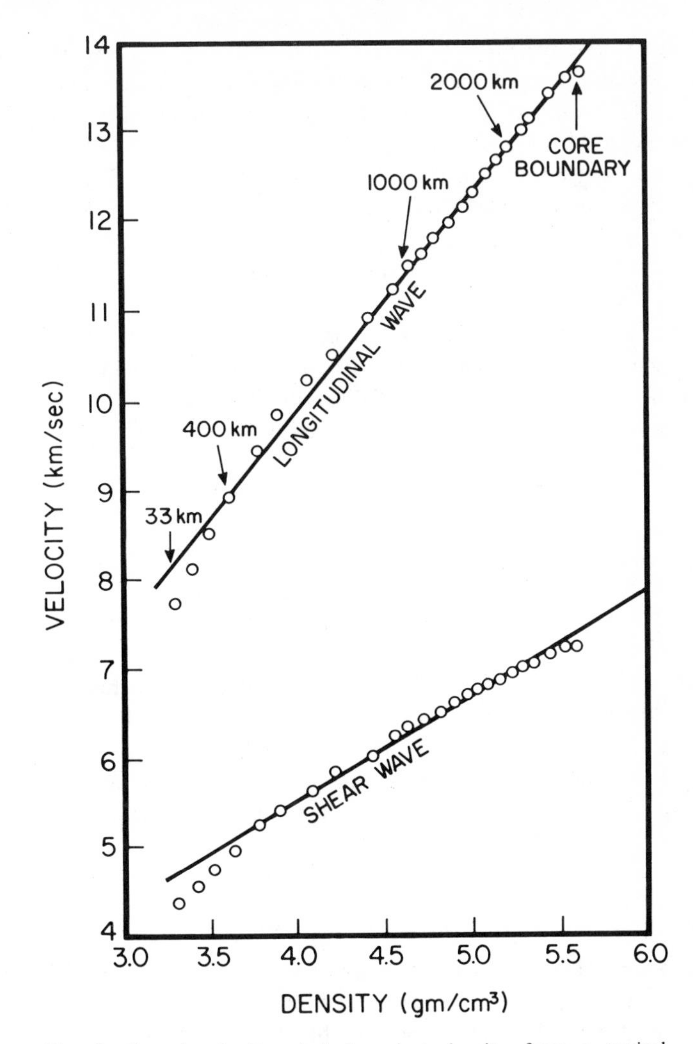

Fig. 1. Sound velocity plotted against density from a typical seismological solution of the mantle.

by

$$K_0 = \frac{2}{9V_0} \frac{A}{r_0} \left[1 + \frac{v''(r_0) r_0}{2v'(r_0)} \right] \tag{3}$$

where the attractive potential is in turn given by $-A/r_0$. Few repulsive potentials yield the result that the term in the brackets is constant. One such function is $v(r) = b/r^n$. On the other hand, the quantity in the brackets of (3) is a slowly varying function of r_0, so that over a limited range of r_0, (2) holds. If the attractive potential $-A/r_0$ is coulombic, then A must be equal to $A_m Z_1 Z_2 e^2$, where A_m is the Madelung constant and $Z_1 Z_2$ is the valence.

The value of A_m is essentially independent of structure, if the attractive energy is defined as $\phi_a = -A_m Z_1 Z_2 e^2/r_0$, so that the variation of A in (3) is essentially found in the valence product $Z_1 Z_2$. Taking the logarithm of (3), we have (Anderson and Nafe, 1965):

$$ln K_0 = ln Z_1 Z_2 - x ln V_0 + \text{constant} \tag{4}$$

where $x = 4/3 \pm \epsilon$, and ϵ is a small number depending upon the exact nature of the repulsive potential $\nu(r)$. Empirically x is very close to 1 for alkali halides, fluorides, and for oxides. This is shown in Fig. 2. For the three classes of ionic solids shown in Fig. 2, we have the rule that

$$\frac{K_0 V_0}{S^2 Z_1 Z_2} = \text{constant} \tag{5}$$

where S is now introduced to account for the fact that the actual attraction between atoms is controlled by SZ, in place of Z, and where S, the ionicity, is less than unity. From Fig. 2, it can be determined that S for the oxides is about 70% of that of the ionicity of the alkali halides (although other definitions give the same value of S for alkali halides and oxides).

The units of V_0 must be consistent with the definitions of bulk modulus and r_0. This means that V_0 is the molar volume of an atomic pair. That is

$$V_0 = \frac{2M/p}{\rho_0} \tag{6}$$

where M/p is the mean atomic weight. If (6) is used for oxides of a more complicated nature, (5) is easily applied to demonstrate the relationship between K_0 and V_0. Figure 3 shows that the linear relationship

$$K_0 \approx \frac{\rho_0}{(M/p)\, r_0} \tag{7}$$

applies to oxides.

It is seen that K_0 does indeed increase linearly with reciprocal volume, as long as the structure is constant and the "average" valence product is constant. Equation (7) applies to a series of oxides in which the mass and the density both change.

Now the essential difference between data plotted in Fig. 2 and Birch's law, is that Birch's law applies to the special case where the mean average mass M/p remains constant. To show this difference, a dashed line is plotted in Fig. 3,

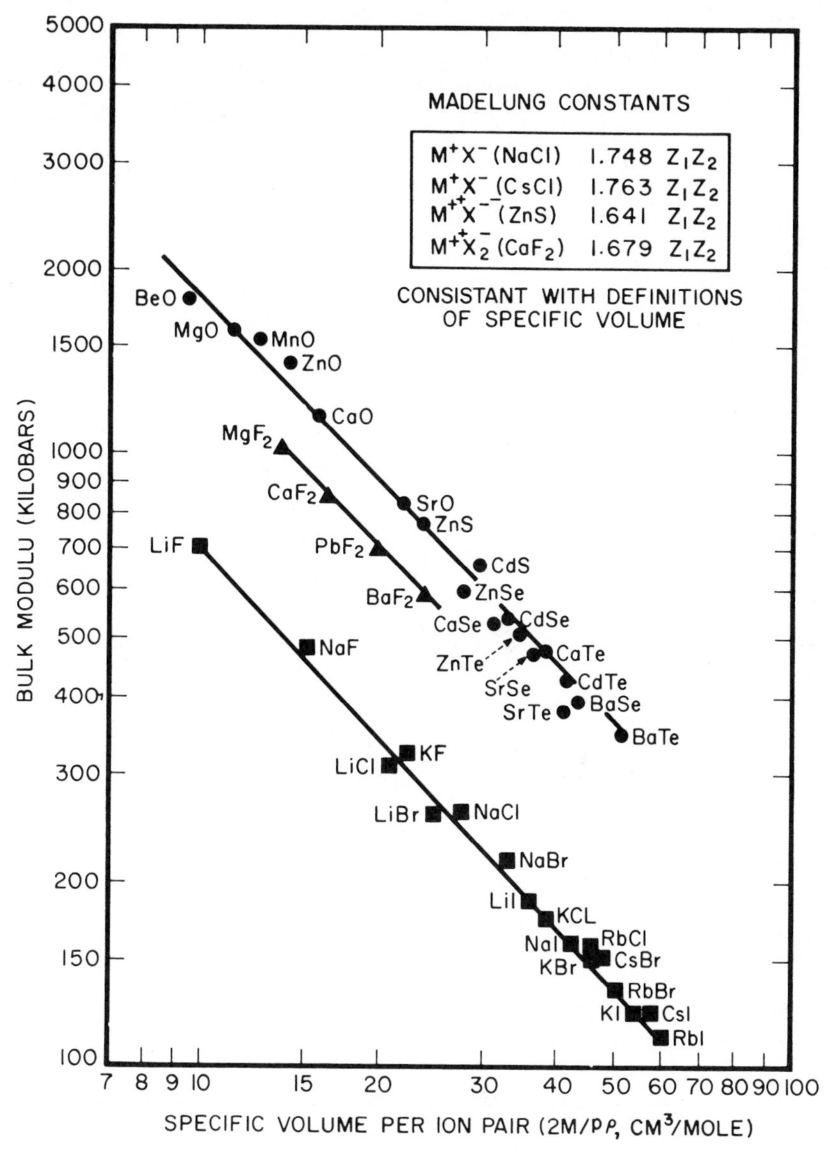

Fig. 2. The bulk modulus versus specific volume for diatomic ionic solids. The slope of the solid lines is -1.

which represents the data of oxides that have nearly constant M/p (very close to 21). The dashed line represents a fourth power law. That is to say, Birch's law, in the dimensions of K_0 and V_0 instead of V_p and ρ_0, is represented closely by $K_0 \approx V_0^{-4}$. On the other hand, oxide data that are plotted in an isostructural series are represented by $K_0 \approx V_0^{-1}$ (Anderson and Soga, 1968).

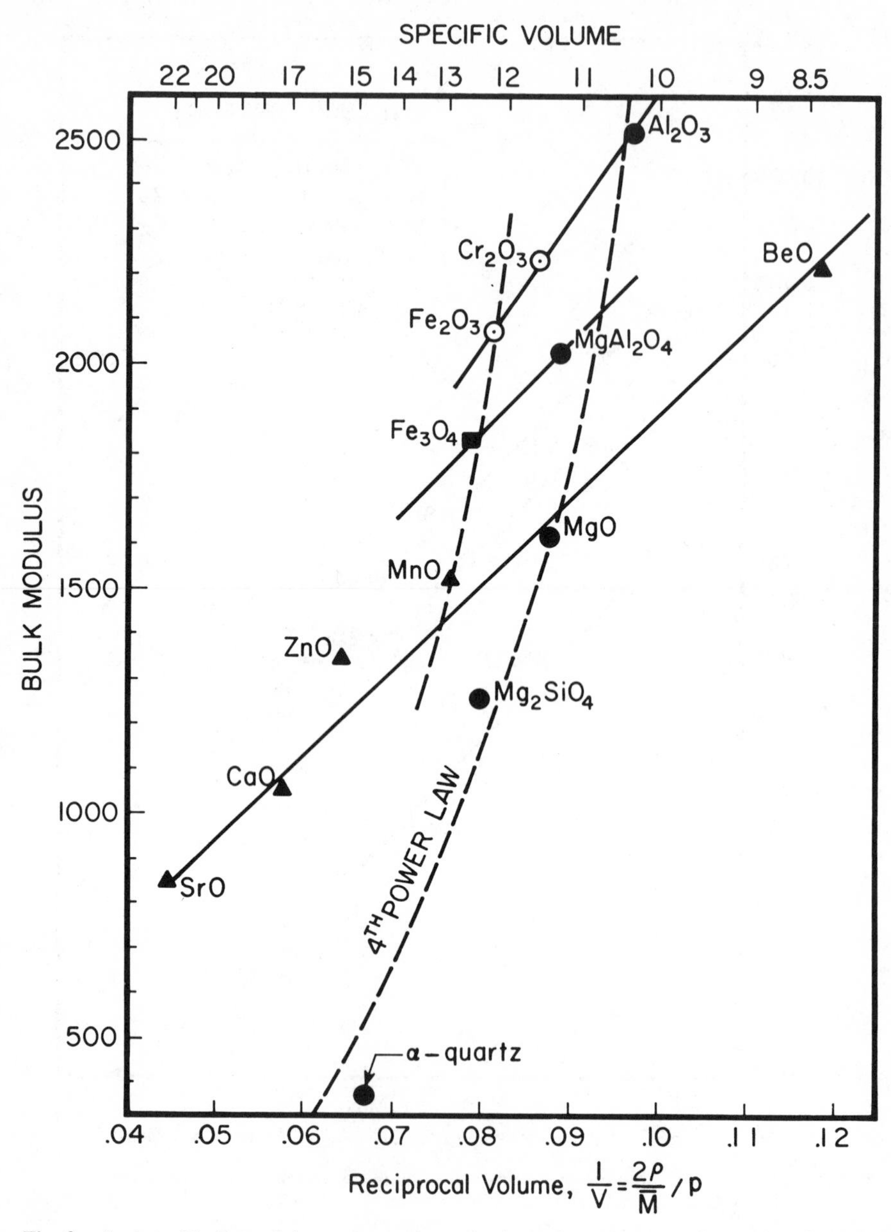

Fig. 3. A plot of bulk modulus versus reciprocal volume for oxides on a linear plot. Oxides with the same structure fall on straight lines.

It is much easier to represent power functions on log-log plots than on linear plots. The distinction between a series of oxides that are isostructurally related and a series of oxides that are related by constant mass is shown schematically in Fig. 4. Data of ionic solids is replotted in Fig. 5 to show the distinction between the two modes. The dashed lines represent sequences at common mean atomic weight. The primary data is listed in Table 1.

A few solids fail to fit this neat picture. For geophysics, the important exception is CaO (Simmons, 1964). While the value of CaO falls on the line for the diatomic solids in Fig. 2, we would expect it to reverse its relative position with ZnO due to its mass. Nevertheless, in spite of exceptions, the patterns shown in Figs. 4 and 5 are remarkable for their simplicity.

If we consider a sequence of oxides with a common value of M/p, then we have for the bulk sound velocity (Anderson and Soga, 1968)

$$V_K = \sqrt{\frac{K_0}{\rho_0}} = \frac{\text{constant}}{(M/p)^2} \rho_0^{1.5} \tag{8}$$

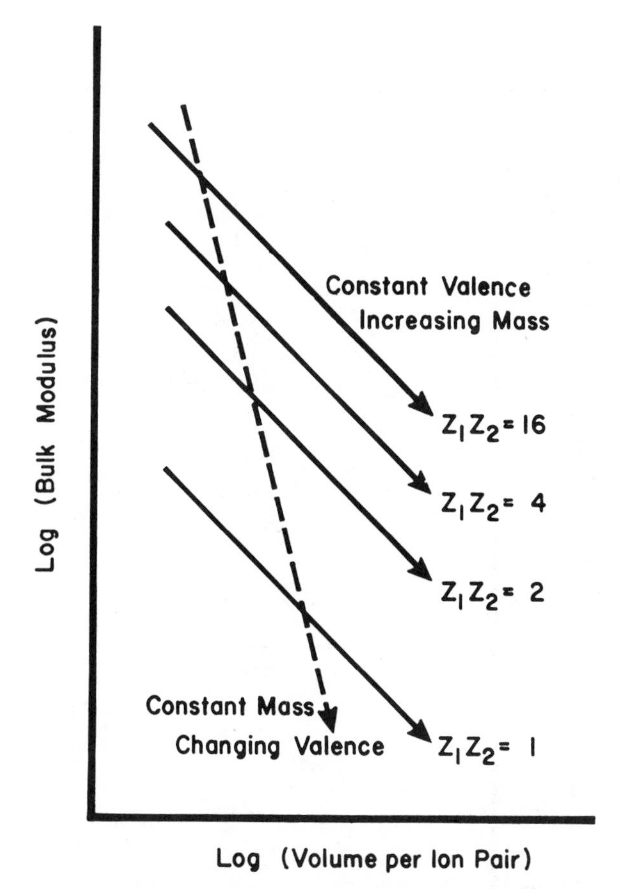

Fig. 4. Schematic plot of ln K_0 – ln V_0 showing the two principles of classifying a sequence of solids. The solid lines are isostructural sequences (slope = – 1). The dashed line is a sequence at constant M/p (slope = – 4).

This rule is empirically indistinguishable from the arbitrary straight line (analogous to Birch's law).

$$V_K = a + b\rho_0$$

in the range of density $\rho_0 = 2.0 - 4.0$.

The seismic parameter $\phi = K_0/\rho_0$, is given from the above by

$$\phi = \frac{\text{constant}}{(M/p)^4}\rho^{3.0}$$

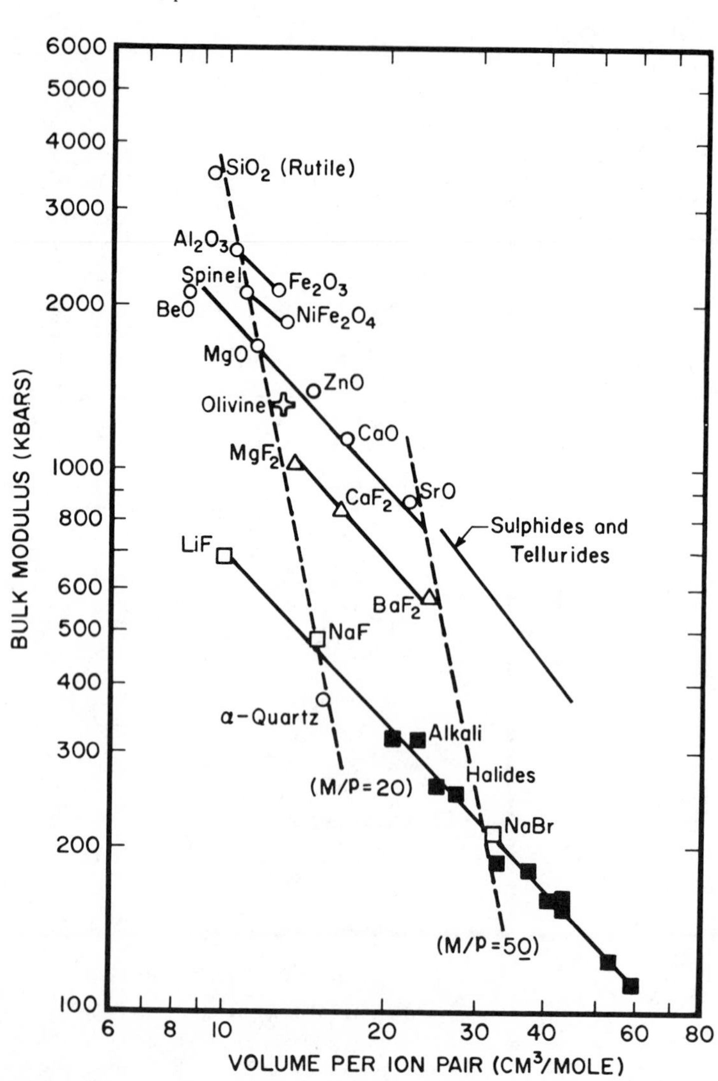

Fig. 5. Elastic constant data illustrating the principle described in Fig. 4.

Table 1. The effect of specific volume ($2M/p\rho$) upon the bulk moduli of oxides and germanates

Solid	Crystal structure	Density ρ (gm/cm^3)	Mean atomic weight	Specific volume $V_0 = 2M/p\rho$ (cm^3/mole)	Bulk modulus K_0 (kb)	$Z_c Z_a$	$K_0 V_0/(2Z_c Z_a e^2)$
Al_2O_3	Corundum	3.972	20.39	10.27	2,521	6	0.0936
Cr_2O_3		5.245	30.40	11.59	2,240	6	0.0938
Fe_2O_3		5.254	31.94	12.16	2,066	6	0.0908
$MgAl_2O_4$	Spinel	3.619	20.36	11.25	2,020	5.33	0.0924
$NiFe_2O_4$		5.313	33.49	12.61	1,823	5.33	0.0935
MgO	Rocksalt	3.583	20.16	11.25	1,622	4	0.0989
CaO		3.285	28.84	17.07	1,059	4	0.0980
NiO		6.810	37.35	10.97	1,900	4	0.1129
MnO		5.45	35.47	13.00	1,540	4	0.1175
SrO		4.66	51.81	22.24	867	4	0.1220
BeO	Wurtzite	3.000	12.51	8.34	2,201	4	0.995
ZnO		5.624	40.68	15.46	1,394	4	0.1168
SiO_2	Rutile	4.280	20.03	9.36	3,600	8	0.913
TiO_2		4.264	26.63	12.49	2,158	8	0.0730
Mg_2SiO_4	Olivine	3.224	20.12	12.48	1,286	8	0.0435
α-SiO_2	α-quartz	2.648	20.03	15.13	377	8	0.0155
GeO_2	Vitreous	3.629	34.86	19.21	239	8	0.0125
GeO_2	Quartz	4.28	34.86	16.29	350	8	0.0154
GeO_2	Rutile	6.277	34.86	11.17	1,950	8	0.0591

Inverting the above, we have the seismic equation of state (Anderson, 1967) in one of its forms:

$$\rho_0 = A\phi^{1/3} \tag{9}$$

which is a particularly useful relationship for finding the density distribution from the seismic velocity. Equations (8) and (9), as well as the fourth power law between K_0 and V_0, are all equivalent forms of Birch's law, except in the former the bulk modulus is used for the elastic constant.

A TOUR AROUND THE PERIODIC TABLE

It would be very satisfying if some fundamental law of ionic bonding could be found which would explain the behavior of oxides and alkali halides, as shown in Fig. 5. The difficulty is that this classification is not unique with ionic solids. In fact, it appears to hold for all elements that solidify.

Let us consider the value of x in the equation

$$K_0 V_0^x = \text{constant} \tag{10}$$

for certain sequences in the periodic table. As a beginning, take the alkali metals sequence, Li, Na, K, Rb, and Cs. The data for K_0 and V_0 are shown in Table 2, and the data plotted in Fig. 6.

The best value of x in (10) is 0.9, but little error is made if $x = 1$ is chosen. Thus, the same empirical law holds for alkali metals as for alkali halides.

Thus, we conclude that the empirical law $K_0 V_0$ = constant is general and does not depend upon the assumption that coulombic electrostatic forces hold atoms together.

We might inquire about the expected behavior for the covalent types of elements and compounds, where the ionic approximation of binding should not hold very well. In other words, what value of x in (10) will characterize the covalent compounds?

In Table 3, we show the data for elements of group 4 and some III-V and

Table 2. Volume-modulus data of the alkali metals

Element	Atomic weight	Density (gm/cm^3)	Specific volume (cm^3/mole)	Bulk modulus (kb)
Li	6.94	0.54	25.6	118
Na	22.96	0.97	47.3	69
K	39.10	0.86	91.0	33
Rb	85.47	1.53	112.0	30
Cs	132.9	1.87	142.0	21

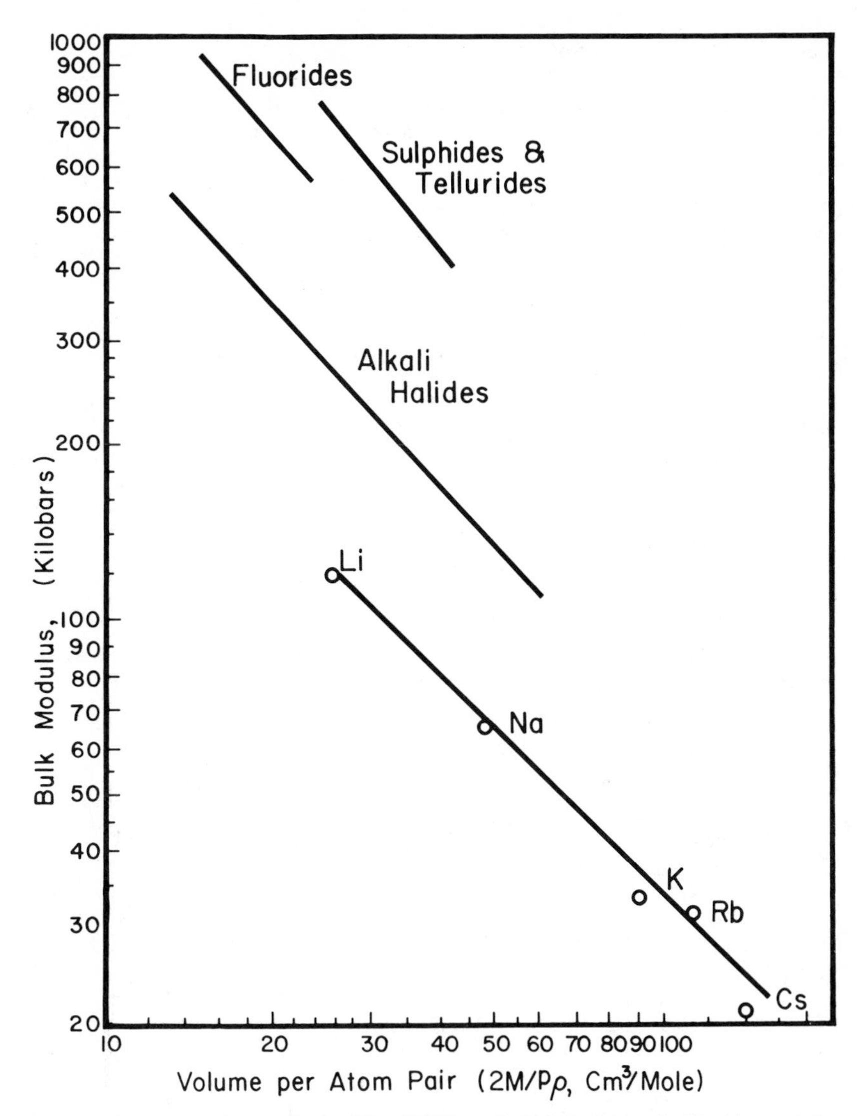

Fig. 6. The bulk modulus and volume relation of the alkali metals. The slope of the metal sequence is parallel to those of ionic solids.

IV-IV compounds. (The data for the elastic constants come from references listed in Appendix A.) These data are plotted in Fig. 7. It turns out that the most consistent value for x is 4/3.

The value 4/3 is not much larger than unity, and is consistent with an assumption of an inverse power law of attraction. We conclude that covalent bonding need not necessarily change the assumptions that the attractive potential varies inversely with r.

Table 3. Elastic constants of covalent solids

	Elastic constants c_{11} (Kb)	c_{12}	c_{44}	Bulk modulus	Shear modulus	Density (gm/cm^3)	$2M/p$
Diamond	10,760	1,250	5,758	4,420	5,334	3.512	6.82
TiC	5,245	980	1,809	2,401	1,929	4.93	12.14
ZrC	4,720	987	1,593	2,231	1,635	6.66	15.30
SiC	4,105	1,643	1,938	2,464	1,900	3.32	12.10
UC	3,200	862	650	1,640	510	13.63	19.2
Si	1,657	639	831	978	665	2.331	24.05
Ge	1,293	485	672	754	557	5.291	27.40
GaAs	1,188	532	594	754	466	5.307	27.25
GaSb	897	412	448	564	342	5.619	34.08
InSb	672	367	302	469	230	5.789	40.89
AlSb	894	442	415	592	325	4.360	34.11
InAs	833	453	396	572	305	5.690	33.30

We might expect that some kind of law, similar to (10) would prevail for the second and third group elements in the periodic table. Table 4 shows the data. Some of the elements are missing (Ra, Ga, Sc, Ac), because of a lack of information on bulk modulus.

We find that the second group elements fall into two subgroups, group 2a,

Table 4. Volume-modulus data of group 2 and 3 of the periodic table

	Atomic number	Element	Atomic weight	Density	Specific volume	Bulk modulus
Group	4	Be	9.01	1.848	9.75	1,023
2b	30	Zn	65.37	7.134	18.33	610
	48	Cd	112.4	8.749	25.69	556
Group	4	Be	9.01	1.848	9.75	1,023
2a	12	Mg	24.32	1.799	26.5	361
	20	Ca	40.08	1.538	52.2	155
	38	Sr	87.62	2.639	66.5	118
	56	Ba	137.34	3.62	75.8	105
Group	5	B	10.82	2.34	9.24	1,820
3a	13	Al	26.98	2.69	20.00	738
	49	In	114.82	7.30	31.42	419
	81	Tl	204.37	11.85	34.5	366
Group	5	B	10.82	2.34	9.24	1,820
3b	39	Y	88.90	4.472	39.5	373
	57	La	138.91	6.15	45.5	248
	71	Lu	175	9.81	35.67	419

Note.–Data from Gschneidner (1964).

composed of Be, Mg, Ca, Sr, Ba, and group 2b, composed of Be, Zn, Cd. We have:

$$K_0 V_0^{1.15} = \text{constant, group } 2a \tag{11}$$

and

$$K_0 V_0^{0.8} = \text{constant, group } 2b \tag{12}$$

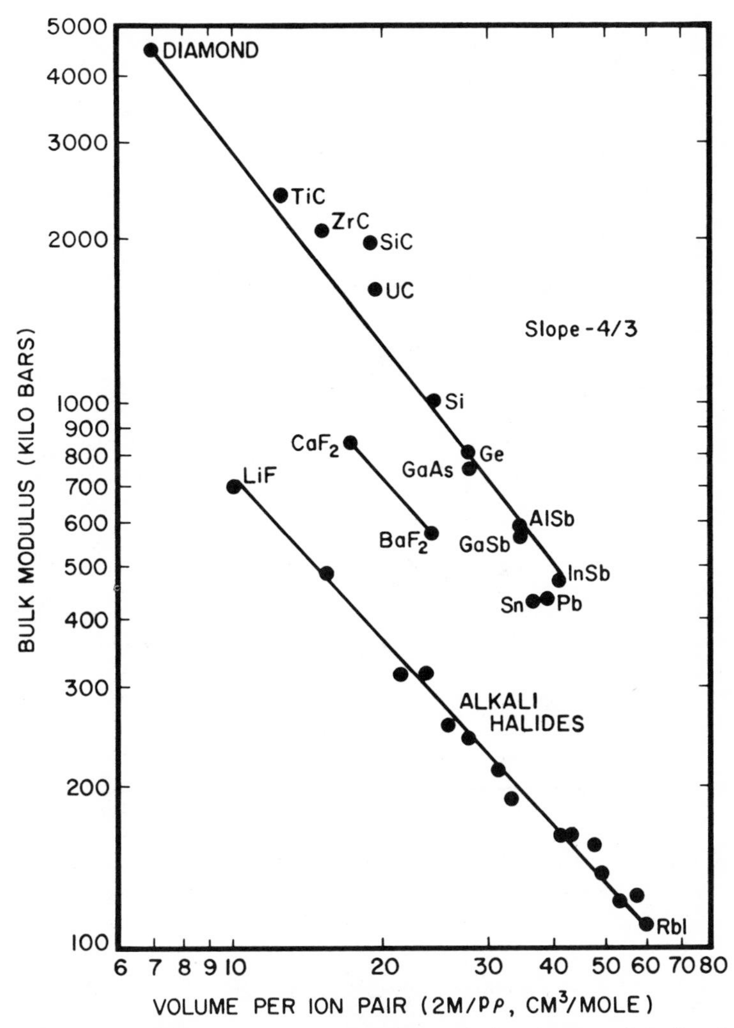

Fig. 7. The bulk modulus of covalent solids, including elements of the fourth and fifth groups, IV-IV compounds and III-V compounds. The slope of the line is – 4/3. Also shown are lines for alkali halides and CaF_2-BaF_2.

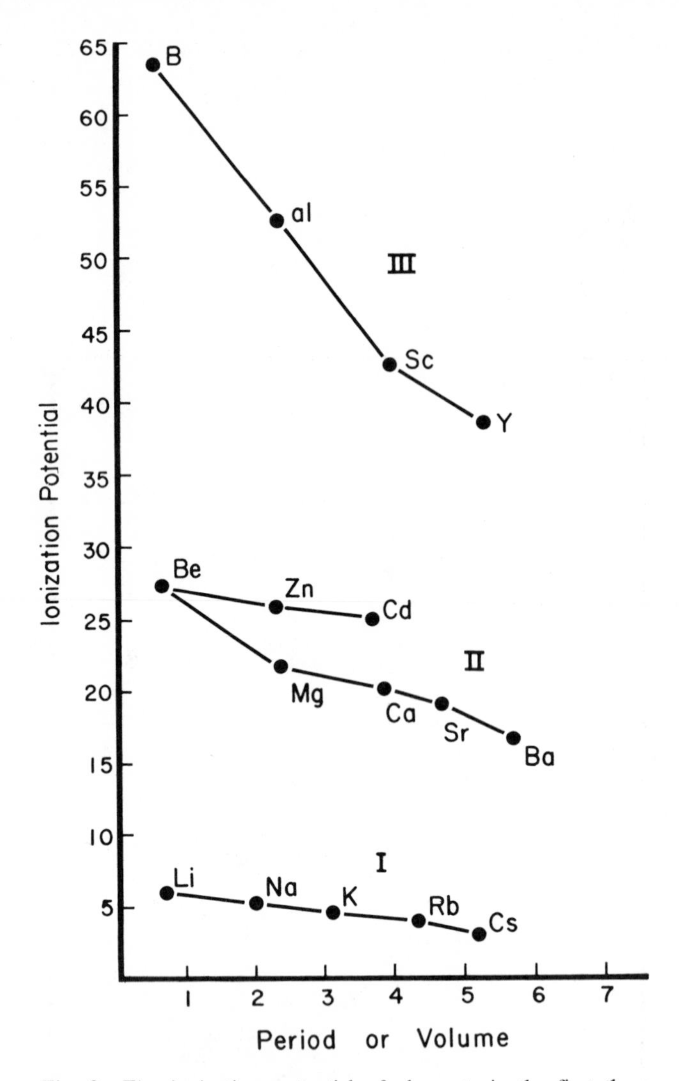

Fig. 8. The ionization potential of elements in the first three groups of the periodic table.

This is consistent with many chemical properties. On the other hand, the third group elements do not fall into two subgroups. They are characterized by the expression

$$K_0 V_0^{1.15} = \text{constant, groups } 3a \text{ and } 3b \tag{13}$$

The variation of bulk modulus with specific volume corresponds to the variation of ionization potential with period number. Going down a column in the periodic table, one finds a steady decrease in ionization potential as shown in Fig. 8, resembling the variation of bulk modulus with volume. We certainly expect the strength of the bond, as measured by its spring constant, to have some relation to the energy required to separate the first electron, or valence electron, from the ground state.

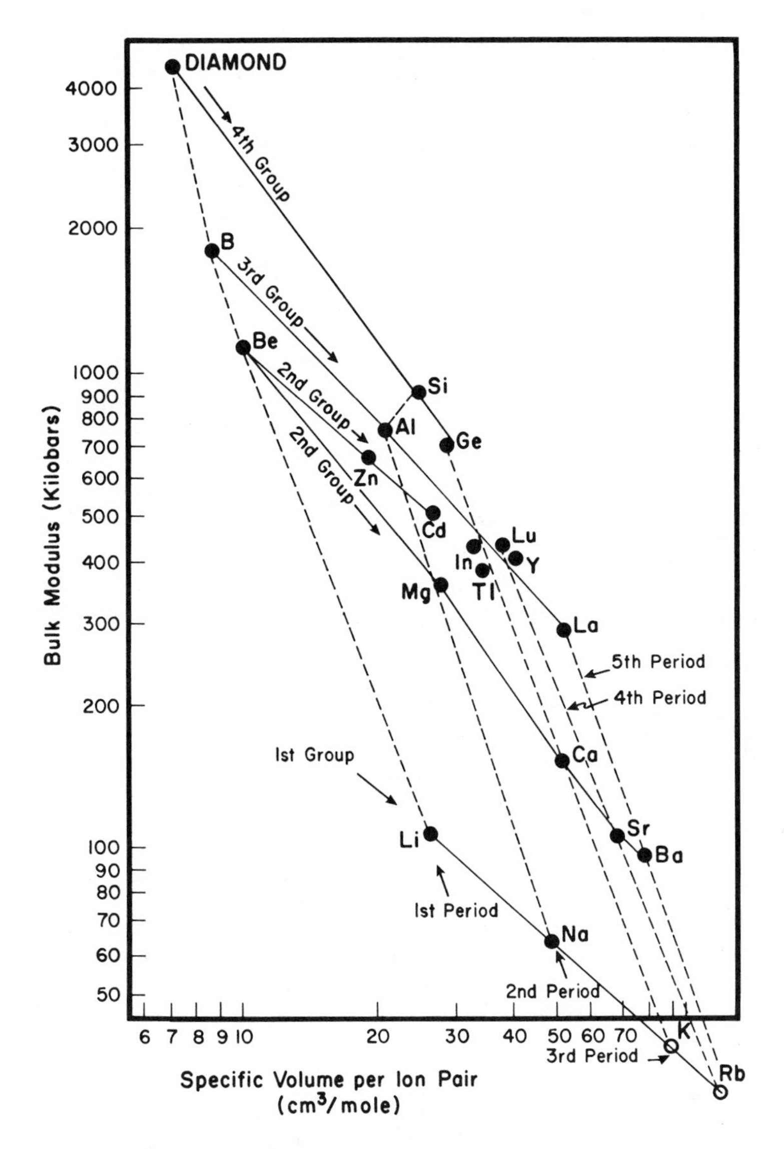

Fig. 9. The variation of bulk modulus with volume for elements in the periodic table. Groups lie on lines with slopes near – 1. Periods lie on lines with slopes near – 4.

We might expect another correlation between bulk modulus and ionization potential as we follow sequence changing groups. In this case, the ionization potential generally increases as the atomic number increases, and this is attributed to the increase in nuclear charge (a higher charge producing a larger attraction on the electron). These effects on individual periods are shown in Fig. 9 as dashed lines.

Data for the first 4 groups are plotted in Fig. 9 as solid lines. Data in the same periods are connected by dashed lines. We find the following correlations

$$
\begin{aligned}
K_0 V_0^3 &= \text{constant, 1st period (Li, Be, B)} \\
K_0 V_0^4 &= \text{constant, 2nd period (Na, Mg, Al)} \\
K_0 V_0^4 &= \text{constant, 3rd period (K, Ca, Ge)} \qquad (14) \\
K_0 V_0^4 &= \text{constant, 4th period (Rb, Sr, Y)} \\
K_0 V_0^{3.5} &= \text{constant, 5th period (Cs, Ba, La)}
\end{aligned}
$$

There are exceptions, and not all of the elements fit into this neat pattern, but the difference between the K_0-V_0 law for groups and for periods is very pronounced. A change in volume along the path of a period is usually accompanied by a great change in bulk modulus.

Proceeding along the path of a group, one finds that the mass changes rapidly while the number of electrons in the outer subshell does not change; that is, the

Table 5. Volume-modulus data of the third long period

Atomic number	Element	Atomic weight	Density	Specific volume	Bulk modulus
70	Yb	173.6	7.02	49.44	135
71	Lu	175	9.81	35.67	419
72	Hf	178.5	13.3	26.8	1,100
73	Ta	181	16.68	21.60	2,040
74	W	183.9	19.3	19.05	3,296
75	Re	186.2	20.5	18.10	3,790
76	Os	190.2	22.5	16.9	?
77	Ir	192.2	22.4	17.1	3,620
78	Pt	195.1	21.3	18.4	2,838
79	Au	196.9	19.3	20.2	1,766
80	Hg	200.6	14.11	28.2	288
81	Tl	204.4	11.85	34.5	366
82	Pb	207.2	11.35	36.5	438
83	Bi	208.9	9.75	42.8	320

Note.–Data from Gschneidner (1964).

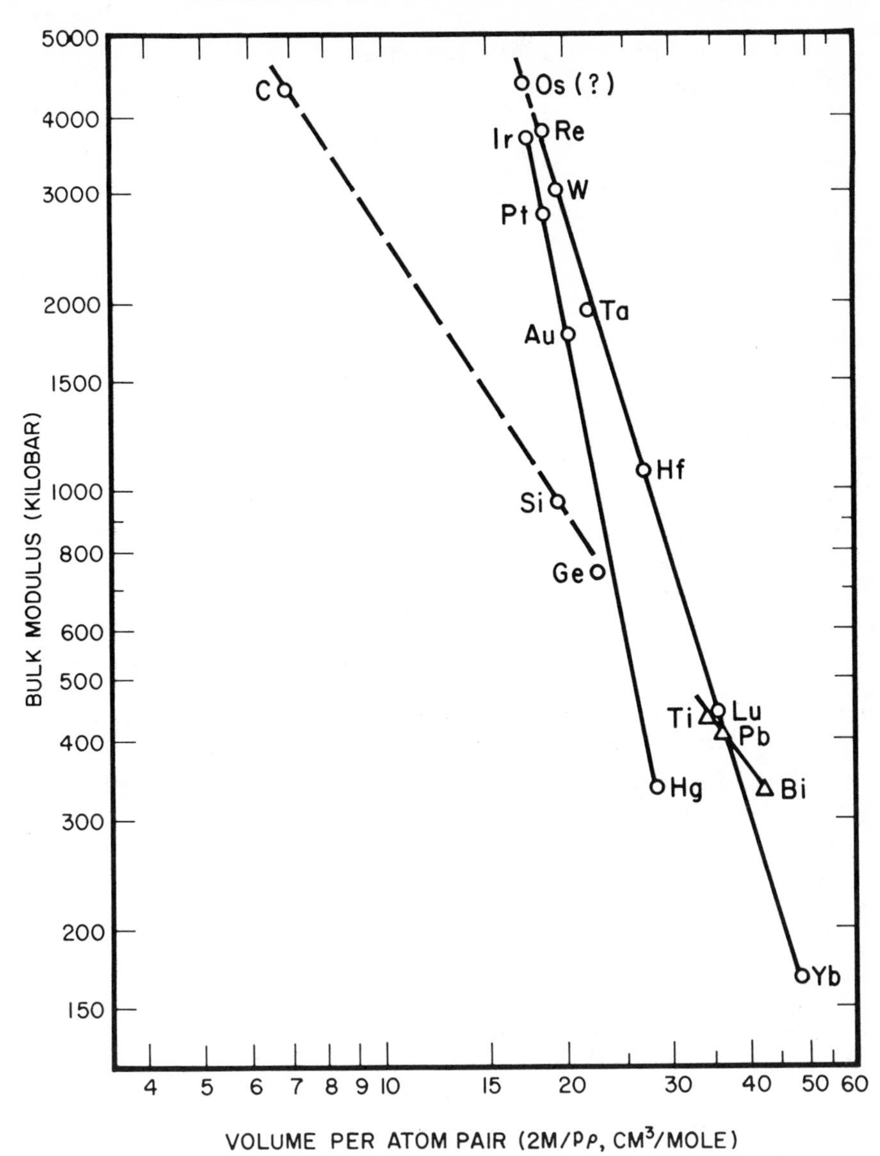

Fig. 10. Bulk modulus versus volume for elements in the third long period. The data lie on two straight lines with large slope (excepting Tl, Pb, and Bi).

valence state does not change. Proceeding along the path of a period, one finds that the valence state changes, but the mass does not change very rapidly. For example, in the sequence, Na, Mg, Al, the electron states are: 1 in the 3s, 2 in the 3s, 2 in the 3s, and 1 in the 3p; the masses are 22.92, 24.32, 26.99.

The fact that the masses do not change rapidly in this sequence is an important

Table 6. Electron structure of the third long period of the chemical elements

Atomic number	Element	Electron 5s 5p 5d 5f	Structure 6s 6p
70	Yb	2, 6, 0, 0	2
71	Lu	2, 6, 1, 0	2
72	Hf	2, 6, 2, 0	2
73	Ta	2, 6, 3, 0	2
74	W	2, 6, 4, 0	2
75	Re	2, 6, 5, 0	2
76	Os	2, 6, 6, 0	2
77	Ir	2, 6, 9, 0	0
78	Pt	2, 6, 9, 0	1
79	Au	2, 6, 10, 0	1
80	Hg	2, 6, 10, 0	2
81	Tl	2, 6, 10, 0	2, 1
82	Pb	2, 6, 10, 0	2, 2
83	Bi	2, 6, 10, 0	2, 3

consideration in view of the constancy of M/p for oxides and silicates. What would be valuable is a sequence where the mass and the outer electronic state do not change significantly, but the volume does. Such a sequence is found in the 3rd long period. The data are shown in Table 5.

Table 7. Volume-modulus data of the second long period

Atomic number	Element	Atomic weight	Density	Specific volume	Bulk modulus
38	Sr	88.62	2.639	66.5	118
39	Y	88.90	4.472	39.5	373
40	Zr	91.22	6.4	28.5	850
41	Nb	92.91	8.58	21.4	1,736
42	Mo	95.50	10.20	18.3	2,779
43	Tc				
44	Ru	101.1	12.2	16.6	3,271
45	Rh	102.2	12.5	16.3	2,758
46	Pd	106.4	12.6	17.4	1,844
47	Ag	107.9	10.53	20.1	1,027
48	Cd	112.4	8.75	25.7	475
49	In	114.82	7.30	31.4	419
50	Sn	118.69	7.31	32.0	553
51	Sb	121.76	6.69	33.3	390
52	Te	127.6	6.24	40.03	235

Note.–Data from Gschneidner (1964).

The data for K_0 versus V_0 are plotted in Fig. 10. These data fall into three sequences: three different straight lines all determined by the electron structures (see Table 6). In the first sequence, ytterbium to osmium, the P electron shell has only two electrons filling the 6s subshell, and electrons are added deep within the 5d electron shell. In this sequence, then, the valence electrons do not change at all; the mass changes by 9%, and the volume changes by 108%. In the next sequence, iridium to mercury, the outer electrons in the 6s subshell change as well as those within the 5d subshell. The mass changes by 4%, and the volume changes by 44%. In the third sequence, thalium to bismuth, additions of electrons occur in the 6p subshells while the 6s subshell remains unchanged. Attention is drawn to the two first sequences, where

$$K_0 V_0^4 = \text{constant (Yb to Os)} \tag{15}$$

$$K_0 V_0^5 = \text{constant (Ir to Hg)} \tag{16}$$

The data for the sequence from Yb to Re lie on a remarkably straight line as shown in the preceding graph. The small change in mass, the relatively larger change in volume, and the constancy of the outer electronic state makes this sequence an idealized model when we compare it to K_0 versus V_0 for oxides and silicates.

We draw attention to the break in the period between Os and Ir, and also between Hg and Tl. We now demonstrate that these breaks are generally found exactly in the middle of the so-called 8th group. This is shown by examination of the data for the 2nd long period in Table 7, plotted in Fig. 11, with electron structures given in Table 8.

A summary of all of the period behavior is given in Table 9. The data for the transition metals and the lathanides are not repeated here.

The situation for the lanthanides is not typical of the other long periods and might be a result of the "lanthanide contraction." Nevertheless, a generality emerges. For a sequence where the mass changes little and the outer electron state remains constant, the relation, $K_0 V_0^x$ = constant, is such that x is a large number, between 3 and 5.

Figures 7, 9, 10, and 11 are plotted with V_0 in units of volume per ion pair. This is done to make the plots comparable to Fig. 2, where the data for diatomic ionic solids are presented. For theories of bonding applicable to the elements, the atomic volume would be more appropriate.

Figure 9 has a very interesting pattern showing the relative relationships of K among the elements. It is characteristic of many relationships that arise from the nature of the periodic table. A plot of the first ionization potential versus atomic volume looks very much like Fig. 9, as well as Figs. 10 and 11. The correlation between the first ionization potential and bulk modulus is pronounced and

Table 8. Electron structure of the second long period of the chemical elements

Atomic number	Element	Electron O	Structure P
38	Sr	2, 6, 0, 0	2
39	Y	2, 6, 1, 0	2
40	Zr	2, 6, 2, 0	2
41	Nb	2, 6, 4, 0	1
42	Mo	2, 6, 5, 0	1
43	Te	2, 6, 6, 0	1
44	Ru	2, 6, 7, 0	1
45	Rh	2, 6, 8, 0	1
46	Pd	2, 6, 10, 0	0
47	Ag	2, 6, 10, 0	1
48	Cd	2, 6, 10, 0	2
49	In	2, 6, 10, 0	2, 1
50	Sn	2, 6, 10, 0	2, 2
51	Sb	2, 6, 10, 0	2, 3
52	Te	2, 6, 10, 0	2, 4

Table 9. Periodic table showing breaks in the plots of K-V curves

1a	2a	3b	4b	5b	6b	7b		8b		1b	2b	3a	4a	5a	6a
													Si		
			22	23	24	25	26	27	28	29	30		32		
			Ti	V	Cr	Mn	Fe	Co	Ni	Cu	Zn		Ge		
37	38	39	40	41	42	43	44	45	46	47	48	49	50	51	52
Rb	Sr	Y	Zr	Nb	Mo	Tc	Ru	Rh	Pd	Ag	Cd	In	Sn	Sb	Te
55	56	57	58	59	60	61	62	63	64	65	66	67	68	69	70
Cs	Ba	La	Ce	Pr	Nd	Pm	Sm	Eu	Gd	Tb	Dy	Ho	Er	Tm	Yb
		71	72	73	74	75	76	77	78	79	80	81	82	83	84
		Lu	Hf	Ta	W	Re	Os	Ir	Pt	Au	Hg	Tl	Pb	Bi	Po

1st break　　　　2nd break

3rd period, left (23-26) K_0V_0 = constant (17)
3rd period, middle (27-30) $K_0V_0^3$ = constant (18)
2nd long period, left (38-44) $K_0V_0^4$ = constant (19)
2nd long period, middle (45-48) $K_0V_0^{4.5}$ = constant (20)
2nd long period, right (49-52) $K_0V_0^{1.3}$ = constant (21)
3rd long period, left (71-76) $K_0V_0^4$ = constant (22)
3rd long period, middle (77-80) $K_0V_0^5$ = constant (23)
3rd long period, right (81-83) $K_0V_0^{1.3}$ = constant (24)

Lanthanides, left (57-62), excepting Ce (no. 58): $K_0V_0^{3.5}$ = constant (25)
Lanthanides, middle (63-66), a cluster at K_0 = 400 Kb, V_0 = 38.
Lanthanides, right (67-70), a cluster at K_0 = 400 Kb, V_0 = 38.

indicates that the power law found between K_0 and V_0 in the periods may arise from the electronic binding that controls the ionization potential.

The relationship between K_0 and V_0, for the elements shown here, is somewhat the same as the relationship between the compressibility and V_0 found by Kempter (1965) who noticed repetitive behavior among the periods.

Another very significant relationship between the bulk modulus and specific volume has been proposed by Knopoff (1965), using an empirical method to extrapolate between finite strain theory and the Thomas-Fermi equation of

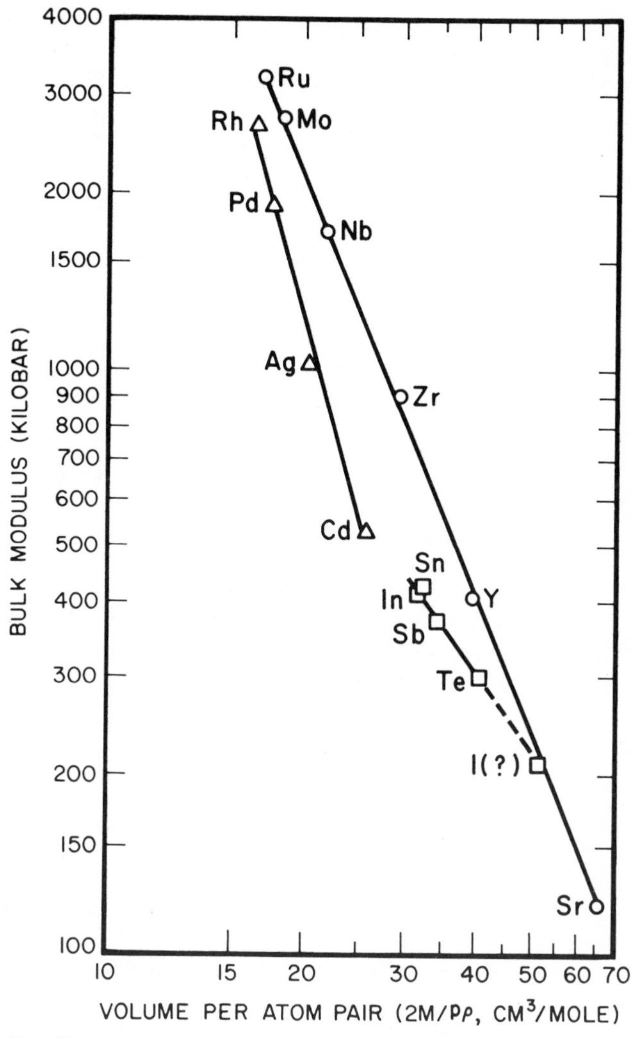

Fig. 11. Bulk modulus versus volume for elements in the second long period. Elements 38 to 48 lie on lines with large slope, near -4.

state. Given the atomic number (or its equivalent for compounds) and the specific volume, K_0 is specified.

VELOCITY-DENSITY RELATIONSHIPS

It was shown that for the elements, the volume variation of the bulk modulus depends upon whether one considers a sequence in a period or in a group. Likewise, the relationship between the bulk sound velocity V_K and density depends upon how one classifies the elements ($V_K = \sqrt{K/\rho}$).

For a sequence composed of elements in a period, the sound velocity increases with density; but for sequences of elements in a group, the sound velocity decreases with density. This is shown in Fig. 12. For comparison purposes, the $V_K - \rho_0$ relationship for a sequence of oxides and silicates is also plotted.

Again, it is demonstrated that the behavior of oxides, with regard to the velocity-density relationship, resembles long periods in the periodic table.

The relationship between V_p and K_0 is given by

$$V_p = \sqrt{\frac{K_0}{\rho_0}}\left[1 + 2\,\frac{(1 - 2\sigma_0)}{(1 + \sigma_0)}\right]^{1/2} \tag{26}$$

Now, applying the – 4 power law, we have (Anderson and Soga, 1968):

$$V_p = \frac{A\rho_0{}^{1.5}}{(M/p)^2}\left[1 + 2\,\frac{(1 - 2\sigma_0)}{(1 + \sigma_0)}\right]^{1/2} \tag{27}$$

The effect of changes in σ_0 upon V_p is less than the effect of ρ_0 and μ/p for rock-forming minerals, so that according to (27) V_p is controlled by ρ_0 and M/p. Now, (27) is indistinguishable from Birch's law, if the density range is restricted to 2 to 4 gm/cm^3. The choice between them is a matter of taste or convenience when an empirical law is constructed to represent measured data. Equation (27), however, has the advantage that the dependence of V_p upon M/p is explicit. For example, at a given density, the value of V_p varies inversely as the square of M/p. For instance, the value of M/p for germanates is 35, in contrast to SiO_2 which is near 20. We would therefore expect, according to (27), that V_p of germanates is 1/3 in value to that of SiO_2 (at constant ρ_0).

The sound velocities of germanates, in three phases with a wide spread in ρ_0, has been measured recently by Soga (1969, 1970). An empirical equation that describes the data of SiO_2 in its various phases, iron-rich silicates, and germanates is

$$V_p = \frac{600\rho_0{}^{1.5}}{(M/p)^2} \tag{28}$$

This is plotted in Fig. 13.

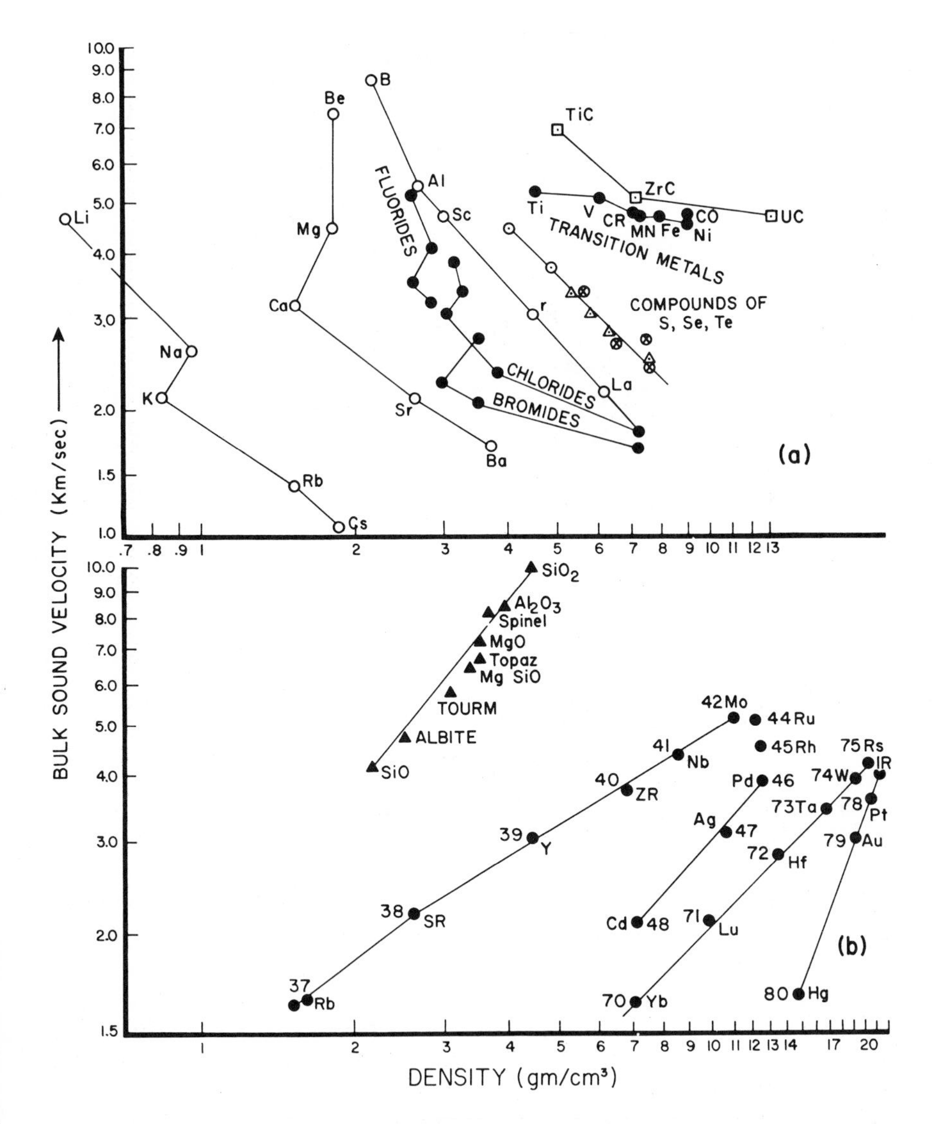

Fig. 12. Bulk sound velocity versus density on a log-log plot. Isostructural sequences (groups) are shown in (*a*) in which the velocity decreases as the density increases. Sequences that have constant mass (periods and a class of oxides) are plotted in (*b*) below, in which the sound velocity increases as the density increases.

The equivalent expression for the shear velocity is

$$V_s = \frac{360 \rho_0{}^{1.5}}{(M/p)^2} \tag{29}$$

This is plotted in Fig. 14.

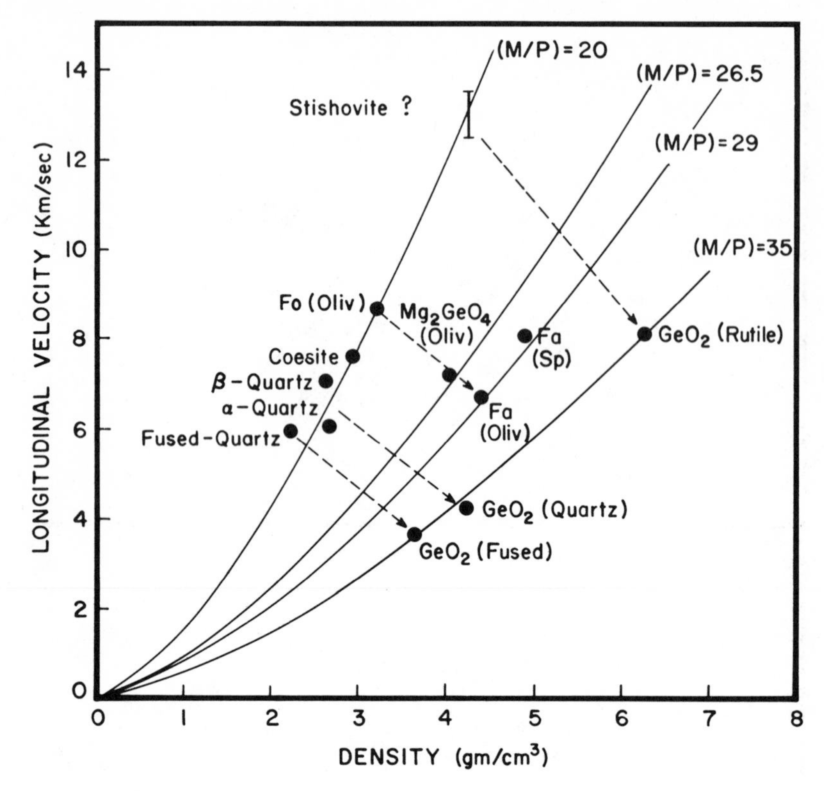

Fig. 13. V_p versus density on a linear plot, according to the equation $V_p = 600\ \rho_0^{1.5}/(M/p)^2$. Dashed lines show isostructural sequences.

Equations (28) and (29) are especially valuable because they permit the determination of V_s for iron-rich compounds by interpolation between silicates and germinates.

The dashed lines in Figs. 13 and 14 show isostructural variation. These lines represent the velocity varying as the inverse square root of M/p. This is the change of sound velocity with density one expects with solid solution substitution. The variation of V_p with density along the dashed lines is equivalent to the variation of V_p with density for groups in the periodic table, as shown in the upper part of Fig. 12. These isostructural lines are equivalent to the solid lines plotted in Fig. 5.

As emphasized, (28) is indistinguishable from a straight line at constant M/p for the obtainable density range. It is interesting to note, however, that (28) requires that the slope, given by b in (1), decreases as M/p increases.

It should be emphasized that the choice of the units of volume is arbitrary, but it does not change the major conclusions. V_0 can be volume per ion pair, as used here, or volume per average atom, or volume per molecule. Relationships of the elastic constants using V_0 in units of volume per molecule were recently used by

Anderson and Anderson (1970) to demonstrate methods of predicting elastic constants of high pressure forms of certain silicates. The systematics between V_p and ρ, and V_s and ρ, in the context of Birch's law but equivalent to Eqs. (28) and (29) has been presented by Liebermann (1970).

DEGENERACIES IN PATTERNS OF ELASTIC CONSTANTS

There is a very practical consequence in the behavior of K_0 versus ρ_0 along the sequence of a group. The value of K for one member of the group at high pressure will be close to the value of another member of the group at zero pressure. This behavior has been formalized by the name "the law of corresponding states" (Anderson, 1967).

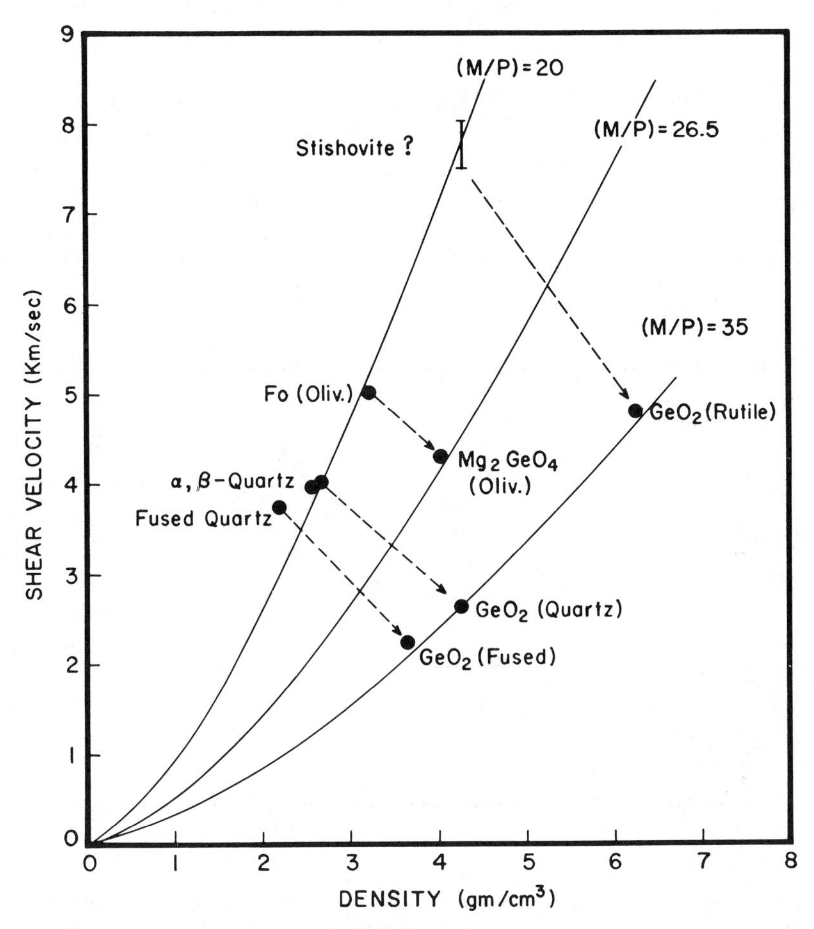

Fig. 14. V_s versus density on a linear plot, according to the equation $V_s = 350\, \rho_0{}^{1.5}/(M/p)^2$. Dashed lines show isostructural sequences.

The point is perhaps best demonstrated by considering plots of sound velocity versus density. In Fig. 15, the variation of V_K versus ρ_0 is plotted for a number of silicates. Further the values of V_K versus ρ are plotted at high pressure, where the data are taken from various shock wave measurements. There is considerable scatter, but the point is made that the velocity solutions overlap when M/p is nearly the same.

In contrast to this behavior, the solutions of V_K versus ρ_0 for the alkali halides tend to spread out as shown in Fig. 16. Rarely does the velocity of one alkali halide at high pressure overlap the velocity of another alkali halide. (There is, of course, a wide spread in M/p.)

The consequence of the degeneracy of solutions for V_s, V_p, and V_K for oxides and silicates is that solutions for the composition of the earth's interior are not unique when based upon seismological evidence. Mixtures of two or more quite different compounds at a particular pressure can yield the same velocity as one given compound at that pressure, provided only that the value of M/p is the same.

Granted that the value of V_K from various oxides tend to overlap the problem posed to geophysics is to determine how this overlap affects seismic solutions of the earth's mantle. Variability of solutions arises because of the uncertainty in M/p. Results of geochemical calculations indicate that M/p should be contained between 21 and 23. Thermodynamic solutions arising from atomic potential functions give relationships between K and V. The plots of V_p and V_s versus ρ_0, shown in Fig. 1, are converted into K versus V in Fig. 17, with two

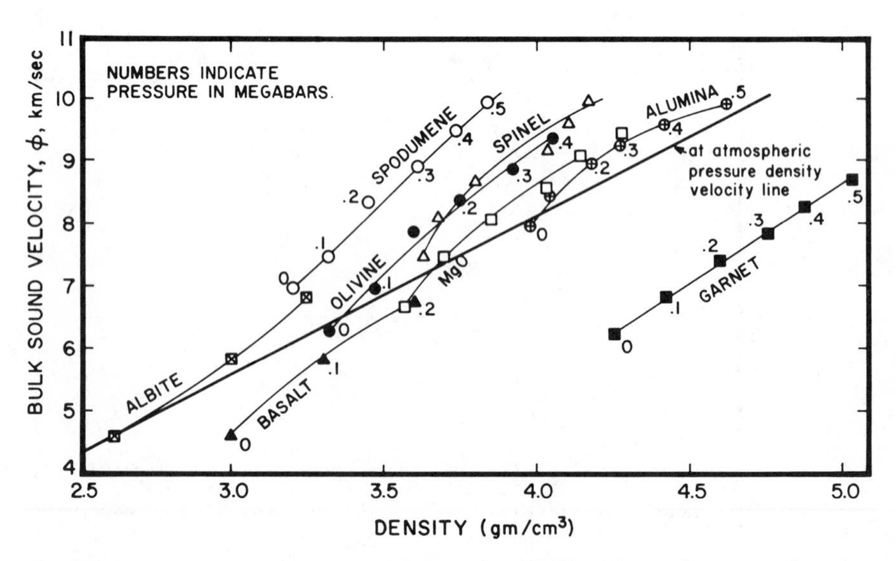

Fig. 15. V_K versus ρ_0 on a linear plot. The variation of V_K with ρ as the pressure increases tends to superimpose on the $V_K - \rho_0$ line if M/p is constant.

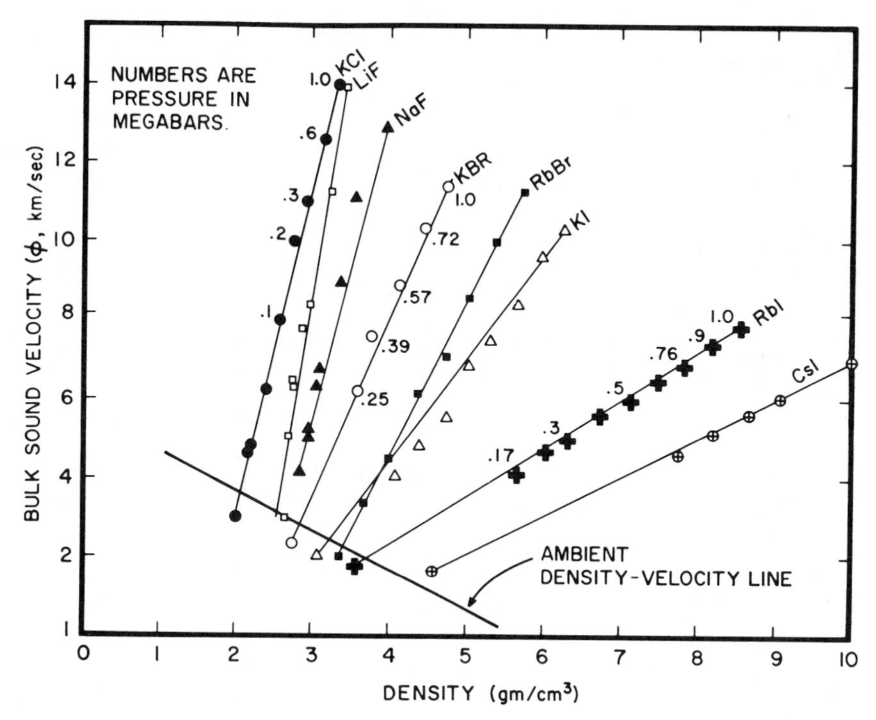

Fig. 16. V_K versus ρ as the pressure increases for alkali halides. These data do not tend to superimpose upon the line showing V_K versus ρ_0 at zero pressure.

values of M/p. Thus, seismic solutions of the earth's mantle are contained within the shaded region which is called the seismic slot.

On this same diagram (Fig. 17), the value of K and V are plotted for MgO, Al_2O_3, olivine, and garnet for several pressures and temperatures. The values of K and V at high pressure are computed from the Birch-Murnaghan equation of state, but the points are printed with large symbols to admit other methods of extrapolation.

These several plotted points illustrate the difficulty. Most solutions of allowable oxides or silicates, at whatever pressure or temperature one may consider, fall within the seismic slot. Thus, solutions to the earth's composition must arise by considering another factor in addition to knowledge of elasticity. Additional factors are usually based upon geochemical methods. For example, the existence of a discontinuity in V_p at a given pressure (not seen in the plot of V_p versus ρ) can be identified with a measurable phase change of a particular compound at that pressure.

A few additional points can be gleaned from Fig. 17. First, the solutions at low volume tend to be close to the M/p = 23 line, while all the solutions at high volume tend to be close to the M/p = 21 line. As a consequence, it can be stated

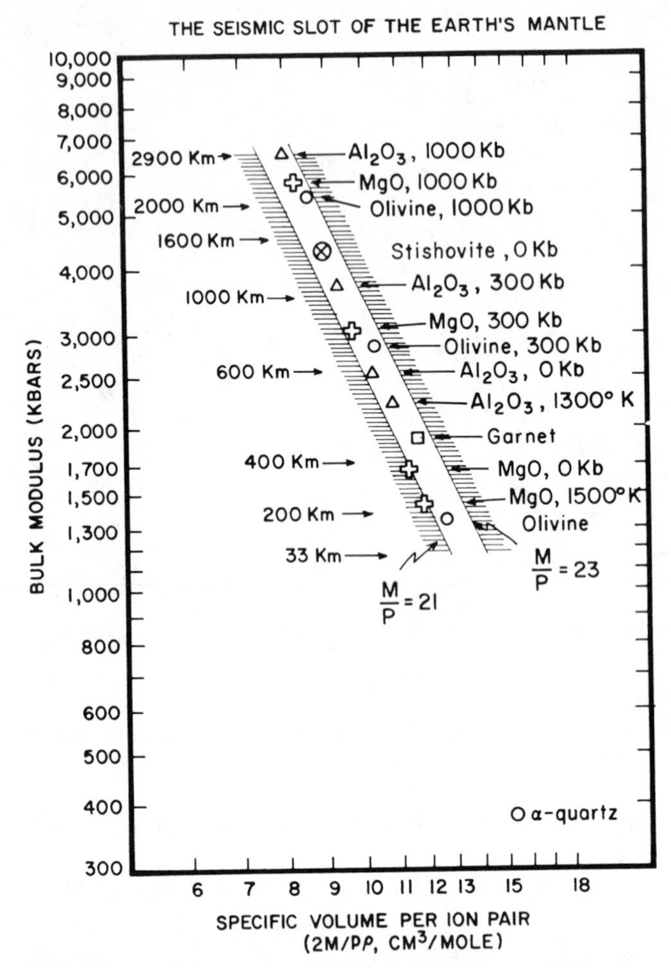

Fig. 17. The seismic slot; the zone between the dashed regions, represents solutions allowed for the earth's mantle with M/p between 21 and 23. Many data of minerals, at various pressures and temperatures, lie in this slot. The seismic slot has a slope of -3, while the data lie on a line with a slope of -4.

that the iron content will increase in the lower mantle over what it is in the upper mantle no matter what oxides are taken for the upper mantle. This same result can be placed in the context of solid-state physics as follows: The slope of the $\ln K - \ln V$ plot for oxides tends to be near -4, while the slope of the $\ln K - \ln V$ plot at constant M/p for the mantle is close to -3. If the amount of iron is slightly increased in a silicate, the slope of the $\ln K$-$\ln V$ plot would be -3.

Second, the specific volumes of all oxides tends to converge to the same value in the lower mantle. This is a consequence of the fact that ρ_0, and K_0, the parameters in an equation of state, are interrelated in oxides. In Fig. 5, we see that for a given value of M/p, the higher the value of ρ_0, the higher the value of K_0. In this case, when one plots ρ versus P, the higher the value of ρ_0, the smaller the value of $(d\rho/dP)_{P \to 0}$. Thus, all trajectories of ρ tend to converge at some pressure, no matter what the initial value of ρ_0 may be (Anderson and Liebermann, 1969). If, however, an isostructural sequence is considered (that is, the density is increased by increasing the iron concentration), two trajectories starting out at different values of ρ_0 will not converge at a higher pressure.

There is some evidence, not complete, that for constant M/p, a higher value of ρ_0 requires with it a lower value of K_0', as shown in Fig. 18. Conversely, in an isostructural sequence, the value of K_0' is independent of ρ_0 (dashed lines in Fig. 18). This evidence would tend to accentuate the convergence of density trajectories described above. The geophysical implication of these results is that the density of the lower mantle is independent of the presence of phase changes in the upper mantle and independent of the density at the crust mantle boundary.

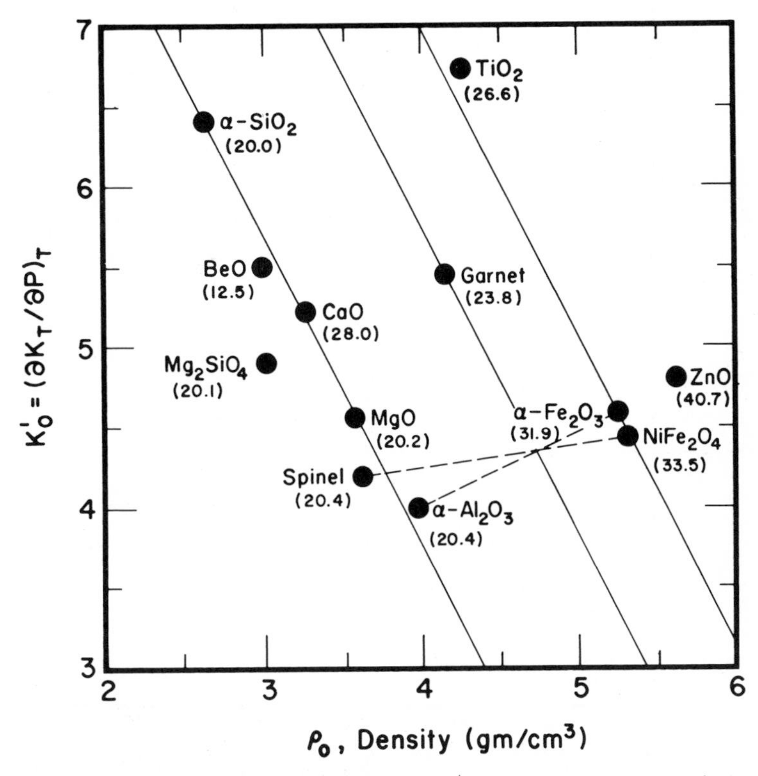

Fig. 18. K_0 versus ρ_0 for oxides (values of M/p are shown in parentheses). At constant M/p, a large density appears to require a smaller value of K_0'. The dashed lines show isostructural variation. Here K_0' is independent of ρ_0.

Another kind of degeneracy arises in extrapolations to high temperatures at ambient pressure. The temperature coefficients are controlled by two anharmonic parameters γ_{th} and δ_s [for definitions and background, see Anderson et al. (1968)]. It has been shown (Anderson, 1966b; Soga et al., 1966) that the extrapolation formula for K versus T is

$$K = K_0 - \frac{\gamma_{th}\delta_s}{V_0}[H(T) - H(0)](1 - x + x^2 + \cdots) \tag{30}$$

where $x = 1/2\ \beta T$, and β is the coefficient of thermal expansivity. The plot of (30) for alumina is shown in Fig. 19. Since $V = V_0(1 + \beta T)$, (30) is easily

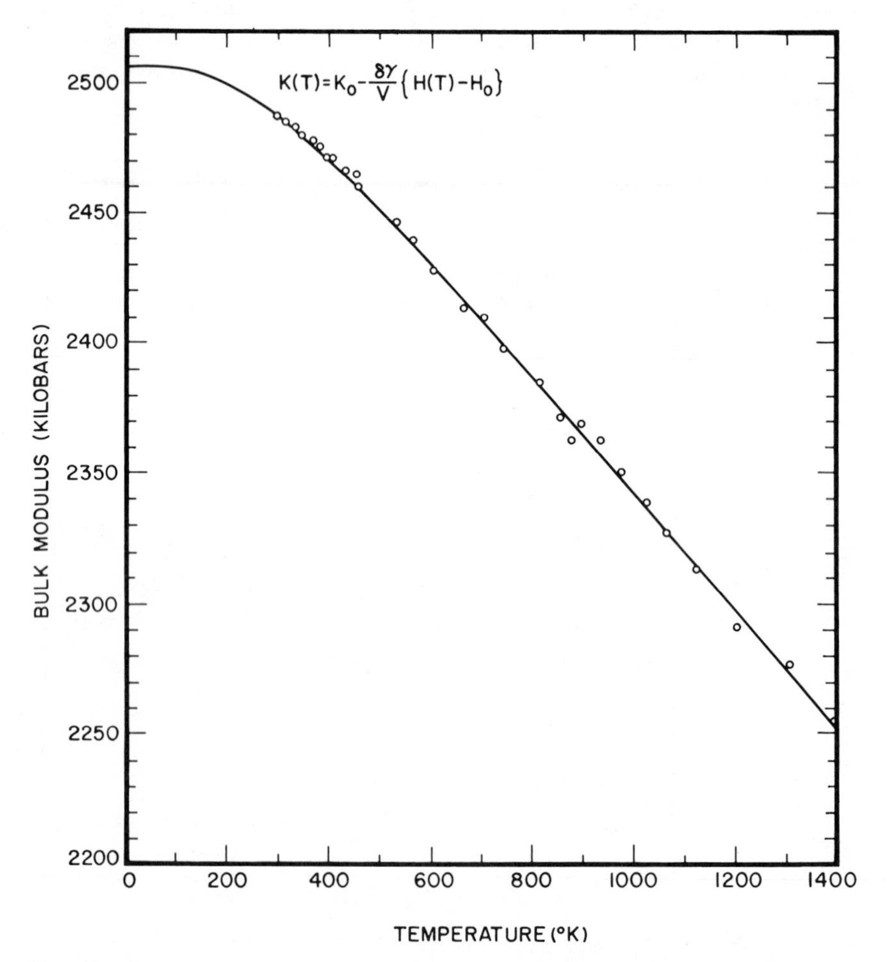

Fig. 19. K versus temperature for Al_2O_3. The solid line is theoretical utilizing the measured enthalpy and two anharmonic parameters.

transformed to

$$\phi = \phi_0 - f(T)$$

where

$$f(T) = \gamma_{th}\delta_s [H(T) - H(0)] \tag{31}$$

and ϕ is the seismic parameter, K/ρ.

It is expected from simple considerations of the atomic potential that $\delta_s + \gamma_{th}$ = constant. This rule is followed approximately, as shown by Fig. 20. The variation in δ_s and γ_{th} is sufficiently small that the product $\delta_s\gamma_{th}$ is close to value 5 for oxides. (The contribution $f(T)$ to ϕ is so small that small errors in $\gamma_{th}\delta_s = 5$ are not important.) This means that a good estimate of the seismic parameter at high temperature can be made if the enthalpy has been measured at that temperature.

LIMITS ON THE VALUE OF K'_0

In Fig. 18, it is demonstrated that experimental values of K'_0 vary from about 4 to 6.5. There are two significant results. First, the value of K'_0 is rarely 4, and this value is not typical even on the average. Second, the value of K'_0 is not ever significantly less than 4.

The fact that K'_0 is rarely 4 means that Eulerian finite strain theory to the second order is not a sufficient approximation to the equation of state. The Birch-Murnaghan equation of state is to third order (Birch, 1952)

$$\frac{P}{K_0} = \frac{3}{2}\left\{\left(\frac{\rho}{\rho_0}\right)^{7/3} - \left(\frac{\rho}{\rho_0}\right)^{5/3}\right\}\left\{1 + \frac{3}{4}(K'_0 - 4)\left[\left(\frac{\rho}{\rho_0}\right)^{2/3} - 1\right]\right\} \tag{33}$$

The term in the second set of braces approaches unity at large strain, only when $K'_0 = 4$, a rare event.

The fact that K'_0 is never much less than 4 can be understood from lattice theory. In the lattice potential

$$\phi = -\frac{A}{r} + \frac{b}{r^n}$$

$$K'_0 = \frac{n+7}{3} \tag{34}$$

The equations for the elastic constants at vanishing pressure are found from

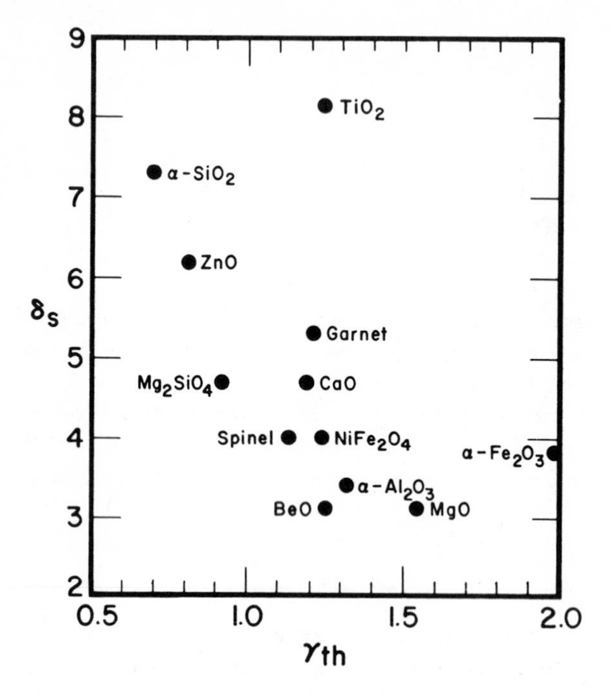

Fig. 20. δ_s versus γ_{th}. The two parameters are required in the equation plotted in Fig. 19. Roughly, the product of δ_s and γ_{th} is a constant. TiO_2 is the major exception.

lattice theory. Using the above potential (Anderson, 1970), the elastic constants can be found in terms of n, and therefore K_0'.

In the following equations for elastic constants, $p = z^2e^2/V_0a_0$ and $c_s = 1/2\,(c_{11} - c_{12})$, where z is the valence, e is the electronic charge, V_0 is the specific volume, and a_0 is the lattice constant. The equations for the elastic constants at zero pressure are

NaCl	CsCl
$c_{11} = p(0.5825\,n - 1.9735)$	$c_{11} = p(0.1942\,n + 1.2287)$
$c_{44} = p(0.6955)$	$c_{44} = p(0.1942\,n - 0.814)$
$c_s = p(0.2912\,n - 1.3345)$	$c_s = p(1.0513)$
$K_0 = p[0.1942\,(n - 1)]$	$K_0 = p[0.1942\,(n - 1)]$

For NaCl, c_s is negative for K_0' less than 3.856, and for CsCl, c_{44} is negative for K_0' below 4.732. If either c_s or c_{44} is negative, the lattice is unstable. These two illustrations depend upon a special model of forces. Nevertheless, they illustrate the point, and reports in which the value of K_0' is less than 3.5 must be viewed with suspicion.

For a given n, K_0 is the same for the two cubic lattices, but the other elastic constants differ from one cubic lattice to the other. NaCl and CsCl share the same crystallographic point group, but are in different crystallographic space groups. Thus, given a common potential between atoms, that is the same K_0', the bulk modulus is common for all members of a point group, but the other elastic constants are common only between members of a space group. (The value of p must be common between two structures to give the same value of K_0.)

LIMITS ON THE VALUE OF $d\mu/dP$ OR $d \ln V_s/d(P/K_0)$ AT $P = 0$

While the value of K_0' must be greater than a given value (roughly 4) for any lattice to be stable, there are lattices in which the dimensionless pressure derivative of the shear velocity, $d \ln V_s/d(P/K_0)$, can be negative. This point is easily demonstrated by using finite strain theory appropriate to elastic isotropy to the first order in strain (Sammis et al., 1970).

To the same degree of approximation that holds for (33), we have (Sammis et al., 1970)

$$\left\{\frac{1}{V_p}\left[\frac{\partial V_p}{\partial (P/K_0)}\right]\right\}_0 = \frac{13\lambda + 14\mu}{6(\lambda + 2\mu)} - \frac{(18l + 4m)}{6(\lambda + 2\mu)} \tag{35}$$

$$\left\{\frac{1}{V_s}\left[\frac{\partial V_s}{\partial (P/K_0)}\right]\right\}_0 = \frac{3\lambda + 6\mu}{6\mu} - \frac{(3m + n)}{6\mu} \tag{36}$$

$$\left(\frac{\partial K}{\partial P}\right)_0 = 4 - \frac{1}{3}\left(\frac{18l + 6m + (2/3)m}{K_0}\right) \tag{37}$$

where l, m, and n are elastic constants of the third order, and λ and μ are isotropic second-order elastic constants.

Given a finite hydrostatic stain, it is not possible to determine l, m, and n individually, but only in the combination (Sammis et al., 1970)

$$18l + 4m = l'$$

$$\frac{1}{2}(3m + n) = m'$$

where l, m, and n are third-order elastic constants only appropriate to elastic isotropy. The important point is that l, m, and n are independent, and while m and n determine the pressure derivative of V_s, m and n *plus* l determine the pressure derivative of K_0.

The shear velocity gradient is zero whenever the following holds between second-order and third-order elastic constants.

$$3m + n = -3(\lambda + 2\mu)$$

Note that when $(dV_s/dP)_0 = 0$, it is possible for dK/dP to be greater, less or equal to 4, and for dV_p/dP to be normal because of the independence of the third-order parameter l. The magnitude and sign of l, m, and n cannot be found from finite strain theory itself, as this theory does not have within its basic formulation any information on the values of the separate derivatives of the free energy. Finite strain theory, however, has information about the relationships between the elastic constants arising from internal consistency arguments. From these relationships, one cannot disallow the possibility that the gradient of the shear velocity will be negative or zero; and further no information that can be derived from P, V, T thermodynamics or P-V measurements can be used to predict, through finite strain theory, what the sign and magnitude of the shear velocity gradient will be. This results from the fact that pressure experiments on V_s only involve two of the three third-order constants in the combination $3m + n$.

The framework of lattice theory permits the possibility of deriving the pressure derivative of the elastic constants, the theory being limited by the knowledge of the details of the potential between atoms. For example, if we consider the potential that leads to (34) to find the elastic constants, we must specify it further to include the effect of nearest neighbors which, for example, may number as many as M. This summation over a total lattice would be

$$\phi = -\frac{A}{r} + Mv(r) \tag{38}$$

For NaCl, $M = 6$; for CsCl, $M = 8$; for ZnS, $M = 4$. The difference in M has a great effect on all elastic constants, except K. Proceeding in the prescribed manner (Anderson and Liebermann, 1969), we can construct Table 10 for the pressure derivatives of the elastic constants for the three structures.

Examination of this table indicates that $(dK/dP)_0$ is the same for all structures. However, dc_s/dP in ZnS is negative for all values of n larger than 1.589 (that is, $K_0' > 2.83$), and therefore is always negative. For CsCl, dc_s/dP is negative for n larger than 13.4 (that is, $K_0' > 6.8$). For NaCl, dc_{44}/dP is negative for n greater than 5.776 (that is, $K_0' > 4.26$). Of course, the accuracy of these predictions depends upon the reliability of (38) as a representation of the lattice potential.

The point, however, is that lattice theory has within itself the capability of predicting the values of elastic constants. The demonstration above serves to

Table 10. The pressure derivatives of the elastic constants derived from the potential given by Eq. (38)

Pressure derivative of	NaCl ($M = 6$)	CsCl ($M = 8$)	ZnS ($M = 4$)
c_{11}	$\frac{(n+3)(n+1) - 17.55}{n-1}$	$\frac{n+3}{3} + \frac{16.53}{n-1}$	$\frac{n+3}{3} + \frac{0.785}{n-1}$
c_{44}	$\frac{5.776 - n}{n-1}$	$\frac{n+3}{3} - \frac{8.264}{n-1}$	
c_s	$\frac{n(n+3) - 18.327}{2(n-1)}$	$\frac{13.396 - n}{n-1}$	$\frac{1.589 - n}{n-1}$
K	$\frac{n+7}{3}$	$\frac{n+7}{3}$	$\frac{n+7}{3}$

underline the point already derived from finite strain theory that knowledge of K_0 and K_0' alone tells us nothing about the magnitude of the shear constants and their pressure derivatives. Further, in order to predict the value of the pressure derivatives of the elastic constants, some information about the symmetry properties of the crystals has to be invoked.

These conclusions hold even for elastic isotropy. For a polycrystalline aggregate of ZnS crystallites, the value of dV_s/dP is always negative. For a polycrystalline aggregate of CsCl crystallites, the value of dV_s/dP is always positive. For a polycrystalline aggregate of NaCl or CaF_2 crystallites, the value of dV_s/dP may be positive or negative depending upon the details of the potential given by (38) (Anderson and Demarest, 1971).

The complete solution of the frequency spectrum of real lattices is now needed in order that the value of the elastic constants can be predicted at very high pressures and very high temperatures. In the meantime some progress has been made (Anderson, 1970; Sammis, 1970). In general, it can be said that the lattices with low coordination will tend to have negative (or small positive) values of dV_s/dP. This is shown in Fig. 21, where a particular solution for $d \ln V_s/d(P/K_0)$ for the three cubic lattices is given on the left. Actual data are plotted on the right. The parameter on the abscissa is the same as the first term on the RHS of (36), $(3\lambda + 6\mu)/6\mu$, and the straight line on Fig. 21 is (36) taking $(3m + n) = 0$.

A very simple form of lattice theory gives the two branches (NaCl and ZnS) in Fig. 21, which come reasonably close to predicting the measured values. There is every reason to believe that improvements in lattice theory will result in the prediction of the measured results with more accuracy.

The results of the last two sections have great relevance to geophysics and geochemistry of the earth's interior, where it is assumed that the elasticity of the

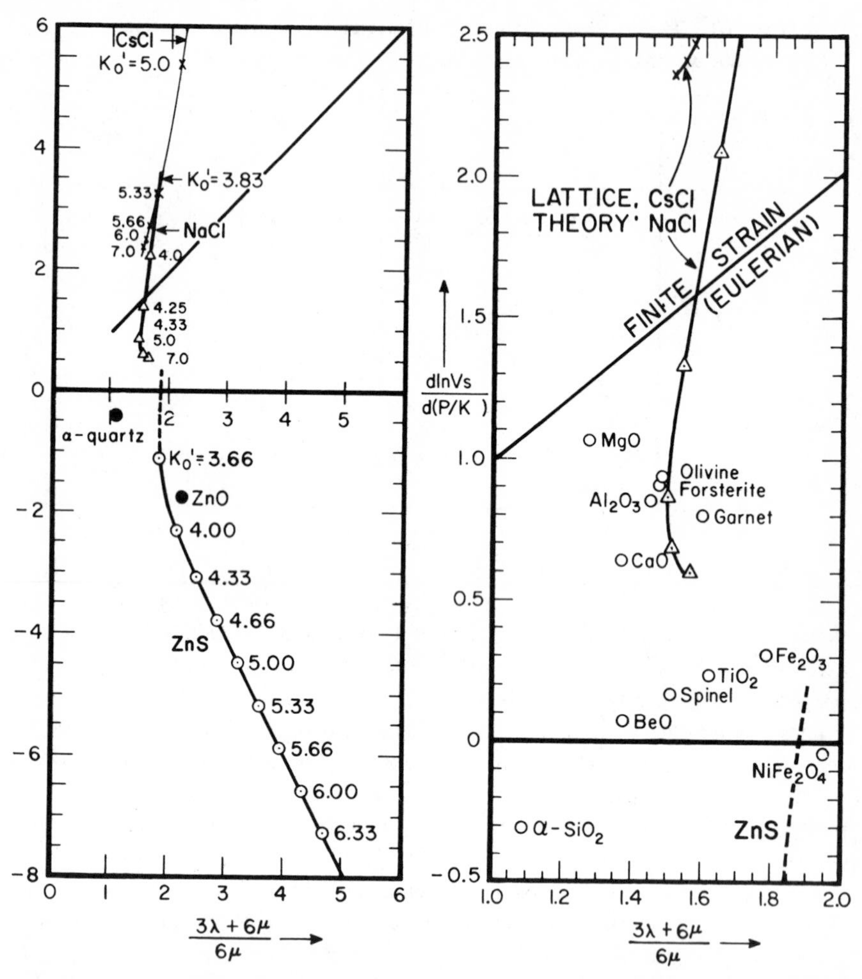

Fig. 21. $d \ln V_s/d(P/K_0)$ versus the parameter $(3\lambda + 6\mu)/6\mu$, The curved branches are special solutions from lattice theory, arising from the central force assumption. The data all fall well below the finite strain solution which ignores third-order constants. Negative values are allowed by the ZnS solution.

earth's interior at any point is isotropic. The earth actually consists of closely packed crystallites, and the crystallites each have anisotropic elastic constants. The seismic waves propagate with a velocity that depends upon the space average of the crystallites' elastic constants along the ray path. From seismic theory, a consistent solution is found, which shows how the isotropic velocity and density varies with depth. In order to relate this seismic solution to a presumed composition model of the planet, one has to find how well the seismic velocity compares with the "averaged" isotropic sound velocity of the presumed composition. In doing this, we are faced with the problem that the magnitude and sign of the pressure derivatives of the shear constants in the "isotropic" material depend greatly upon the crystal structure. It is not enough to say that a material is isotropic in order to investigate the elasticity of matter at high pressure; one must also specify the details of the crystallographic structures in the crystallites of the aggregate if the behavior of V_s at high pressure is to be predicted.

APPENDIX A

References to Elastic Constant Data

All elements, Gschneidner (1964)
Alkali halides, Anderson and Nafe (1965)
TiC, Chang and Graham (1966)
ZrC, *ibid.*
SiC, Kucher (1963)
UC, Graham et al. (1963)
GaAs, Garland and Park (1962)
GaSb, Einspruch and Manning (1963)
InSb, Potter (1956)
AlSb, Bolef and Menes (1960)
InAs, Gerlich (1963)
Al_2O_3, Anderson et al. (1968)
Cr_2O_3, Schreiber (1969)
Fe_2O_3, Anderson et al. (1968)
$MgAl_2O_3$, *ibid.*
$NiFe_2O_3$, *ibid.*
MgO, *ibid.*
CaO, *ibid.*
BeO, *ibid.*
ZnO, *ibid.*
SiO_2, *ibid.*
SrO, Johnston et al. (1970)
MnO, Oliver (1970)
Mg_2SiO_4, Kumazawa & Anderson (1969)
TiO_2, Manghnani (1969)
GeO_2 (glass), Soga (1969)
GeO_2 (quartz), Soga (1969)
GeO_2 (rutile), Soga (1970)

ACKNOWLEDGMENTS

I greatly appreciate the opportunity of reading advance manuscripts of Charles Sammis, Leon Thomsen, and R. Liebermann.

This work was sponsored by NSF Grant No. 16082.

REFERENCES

Anderson, D. L., A seismic equation of state, Geophys. J. Astron. Soc., 9-30, 1967.

Anderson, D. L., and Anderson, O. L., The bulk modulus-volume relationship, J. Geophys. Res., **75,** 3494-3500, 1970.

Anderson, O. L., A proposed law of corresponding states for oxide compounds, J. Geophys. Res., **71,** 4963-4971, 1966a.

______, Derivation of Wachtman's equation for the temperature dependence of elastic moduli of oxide compounds, Phys. Rev., **144,** 553-557, 1966b.

______, Elastic constants of the central force model for three cubic structures: Pressure derivatives and equations of state, J. Geophys. Res., **75,** 2719-2790, 1970.

Anderson, O. L., and Demarest, H. H., Elastic constants of the central force model for cubic structures: Polycrystallize aggregates and instabilities, J. Geophys. Res., **76,** 1349-1369, 1971.

Anderson, O. L., and Nafe, J. F., The bulk-modulus relationship for oxide compounds and related geophysical problems, J. Geophys. Res., **70,** 3951-3963, 1965.

Anderson, O. L., and Liebermann, R., Elastic constants of oxide compounds used to estimate the properties of the earth's interiors, *in* Application of Modern Physics to the Earth and Planetary Interiors, pp. 425-448, edited by S. K. Runcorn, John Wiley & Sons, New York, 1969.

Anderson, O. L., and Soga, N., A restriction to the law of corresponding states, J. Geophys. Res., **72,** 5754-5757, 1968.

Anderson, O. L., Schreiber, E., Liebermann, R. C., and Soga, N., Some elastic constant data on minerals relevant to geophysics, Rev. Geophys., **6,** 491-524, 1968.

Birch, F., The velocity of compressional waves in rocks to 10 kilobars, 2, J. Geophys. Res., **66,** 2199-2224, 1961.

______, Elasticity and composition of the earth's interior, J. Geophys. Res., **57,** 227-286, 1952.

Bolef, D. I., and Menes, M., Elastic constants of single crystal aluminum antimonide, J. Appl. Phys., **31,** 1426-1427, 1960.

Bridgman, P. W., The compressibility of thirty metals as a function of pressure and temperature, Proc. Am. Acad. Arts Sci., **58,** 165-242, 1923.

Chang, R., and Graham, L. T., Low-temperature elastic properties of ZrC and TiC, J. Appl. Phys., **37,** 3778-3783, 1966.

Cline, C. F., Dunnegin, H. L., and Henderson, G. W., Elastic constants of hexagonal BeO, ZnS, and CdSe, J. Appl. Phys., **38,** 1944-1948, 1967.

Einspruch, N. G., and Manning, R. J., Elastic constants of compound semiconductors–ZnS, PbTe, GaSb, J. Acoust. Soc. Am., **35,** 215-216, 1963.

Garland, C. W., and Park, K. C., Low temperature elastic constants of gallium arsenide, J. Appl. Phys., **33,** 759-760, 1962.

Gerlich, D., Elastic constants of single-crystal indium arsenide, J. Appl. Phys., **34,** 2915, 1963.

Graham, L. J., Nadler, H., and Chang, R., Elastic constants of UC, J. Appl. Phys., **34,** 1572-1573, 1963.

Gschneidner, K. A., Physical properties and interrelationships of metallic and semimetalic elements, *in* Solid State Physics, vol. 16, edited by F. Seitz and D. Turnbull, Academic Press, New York, 1964.

Johnston, D. L., Thrasher, P. H., and Kearney, R. J., Elastic constants of SrO, J. Appl. Phys., **41,** 427-428, 1970.

Kempter, D. P., Isoperiodic variation of compressibilities of the elements, Phys. Stat. Sol., **8,** 161-172, 1965.

Knopoff, L., Approximate compressibility of elements and compounds, Phys. Rev., A1445-A1447, 1965.

Kucher, T. I., Concerning the similarity between the characteristic frequency-dispersion curves of diamond type crystals, Soviet Phys.–Solid State, English transl., **4,** 1747-1752, 1963.

Kumazawa, M., and Anderson, O. L., Elastic moduli, pressure derivatives and temperature derivatives of single-crystal olivine and single-crystal forsterite, J. Geophys. Res., **74,** 5961-5972, 1969.

Liebermann, R. C., Velocity-density systematics in the olivine and spinel phases of the Mg_2SiO_4-Fe_2SiO_4 system, J. Geophys. Res., **75**(20), 1970.

Manghnani, M. H., Elastic constants of single-crystal rutile under pressures to 7.5 kilobars, J. Geophys. Res., **74,** 4317-4329, 1969.

Nafe, J. E., and Drake, C. L., Variation with depth in shallow and deep water marine sediments of porosity, density, and the velocities of compressional and shear waves, Geophysics, **22,** 523-552, 1957.

Oliver, D. W., The elastic moduli of MnO, J. Appl. Phys., **41,** 427, 1970.

Potter, R. F., Elastic moduli of indium antimonide, Phys. Rev., **103,** 47-50, 1956.

Sammis, C. G., The pressure dependence of the elastic constants in the NaCl and spinel structure from a lattice model, Geophys. J. Roy. Astron. Soc., **19,** 285-297, 1970.

Sammis, C., Anderson, D. L., Jordan, T., A note on the application of finite strain theory to ultrasonic and seismological data, J. Geophys. Res., **75,** 4478-4480, 1970.

Schreiber, E., The effect of solid solutions upon the bulk modulus and its pressure derivative: Implications for equations of state, Earth Planet. Sci. Letters, **7,** 137-140, 1969.

Simmons, G., Velocity of shear waves in rocks to 10 kilobars, J. Geophys. Res., **69,** 1123-1130, 1964.

Soga, N., Pressure derivatives of the elastic constants of vitreous germania at 25°, – 78.5°, and – 195.8°, J. Appl. Phys., **40,** 3382-3385, 1969.

______, Sound velocities of germanites compounds and its relation to the law of corresponding states, J. Geophys. Res., submitted, 1970.

Soga, N., Schreiber, E., and Anderson, O. L., Estimation of bulk modulus and sound velocities of oxides at very high temperatures, J. Geophys. Res., **71,** 5315-5320, 1966.

Thomsen, L., On the fourth-order anharmonic equation of state of solids, J. Chem. Phys. Solids, **31,** 2003-2016, 1970.

22
THE MELTING OF METALS

by **L. Knopoff and J. N. Shapiro**

A brief review of the Lindemann law is presented, including the improvements of Lennard-Jones and Devonshire, and Born. Lattice dynamics are used to reformulate the Lindemann criterion in terms of the frequency spectrum. Results are then presented for some face- and body-centered metals, using a central-force two-constant model. The initial slopes of the melting curves are calculated from third-order elastic constants. Improvements to the theory are suggested by considering the correlation present in the liquid state. Mercury and water are used as examples.

INTRODUCTION

In the year 1910, Lindemann (1910) gave a model for melting which, although elementary, has remained an active part of the extensive literature on melting since that time. Lindemann assumed that all the atoms in a solid were thermally excited simple harmonic oscillators, all presumed to be oscillating with the same frequency, under the Einstein approximation for the calculation of specific heat. Simply, then, at melting, one may equate the thermal energy imparted to each oscillator, $k_B T/2$ to the kinetic energy of each oscillator, $m\omega^2(\delta a)^2/2$ with $\omega = 2\pi\nu$ the angular frequency, m the mass of the oscillator, a the nearest neighbor spacing, and δ a dimensionless number, $0 < \delta < 1$. The amplitude of the oscillations is thus δa. Boltzmann's constant is k_B.

$$\frac{1}{2} k_B T_m = \frac{1}{2} m\omega^2 (\delta a)^2 \tag{1}$$

On Lindemann's model, the material dissociates when the energy of thermal oscillation is sufficiently great to cause neighboring atoms to collide; this is presumed to take place when the amplitude of the oscillations is some critical fraction δ of the nearest neighbor distance. If the nearest neighbor distance is expressed in terms of the specific volume V_m taken at melting, we get the formula, in the form given by Lindemann,

$$T_m = c_1 M\omega^2 V_m^{2/3} \tag{2}$$

where c_1 is supposed to be a universal constant, M is the molecular weight and ω, V_m, T_m are taken at the melting point.

Gilvarry (1956) used the Debye approximation for the spectrum to estimate the frequency ω and obtained the relation

$$T_m = c_2 K_m V_m f(\sigma) \tag{3}$$

where c_2 is a constant related to c_1, K_m is the acoustically measured bulk modulus at melting, and $f(\sigma)$ is a complicated but slowly varying function of Poisson's ratio, designed to take into account the transverse and longitudinal wave contributions to the vibrational spectrum. The Gilvarry version of the Lindemann relation could give only approximately constant values of c_2 because the emphasis is placed on the measurement of the properties (K, σ) at acoustic frequencies rather than in the neighborhood of the Debye frequency; the properties of the shear modes of wave propagation are a much more significant determinant of (3) than are the bulk modes (Knopoff and Shapiro, 1969).

It is of further interest to note a comment due to Lennard-Jones and Devonshire (1939a; hereafter abbreviated L-JD) in reference to (2): "It is to be observed, however, that this formula is not so much a formula for the melting temperature as a correlation between two observables T_m and ν, both of which should in a more fundamental theory be expressed in terms of interatomic forces."

The discussion of theories of melting based upon Lindemann's law therefore has three parts: (1) the prediction of the melting temperature from some criterion of melting expressed in terms of a model of interatomic forces; (2) the empirical determination of whether Lindemann's law is indeed universal; and (3) the determination of the appropriate frequency to be used in (1) and (2). We shall show below that the second and third problems are intimately related. We can only give a partial answer to each of these problems.

MELTING TEMPERATURES

The first problem, that of the prediction of the melting temperature, is in principle truly separable from the other two since it invokes only the calculation of a macroscopic, measurable quantity from some postulate of the nature of the intracrystalline potential and some melting criterion. This problem has been studied by a number of authors. We cite two of the most important discussions of the problem, namely those due to Born (1939) and to L-JD (1939a, b). Born establishes the criterion that the crystal should have zero shear modulus in the molten state, i.e., $c_{44} = 0$; he is not concerned whether the nonsolid state is liquid or gaseous. L-JD propose, as an alternate condition, the criterion by which a material passes from a highly ordered state to a partially disordered one. L-JD recognize that there *is* short-range order in a liquid and adjust their statistical mechanics to take into account the probability that the material has some order in the molten state. The theory of L-JD yields the volume change upon melting, the coefficients of expansion in the liquid and solid states, and the entropy change upon melting. After adjusting an empirical parameter, L-JD obtain values for the melting temperatures of Ne, A, N_2, CO, CH_4, to within 10%. Born does not predict melting temperatures.

Both theories apply to the melting of nonionic, nonmetallic solids. Both theories take the same two-body potential, namely

$$\phi(r) = \frac{\alpha}{r^{12}} - \frac{\beta}{r^6} \tag{4}$$

and predict that a sharp melting point should exist. Both theories show that the Lindemann relation derives from the models. It is presumed that the two theories give different melting temperatures.

What is truly important about these theories is that, in both cases, the Lindemann relation is derived for any two-body potential $\phi(r)$. Both theories use the harmonic oscillator approximation, which means that the specific form of $\phi(r)$ is not important; the local intracrystalline potential for a single particle is a quadratic form, which is the sum of the second-order terms in the Taylor expansions of $\phi(r)$ for all the neighboring particles, the expansion taken about the equilibrium position of each neighbor. We may assume that a change in the form of $\phi(r)$ will change the size of the constant in the Lindemann law: δ, c_1, c_2 in its various forms. If the Lindemann parameter is indeed a universal constant, it must be a determinant in obtaining the correct function $\phi(r)$, as is the melting temperature itself. At best the values of δ and T_m can be considered to be constraints on the classes of allowable functions $\phi(r)$, because of the harmonic oscillator approximation.

The calculation of the melting temperature has not yet been carried out for metals or for ionic crystals. In the case of metals, Born has noted, "Metals need a special treatment as the forces arising from the free electrons are not central." In the discussion that follows, we shall take into account the effects of free electrons in metals.

It is at this point that we leave the first problem, confident at least that there exists a machinery for handling the problem of the calculation of the melting temperature for some solids and hopeful that this can be extended to ionic and other nonmetallic solids. The reconciliation of the several competitive melting criteria in the hope that they all lead to a common melting temperature is still unresolved. But the criteria thus far discussed *do* give Lindemann's law as a consequence. If it is not true in nature that δ is a universal constant, then we may have to return to the first problem and investigate modifications of the theories of the determination of the melting temperature.

THE DETERMINATION OF THE LINDEMANN PARAMETER

Using Debye's method of calculating the frequency ω, that is using Eq. (3), Gilvarry obtained values of δ for the metals as shown in Table 1. It is seen that the results cannot be taken as conclusive that Lindemann's law is universal. Before rejecting the constructions of Lindemann's theory and those of the theory of melting consistent with it, we investigate the third problem, namely that of determining the frequency ω appropriate to Eq. (1).

The simple Lindemann-Einstein theory assumes that all the atoms in the lattice vibrate with the same frequency. That is, of course, not valid. N atoms in an

Table 1. Gilvarry's values for the Lindemann parameter (*nnd* = nearest neighbor distance)

Element	Structure	δ_{nnd}
Cs	bcc	.070
Rb	bcc	.065
K	bcc	.077
Na	bcc	.067
Li	bcc	.079
Pb	fcc	.077
Al	fcc	.066
Ag	fcc	.082
Cu	fcc	.079
Fe	bcc (trans. below melting)	.079
Ni	fcc	.085
Pd	fcc	.074
Pt	fcc	.082
Zn	hcp	.068
Mg	hcp	.074
Hg	rhomb	.045
Sn	tetra.	.046
Bi	rhomb	.048
Sb	rhomb	.051

m-dimensional lattice vibrate with mN degrees of freedom in mN modes. The atoms are all coupled, one to the other, and so the system cannot be considered to vibrate monochromatically; indeed, the monochromatic condition would arise only if there were no coupling between atoms. In a one-dimensional regular lattice, there is no degeneracy among the N modes. In a three-dimensional regular lattice, there is some degeneracy among the $3N$ eigenfrequencies. The modification of (1), which takes into account appropriately the fact that the atoms in the lattice are vibrating with different frequencies, was given by Blackman (1937) and is

$$\delta^2 = \frac{k_B T_m}{ma^2} \int_0^{\omega_{max}} \frac{g(\omega)\, d\omega}{\omega^2} \tag{5}$$

where $g(\omega)d\omega$ is the fractional number of vibrational frequencies in the frequency interval between ω and $\omega + d\omega$. We call $g(\omega)$ the spectrum. The spectrum is normalized by

$$\int_0^{\omega_{max}} g(\omega)\, d\omega = 1 \tag{6}$$

A succinct derivation of (5) is given by Pines (1963).

If $g(\omega) \sim \omega^n$, (5) becomes

$$\delta^2 = \frac{k_B T_m}{ma^2} \frac{(n+1)}{(n-1)} \frac{1}{\omega_{max}^2} \tag{7}$$

The case $n = 2$ gives the spectrum according to Debye's model. The result (5) is also consistent with the Einstein spectrum,

$$g(\omega) = \delta(\omega - \omega_E)$$

These lead directly to (1), if $\omega = \omega_E = \omega_{max} 3^{-1/2}$ ($n = 2$). Thus, we can get the original Lindemann relation by the use of either the Einstein or Debye estimates of the spectrum. For any other spectral distribution $g(\omega)$, it is clear that (5) and (6) yield

$$\delta^2 = \frac{k_B T_m}{ma^2 \omega_{max}^2 s} \tag{8}$$

where s is a pure number. It is our contention that the Lindemann Law can neither be demonstrated to hold nor to be violated in nature until s is evaluated separately for each material from the proper spectrum for that material.

We propose to calculate $g(\omega)$ by using a two-body potential function $\phi(r)$ which does not have a significant long-range part, although we recognize that the long-range component is important in preserving the long-range order of the solid. This is the same approximation made by Born, L-JD, and others. Instead of specifying the form of $\phi(r)$, we propose to evaluate the quadratic potential at a lattice site in terms of the nearest and first next-nearest neighbors and assume that the potential falls off so rapidly beyond the second neighbors that the effect of the more distant atoms can be neglected. Thus, our local potential is

$$\psi = \psi_0 + \psi_2 r^2 \tag{9}$$

The problem is to calculate ψ_2.

Fuchs (1935, 1936) has given a method that circumvents the question of noncentral forces resulting from the free electrons in a metal. Fuchs' relations are

$$\begin{aligned} c_{11} - c_{12} &= c'_{11} - c'_{12} \\ c_{44} &= c'_{44} \end{aligned} \tag{10}$$

Namely, for *bcc* and *fcc* metals, the shear elastic constants of a hypothetical metal with the electronic effects removed can be derived from the properties of the real metal through (10). With only two degrees of freedom in (10), we can calculate the two force constants (α_1, α_2) resulting from the nearest and first next-nearest neighbors according to the expressions

$$\begin{array}{ll} \textit{fcc} & \textit{bcc} \\ \dfrac{\alpha_1}{a} = c'_{44} & \dfrac{2\alpha_1}{3a} = c'_{44} \\ \dfrac{4\alpha_2}{a} = c'_{11} - c'_{12} - c'_{44} & \dfrac{2\alpha_2}{a} = c'_{11} - c'_{12} \end{array} \tag{11}$$

for the two cubic structures, where a is the lattice parameter. We express the condition that the forces beyond the two nearest neighbors be small by imposing the requirement $|\alpha_2/\alpha_1| \ll 1$. In practice, we take $|\alpha_2/\alpha_1| < 1/4$. Values of α_2/α_1 at room temperature are listed in Table 2 for a number of metals. The values of the elastic constants at the melting point are, in general, obtained by

Table 2. Ratio α_2/α_1 for cubic elements (at $T = 300^\circ$ K unless otherwise noted)

Element (bcc)	$\frac{1}{3}\frac{c_{11} - c_{12}}{c_{44}} = \frac{\alpha_2}{\alpha_1}$
Cr	.93
*Cs [78°K]	.092
Fe	.28
*K	.11
*Li	.076
Mo	.73
*Na	.093
Nb	1.3
*Rb [78°K]	.088
Ta	.42
V	.85
W	.66
Element (fcc)	$\frac{1}{4}\frac{c_{11} - c_{12} - c_{44}}{c_{44}} = \frac{\alpha_2}{\alpha_1}$
*Ag	- .082
*Al	+ .16
*Au	- .074
*Cu	- .094
Ni	- .045
Pb	- .13
Pd	- .072
Th	- .11
Ir [0°K]	+ .08
Pt	+ .06

extrapolation of data taken at room temperature and below. The materials used in this study are starred in Table 2; the others have been rejected for one or more of the following reasons: (1) inappropriate values of α_2/α_1, (2) phase changes between room conditions and melting, (3) inapplicability of linear extrapolation conditions resulting from curvature of the dependence of the elastic coefficients on temperature, or (4) insufficient data.

It should be noted that the model in which the elastic properties of the metal are attributed to the two nearest neighbors is in reality a lumped-parameter model. The effect of all the lattice points is presumed to be concentrated in these two "springs"; we do not state that the force constants α_1 and α_2 are the real values for these two springs.

With the force constants for each metal in hand, we can now solve for the spectrum $g(\omega)$. The one-dimensional version of this problem is shown in Fig. 1. For a large, finite lattice, the spectrum is discrete but dense and may be

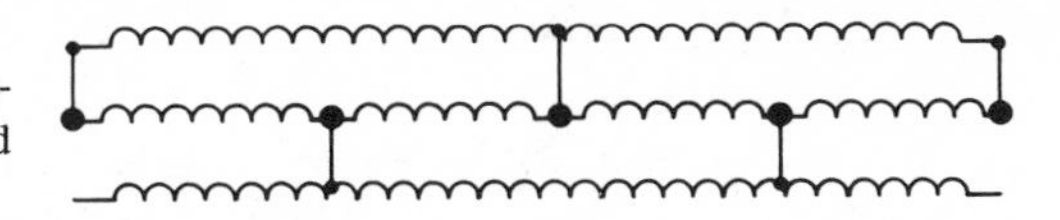

Fig. 1. Schematic diagram of a one-dimensional lattice with nearest and next-nearest neighbor interactions.

considered to be continuous (thereby giving meaning to the quantity $d\omega$). The lattice of Fig. 1 is easily solved for $g(\omega)$ by usual methods (Brillouin, 1946). The corresponding three-dimensional problem is made more complicated only by the geometry. Shapiro (1970) has modified the moment-trace method of Montroll (Maradudin et al., 1963) to obtain the moments of the spectrum, including the inverse second moment, by a powerful arithmetic device. For the purposes of possible academic interest, we show $g(\omega)$ for an *fcc* lattice with two different values of α_2/α_1 (Fig. 2).

The final result is

$$\delta^2 = \frac{k_B T_m}{3c_{44}a^3} f\left(\frac{\alpha_2}{\alpha_1}\right) \tag{12}$$

The functions f_{fcc} and f_{bcc} are plotted in Fig. 3, thus completing our task.

A correction to these quantities can be made to take into account the fact that the neighboring atoms are not stationary in space but are themselves vibrating. The method is outlined by Overton (1963). The corrections are, in general, small except for the elements of low atomic weight.

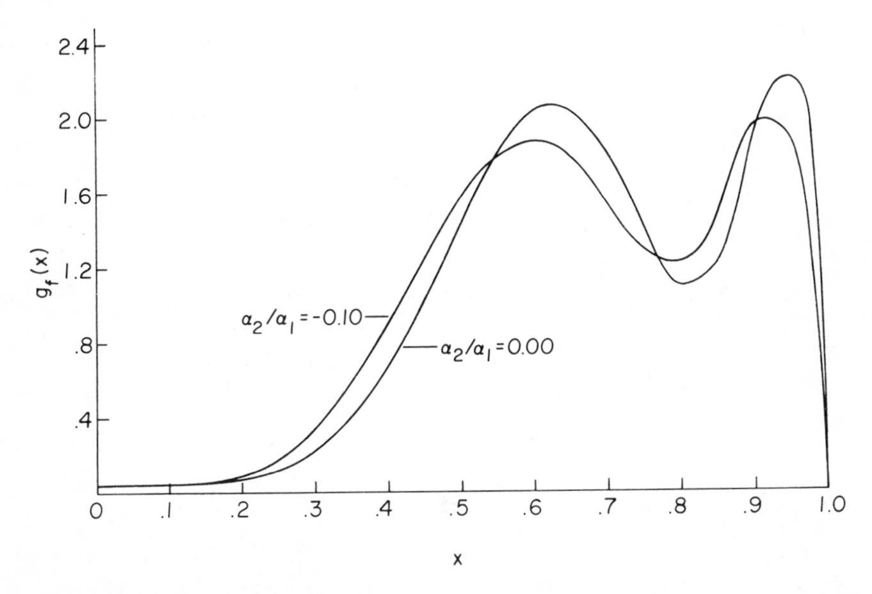

Fig. 2. Spectrum for an *fcc* lattice with $\alpha_2/\alpha_1 = 0.95$ and 1.0.

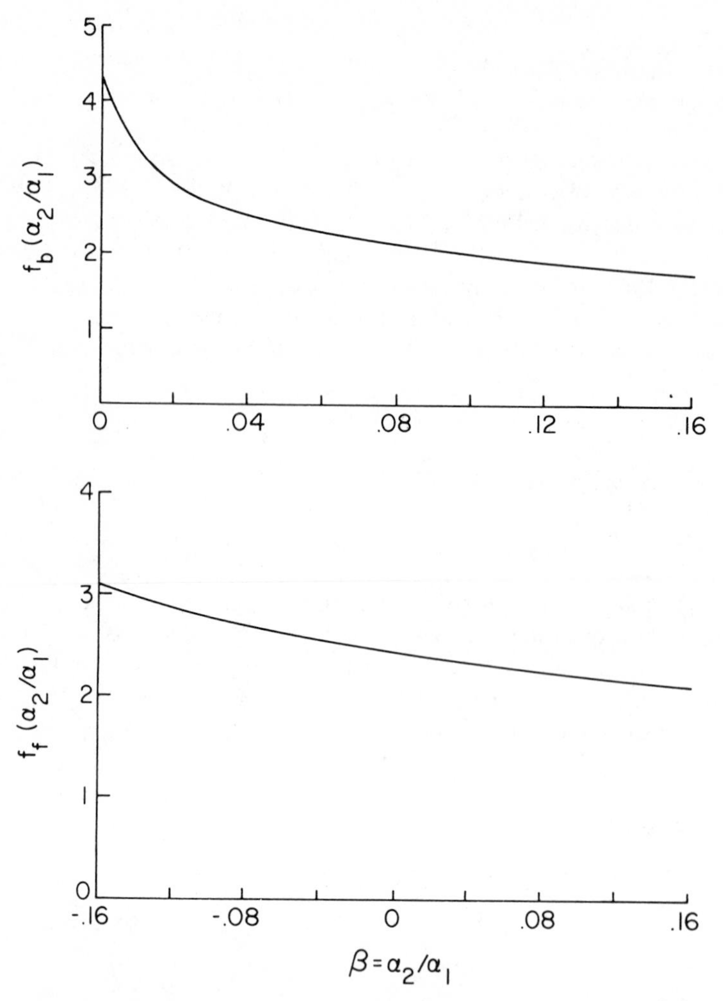

Fig. 3. The functions f_{fcc} and f_{bcc} for ranges of α_2/α_1 of interest.

The results for this calculation are shown in Table 3. It is seen that the values of δ are much more consistent among the entries than any estimates heretofore. We conclude that the Lindemann postulate is indeed a "law of corresponding states," but only for metals of the same crystallographic class, at least at one atmosphere (Shapiro, 1970).

SLOPES OF THE MELTING CURVES

There is no obvious reason that this demonstration should not be extended to elevated pressures. In this case, we may assume that δ is the appropriate constant determined above and solve (12) for T_m at some elevated pressure. We expand the right-hand side of (12) in a Taylor series; the terms of higher order involve derivatives *along the melting curve* of the quantities c_{44}, V_m, and α_2/α_1. In terms of isothermal and isobaric derivatives, the most important terms are

$$\left.\frac{\partial}{\partial P}\right|_T, \quad \left.\frac{\partial}{\partial T}\right|_P, \quad \text{and} \quad \frac{\partial^2}{\partial P \partial T}$$

to low orders. Unfortunately, data for the partial derivatives, and especially for

$$\frac{\partial^2 c_{44}}{\partial P \partial T}, \quad \frac{\partial^2 c_{12}}{\partial P \partial T}, \quad \frac{\partial^2 c_{11}}{\partial P \partial T}$$

are not plentifully available for the metals. For those that are available, we extrapolate by the usual linear procedures to predict the properties of the metals at elevated pressures at the melting point. Comparison is made in Table 4 with the observations of Newton et al. (1962), Kennedy and Newton (1963), Jayaraman et al. (1963), and Cohen et al. (1966) for the slopes of the melting curves. In almost every case for which experimental data on the derivatives of elastic constants are available, our empirical slopes are systematically less than the experimental slopes.

We may offer at least two explanations for this systematic difference. Our easiest course is to place upon the experimentalists the burden of the problem of making the two sets of numbers equal; the measurement of $\partial c_{44}/\partial P$, etc. is admittedly difficult and that of the derivative of this quantity with temperature is even more difficult.

Table 3. Theoretical Lindemann parameters using lattice dynamics

	Element	δ
b.c.c.		
	Li	.118
	Na	.108
	K	.109
	Rb	.110
	Cs	.106
f.c.c.		
	Al	.092
	Cu	.093
	Ag	.090
	Au	.097

Table 4. Ratio dT_M/dP for seven cubic metals

Element	dT_M/dP	
	Theoretical	Experimental
Na	4.94[a]	8.0[f]
K	9.74[a]	15.3[f]
Al	5.94[a] 4.86[b]	5.9[g]
Cu	3.02[a] 3.13[c]	3.95[h]
Ag	4.73[a] 6.17[c]	5.87[h]
Au	4.46[a] 3.72[c]	6.12[h]
Pb	5.03[d] 5.89[e]	6.5[i]

[a]Barsch and Chang (1967).
[b]Thomas, Jr. (1968).
[c]Hiki and Granato (1966).
[d]Miller and Schuele (1969).
[e]Data on pressure derivatives taken from (d) and extrapolated to melting.
[f]Newton et al. (1962).
[g]Jayaraman et al. (1963).
[h]Cohen et al. (1966).
[i]Kennedy and Newton (1963).

It would be honest, however, to admit that certain theoretical difficulties are also connected with the problem of the calculation of dT_M/dP at zero pressure, or, what is equivalent in our approximation, T_M at slightly elevated pressure. In the L-JD model of melting, the melting is presumed to take place on the basis of a transition from the highly correlated solid state to that of a poorly correlated liquid. We wish to argue that the correlation of a liquid probably increases with pressure and that one will find a dependence of the Lindemann parameter δ on the correlation of the liquid. Qualitative arguments can be made, at present, that the melting temperature at one atmosphere, predicted by the L-JD model, rises if it is required that the solid dissociate into slightly ordered clusters of atoms instead of into a totally disordered fluid. The usual Lindemann relation presumes dislocation of the solid without reference to the precise state, liquid or gaseous, of the nonsolid phase. It is our contention that the properties of the liquid must be taken into account and, when this is done for an increase in correlation, the theoretical values of dT_M/dP will be more in accord with experiment. Regrettably, the quantitative evaluation of the dependence of δ

upon correlation in the liquid has not yet been carried out. We now make qualitative arguments that the correlation in liquids increases with increasing pressure.

CORRELATION IN LIQUIDS

What distinguishes a liquid from a gas is the spatial correlation; a liquid has short-range correlation, while a gas does not. We define the spatial density correlation function as the integral

$$f(r) = \langle \int \rho(r, \theta, \phi)\, d\Omega \rangle$$

in a noninertial coordinate system with origin fixed in some particle and where the average is taken over all particles and over a long time. We use this coordinate system to avoid the complicating feature of the motion of the reference particles in an inertial frame. This quantity, properly normalized, gives the probability that we find mass at a distance r from any particle. There will be a strong self-correlation effect at short range because of the mass distribution in the reference particle. This self term is usually subtracted from $f(r)$ to leave the function $g(r)$, the density correlation with neighbors. For a liquid, the function $g(r)$ rises from zero at $r = 0$, usually executes a few damped oscillations and finally settles to a constant value indicating that there is a uniform probability of finding another particle at large distance from any other. The correlation function for liquids at one atmosphere is usually obtained experimentally by x-ray scattering or neutron diffraction experiments (Egelstaff, 1967; March, 1968; Ubbelohde, 1965). For gases, the correlation function rises from zero to a constant value without oscillations. Typically, two or three oscillations can be seen in the correlation function for liquids.

Ubbelohde calls these correlated micro-units of liquids *clusters.* We presume that the "size" of these clusters increases if the number of oscillations in $g(r)$ increases. Experiments show that the local structure of such clusters in the melt, at temperatures just above the melting point, is the same as that just below the melting point and with the same lattice dimensions. Thus, we conclude that the volume expansion of the material upon melting is a result of the disorder among the clusters and not of an intrinsic change in lattice dimensions. We assume that these clusters have a locally rigid structure. As a consequence, we propose to investigate the vibration spectrum of these clusters to learn something about their correlation.

A quantity that is a useful indicator of the properties of the spectrum is the Grüneisen parameter $\Gamma(V, T)$, which we define to be the dimensionless combination of thermodynamic derivatives

$$\Gamma(V,T) = \frac{K_s \alpha V}{C_p} \tag{15}$$

where K_s is the adiabatic bulk modulus, α is the thermal coefficient of expansion, V is the specific volume, and C_p is the specific heat at constant pressure. In the case of a large crystalline solid of finite size, Grüneisen (1912) obtained the equation

$$\Delta P = \frac{\gamma U}{V} \tag{16}$$

where ΔP is the pressure excess required to keep the solid confined to the same specific volume V when the temperature is raised from zero to T, and U is the internal energy. In this case, namely for a large metallic crystal, $\gamma = \Gamma$ and is independent of temperature. It is easy to show that, if Γ is independent of T, (16) follows from (15) and also, if (16) holds for Γ independent of T, (15) follows. If the Debye model of the spectrum of a solid is assumed, $\gamma = \gamma_D$ with

$$\gamma_D = -\frac{d \log \omega}{d \log V} \tag{17}$$

Shapiro and Knopoff (1970) have shown how to calculate γ for a more realistic model, namely for cubic crystals using models considered above.

For liquids, we can evaluate Γ from (15) from direct measurements of the velocity of sound C_s,

$$\Gamma = \frac{C_S^2 \alpha}{C_p} \tag{18}$$

Because of its high compressibility, water is the liquid with the largest span of volumes for which experimental data are available for the calculation of Γ. Water is also the liquid with the most precise (P, V, T) data for the taking of thermodynamic derivatives. $\Gamma(V, T)$ for water is shown in Fig. 4 (Knopoff and Shapiro, 1970) using data of Vedam and Holton (1968). $\Gamma(V, T)$ can be drawn for Hg from the data of Davis and Gordon (1967) (Fig. 5).

What is most remarkable about the behavior of Γ for both H_2O and Hg is that it is a decreasing function of V. For solids, γ is an increasing function of V (Fig. 4); this fact is often used in making reductions of shock wave observations from Hugoniot to isothermal equations of state. Despite our earlier criticism of Debye's theory of the spectrum, it gives a satisfactory *qualitative* description of

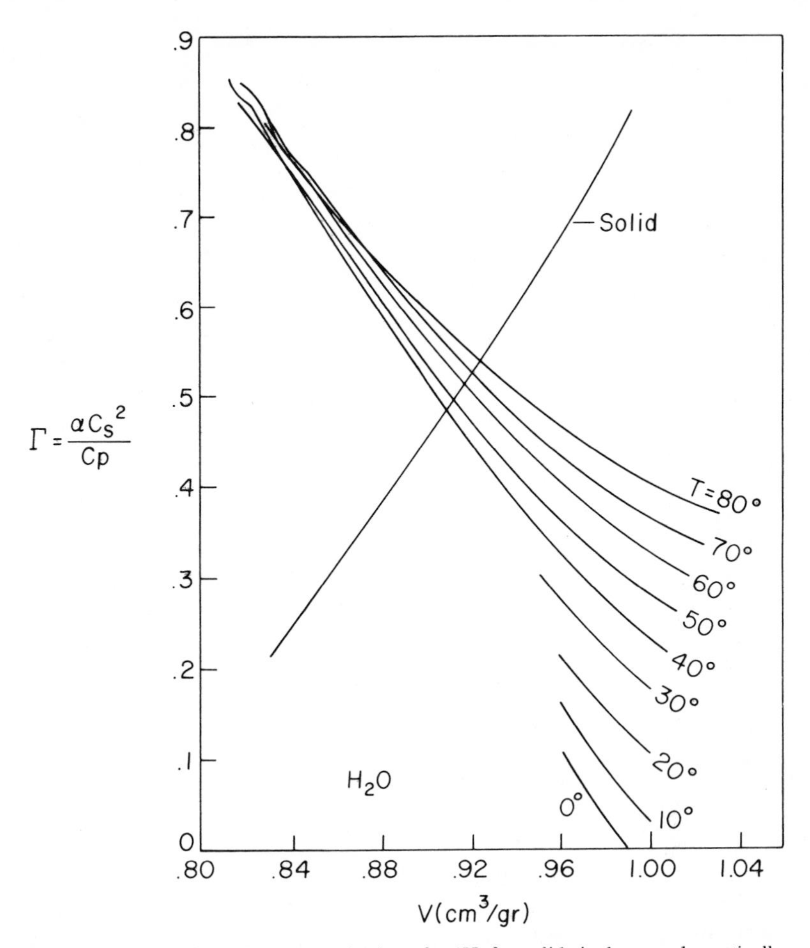

Fig. 4. $\Gamma(V, T)$ for water. The behavior of $\gamma(V)$ for solids is shown schematically.

the vibrational behavior of a crystal. Knopoff and Shapiro (1970) have shown that, if a postulate is made that the size of clusters increases with increasing pressure, and if the clusters are considered as microscopic *solid* units, γ is not independent of temperature, and is

$$\gamma = \gamma_D + \xi(T, V) \frac{d \log N(V)}{d \log V} \tag{19}$$

where ξ is a known function. The additional term, over and above the term γ_D for the almost-infinite solid, arises from the dispersion relation. If the rate of growth of a cluster $N(V)$ is stronger than the power law, $N \sim V^s$, $(s < 0)$, e.g., $N/N_0 \sim \exp(V/V_0)^p$, $(p < 0)$, then $\gamma(V)$ for a finite lattice may have the slope $d\gamma/dV|_T$ reversed from that for an infinite solid. In one dimension, for $N > 10$,

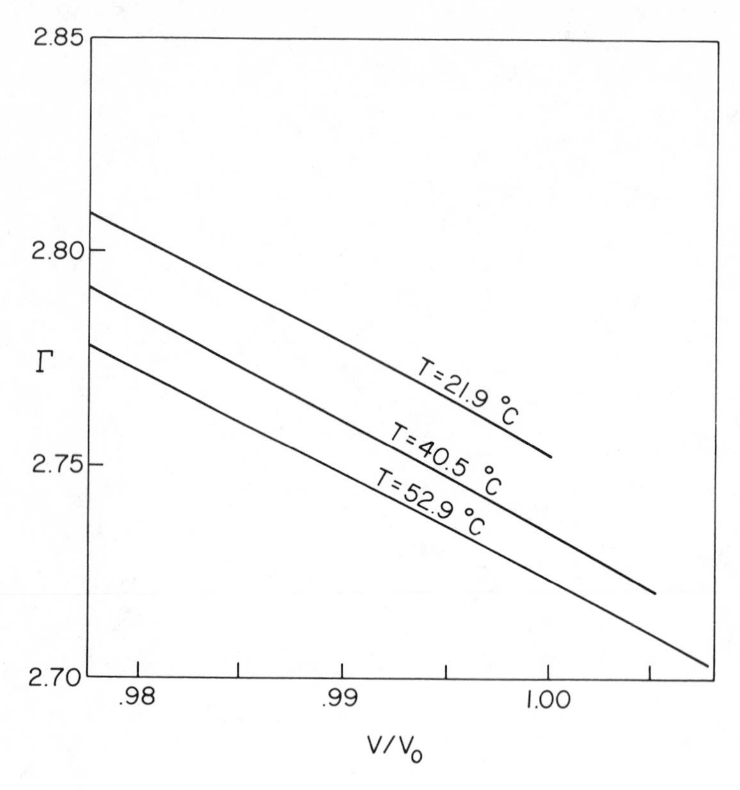

Fig. 5. $\Gamma(V, T)$ for Hg.

approximately, ξ is substantially independent of V. The one-dimensional theory predicts γ is an increasing function of T at constant volume; this is the case for H_2O but not for Hg! This difficulty may be a result of the different three-dimensional lattice structures of the two liquids, but we have not yet explored this point.

Thus, if this model is valid, the size of liquid clusters increases with increasing compression, i.e., we expect more oscillations in the correlation function $g(r)$ with increasing pressure. In view of our earlier speculation, this may indicate that our estimates of dT_M/dP are too low. Hence, we may postulate an additional constraint to the statement of Lindemann's law: this law holds for metals of the same crystallographic class, for those metals whose liquid phases at one atmosphere have low correlation numbers.

Is it possible to find a pressure sufficiently high that the growth of clusters has advanced to such a state that the clusters are immobilized into a granular solid? This may be, but we have no evidence bearing on this problem. Molten iron at 1.5 to 2.5 megabars certainly does not transmit S waves, but perhaps the inner core is such a phase. The observations on water for $V < 0.84$ cm^3, which show that Γ is independent of T, should not be interpreted in this way, since $\Gamma(V)$ still has the reversed slope from that of a solid in this range. Can ionic solids, such as the silicates, under high pressures exhibit quasi-solid behavior in the molten state? If this speculation is correct, then melting ceases to be a property of matter at sufficiently high pressures and thus renders trivial the old controversy of the value of $T_m(V)$ as V approaches zero. This notion is not new; indeed it had been considered and rejected by P. W. Bridgman.

EPILOGUE

Where does the future in this type of research lie? We think that investigation in the *near* future must study (1) whether Lindemann's law is applicable for ionic solids, so that we can start to study the behavior of mantle materials, (2) what the quantitative nature of melting is at high pressures, (3) the estimation of the importance of anharmonic terms in the frequency spectrum, in view of our rather surprising result that the Lindemann law holds to good accuracy without invoking these terms, (4) the determination of the class of two-body potentials that give accurate predictions of both the melting temperature and the Lindemann parameter δ, and (5) the demonstration of the equivalence of the various criteria for melting, although we think this problem should have the lowest priority. To determine theoretical melting conditions at pressures greater than one atmosphere, we shall need experimental data in both the liquid and solid states of the elastic properties of materials near the melting point; we herewith enter our plea for the execution of these crucial measurements. Finally, with regard to the theoretical program we have outlined, we would urge, on the basis of the experience described above, that these studies proceed slowly and that little of the older work be taken as valid–including this one.

REFERENCES

Barsch, G. R., and Chang, Z. P., Adiabatic, isothermal and intermediate pressure derivatives of the elastic constants for cubic symmetry, II, Phys. Stat. Sol., **19**, 139-151, 1967.

Blackman, M., The effect of temperature on the reflexion of X-rays, Proc. Cambridge Phil. Soc., **33**, 380-384, 1937.

Born, M., Thermodynamics of crystals and melting, J. Chem. Phys., **7**, 591-603, 1939.

Brillouin, L., Wave Propagation in Periodic Structures, McGraw-Hill, New York, 1946.

Cohen, L. H., Klement, W., Jr., and Kennedy, G. C., Melting of copper, silver and gold at high pressures, Phys. Rev., **145**, 519-525, 1966.

Davis, L. A., and Gordon, R. B., Compression of mercury at high pressure, J. Chem. Phys., **46**, 2650-2660, 1967.

Egelstaff, P. A., An Introduction to the Liquid State, Academic Press, New York, 1967.

Fuchs, K., A quantum mechanical calculation of the elastic constants of monovalent metals, Proc. Roy. Soc., London, Ser. A, **153**, 622-639, 1935.

______, The elastic constants and specific heats of the alkali metals, Proc. Roy. Soc., London, Ser. A, **157**, 444-450, 1936.

Gilvarry, J. J., The Lindemann and Grüneisen laws, Phys. Rev., **102**, 308-316, 1956.

Grüneisen, E., Theorie des festen Zustandes einatomiger Elemente, Ann. Physik, **39**, 257-306, 1912.

Hiki, Y., and Granato, A. V., Anharmonicity in noble metals; higher order elastic constants, Phys. Rev., **144,** 411-419, 1966.

Jayaraman, A., Klement, W., Jr., Newton, R. C., and Kennedy, G. C., Fusion curves and polymorphic transitions of the group III elements, aluminium, gallium, indium, and thallium at high pressures, J. Phys. Chem. Solids, **24,** 7-18, 1963.

Kennedy, G. C., and Newton, R. C., *in* Solids Under Pressure, p. 171, McGraw-Hill, New York, 1963.

Knopoff, L., and Shapiro, J. N., Comments on the interrelationships between Grüneisen's parameter and shock and isothermal equations of state, J. Geophys. Res., **74,** 1439-1450, 1969.

______, A pseudo-Grüneisen parameter for liquids, Phys. Rev. B, **1**, 3893-3895, 1970.

Lennard-Jones, J. E., and Devonshire, A. F., Critical and cooperative phenomena III, A theory of melting and the structure of liquids, Proc. Roy. Soc., London, Ser. A, **169,** 317-338, 1939a.

_____, Critical and cooperative phenomena IV, A theory of disorder in solids and liquids and the process of melting, Proc. Roy. Soc., London, Ser. A, **170,** 464-484, 1939b.

Lindemann, F. A., Molecular frequencies, Phys. Zeitschr., **11,** 609-612, 1910.

Maradudin, A. A., Montroll, E. W., and Weiss, G. H., Theory of Lattice Dynamics in the Harmonic Approximation, Academic Press, New York, 1963.

March, N. H., Liquid Metals, Pergamon Press, New York, 1968.

Miller, R. A., and Schuele, D. E., The pressure derivatives of the elastic constants of lead, J. Phys. Chem. Solids, **30,** 589-600, 1969.

Newton, R. C., Jayaraman, A., and Kennedy, G. C., The fusion curves of the alkali metals up to 50 kilobars, J. Geophys. Res., **67,** 2559-2566, 1962.

Overton, W. C., FCC elastic constants for arbitrary central forces by the method of long waves, J. Chem. Phys., **38,** 2482-2493, 1963.

Pines, D., Elementary Excitations in Solids, W. A. Benjamin, New York, 1963.

Shapiro, J. N., Lindemann law and lattice dynamics, Phys. Rev. B, **1**, 3982-3989, 1970.

Shapiro, J. N., and Knopoff, L., Grüneisen parameter for Bornvon Kármán lattices, Phys. Rev. B, **1**, 3990-3992, 1970.

Thomas, J. F., Jr., Third-order elastic constants of aluminum, Phys. Rev., **175,** 955-962, 1968.

Ubbelohde, A. R., Melting and Crystal Structure, Oxford Univ. Press, 1965.

Vedam, R., and Holton, G., Specific volumes of water at high pressures, obtained from ultrasonic-propagation measurements, J. Acoust. Soc. Am., **43,** 108-116, 1968.

23
STRENGTH OF METAMORPHOSED GRAYWACKE AND OTHER ROCKS

by **Eugene C. Robertson***

The strengths of mantle rocks may reach 100 bars, even for stresses of long duration; the minimum strength may be less than 0.01 bar. The strengths of crustal rocks are better known from experiments and range from 1 bar to more than 20 kb, depending on the pressure-temperature-time-composition conditions.

Deformation experiments on an isochemical suite of rocks of graywacke composition show that confining pressure (to 6 kb) and decrease in porosity (from 35 to 1%) strengthens the rocks but that they are weakened by applying pore water under pressure (to 4 kb), by heating (to 450°C) and by decreasing the strain rate (to 10^{-6} sec^{-1}). Metamorphism strengthens the rocks of this chemically equivalent suite in two stages: low grade by compaction and zeolite cementation of the graywacke, and intermediate grade by recrystallization and induration of the zeolite facies to the blueschist facies and to the greenschist facies. The experimental data indicate that the strength of the graywacke might sustain a tectonic overpressure sufficient to form the blueschist facies, but weakening by pore-water pressure or by interbedded shale could counteract this effect.

*Publication authorized by the Director, U.S. Geological Survey.

INTRODUCTION

The strengths of a suite of rocks of graywacke composition were studied experimentally, and these new results are discussed in relation to general concepts of rock strength and metamorphic grade. Considerable attention is also given to data from other sources, principally experimental ones, on the strength of rocks in the crust and mantle, as affected by the pressure, temperature, and chemical conditions. Although experiments on small samples cannot be directly extrapolated to large bodies of rock in the earth because of the larger scale heterogeneities and structural complexities, they give insight into possible effects of rock deformation in the earth. One intention of this paper is to place the new results in a framework that will elucidate the factors affecting rock strength and be informative to readers unfamiliar with the concepts of rock deformation.

Ultimately, the geologic structures developed in the relatively strong and brittle rocks in the crust must be related to the movements of the relatively weak rocks deeper in the earth, and conjectures on the structural mechanisms of that relationship are receiving much attention at present. Elsewhere in this volume, Gilluly, Kaula, McKenzie, and Menard discuss tectonic features associated with the hypothesis of plate tectonics, and Griggs discusses this hypothesis in regard to the way in which strong crustal rocks interact with weak rocks in the mantle to produce deep-focus earthquakes.

The estimates of the strength of the mantle of the earth made by Francis Birch (1964) are still useful approximations. (The strength of the mantle may be described as the perceptible resistance to continuous deformation of long duration, or more specifically, as the stress difference producing some minimum observable strain by creep perhaps at a rate of 10^{-11} sec^{-1} or more.) Birch used data from geodesy, gravity, and geology. From satellite observations, the equipotential surface is found to show significant departures from a geoid calculated for an assumed hydrostatic equilibrium. Birch (1964) pointed out that these broad departures from the equilibrium figure indicate nonhydrostatic stresses of the order of 100 bars in the earth's mantle. A stress difference of 10 to 100 bars is also indicated by the lack of correspondence between geoidal height and topography of continental size. The observations of gravity show that most of the solid earth's surface is within a few hundred meters of the elevation it would have if the crust were floating on a mantle of high density but low strength, and according to Birch, this lack of compensation over long time periods suggests that the yield point in the upper mantle must be of the order of 100 bars. Birch (1964), however, found evidence for greater strength of rocks near the earth's surface on a smaller scale. The strength of the crust and upper mantle may be a few thousand bars under high mountains, for example under the island of Hawaii, and under the Green Mountain anticlinorium where a positive gravity anomaly of 50 mgal may have persisted for 100 m.y.

A comprehensive review of the mechanical properties of rocks in the mantle, analyzed from many points of view, is given in the proceedings of a symposium on nonelastic processes in the mantle (Tozer, 1967); the treatment of the phenomena of creep, plasticity, viscosity, attenuation, rigidity, tidal damping, fracturing, and convection is theoretical for the most part, as the experimental data and field observations bearing on these phenomena are quite meagre. The experimental data on the strength of mantle rocks were recently reviewed by Raleigh and Kirby (1970). Weertman (1970) gives a recent theoretical analysis of creep mechansims in ultramafic rocks under the conditions of the upper mantle.

Results of these studies suggest a range in strength of mantle rocks from 10^{-2} to 10^{+2} bars; this range is low enough to satisfy the theoretical models of convection in the upper mantle and to permit sea floor spreading. As implied above, however, strength can vary with time of application of stress; in most models, steady-state creep or viscoelastic behavior is assumed, whereas most of the available creep experiments show an exponential strain-time relationship and work-hardening in silicate rocks. Much more experimental work needs to be done on deformation mechanisms and rates of mantle rocks under the actual pressure and temperature conditions of the mantle in order to specify such processes as creep, damping, and convection.

Data from experiments on rock deformation provide estimates of the strength of particular rock specimens, in contrast to the strength estimates for large rock masses made from data on mass distribution and the figure of the earth. Although the rocks used in the laboratory tests are taken from the earth's surface and so represent only a very small part of the earth's volume, their mechanical properties can be studied under conditions similar to those of the earth's interior. *Rock strength in laboratory experiments may be defined as the maximum stress difference at fracture, or if fracture does not occur, the greatest stress difference producing an observable steady-state creep.*

Innumerable investigations have been made of the strengths of rocks at room temperature, many of which are described in the proceedings volumes of the many Rock Mechanics symposia. Examples of a few representative investigations of the effects of other conditions on strength can be listed: studies at higher temperature by Griggs et al. (1960) and Handin and Hager (1958); studies of strain rate at high temperature by Heard (1963) and Heard and Carter (1968); studies of pore pressure by Handin et al. (1963) and by Brace and Martin (1968). A complete listing of references for experimental research on the mechanical properties of rocks from 1870 through 1969 is given in Gralewska (1969).

An interesting experimental manifestation of rock strength has been observed by Brace and Byerlee (1966) in which there is a cyclic build-up of strength across a locked fracture surface followed by sudden sliding displacement, a behavior called "stick-slip." This behavior is similar to the repetitive motions of shallow crustal earthquakes and their aftershocks. At low temperatures and low confining pressures, stick-slip was observed in experiments on fresh granite, whereas stable sliding occurs as a post-fracture behavior in altered gabbro, dunite, and granite (Byerlee and Brace, 1969). If the temperature is high enough, stable sliding is also found in fresh granite and gabbro, which may account for the depth limit of 15 km for most California earthquakes, with stable sliding occurring at greater depths because of the higher temperatures there (Brace and Byerlee, 1970).

In the studies reported to date, particular rock types or rocks from specific localities have been used in deformation experiments. No systematic work has been done on differences in mechanical properties in rocks of constant chemical composition but varying metamorphic grade with respect to consolidation, lithification, and changes in mineral assemblages. An increase in strength and rigidity with increase in solidity and induration would certainly be anticipated, but experiments have been needed to show in detail how improvement of the intergranular bonding affects strength. This paper is a report of such an investigation of a nearly isochemical suite of rocks at room temperature and under moderate confining pressure. Some unpublished as well as published data on one member of the rock suite were supplied by W. F. Brace and his coworkers from MIT.

SAMPLE COMPOSITION AND METAMORPHIC GRADE

The suite of eight types of rocks studied is a collection of metamorphosed graywackes from the Franciscan Formation and similar rocks from California, Oregon, and Nevada. The chemical composition of all rock types is similar, each containing 65 to 70% SiO_2 (ignoring $CaCO_3$), for example; but the petrographies are quite different. The rocks range in porosity and metamorphic grade from loosely consolidated sediment of the zeolite facies to a dense jadeite-bearing rock of the blueschist facies, and to a massive metagraywacke with albite and clinozoisite of the greenschist facies. The chemical analyses, modal compositions, calculated blueschist compositions, densities, porosities, and localities of the rocks studied are listed in Table 1.

Zeolite Facies

Rocks of the zeolite facies, samples 22, 63, and 64 (Table 1) were chosen as very low grade metamorphic equivalents of the Franciscan Formation, as was sample B24f, which is a rhyolite tuff with a similar chemical composition and composed of more than half zeolite minerals. (Detrital orthoclase is incorporated with albite in the modes in Table 1 for samples 22, 63, 64, B24f, 52, and 1; no attempt was made to identify and count it in the thin sections. In these same samples, the brown hydromicas, including stilpnomelane and detrital biotite, have been combined as biotite.) The diagenesis of the porous sediments to zeolite-bearing rocks was principally a near-surface compaction, a removal of water from the original sediment, and low-grade metamorphism. Judging from the data of Athy (1930) and assuming that overlying load of rock and slightly raised geotherms suffice in the diagenesis, the sediment would be compacted to 35% porosity at about 0.3-km depth, and to 2.5% porosity at about 3-km depth.

Table 1. Chemical analyses, modes, and calculated blueschist compositions of a suite of siliceous rocks[1]

Chemical analyses[2]

Oxide	(Weight percent)							
	B24f	22	63	64	52	1	8	14
SiO_2	65.0	66.1	55.8	59.3	69.0	65.5	61.1	69.7
Al_2O_3	13.4	13.9	9.9	10.8	14.3	15.2	12.3	12.7
Fe_2O_3	1.9	2.4	1.4	.80	.89	.52	1.6	1.3
FeO	.3	1.4	3.6	4.0	2.1	4.8	2.3	2.5
MgO	1.2	1.9	3.8	3.9	1.6	2.3	2.3	2.5
CaO	3.7	4.0	9.4	6.5	2.5	1.1	7.5	2.6
Na_2O	1.4	2.5	2.1	2.2	3.5	3.5	2.4	1.7
K_2O	1.5	1.9	.98	1.0	2.5	2.3	1.8	2.0
H_2O	10.6	2.6	3.1	3.1	2.3	2.4	2.8	3.6
TiO_2	.3	.60	.50	.54	.46	1.0	.54	.5
P_2O_5	.03	.11	.09	.09	.13	.20	.11	.11
MnO	.04	.08	.75	.56	.03	.12	.06	.05
CO_2	.16	1.5	7.5	5.5	.40	$< .05$	5.1	.42
Total	99.5	99.0	98.9	98.3	99.7	99.0	99.9	99.7

Mode

Mineral[3]	(Volume percent)							
	B24f	22	63	64	52	1	8	14
lim	1	2	1	1				
il-mt-rut	1	1	1	1	1	1	1	1
sph-ap						1	1	1
mv		10	2	2	12	13	15	15
bio	4	3			7	2		
czo					7	2		
ct		3	16	13	1			
arag							11	1
zeol	54	22	22	21				
law							2	9
glauc							6	
clt		5	16	19	5	13	9	10
ab	12	18	11	12	29	30		
jd							12	11
qz	28	36	31	31	38	38	43	52
Total	100	100	100	100	100	100	100	100

Table 1. Chemical analyses, modes, and calculated blueschist compositions of a suite of siliceous rocks (continued)

Calculated blueschist composition

Mineral[3]	(Volume percent)							
	B24f	22	63	64	52	1	8	14
misc.	2	3	2	2	2	2	2	2
bio					7			
mv	13	16	9	8	13	19	15	15
arag		3	16	12	1		11	1
law	18	8			9	2	2	9
clt		5	15	18		8	9	11
glauc	9	4	3		3	14	6	
jd	8	14	12	14	20	16	12	10
qz	50	47	43	46	45	39	43	52
Total	100	100	100	100	100	100	100	100
Density (gm/cm^3)	1.59	2.42	2.45	2.60	2.69	2.66	2.80	2.82
Porosity %	35	9.0	7.5	2.6	0.6	1.0	0.6	0.6

[1]*Localities of samples*: No. 22–Valley sequence (Knoxville Formation), near San Luis Reservoir, Calif. No. 63, 64–Umpqua Formation, Curry County, Oregon. No. 52–Franciscan Formation, San Rafael, Calif. No. 1–Franciscan Formation, San Bruno Mountain, Calif. No. 8, 14–Franciscan Formation, Pacheco Pass, Calif. No. B24f–Oak Spring Formation (of former usage), Tos_4 bed, Nevada Test Site, Nevada. Collectors of samples: B24f, G. V. Keller; other samples, R. G. Coleman and E. C. Robertson.

[2]Chemical analyses of 22, 63, 64, 52, 1, 8 and 14 performed by P. Elmore, L. Artis, J. Glenn, G. Chloe, H. Smith, J. Kelsey, U.S. Geological Survey. Chemical analysis of B24f by M. Balazs, P. Elmore, S. Botts, M. Mack, D. Powers, P. Bamett, K. Hazel, R. Havens, U.S. Geological Survey.

[3]Abbreviations: ab–albite, ac–acmite, and–andalusite, ap–apatite, arag–aragonite, bio–biotite, ct–calcite, clt–chlorite, czo–clinozoisite, cord–cordierite, di–diopside, flsp–feldspar, glauc–glaucophane, il–ilmenite, jd–jadeite, ky–kyanite, laum–laumontite, law–lawsonite, lim–limonite, mt–magnetite, mv–muscovite, or–orthoclase, pyroph–pyrophyllite, px–pyroxene, qz–quartz, rut–rutile, sph–sphene, staur–staurolite, zeol–zeolite, zo–zoisite; misc–miscellaneous.

Normally, at 3 km the pressure would be about 1 kb and the temperature about 100°C; however, the field of stability of the zeolite facies might extend to 3-kb pressure and to 200°C, as shown in Fig. 1; Madsen and Murata (1970) estimate that the laumontite found in sandstone in the California Coast Ranges formed at 100 to 240°C and at 2- to 6-km depth. An upper pressure limit is placed approximately by the steep laumontite-lawsonite phase boundary (Thompson, 1970). A photomicrograph in Fig. 2a shows the petrography, texture, and consolidation of sample 63.

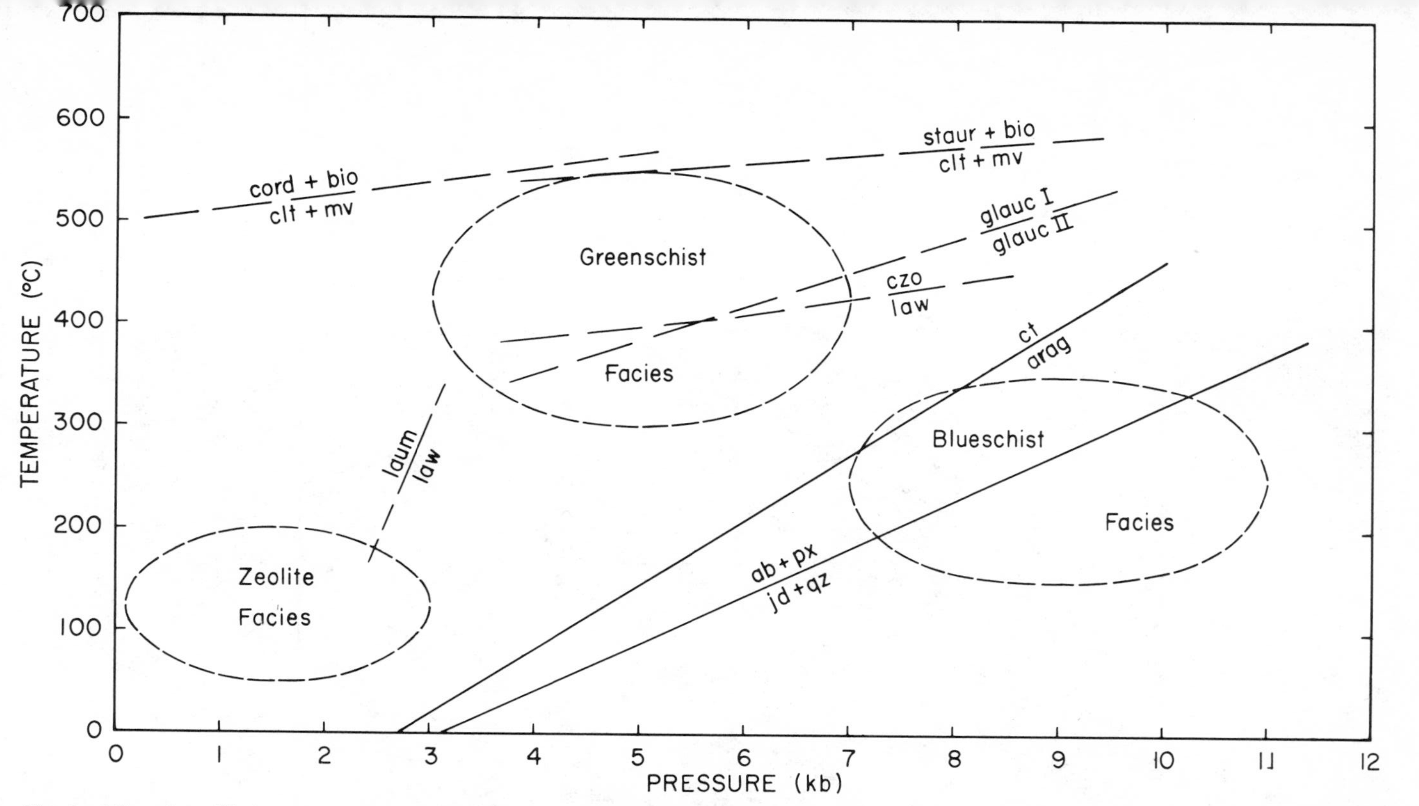

Fig. 1. Mineral equilibria and metamorphic facies of graywacke on the pressure-temperature field; for abbreviations, see footnote to Table 1: ab + px = ($jd_{82} - ac_{14} - di_4$) + qz, from Newton and Smith (1967); ct = arag, from Boettcher and Wyllie (1968); laum = law + 2 qz + 2 water, from Thompson (1970); 4 law + 2 qz = 2 zo + pyroph + 6 water, from Nitsch, cited in Winkler (1970), and an + water = zo + ky + qz, from Newton and Kennedy (1963); glauc I = glauc II, from Ernst (1961, 1963); (clt + mv + qz) = (cord + bio + and + water), from Hirschberg and Winkler (1968); (clt + mv) = (staur + bio + qz + water), from Horschek (1969).

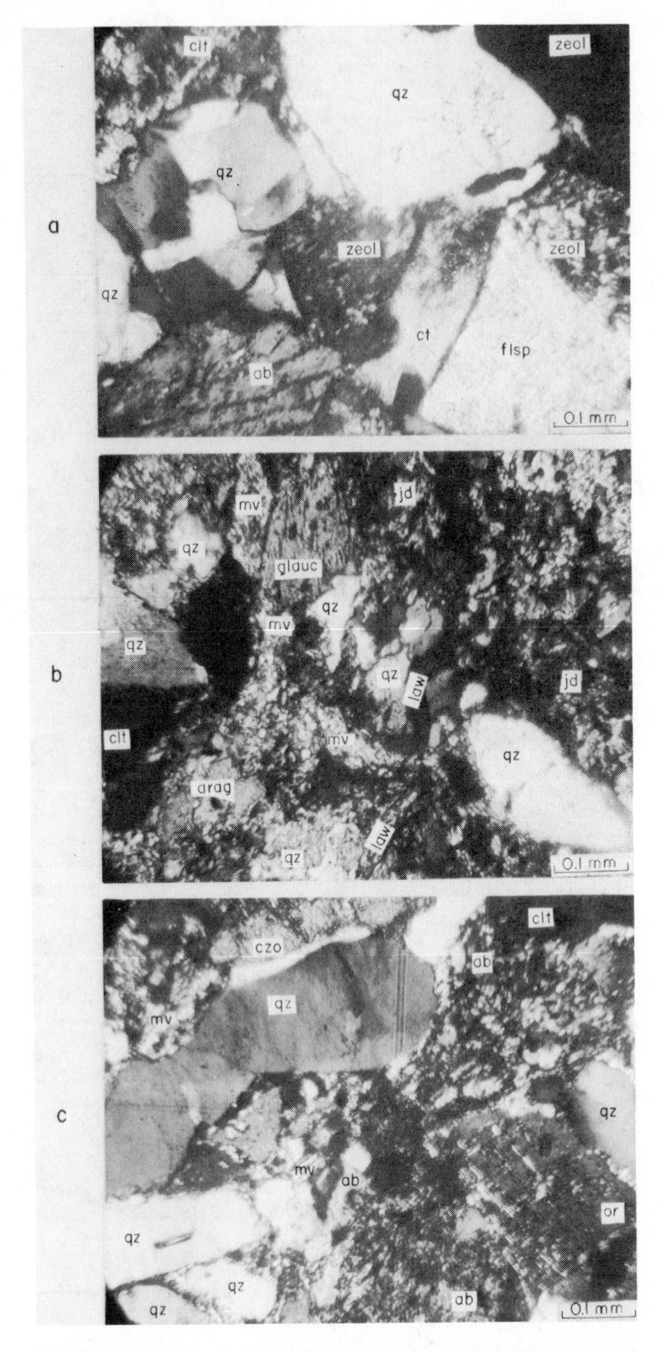

Fig. 2. Photomicrographs of graywacke in thin sections under light with crossed polarization: (*a*) zeolite facies, sample 63, (*b*) blueschist facies, sample 8, (*c*) greenschist facies, sample 52. (For abbreviations, see footnote to Table 1.)

Blueschist Facies

The relatively low-temperature, high-pressure mineral assemblage in the blueschist facies of the Franciscan Formation is represented in this study by samples 8 and 14 in Table 1; the diagnostic minerals observed include jadeite, lawsonite, glaucophane, and aragonite. The texture and petrography of sample 8 is shown in Fig. 2b. The pertinent phase boundaries shown in Fig. 1 suggest that the conditions of metamorphism might have ranged from 7 to 11 kb and from 150 to 350°C; the point, 10 kb and 300°C, may be chosen to typify the conditions of formation for an assemblage containing jadeite but not albite in equilibrium. By study of oxygen isotopes, Taylor and Coleman (1968) determined the temperature of formation of low-temperature blueschist to range from 270 to 315°C.

Greenschist Facies

At somewhat higher temperature and lower pressure, a greenschist mineral assemblage is stable in the Franciscan metagraywacke, as in sample 52 and 1 in Table 1. The diagnostic minerals observed are albite, chlorite, and clinozoisite (not epidote, as in most greenschists). A photomicrograph of sample 52 in thin section is shown in Fig. 2c. The clinozoisite-lawsonite boundary in Fig. 1 is an upper limit for lawsonite to be stable, but not a lower limit for clinozoisite to form, although the lower limit is probably less than 100°C below. An upper temperature limit for greenschist is provided by the boundary of the chlorite stability field, as shown in Fig. 1. The range of conditions of metamorphism for the greenschist facies might be from 3 to 7 kb and from 300 to 550°C, as suggested by the stability fields in Fig. 1; the point, 5 kb and 350°C, is chosen to typify the formation conditions of the equilibrium assemblage.

Discussion

The identification of the clinozoisite in samples 52 and 1 has been checked, but apparently the occurrence is rare; pumpellyite or prehnite are usually observed instead (R. G. Coleman, 1971, and M. C. Blake, Jr., 1971, personal communications). Presumably, the body of Franciscan Formation cropping out just south and just north of San Francisco was subjected to somewhat higher temperature than most of the other bodies. An intermediate field of pumpellyite-prehnite facies probably should be shown between the other three fields in Fig. 1 to characterize fully the metamorphism of the Franciscan Formation; no samples of this facies were obtained for this study. It is not likely that the greenschist facies and the blueschist facies formed by further metamorphism of the other, but probably formed directly from the zeolite facies, possibly passing through the pumpellyite-prehnite facies. As might be expected, in some places (e.g., the Diablo Range), the high-pressure minerals occur as mixed, nonequilibrium assemblages with low-pressure minerals; one such rock was studied by Brace et al. (1970).

The close equivalence of the members of the suite of rocks studied here is shown in Table 1 by the similarity of their mineral compositions as calculated for the blueschist facies; therefore, despite some small differences, their chemical composition is assumed to be essentially constant, thus removing that variable from consideration. Two aspects of the effect of metamorphism on the strength of this suite of rocks present themselves, decrease in porosity and change in petrography, and results of experiments on both properties are described.

APPARATUS AND METHOD

The apparatus used was a simple ram and pressure cylinder with an auxiliary hydraulic system for building up confining pressure. Experiments were all made at room temperature. This and the more complex types of high pressure apparatus, such as that used by Brace et al. (1970), are briefly described by Heard (1968, Figs. 17, 18, 22).

The hydrostatic pressure of the surrounding fluid on test samples is called confining pressure, P. In the earth, the confining pressure may be considered to be equal to the weight of the superincumbent load of rock. In my experiments, all tests were in compression. A compressive load was applied to the flat ends of a rock cylinder, producing axial stress in the cylinder higher than the lateral hydrostatic pressure. Axial loading was increased until failure occurred by fracture, and the difference between the axial stress and the lateral confining pressure, the stress difference, at failure is taken as the strength, S, of the rock sample. In some rocks, like the zeolite-bearing ones, some nonelastic or plastic deformation may precede fracture.

Test samples in my experiments were cylinders of two sizes, 1.2-cm diameter by 3.8-cm length, and 2.5-cm diameter by 6.2-cm length. In the room temperature tests, impervious rubber jackets were placed around each sample; Brace et al. (1970) used copper jackets in their heated runs. Pressure and stress measurements are estimated to be reliable to about 5% of the measurement; strain measurements are reliable to about 0.1% strain. The details of methods of application of pressure, load, and heat are given in the papers referred to in the introduction. The properties and conditions affecting strength are considered below essentially in order of increasing complexity.

EXPERIMENTAL RESULTS

Porosity

The strength increase of the graywacke samples with decrease in intergranular porosity is shown by the order of the curves from bottom to top in Fig. 3. The maximum scatter in the test results is shown by the points plotted for sample 63; the other curves are better defined. The inverse relation between strength and porosity is shown in Fig. 4 at one confining pressure, P = 2 kb; the maximum change in strength occurs for samples with intergranular porosities below 10%. The strength increase is not closely determined by the results, but the trend is clear.

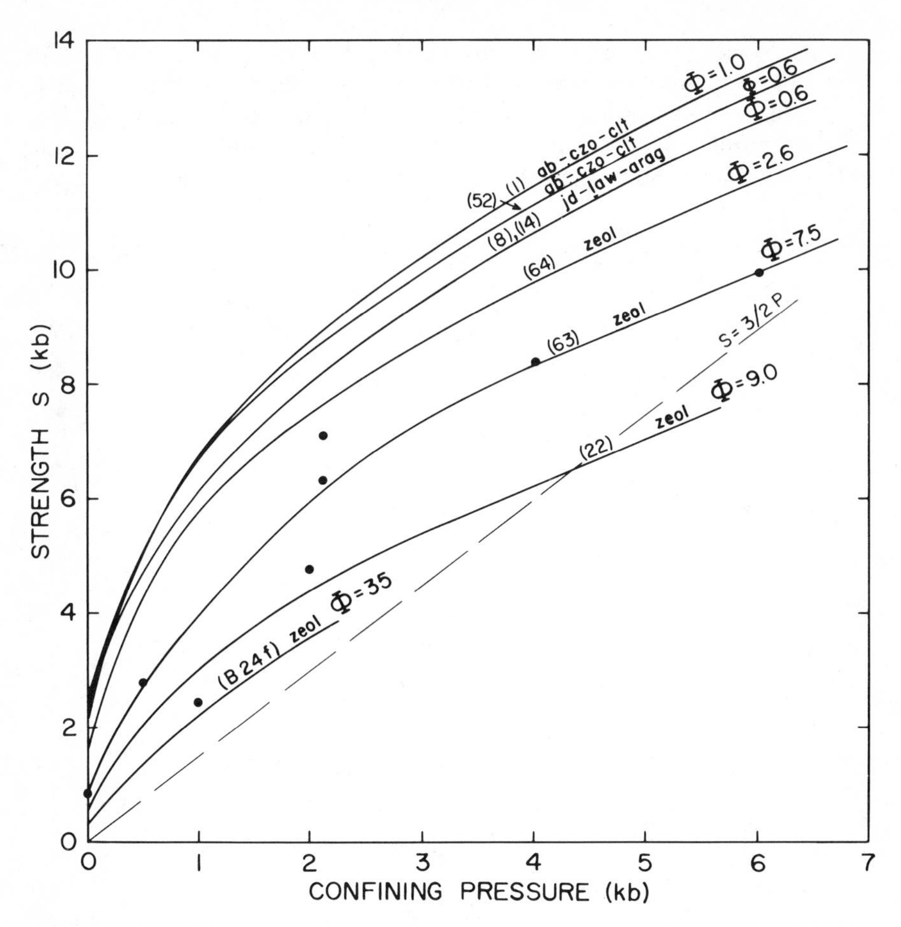

Fig. 3. Effect on the compressive strength of graywacke samples of porosity, ϕ, confining pressure, and metamorphic grade. The samples were tested dry at room temperature. Points are for sample 63 and show the maximum scatter.

As might be expected, the porosity resulting from fractures and joints weakens rock below its strength in the unbroken state. Two studies may be cited: Byerlee (1967) with a study of friction across rock surfaces and Donath (1961) with a study of anisotropy in strength due to slaty cleavage.

Confining Pressure

One of the earliest, most important, and best authenticated results of experiments on rock deformation was the finding of a very marked strengthening of rocks with increasing confining pressure (Mogi, 1966). The rise of each curve in Fig. 3 shows the increase of strength with confining pressure for each sample, the greatest rate of increase occurring in the first 2 kb of confining pressure. This direct relation between compressive strength and confining pressure applies to all rocks; the curves in Fig. 5 (after Mogi, 1966) show the effect for a variety of rocks; for example, for granite, basalt, and quartzite, the strength under 1-kb confining pressure is three times the 1-atm value (Fig. 5).

Quartzite, single crystal quartz, and quartz glass have been studied in strength tests to very high confining pressures, as shown in Fig. 6. The mineral, the rock,

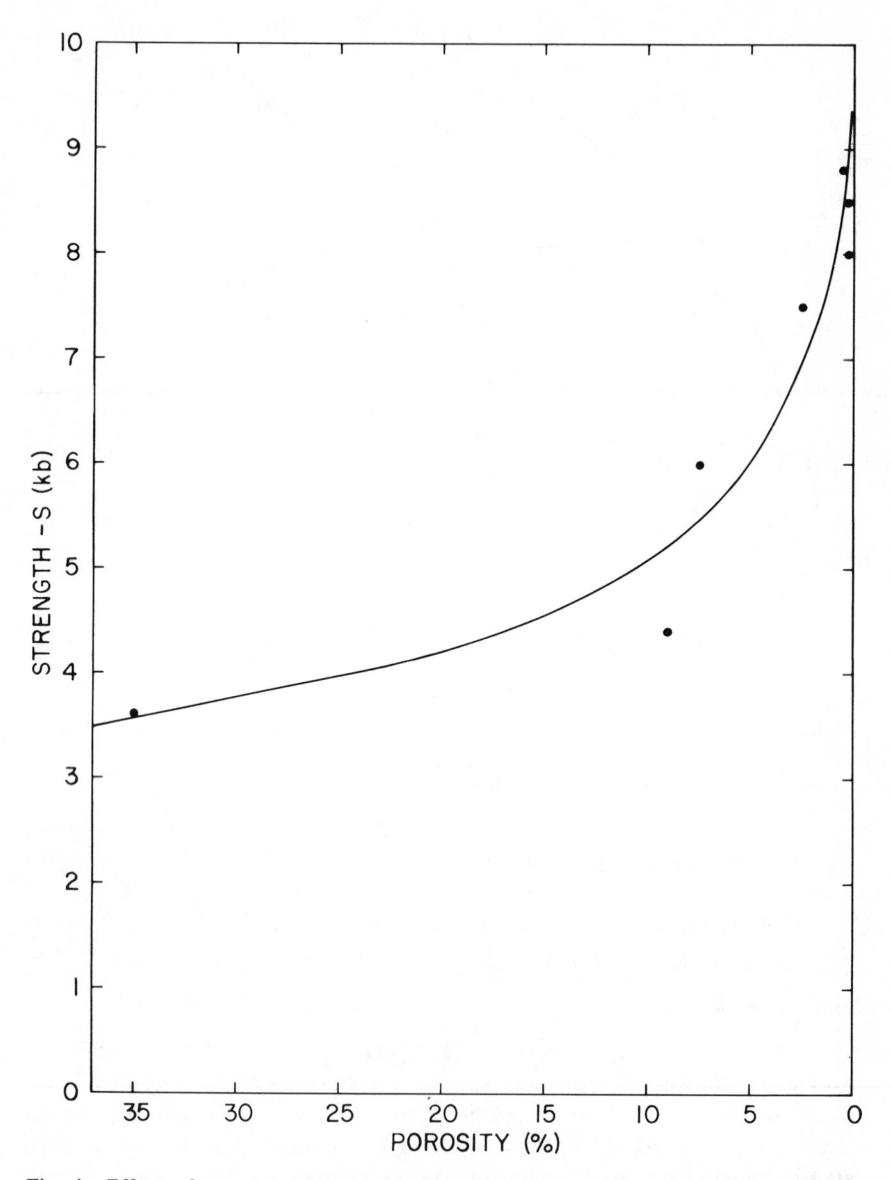

Fig. 4. Effect of porosity on strength of samples of graywacke suite under confining pressure of 2 kb and at room temperature. (Porosity scale is inverted.)

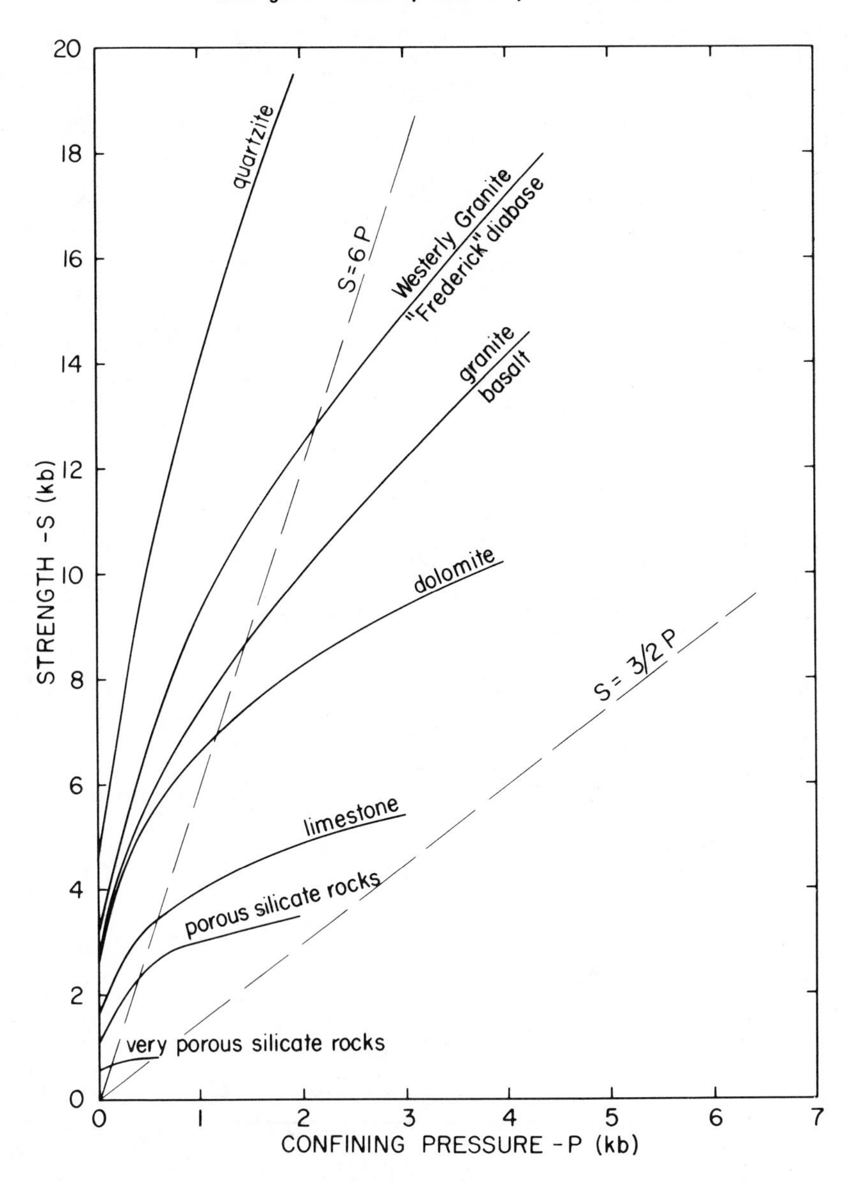

Fig. 5. Increase of compressive strength with confining pressure for various rocks (after Mogi, 1966).

and even the glass of silica can attain phenomenal strengths if the surrounding pressure is high enough; furthermore, as shown by two points for quartzite, heating to 500°C does not weaken the rock appreciably.

Confining pressure also increases friction strength. Byerlee's investigation (1967) of the shear strength across an existing saw-cut or fracture surface in

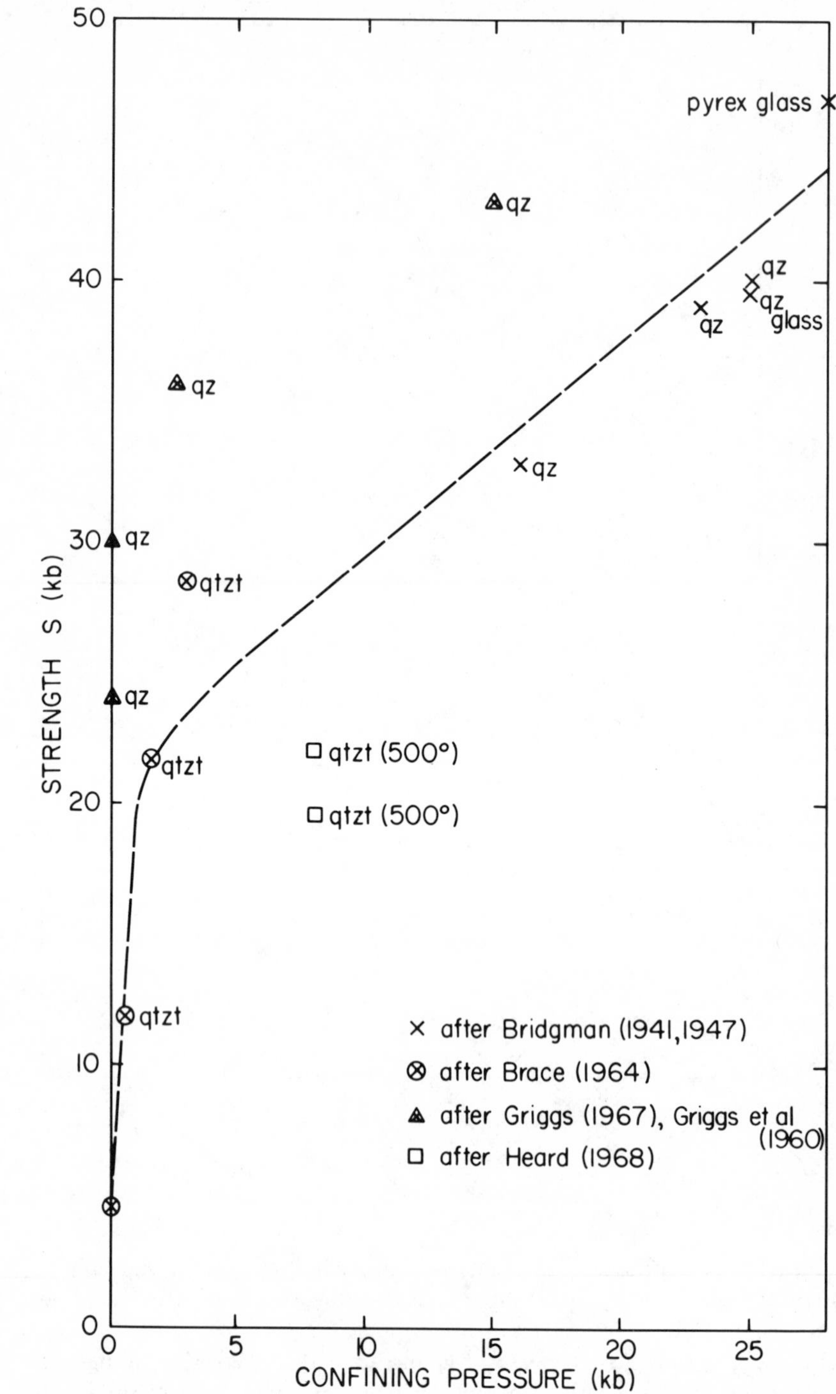

Fig. 6. Strength of quartz, quartzite, and quartz glass as affected by confining pressure to 28 kb.

Westerly Granite revealed an increase in friction from half the fracture strength at P = 1 kb to equal in amount at P = 11 kb. Byerlee (1968) also found that the strengths of powdered quartz and granite increase markedly with confining pressure; Byerlee and Brace (1969) observed that the strength of *powdered* Westerly Granite is 11 kb at P = 8 kb, compared with the strength of virgin granite of 22 kb at that pressure.

An explanation of the strengthening of silicate rocks by confining pressure may be that in reducing the initial porosity, both intergranular and fracture porosity, the pressure improves the interlocking and bonding between grains.

Water in the Pores and in the Crystal Structure

One of the important conditions that reduces the strength of a sedimentary rock is the mechanical effect of pressure of water in the pores. In one series of tests, samples from block 64 were saturated with water and compressed in two ways. In one way, an open chamber in the adjacent steel anvil plug provided an exit at one end of the sample so that water could be squeezed out as the sample was compressed; in the second way without such a chamber, as the sample was compressed, the water came under the same pressure as the confining pressure. In the latter case, the rock was extremely weak, as shown in Fig. 7. In the case

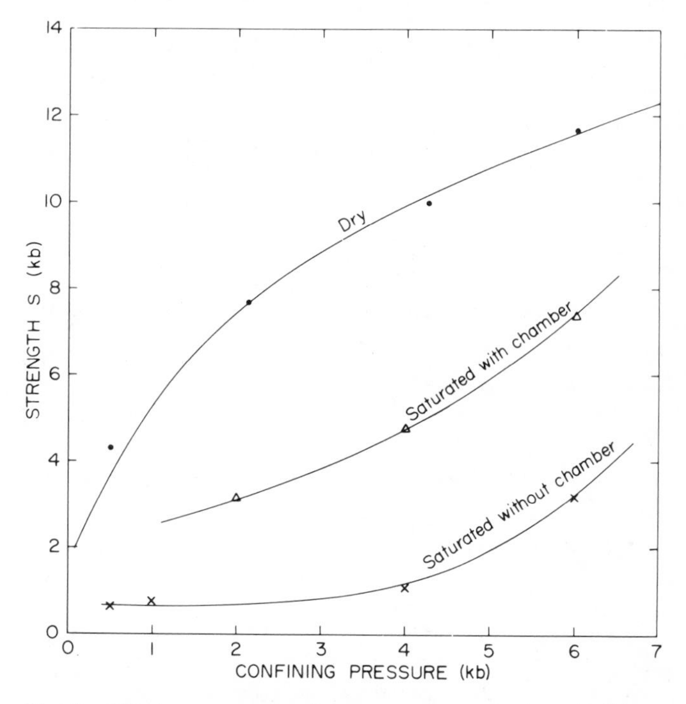

Fig. 7. Effect on strength of a porous graywacke (sample 64) of water in the pores under pressure, and of water vented to an open chamber, at room temperature.

of the vented system, the water was under a reduced pressure in the pores, and presumably under a condition similar to a sedimentary rock in the earth being compacted, water being squeezed from the pores. (Although the time constant for pore pressure should be much greater in the earth, the loading time is presumably also much greater.) The dry samples had a much higher strength. Similar results, but not as pronounced, were obtained on quartzitic shale (porosity of 0.3%) and quartzitic sandstone (porosity of 15%) by Colback and Wiid (1965).

In Fig. 8 similar curves are shown for specimens from the entire graywacke suite. The rock samples were saturated with water and tested so the water could vent to the open chamber. The weakening resulting from the pore pressure again is evident, and it is more pronounced the greater the porosity; in samples of the two dense facies, however, the effect is small. The weakening is probably a result of building up of local pore pressures rather than of lubrication between grains (W. F. Brace, oral communication, 1970).

A series of tests was made by Brace et al. (1970) to determine the effect of pore water under pressure on samples of a mixed-facies rock (jadeite with albite) of Franciscan graywacke from Pacheco Pass, Calif.; the results are shown in Fig. 9. A downward trend is displayed by the points, but the relationship is uncertain. The curves in Fig. 9 were drawn to simulate the more complete results

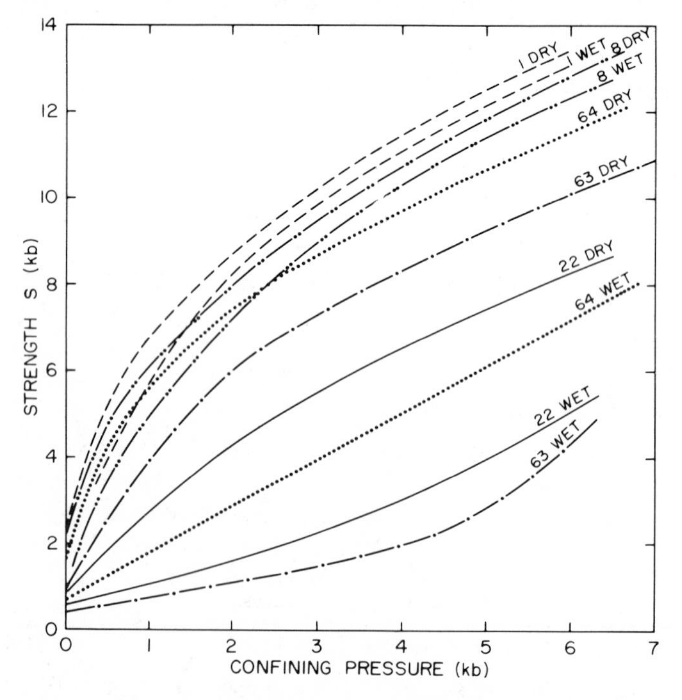

Fig. 8. Effect on strength of a suite of graywacke samples of water saturating the pores but vented to an open chamber.

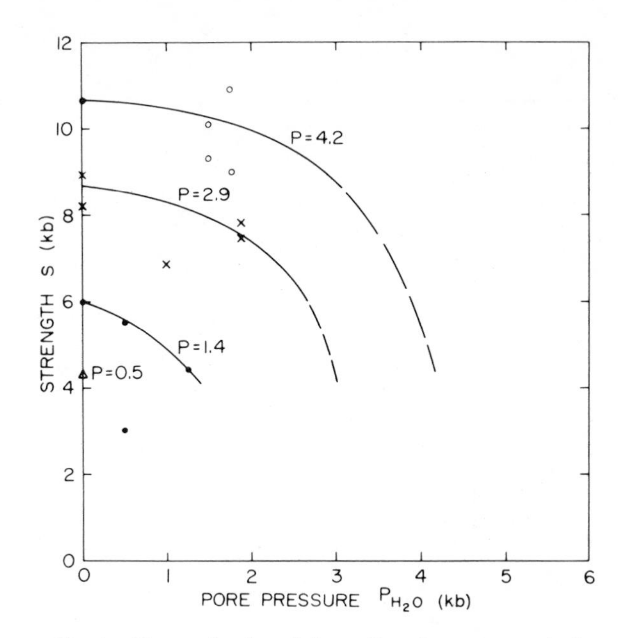

Fig. 9. The weakening of dense Franciscan graywacke by controlled pore-water pressure, at three confining pressures (P) at room temperature (after Brace et al., 1970).

of Serdengecti and Boozer (1961) for Carthage marble, shown in Fig. 10. They tested the marble at pore water pressures up to the confining pressure at three pressure levels. From their results, the conclusion may be drawn that the strength of the rock is not reduced more than 20% until the pore pressure is greater than about three-fourths of the confining pressure; the strength decreases rapidly with higher pore pressures. The ultimate strength of the rock at pore pressure of water equal to the confining pressure is essentially that of the unconfined rock sample.

Another way in which water could weaken quartz-rich rocks like graywacke was found by Griggs (1967). Synthetic quartz crystals have about 0.1% water distributed through the crystal structure, and because of this water, the crystals, when heated above 400°C, exhibit an extreme weakening, as shown in Fig. 11. This *hydrolytic weakening,* as it is called by Griggs, has been observed neither in natural quartz crystals (Fig. 6) nor in natural crystals of other minerals; however, Griggs has been able to force water into the crystal structure of natural quartz, olivine, and pyroxene crystals at 1000°C and 20 kb so that subsequently they showed the hydrolytic weakening.

Temperature

If water is present in the crystal structure of mineral grains in a rock being heated, the rock strength may be markedly reduced at some relatively low

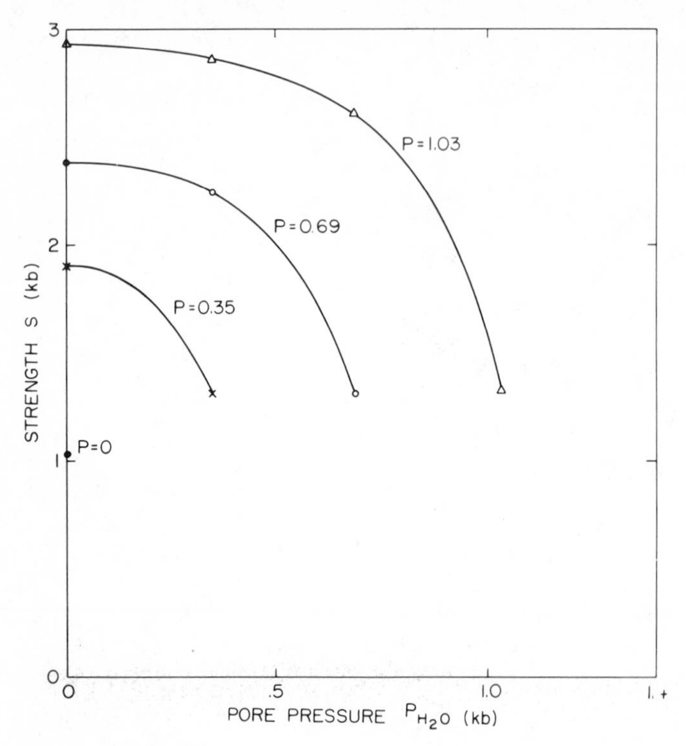

Fig. 10. Pore water pressure and the strength of "Carthage" marble, at three confining pressures (*P*) at room temperature (after Serdengectic and Boozer, 1961).

temperature by breakdown of the grains. In hydrolytic weakening the water is an impurity in the crystalline quartz, whereas the weakening of serpentinite at 300 to 500°C found in strength tests by Raleigh and Paterson (1965) is a result of the water bound in the mineral structure. Similarly the zeolite mineral, laumontite, breaks down at about 300°C at room pressure (Thompson, 1970), and although strength tests with heating have not been made on zeolite-bearing rocks, it is almost certain that the heating would weaken the rock greatly as the zeolite grains break down. The pressure effect on the breakdown temperature of serpentinite and laumontite apparently is a small increase, about 50°C for 5 kb.

The high-pressure minerals in the blueschist facies, jadeite, lawsonite, aragonite, and glaucophane, are metastable at room conditions, and they each break down to stable phases on heating at atmospheric pressure. Although the effect of temperature on strength is complicated when accompanied by these transformations, the samples of the mixed-facies rock of the Franciscan Formation studied by Brace et al. (1970) were not much weakened when heated dry to temperatures below 450°C, as shown in Fig. 12. In these laboratory tests, breakdown of the metastable minerals would not be significant because of the short duration of the experiments. The decrease in fracture strength of this rock

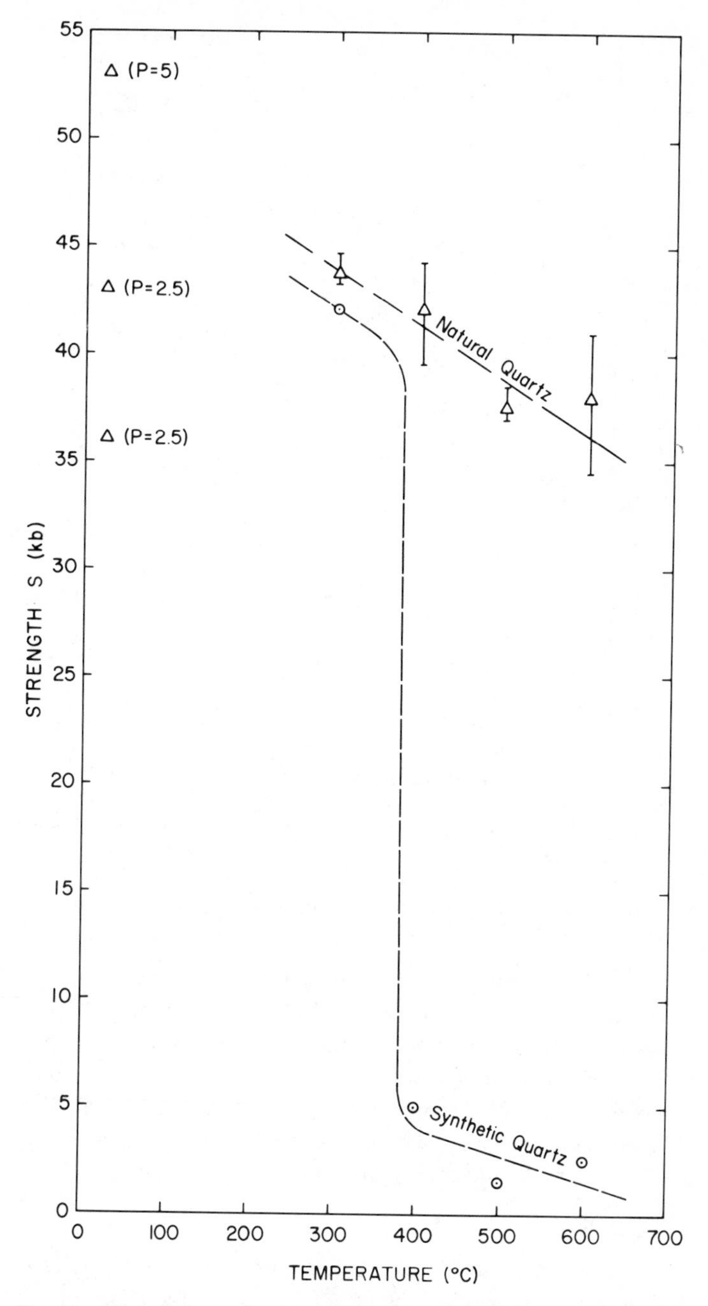

Fig. 11. Hydrolytic weakening of *synthetic* quartz crystals (circles) as a function of temperature compared with the strength decrease in *natural* quartz crystals (triangles) resulting from heating; both sets of samples were tested in compression under 15-kb confining pressure, except for three as marked (after Griggs, 1967, and Griggs et al., 1960).

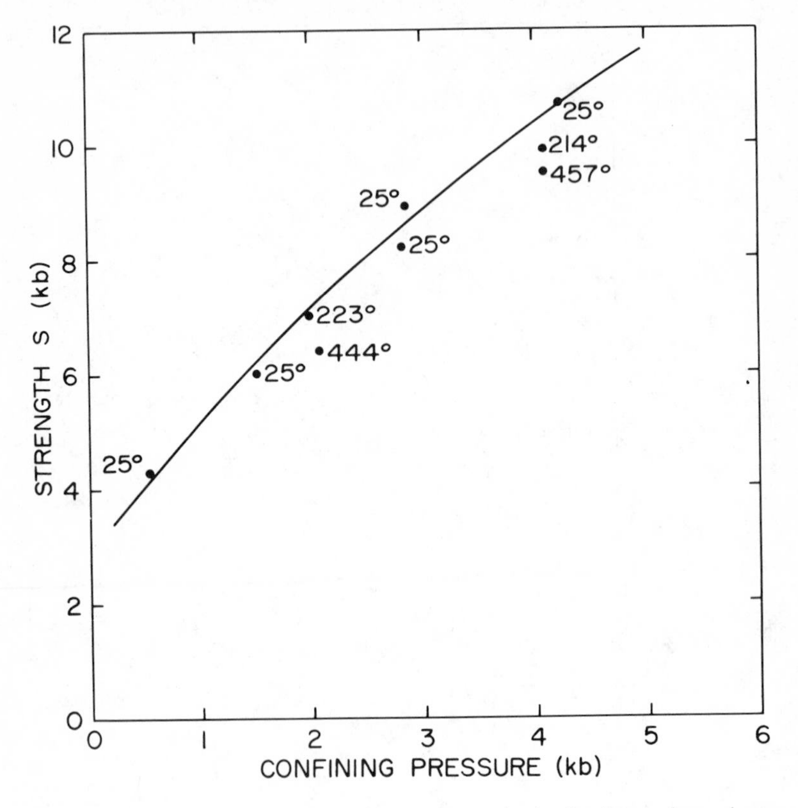

Fig. 12. Weakening of dense Franciscan graywacke by heating under various confining pressures. Samples run at 25°C were tested at a strain rate of 10^{-6} sec^{-1} (shown by solid line); samples at about 220°C and about 450°C were tested at 10^{-4} sec^{-1} (from Brace et al., 1970).

(Fig. 12) is about 10% by heating at 450°C and even less at 220°C. A somewhat greater weakening is observed on heating granite samples, as shown by the results of Griggs et al. (1960) in Fig. 13; these results are presented by way of comparison with graywacke and to show the effect on strength of higher temperatures. It may be suggested that differential thermal expansion of the anisotropic mineral grains in these rocks weakens the intergranular bonds and thus lowers the rock strength.

Strain Rate

In the same series of experiments of Brace et al. (1970) on samples of Franciscan graywacke, the effect of varying the change in strain with time was also studied. Extrapolating slightly from their results to temperatures of 25 and 420°C and to $P = 2$ kb, it is found that there is a decrease of 5 to 10% in the strength of the graywacke at fracture because of decrease in strain rate from 10^{-4} sec^{-1} to 10^{-6} sec^{-1}; these results are shown by the two curves marked "gwke" in Fig. 14. If the confining pressure were increased, the curves would be displaced upwards, but the weakening effect of slowing the strain rate would be about the same.

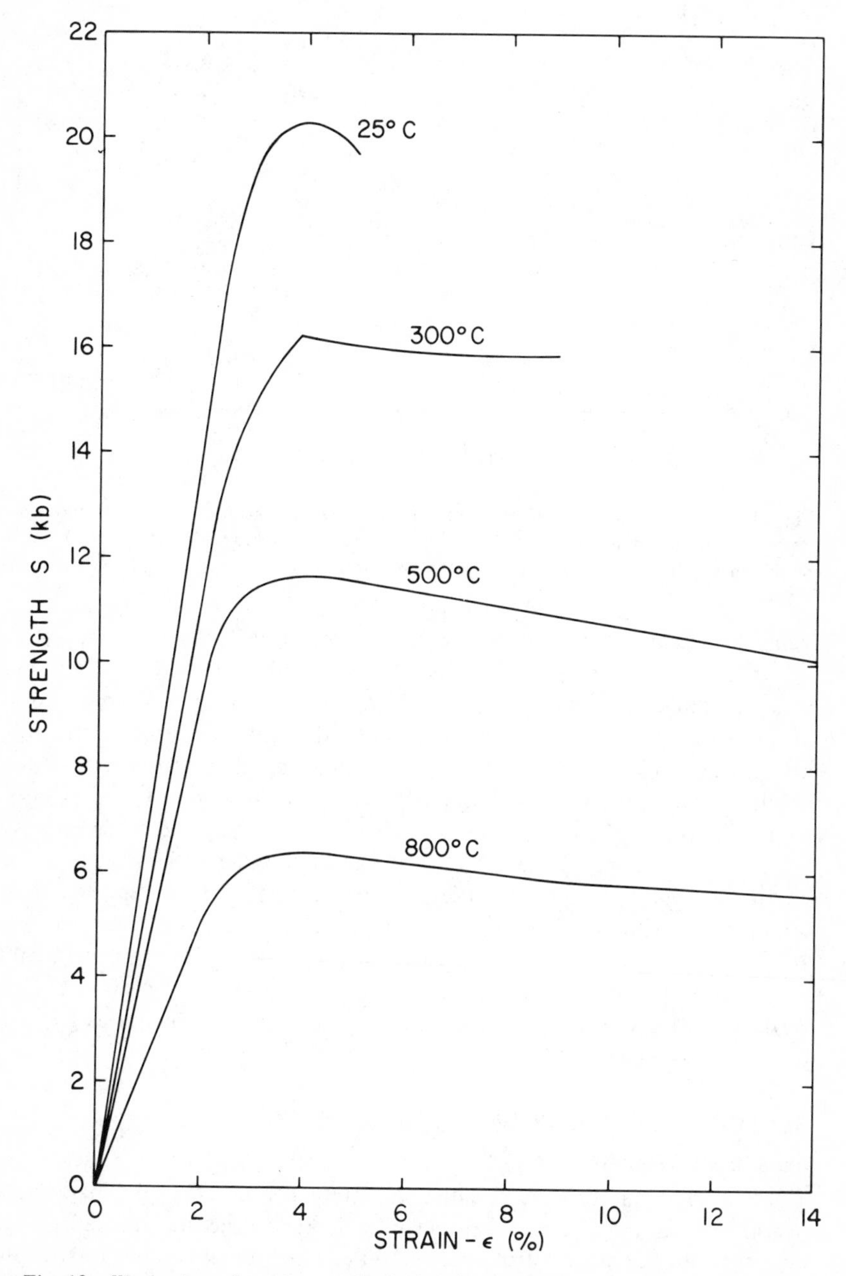

Fig. 13. Weakening of granite samples by heating during compression tests under 5-kb confining pressure at a strain rate of 10^{-4} $\sec^{-1}$ (from Griggs et al., 1960).

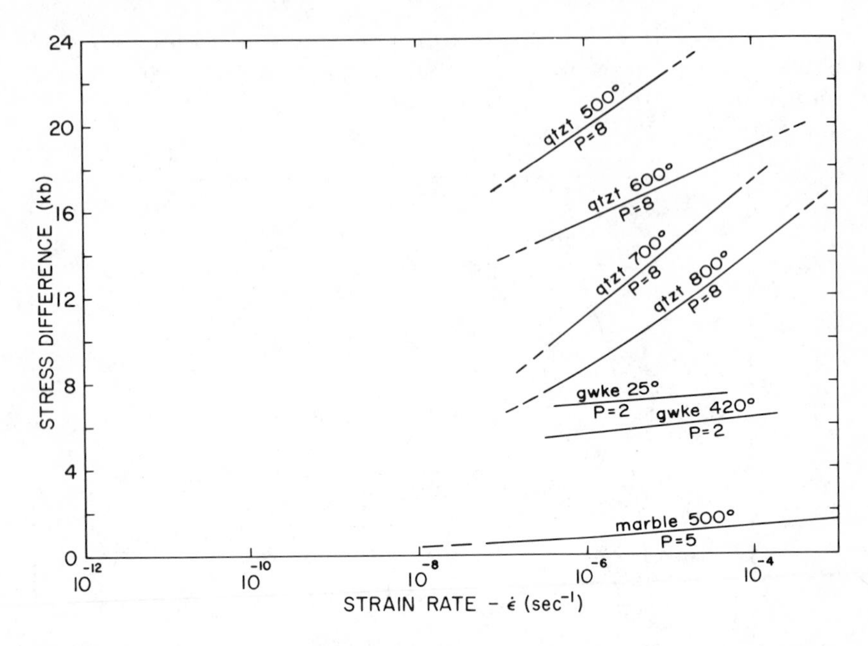

Fig. 14. Effect of varying the strain rate at various temperatures in compression tests on graywacke (after Brace et al., 1970), quartzite (after Heard and Carter, 1968), and marble (after Heard, 1963). (*P* is confining pressure.)

Similar weakening by decreasing the strain rate at constant temperature is shown in Fig. 14 from the results of Heard and Carter (1968) on Simpson orthoquartzite. The data are not consistent, as shown by the lack of uniformity in slopes of the lines; however, it is apparent that at temperatures above 500°C, the strength of quartzite is markedly decreased by slowing the strain rate. As another comparison with the graywacke results, a curve from Heard (1963) for "Yule" marble at 500°C is also shown in Fig. 14; the slope of the line is the same as those for graywacke, although the absolute value of the strength is much less for marble under the conditions shown. Donath (1966) reported that he found no systematic decrease in strength with decreasing strain rate from 10^{-3} sec^{-1} to 10^{-7} sec^{-1} on sandstone and other rock samples, at room temperature for $P = 2$ kb. In general, at temperatures below 400°C, the weakening of siliceous rocks by decreasing the strain rate is not as great as at temperatures above 400°C.

GEOLOGICAL IMPLICATIONS

Experimental Results

The strength in rocks is sensitive to internal structure. The character and relationships of the mineral grains make up the rock structure, and depending on how they influence this structure, certain environmental conditions strengthen the rocks and others weaken it. Confining pressure stengthens graywacke and other rocks, as seen in Figs. 3, 5, and 6. Apparently, continued cycling of stress from low to high levels, a work hardening, increases rock strength (Donath, 1968; Robertson, 1955); thus strain history may also be important. Finally, metamorphic effects, including compaction and diagenesis, have been found to improve the strength of graywacke and other rocks, as discussed below. On the other hand, weakening of rocks occurs with structural damage to the mineral grains and intergranular bonds by heating (Figs. 12, 13), by the presence of water in the pores (Figs. 7, 8) or in the crystal structure (Fig. 11), especially if the water is under pressure (Figs. 9, 10), and by slowing the strain rate (Fig. 14). The effect of each of these elements on strength of rocks has been described briefly in this paper and at length by the previous investigators cited.

Other factors being equal, increase in strength of rocks may be said to occur when intergranular bonding improves. Reduction in porosity by compaction and the accompanying rearrangement of grains during diagenesis of the zeolite facies of the graywacke is more important in improving the strength of rocks of this facies than the gross mineral assemblage (Fig. 4). With compaction during diagenesis, the zeolites form a cement between grains, the large quartz and feldspar grains interlock more (Fig. 2a), and the strength increases; this occurs without any significant change in mineralogy, for example, in the content of the cementing zeolite minerals in samples 22, 63, and 64. Calcite would be expected to weaken the rock (see limestone curve in Fig. 5), but the calcite content apparently does not affect the strength much; for example, compare sample 22 with samples 63 and 64 in Table 1 and Fig. 3. Strength increase with diminishing porosity and with recrystallization and induration during diagenesis or higher grade metamorphism has been observed in a general way in unsystematic tests made on other sedimentary rocks by previous workers.

With increase in temperature and pressure through the field of the zeolite facies into the other fields (Fig. 1), recrystallization and induration occur, the porosity is reduced, and the strength increases. The fairly dense zeolite rock, sample 64 (porosity, 2.6%), is nearly as strong as the blueschist samples 8 and 14 (porosity, 1%) (Fig. 3), a point to be referred to later. The high-pressure blueschist facies, formed at about 10 kb and 300°C, appears to be slightly weaker than the relatively higher temperature, lower pressure greenschist facies, formed at about 5 kb and 350°C, possibly because the higher temperature produced greater recrystallization and grain bonding; however, changes in grain structure are not at all obvious in the photomicrographs in Fig. 2. The preservation of the original clastic grain structure of the sediment without crushing (Fig. 2) through metamorphism under the cited pressures and temperatures, despite mineral phase changes, implies that the interlocking grains have great strength. Other sedimentary rocks might be expected to behave in a similar way.

Application of these experimental results may be made to metagraywacke in the earth and to other rocks in place, the certainty depending on knowledge of the prevailing conditions. In the earth, however, account must be taken of inhomogeneities of a larger scale than sample size.

A convenient fiction often assumed in mathematical models of the earth is that the medium of interest is a homogeneous continuum, but rock in the earth is far from being homogeneous. Geological variety is obvious, ranging from siltstone to dunite, and aside from the environmental conditions, each problem involving rock strength must consider the local rock types, their mechanical properties, and the spatial relationships of discontinuities between rock masses; of the latter, joints and faults are the most important. To some extent, the results of rock deformation experiments on samples can be extrapolated to jointed and faulted rock masses in the crust of the earth. If the confining pressure is high enough, friction strength becomes equal to fracture strength, as shown by the experiments of Byerlee (1967) on saw-cut and fracture surfaces in granite. A similar strengthening at high confining pressure was found by Byerlee and Brace (1969) on powdered granite. Assuming that these results apply to most other rocks, fracture and breccia zones would weaken the rock at shallow depths, but the strength across such discontinuities would equal that of the unbroken rock below some minimum depth, and would approach it at shallower depths, as indicated by Byerlee's results (1967).

Estimating Strength of Crustal Rocks

The strength curves in Fig. 3 and 5 are approximately parabolic, but each rock type has a different curve. It would be useful to have a formula providing a rough approximation for the strength of rocks, applicable to massive rocks in the shallow crust. We might seek a general expression by using experiments performed under room conditions, at low confining pressures; this is permissible if we assume the rocks to be unfractured, with low pore-water content, at temperatures below 400°C, and weakened only slightly by slow, geologic strain rates. An approximation indicated in a paper by Birch (1955) was that the strength of crustal rocks is constant at 6 kb, and although this value would be roughly applicable to dense rocks below 2 km (Fig. 5), the increase in strength resulting from confining pressure is significant and should be added. In another formula (Robertson, 1955), based on results of various tests at low confining pressure, the maximum shearing stress (that is, one-half the strength) was found to be approximately equal to the mean stress; in terms of the present compression experiments, this relation is equivalent to $S = 6P$ (shown in Fig. 5). This formula gives the approximate range of strength for massive rocks in the pressure range, 0.5 kb $< P <$ 3 kb; it provides reasonable values for rocks in the shallow crust, perhaps from 1- to 10-km range in depth.

As remarked above, porous silicate rocks have relatively low strength (Fig. 5). In fact, rather high ductility was observed in the laboratory tests on sample B24f before failure by fracture; this behavior occurs probably because the zeolite grains, composing half the rock, are relatively weak and plastic. Based on the stress-strain curves and the sample-coherence of these porous rocks, a nearly ideal plasticity may be postulated to apply to them, and the simple approximation may be made that strength is equal to the mean stress. For compression experiments, this model is represented by the line, $S = 3/2\,P$ (shown in Figs. 3 and 5), which seems to be a lower limit even for porous rocks.

These two formulas may apply roughly as limiting cases, but as might be expected, there is no single approximation for rock strength that will apply to all rocks in the crust. Each rock mass and its geologic conditions must be considered individually.

Pressure in the Metamorphism of the Blueschist Facies

Many comprehensive discussions have been published of the pressure-temperature conditions during the reconstitution of the graywacke to the blueschist facies of the Franciscan Formation in California, and a good recent review of the previous work is given by Ernst (1971). A controversy has arisen over the nature of the geologic process by which the high pressure was attained in forming the blueschist rocks. Coleman and Lee (1962) first suggested that tectonic overpressure could produce the required mean pressure, and Ernst (1971) has maintained that deep burial was required. (Tectonic overpressure is merely the excess of lateral pressure in the earth over lithostatic, vertical, load pressure.) The geologic evidence adduced is not fully compelling, so the results of experiments of Brace et al. (1970) on the strength of some Franciscan samples shed new light on the question. The strength of the graywacke is an important consideration because it determines the amount of tectonic overpressure it will withstand during metamorphism. Brace and his collaborators concluded that the Franciscan rocks would be too weak to withstand the necessary overpressure, and so they came down on the side of deep burial. Their results and the new experimental results presented in this paper can be assessed for another interpretation of the problem.

Most of the experimental data of Brace et al. (1970) have been reproduced and commented upon in the preceding sections. The strengthening of graywacke by increasing confining (i.e., lithostatic) pressure has just been described, but the weakening conditions bear repeating. Heating the graywacke to 300°C would reduce its strength 10% or less (Fig. 12). Decreasing the strain rate from laboratory rates (about 10^{-4} sec^{-1}) to geologic rates (10^{-13} sec^{-1}, Savage and Burford, 1970) would weaken the graywacke by about 20% (Fig. 14). The effects of these two conditions are important, but they are not as crucial as raising the pore pressure.

The effect of pore pressure is large in porous rocks, especially when it is close to the confining pressure, as shown in Figs. 7, 8, and 9. The detailed arguments for high pore pressure during metamorphism in Brace et al. (1970) and Taylor and Coleman (1968) are strong, but do not seem incontrovertible. Some of the results of this study suggest that pore-water pressures could have been low and that the relatively high strengths of nearly dry rock are therefore applicable. As an example, we may consider sample 64 (in the zeolite facies) whose strength is close to that of the blueschist samples 8 and 14 (Fig. 3). At its porosity of 2.6%, even if reduced by 4-kb overburden pressure, the estimated permeability of 10^{-4} darcy (extrapolated on the basis of data from Fraser, 1935, and Fatt, 1953) would be more than adequate to leak off pore water, as the temperature increased from 100 to 300°C. (Water needed for crystal structural positions in the blueschist minerals would be available by diffusion from the zeolite minerals, as indicated in the calculated blueschist compositions in Table 1.)

The experiments made in studying the effect of pore pressure on strength must be carefully performed. If the water in the pores is injected under a pressure near the confining pressure, it will weaken the rock sample, as shown in Fig. 7 for saturated samples tested without an adjacent chamber. Apparently, the mineral grains are forced apart, and the sample weakened thereby. This may be quite important for rocks of low porosity, requiring long pretest pumping times to saturate the samples. The dilatancy accompanying deformation of the samples (Brace and Martin, 1968) increases permeability, which would allow water to leak off. Therefore, interpretation of pore pressure experiments must be done carefully in applying them to rocks in the earth.

At a confining pressure of 5 kb, the strength of sample 64 would be about 11 kb, and after reducing it 25% for heat and strain rate effects, and assuming the principal lateral stresses to be equal, the effective mean pressure on the rock would be about 10 kb, enough to produce the high pressure minerals of the blueschist rocks. In other words, the confining pressure of 5 kb would be equal to the rock weight at about 15 km, and the strength at that depth could withstand the 8-kb stress difference required to produce the mean pressure of 10 kb needed to form the high pressure minerals. The strength at a shallower depth, perhaps 10 km, would be sufficient if the maximum stress rather than the mean stress determines the pressure at which mineral phase changes occur. The 5-kb mean pressure for the greenschist facies can be produced easily in the shallow crust.

The remaining important weakening factor described by Brace et al. (1970) is that the strength of the graywacke mass would be reduced by the presence of weaker interbedded shale. The answer to the importance of the shale in determining the permissible mean stress requires further geologic field study. Detailed mapping indicates that the shale may not be as ubiquitous or continuous in the Franciscan Formation as thought heretofore (W. R. Cotton, 1970, oral communication), and local bodies of blueschist rocks may be autonomous and strong. We may conclude that under favorable conditions of pore pressure and occurrence of shale, the graywacke could be strong enough to resist crushing and permit the high-pressure minerals to form in the upper crust.

ACKNOWLEDGMENTS

I am indebted to Robert G. Coleman for help in collecting field specimens and in providing geological data. This paper would not have appeared without the help of Rudolph Raspet and R. W. Werre, who performed many of the experiments and helped develop and build the apparatus; I acknowledge their support with many thanks. A cogent review of the initial version of the manuscript by James F. Hays materially aided me in this presentation; Herbert R. Shaw also greatly helped me by his review. Finally, I am grateful to William F. Brace (who was supposed to write this paper, and should have) for supplying unpublished data, apparatus, quarters and rations, and encouragement.

Note added in proof: J. S. Huebner (oral communication, 1971) has just determined the unit cell parameters and thus fully identified clinozoisite in my sample 52; however, the amount separated and measured was < 1%, and the finegrained remainder of the estimated 7% in the rock may be pumpellyite. The clinozoisite phenocrysts are still considered to be metamorphic and not detrital in origin from petrographic observations, and so the rock can be left in the greenschist facies.

REFERENCES

Athy, L. F., Density, porosity, and compaction of sedimentary rocks, Am. Assoc. Petrol. Geol. Bull. **14**, 1-24, 1930.

Birch, F., Physics of the crust, *in* The Crust of the Earth, edited by A. Poldervaart, pp. 101-118, Geol. Soc. Amer. Spec. Paper 62, 1955.

______, Megageological considerations in rock mechanics, *in* State of Stress in the Earth's Crust, edited by W. R. Judd, pp. 55-80, Am. Elsevier Publ. Co., New York, 1964.

Boettcher, A. L., and Wyllie, P. J., The calcite-aragonite transition measured in the system CaO-CO_2-H_2O, J. Geology, **76**, 314-330, 1968.

Brace, W. F., Brittle fracture of rocks, *in* State of Stress in the Earth's Crust, edited by W. R. Judd, pp. 111-173, Am. Elsevier Publ. Co., New York, 1964.

Brace, W. F., and Byerlee, J. D., Stick-slip as a mechanism for earthquakes, Science, **153**, 990-992, 1966.

______, California earthquakes: Why only shallow focus? Science, **168**, 1573-1575, 1970.

Brace, W. F., and Martin, R. J., III, A test of the law of effective stress for crystalline rocks of low porosity, Int. J. Rock Mech. Min. Science, **5**, 415-426, 1968.

Brace, W. F., Ernst, W. G., and Kallberg, R. A., An experimental study of tectonic overpressure in Franciscan rocks, Geol. Soc. Am. Bull., **81**, 1325-1338, 1970.

Bridgman, P. W., Explanations toward the limit of utilizable pressures, J. Appl. Phys., **12**, 461-469, 1941.

______, The effect of hydrostatic pressure on the fracture of brittle substances, J. Appl. Phys., **18**, 246-258, 1947.

Byerlee, J. D., Frictional characteristics of granite under high confining pressure, J. Geophys. Res., **72**, 3639-3648, 1967.

______, Deformational characteristics of rock powders, Am. Geophys. Union Trans., **49**, 314, 1968.

Byerlee, J. D., and Brace, W. F., High-pressure mechanical instability in rocks, Science, **164**, 713-715, 1969.

Colback, P. S. B., and Wiid, B. L., The influence of moisture content on the compressive strength of rock, Rock Mech. Sympos., Dept. Mines Tech. Surveys, Mines Branch, Ottawa, 65-83, 1965.

Coleman, R. G., and Lee, D. E., Metamorphic aragonite in the glaucophane schists of Cazadero, California, Am. J. Sci., **260**, 577-595, 1962.

Donath, F. A., Experimental study of shear failure in anisotropic rocks, Geol. Soc. Amer. Bull., **72**, 985-990, 1961.

______, Experimental support for high pressure at shallow depth, Geol. Soc. Amer. Spec. Paper 101, 55-56, 1966.

______, Role of experimental rock deformation in dynamic structure geology, *in* Rock Mechanics Seminar, edited by R. E. Riecker, pp. 355-437, Clearinghouse for Federal Scientific and Technical Information, Dept. of Commerce, Washington, D.C., 1968.

Ernst, W. G., Stability relations of glaucophane, Am. J. Sci., **259**, 735-765, 1961.

______, Polymorphism in alkali amphiboles, Am. Mineralogist, **48**, 241-260, 1963.

______, Do mineral parageneses reflect unusually high-pressure conditions of Franciscan metamorphism? Am. J. Sci., **270**, 81-108, 1971.

Fatt, I., The effect of overburden pressure on relative permeability, J. Petrol. Tech., sec. 1, 15-16, 1953.

Fraser, H. J., Experimental study of the porosity and permeability of clastic sediments, J. Geology, **43**, 910-1010, 1935.

Gralewska, A., ed., KWIC index of rock mechanics literature, vol. 1 and 2, Amer. Inst. Min. Met. Petroleum Engr., New York, 863 pp., 1969.

Griggs, D. T., Hydrolytic weakening of quartz and other silicates, Geophys. J., **14**, 19-31, 1967.

Griggs, D. T., Turner, F. J., and Heard, H. C., Deformation of rocks at 500°C to 800°C, *in* Rock Deformation, pp. 39-104, Geol. Soc. Amer. Memoir 79, 1960.

Handin, J., and Hager, R. V., Experimental deformation of sedimentary rocks under confining pressure; tests at high temperature, Am. Assoc. Petrol. Geol. Bull., **42**, 2892-2934, 1958.

Handin, John, Hager, R. V., Jr., Friedman, M., and Feather, J. N., Experimental deformation of sedimentary rocks under confining pressure: Pore pressure tests, Am. Assoc. Petrol. Geol. Bull., **47**, 717-755, 1963.

Heard, H. C., Effect of large changes in strain rate in the experimental deformation of Yule marble, J. Geol., **71**, 162-195, 1963.

______, Experimental deformation of rocks and the problem of extrapolation to nature, *in* Rock Mechanics Seminar, edited by R. E. Riecker, **2**, 439-508, Clearinghouse for Federal Scientific and Technical Information, Dept. of Commerce, Washington, D.C., 1968.

Heard, H. C., and Carter, N. L., Experimentally induced "natural" intragranular flow in quartz and quartzite, Am. J. Sci., **266**, 1-42, 1968.

Hirschberg, A., and Winkler, H. G. F., Stabilitätsbeziehungen zwischen Chlorit, Cordierit and Almandin bei der metamorphose, Contrib. Miner. Petrologie, **18**, 17-42, 1968.

Hoschek, G., The stability of staurolite and chloritoid, Contrib. Miner. Petrologie, **22**, 208-232, 1969.

Madsen, B. M., and Murata, K. J., Occurrence of laumontite in Tertiary sandstones of the central Coast Ranges, California, U.S. Geol. Survey Prof. Paper 700-D, 188-195, 1970.

Mogi, K., Pressure dependence of rock strength and transition from brittle fracture to ductile flow, Earthquake Res. Inst. Bull., **44**, 215-232, 1966.

Newton, R. C., and Smith, J. V., Investigations concerning the breakdown of albite at depth in the earth, J. Geol., **75**, 268-286, 1967.

Newton, R. C., and Kennedy, G. C., Some equilibrium reactions in the join $CaAl_2Si_2O_6$-H_2O, J. Geophys. Res., **68**, 2967-2983, 1963.

Raleigh, C. B., and Paterson, M. S., Experimental information of serpentinite and its tectonic implications, J. Geophys. Res., **70**, 3965-3985, 1965.

Raleigh, C. B., and Kirby, S. H., Creep in the upper mantle, *in* The Mineralogy and Petrology of the Upper Mantle, edited by B. A. Morgan, pp. 113-121, Mineral. Soc. Amer. Spec. Paper 3, 1970.

Robertson, E. C., Experimental study of the strength of rocks, Geol. Soc. Am. Bull., **66**, 1275-1314, 1955.

Savage, J. C., and Burford, R. O., Accumulation of tectonic strain in California, Seism. Soc. Am. Bull., **60**, 1877-1896, 1970.

Serdengecti, S., and Boozer, G. D., The effects of strain rate and temperature on the behavior of rocks subjected to triaxial compression, Penn. State Univ., Mineral Industries Experiment Station Bull., **76**, 83-97, 1961.

Taylor, H. P., Jr., and Coleman, R. G., O^{18}/O^{16} ratios of coexisting minerals in glaucophane-bearing metamorphic rocks, Geol. Soc. Am. Bull., **79**, 1727-1756, 1968.

Thompson, A. B., Laumontite equilibria and the zeolite facies, Am. J. Science, **269**, 267-275, 1970.

Tozer, D. C., ed., Non-elastic processes in the mantle, Roy. Astron. Soc. Geophys. J., **14**, 450 p., 1967.

Weertman, J., The creep of the earth's mantle, Rev. Geophys. Space Phys., **8**, 145-168, 1970.

Winkler, H. G. F., Abolition of metamorphic facies, introduction of the four divisions of metamorphic stage, and of a classification based on isograds in common rocks, Neues Jahrb. Mineral., **5**, 189-248, 1970.

AUTHOR INDEX

References are indexed for the pages on which an author's work is cited and are arranged alphabetically according to each author's last name. Numbers in **bold face** indicate the inclusive pages of chapters in this book. A number in *italics* indicates the page of an "et al." citation.

TOPIC INDEX

This index is designed to show what broad topics are covered in this book and thereby to be a reader's guide rather than a detailed classification, as would be found in a comprehensive textbook. The reader can determine where a topic of interest is discussed and its page location by inspection of the subjects listed under the key-word headings. These first-order headings are cross-referenced to keep duplications of the topic entries to a minimum. In addition, this index is short enough that a rapid perusal of it will show the scope of the topics described in the book.